Beiträge zum ausländischen und internationalen Privatrecht

151

Herausgegeben vom

Max-Planck-Institut für ausländisches
und internationales Privatrecht

Direktorium:

Holger Fleischer, Ralf Michaels, Anne Röthel

Global Legal Pluralism and Rights of Nature

Edited by

Daniel Bonilla Maldonado and
Ralf Michaels

Mohr Siebeck

Daniel Bonilla Maldonado is Full Professor of Law at the Universidad de los Andes, Bogotá.

Ralf Michaels is Director at the Max Planck Institute for Comparative and International Private Law in Hamburg; Chair of Global Law, Queen Mary University in London; Professor of Law, University of Hamburg.

Open Access funding provided by Max Planck Society.

ISBN 978-3-16-200402-4 / eISBN 978-3-16-200403-1
DOI 10.1628/ 978-3-16-200403-1

ISSN 0340-6709 / eISSN 2568-6577
(Beiträge zum ausländischen und internationalen Privatrecht)

The Deutsche Nationalbibliothek lists this publication in the Deutsche Nationalbibliographie; detailed bibliographic data arc availablc at *https://dnb.dnb.de.*

Published by Mohr Siebeck Tübingen 2026.

Printed on non-aging paper.

Mohr Siebeck GmbH & Co. KG, Wilhelmstraße 18, 72074 Tübingen, Germany
www.mohrsiebeck.com, info@mohrsiebeck.com

Introduction

I.

Rights of nature (RoN) have moved, in record time, from being a slightly eccentric niche topic to a mainstream field of research. By now there are tens of thousands of publications dedicated to the topic. Given that RoN spread through a process of rapid legal transplantation, it is not surprising that much of that literature is, to some extent, comparative in nature. Yet another comparative law publication on the subject is therefore in need of justification, lest it merely repeat what others have produced before. We believe this justification lies in a number of characteristics that make this project unique.

First, we link rights of nature with discussions on global legal pluralism (GLP). This provides a novel and, as it turns out, unusually productive perspective on both phenomena. This means that we are interested in interrelations between legal orders, both horizontally (country to country) and vertically (between local municipalities and states and international law), between official and unofficial laws, between South and North, and between South and South. And it means that we can connect two discourses that each emerge, in different and complicated ways, from North-South legal relations.

Second, this project resulted from the proceedings of the International Association of Comparative Law. This meant that we were able to bring together expertise of numerous authors from around the world who provided us with detailed information about their respective legal systems, thereby enabling a depth of information that a single researcher could not produce. The number of country reports that this book includes, together with a number of special reports, sets it apart from most comparative analyses that focus on only a few countries.

It is not solely the breadth of countries under scrutiny that sets this project apart. Many comparative studies focus on the same few countries – Ecuador, Bolivia, Columbia, New Zealand, perhaps India – that have adopted RoN. Other than them, thirdly, our list of countries includes a good number of 'negative' cases – legal orders that have not adopted RoN. The failure to adopt RoN in those countries is crucial for a full understanding of the global situation of RoN and of the conditions for their success, and their inclusion thus deepens the analysis considerably.[1]

[1] See Martha-Cecilia Dietrich and others, 'Stories of Systemic Failure? Landscaping the Rights of Nature in Europe' (2024) 8 JLA 46–76.

Fourthly, unlike studies that concentrate on positive law, we bring together formal law, socioeconomic and cultural conditions, and analysis of the movements behind the adoption of RoN. Individually, each of these fields is well covered in the literature – formal law mainly by lawyers, socioeconomic and cultural conditions by anthropologists, movement analysis by sociologists and political scientists. Bringing them together offers more insights than the mere sum of their parts. RoN are fascinating in no small part because of the way in which social movements translated normativities into formal law, and by the alterations that these normativities assume in the process. GLP is characterized not only by static interrelations but by processes of engagement. We hope this study can shed some light on these translations and processes.

Finally, although North/South relations do not define our entire analysis, the way in which they permeate these debates made it necessary to address these relations in the work. North/South relations easily invite hierarchies. In some of those the North stands on top – Northerners are experts and Southerners informants,[2] or Northerners are sophisticated and Southerners are 'legal barbarians'.[3] In some of these the opposite is true – Southerners are idealized as 'noble savages' who live in accordance with nature, unlike Northerners who can only destroy. None of these descriptions is fully accurate in either its juxtaposition of North and South or in its lumping together significantly different approaches under each label. RoN, viewed through the lens of GLP, paint a more differentiated picture.

II.

The contributions to this book combine two general reports, five special reports, and 20 country reports from a theme of the 2022 Conference of the International Association of Comparative Law in Paraguay, for which the two of us served as general rapporteurs. Among the special reports, Daniel Bonilla's provides a categorization of poietic, mimetic and resistance patterns that also serves as structure for this volume; he also analyzes the political economy of legal knowledge and the various transitions and cross-fertilizations RoN undertake.[4] Ralf Michaels addresses GLP and RoN each as internally hybrid, such that their combination becomes a hybrid of hybrids, before surveying the responses in the country reports to the questionnaire.[5]

[2] See Annelise Riles, 'From Comparison to Collaboration: Experiments with a New Scholarly and Political Form' (2015) 78(1/2) L&CP 147–183.

[3] Daniel Bonilla Maldonado, *Legal Barbarians: Identity, Modern Comparative Law and the Global South* (CUP 2020).

[4] Daniel Bonilla Maldonado, 'The Globalization of Rights of Nature. Global Legal Pluralism and the Rights of Nature' in this volume, 3–46.

[5] Ralf Michaels, 'A Hybrid of Hybrids. Global Legal Pluralism and Rights of Nature' in this volume, 47–94.

The five special reports address the topic in a general fashion. Mihnea Tananescu, writing about theoretical foundations, emphasizes different origins and different conceptualizations of RoN, discarding the myth of a uniform concept.[6] Mara Tignino sketches the role of RoN in international law with a specific eye to the difference between an anthropocentric right *to* nature and ecocentric rights *of* nature.[7] Dirk Hanschel and Annette Mehlhorn demonstrate the fruitfulness of combining anthropological and legal perspectives on RoN.[8] Mari Margil provides an insider perspective on the role of NGOs for RoN, particularly within the United States.[9] The struggle to give rights to the Magpie River and other rivers in Canada is explained by Yenny Vega Cárdenas and Uapukun Metokosho.[10]

Among the country reports, we follow the structure proposed by Bonilla of poietic, mimetic, and resistant legal orders[11] – with the caveat, explained in more detail below, that this categorization is necessarily simpler than reality. As poietic countries – those that introduced RoN first – we characterize the legal orders of Ecuador, Bolivia, and New Zealand. Gabriela Espinoza shows how RoN were first established and then, over time, consolidated in Ecuador.[12] Clemens Gregor Barié, in his report on Bolivia, demonstrates the much more arduous path towards the recognition of RoN in Bolivia, as well as the obstacles they face.[13] Catherine Iorns outlines the very different foundations and structure of RoN in Aotearoa New Zealand, as laid out in treaties and laws stipulating guardianship for local Maori.[14]

Mimesis of the idea of RoN happened in several legal orders, of which we discuss a significant number. Tania Luna Blanco and David de la Torre Vargas present the successful adoption of RoN in Colombia, where they were recognized primarily by the courts.[15] India's path towards ecocentrism and the recognition

---

[6] Minhea Tănăsescu, 'Theoretical Foundations of the Rights of Nature. Origins and Conceptual Characteristics' in this volume, 97–130.

[7] Mara Tignino, 'International Law. Rights of Nature and Rights to Nature: Emerging Trends' in this volume, 131–143.

[8] Dirk Hanschel and Annette Mehlhorn, 'Law and Anthropology. An Interdisciplinary Perspective on Rights of Natur' in this volume, 145–164.

[9] Mari Margil, 'NGOs in the United States. Birth of a Movement in this volume', 165–175.

[10] Yenny Vega Cárdenas and Uapukun Mestokosho, 'Recognizing the Legal Personality of a River. The Magpie River/*Mutehekau Shipu* in Canada' in this volume, 177–207.

[11] Bonilla Maldonado, 'The Globalization of Rights of Nature' (n 4) II.

[12] Gabriela Espinoza, 'Ecuador. The Nature of the Rights of Nature' in this volume, 211–246.

[13] Cletus Gregor Barié, 'Bolivia. The Rights of Mother Earth: Fifteen Years of Solitude' in this volume, 247–282.

[14] Catherine Iorns, 'Aotearoa New Zealand. Protecting Nature through Upholding Indigenous Rights and Human Responsibilities' in this volume, 283–316.

[15] Tania Luna Blanco and David de la Torre Vargas, 'Colombia. Graffiti on the Colombian Legal Architecture: The Emergence of Non-human Subjects and the Rights of Nature' in this volume, 319–340.

of RoN is explained by Manjeri Subin Sunder Raj.[16] Tajudeen Sanni describes Uganda as the first country on the African continent that introduced RoN, closely following the Ecuadorian model.[17] South Africa, presented by Louis J Kotzé, is another country that can be characterized as a mimetic legal order.[18] Several other countries that could be included in this group – Spain, Panama, and Peru, to name just three – have no country reports but are taken account of in our general reports.[19]

The largest portion of legal systems in the world can be counted as resisters, at least insofar as they do not officially recognize RoN. Their study is of interest regardless.[20] Our book contains no fewer than thirteen reports on countries that do not recognize RoN – though in most of them they are discussed. Antonio Carlos Wolkmer and Debora Ferrazzo consider national rejection and local adoption of RoN in Brazil and the potential that the constitutional right to an ecologically balanced environment may lead to broader recognition.[21] Stéphanie Roy, Gaële Gidrol-Mistral, and Alexandra Popovici focus on Canada, with special attention to Québec; they suggest a particularly interesting way of recognizing RoN, effectively as *res communis* under the Code Civil of Québec.[22] Jacques deLisle provides a particularly interesting perspective on a country rarely discussed in RoN literature – China – and describes the reasons why RoN have a particularly difficult stand there.[23] Konstantinos A Rokas' analysis of Cyprus shows how small countries sometimes almost entirely stand outside of discourses like the one on

---

[16] Manjeri Subin Sunder Raj, 'India. Sowing the Seeds of "Nature Rights"' in this volume, 341–352.

[17] Sanni Tajudeen, 'Uganda: The First African Country to Recognize Rights of Nature' in this volume, 353–366.

[18] Louis J Kotzé, 'South Africa. Protecting Nature through an Anthropocentric-Oriented Rights Paradigm' in this volume, 367–374.

[19] On the Mar Menor in Spain, see, eg, Marie-Christine Fuchs, 'Rights of Nature Reach Europe' (*Verfassungsblog*, 24 February 2023) <https://verfassungsblog.de/rights-of-nature-reach-europe/>; José Francisco Alenza García, 'Sobre los aparentes derechos reconocidos al Mar Menor', (2025) 226 Revista de administración pública 155–186. On Panama, see Ley 287 del 24 de febrero de 2022 <https://www.gacetaoficial.gob.pa/pdfTemp/29484_A/90384.pdf>; Corte Suprema de Justicia 27 Nov 2023, GO N° 29922. On Peru, see STC Exp N° 00010-2022-0-1901 <https://iuslatin.pe/historica-sentencia-rio-maranon-y-sus-afluentes-son-declarados-sujetos-derechos-expediente-00010-2022-0-1901/>; Mariano Andres Bustamante Jimenez, 'Los Ríos como Titulares de Derechos en el Perú' (2025) 18 YachaQ: Revista de Derecho 75–89 <https://doi.org/10.51343/yq.vi18.1673>.

[20] See Dietrich and others (n 1).

[21] Antonio Carlos Wolkmer and Debora Ferrazzo, 'Brazil. Isolated Legislative and Judicial Inroads in a Dependent Capitalism Landscape' in this volume, 377–392.

[22] Stéphanie Roy, Gaële Gidrol-Mistral and Alexandra Popovici, 'Québec (and Canada). Nature, Culture, Rupture: Tools for Environmental Protection in a Liberal Anthropocentric Legal System' in this volume, 573–598.

[23] Jacques deLisle, 'China. Limited Prospects for the Rights of Nature in an Inhospitable Legal-Political Environment' in this volume, 393–425.

RoN – and instead move in other ways towards more ecological law.[24] Jakub Harašta shows that the Czech Republic takes a decidedly anthropocentric approach to environmental protection and thereby rejects RoN quite firmly, despite early musings in Czech scholarship.[25] Yvette Lind shows how a Northern country like Denmark could potentially introduce RoN based on the input of local and indigenous communities.[26] Elena Ewering, Janina Reimann, and Tore Vetter give a detailed perspective on the reasons for the failure, so far, of RoN in Germany, recent court decisions to the contrary notwithstanding.[27] Domenico di Micco and Michele Graziadei place RoN in a broader history of Italian law and ecology and point to public parks as interesting substitutes for RoN.[28] Keisuke Abe reports on environmental protection in Japan.[29] Philip Paiement writes about the Netherlands as a country that does not yet recognize RoN but that has the potential, with several ongoing initiatives.[30] Sören Koch, Esmeralda Colombo, and Catalina Vallejo Piedrahíta discuss the potential of rights of nature in Norway – especially on the basis of Sámi legal culture – along with potential alternatives for protecting nature.[31] Romania is, as Diana Botău shows in her report, another 'unfertile ground' for RoN.[32] Finally, Laura Affolter and Sian Affolter show how a lively debate in Switzerland has been spurred by both local initiatives and ideas borrowed from afar.[33]

III.

This book has been a long time in the making. Several unforeseen events, including a publisher's withdrawal, delayed publication. We are especially grateful to all contributors for their texts, for their willingness to adapt – sometimes sig-

---

[24] Konstantinos A Rokas, 'Cyprus. Constitutional Adoption of a Right to Nature, but No Formal Recognition of Nature's Rights' in this volume, 427–440.

[25] Jakub Harašta, 'Czech Republic. An Anthropocentric Approach to Environmental Protection' in this volume, 441–454.

[26] Yvette Lind, 'Denmark. Could Danish Human- and Animal-Rights based Environmental Protection Provide Rights to Nature?' in this volume, 455–462.

[27] Elena Ewering, Janina Reimann and Tore Vetter, 'Germany. Rights of Nature: A Paradigm Change for the German Legal System' in this volume, 463–491.

[28] Domenico di Micco and Michele Graziadei, 'Italy. Establishing the Rights of Nature in Italy – Between Tradition and Change' in this volume, 493–514.

[29] Keisuke Abe, 'Japan. Bringing the Alternative View to Fruition' in this volume, 515–525.

[30] Phillip Paiement, 'Netherlands. Rights of Nature, Globalization, and Legal Pluralism' in this volume, 527–538.

[31] Sören Koch, Esmeralda Colombo and Catalina Vallejo Piedrahíta, 'Norway. Rights of Nature in the Norwegian Legal Culture: To Be or Not to Be?' in this volume, 539–571.

[32] Diana Botău, 'Romania. Unfertile Ground for the Rights of Nature' in this volume, 599–609.

[33] Laura Affolter and Sian Affolter, 'Switzerland. Mapping the Scene' in this volume, 611–635.

nificantly – to our requests, and for their patience with publication of this book. Some of the contributions were previously published in Spanish as part of an earlier special issue that we were privileged to coedit with Patricia Zalamea Fajardo.[34]

Special thanks go to Chiara Goetzke and Dharmita Prasad who, at different stages, supported the project at the Max Planck Institute. The editorial services team at the Institute was, as always, invaluable in turning the many texts into a book. Our thanks go to Christian Eckl, Michael Friedman, Janina Jentz, and Anja Rosenthal for the careful editing and typesetting. The Max Planck Digital Library makes it possible for the book to appear open access, a crucial element for the global discourse on topics like ours.

Bogotá/Colombia and Hamburg/Germany, December 2025

Daniel Bonilla Maldonado
Ralf Michaels

---

[34] Daniel Bonilla Maldonado, Ralf Michaels and Patricia Zalamea Fajardo (eds), 'Los derechos de la naturaleza: diálogos entre el derecho y las artes' (2022) 4 Naturaleza y Sociedad: Desafíos Medioambientales 1.

Content

III. Poietic Legal Orders

IV. Mimetic Legal Orders

V. Resisting Legal Orders

Appendix

Contributors

Keisuke Mark Abe, Prof Dr, LL M (Harvard)
Professor at Waseda University Faculty of Law

Laura Affolter, Dr
Senior researcher at the Humanities Center for Advanced Studies "Futures of Sustainability" at the University of Hamburg; Associate Researcher at the Institute of Social Anthropology at the University of Bern

Sian Affolter, Dr
Senior researcher at the Institute for European Law at the University of Fribourg

Cletus Gregor Barié, Phd Social Anthropology (Vrije Universiteit Amsterdam)
Senior Consultant at Deutsche Welle Akademie (Bonn/Berlin); independent advisor on indigenous rights and conflict transformation

Tania Luna Blanco, Prof Dr, LL M
Professor and researcher at the Department of Public Law, and Director of the Doctoral Program in Legal Sciences and the Public Interest Law and Litigation Clinic at the Faculty of Legal Sciences, Pontificia Universidad Javeriana (Bogotá, Colombia)

Daniel Bonilla Maldonado, Prof Dr
Full Professor of Law, Universidad de los Andes, Bogotá

Diana Botău, PhD ('Babeș-Bolyai' University)
Associate Professor and Vice Dean of the Faculty of Law, 'Babeș-Bolyai' University; lawyer and member of Cluj Bar

Esmeralda Colombo, Esq. Dr, LL M (Columbia)
Marie Skłodowska-Curie Postdoctoral Fellow at the RFF-CMCC European Institute on Economics and the Environment (EIEE)

David de la Torre Vargas, Prof Ph D
University Professor at Universidad del Cauca, Universidad Surcolombiana, and Universidad Cooperativa de Colombia; currently serving as Judicial Procurator 153 for Administrative Affairs

Jacques deLisle, Prof
Stephen A. Cozen Professor of Law, Professor of Political Science, and Director of the Center for the Study of Contemporary China, University of Pennsylvania; Chair, Asia Program, Foreign Policy Research Institute

Domenico di Micco, Dr
Senior Lecturer of Comparative Law; Director of ISAIDAT, Subalpine Research Centre for Transnational Studies, Turin

Gabriela Espinoza, PhD (Los Andes University), LL M (Yale University)
Senior Judicial Advisor of the Ecuadorian Constitutional Court

Elena Ewering
Legal trainee (Rechtsreferendarin) at the Oberlandesgericht Hamm; former Research Assistant at the Kassel Institute for Sustainability of the University of Kassel, Germany; part of the research project 'Nature as a Legal Entity', funded by the German Research Foundation

Debora Ferrazzo, Prof Dra, (UFPR), Brazil
PhD in Law from the Federal University of Paraná (UFPR); professor of the Postgraduate Program in Law at the University of the Extreme South of Santa Catarina (UNESC), and professor of 'Rights of Nature and Animal Rights' in the Law School; lawyer

Gaële Gidrol-Mistral, Prof
Professeuse, Département des sciences juridiques, Université du Québec à Montréal; Directrice, Groupe de Réflexion en Droit Privé

Michele Graziadei, Prof
Full Professor of Comparative Law; Past President of SIRD, Italian Society of Comparative Law; Fellow of Accademia delle Scienze di Torino

Dirk Hanschel, Prof Dr MCL (Mannheim/Adelaide)
Professor at the University of Halle, Germany, Chair of German, European and International Public Law; from 2019 until 2025, Fellow at the Max Planck Institute for Social Anthropology, Halle (Saale)

Jakub Harašta, PhD
Assistant Professor at the Institute of Law and Technology, Faculty of Law, Masaryk University

Catherine J Iorns, Prof, BA, LLB (Well), LL M (Yale)
Professor of Law, Victoria University of Wellington

Sören Koch, Prof Dr, PhD (Bergen)
Professor of Law, President of the Norwegian Association of Comparative Law, Member of the European Law Institute and of the International Academy of Comparative Law

Louis J Kotzé, Prof Dr Dr
Chair of the Law Group and Professor of Law, Wageningen University and Research, the Netherlands

Yvette Lind, Prof Dr
Professor of Law at BI Norwegian Business School; Head of BI Tax and Fiscal Policy Centre at BI Norwegian Business School

Mari Margil, MPP (Harvard Kennedy School of Government)
Executive Director, Center for Democratic and Environmental Rights

Annette Mehlhorn, Dr
Postdoctoral researcher in the project 'Rights of Nature without Biocultural rights? – Investigating the conflicting dimensions of ecocentrism', Max Planck Fellow Group 'Environmental Rights in Cultural Context' (ERCC), Max Planck Institute for Social Anthropology, Halle (Saale)

Uapukun Mestokosho
Young leader from the Innu community of Ekuanitshit, located on the North Shore of Quebec, Canada; considered guardian of the Land and Waters, emerging filmmaker and former candidate for Quebec Solidaire political party in the provincial electoral district of Duplessis

Ralf Michaels, Prof Dr, LL M (Cambridge)
Director at the Max Planck Institute for Comparative and International Private Law in Hamburg; Chair of Global Law at the Queen Mary University in London; Professor at the University of Hamburg

Phillip Paiement, Prof Dr
Professor of Law and Governance in the Anthropocene at Tilburg University, the Netherlands

Alexandra Popovici, Prof
Professeuse et directrice des programmes de common law et droit transnational, Faculté de droit, Université de Sherbrooke

Manjeri Subin Sunder Raj, Dr PhD, LL M, BA, LLB (Hons)
Associate Professor of Law, CHRIST University; expert member, Harmony with Nature, United Nations

Janina Reimann
Research Assistant and PhD candidate in the Section 'Just Transitions' at the Kassel Institute for Sustainability of the University of Kassel, Germany; part of the research project 'Nature as a Legal Entity', funded by the German Research Foundation

Konstantinos Rokas, Assist. Prof Dr, LL M (National and Kapodistrian University of Athens), LL M (University Paris II Panthéon-Assas)
Currently Assistant Professor of Private International Law and of Law of International Transactions, Aristotle University of Thessaloniki Faculty of Law; formerly Lecturer, University of Nicosia, Cyprus

Stéphanie Roy, Prof, LLD (ULaval), LL M (McGill)
Assistant professor at the University of Sherbrooke

Tajudeen Sanni, Dr
Associate Professor and Head of the Center for Transformative Environment Law, Faculty of Shariah and Law, Villa College, Maldives; Uganda Pentecostal University

Mihnea Tănăsescu, Prof Dr
Assistant Professor, National Fund for Scientific Research (FNRS) and University of Mons, Belgium

Mara Tignino, PhD (Geneva Graduate Institute), HDR (University of Jean Moulin Lyon 3)
Senior Lecturer at the Faculty of Law and the Global Studies Institute, University of Geneva, and Senior International Law Specialist, Geneva Water Hub

Catalina Vallejo Piedrahíta, PhD
Professor of Law at Eafit University; country case researcher for the Riverine Rights Project (OsloMet); PI Ecological Transitional Justice Project (Eafit University); member of LawTransform, Norway

Yenny Vega Cárdenas, PhD
Water Law expert, Attorney at Law, Quebec's Bar (Canada); Research Advisor at the University of Sherbrooke; President of the International Observatory on Nature's Rights; expert member, Harmony with Nature, United Nations

Tore Vetter
Administrative Lawyer in Hamburg

Antonio Carlos Wolkmer, Prof Dr, UFSC (Brazil)
PhD in Law; Emeritus Professor and Retired Titular Professor at Universidade Federal de Santa Catarina; Professor of Postgraduate Law Courses (Master's and Doctorate) at Universidade La Salle and others

Abbreviations

AASTGB	Advances in Animal Science, Theriogenology, Genetics and Breeding
ABA J	American Bar Association Journal
ABT	Authority for Social Control and Inspection of Forests and Lands
ACHPR	African Commission on Human and Peoples' Rights
ADI	Ação Direta de Inconstitucionalidade [Brazil]
AELA	Australian Earth Laws Alliance
AER	Australian Environment Review
AFRICE	African Institute of Culture and Ecology
AIDA	Interamerican Association for Environmental Defense
AIR	All India Reporter
AJAM	Jurisdictional Administrative Mining Authority
AJCL	American Journal of Comparative Law
AJIL	American Journal of International Law
Ala L Rev	Alabama Law Review
ANAPO	Asociación de Productores de Oleaginosas y Trigo
ANARDE	Advocates for Natural Resources and Development
Annu Rev Law Soc Sci	Annual Review of Law and Social Science
APA	Administrative Procedure Act
APJEL	Asia Pacific Journal of Environmental Law
APuZ	Aus Politik und Zeitgeschichte
Arch phil droit	Archives de philosophie du droit
BEK	bekendtgørelse (ministerial decree laying down more detailed rules for the implementation of a law) [Denmark]
BG	Bundesgericht
BGE	Entscheidungen des Schweizerischen Bundesgerichts
BNatSchG	Bundesnaturschutzgesetz
BPAs	Bisphenol A
BT-Drs.	Bundestagsdrucksache
Buff Env LJ	Buffalo Environmental Law Journal
BVerfG	Bundesverfassungsgericht (Federal Constitutional Court) [Germany]
BVerwG	Bundesverwaltungsgericht
CAINCO	Cámara de Industria, Comercio, Servicios y Turismo de Santa Cruz
Can J L Soc	Canadian Journal of Law and Society
CAO	Cámara Agropecuaria del Oriente
CBD	United Nations Convention on Biological Diversity
CC	Constitutional Court

CCQ	Civil Code of Québec
CDER	Center for Democratic and Environmental Rights
CEBEM	Centro Boliviano de Estudios Multidisciplinarios
CEDIB	Centro de Documentación e Información Bolivia
CEDLA	Centro de Estudios para el Desarallo Laboral y Agrario
CEJIS	Centro de Estudios Jurídicos e Investigación Social
CEJIS	Centro de Estudios Jurídicos e Investigación Legal
CELDF	Community Environmental Legal Defense Fund
CFREU	Charter of Fundamental Rights of the European Union
CICAJ	Centro de Investigación, Capacitación y Asesoría Jurídica (Pontificia Universidad Católica del Perú)
CIDOB	Confederación de Pueblos Indígenas Del Oriente Boliviano
CIPCA	Centro de Investigación y Promoción del Campesinado
CJEU	Court of Justice of the European Union
CJLACS	Canadian Journal of Latin American and Caribbean Studies/ Revue canadienne des études latino-américaines et caraïbes
CLJ	Cambridge Law Journal
CLR	Cyprus Law Reports
CODEPANAL	Committee for the Defence of National Heritage
Comp Pol Stud	Comparative Political Studies
CONAIE	Confederation of Indigenous Nationalities of Ecuador
CONALCAM	Coordinadora Nacional por el Cambio
CONAMAQ	Consejo Nacional de Ayllus y Markas del Qallasuyu
CONTIOCAP	National Coordinator for the Defence of Indigenous Peasant Territories and Protected Areas of Bolivia
COP21	2015 Paris Agreement on Climate Change
CPEM-B	Central de los Pueblos Étnicos Mojeños de Beni
CQLR	Québec Laws and Regulations
CSUTB	Confederación Sindical Única de Trabajadores Campesinos de Bolivia
CUP	Cambridge University Press
DIIS	Danish Institute for International Studies
DÖV	Die öffentliche Verwaltung
DPSP	Directive Principle of State Policy [India]
Duke UP	Duke University Press
EAJENR	East African Journal of Environment and Natural Resources
EALS	East Africa Law Society
ECHR	European Convention on Human Rights
ECLAC	Economic Commission for Latin America
Ecology LQ	Ecology Law Quarterly
ECtHR	European Court of Human Rights
EESC	European Economic and Social Committee
eg	exempli gratia
ELJ	European Law Journal
ELR	Environmental Law Reporter
ENC	Environment Court [New Zealand]
EnvAct	Act on the Environment
EPA	Environmental Protection Act

EPI	Environmental Performance Index
EPL	Environmental Policy and Law
ERLACS	European Review of Latin American and Caribbean Studies
ERPL	European Review of Private Law
EU	European Union
EurUP	Zeitschrift für Europäisches Umwelt- und Planungsrecht
FAN	Fundación Amigos de la Naturaleza
FCC	Federal Constitutional Court [Germany]
FEGA-SACRUZ	Federación Departamental de Ganaderos de Santa Cruz
FNQLSDI	First Nations of Quebec and Labrador Sustainable Development Institute
FONABOSQUE	National Forestry Development Fund
Fordham Env L Rev	Fordham Environmental Law Review
GARN	Global Alliance for the Rights of Nature
Geo Envt'l L Rev	Georgetown Environmental Law Review
Geo Int'l Envt'l L Rev	Georgetown International Environmental Law Review
Ger YBIL	German Yearbook of International Law
GLJ	German Law Journal
GLR	Griffith Law Review
GO	Gaceta Oficial
Harv Envt'l L Rev	Harvard Environmental Law Review
Harv Int LJ	Harvard International Law Journal
Harv UP	Harvard University Press
IACHR	Inter-American Court of Human Rights
IBAMA	Institute for the Environment and Renewable Natural Resources
IJGLS	Indiana Journal of Global Legal Studies
IJPL	International Journal of Procedural Law
ILM	International Legal Materials
INRA	National Agrarian Reform Institute
Int JL Con	International Journal of Law in Context
IPBES	Intergovernmental Science-Policy Platform on Biodiversity and Ecosystem Services
IPSP	Political Instrument for the Sovereignty of the People
IUCN	International Union for Conservation of Nature
IVN	Instituut voor Natuureducatie en Duurzaamheid
J Legal Plur	Journal of Legal Pluralism and Unofficial Law
JCL	Journal of Comparative Law
JED	Journal of Environment and Development
JEEPL	Journal for European Environmental and Planning Law
JEL	Journal of Environmental Law
JELP	Journal of Environmental Law and Policy
JELS	Journal of Empirical Legal Studies
JEM	Journal of Environmental Management
JEP	Special Jurisdiction for Peace

JHRE	Journal of Human Rights and the Environment
JL Soc	Journal of Law and Society
JLA	Journal of Legal Anthropology
JMS	Journal of Mountain Science
JURA	Juristische Ausbildung
JuS	Juristische Schulung
KJ	Kritische Justiz
KritV	Kritische Vierteljahresschrift für Gesetzgebung und Rechtswissenschaft
L&CP	Law & Contemporary Problems
LACES	Latin American and Caribbean Ethnic Studies
LBK	Statutory instrument acting as a compilation of a laws or statute with subsequent amendments [Denmark]
LCH	Law, Culture and the Humanities
LEAD J	Law, Environment and Development Journal
Leiden JIL	Leiden Journal of International Law
LJ	Law Journal
LLC	Limited Liability Company
MAS	Movement Towards Socialism
McGill LJ	McGill Law Journal
MDG	Miljøparti de Grønne [Norway]
Mich J Int L	Michigan Journal of International Law
MJ	Maastricht Journal of European and Comparative Law
MLR	Modern Law Review
MLRA	Marine Living Resources Act
MMAYA	Ministry of Environment and Water
NAD-DIS	Nuovi Autoritarismi e Democrazie: Diritto, Istituzioni, Società
NAPE	National Association of Professional Environmentalists
Nat Med	Nature Medicine
NCHA	Federal Act on the Protection of Nature and Cultural Heritage [Switzerland]
NDA	Nature Diversity Act
NEA	National Environment Act No 5 of 2019 [Uganda]
NEM:BA	National Environmental Management: Biodiversity Act 10 of 2004
NEM:PAA	National Environmental Management: Protected Areas Act 57 of 2003
NEMA	National Environmental Management Act 107 of 1998 [South Africa]; National Environment Management Authority [Uganda]
NGO	Non-governmental organization
NH Rev Stat	New Hampshire Revised Statutes
NHPA	National Historic Preservation Act
NJW-RR	Neue Juristische Wochenschrift, Rechtsprechungs-Report
NRDC	Natural Resources Defense Council
NVwZ	Neue Zeitschrift für Verwaltungsrecht
NZEnvC	New Zealand Environmental Court

NZHC	New Zealand High Court
NZLR	New Zealand Law Reports
NZRMA	New Zealand Resource Management Act Court
NZTPA	New Zealand Town Planning Appeal Authority
OAU	Organization of African Unity
OIDN	International Observatory of the Rights of Nature
OJLS	Oxford Journal of Legal Studies
OUP	Oxford University Press
P&H	Punjab and Haryana [India]
Pace Envtl L Rev.	Pace Environmental Law Review
PCA	Pollution Control Act
PCBs	Polychlorinated biphenyls
Penn St JL & Int'l Aff	Penn State Journal of Law and International Affairs
PFAs	Per- and polyfluoroalkyl substances
PIL	Public Interest Litigation [India]
PNCC	National Climate Change Program
PoLAR	Political and Legal Anthropology Review
PUP	Princeton University Press
QCCA	Court of Appeals of Quebec
QCCS	Superior Court of Quebec
RabelsZ	Rabels Zeitschrift für ausländisches und internationales Privatrecht
RCM	Regional County Municipality
RECIEL	Review of European Community and International Environmental Law
RGD	Revue générale de droit
RJEUL	Revue juridique des étudiants et étudiantes de l'Université Laval
RJPA	Romanian Journal of Public Affairs
RLRQ	Recueil des lois et des règlements du Québec
RMA	Resource Management Act 1991 [New Zealand]
RoC	Republic of Cyprus
RoN	Rights of Nature
RTP	Radio Televisión Popular
S Cross U L Rev	Southern Cross University Law Review
SAC	Supreme Administrative Court
SCal LRev	Southern California Law Review
SCC	Supreme Court of Canada; Supreme Court Cases [India]
SCR	Supreme Court Reports [Canada]
SERNAP	National Protected Areas Service
SNST	sacred natural sites and territories
STC Exp	Sentencia del Tribunal Constitucional, Expediente
STF	Supreme Federal Court
Sy J Int L	Syracuse Journal of International Law

TEL	Transnational Environmental Law
TFI	Tribunal de première instance
TIPNIS	Isiboró Sécure National Park and Indigenous Territory
TLCP	Transnational Law and Contemporary Problems
Trans Inst Br Geogr	Transactions of the Institute of British Geographers
TRNC	Turkish Republic of Northern Cyprus
Tulane Env LJ	Tulane Environmental Law Journal
U Ark Little Rock L Rev	University of Arkansas at Little Rock Law Review
U Ill L Rev	University of Illinois Law Review
U Pa Asian L Rev	University of Pennsylvania Asian Law Review
U Pa JIL	University of Pennsylvania Journal of International Law
UCP	University of Chicago Press
UDHR	Universal Declaration of Human Rights
UKPC	United Kingdom Privy Council
ULR	Utah Law Review
UmwRG	Umweltrechtsbehelfsgesetz
UN	United Nations
UNDP	United Nations Development Program
UNEP-WCMC	United Nations Environment Program World Conservation Monitoring Center
UNESCO	United Nations Educational, Scientific and Cultural Organization
UNGA	United Nations General Assembly
UNHRC	United Nations Human Rights Council
UNITAS	Unión Nacional de Instituciones para el Trabajo de Acción Social
UP	University Press
US	United States
USA	United States of America
UW Press	University of Wisconsin Press
Va L Rev	Virginia Law Review
Val U L Rev	Valparaiso University Law Review
VRÜ	Verfassung und Recht in Übersee
Vt JEL	Vermont Journal of Environmental Law
Vt L Rev	Vermont Law Review
VwGO	Verwaltungsgerichtsordnung
Wai	Waitangi Tribunal's Wai Numbering System
Wash Int LJ	Washington International Law Journal
WCC	World Conservation Congress; Western Cape High Court, Cape Town
Widener L Rev	Widener Law Review
Wis L Rev	Wisconsin Law Review
ZaöRV	Zeitschrift für ausländisches öffentliches Recht und Völkerrecht
ZAWCHC	High Court of South Africa Western Cape
ZfRSoz	Zeitschrift für Rechtssoziologie – The German Journal of Law and Society
ZUR	Zeitschrift für Umweltrecht

I. General Reports

The Globalization of Rights of Nature

Global Legal Pluralism and the Rights of Nature

Daniel Bonilla Maldonado[*]

I. Introduction

Nature is a subject of rights. The legal systems of countries as dissimilar as Ecuador, Bolivia, New Zealand, the United States, and Uganda have recognized this new legal subject and have granted nature rights such as the right to life, the right to the functioning of its vital cycles, and the right to restoration of its processes.[1] Nature as a subject of rights therefore opposes the nature-object perspective that

[*] I would like to thank Ralf Michaels for the many useful and insightful comments that he made to previous versions of this article. I would also like to thank Mihnea Tănăsescu, Cletus Barié, Mari Margil, and Mara Tignino for their valuable comments. I presented earlier or shorter versions of the article at the Max Planck Institute for Comparative and Private International Law, the Legal Heterodoxies Workshop organized by New York University, Universidad de Buenos Aires, the University of Brescia School of Law, the Max Planck Institute for Legal History and Legal Theory, and the International Academy of Comparative Law Ankara Roundtable on Environment, Climate Change and Constitutionalism. I would like to thank the students and professors that participated in these events. Their comments, questions, and critiques were very useful in strengthening the article.

This manuscript was finalized in January 2024. The assessment of literature and case law, including internet materials, reflects the legal situation at that time.

[1] Mihnea Tănăsescu, 'Rights of Nature, Legal Personality, and Indigenous Philosophies' (2020) 9 TEL 429.

has historically been dominant in the West. Rights of nature challenge the notion of nature as an object that humans should describe, understand, and master; as an instrument that exists solely to satisfy human needs; and as property that humans can exploit without limit. Rights of nature thus challenge the objectification of nature, its absolute instrumentalization, and the anthropocentrism and extractivism with which science, religion, economics, and law in the West have typically described, evaluated, and interacted with nature.[2] As regards the rights of nature, these ideas have not only had discursive consequences, they have also had practical consequences. These ideas are some of the main causes of the contemporary global environmental crisis. The rights of nature therefore emerge as a discursive and practical model that aims to rethink the concept of nature, as well as the relationship between nature and humans, so that we can effectively address the radical environmental degradation we are currently experiencing.[3]

The rights of nature are now a global discursive and practical pattern.[4] Between 2006, the year in which the rights of nature took precise legal form, and June 2021, we can find 409 initiatives related to the rights of nature around the world. Of these initiatives, 66.5% recognize rights of nature as a whole, 17.8% the rights of rivers or other aquatic ecosystems, and 10.5% the rights of animals.[5] Thirty-nine countries account for 90% of all initiatives.[6] However, the vast majority of the initiatives, 80%, were submitted in the Americas, and 91% of the legal texts were written in English or Spanish. The biggest group among all initiatives, 38%, is made up of municipal legal provisions.[7]

We can divide this global discursive and practical pattern into three parts. First, there are the paradigmatic articulations of the discourse. The prototypical models of the rights of nature emerge from two countries in the Global South, Ecuador and Bolivia, and from a peripheral country in the Global North, New Zealand. Second, there are discursive and practical patterns that reproduce the conceptual structures of the prototypical discourse, although they may vary or complement them, such as the perspectives on the rights of nature that have been embraced by Argentina, Panama, India, Pakistan, and the international human rights system.

---

[2] Michel Serres, *The Natural Contract* (Michigan UP 1995); and Bruno Latour, *Politics of Nature* (Harv UP 2004).

[3] Rafi Youatt, 'Personhood and the Rights of Nature: The New Subjects of Contemporary Earth Politics' (2017) 11 International Political Sociology 39, 43–45.

[4] Craig M Kauffman and Pamela L Martin, 'Constructing Rights of Nature Norms in the US, Ecuador, and New Zealand' (2018) 18(1) Global Environmental Politics 43, 43–44.

[5] Alex Putzer and others, 'Putting the rights of nature on the map. A quantitative analysis of rights of nature initiatives across the world' (2022) 18(1) Journal of Maps 89, 92. 66% of the initiatives offer a specific institutional design for a materialization of the rights of Mother Earth, while 33% of these initiatives are silent on the in terms of the necessary institutional structures.

[6] Putzer and others (n 5) 90.

[7] Putzer and others (n 5) 91.

Finally, we have the discourse and practices that have opposed the rights-of-nature discourse, as encountered in most European countries.[8]

The discursive and practical pattern of rights of nature has been analyzed from a variety of academic perspectives. The rights of nature are an object of study that has firmly positioned itself in areas as diverse as constitutional law, environmental law, legal and political theory, and the social sciences. Constitutional law scholars have examined the contents of political charters that, like the Ecuadorian or Bolivian ones, recognize nature as a subject of rights or that recognize principles, such as good living, that intersect with this type of subject of rights.[9] Environmental law scholars have analyzed the tensions that exist between prevailing environmental regimes and the rights of nature.[10] Legal or political theorists have been concerned with studying the conceptual architecture of these rights or their foundations.[11] Finally, social scientists are beginning to investigate the efficacy of the rights-of-nature discourse.[12]

---

[8] Most countries in Asia and Africa have been indifferent to the rights of nature. In the vast majority of the countries that make up these continents, rights of nature have not been an object of systematic and continuous public discussion, and there have been no rights of nature initiatives introduced. Putzer and others have produced the most complete quantitative analysis of rights of nature presently available. In Map 2 of their article they offer a powerful image of the regions and countries where rights of nature initiatives have been introduced. In grey are the countries and regions where no rights of nature initiative have been introduced. Africa and Asia are almost all grey. See the map at Putzer and others (n 5) 92. Also see, Oluwabusayo Temitope Wuraola, 'The Legal Rights of Natural Entities: African Approaches to the Recognition of Rights of Nature' in Michael Addaney and Ademola Oluborode Jegede (eds), *Human Rights and the Environment under African Union Law* (Palgrave Macmillan 2020). The case of South Africa is particularly notable in that some of its scholars and activists, such as Cormac Cullinan, have been very influential in articulating the foundations of the rights of nature, as well as promoting them internationally. However, South Africa has not legally recognized nature as a subject, and there has been no proposal for such a thing to happen. See Louis J Kotzé, 'South Africa. Protecting Nature through an Anthropocentric-Oriented Rights Paradigm' in this volume, 367ff. Cullinan has written some well-known and well-discussed books on rights of nature. See, for example, Cormac Cullinan, *Wild Law: A Manifesto for Earth Justice, Green Books* (Chelsea Green Publishing 2011).

[9] Daniel Bonilla Maldonado, 'El constitucionalismo radical ambiental y la diversidad cultural en América Latina. Los derechos de la naturaleza y el buen vivir en Ecuador y Bolivia' (2018) 42 Revista Derecho del Estado 3.

[10] Eduardo Gudynas, 'La ecología política del giro biocéntrico en la nueva Constitución de Ecuador' (2009) 32 Revista de Estudios Sociales 34.

[11] Minheau Tănăsescu, *Environment, Political Representation, and the Challenge of Rights: Speaking for Nature* (Palgrave Macmillan 2016).

[12] María Valeria Berros, 'Challenges for the Implementation of the Rights of Nature: Ecuador and Bolivia as the First Instances of an Expanding Movement' (2021) 48 Latin American Perspectives 192; Craig M Kauffman and Pamela L Martin, 'Can Rights of Nature Make Development More Sustainable? Why Some Ecuadorian Lawsuits Succeed and Others Fail' (2017) 92 World Development 130.

This article aims to contribute to the understanding of the rights of nature from a different perspective: comparative law. More precisely, this article aims to contribute to the description, analysis, and comparison of the discursive patterns that convey and underpin the rights of nature. It also aims to analyze these discursive patterns from the perspective of global legal pluralism.[13] To meet these objectives, I divide this article into two main parts.

In the first part (II.), I describe, analyze, and compare three types of discourse related to the rights of nature: Initially, I examine their prototypical models, that is, the models emerging in Bolivia, Ecuador, and New Zealand. Then, I study discourse on the rights of nature that reproduce the conceptual structures of the paradigmatic models, including those emerging in the international human rights system and the legal systems of Colombia and India. Finally, I explore discourse that resists rights of nature, such as literature emerging in the legal systems of much of Western Europe. This section of the article therefore divides discursive patterns related to the rights of nature into poietic, mimetic, and resistance patterns.

In the second and longer part of this article (III.), I analyze rights of nature from three perspectives that are central to contemporary comparative law: the political economy of legal knowledge, legal pluralism, and explanatory theories of legal change. These perspectives revolve around three questions that are of particular interest for a full understanding of rights of nature. The political economy of legal knowledge investigates the discursive and practical patterns that determine where legal knowledge is (and should be) created and who can (and should) create, use, and transfer it.[14] Legal pluralism is concerned with the description and examination of the characteristics and interactions between legal systems within a single state or between national legal systems and international law.[15] Explanatory theories of legal change account for the variables and processes that motivate transformations of legal orders.[16] I therefore divide this second part of the report into three sections.

In the first section (III. 1), I explore how rights of nature challenge the dominant political economy of legal knowledge. Rights of nature are culturally hybrid, as well as epistemologically, politically, and legally heterodox. Rights of nature, and particularly their paradigmatic forms, have been articulated in countries which have historically been considered poor or marginal to the creation of original legal knowledge and which draw on knowledge – such as that of indige-

[13] Ralf Michaels, 'Global Legal Pluralism' (2009) 5 Annual Review of Law and Social Science 243.

[14] Daniel Bonilla Maldonado, 'Economía política del conocimiento jurídico' (2015) 2 Brazilian Journal of Empirical Legal Studies 26; Daniel Bonilla Maldonado, 'The Political Economy of Legal Knowledge', in Daniel Bonilla and Colin Crawford (eds), *Constitutionalism in the Americas* (Elgar Publishing 2018).

[15] John Griffiths, 'What is Legal Pluralism?' (1986) 24 J Legal Plur 1.

[16] Alan Watson, 'Comparative Law and Legal Change' (1978) 37 CLJ 313.

nous peoples – that has usually been excluded from state law-making processes. Likewise, the contents of the rights of nature question some of the central premises of modern law, eg who can be a subject of rights and what rights can be granted to non-human entities.

In the second section (III.2), I analyse the rights of nature from the perspective of external legal pluralism.[17] More precisely, I explore the interactions that have taken place between international law and domestic law with respect to the rights of nature. We cannot accurately describe and analyze rights of nature using the conceptual lenses through which external legal pluralism is typically examined. These dominant perspectives generally focus on the impact of international law on national rights and the ways in which these legal systems transform, use or, oppose international law.[18] Occasionally, these perspectives also examine how national legal orders influence international law, although these analyses generally focus on how a few legal systems in the Global North – in the US, French, or German systems, for example – have influenced international law.[19] Rights of nature are not a consequence of the impact of international law on national law. To the contrary, the legal systems of countries of the Global South, such as Ecuador and Bolivia, or of peripheral countries of the Global North, such as New Zealand, have influenced and partially transformed international law. In the third and final section (III.3), I study how national legal systems have included rights of nature within their structures through a process of cross-fertilization that involves not only North-South processes of knowledge exchange, but also heterodox processes of South-South and South-North exchange. Section IV summarizes my findings in a brief conclusion.

I base the theses offered in this article on twenty-two national reports and six special reports that were written for the 2022 Asunción Conference of the International Academy of Comparative Law.[20] The two types of reports are a consequence of a set of questions that Ralf Michaels and I drafted in our capacity as general rapporteurs for the project. This set of questions touches on central dimensions of the discourse and practice regarding rights of nature, eg their con-

---

[17] Daniel Bonilla Maldonado and Libardo José Ariza, 'El Pluralismo jurídico: contribuciones, debilidades y retos de un concepto polémico', in Brian Tamanaha and others (eds), *Pluralismo Jurídico* (Siglo del Hombre Editores 2007).

[18] Sally Engle Merry, 'Human Rights and Global Legal Pluralism: Reciprocity and Disjuncture' in Franz von Benda-Beckmann, Keebet von Benda-Beckmann and Anne Griffiths (eds), *Mobile People, Mobile Law* (Ashgate Publishing 2005); Sally Engle Merry, 'Anthropology and International Law' (2006) 35 Annual Review of Anthropology 99; Sally Engle Merry, 'From Law and Colonialism to Law and Globalization' (2006) 28 Law and Social Inquiry 569; Helen Quane, 'Legal Pluralism and International Human Rights Law: Inherently Incompatible, Mutually Reinforcing or Something in Between?' (2013) 33 OJLS 675.

[19] Anthony Clark Arend, 'The Evolution of International Law', in JR McNeill and Kenneth Pomeranz (eds), *The Cambridge World History* (CUP 2015).

[20] Twenty-eight reports were written by national rapporteurs. However, not all of these reports are included in this book.

tent, the institutions that have recognized them, their foundations, their effectiveness, and their relationship with international law.[21] The national reports, as their name suggests, focus on the various aspects that make up rights of nature in a particular legal system. The special reports, by contrast, focus on specific issues that we, as general rapporteurs for the project, consider key to a full understanding of rights of nature, including their philosophical underpinnings and their tensions with international law. The theses that I offer in this article also draw on other primary and secondary sources to complement the empirical and theoretical information provided in the national and special reports. Finally, the theses I present in this article are theoretically informed. The questions that guided the national and special reports, as well as the conceptual lenses that allowed for the interpretation of the empirical information they (and other sources) offer, fall within the following theoretical approaches: legal epistemology and epistemic justice; legal Thepluralism; and comparative explanatory theories of legal change.

II. Mapping the Global Legal Discourse on Rights of Nature

We can divide the global discourse on rights of nature into three types of patterns: poietic, mimetic, and resistance. Poietic patterns are those that create the constitutive elements of the discourse and those that articulate the structural components of the global discourse on rights of nature.[22] These poietic patterns emerge in the legal systems of Ecuador, Bolivia, and New Zealand.[23] The rights of nature literature widely considers the models constructed in these three countries as prototypical.[24] Mimetic patterns are those that reproduce the structures of poietic pat-

---

[21] The questionnaire is reproduced at the end of this book (p 639ff). It covers the following eight dimensions of rights of nature: existence and contents; legal and political operators; enforcement; theoretical sources and foundations; relationship between national legal systems; international law; efficacy; and consequences.

[22] Poieisis comes from the Greek term ποιεῖν, which means to create or to make.

[23] Gabriela Espinoza, 'Ecuador. The Nature of the Rights of Nature' in this volume, 211ff.; Cletus Gregor Barié, 'Bolivia. The Rights of Mother Earth: Fifteen Years of Solitude' in this volume, 247ff.; and Catherine Iorns, 'Aotearoa New Zealand. Protecting Nature through Upholding Indigenous Rights and Human Responsibilities' in this volume, 283ff.. The Tamaqua Borough Ordinance no 612 of 2006 (Pennsylvania, United States) was the first legal norm that conceptualized nature as a subject of rights. This legal norm influenced the drafting of the Ecuadorian constitutional norms on the rights of nature. However, this ordinance did not become a central element of the prototypical discourse on the rights of nature. I elaborate on these arguments in this article's section on external legal pluralism.

[24] See, on the one hand, the reports on these three countries in this volume: Espinoza, 'Ecuador' (n 23); Barié, 'Bolivia' (n 23); and Iorns, 'Aotearoa New Zealand' (n 23). On the other hand, see the reports from other parts of the world where the discourse on rights of nature articulated in Ecuador, Bolivia, or New Zealand is recognized as central in the global discourse on rights of nature; in this volume Mara Tignino, 'International Law. Rights of Nature and Rights to Nature: Emerging Trends' in this volume, 131ff.; Dirk Hanschel and Annette Mehlhorn,

terns, although sometimes these are partially transformed or some new contingent components are added to them. Countries in five continents have articulated mimetic patterns: in Africa, Uganda, and Nigeria; in the Americas, Colombia, Panama, Argentina, and Canada, for example; in Asia, India, Bangladesh, and Pakistan; in Europe, Spain; and in Oceania, Australia.[25] Finally, patterns of resistance are those that oppose the recognition of the rights of nature. In Europe, for example, up until 2022 no country had legally recognized rights of nature, and some social, political, or academic sectors opposed them actively. The exceptional and recent recognition in Spain of the Mar Menor as a subject of rights confirms this rule.[26]

1. Poietic Pattern

The poietic discourse on the rights of nature revolves around the following three components: subject, rights, and representation. The prototypical discourse constructed in Ecuador, Bolivia, and New Zealand creates a new legal subject, to whom rights are extended and whose representation is given to third parties, individuals, or institutions. Poietic patterns emerge with the following legal instru-

---

'Law and Anthropology. An Interdisciplinary Perspective on Rights of Nature' in this volume, 145ff.; Mari Margil, 'NGOs in the United States. Birth of a Movement' in this volume, 165ff.; Yenny Vega Cárdenas and Uapukun Mestokosho, 'Recognizing the Legal Personality of a River. The Magpie River/*Mutehekau Shipu* in Canada' in this volume, 177ff.; and Minhea Tănăsescu, 'Theoretical Foundations of the Rights of Nature. Origins and Conceptual Characteristics' in this volume, 97ff.; Domenico di Micco and Michele Graziadei, 'Italy. Establishing the Rights of Nature in Italy: Between Tradition and Change' in this volume, 493ff.; Sören Koch, Esmeralda Colombo and Catalina Vallejo Piedrahíta, 'Norway. Rights of Nature in the Norwegian Legal Culture: To Be or Not to Be?' in this volume, 539ff.; Laura Affolter and Sian Affolter, 'Switzerland. Mapping the Scene' in this volume, 611ff.; Phillip Paiement, 'Netherlands. Rights of Nature, Globalization, and Legal Pluralis' in this volume, 527ff.; Yvette Lind, 'Denmark. Could Danish Human- and Animal-Rights based Environmental Protection Provide Rights to Nature?' in this volume, 455ff.; Jakub Harašta, 'Czech Republic. An Anthropocentric Approach to Environmental Protection' in this volume, 441ff.; Stéphanie Roy, Gaële Gidrol-Mistral and Alexandra Popovici, 'Québec (and Canada). Nature, Culture, Rupture: Tools for Environmental Protection in a Liberal Anthropocentric Legal System' in this volume, 573ff.; Tania Luna Blanco and David de la Torre Vargas, 'Colombia. Graffiti on the Colombian Legal Architecture: The Emergence of Non-human Subjects and the Rights of Nature' in this volume, 319ff.. See also Stefan Grundmann, 'Globalisation et pluralisme juridique – Globalisation and Legal Pluralism. Orders of Pluralism and Rights' in Martin Schmidt-Kessel (ed), *German National Reports on the 21st International Congress of Comparative Law* (Mohr Siebeck 2022).

[25] See, for example in this volume, Tajudeen Sanni, 'Uganda. The First African Country to Recognize Rights of Nature' in this volume, 353ff.; Margil, 'NGOs' (n 24); Blanco and de la Torre, 'Colombia' (n 24); Congreso de la República de Panamá, Ley No 287 de 24 de Febrero de 2022 <https://www.gacetaoficial.gob.pa/pdfTemp/29484_A/90384.pdf>; and on Argentina, Fernando Solanas, Proyecto de ley nacional S-2506/15-2015 (2015); Fernando Solanas, Proyecto de ley nacional S-0615/19-2019 (2019); Fernando Solanas, Proyecto de ley nacional 6118-D-2020 (2020).

[26] Ley 19/2022, de 30 de septiembre, para el reconocimiento de personalidad jurídica a la laguna del Mar Menor y su cuenca, <https://www.boe.es/eli/es/l/2022/09/30/19>.

ments: the Ecuadorian Constitution of 2008,[27] the Bolivian Mother Earth Laws 71 of 2010[28] and 300 of 2012,[29] and the New Zealand Te Urewera Act of 2014 and Te Awa Tupua Act of 2017.[30] In these legal norms a new subject is created:

---

[27] The key components of the rights of nature discourse in Ecuador appear in the Preamble and art 71 of the 2008 Constitution. Preamble: 'Celebrating nature, Pacha Mama, of which we are a part and which is vital to our existence'; Article 71: 'Nature, or Pacha Mama, where life is reproduced and occurs, has the right to integral respect for its existence and for the maintenance and regeneration of its life cycles, structure, functions and evolutionary processes. All persons, communities, peoples and nations can call upon public authorities to enforce the rights of nature. To enforce and interpret these rights, the principles set forth in the Constitution shall be observed, as appropriate. The State shall give incentives to natural persons and legal entities and to communities to protect nature and to promote respect for all the elements comprising an ecosystem.' – <https://pdba.georgetown.edu/Constitutions/Ecuador/english08.html>. See also arts 72, 73, and 74.

[28] Law 71 of 2010 (Bolivia) – 'Article 1. (Object). The purpose of this Law is to recognize the rights of Mother Earth, as well as the obligations and duties of the Plurinational State and society to guarantee the respect of these rights.'

'Article 3. (Mother Earth). Mother Earth is the dynamic living system formed by the indivisible community of all life systems and living beings, interrelated, interdependent and complementary, which share a common destiny. Mother Earth is considered sacred, from the cosmovisions of the indigenous native peasant nations and peoples.'

'Article 5. (Legal Status of Mother Earth). For the purposes of the protection and tutelage of its rights, Mother Earth adopts the character of a collective subject of public interest. Mother Earth and all its components, including human communities, are holders of all the inherent rights recognized in this Law. The application of the rights of Mother Earth shall take into account the specificities and particularities of its various components. The rights established in this Law do not limit the existence of other rights of Mother Earth.' (Author's translations.)

[29] Law 300 of 2012 (Bolivia) – 'Article 5. (Definitions). For the purposes of this Law, the following definitions shall apply: 1) Mother Earth. It is the dynamic living system formed by the indivisible community of all life systems and living beings, interrelated, interdependent and complementary, which share a common destiny. Mother Earth is considered sacred; she nourishes and is the home that contains, sustains and reproduces all living beings, ecosystems, biodiversity, organic societies and individuals that compose her.'

'Article 4. (Principles).

1) Compatibility and Complementarity of Rights, Obligations and Duties.

One right cannot materialize without the others or cannot be above the others, implying the interdependence and mutual support of the following rights:

a) Rights of Mother Earth as a collective subject of public interest.' (Author's translations.)

[30] Te Awa Tupua (Whanganui River Claims Settlement) Act 2017 (New Zealand), <https://legislation.govt.nz/act/public/2017/0007/latest/DLM6831458.html>:

'Subpart 2 –Te Awa Tupua

Te Awa Tupua and Tupua te Kawa

12 Te Awa Tupua recognition

Te Awa Tupua is an indivisible and living whole, comprising the Whanganui river from the mountains to the sea, incorporating all its physical and metaphysical aspects.

13 Tupua te Kawa

Tupua te Kawa comprises the intrinsic values that represent the essence of Te Awa Tupua, namely – *Ko Te Kawa Tuatahi*

nature. In all three countries this new subject is structured around the following elements: nature is (i) a juridically autonomous subject; (ii) culturally hybrid; (iii) sacred and eternal; and (iv) conceived as a space that is both inhabited and constituted by a set of interdependent organic and inorganic entities. In these countries, however, what is conceived of as 'nature' has some variations.[31] In Ecuador and Bolivia, the new subject is nature as a whole, while in New Zealand the new subject is identified with what used to be a natural park, as well as with a river.[32]

The new legal subject has a life and interests of its own; it is a subject that is different from human beings, who also compose it, and a subject that acts according to its own dynamics and interests. This subject, moreover, is a rights-holder. In the case of Bolivia and Ecuador, nature is entitled to such rights as the right to life, to its vital cycles, and to have them reestablished in case they are disturbed.[33] In the case of New Zealand, parliament recognized Te Urewera and the Whanganui River as having the same rights as any other legal subject in the legal system.[34] This subject is also constructed from the intertwining of indigenous religious traditions (Andean and Māori) and modern Western legal traditions.[35] Indigenous peoples imagine nature as a mother, as the vital energy from which life emerges. Mother Nature has agency. Mother Nature has interests. Mother Nature is potentially responsible for her actions.[36] However, Pachamama and Papatūānuku intersect with the concepts of 'subject of rights' and 'rights' that are central to Western

---

(a) *Ko te Awa te mātāpuna o te ora*: the River is the source of spiritual and physical sustenance: Te Awa Tupua is a spiritual and physical entity that supports and sustains both life and natural resources within the Whanganui River and the health and well-being of the iwi, hapū and other communities of the River.

[...]

14 Te Awa Tupua declared to be legal person

(1) Te Awa Tupua is a legal person and has all the rights, powers, duties and responsibilities of a legal person.'

[31] Daniel Bonilla Maldonado, 'Los derechos de la naturaleza: su arquitectura conceptual' (2022) 4 Naturaleza y Sociedad: Desafíos Medioambientales 70.

[32] Constitution of 2008 (Ecuador), Preamble, arts 71, 72, 73, 74; Law 71 of 2010 of the Estado Plurinacional de Bolivia, arts 1, 3, 5; Law 300 of 2012 (Bolivia), art 5; and Te Awa Tupua Act 2017 (New Zealand), arts 12, 13, 14.

[33] Right to life: Law 71 of 2010 (Bolivia), art 7; right to restore its cycles: Constitution of the Republic of Ecuador of 2008, arts 71 and 72; right not be contaminated and to equilibrium: Law 71 of 2010 (Bolivia), art 7.

[34] Te Urewera Act 2014, art 11(1) and Te Awa Tupua Act 2017 (New Zealand), art 14(1).

[35] This argument is explored in more detail in Section III.1.a), below. It is important to consider the critiques offered against rights of nature, among others, that they essentialize indigenous communities; that they do not really represent the religious views of indigenous communities; that they homogenize the very diverse cultures of indigenous peoples; and that they can be used against indigenous groups' rights, for example, when rights of nature are used to restrict self-government rights. See, for example, Hanschel and Mehlhorn, 'Law and Anthropology' (n 24); and Tănăsescu, 'Theoretical Foundations' (n 24).

[36] Minhea Tănăsescu, *Understanding the Rights of Nature: A Critical Introduction* (Transcript Verlag 2022) 62–69, 87–94.

legal modernity. Similarly, following the traditions of the Andean and Māori indigenous peoples, this new legal subject is imagined as a deity that escapes the finite time of humans.[37] The Ecuadorian Constitution, the Bolivian Mother Earth laws, and the New Zealand Te Urewera and Te Awa Tupua laws recognize the religious dimension of Mother Earth for indigenous Andean and Māori peoples.[38] This sacralized subject is also understood as a space that forms an internally heterogeneous whole whose components are interdependent and interact continuously.[39] Nature is imagined as both a subject and a space that is inhabited by all the organic and inorganic entities that exist around the world. Humans, moreover, are not separate from, opposed to, or destined to dominate nature; humans are nature and do not occupy a privileged place in the ecosystemic whole. Humans, though, need the other elements that make up nature to survive and have a full life. Consequently, nature is not understood as a being that must be conserved, that cannot be exploited in any way. Humans can exploit the other elements that make up nature to satisfy their needs and desires. Yet they cannot do so in a way that disrupts the ecosystemic balance or that ignores principles such as reciprocity and interdependence.

The nature-subject, nonetheless, cannot express its will by itself. It does not have a voice that humans can hear. The poietic discursive patterns of the rights of nature therefore create a system of representation for protecting the interests and rights of nature.[40] In the case of Ecuador, any person can represent nature to defend its rights;[41] in Bolivia, the Defensoría de la Madre Tierra, in particular, and the state, in general, should represent nature and protect its rights,[42] in New Zealand, parliament created two legal entities to represent the interests and rights of Te Urewera and the Whanganui River.[43] Humans (or their institutions), thus, have a duty to interpret and protect nature. The rights of nature (and their representatives), however, are in tension with other rights that are also part of the legal systems in which they are located, for example, the right to a healthy environment, the right to equality, and the right to sustainable development. These rights protect nature by appealing to an anthropocentric perspective (nature is defended to safeguard the interests of human beings) or they conceptualize nature as a resource that should be exploited to generate wealth that can be redistributed in the political community. Hence, Bolivia and Ecuador are currently described as

[37] Bonilla Maldonado (n 31).

[38] Constitution of 2008 (Ecuador), Preamble, arts 71, 72, 73, 74; Law 71 of 2012 of the Estado Plurinacional de Bolivia, arts 1, 3, 5; Law 300 of 2012 (Bolivia), art 5; and Te Awa Tupua Act 2017 (New Zealand), arts 12, 13, 14.

[39] Constitution of 2008 (Ecuador), Preamble, arts 71, 72, 73, 74; Law 71 of 2012 of the Estado Plurinacional de Bolivia, arts 1, 3, 5; Law 300 of 2012 (Bolivia), art 5; and Te Awa Tupua Act 2017 (New Zealand), arts 12, 13, 14.

[40] Tănăsescu, 'Theoretical Foundations' (n 24).

[41] Constitution of 2008 (Ecuador), Preamble.

[42] Law 71 of 2010 (Bolivia), art 10.

[43] Te Urewera Act 2014, part 2 and Te Awa Tupua Act 2017 (New Zealand), subpart 3.

neo-extractivist countries or countries committed to 'social extractivism'.[44] Similarly, in New Zealand there are questions about how to balance rights of nature and environmental laws, such as, the Resource Management Act of 1991, the statute 'that establishes New Zealand's legislative framework for decision-making and planning of the use of the environment'.[45]

2. Mimetic Patterns

The mimetic discourse that has emerged across the globe reproduces the poietic discourse on the rights of nature. The former iterates the structure of subject, rights, and representation that constitutes the spine of the latter.[46] This is the case for the rights of nature discourse articulated in countries such as Colombia, Canada, India, Spain, and Uganda. In this mimetic discourse, nature is recognized as a rights-holder subject that must be represented by particular persons or institutions. However, some of these discursive patterns partially transform or complement the prototypical nature-subject discourse. Three of these modifications are particularly relevant for understanding the global discourse on the rights of nature. On the one hand, we can note a secularization of the foundations of the nature-subject or a variation in its religious foundations. In countries such as Colombia, Panama, Spain, and the United States, legal recognition of the nature-subject does not appeal to religion, although in these countries legal and political operators accept that the prototypical discourse has as its source the religious traditions of some indigenous peoples.[47] Nature, in mimetic discourse, remains sacred. However, the sacredness of nature is reinterpreted as deserving exceptional respect given its inherent value, and, therefore, it is the duty of human beings not to harm it. Thus, for example, in the Atrato River opinion the Colombian Constitutional Court bases its declaration that the river is a subject of rights on what it called biocultural rights, through which the Court connects nature with cultural minorities.[48] The Court does not explicitly allude to the cultures of indigenous peoples in support of its decision to recognize rights held by the river.

---

44 Espinoza, 'Ecuador' (n 23); Barié, 'Bolivia' (n 23).

45 See Iorns, 'Aotearoa New Zealand' (n 23).

46 See, for example, Yenny Vega Cárdenas and Uapukun Mestokosho, 'Recognizing the Legal Personality of a River' (n 24); Roy, Gidrol-Mistral and Popovici, 'Québec (and Canada)' (n 24); Sanni, 'Uganda' (n 25); Blanco and de la Torre, 'Colombia' (n 24); Manjeri Subin Sunder Raj, 'India Sowing the Seeds of "Nature Rights"' in this volume, 341ff.; and Ley 19/2022, de 30 de septiembre, para el reconocimiento de personalidad jurídica a la laguna del Mar Menor y su cuenca, <https://www.boe.es/eli/es/l/2022/09/30/19>.

47 Margil, 'NGOs' (n 24); Blanco and de la Torre, 'Colombia' (n 24); Ley No 287 de 24 de Febrero de 2022, <https://www.gacetaoficial.gob.pa/pdfTemp/29484_A/90384.pdf>; and Ley 19/2022, de 30 de septiembre, para el reconocimiento de personalidad jurídica a la laguna del Mar Menor y su cuenca, <https://www.boe.es/eli/es/l/2022/09/30/19>.

48 Corte Constitucional de Colombia, Sentencia T-622 de 2016; and Blanco and de la Torre, 'Colombia' (n 24).

In the case of India, the recognition of the Ganges River or some glaciers as subjects of rights was based on Hinduism; the High Court of Uttarkaband turned to sacred writings of the majority religion in India to support the idea of the nature-subject.[49]

On the other hand, some mimetic discourse considers that the protection of nature can (and should) be achieved by using both rights of nature and environmental law.[50] Although conceptually these two types of discourse are incompatible, the former is based on an ecocentric perspective while the latter is based on an anthropocentric point of view; in political and legal practice, the argument goes, they can (and should) be used at the same time to address the serious environmental problems that human beings face today. The clearest example of this strategy for the protection of nature is the argumentation offered by the Colombian Constitutional Court in the Atrato River ruling.[51] In this judgment, the Court recognizes the river as a subject of rights and creates the institution of the 'river guardians' to protect it, following the structures of the prototypical discursive pattern. However, at the same time, the Court uses national and international environmental law to defend the new subject of rights.[52] The Court considers that nature can (and should) be protected for the inherent value it holds, but also for the instrumental value it has for human beings. We can also find this same reasoning in some interpretations of the international human rights system.[53]

Finally, the mimetic discourse that emerges in international law is beginning to vary the transfer and reinterpretation processes of poietic discourse. In the emergence and consolidation of prototypical discourse, it travelled between national legal systems or moved from national legal systems to international law.[54] Today, however, the international law discourse promoting the rights of nature is beginning to develop a process of top-down diffusion, moving from international law to national law.[55] This new process adds a new layer of legitimacy to the legal discourse on the rights of nature (benefiting from the legitimacy that international law enjoys in many countries), helps to entrench it as a global discourse, and contributes to weakening the objections that resistance discourse has raised against the idea of the nature-subject.

---

49 Raj, 'India' (n 46); and *Salim v State of Uttarakhand*, Writ Petition (PIL) no 126 of 2014 (5 December 2016 and 20 March 2017) <https://www.elaw.org/salim-v-state-uttarakhand-writ-petition-pil-no126-2014-december-5-2016-and-march-20-2017>.

50 Blanco and de la Torre, 'Colombia' (n 24) and Tignino, 'International Law' (n 24).

51 Corte Constitucional de Colombia, Sentencia T-622 de 2016.

52 Tignino, 'International Law' (n 24).

53 Tignino, 'International Law' (n 24).

54 See Section III, below.

55 Graziadei and Di Micco, 'Italy' (n 24) and Affolter and Affolter, 'Switzerland' (n 24).

3. Patterns of Resistance

Resistance discourse is that which explicitly or implicitly opposes rights of nature in the legal systems in which it circulates. This does not mean, of course, that in the legal systems where resistance discourse emerges or moves there are no rights of nature or there are no movements promoting or examining rights of nature. In Europe, for example, academia and social organizations have extensively discussed or promoted rights of nature.[56] Resistance discourse opposes rights of nature for the following reasons.

First, such discourse considers that the nature-subject is premised on a re-enchantment of the world –a retreat from the processes of rationalization, secularization, and scientization of the natural world that shape the modern project. This perspective identifies rights of nature with an animism that intersects with some religious perspectives rightly displaced by modernity.

Second, the rights of nature call into question central elements of the grammar of modern law. Recognizing nature as a subject controverts the idea that, as a general rule, only those individuals who have agency, who are autonomous, rational, and therefore responsible, can be considered subjects of rights. For the discursive resistance patterns, the rights of nature do not fit into the modern legal machinery that allows its legal systems to function efficiently.[57]

Third, modern legal systems already have the necessary tools to confront the environmental problems that are currently degrading ecosystems all over the world. National and international environmental laws are powerful instruments to confront the environmental degradation that characterizes the Anthropocene. The challenge lies in the application of existing environmental legal frameworks, not in the creation of new frameworks that contradict environmental law's premises and conclusions.[58]

---

[56] See, for example, Elena Ewering, Janina Reimann and Tore Vetter, 'Germany. Rights of Nature: A Paradigm Change for the German Legal System' in this volume, 463ff.; Graziadei and Di Micco, 'Italy' (n 24); Koch, Vallejo Piedrahita and Colombo, 'Norway' (n 24); Affolter and Affolter, 'Switzerland' (n 24); Paiement, 'Netherland' (n 24); Lind, 'Denmark' (n 24); and Harašta, 'Czech Republic' (n 24). Two recent cases are also useful to show how civil society in Europe is mobilizing for the recognition of rights of nature. First, the case of the Tavignano River in Corsica, see Le Monde, 'The Fight for Corsica's "Living" River' (*Le Monde*, 25 July 2022) <https://www.lemonde.fr/en/m-le-mag/article/2022/07/25/the-fight-for-corsica-s-living-river_5991334_117.html>; and second, the case of the Rhone river in Switzerland and France, see 'La Métropole et la Ville de Lyon veulent que le fleuve Rhône ait une personnalité juridique' (*ActuLyon*, 9 September 2021) <https://actu.fr/auvergne-rhone-alpes/lyon_69123/la-metropole-et-la-ville-de-lyon-veulent-que-le-fleuve-rhone-ait-une-personnalite-juridique_44767772.html> and <https://www.appeldurhone.org/>.

[57] Graziadei and Di Micco, 'Italy' (n 24); and Koch, Vallejo Piedrahita and Colombo, 'Norway' (n 24).

[58] Affolter and Affolter, 'Switzerland' (n 24); Attila (n 56); Paiement, 'Netherland' (n 24); Lind, 'Denmark' (n 24); and Harašta, 'Czech Republic' (n 24).

Fourth, the rights of nature go against the dominant geopolitics of legal knowledge. The premises from which this geopolitics starts indicate that European culture is the source of the central legal families in the history of law, civil law, and common law. Non-European legal systems, with a few exceptions such as the United States, are merely minor iterations of European legal systems. The implicit premise of some resistance discourse is thus that the Global South would have nothing to teach the Global North on legal matters.[59]

Fifth, and last, rights of nature are an obstacle for development and redistribution policies. They are a tool for unjustly limiting the exploitation of natural resources that are central for economic growth, a key component of development. Rights of nature, furthermore, would eliminate or radically limit extractive activities that are the sources of many jobs and tax revenue that is essential for funding state's social programs.[60]

III. Global Legal Pluralism

The global discourse on rights of nature has emerged, diffused, and transformed itself through processes connected to global legal pluralism.[61] In this section, I will examine the rights-of-nature discourse from the following dimensions: the political economy of legal knowledge, external legal pluralism, and cross-fertilization. The analysis presented below revolves around the prototypical models of rights of nature. However, where relevant, I will also include mimetic discourse on the rights of nature in my analysis.

1. The Political Economy of Legal Knowledge and the Rights of Nature

Rights of nature challenge the dominant political economy of legal knowledge in that they are (i) culturally hybrid; (ii) epistemologically subversive; and (iii) politically and legally emancipatory.

a) The rights of nature and cultural hybridity

The paradigmatic models of the rights of nature, namely the Bolivian, Ecuadorian, and New Zealand models, challenge the political economy of legal knowledge in that they are culturally hybrid.[62] These models combine the vocabulary and grammar of modern law with the religions of indigenous Andean and Māori peo-

---

[59] Bonilla Maldonado, 'The Political Economy' (n 14); and Daniel Bonilla Maldonado, 'Introduction' in Daniel Bonilla Maldonado (ed), *Constitutionalism of the Global South* (CUP 2013).

[60] See, for example, Koch, Vallejo Piedrahita and Colombo, 'Norway' (n 24) and Barié, 'Bolivia' (n 23).

[61] Keebet von Benda-Beckmann and Bertram Turner, 'Legal Pluralism, Social Theory, and the State' (2018) 50 J Legal Plur 255.

[62] Daniel Bonilla Maldonado, *Legal Barbarians: Identity, Modern Comparative Law and the Global South* (CUP 2021) 158–167.

ples. They are modern law and they are indigenous law; they simultaneously 'are' and 'are not'. The syncretism of these models calls into question the following central ideas of the dominant political economy of legal knowledge:[63] (i) there are rich cultures that can create legal knowledge and poor cultures that cannot; (ii) the Global North has rich cultures that enable the production of law while the Global South has weak cultures that do not enable the production of real law, only morality or politics; (iii) within the Global North there are some cultures that are at the origin of Western law – eg French, British, German – and other cultures that, for historical reasons (imperialism and settlement colonization, for example), simply reproduce those cultural and legal matrices, eg Australia and New Zealand; (iv) Global North law is the only (or the most powerful) legal standard and can be transferred globally; (v) Global South countries must transplant this single global legal standard to their jurisdictions if they want to construct true law; and (v) when the Global North law that has been either reproduced in the lesser cultures of the Global North or transplanted to the Global South mixes with local culture, it becomes contaminated. The purity of true law is tainted; it loses clarity and precision; and it becomes a culturally-illegitimate offspring that is conceptually and practically ineffective.[64]

The paradigmatic models of the rights of nature conceive of nature as an autonomous legal subject that has a set of rights.[65] This new legal subject thus draws conceptually on the grammar and vocabulary of modern law. Categories such as 'rights', 'subject of rights', 'agency', 'autonomy', and 'interests' are central to the construction of the nature-subject. These categories, moreover, are put into action following the principles and rules that regulate their uses, eg only subjects with agency or self-interest can be rights-holders. At the same time, however, the paradigmatic models of the rights of nature distance themselves from the grammar and vocabulary of modern law. As seen above, drawing on indigenous (Andean or Māori) religious, political, and legal traditions, these models anthropomorphize, sacralize, and interpret nature as a biocentric and ecosystemic conceptual space.[66]

The subject of rights that prototypical models of the rights of nature create, furthermore, must be interpreted in light of other legal principles and rules that are influenced by or recognize indigenous cultural traditions in order to be properly understood and applied. This set of rights and principles puts modern law even more strongly into question. In the Ecuadorian and Bolivian cases, this is shown

[63] Bonilla Maldonado, 'The Political Economy' (n 14); and Bonilla Maldonado, 'Introduction' (n 59).

[64] Bonilla Maldonado (n 62) chs 2, 3, and 4.

[65] Constitution of 2008 (Ecuador), Preamble, arts 71, 72, 73, 74; Law 71 of 2010 (Bolivia), arts 1, 3, 5; Law 300 of 2012 (Bolivia), art 5; and Te Awa Tupua Act 2017 (New Zealand), arts 12, 13, 14.

[66] Bonilla Maldonado (n 31).

by the principles of plurinationality, interculturality, and good living.[67] The first recognizes that each of these states is constituted not by a single culturally-homogeneous nation, as the classical nation-state model advocates, but by several nations that have cultural and legal traditions different from those upheld by state law and the dominant white/mestizo culture. The second principle prescribes that the basic structure of the state must be built through the continuous dialogue of all its constituent cultures. The cultures that constitute each state should not be understood as closed and isolated systems that repel or only occasionally interact with each other. The third principle shaped by the principles of relationality, interdependence, reciprocity, and balance apply not only among human beings, but also between them and the other components of nature.[68]

In the New Zealand case, the rights of the Whanganui River intersect with four values explicitly recognized by the Crown-Māori agreement on the Whanganui River basin and the Te Awa Tupua Act ratifying the agreement (the values are analogous to those intersecting with rights conferred upon the North Island Forest in the Te Urewera Act):[69] (i) 'Ko te Awa te mātāpuna o te ora: the River is a source of physical and spiritual sustenance'. This first value prescribes that Te Awa Tupua (the Whanganui River basin) is a physical and spiritual entity that ensures the sustenance of all natural resources and peoples inhabiting its banks; (ii) 'E rere kau mai te Awa nui mai te Kahui Maunga ki Tangaroa: the great River flows from the mountains to the sea.' This second value states that the Whanganui River basin is a living, indivisible whole from its source to its mouth; (iii) 'Ko au te Awa, ko te Awa ko au: I am the River and the River is me.' This third value affirms that the Māori people are interconnected and at the same time responsible for the Whanganui River basin; and (iv) 'Ngā manga iti, ngā manga nui e honohono kau ana, ka tupu hei Awa Tupua: the small and large streams that flow into one another form one River.' The latter value provides that the Whanganui River basin is a whole made up of many elements that aim to maintain and promote its wellbeing.

The rights of nature, in short, are a hybrid legal product that breaks with central rules and principles of the dominant political economy of legal knowledge. Rights of nature, first, demonstrate the cultural diversity that exists in contem-

[67] Constitution of 2008 (Ecuador), art 1; Constitution of 2009 (Bolivia) art 1. Also see Raúl Llasag Fernández, 'El sumak kawsay y sus restricciones constitucionales' (2009) 12 Revista de Derecho UASB 113.

[68] Carolina Silva Portero, '¿Qué es el buen vivir en la Constitución?' in Ramiro Ávila Santamaría (ed), *La Constitución del 2008 en el contexto andino. Análisis desde la doctrina y el derecho comparado* (Ministerio de Justicia y Derechos Humanos 2008); Josef Estermann, *Filosofía andina. Estudio intercultural de la sabiduría autóctona andina* (Abya-Yala 1989).

[69] Te Awa Tupua Act 2017 (New Zealand), art 13. Also see, Mike Boyes, 'Re-envisioning Nature from a New Zealand Māori perspective' in *European Institute for Outdoor Adventure Education and Experiential Learning Conference* (Ratece-Planica, Slovenia 2010) 1–7; and Catherina Iorns Magallanes, 'Māori Cultural Rights in Aotearoa New Zealand: Protecting the Cosmology That Protects the Environment' (2015) 21 Widener L Rev 273.

porary states. The rights of nature make explicit that cultural homogeneity within states is not a descriptive but a normative argument that generally leads to the marginalization of minority cultures.[70] Second, it makes clear the richness of non-Western cultures and their capacity to create valuable legal knowledge.[71] Third, it shows that this legal knowledge can be useful to other cultural communities and, therefore, that it can be exportable or used productively within the states in which it emerges. Fourth, it illuminates precisely the ways in which cultural hybridity is not a source of pollution but of legal creation and innovation.[72] Fifth, and finally, it blurs the distinction between original law and copied law that forms the basis of much of the dominant political economy of legal knowledge.[73]

b) The rights of nature and epistemic justice

The rights of nature are epistemically subversive because they are articulated and grounded by appealing to two sources of knowledge that have traditionally been considered poor for the generation of legal products.[74] The rights of nature face a dual historical epistemic injustice. The first is the exclusion or marginalization of the three legal systems that have created the prototypical models of the rights of nature – Bolivia, Ecuador, and New Zealand – from the global legal dialogue. These legal systems have been excluded or marginalized insofar as they have been qualified, generally a priori, as poor spaces for the creation of original legal products. The legal systems of these three countries, consequently, are not considered valuable objects of study by the transnational legal academy, their legal institutions are regarded as mere minor iterations of the law that has been creat-

[70] On rights of nature, cultural homogeneity, and the nation state, see Daniel Bonilla Maldonado, 'Beyond the State, Can State Law Survive the Twenty-First Century?' in Thomas Duve and Tamar Herzog (eds), *The History of Latin American Law* (CUP 2023) and Daniel Bonilla Maldonado, 'El constitucionalismo radical ambiental y la diversidad cultural en América Latina. Los derechos de la naturaleza y el buen vivir en Ecuador y Bolivia' (2018) 42 Revista Derecho del Estado 9. On cultural homogeneity and the nation state, see, Paul Hirst and Grahame Thompson, 'Globalization and the Future of the Nation State' (1995) 24 Economy and Society 408; Hans-Joachim König, 'Discursos de identidad, Estado-nación y ciudadanía en América Latina: viejos problemas, nuevos enfoques y dimensiones' (2005) 11 Historia y sociedad 18.

[71] Karl S Zimmerer, 'The Indigenous Andean Concept of "Kawsay", the Politics of Knowledge and Development, and the Borderlands of Environmental Sustainability in Latin America' (2012) 127 PMLA 600; Catherine Walsh, 'Afro and Indigenous Life– Visions in/and Politics. (De)colonial Perspectives in Bolivia and Ecuador' (2011) 18 Revista de estudios bolivarianos 49; Victoria Haidar and Valeria Berros, 'Entre el sumak kawsay y la "vida en armonía con la naturaleza": disputas en la circulación y traducción de perspectivas respecto de la regulación de la cuestión ecológica en el espacio global' (2015) 32 Revista THEOMAI 128; and Bonilla Maldonado (n 62) ch 5.

[72] Daniel Bonilla Maldonado, 'El constitucionalismo radical ambiental y la diversidad cultural en América Latina. Los derechos de la naturaleza y el buen vivir en Ecuador y Bolivia' (2018) 42 Revista Derecho del Estado 3.

[73] Bonilla Maldonado (n 62) ch 5.

[74] Miranda Fricker, *Epistemic Injustice: Power and the Ethics of Knowing* (OUP 2009).

ed in other legal systems, and their academic output is ignored or undervalued by the international nodes of legal research.[75] The second epistemic injustice with which the rights of nature struggle is the exclusion or marginalization of indigenous legal, moral, political, and religious knowledge from the processes of legal knowledge creation in modern states.[76]

This subsection of the article thus focuses on showing how the political economy of legal knowledge has closed its doors to two peripheral legal voices. The previous subsection, by contrast, focuses on the cultural hybridity of the rights of nature to show how the normative patterns that determine who can create, transfer, and legitimately use legal knowledge negatively evaluate non-Western cultures, reject their capability of creating legal knowledge, and disqualify any syncretic legal product. Yet the two sections are closely linked and they complement and nourish each other.

Countries of the Global South or peripheral countries of the Global North have created the paradigmatic formulations of the rights of nature.[77] Ecuador is the only country in the world that has recognized rights of nature constitutionally.[78] Bolivia was the first country to recognize the rights of nature in two laws of national scope that develop aspects of the Constitution that intersect with these types of rights: the recognition of Pachamama as a deity and the principle of good living.[79] New Zealand, finally, is the first country to enact two national laws related to the rights of nature that derive from the interpretation to be given to a country's founding treaty (the Treaty of Waitangi).[80] All three countries also offer an institutional design for the implementation of the rights of nature.[81] The countries of the Global North, by contrast, have not articulated a discourse on the rights of nature, and they have not included such rights in their legal systems or have only done so in local spaces that have little political, legal, or economic relevance.[82] Australia and Spain are the only exceptions to this rule.[83] In Africa, only Uganda and Nigeria have conferred rights to nature, and in Asia only India, Pakistan, and Bangladesh have done so.[84]

---

[75] Fricker (n 74) ch 1.

[76] Fricker (n 74) ch 1. This does not mean, of course, that the discourse on rights of nature has been constructed solely by indigenous peoples' traditions. The works of authors like Christopher Stone, Cormac Cullinan, Thomas Berry, and Godofredo Stutzin have also been influential in this process.

[77] See Section II, above.

[78] Constitution of 2008 (Ecuador), Preamble, arts 71, 72, 73, 74.

[79] Law 71 of 2010 (Bolivia), arts 1, 3, 5; Law 300 of 2012 (Bolivia), art 5.

[80] Te Awa Tupua Act 2017 (New Zealand), arts 12, 13, 14.

[81] See Section II, above.

[82] Ibid.

[83] Erin O'Donnell and Julia Talbot-Jones, 'Creating Legal Rights for Rivers: Lessons from Australia, New Zealand, and India' (2018) 23(1) Ecology and Society.

[84] O'Donnell and Talbot-Jones (n 83). The decision made by the Uttarakhand High Court in India has presently been stayed pending appeal to the Supreme Court.

Dominant legal epistemology usually imagines Ecuador, Bolivia, and New Zealand, however, as poor spaces for the creation of legal knowledge and as mimetic – and not poietic – countries in legal matters.[85] The dominant political economy of legal knowledge has historically considered the legal systems of Ecuador, Bolivia, and New Zealand to be offspring systems that reproduce the premises and contents of the parent systems with varying degrees of precision.[86] Ecuador and Bolivia, which belong to the civil tradition, reproduce the premises and fundamental contents of the French and German legal systems. New Zealand, which belongs to the common law tradition, reproduces the legal system of Great Britain. In the prevailing understanding of 20th century comparative law literature, largely focused on legal families, the Ecuadorian, Bolivian, and New Zealand legal systems are iterations of other legal systems. In this line of argumentation, it is necessary to examine the parent legal systems, and not the offspring, to understand the structures of the two families that make up Western law.[87] The legal systems of these three countries, therefore, are not relevant objects of study and do not generate original legal products of quality. The legal voices of these three countries are therefore excluded from or marginalized in the global legal dialogue.

In this case, however, Bolivia, Ecuador, and New Zealand have constructed a legal product that has been interpreted as an instrument that can help address the radical environmental deterioration being experienced around the world.[88] Politicians, academics, and activists, among others, have interpreted this legal product as one that can (and should) travel from the peripheral legal spaces in which it was created to the rest of the world, including the Global North. The 2008 Ecuadorian Constitution of 2008, the 2010 and 2012 Bolivian Mother Earth laws, and the 2014 Te Urewera and 2017 Te Awa Tupua Acts in New Zealand are exportable legal products. The contents of these legal artefacts could contribute to a rethinking of the relationships between humans, nature, law, and economy that have largely been the cause of the environmental crisis that the world is experi-

85 Mariana Pargendler, 'The Rise and Decline of Legal Families' (2012) 60 AJCL 1043.

86 R Daniel Kelemen and Eric C Sibbitt, 'The Globalization of American Law' (2004) 58 International Organization 103; John Henry Merryman, 'Comparative Law and Social Change. On the Origins, Style, Decline and Revival of the Law and Development Movement' (1977) 25 AJCL 457, 484–489; Kerry Rittich, 'The Future of Law and Development: Second-Generation Reforms and the Incorporation of the Social' in David M Trubek and Alvaro Santos (eds), *The New Law and Economic Development: A Critical Appraisal*, (CUP 2006); and Bonilla Maldonado, 'Introduction' (n 59).

87 David S Clark, 'The Idea of the Civil Law Tradition', in John Henry Merryman and David S Clark (eds), *Comparative and Private International Law* (Duncker and Humblot 1990); and John R Schmidhauser, 'Legal Imperialism: Its Enduring Impact on Colonial and Post-Colonial Judicial Systems' (1992) 13 International Political Science Review/Revue Internationale de Science Politique 321.

88 Eduardo Gudynas, 'La ecología política del giro biocéntrico en la nueva Constitución de Ecuador' (2009) 32 Revista de Estudios Sociales 34.

encing in the 21st century.[89] The Ecuadorian and Bolivian *legal barbarians*, or the New Zealand peripheral subjects of rights, are shown as autonomous legal subjects who have the capacity to create original juridical knowledge. Likewise, they are shown to be subjects whose identities are a consequence of the internally-heterogeneous and strong cultures which have constructed them and that they simultaneously help to construct.

The rights of nature are also epistemically innovative because they interweave the contents and interpretations of the state legal systems of Ecuador, Bolivia, and New Zealand with the religious traditions of the indigenous peoples that make up these three countries.[90] The knowledge of Andean, Aymara, and Quechua indigenous peoples, among others, and the knowledge of the Māori people, are included in state legal systems through the rights of nature. The state legal system hears and includes the voices of these cultural minorities. As indicated in the previous section, through the rights of nature, the Ecuadorian, Bolivian, and New Zealand legal systems incorporate into their structures central ideas from the religious and cultural traditions of the indigenous Andean and Māori peoples.[91]

This indigenous knowledge, it is important to note, is not incorporated into a part of the formal legal system that is applicable only to indigenous peoples. It is not a mechanism to enrich or complexify indigenous jurisdictions or indigenous peoples' self-governance rights. This knowledge is included in legal norms, ie the Constitution or statutes, that apply to all citizens.[92] The recognition of cultural diversity through the rights of nature and the recognition of the value of the historically-marginalized knowledge of indigenous peoples, is therefore transformative. Indigenous peoples' knowledge does not 'go up' to the legal order to 'come down' again to their own political and legal spaces, as norms to be applied only within the indigenous jurisdiction, for example. This knowledge 'goes up' into the legal system and 'comes down' as norms to be applied to all members of the political community. Furthermore, this knowledge, already intertwined with state law, as mentioned in the previous subsection, is reinterpreted as valuable knowledge that can be used to solve global, regional, or other states' prob-

[89] Eduardo Gudynas, 'La dimensión ecológica del buen vivir: entre el fantasma de la modernidad y el desafío biocéntrico' (2009) 4 Revista Obets 49; and Sarah A Radcliffe, 'Development for a Postneoliberal Era? Sumak Kawsay, Living Well and the Limits to Decolonisation in Ecuador' (2012) 43 Geoforum 241; Stefan Knauß, 'Conceptualizing Human Stewardship in the Anthropocene: The Rights of Nature in Ecuador, New Zealand and India' (2018) 31 Journal of Agricultural and Environmental Ethics 703.

[90] Cletus Gregor Barié, 'Nuevas narrativas constitucionales en Bolivia y Ecuador: el buen vivir y los derechos de la naturaleza' (2014) 59 Latinoamérica 9; Gary Brierley and others, 'A Geomorphic Perspective on the Rights of the River in Aotearoa New Zealand' (2019) 35 River Res Applic 1640.

[91] See Section III. 1. a), above.

[92] Karl S Zimmerer, 'Environmental Governance Through "Speaking Like an Indigenous State" and Respatializing Resources: Ethical Livelihood Concepts in Bolivia as Versatility or Verisimilitude?' (2015) 64 Geoforum 314.

lems.[93] Rights of nature, a hybrid legal product – a mixture of a child legal system and indigenous knowledge – is assessed as a relevant transnational academic object of study and as a legal product translatable to other state jurisdictions.[94]

The rights of nature are thus a materialization of epistemic justice. The voices of entire peoples that the law had disregarded and dismissed as primitive or barbaric are now heard to create legal knowledge. The voices of indigenous peoples, in the best interpretations of the rights of nature, are not exoticized or romanticized. Rather, they are understood as sources of knowledge that offer innovative ways of thinking about the relationship between humans, nature, law, and economics; they are sources of knowledge that can enrich thinking about these issues anywhere in the world.[95]

c) The rights of nature, epistemic justice, and political emancipation

The rights of nature are also epistemically innovative because (i) they challenge some of the assumptions of modern Western law and (ii) because they are created through heterodox and emancipatory processes of legal knowledge generation. The rights of nature challenge the idea that only human beings can be rights-holders and, correspondingly, that subjective rights can be assigned only to human beings. Modern Western law, founded on liberalism, considers as a rule that only moral, autonomous, and rational agents can be holders of subjective rights.[96] Legal persons, minors, and some people with severe cognitive or physical disabilities constitute the exceptions that confirm the rule.[97] For this legal tradition, nature, a thing over which rights are vested, particularly the right to property, could never be the holder of subjective rights. However, the rights of nature recognize

[93] See Section III. 1.a), above, on legal hybridity.

[94] Minhea Tănăsescu, 'Rights of Nature, Legal Personality, and Indigenous Philosophies' (2020) 9 TEL 429.

[95] The most recent quantitative analysis on the subject shows that 18.6% of the initiatives related to the rights of nature refer to or are based on indigenous beliefs; only 2.4% come directly from indigenous peoples. Putzer and others (n 5) 92. However, this quantitative analysis obscures the fact that even if the initiatives do not explicitly mention indigenous traditions, many of them may have been influenced by them. CELDF, which has directly and explicitly influenced the drafting of most local legal norms on the rights of nature in the US, explicitly acknowledges the intersection between rights of nature and indigenous traditions. The other major quantitative analysis, which precedes this one, does not address the issue from a numerical perspective, although it does emphasize the Ecuadorian, Bolivian, and New Zealand models in its analysis. Kauffman and Martin (n 4).

[96] Seth Epstein and others 'Liberalism and Rights of Nature: A Comparative Legal and Historical Perspective' (2022) Law, Culture and the Humanities <https://journals.sagepub.com/doi/10.1177/17438721211065735>. Also see Patricia Williams, 'On Being the Object of Property' in Katherine Bartlett and Rosanne Kennedy (eds), *Feminist Legal Theory: Readings in Law and Gender* (Westview Press 1991) 165 (examining the type of subject presupposed by classical liberalism).

[97] Elvia Arcelia Quintana Adriano, 'The Natural Person, Legal Entity or Juridical Person and Juridical Personality' (2015) 4 Penn St JL & Int'l Aff 363.

that *nature* is a non-human agent that can hold rights and, therefore, also obligations. This new legal subject has no voice of its own. However, human beings can represent its voice, as managers do with respect to legal persons, or as parents and guardians represent minors or persons with particularly-serious disabilities.[98] While there may be doubts about how the voice of nature can be heard, or how to interpret it, the attempt to do so implies an important shift away from the anthropocentric perspectives constitutive of modern Western law that prescribe that the only voice to be heard is that of human beings: the center of the universe.[99]

The rights of nature are also epistemically emancipatory because they are a consequence of constitutional or legal processes involving actors who have historically been at the margins of both the political community and state law-making processes.[100] Indigenous peoples' voices do not just enter indirectly, through mediators from the majority culture, into the state legal and political system to construct the rights of nature. Indigenous peoples are themselves the ones who present and defend their traditions about nature.

In Ecuador, the national Constituent Assembly had indigenous assembly members who promoted the recognition of the rights of nature as part of the agenda of the political movements they represented.[101] Similarly, indigenous organizations put pressure on the Constituent Assembly to recognize the rights of nature, and indigenous organizations were part of the political movement that brought President Rafael Correa to power, who later convened the national Constituent Assembly.[102] In Bolivia, the government of Evo Morales, an indigenous Aymara, as well as some indigenous movements and parties, promoted the recognition of Pachamama and the principle of *buen vivir* in the 2009 Constitution. Likewise, these actors promoted the recognition of the rights of nature in the laws of Mother Earth.[103] Cultural minorities in Bolivia and Ecuador managed to position them-

---

98 Tănăsescu (n 11).

99 Susana Borrás, 'New Transitions from Human Rights to the Environment to the Rights of Nature' (2016) 5 TEL 113; and Eduardi Gudynas, 'Buen Vivir: Today's tomorrow' (2011) 54 Development 441.

100 Barié (n 90) 10; Paola Villavicencio Calzadilla and Louis J Kotzé. 'Living in Harmony with Nature? A Critical Appraisal of the Rights of Mother Earth in Bolivia' (2018) 7 TEL 397; Iorns Magallanes (n 69).

101 Minhea Tănăsescu, 'The Rights of Nature in Ecuador: The Making of an Idea' (2013) 70 International Journal of Environmental Studies 6.

102 This is not to say, of course, that all indigenous (or non-indigenous) organizations and parties agreed with or promoted the rights of nature equally strongly. The disputes between the Correa government and the indigenous movements that supported them are well known, as are the conflicts between the different sectors that made up the Constituent Assembly – liberal, conservative, and left – regarding the value of the rights of nature. Tănăsescu (n 101).

103 Jorge Lazarte, 'Plurinacionalismo y multiculturalismo en la Asamblea Constituyente de Bolivia' (2009) 33 Revista Internacional de Filosofía Política 71.

selves politically in these processes of legal creation and were able to make their voices heard effectively.[104]

In New Zealand, the recognition of the Whanganui River as a subject of rights is a consequence of a long political process that ended with the Te Awa Tupua agreement, which reinterpreted the founding treaty of New Zealand: the Treaty of Waitangi.[105] The Māori people, the people who inhabited the lands that are now New Zealand before the arrival of Europeans in Oceania, have lived on the margins of the political community. However, in this case, they reached an agreement with the government to change the dominant understandings of the treaty that created the New Zealand political community. Both the process and its conclusions challenge the historical status quo in New Zealand for the following reasons: the negotiation that ends with the agreement is between the Māori people and the New Zealand government, two peoples who in this instance negotiated horizontally;[106] the agreement conferring rights to the Whanganui River is a syncretic product in which indigenous traditions and those of the dominant culture intersect;[107] the process of creating a statute in parliament – in this case the ratification of the treaty – was attended for the first time by Māori leaders;[108] and the legal entity representing the Whanganui River is composed equally of representatives of the parties creating the treaty.[109] New Zealand, like Ecuador and Bolivia, is a culturally-diverse country that has historically discriminated against their

[104] Marc Becker, 'Correa, Indigenous movements, and the Writing of a New Constitution in Ecuador' (2011) 38 Latin American Perspectives 48. In both countries, however, the rights of nature were (and still are) in tension with the principle of material equality and the redistribution objectives it demands; they were (are) promoted by a notable segment of those leftist and indigenist movements that have promoted the rights of nature, such as the MAS and Alianza PAIS. Material justice, in the interpretation of these movements, demands that the governments of economically-developing countries like Ecuador and Bolivia make intensive use of natural resources. The extractivism that these movements continue to defend politically and legally is in conflict with the rights of nature. Rickard Lalander, 'Entre el ecocentrismo y el pragmatismo ambiental: Consideraciones inductivas sobre desarrollo, extractivismo y los derechos de la naturaleza en Bolivia y Ecuador' (2015) 6 Revista Chilena de Derecho y Ciencia Política 118; Rickard Lalander, 'Rights of Nature and the Indigenous Peoples in Bolivia and Ecuador: a Straitjacket for Progressive Development Politics?' (2014) 3 Revista Iberoamericana de Estudios de Desarrollo 148.

[105] Craig M Kauffman, 'Managing People for the Benefit of the Land: Practicing Earth Jurisprudence in Te Urewera, New Zealand' (2020) 27 Interdisciplinary Studies in Literature and Environment 578.

[106] Kauffman (n 105).

[107] Katherine Sanders, 'Beyond Human Ownership? Property, Power and Legal Personality for Nature in Aotearoa New Zealand' (2017) 30 JEL 1.

[108] 'Innovative Bill Protects Whanganui River with Legal Personhood' (New Zealand Parliament, 28 March 2017) <https://www.parliament.nz/en/get-involved/features/innovative-bill-protects-whanganui-river-with-legal-personhood/>.

[109] Catherine Iorns Magallanes, 'Nature as an Ancestor: Two Examples of Legal Personality for Nature in New Zealand' (2015) Special Issue 22 VertigO – La revue électronique en sciences de l'environnement, <https://doi.org/10.4000/vertigo.16199>.

cultural minorities. The rights of nature show that this inequality of power and these injustices are not inevitable or necessary.

2. External Legal Pluralism and the Rights of Nature

Legal pluralism offers a set of conceptual lenses through which to describe, analyze, and evaluate the heterogeneity of normative systems that exist within and outside states, as well as their interactions.[110] This heterogeneity, which is not always normatively defensible, challenges the assumptions underlying the legal monism that is dominant in modern law whereby states have (and should have) an autonomous, unitary, and hierarchical legal order that reflects the ethos of the nation.[111]

The rights of nature have created a form of external legal pluralism that has not been regularly examined. The paradigmatic models of the rights of nature are legal products that have emerged in peripheral national legal systems and that have influenced and partially transformed international law. This bottom-up legal movement, from the legal periphery to the center of international law, is concretized in the following two ways: *first*, the rights of nature severely question numerous core aspects of international environmental law.[112] The national law establishing rights of nature are an irritant to the international environmental legal system discursively. National legal systems promoting rights of nature attack the premises and foundations of international environmental law. This first type of interaction between national rights and international law is fundamental for the latter's subsequent transformation of itself, opening spaces for the rights of nature within its structures.

The rights of nature are grounded in a biocentric perspective opposing the anthropocentric view that justifies mainstream international environmental law.[113] The rights of nature see nature as a living being and as an internally diverse whole that is made up of interrelated and interdependent components. From this perspective, humans are not at the center of nature, even though they are nature. By contrast, mainstream international environmental law is articulated and founded around the idea that humans are at the heart of nature. *Nature* must be protected because this benefits the interests or enables a realization of the rights of individuals. The principles of precaution, prevention, sustainability, environmental justice, or intergenerational environmental justice, among others, are usually in-

---

[110] Ralf Michaels, 'Globalisation and Law: Law Beyond the State', in Reza Banakar and Max Travers (eds), *Law and Society Theory* (Hart Publishing 2013) 294.

[111] Michaels (n 13) 258.

[112] Helen Dancer, 'Harmony with Nature: Towards a New Deep Legal Pluralism' (2021) 53 J Legal Plur 21.

[113] Louis J Kotzé and Duncan French, 'The Anthropocentric Ontology of International Environmental Law and the Sustainable Development Goals: Towards an Ecocentric Rule of Law in the Anthropocene' (2018) 7 Global Journal of Comparative Law 5.

terpreted in an anthropocentric way. Nature must be protected by appealing to these principles because it is useful for humanity.[114]

We should not forget, however, that the rights of nature emerged in a context where international environmental law, as well as state environmental law, had already opened a space, albeit anthropocentric, for the legal protection of nature. An important percentage of the initiatives on rights of nature that have been presented around the world, 41.1%, include in their rationale a questioning of capitalism, an economic model that is continually in tension with national and international environmental law, and 39.4% of these initiatives include the right to a healthy environment as a justification.[115] The tensions between international environmental law and the rights of nature, moreover, do not imply that the rights of nature and international environmental law cannot be used as concomitant legal strategies. Even if their rationales are incompatible, the two types of rights can be used in practice to protect nature.[116]

Second, the rights of nature have opened up spaces in international law. The UN's Harmony with Nature program,[117] the creation of international Mother Earth tribunals,[118] the UN's designation of 22 April as International Mother Earth Day,[119] the proclamation of the Universal Declaration of the Rights of Nature,[120] and the recognition of the rights of nature by the Inter-American Court of Human Rights[121] are evidence of the space that the rights of nature already occupy in international law. The soft and hard international law on the rights of nature is a consequence, mainly, of the following three factors: (i) the actions of national and international non-governmental organizations; (ii) the strategies of the Bolivian and Ecuadorian governments to position the rights of nature regionally and globally; and (iii) the work of academics from different disciplines that promotes the rights of nature. I will now trace the central steps that national legal models on the rights of nature took to transform international law; the central contents of this new international law will be outlined as well. This analysis is key in understanding the emergence and characteristics of the external legal pluralism that arose with the rights of nature.

---

[114] Kotzé and French (n 113); and Klaus Bosselmann, 'Environmental and Human Rights in Ethical Context' in Anna Grear and Louis J Kotzé (eds), *Research Handbook on Human Rights and the Environment* (Edward Elgar 2015) 531–549.

[115] Putzer and others (n 5) 92.

[116] See Section II, above, and Corte Constitucional de Colombia, Sentencia T-622 de 2016.

[117] <http://www.harmonywithnatureun.org/>.

[118] <https://www.garn.org/rights-of-nature-tribunal/>.

[119] <https://www.un.org/en/observances/earth-day>.

[120] <https://www.un.org/en/about-us/universal-declaration-of-human-rights>.

[121] Inter-American Court of Human Rights, advisory opinion, 15 November 2017, OC-23/17, serie A, no 23; and Inter-American Court of Human Rights 6 February 2020, *case of the indigenous communities' members of the Lhaka Honhat (our land) Association v Argentina*, serie C, no 400.

a) The rights of nature and NGOs: The transition from the national to the international level

The formal recognition of the rights of nature is taking shape in the 21st century. Two international non-governmental organizations have played a central role in their recognition and transnationalization, although many more are active in the networks that have promoted these type of rights:[122] the Community Environmental Legal Defence Fund (CELDF)[123] and the Global Alliance for the Rights of Nature (GARN).[124] CELDF, a US-based environmental non-governmental organization that promotes participatory democracy,[125] was a key player in the process that led to the enactment of the world's first legal norm on the rights of nature and played a notable role in the recognition of these rights in the 2008 Ecuadorian Constitution.[126] CELDF contributed to the initial construction of the rights of nature discourse that would later globalize and influence international law.

The first legal norm that recognized the rights of nature globally was the ordinance issued in Tamaqua Borough, Pennsylvania, United States, in 2006. This ordinance recognizes that 'natural communities and ecosystems should be considered as persons' for all matters related to the protection of their civil rights. The purpose of the ordinance is to stop companies from mismanaging wastewater in the area. One of the people who was to become the village supervisor, Cathy Miorelli, participated in one of CELDF's workshops on nature protection and local democracy.[127] Consequently, when she took office, she invited CELDF to advise the town on how to address the environmental problems that were nega-

[122] See, for example, the inventory provided by Craig M Kauffman and Pamela L Martin, *The Politics of Rights of Nature: Strategies for Building a More Sustainable Future* (Massachusetts Institute of Technology Press 2021) 30: 'The Gaia Foundation similarly trains civil society and community leaders in Africa, and the Earth Law Center and others work with local and national governments around the world to draft RoN laws. The Australian Earth Laws Alliance, Earth Law Center, and other organisations train lawyers and judges to advance RoN through the courts. Other groups, like Movement Rights and the Indigenous Environmental Action Network, promote RoN among Indigenous and other grassroots activists. The Scottish organisation Nature's Rights, The Center for Democratic and Environmental Rights, the Ecological Law and Governance Association, and other organisations hold regular webinars to raise public awareness. End Ecocide on Earth and others are working to amend the Rome Statute to add ecocide (i.e., serious violations of RoN) to the list of crimes that could be brought to the International Criminal Court'.

[123] <https://celdf.org/>.

[124] <https://www.garn.org/>.

[125] 'CELDF is assisting civil society, indigenous peoples, communities, and governments to advance laws and policies for the protection of Nature and the environment. This includes providing legislative and policy drafting, legal research, public engagement and education, ongoing support during global crises, and trainings'. See <https://celdf.org/> and <https://celdf.org/rights-of-nature/>.

[126] CELDF, 'Press Release: Ecuador Approves New Constitution: Voters Approve Rights of Nature' (28 September 2008) <https://celdf.org/2008/09/press-release-ecuador-approves-new-constitution-voters-approve-rights-of-nature/>.

[127] Kauffman and Martin (n 122) 53–54.

tively affecting it. The town drafted the ordinance with CELDF's ideas about the rights of nature and the place of local democracy in the American political system in mind.[128]

However, this norm has not proven as consequential as the recognition of these rights in the 2008 Ecuadorian Constitution, the 2010 and 2012 Bolivian Mother Earth laws, and the 2014 and 2017 New Zealand Te Ure Wera and Te Awa Tupua laws. The US ordinance is applicable only in the jurisdiction of this single town in the State of Pennsylvania, and as it has not been challenged before any court, its contents have not been clarified by judges; the authorities that administer the town have not refined it, either.[129] Conceptually, as compared to the Ecuadorian, Bolivian, and New Zealand norms, this ordinance is more limited and less innovative: it only indicates that 'natural communities and ecosystems' are the subjects of rights, affording these communities and ecosystems the civil rights to which all US citizens are entitled.[130] It does not grant nature any particular rights. The grounds on which it relies in establishing these new legal subjects, moreover, are anthropocentric arguments: the protection of the rights of the town's inhabitants and participatory democracy.[131] Humans are the central concern of the ordinance: nature is protected for their welfare. The ordinance also emerges in a context where conflicts between municipalities, states, and the federal government over environmental issues are a constant. The ordinance is a manner of interpreting US federalism that empowers municipalities vis-à-vis the state and federal governments. This first rule has had multiple iterations in the United States. Currently, 155 localities have issued or discussed regulations affording rights to nature; their contents are remarkably similar since in most instances CELDF advised on their drafting.[132]

After this first achievement in favor of the rights of nature, the Ecuadorian government invited CELDF to advise the Constituent Assembly.[133] In this pro-

---

[128] CELDF, 'Tamaqua Borough, Pennsylvania' (31 August 2015) <https://celdf.org/2015/08/tamaqua-borough/>; and 'The Legal Defense Fund has assisted communities in the U.S. – in Pennsylvania, New Hampshire, and Virginia – to draft and adopt first-in-the-nation laws that change the status of ecosystems from being regarded as *property* under the law to being recognised as *rights-bearing entities*.' CELDF (n 126) (emphasis in original).

[129] Kauffman and Martin (n 122) ch 3.

[130] Tamaqua Borough, 2006, Ordinance no 612, s 7.6: 'Borough residents, natural communities, and ecosystems shall be considered to be "persons" for purposes of the enforcement of the civil rights of those residents, natural communities, and ecosystems'.

[131] Tamaqua Borough, 2006, Ordinance no 612, s 3 – Findings and Purposes.

[132] Putzer and others (n 5) 92. Huneeus identifies 56 approved initiatives on rights of nature in the United States. Alexandra Valeria Huneeus, 'The Legal Struggle for Rights of Nature in the United States' (2022) 1 Wis L Rev 134; Tănăsescu (n 11) 111 ('What I want to point out here is that, from 2008 onward, US municipalities have passed strikingly similar ordinances, as a result of the central role of CELDF and the overarching representation of nature through rights, which carries its inherent elements through various cases').

[133] 'In 2008, CELDF was invited to meet with the Ecuador Constituent Assembly as they drafted a new constitution. We assisted the Assembly with constitutional provisions regarding

cess, CELFD explicitly pointed out the limits of traditional anthropocentric environmental law and promoted the intrinsic value of nature. CELDF's efforts have continued in many other countries of the Global South and North, including India, Nepal, Colombia, Cameroon, Australia, Wales, and England.[134] CELDF is a global player in the promotion of the rights of nature. As of January 2021, public officials, state institutions, activists, and social organizations had introduced 409 initiatives on the rights of nature around the world.[135] The introduction of these initiatives has been fundamental in the transformation of international environmental law. The critical mass created from below has created pressure for a modification of the anthropocentric protection of nature.

GARN was created in 2010 in Patate, Ecuador, with the aim of promoting the rights of nature worldwide.[136] GARN is currently composed of 1,400 members from ninety-nine countries.[137] GARN's main achievements with regard to the pro-

Nature's defense. Ecuadorians adopted their new constitution by an overwhelming majority, making Ecuador the first country in the world to recognise the Rights of Nature in its constitution.' <https://celdf.org/rights-of-nature/>; and 'Over the past year, the Legal Defense Fund was invited to assist the Ecuadorian Constituent Assembly to develop and draft provisions for the new constitution to put ecosystem rights directly into the Ecuadorian constitution. The elected Delegates to the Constituent Assembly requested that the Legal Defense Fund draft language based on ordinances developed and adopted by municipalities in the U.S.'; 'The Pachamama Alliance, with offices in San Francisco and Quito, played a key role in facilitating the Legal Defense Fund's involvement in the drafting of Ecuador's new constitution.' CELDF (n 126); see also, CELDF, 'Blog: A Decade of Legality: Rights of Nature Coming of Age' (21 September 2018) <https://celdf.org/2018/09/blog-a-decade-of-legality-rights-of-nature-coming-of-age/>.

[134] 'Following Ecuador's constitutional process, people and civil society in other countries learned about the movement. Today, CELDF's International Center for the Rights of Nature is working in India, Nepal, Australia, Cameroon, Colombia, the United States, and other countries. We assist people, communities, and groups in advancing the protection of Nature's rights. Over the past several years, CELDF has also traveled to Nepal, India, Colombia, and Australia. We have presented on Nature rights initiatives and help people and civil society launch related campaigns'. <https://celdf.org/rights-of-nature/> and CELDF, 'Green Party of England and Wales Adopts Rights of Nature Policy: CELDF Assisted in Drafting New Rights-Based Plank' (29 February 2016) <https://celdf.org/2016/02/green-party-of-england-and-wales-adopts-rights-of-nature-policy-celdf-assisted-in-drafting-new-rights-based-plank/>.

[135] Putzer and others (n 5) 90.

[136] <https://www.garn.org/rights-of-nature-timeline/>; and 'The Global Alliance for the Rights of Nature (GARN) establishes that in order to ensure an environmentally sustainable future, humans must reorient themselves from an exploitative and ultimately self- destructive relationship with nature, to one that honors the deep interrelationship of all life and contributes to the health and integrity of the natural environment. An essential step in achieving this is to create a system of jurisprudence that sees and treats nature as a fundamental, rights bearing entity and not as mere property to be exploited at will. Breaking out of the human-centered limitations of our current legal systems by recognising, respecting and enforcing Rights of Nature is one of the most transformative and highly leveraged actions that humanity can take to create a sustainable future for all.' – <https://www.garn.org/>.

[137] <https://www.garn.org/>.

motion of the rights of nature in the international context are twofold: first, it has contributed notably to the global diffusion of the Universal Declaration of the Rights of Mother Earth. Cormac Cullinan, one of GARN's founders, played a central role in the drafting of the Universal Declaration;[138] second, the organization created the International Mother Earth Tribunals and is in charge of its administrative secretariat. Alberto Acosta, the vice-president of the Constituent Assembly of Ecuador, who later became a member of GARN, was one of its main promoters and has actively participated in the implementation of these tribunals.[139]

b) The Universal Declaration and the International Mother Earth Tribunals

The Universal Declaration of Mother Earth was drafted in 2010 with the Universal Declaration of Human Rights as a reference, and Cormac Cullinan, as mentioned above, one of the founders of GARN, was one of its main articulators.[140] The Declaration, which was drafted during the World People's Conference on Climate Change and the Rights of Mother Earth held in Cochabamba, Bolivia, urges the United Nations General Assembly to adopt it, although so far this has not happened. The contents of the Declaration are structured around the following elements: (i) the environmental crisis facing humanity; (ii) capitalism as one of the central causes of this crisis; (iii) nature as a maternal entity constituting a whole that is composed of organic and inorganic elements that are interrelated and interdependent; (iv) the importance of recognizing nature as a subject of rights that has a life of its own; (v) the urgency of recognizing that nature has a particular set of rights such as the right to life, the right to be respected, the right to its identity and integrity, and the right to the regeneration of its vital cycles; (vi) the need to recognize the rights of Mother Earth as a means to materialize human rights; and (vii) the obligations of humans with respect to Mother Earth.[141] The influence of the Ecuadorian and Bolivian models on the Declaration – the New Zealand model had not yet been articulated when the Declaration was drafted – are evident. The hybrid nature of the rights-of-nature discourse (modern law and indigenous knowledge) and the anthropomorphization, sacralization, and interpretation of nature as a biocentric and ecosystemic space form the backbone of the Declaration, as is the case with the Ecuadorian and Bolivian models of the rights of nature.

---

[138] GARN, 'Cormac Cullinan SAB Environmentalist of the Year (2012)' (23 October 2012) <https://www.garn.org/cullinan-environmentalist-of-the-year-2012/>. The Declaration was drafted in April of 2010 and GARN was founded in September of 2010.

[139] <https://www.rightsofnaturetribunal.org/about-us/>; and Alberto Acosta, 'Toward the Universal Declaration of Rights of Nature: Thoughts for Action' (2010) 24 AFESE Journal.

[140] Dancer (n 112) 37.

[141] GARN, 'Universal Declaration for the Rights of Mother Earth of 22 April 2010' <https://www.garn.org/universal-declaration/>.

The International Mother Earth Tribunals were created with the International War Crimes Tribunal and also the Permanent Peoples' Tribunal[142] in mind. To date, GARN has organized five international tribunals (Quito, 2014; Lima, 2014; Paris, 2015; Bonn, 2017; and Glasgow 2021), two regional tribunals (Chile, 2019 and Europe, 2020), and eight local tribunals (Quito, twice in 2014; San Francisco, twice in 2014 and 2016; Brisbane, 2014; Bemidji Ponca Nation – Oklahoma, Espiritu Santo; Queensland and Stockholm).[143] We can classify the objectives of these international tribunals as follows: On the one hand, they have democratic and representative purposes.[144] The tribunals want to become a space for citizens and indigenous communities allowing them to make their voices heard about environmental damage caused by governments and corporations, as well as to advance solutions to this damage. On the other hand, the tribunals pursue discursive ends.[145] These tribunals seek to rethink the theoretical frameworks within which human-nature relations are interpreted and environmental problems are understood and solved. Ultimately, these tribunals seek to achieve normative goals.[146] More precisely, they seek to generate alternatives for addressing environmental problems and to educate citizens and governments on these same issues.

The theoretical framework within which the goals pursued by the tribunals are embedded is that of the rights of nature. The tribunals are committed to the idea that nature is an autonomous subject with particular rights and that human interests have the same value as the interests of nature. Again, the intersection between the paradigmatic models of the rights of nature and the institutions created by GARN are evident. The purposes of GARN, the purposes of the Universal Declaration of the Rights of Nature, which GARN was instrumental in drafting, and those purposes that the International Tribunals of Mother Earth aim to realize are aligned.

c) Bolivia, Ecuador, and the international rights of nature

The government of Ecuador and, with particular intensity, the government of Bolivia, have acted explicitly and directly to position the rights of nature internationally.[147] Rights of nature have been an important and strategic component of these two countries' foreign policy. The government of Bolivia, during the presi-

[142] <https://www.rightsofnaturetribunal.org/about-us/> and <https://www.rightsofnaturetribunal.org/>.

[143] <https://www.garn.org/rights-of-nature-tribunal/> and <https://www.rightsofnaturetribunal.org/>; <https://www.rightsofnaturetribunal.org/about-us/>.

[144] <https://www.rightsofnaturetribunal.org/> and <https://www.rightsofnaturetribunal.org/about-us/>.

[145] Ibid and Michelle Maloney, 'Building an Alternative Jurisprudence for the Earth: the International Rights of Nature Tribunal' (2016) 41 Vt L Rev 129.

[146] Ibid.

[147] Kauffman and Martin (n 122) ch 2.

dency of Evo Morales, played a central role in the promulgation of the UN declaration recognizing 22 April as International Mother Earth Day. The recognition of International Mother Earth Day in 2009 was made at the request of the Morales government.[148] During his address to the General Assembly that year, the Bolivian president indicated that he hoped the 21st century would be the century of Mother Earth and urged the UN to draft and adopt a universal declaration of the rights of Mother Earth.[149] The General Assembly did not accept Morales' invitation. Bolivia, in response to the poor results of the Copenhagen Climate Change Conference in 2009, organized the World People's Conference on Climate Change and the Rights of Mother Earth in Cochabamba.[150] During this Conference, which took place in 2010, the Universal Declaration of the Rights of Mother Earth was drafted.[151] However, the greatest influence that Bolivia had in its campaign to position the rights of nature in the international context was the creation by the UN – at Bolivia's request and with its provision of partial funding – of the Harmony with Nature program.[152] In the context of this program, the UN has, since 2009, enacted twelve resolutions promoting the rights of nature,[153] organized eleven interactive dialogues on the rights of nature among the member countries of the General Assembly,[154] and created the committee of experts on the rights of nature;[155] further, the secretary general of the organization has drafted eleven reports related to these rights.[156]

---

[148] United Nations, 'General Assembly Proclaims 22 April "International Mother Earth Day" Adopting by Consensus Bolivia-led Resolution' (22 April 2009) <https://press.un.org/en/2009/ga10823.doc.htm>; United Nations, 'General Assembly to Consider Designating 22 April as International Mother Earth Day' (22 April 2009) <https://press.un.org/en/2009/ga10822.doc.htm>; David R Boyd, *The Rights of Nature: A Legal Revolution that Could Save the World* (ECW Press 2017) 258–259; and GARN, 'Bolivia's Leadership' (19 January 2012) <https://www.garn.org/bolivia-experience/>.

[149] United Nations (n 148); Boyd (n 148) 259.

[150] Boyd (n 148) 258; Fuad Rahmat, 'Beyond Copenhagen? World People's Conference on Climate Change and the Rights of Mother Earth, Cochabamba, Bolivia, 19–22 April 2010' (2010) 162 Radical Philosopy 63 <https://www.radicalphilosophyarchive.com/conference-report/beyond-copenhagen/>.

[151] Kauffman and Martin (n 122) 33; GARN, 'Bolivia's Leadership' (19 January 2012) <https://www.garn.org/bolivia-experience/>.

[152] 'Agreement between the Government of the Plurinational State of Bolivia and the Unites Nations Department of Economic and Social Addairs' <http://files.harmonywithnatureun.org/uploads/upload49.pdf>.

[153] <http://www.harmonywithnatureun.org/chronology/> and <http://www.harmonywithnatureun.org/unDocs/>.

[154] <http://www.harmonywithnatureun.org/dialogues/>.

[155] <http://www.harmonywithnatureun.org/welcome/> and <http://www.harmonywithnatureun.org/experts/>.

[156] <http://www.harmonywithnatureun.org/unDocs/>.

The impact that the rights of nature have had on the UN can also be seen in Resolution 66/288 issued by the UN Conference on Sustainable Development.[157] In this 2012 resolution, the UN recognizes the relationship between intergenerational environmental justice and harmony with nature.[158] This resolution, mainly because of the efforts of Ecuador and Bolivia, recognizes Mother Earth, notes the recognition of the rights of nature by several states, and highlights the importance of living in harmony with nature.[159] Ecuador had already obtained the support of the countries of the Community of Latin American and Caribbean States (CELALC) at the meeting held by this organization in Quito, ensuring that the possibility of adopting a Universal Declaration of the Rights of Nature would be discussed at Rio+22.[160] This impact from advancing rights of nature can be further seen in the final report of the Rio+22 Conference held in South Africa, where the UN tries to strike a balance between sustainable development and living in harmony with nature.[161] Similarly, we can see such influence in the Declaration and Plan of Action of Santa Cruz de la Sierra – Towards a New World Order for Living Well.[162] In this Declaration, leaders of the Group of 77, a group of developing countries, recognize the importance of living in harmony with nature, note the recognition that several countries have made of the rights of nature, and emphasize the need to respect it.[163] We can also see this impact in the UN Sustainable Development Goals produced in 2015.[164] In Goal 12.8, the UN recognizes the importance of citizens being aware of and having access to all relevant information so that development and their lifestyles are in harmony with nature.[165]

Finally, and very importantly because it is an instance of recognition within hard international law, the Inter-American Court of Human Rights has recognized the rights of nature in both its advisory opinion OC-23/17 (requested by Colombia in relation to its pursuit of infrastructure projects)[166] as well as in its ruling in

[157] United Nations General Assembly, 'Resolution 66/288 adopted by the General Assembly on 27 July 2012' <https://www.un.org/en/development/desa/population/migration/generalassembly/docs/globalcompact/A_RES_66_288.pdf>.

[158] United Nations General Assembly (n 157) art 86.

[159] United Nations General Assembly (n 157) art 39: 'We recognise that planet Earth and its ecosystems are our home and that "Mother Earth" is a common expression in a number of countries and regions, and we note that some countries recognise the rights of nature in the context of the promotion of sustainable development'; art 40: 'We call for holistic and integrated approaches to sustainable development that will guide humanity to live in harmony with nature and lead to efforts to restore the health and integrity of the Earth's ecosystem'.

[160] Kauffman and Martin (n 122) 35.

[161] Principle 1 and art 4.11 <https://www.un.org/en/conferences/environment/rio1992>.

[162] United Nations General Assembly, 'Session 68/948 of 7 July 2024' <https://www.g77.org/doc/A-68-948(E).pdf>.

[163] United Nations General Assembly (n 162) paras 7, 14, 30, 31, 40, 51, 210, 211.

[164] <https://sdgs.un.org/es/goals>.

[165] <https://indicators.report/goals/goal-12/>.

[166] Inter-American Court of Human Rights, advisory opinion, 15 November 2017, OC-23/17, serie A, no 23.

the *Case of the Indigenous Communities of the Lhaka Honhat (Our Land) Association v Argentina.*[167] In this regard, the Inter-American Court included a paragraph in the advisory opinion that was later reproduced in the aforementioned judgment:

> 'This Court considers it important to emphasise that the right to a healthy environment as an autonomous right, unlike other rights, protects the components of the environment, such as forests, rivers, seas and others, as juridical interests in themselves, even in the absence of certainty or evidence of risk to individual persons. It is a matter of protecting nature and the environment not only because of its connection with a utility for human beings or because of the effects that its degradation could cause on other rights of persons, such as health, life or personal integrity, but also because of its importance for the other living organisms with which the planet is shared, also deserving of protection in themselves. [...] the right to a healthy environment as an autonomous right is different from the environmental content that arises from the protection of other rights, such as the right to life or the right to personal integrity.'

Neither the advisory opinion nor the Lhaka Honhat ruling identifies any of the sources or influences that led to the Court's recognition of rights of nature, and it can be observed that the Inter-American Convention on Human Rights does not recognize nature as a subject of rights. However, the Court has been in dialogue with the supreme or constitutional courts of the countries of the region for many years.[168] The judgments of the Inter-American Court have influenced the courts of countries such as Colombia, Ecuador, and Bolivia. However, the jurisprudence of these national courts has also influenced the jurisprudence of the Inter-American Court.[169] Furthermore, the Inter-American Court, as the head of the Inter-American human rights system, is familiar with the basic structures of the legal systems of the Latin American countries that are party to the Convention.

d) The epistemic community promoting the rights of nature

Finally, the impact that the rights of nature have had on international law would not have been possible without the creation of an epistemic community to de-

---

[167] Inter-American Court of Human Rights 6 February 2020, *Case of the Indigenous Communities' Members of the Lhaka Honhat (Our Land) Association v Argentina*, serie C, no 400.

[168] René Urueña, 'Luchas Locales, Cortes Internacionales. Una exploración De La protección Multinivel De Los Derechos Humanos En América Latina' (2013) 30 Revista Derecho Del Estado 301.

[169] Supreme Court of Argentina, Causa 17.768, *causa Recurso de Hecho: Simón, Julio Héctor s/Privación de la libertad y otros*; Leonardo Alberto Franco, 'Recepción de la jurisprudencia interamericana en el ordenamiento jurídico argentino', in Sergio Ramírez García and Mireya Castañeda Hernández (eds), *Recepción nacional del derecho internacional de los derechos humanos y admisión de la competencia contenciosa de la Corte Interamericana* (UNAM 2009); Tribunal Constitucional del Perú 7 July 2006, no 0174-2006-PHC/TC case *John McCarter*; Tribunal Constitucional del Perú 7 July 2006, no 8817-2005-PHC/TC case *César Alfonso Ausin de Irruarízaga*; Tribunal Constitucional del Perú 29 November 2005, no 4587-2004-AA/TC case *Santiago Martín Rivas*.

scribe, analyze, critique, and promote them.[170] An enormous number of scholarly articles has followed the seminal work of authors such as Aldo Leopold,[171] Christopher Stone,[172] Godefroy Stutzin,[173] Thomas Berry,[174] and Cormac Cullinan[175] on the rights of nature. A search on 'derechos de la naturaleza' and 'rights of nature' in Google Scholar yields 21,100 publications on the subject in Spanish and 15,800 in English. In HeinOnline, 2,079 texts appear when searching for the words 'rights of nature', and in JSTOR 555 articles appear when entering these same words in the search engine. Over the last fifteen years, a network of scholars studying these rights has developed, and the corresponding body of scholarship is profoundly heterogeneous.[176] There is immense diversity in the field with respect to the disciplines in which these scholars work (eg law, ecology, anthropology, sociology, and political science); the theoretical approaches to which they are committed (eg feminism, Marxism, postcolonial studies, liberal political philosophy, and decolonial perspectives); the political perspectives they advocate for (eg socialism or liberalism); and the methods they use to advance their research (quantitative methods, qualitative methods, or purely theoretical analysis).

Nonetheless, this network of scholars has constructed a set of common discursive and practical patterns with respect to the rights of nature.[177] This academic and political knowledge and ways in which research is conducted have legitimized the rights of nature politically and as an object of study. This network of scholars recognizes the value of the rights of nature in understanding and potentially solving acute social problems of the contemporary world, such as climate

---

[170] Peter Haas, 'Epistemic Communities', in Lavanya Rajamani and Jacqueline Peel (eds), *The Oxford Handbook of International Environmental Law* (2nd edn, OUP 2021); Carolina Valladares and Rutgerd Boelens, 'Extractivism and the Rights of Nature: Governmentality, "Convenient Communities" and Epistemic Pacts in Ecuador' (2017) 26 Environmental Politics 1015.

[171] Aldo Leopold, *A Sand County Almanac* and *Sketches Here and There* (OUP 1949).

[172] Christopher Stone, *Should Trees Have Standing? And Other Essays of Law, Morals, and the Environment* (OUP 2010).

[173] Godofredo Stutzin, 'Should We Recognize Nature's Claim to Legal Rights? An Essay' (1976) 2 EPL 129.

[174] Thomas Berry, 'Rights of Earth: We Need a New Legal Framework Which Recognises the Rights of All Living Beings' in Peter Burdon (ed), *Exploring Wild Law: The Philosophy of Earth Jurisprudence* (Wakefield Press 2011).

[175] Cromac Cullinan, *Wild Law: A Manifesto for Earth Justice* (Green Books 2011).

[176] See, for example, the network of experts created by the Harmony with Nature program. The program states that '[the Network] gathers leading experts at the cutting-edge of the natural and social sciences. Their disciplines encompass physics, biology, ecology, economics, sociology, law, ethics, spirituality, anthropology, philosophy, medicine, linguistics and more. The work of these experts advances an Earth-centered worldview in which human-Earth relationships are balanced within the greater web of life to live in Harmony with Nature.' <http://www.harmonywithnatureun.org/welcome/>.

[177] Cormac Cullinan, 'A History of Wild Law' in Peter Burdon (ed), *Exploring Wild Law: The Philosophy of Earth Jurisprudence* (Wakefield Press 2011).

change, ecosystem degradation, and consumerism. Equally, these common patterns have allowed rights of nature to be known by citizens, by those with political power, and by those serving in a law-making capacity in both national and international contexts. This network of authors also recognizes that the environmental problems facing the world cannot be addressed solely from a national perspective. The solution to these problems has an unavoidable international dimension. National and international law must therefore join forces to confront these common problems.[178]

The impact of the emerging rights of nature in national legal systems on international law has led the UN's Harmony with Nature program to create its own network of experts.[179] The macro-objective of this international network, which includes academics, activists, and practitioners, is as follows:

> 'The Knowledge Network seeks to increase the tools and resources rooted in the human-Earth interconnection available to inform policy makers and urge societies on all continents to reconsider how they interact with the natural world. It offers an alternative approach to current thinking and practice that can guide humanity towards a more holistic, peaceful and balanced future.'[180]

The Network is composed of 185 researchers and eighteen young professionals from around the world. Of these researchers, 25.4% are South American; 22.8% are North American; 28.1% are European; 10.2% are Asian; 8.5% are African; and 5% are from Oceania. Several of the experts who have links to non-governmental organizations, such as GARN, and who have produced publications on the rights of nature are part of this Knowledge Network, including Cormac Cullinan, Ramiro Avila, Natalia Greene, Valerie Cabanes, Fiona Wilton, Lisa Mead, Yenny Vega, and Linda Sheehan. Some of the Network's experts have also been part of the eleven Interactive Dialogues among the countries that are part of the United Nations General Assembly, thus broadening their influence both in this international organization and in the legal systems that comprise it.[181]

The General Assembly decided in 2010 to organize the first Interactive Dialogue to commemorate International Mother Earth Day.[182] This first dialogue 'discussed ways to promote a holistic approach to sustainable development in Harmony with Nature and shared national experiences on criteria and indicators for measuring sustainable development in harmony with nature'.[183] The other ten

---

178 <http://www.harmonywithnatureun.org/ejInputs/> and <http://www.harmonywithnatureun.org/references/>.

179 <http://www.harmonywithnatureun.org/experts/> and <http://www.harmonywithnatureun.org/welcome/>.

180 <http://www.harmonywithnatureun.org/welcome/>.

181 <http://www.harmonywithnatureun.org/experts/> and <http://www.harmonywithnatureun.org/statistics/>.

182 <http://www.harmonywithnatureun.org/dialogue/L!XKqcMUU5QAyQJV0!JHfMS+8MMM4vIhC%20+nstpTTbWKF7wkqBqTA9984zP+iaLIGK0aRG1wG661ZaoBacjiBPQ==/>.

183 <http://www.harmonywithnatureun.org/chronology/>.

Dialogues organized by the Harmony with Nature program have continued to discuss issues related to the rights of nature, such as scientific research on the impacts of human activities on the functioning of the earth and economic approaches that could contribute to advancing the construction of an ethical relationship between humans and nature.[184] The Network therefore accepts as premises those found in the approaches of Ecuador, Bolivia, and New Zealand, the paradigmatic models on rights of nature. In this regard, the Network adds,

> '[This] is an online platform of practitioners, scholars and researchers dedicated to strengthening interdisciplinary collaborations to promote a non-anthropocentric or Earth-centered worldview, also called Earth Jurisprudence. At the core of this worldview is the recognition of the intrinsic value of Nature and of human-Earth relationships that are symbiotic, interconnected and subject to the natural laws of the Universe […] Many worldviews of indigenous peoples, ancient spiritual and philosophical traditions around the world have long recognised the symbiotic nature of the human-Earth relationship. It is these deep spiritual and philosophical relationships that underpin and inspire the work of this Knowledge Network in both theory and practice.'[185]

In sum, the rights of nature as well as their articulation, interpretation, and materialization do not emerge and consolidate in international law before then proceeding to influence national law. To the contrary, paradigmatic models of the rights of nature emerged in countries of the Global South and a peripheral country of the Global North. These legal systems dialogue with, irritate, and transform international law through a political and legal process progressing from the bottom (the legal periphery) to the top (international law).

3. *Cross-Fertilization and the Rights of Nature*

The emergence and transnationalization of rights of nature calls into question explanatory theories of legal change in comparative law.[186] The emergence and denationalization of rights of nature is not a consequence of changes in the monolithic and homogenous cultural or economic structures for which they are supposedly the epiphenomenon; they are not a consequence of one-way transfers between two countries, usually from the Global North to the Global South; and they are not an effect of legal transfers that travel from the local to the global and then to new local spaces that adapt and transform them. Rather, the following internal and external processes can explain the emergence and transnationalization of the rights of nature.

The paradigmatic models of the rights of nature, the Ecuadorian, Bolivian, and New Zealand models, emerged as a consequence of political processes that rec-

---

[184] <http://www.harmonywithnatureun.org/chronology/>.

[185] <http://www.harmonywithnatureun.org/welcome/>.

[186] Alan Watson, *Society and Legal Change* (Scottish Academic Press 1977); Alan Watson, *The Evolution of Law* (John Hopkins UP 1985); Alan Watson, 'Aspects of reception of law' (1996) 44 AJCL 335; and George Mousourakis, 'Comparative Law, Legal Transplants and Legal Change' in George Mousourakis (ed), *Comparative Law and Legal Traditions* (Springer 2019).

ognize the multicultural character that these three states have had since their inception.[187] This recognition is mainly a consequence of the agency of indigenous political movements and social organizations that, in alliance with progressive social movements, have managed to position their agendas at the center of the political community.[188] It is also the consequence of the formal or informal recognition of the principles of plurinationality and interculturality.[189] The Ecuadorian Constitution of 2008, the Bolivian Mother Earth laws of 2010 and 2012, and the New Zealand Te Urewera and Te Awa Tupua laws of 2014 and 2017 emerged as a consequence of the recognition that these three countries are not monocultural; that the cultures that compose them must be valued; and that dialogue between these cultures must be the way to build the state they share.[190] The paradigmatic models of the rights of nature, therefore, did not arise because of a change in the unique and homogenous cultures that supposedly constitute these three states, nor are they a consequence of a change in the economic structure of these three countries.

The transnationalization of these prototypical models of the rights of nature has primarily been a process of cross-fertilization involving state and non-state actors, as well as peripheral North-South, South-South, and South-North dialogues and mutual learning. This cross-fertilization has occurred between non-governmental organizations and primary legislatures; between courts; between primary or secondary legislators and courts; and between legislators. The following are globally recognized cases that support these arguments.

a) Cross-fertilization between the peripheral north and the Global South

The emergence of paradigmatic models of the rights of nature is partly a consequence of a dialogue between the peripheral North and the Global South. This dialogue was between a US non-governmental organization and the state of Ecuador.[191] The dialogue was not between the state of Ecuador and any institution of the US legal system. As discussed above, Ecuador invited CELDF to advise the Constituent Assembly on issues related to the rights of nature. CELDF had played a central role in the drafting of the Tamaqua ordinance and continued to do so in

---

[187] Catherine Walsh, *Interculturalidad, Estado, Sociedad. Luchas (de)coloniales de nuestra época* (Universidad Andina Simón Bolívar 2009).

[188] Jorge Lazarte, 'Plurinacionalismo y multiculturalismo en la Asamblea Constituyente de Bolivia' (2009) 33 Revista Internacional de Filosofía Política 71; Marc Becker, 'Correa, Indigenous Movements, and the Writing of a New Constitution in Ecuador' (2011) 38 Latin American Perspectives; and Iorns Magallanes (n 69).

[189] CatherineWalsh, '(De)Construir la interculturalidad. Consideraciones críticas desde la política, la colonialidad y los movimientos indígenas y negros en el Ecuador' in Norma Fuller (ed), *Interculturalidad y Política* (Red de Apoyo de las Ciencias Sociales 2002).

[190] Ramiro Fernando Ávila Santamaría, *El neoconstitucionalismo transformador el estado y el derecho en la constitución de 2008* (Universidad Simón Boliva, Abya-Yala 2011) chs 9–10.

[191] CELDF (n 126); also see CELDF (n 133).

other US cities and towns that subsequently recognized rights of nature, including Toledo, Ohio; Grant Township, Pennsylvania; and Denver, Colorado.[192] But the Ecuadorian Constituent Assembly intertwined CELDF's political experience and knowledge with the local political experiences and knowledge of Andean indigenous peoples. CELDF subsequently incorporated arguments stemming from the knowledge of Andean indigenous peoples among those it uses to promote the rights of nature around the world.[193] In this regard, CELDF states,

> 'CELDF recognises that the Rights of Nature are a concept deeply rooted in traditional indigenous knowledge. This timeline reflects an evolution of Western law and philosophy that is influenced by indigenous cosmovisions, as well as by indigenous-led efforts around the world'.[194]

The cross-fertilization between the peripheral North and the South is also evident in the dialogue that has taken place between the New Zealand legislature, the Colombian Constitutional Court, and the Uttarakhand High Court in India. In the Atrato River judgment, the Colombian Constitutional Court mentioned Te Awa Tupua as one of its sources and took the concept of river guardians from the agreement to apply it in the Colombian context.[195] And just as the New Zealand legislator created a juridical person to represent the Whanganui River, the Colombian court created the figure of the guardians of the Atrato River to achieve this objective. Delegates from the government and the Māori people represent the Whanganui River; riparian communities and grassroots organizations that inhabit the riverbanks, as well as government officials, represent the Atrato River.[196] The Uttarakhand High Court, after recognizing the relevant glaciers as subjects of rights, and following the New Zealand model,[197] appointed a committee of three

---

[192] CELDF, 'Tamaqua Borough, Pennsylvania' (31 August 2015) <https://celdf.org/2015/08/tamaqua-borough/>; and CELDF (n 126).

[193] CELDF also indicates that 'Aboriginal nations and communities retain sovereignty and knowledge over the natural ecosystems they have evolved with. To recognise and follow the natural laws of nature necessitates the elevation particularly of the Traditional Knowledge of local indigenous communities.' <https://celdf.org/rights-of-nature/>.

[194] <https://celdf.org/rights-of-nature/timeline/>.

[195] Corte Constitucional de Colombia, Sentencia T-622 de 2016. 'The most recent case of protection of the rights of nature occurred in New Zealand in 2012. It is an agreement with the force of law signed between the New Zealand government and the Māori Aboriginal community of the Whanganui River, protectors of the river – the third largest in the country – for hundreds of years. In the agreement, which is now law – Te Urewera Act 2014 –, the New Zealand government recognises the river as having the same legal status as a person or corporation, i.e. it grants it the right to exist, to be maintained, and to thrive, along with the obligation to be respected. Under this design, as a legal person or rights-bearing entity, the river has two legal guardians: the New Zealand government and the Whanganui River community, which together form a stewardship and protection committee. […] Similarly, the government of New Zealand recently recognised the Whanganui River as a subject of rights'. <https://www.corteconstitucional.gov.co/relatoria/2016/t-622-16.htm>.

[196] Corte Constitucional de Colombia, Sentencia T-622 de 2016.

[197] *Lalit Miglani v State of Uttarakhand and others* (Writ Petition (PIL) no 140 of 2015): The Court states that '[t]he New Zealand Parliament has enacted "Te 42 Urewera Act 2014"

civil servants to serve as their guardians. The Uttarakhand High Court did not, of course, base its decision on Andean indigenous religions, but on Hinduism.[198] The Court makes clear in this judgment how, for Hinduism (as in an earlier judgment declaring the Ganges and Yamuna rivers to be subjects of rights[199]), the Ganges and Yamuna river basins are sacred.[200] By appealing to a non-Western religion, the Indian court grounds the personification, sacralization, hybridity, and ecosystemic character of nature, just as has been done by the Ecuadorian, Bolivian, and New Zealand legal norms that recognize the rights of nature.

b) South-south cross-fertilization

The exchanges between the Colombian Constitutional Court, the Ecuadorian Constituent Assembly, the Bolivian legislature, and the Ecuadorian Constitutional Court confirm South-South cross-fertilization. This exchange began as a dialogue between the judicial branch and the primary constituent but later became a dialogue between supreme courts. In the Atrato River ruling, the Colombian Constitutional Court cites the rights of nature recognized by the Ecuadorian Constitution and the Bolivian legal system as one of the sources from which it draws.[201]

---

whereby the "Urewera National Park" has been given the [status of a] legal entity under Section 11 of the Act. The purpose of the Act is to preserve, as far as possible, Te Urewera in its natural state, the indigenous ecological systems, biodiversity and its historical cultural heritage.' <https://indiankanoon.org/doc/92201770/>.

198 *Salim v State of Uttarakhand*, Writ Petition (PIL) no 126 of 2014 (5 December 2016 and 20 March 2017) <https://www.elaw.org/salim-v-state-uttarakhand-writ-petition-pil-no126-2014-december-5-2016-and-march-20-2017> The court states that, 'Rivers Ganges and Yamuna are worshipped by Hindus. These rivers are very sacred and revered. The Hindus have a deep spiritual connection with Rivers Ganges and Yamuna. According to Hindu beliefs, a dip in River Ganga can wash away all the sins. The Ganga is also called "Ganga Maa". It finds mentioned in ancient Hindu scriptures including "Rigveda". The river Ganga originates from Gaumukh Glacier and River Yamuna originates from Yamnotri Glacier. […] All the Hindus have deep Astha in rivers Ganga and Yamuna and they collectively connect with these rivers. Rivers Ganga and Yamuna are central to the existence of half of Indian population and their health and well-being. The rivers have provided both physical and spiritual sustenance to all of us from time immemorial. Rivers Ganga and Yamuna have spiritual and physical sustenance. They support and assist both the life and natural resources and health and well-being of the entire community. Rivers Ganga and Yamuna are breathing, living and sustaining the communities from mountains to sea'.

199 *Salim v State of Uttarakhand*, Writ Petition (PIL) no 126 of 2014.

200 Erin L O'Donnell, 'At the Intersection of the Sacred and the Legal: Rights for Nature in Uttarakhand, India' (2017) 30 JEL 135–144.

201 Corte Constitucional de Colombia, Sentencia T-622 de 2016, M.P. Jorge Iván Palacio. 'In accordance with the above, the sentence in question states that "it is admissible for the Court to sustain that the heterogeneous approaches to environmental protection find support in the provisions of the 1991 Charter. The paradigm to which the so-called 'Ecological Constitution' leads us, because it corresponds to a dynamic and open instrument, supported by a system of evidence and collective representations, implies for contemporary society to take ecosystems and natural communities seriously, advancing towards a juridical approach that is more com-

This ruling is the one at the start of the Colombian jurisprudential line on the rights of nature, and it is a decision that has been widely discussed in specialized global literature.[202] The Colombian Court does not, however, base its ruling on Andean indigenous religions, as the Ecuadorian Constitutional Assembly does, but rather on biocultural rights, which include the cultural rights held by Colombia's indigenous peoples according to the 1991 Constitution.[203] After 2009, the Ecuadorian Constitutional Court issued fifty rulings on the rights of nature.[204] The earlier Colombian Constitutional Court rulings on cultural diversity were one of the sources that nurtured this jurisprudence.[205]

Argentine, Peruvian, and Panamanian legislators' uses of the Ecuadorian and Bolivian experiences also confirm South-South cross-fertilization. Bill S-2506/15 presented by Argentine Congressmen Fernando E Solanas and Rubén H Giustiniani proposes to recognize the rights of nature.[206] In its explanation, the bill recognizes the influence that the law of Ecuador and Bolivia had in the drafting of the proposed law. In this regard, the bill says:

> 'This debate is circulating around the world after having been put on the political agenda by the new Constitution of Ecuador in 2008 (from which – together with Bolivian legislation – we draw inspiration for this bill) which already in its preamble states […].'[207]

This same bill was presented again in 2019 and 2020. The 2020 bill also explicitly acknowledges the influence of GARN in its drafting. In this regard, the bill states:

---

mitted to them, as goods that are in themselves the object of guarantee and protection." In addition, it can be said that the Constitutions of Ecuador (2008) and Bolivia (2009) have adopted this approach to environmental protection. Similarly, the government of New Zealand recently recognized the Whanganui River as a subject of rights' (italics in the original).

[202] Iván Vargas Chaves and others. 'Recognizing the Rights of Nature in Colombia: the Atrato River Case' (2020) 17 Jurídicas 13; Paola Calzadilla Villiavicencio, 'A Paradigm Shift in Courts' View on Nature: The Atrato River and Amazon Basin Cases in Colombia' (2019) 15 LEAD J.; Craig M Kauffman and Pamela L Martin, 'When Rivers Have Rights: Case Comparisons of New Zealand, Colombia, and India' (2018) 4 International Studies Association Annual Conference, San Francisco; Elizabeth Macpherson and Felipe Clavijo Ospina, 'The Pluralism of River Rights in Aotearoa New Zealand and Colombia' (2018) 25(6) The Journal of Water Law 283.

[203] Corte Constitucional de Colombia, Sentencia T-622 de 2016; Elizabeth Macpherson and others, 'Constitutional Law, Ecosystems, and Indigenous Peoples in Colombia: Biocultural Rights and Legal Subjects' (2020) 9 TEL 521.

[204] Espinoza, 'Ecuador' (n 23).

[205] Espinoza, 'Ecuador' (n 23).

[206] Fernando Solanas, Proyecto de ley nacional S-2506/15-2015 (2015); Fernando Solanas, Proyecto de ley nacional S-0615/19-2019 (2019); Fernando Solanas, Proyecto de ley nacional 6118-D-2020 (2020).

[207] Fernando Solanas, Proyecto de ley nacional S-2506/15-2015 (2015).

'Likewise, this bill includes the contributions of the Global Alliance for the Rights of Nature (GARN), in the framework of a campaign on the rights of nature that has, among other objectives, to promote legislation that recognises the rights of nature in the Argentine Republic'.[208]

Draft Law 6957 of 2020 in Peru also proposes a recognition of the rights of nature.[209] After stating in the grounds for the law that '[d]iverse countries around the world have recognized, within their domestic regulatory framework, that nature, ecosystems, rivers, species, etc. have rights as living entities and as subjects of rights',[210] the bill names Ecuador and Bolivia as its first two references. However, the bill also mentions other countries from the Global South (Mexico, Brazil, Argentina, Colombia, and Uganda) as well as some from the Global North (the United States, New Zealand, and Australia) as sources for its contents.[211] Finally, Panama recognized nature as a subject of rights and granted it a broad set of specific rights in Law No 287 of 24 February 2022.[212] The law cites the Constitution and the jurisprudence of the Constitutional Court of Ecuador, the Bolivian legal system, and the jurisprudence of the Colombian Constitutional Court as sources for its contents. The bill also refers to the experiences of the

[208] In Argentina, see also the proposal for the recognition of the rights of nature in the city of Santa Fe and the proposal of Rosario's Municipal Council to the Argentine Congress to recognize the Paraná River as a subject of rights. The draft ordinance of the city of Santa Fe recognizes the rights of nature in the context of the city's struggle against the use of glyphosate. In the rationale supporting the draft ordinance, much of the federal bill is reproduced. In some of these paragraphs, the influence of GARN and Ecuador in the drafting of the proposal is acknowledged. The draft ordinance also recognizes the importance of the fact that the UN Harmony with Nature program page referenced the first proposal and its approval. This is an example of the relevance that the UN program on the rights of nature is already beginning to have in state legal systems. In this regard, the proposed ordinance states: 'This legislative milestone was the motivation, among other recognitions, for the City Council of the city of Santa Fe being referenced by Harmony With Nature, of the United Nations Organisation'. <http://files.harmonywithnatureun.org/uploads/upload1038.pdf>; The Municipal Council of Rosario supports in the national level a proposal made by the Cátedra del Agua of Universidad Nacional de Rosario and some of Rosario's unions. The Declaration of Rosario's Municipal Council states '[t]hat the most important of the antecedents [of the proposed recognition of rights to the Paraná River] is the Constitution of the Republic of Ecuador sanctioned in 2008, which recognised nature as a subject of rights'. The Declaration also mentions the jurisprudence of both the Colombian Constitutional Court and the High Court of Uttarkaband as well as New Zealand law. Expte. 255.740-F-2020, C.M.-Consejo Municipal de Rosario, 09 de noviembre ds e 2020, <http://files.harmonywithnatureun.org/uploads/upload1049.pdf>.

[209] Congreso de la República del Perú, Proyecto de Ley 06957-CR <https://leyes.congreso.gob.pe/Documentos/2016_2021/Dictamenes/Proyectos_de_Ley/06957DC19MAY20210517.pdf >.

[210] Congreso de la República del Perú, Proyecto de Ley 06957-CR, Exposición de Motivos, s 3.

[211] Ibid.

[212] Congreso de la República de Panamá, Ley No 287 de 24 de Febrero de 2022 <https://www.gacetaoficial.gob.pa/pdfTemp/29484_A/90384.pdf>.

United States, Brazil, Mexico, New Zealand, Australia, and Uganda with respect to the rights of nature.[213]

c) South-north cross-fertilization

As regards the rights of nature and their paradigmatic models, the direct impact of peripheral legal systems on regimes of the Global North is only now beginning to be identified. However, there are already two cases where the South-North influence has been relevant for the creation of rights of nature. A first example can be found in the ordinances that CELDF helped to draft after its work in the Constituent Assembly of Ecuador. In these processes in the United States as well as in Australia, England, and Wales, CELDF has incorporated components of the Ecuadorian, Bolivian, and New Zealand models.[214] CELDF, as described above, has explicitly acknowledged the influence of indigenous knowledge in the way it understands and grounds the rights of nature.

Second, South-North influence has been relevant in the proposed recognition of the St. Lawrence River as a subject of rights in Quebec, Canada, as was presented to the federal parliament,[215] and in the recognition of the Magpie River as a subject of rights by two Innui indigenous peoples.[216] These were both processes driven by the Rights of Nature Observatory, a Montreal-based non-governmental organization that promotes the rights of nature.[217] The Observatory contributed to the drafting of both the legal norms that seek to recognize the St. Lawrence

---

[213] The bill states that '[t]he concept of rights of nature has been developed in various jurisdictions, through laws and court decisions recognising nature as a subject with inalienable rights at various levels of government around the world, including Ecuador, Bolivia, Colombia, India, Mexico, New Zealand, and the United States, in California. Laws are developing further in Argentina, Brazil, Cameroon, Nepal, and elsewhere.' Ibid. The press release with which Leatherback Project, Earth Law Center, and Rights for Nature announced and celebrated the Panamanian Congress's approval of a bill they helped construct also notes that 'Panama now joins a number of countries that recognise the Rights of Nature at the national level. The list also includes Bolivia, which passed the Law of the Rights of Mother Earth (071) in 2010 and Law of Mother Earth and Integral Development for Living Well (300) in 2012, and Ecuador, which codified the rights of nature (Pachamama) into its Constitution in 2008. Uganda also has a national rights of nature provision that applies only to designated areas, and Chile is considering incorporating rights of nature into its new Constitution.' at <http://files.harmonywithnatureun.org/uploads/upload1169.pdf>.

[214] <https://celdf.org/rights-of-nature/timeline/> and CELDF (n 134).

[215] House of Commons of Canada, Bill C-271 – St. Lawrence River Capacity and Protection Act. at <http://files.harmonywithnatureun.org/uploads/upload1277.pdf>.

[216] Conseil Des Innu Ekuannishit, No consecutif 919-082, No de Dossier, 919-01-18, <http://files.harmonywithnatureun.org/uploads/upload1072.pdf>; and Province de Québec – Municipalité Régionale de Comté de Minganie, Résolution no 025-21, Reconnaissance de la personnalité juridique et des droitsde la rivière Magpie – Mutehekau Shipu at <http://files.harmonywithnatureun.org/uploads/upload1069.pdf>.

[217] 'We promote the recognition of the rights of Nature in Quebec and Canada, and contribute to their recognition worldwide.' at <https://observatoirenature.org/observatorio/en/home/>.

River as a subject of rights and to the two identically worded norms issued by the Ekuanitshit Innu Council and the Minganie MRC that serve the same purpose with respect to the Magpie River.[218] The Observatory mentions the Colombian ruling on the Atrato River, the Indian ruling recognizing the Ganges and Yamuna rivers, the Bangladeshi ruling recognizing all rivers in Bangladesh as subjects of rights, the Ecuadorian Constitution, the Bolivian Mother Earth laws, and the New Zealand laws recognizing a river and a forest as subjects of rights as antecedents to the proposal to recognize the St. Lawrence River as a subject of rights.[219] It also recognizes the case of the Atrato River as one of the influences on the norms on the Magpie River. However, it also mentions the New Zealand case[220] and the resolutions of some US indigenous peoples that recognize river rights.[221]

IV. Conclusion

The rights of nature are a global discourse today. This global discourse constructs a new subject of rights – nature – and it confers a particular series of rights to nature, including the right to life and the right not to be contaminated. The discur-

---

[218] 'The recognition of the Magpie River has inspired other initiatives across the country to embrace the rights of Nature movement. Some of them are the St. Lawrence River project led by the International Observatory of Nature's Rights, which seeks to recognize the personhood and rights of the river. The Observatory has started a conversation with political parties in Quebec and in Canada at federal level, also with many Indigenous territories and municipalities that the St. Lawrence flows through'. Also, '[t]he International Observatory of the Rights of Nature (IOIRN) was deeply involved in the drafting of the resolutions that were adopted in the case of the Magpie River in Canada'; and 'the team of the International Observatory of the Rights of Nature, which contributed to the drafting of the two parallel resolutions [on the Magpie River], is grateful for the valuable contribution of the Observatory's research associates, experts in the rights of nature and indigenous law […]'. Vega Cárdenas and Mestokosho, 'Recognizing the Legal Personality of a River' (n 24) 190ff.

[219] Vega Cárdenas and Mestokosho, 'Recognizing the Legal Personality of a River' (n 24) 190ff.

[220] 'Innu youth were profoundly inspired by the Māori negotiations with the government of New Zealand, which spanned decades and ultimately gifted us with the fruit of legal personality for the Whanganui River. Links were created between Indigenous young leaders around the world, particularly at the second *Mni Ki Wakan*: World Indigenous Peoples Decade of Water Summit.' Vega Cárdenas and Mestokosho, 'Recognizing the Legal Personality of a River' (n 24) 190ff.

[221] 'The OIDN acknowledges that the rights of the River as recognised in the resolutions were inspired by the Universal Declaration of the Rights of Rivers written by the Earth Law Center, the *Te Awa Tupua* (Whanganui River Claims Settlement Act) that was adopted in New Zealand recognising the rights of the Whanganui River, the Colombian Constitutional Court decision C-622/2016 in which the *Atrato* river was recognised as a legal entity with rights, and some Indigenous resolutions adopted in the United States to grant rights to rivers such as the Snake and Klamath River, among others.' Vega Cárdenas and Mestokosho, 'Recognizing the Legal Personality of a River' (n 24) 190ff.

sive pattern on the rights of nature has three components: poietic discourse, mimetic discourse, and discourse on resistance/indifference. The first articulates the central discourse patterns on rights of nature, that is, the seminal discursive structures on the rights of nature (Bolivia, Ecuador, and New Zealand). The second collects discursive patterns that reproduce, albeit with some variations, the prototypical models on the rights of nature. The third offers discursive patterns that resist or ignore rights of nature. However, we cannot fully understand the rights of nature without examining them from the perspective of comparative law. The rights of nature intersect with three central themes in contemporary comparative law: the political economy of legal knowledge, legal pluralism, and cross-fertilization.

The rights of nature challenge the dominant political economy of legal knowledge through the culturally-hybrid character of their paradigmatic models, and they call into question the contexts of legal knowledge production from which they originate and the political processes that led to their emergence. Similarly, these paradigmatic models challenge traditional views of legal pluralism. The rights of nature are a discourse and a practice that constructs an external legal pluralism from the periphery, in contrast to traditional views that focus on internal legal pluralism or that assume a vertical, top-down perspective between international law and national rights. Finally, the rights of nature explain legal change through heterodox political processes that question the traditional relationship between culture and law, as well as the direction taken by the dialogues and mutual learning that develop between legal systems. The rights of nature open promising doors for rethinking the way we conceive of nature and the relationship that human beings have with it. However, the rights of nature are still a young legal product. The rights of nature as well as their foundations, structures, and institutions are still evolving, as are the consequences they may have for the world.

A Hybrid of Hybrids

Global Legal Pluralism and Rights of Nature

Ralf Michaels[*]

I. Combinations[1]

Combining global legal pluralism (GLP) and rights of nature (RoN) in a single study may be unexpected at first.[2] Not fully, perhaps: to the extent that RoN emerge

---

[*] The manuscript was finalized and all internet sources were last accessed in October 2025. The assessment of literature, case law, and other materials reflects the legal situation at that time. Thanks to Michael Friedman and Dharmita Prasad for suggestions and language edits.

[1] My assumption is that no general introduction is needed for either legal pluralism or rights of nature as both are well established. For an introduction to the former, see, eg, Brian Z. Tamanaha, *Legal Pluralism Explained* (OUP 2021). For an introduction to the latter, see, eg, Mihnea Tănăsescu, *Understanding the Rights of Nature – A Critical Introduction* (transcript 2022).

[2] Though it is not unprecedented. See, eg, Helen Dancer, 'Harmony with Nature: towards a new deep legal pluralism' (2020) 53 J Legal Plur 21; Manas Divetia and Arjun Chaudhary, 'Legal Pluralism: Re-Engaging the Narrative to Solve Global Problems' (2023) 3 Indian J Integrated Rsch L 1; Elizabeth J Macpherson and Felipe C Ospina 'The Pluralism of River Rights in Aotearoa, New Zealand and Colombia' (2019) 25 J Water Law 283, 289–293. See also Minhea

sfrom indigenous communities,[3] they meet the interest of students of legal pluralism in unofficial law. But generally, one might think that as a very local phenomenon – rivers, mountains, and forests do, after all, exist at and as places – RoN has very little do with GLP as an attempt to understand law in the world at large. At least one might not expect to find specific insights beyond the relative triviality that GLP, in focusing on all legal orders in the world, will somehow also need to make place for RoN.[4]

Such skepticism would be unwarranted: the combination is unexpectedly generative in more than one way. Viewing GLP through the lens of RoN is productive, as is viewing RoN through the lens of GLP. The result is not a mere amalgam but an actual hybrid of disciplines, methods, projects that are in fruitful tension with each other.[5] Moreover, it is a hybrid of hybrids, since closer analysis reveals that both of the traditions coming together within this field – GLP and RoN – are themselves also hybrids. GLP is a hybrid between globalization and legal pluralism, and depending on which is prior, exists either as globalized legal pluralism or as pluralized global law. RON, on the other hand, is a hybrid between nature and rights, and exists as nature-based or as rights-based RON. Globalied legal pluralism nicely aligns with nature-based RON, whereas pluralized legal globalization aligns with rights-based approaches to RoN. And although it would be too simplistic to allocate the first approach to the Global South and the second to the Global North (not least because these categories themselves are somewhat questionable), we find, nonetheless, that only the former combination is truly paradigm shifting, whereas the latter is more compatible with Global North concepts of law precisely because it does not shift paradigms.

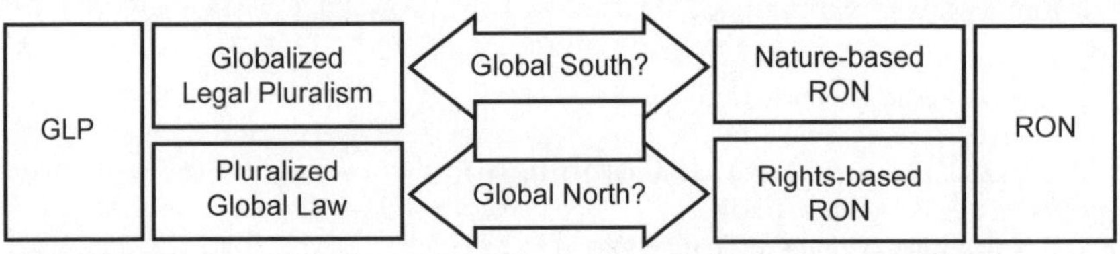

Tănăsescu, 'Theoretical Foundations of the Rights of Nature. Origins and Conceptual Characteristics' in this volume, 105–110, I.4.

[3] For doubts as to this genealogy, see Tănăsescu, 'Theoretical Foundations' (n 2) 105ff.; see also II.6.a), below.

[4] See Tănăsescu, 'Theoretical Foundations' (n 2) 105 ('rights of nature […] is but a subconcern of a much wider movement for the recognition of diverse legal traditions.')

[5] Daniel Bonilla Maldonado, 'Global Legal Pluralism and the Rights of Nature' in this volume, 16–19.

1. Global Legal Pluralism: A Hybrid of Legal Pluralism and Globalization

Global legal pluralism is a hybrid in that the field emerges from the confluence of two different traditions.[6] On the one hand we have *globalized legal pluralism*, emerging from the (predominantly local) tradition of legal pluralism that becomes globalized. Around the turn of the century, scholars began scaling up originally localized studies of normative plurality, typically in a colonial or postcolonial context and mostly as developed by anthropologists,[7] to a transnational and global level.[8] Without giving up anthropology's traditional attention to local contexts, these scholars noted interdependences between these local contexts and international law and institutions, or indeed between different localities. On the other hand we find the tradition of *pluralized global law*. In this strand, scholars interested in the globalization and transnationalization of law developed an interest in legal pluralism as a way to account for the possibility of non-state law and the plurality of laws on the global sphere.[9] If both traditions thus seem to meet, a full merger does not happen.

a) Convergence

That is not to say that these two traditions do not converge at all. They share a number of ideas. One is the opposition against traditional understandings of law that are monist in nature, that understand law as exclusively the domain of the state, and that downplay or even deny the plurality of, and interaction among, various legal orders. Legal pluralism began as a field explicitly challenging this monism, emphasizing the existence of non-state laws, in particular religious or customary laws.[10] Similarly, students of globalization and law had to grapple with

---

[6] Ralf Michaels, 'Global Legal Pluralism' (2009) 5 Annu Rev Law Soc Sci 243, 244. For other surveys, see, eg, Paul Schiff Berman, 'Understanding Global Legal Pluralism: From Local to Global, from Descriptive to Normative' in Paul Schiff Berman (ed), *Oxford Handbook of Global Legal Pluralism* (OUP 2020) 1–35; Tamanaha (n 1) ch 4.

[7] Notably, not all intellectual traditions of legal pluralism emerge from such contexts.

[8] Keebet von Benda-Beckmann, 'Transnational dimensions of legal pluralism' in Wolfgang Fikentscher (ed), *Begegnung und Konflikt – eine kulturanthropologische Bestandsaufnahme* (Verlag der Bayerischen Akademie der Wissenschaften 2001) 33–49; Keebet von Benda-Beckmann and Volkmar Gessner, 'Rechtspluralismus und globale soziale Bewegungen' (2002) 23 ZfRSoz 277–306; Sally Engle Merry, 'International Law and Sociolegal Scholarship: Toward a Spatial Global Legal Pluralism' (2007) 41 Studies in Law, Politics and Society 149–168; Boaventura de Sousa Santos, *Towards a New Legal Common Sense* (CUP 2002) 92. On this trajectory, see also Paul Schiff Berman, 'The New Legal Pluralism', (2009) 5 Annu Rev Law Soc Sci 225–242; Keebet von Benda-Beckmann and Bertram Turner, 'Anthropological Roots of Global Legal Pluralism' in Berman (n 6) 66–141.

[9] Gunther Teubner, 'Global Bukowina' in Gunther Teubner (ed), *Global Law Without a State* (Brookfield: Dartmouth 1997) 3–28; Francis Snyder, 'Governing Economic Globalisation: Global Legal Pluralism and European Law' (1999) 5 ELJ 334–374.

[10] Legal pluralism describes a situation in which two or more laws (or normative orders) coexist in (or are obeyed by) one social field (or a population or an individual). Michaels (n 6)

non-state laws, among them most obviously international law, but also transnational normative systems like the lex mercatoria.[11]

That recognition of non-state law has implications. Both legal pluralism and global law assume that a plurality of laws exists and that there are interactions between them. From the perspective of legal pluralism and sociolegal studies, Boaventura de Sousa Santos coined the influential term 'interlegality' for these interactions.[12] From the perspective of globalization, the existence of a plurality of laws that interact is not novel – in fact, an entire field deals with this issue, namely conflict of laws.[13] Under the heading of global legal pluralism, however, the concept of interlegality has been freely adopted here as well.[14]

Moreover, there is agreement that legal pluralism on a local level is connected to global and transnational interactions and vice versa – even if legal pluralists and scholars of globalization reach that connection from opposite ends. Local situations of legal pluralism have often been influenced by transnational and global dynamics, beginning with colonization, up to the present stage of late globalization. Global law on the other side has always had to account for local pluralities, as becomes visible, for example, in the attention paid by international law to minority rights and indigenous communities. These traditions converge in fruitful ways, bringing their differences in experience and focus to the table.

b) Differences

Beneath this convergence, however, the differences between the two traditions are not eradicated. Instead, they are carried into the field of GLP, making it a hybrid. These two traditions rest, largely, on different *disciplinary traditions* – legal pluralism was a project primarily for anthropologists and sociologists, whereas the globalization of law occupied mainly lawyers and legal theorists. They rep-

---

245, expanding on Sally Engle Merry, 'Legal Pluralism' (1988) 22 Law Soc Rev 869, 870. There is disagreement about the appropriateness of referring to normativities other than state law as law; for one powerful counterposition, see Simon Roberts, 'Against Legal Pluralism – Some Reflections on the Contemporary Enlargement of the Legal Domain' (1998) 30 J Legal Plur 95–106; Simon Roberts, 'After Government? On Representing Law Without the State' (2005) 68 MLR 1. Such concerns led one prominent early proponent of legal pluralism to abandon the concepts of legal pluralism in favor of normative pluralism: John Griffiths, 'The Idea of Sociology of Law and its Relation to Law and to Sociology' in Michael Freeman (ed), *Law and Sociology: Current Legal Issues*, vol 8 (OUP 2006) 49, 63–64.

[11] Teubner (n 9).

[12] Boaventura de Sousa Santos, 'Law: a Map of Misreading. Toward a Postmodern Conception of Law' (1987) 14 JL Soc 279–302; on the history of the concept, see, eg, Rainer Nickel, 'Interlegalidad' (2015) 8 Eunomia 205–211.

[13] Didier Boden, 'Le pluralisme juridique en droit international privé' (2005) 49 Arch phil droit 275–316; Ralf Michaels, 'Global Legal Pluralism and Conflict of Laws' in Berman (n 6) 629–648.

[14] Eg Jan Klabbers and Gianluigi Palombella (eds), *The Challenge of Inter-Legality* (CUP 2019).

resent different *methodologies* – thick description on the one hand; abstraction, normativity, and ordering on the other. Their *model cases* are different – local encounters between different communities on the one hand; the global interaction of various state laws and treaties and institutions on the other. They tend to represent different *attitudes* – a fascination with plurality and porous boundaries on the one hand; a concern with ultimate disorder on the other. And they invite different *projects* – emancipation from the global dominance of liberal, state-based, market-oriented law on the one hand; the organization and ultimately ordered maintenance of such plurality on the other.

(1) GLP as globalized legal pluralism

As regards GLP as globalized legal pluralism, its *disciplinary origins* lie in anthropology and sociology. This is not to say that legal pluralism has not also always been a concern for lawyers. Santi Romano's studies of legal pluralism and Eugen Ehrlich's account of living law (if we count this as a work of legal pluralism) were the work of lawyers, as were many of the accounts of the legal plurality characterizing the European middle ages.[15] Similarly, legal pluralism serves as the title for lawyers' attempts to understand and organize the interrelations between European and indigenous laws in existing and prior colonial contexts.[16] However, there is a distinct school of legal pluralism that emerges from the interest of anthropologists and sociologists in non-European normative orders, especially in Africa and Asia, and from the concrete and local ways in which they interact with each other and, more importantly, the European law of the colonizers.

Anthropologists and sociologist are, by disciplinary training, interested in practices more than legal rules.[17] Their *methodological approach* to law is, in the famous words of Clifford Geertz, one of 'thick description' rather than the abstraction typically found in legal rules and judicial argument.[18] What they are inter-

[15] Santi Romano, *The Legal Order* (Routledge 2017, first published in Italian 1917/18); Eugen Ehrlich, *Fundamental Principles of the Sociology of Law* (Ayer Co Pub 2001, first published in German 1913). For the history of legal pluralism, see also Emanuele Conte, 'Legal Pluralism from History to Theory and Back: Otto von Gierke, Santi Romano, and Francesco Calasso on Medieval Institutions' (2024) 42 Law and History Review 169–180. See, generally, Tamanaha (n 1); Ralf Seinecke, *Das Recht des Rechtspluralismus* (Mohr Siebeck 2015).

[16] MB Hooker, *Legal Pluralism: An Introduction to Colonial and Neo-colonial Laws* (OUP 1976).

[17] In fact, their disinterest in – and indeed suspicion of – legal rules and the operations of the law and its officials sometimes goes so far as to create an actual gap of relevant attention; see Ralf Michaels and Annelise Riles, 'Law as Technique' in Marie-Claire Foblets and others (eds), *Oxford Handbook of Law and Anthropology* (OUP 2022) 860–878.

[18] Clifford Geertz, 'Thick Description: Toward an Interpretive Theory of Culture' in Clifford Geertz, *The Interpretation of Cultures: Selected Essays* (Basic Books 1973) 3–30; Kathrin Seidel, 'Comparative Methods in Legal Anthropology: Thick Comparison Through (Cultural) Translation' in Maurice Adams and Mark van Hoecke (eds), *Comparative Methods in Law, Humanities and Social Sciences* (Edward Elgar 2021) 96–120.

ested in is how plural laws operate in concrete societies – how and by whom they are made, whose interests they serve, how they operate between legal officials and subjects of the law, what general attitudes individuals have towards plural legal rules. Interlegality is here a largely descriptive term for a situation in the in-between of laws that is not – and cannot be – resolved, but also need not be deplored. This methodological inheritance is then transferred to the transnational and global sphere and applied there.[19]

We see this attitude also in the *model cases* addressed in this tradition, which tend to be, by and large, localized even when they introduce global elements, eg the invocation ('vernacularization') of human rights by Indian women[20] or the local variants of Islamic law.[21] International law and religious law may be international or even global, and serve as a discursive connection between different localities. But the interest remains, very much, in the local. And that interest translates into *attitudes* that remain, by and large, rather critical of globalization, at least insofar as it is understood as global universalization of norms, laws, and market conditions. Localized customary law is viewed as being under threat from the forces of globalization, often viewed as extensions of the state.

(2) GLP as pluralized legal globalization

Looking at this disciplinary inheritance through these categories helps clarify the differences existing within the seemingly uniform field of GLP, because the other tradition of GLP, that coming from globalization, looks very different. *Disciplinarily*, legal globalization's roots of analyses lie not in anthropology and sociology but in legal disciplines like public international law, comparative law, transnational commercial law, legal theory, and other disciplines having a close affinity like international relations and economics.[22] Even where concepts like pluralism and interlegality are borrowed from anthropology, they are typically translated into the rationality of these disciplines.

Unsurprisingly, these different methodological roots lead to different *methodologies*. We find less attention to developments within specific spaces and more attempts to explain the world of plural law in its entirety. Instead of close ethnographic attention to detail, students of pluralized global law tend to strive for abstraction and generalization. And their focus lies less on practices and more on legal rules, norms, and institutions.

---

[19] Sally Engle Merry, 'An Anthropological Perspective on Legal Pluralism', in Berman (n 6) 169–185.

[20] Peggy Levitt and Sally Engle Merry, 'Vernacularization on the Ground: Local Uses of Global Women's Rights in Peru, China, India and the United States' (2009) 9 Global Networks 441–461.

[21] von Benda-Beckmann (n 8) 41ff.

[22] Ralf Michaels, 'Welche Globalisierung für das Recht? Welches Recht für die Globalisierung?' (2005) 69 RabelsZ 525, 526ff.

The *model cases* of pluralized legal globalization are not local norms. Instead, pluralized global law often starts from interactions between traditional legal orders, in particular between state laws, between state law and international law, or between national and supranational courts. Where attention is paid to non-state laws outside supranational and international law, the non-state laws in question are typically those emerging from the globalized economy – be it lex mercatoria and its close cousins having similarly Latinized names (lex sportiva etc.) or be it a presumed autonomous law of the internet. These are not normativities emerging from closely-knit communities – to the contrary, they are laws presumably needed precisely in the absence of such communities.

And the *projects* emerging from pluralized legal globalization are specific to these underlying traditions and methodologies, too. Here, plurality and its ensuing disorder is more often seen as a problem than as an asset. Consequently, projects aim at 'ordering pluralism', typically by legal rules and institutions, and thus rely on a strongly normative perspective.

Table 1: Global Legal Pluralism

	Globalized Legal Pluralism	*Pluralized Global Law*
Similarities	Acceptance of non-state law Plurality of, and interactions between, laws ('interlegality') Connections between global and local	
Disciplinary traditions	Anthropology, Sociology	Law, International Relations
Methods	Thick description	Normativity
Model cases	Local communities	State interactions
Attitudes	Plurality as a benefit	Plurality as a problem
Projects	Emancipation from dominance	Ordering of pluralism

2. *Rights of Nature: A Hybrid between Nature and Rights*

It is perhaps less obvious but no less important that the global discourse on rights of nature also lies at the convergence of (at least) two very different traditions, with the consequence that what may seem like a broadly common discourse is actually also a hybrid that displays a set of both commonalities and internal frictions. Here, one tradition begins with ideas of nature and its subjectivity. In this tradition, providing nature with rights is only a second step, a consequence. The other tradition, by contrast, starts with the idea of rights and expands the set of possible rightsholders to include nature. We can call the first tradition a nature-based approach and the second a rights-based approach to RoN. Distinguishing them in this way requires working with ideal types. In reality, many intermediate types exist, and the political character of RoN means that existing differences are frequently downplayed in favor of building effective alliances across traditions. Nonetheless, separating them is instructive.

a) Convergence

That this internal friction within RoN is not immediately visible is due in large part to the fact that the field, despite its remarkable and remarkably rapid rise, remains a recent minority approach directed against an overwhelmingly established paradigm of law. In that paradigm, a strict separation exists between the subjects and objects of law. Subjects are natural persons and legal entities; they are capable of holding rights. Neither nature itself nor specific emanations – forests, animals, lakes – are considered such a person. As a consequence, nature cannot hold rights but only be the object of rights. At best, what can exist are rights *to* nature – rights (of humans) in which nature remains an object.[23]

Against this orthodoxy, RoN, in both named traditions, emphasizes that nature can be a subject and a rightsholder – whether that rightsholder is nature as such or whether it be distinct entities is insofar secondary. Importantly, RoN insists that these rights are not only moral or metaphorical but legal rights that can be enforced in courts of law and that can trump the interests and rights of humans or corporations. RoN can rely on positive law where such rights have been implemented in constitutions, treaties, statutes, or judge-made law. But RoN also relies on ontological, deontological, and instrumental arguments for why such rights are desirable. Ontologically, subjectivity is not restricted to humans. Deontologically, nature ought to be protected as nature. Instrumentally, its protection contributes to sustainability, and environmental protection and thus benefits humans and the planet.

As a consequence, RoN discourse is mostly about policy and law reform and not theory. Of course, RoN are also of considerable theoretical interest, but even scholars interested in their theory seldom detach that theoretical interest from normative considerations. Largely, RoN proponents work actively towards the implementation of rights of nature within legal orders, and towards strengthening such rights where they exist. Because these policy interests align, broadly, among these traditions, the differences do not matter very much compared to the advantages of an alliance.

Finally, we find that the discourse on RoN continues to revolve around the first countries in which RoN were recognized, first and foremost Ecuador, Bolivia, and New Zealand, but also Colombia. From the perspective of comparative law this is remarkable, because at least the Latin American countries in this list are

[23] Fátima Alves and others, 'The Rights of Nature and The human Right to Nature: An Overview of the European Legal System and Challenges for the Ecological Transition' (2023) 11 Frontiers in Environmental Science 1175143, <https://doi.org/10.3389/fenvs.2023.1175143>; see also Mara Tignino, 'International Law. Rights of Nature and Rights to Nature: Emerging Trends' in this volume, 137–142. On rights to nature in Cyprus since a recent constitutional amendment, see Konstantinos A Rokas, 'Cyprus. Constitutional Adoption of a Right to Nature, but No Formal Recognition of Nature's Rights' in this volume, 429–432.

frequently viewed as having inferior laws, with the consequence that comparative lawyers' interest is traditionally in their failings, not in what they can model.[24]

b) Differences

As is the case within GLP, there are, however, important differences between different traditions of RoN.[25]

(1) Nature-based approaches to RoN

Let me begin with nature-based approaches to RoN. Here, the *disciplinary origin* is not legal but ontological, frequently emerging from non-European, indigenous ontologies. In these ontologies, nature is a subject.[26] There are different variants of this ontological staring point. In one, subjectivity is multiplied in the sense that everything in the world, whether living or not living, has such subjectivity. In a second one, there is only one real subject: Pachamama for example. In a third one, the very idea of subjectivity is rejected because it suggests a separation of the subject from the surrounding world. Either way, these ontologies suggest a dramatic difference from European ontologies that rest heavily on the separation of man and environment, or on subject and object, and the idea that man should subdue the earth (Genesis 1:28).

This ontological approach invites a specific *methodological* corpus. It becomes necessary to transcend the difference between subject and object. Humans and nature can no longer be seen as separated because humans are part of nature, and human violence against nature thereby becomes self-violation at the same time. The implications of this are obviously quite profound in general, but also in particular for the law. Subjective rights are usually rights that imply both objects and a separation from others: they create legal relations with others regarding certain objects. The idea of rights must stand in tension to an ontology in which neither objects nor separations exist in the same way. The form of RoN is then a compromise, a tactical move to ensure the subjectivity of nature.

The need for such a tactical move arose first, not surprisingly, in countries in the Global South. Here, colonial and postcolonial extractive policies had devastating consequences, material and epistemic. Materially, nature was and is being

---

[24] See, eg, Jorge Esquirol, *Ruling the Law: Legitimacy and Failure in Latin American Legal Systems* (CUP 2019); Daniel Maldonado Bonilla, *Legal Barbarians: Identity, Modern Comparative Law and the Global South* (CUP 2020); Lena Salaymeh and Ralf Michaels, 'Decolonial Comparative Law: A Conceptual Beginning' (2022) 86 RabelsZ 166–188; Kevin E Davis and Mariana Pargendler (eds), *Legal Heterodoxy in the Global South* (CUP 2025).

[25] See Ariel Rawson and Becky Mansfield, 'Producing juridical knowledge: "Rights of Nature" or the Naturalization of Rights?' (2018) 1 Environment and Planning E: Nature and Space 99–119.

[26] Eric Katz, *Nature as Subject: Human Obligation and Natural Community* (Rowman and Littlefield 1996); Constitutional Court of Colombia 10 November 2016, T-622/16, *The Atrato River Case*; Bonilla, 'The Globalization of Rights of Nature' (n 5) 3.

destroyed at rapid pace. Epistemically, European universalism came to replace local ontologies. It is not surprising, therefore, that RoN first arose, properly, in Latin America, as the fruit of local movements resisting colonialism. And it is not surprising that these movements serve as the *model case* for the RoN discourse worldwide. In this sense, RoN is a decolonial project.[27] Although RoN borrow the legal form of rights from European law, their *attitude* is therefore nothing less than revolutionary. The aim of nature-based RoN is – indeed must be – to challenge the existing paradigm of law. Whether that means establishing a different paradigm or overcoming law altogether is of comparatively minor importance.

(2) Rights-based approaches to RoN

Right-based approaches differ from nature-based approaches in all these regards. Their *origin* lies in European traditions of law, more specifically in the invention of subjective rights.[28] The origin of this idea lies in a confluence of a religious emphasis on the individual and a doctrinal development in which ius in an objective sense (law) became separated from ius in a subjective sense (right). Such rights originally belonged only to humans, more precisely males with a certain status (and often as representatives of households). Those holding no rights were objects of such rights – unless they could be emancipated. It took time until the ability to hold rights was expanded to children, women, and enslaved people. It also took time until this ability was granted to non-humans, like states and corporations. And developments are still in flux.

Within this tradition, granting rights to nature does not flow as naturally as is sometimes claimed. Traditionally, rights are given to non-humans only if the recipients are either human-like (such as animals or robots) or represent humans behind them (such as states or corporations). Christopher Stone, in his seminal 1972 article, indeed drew on this trajectory.[29] Rivers and mountains are neither,

[27] Ariella Kely Besing Motter and Miguel Etinger de Araujo Junior, 'From the Modern Individual to Nature: Decolonial Perspectives of the Concept of "Subject of Rights"' (2024) 15 Revista Catalana de Dret Ambiental 1, 8ff; Maria Andrea Nardi, Torsten Krause and Fariborz Zelli, 'Diverse Understandings and Values of Nature at the Peace–Environment Nexus: A Critical Analysis and Policy Implications Towards Decolonial Peace' (2024) 29 Ecology and Society 41.

[28] A considerable debate exists about the origin of subjective rights – their arising as early as Greek antiquity, or as late as with Ockham. This is not relevant here. See, from the voluminous literature, Richard Tuck, *Natural Rights Theories: Their Origin and Development* (CUP 1981); Annabel S Brett, *Liberty, Right and Nature: Individual Rights in Later Scholastic Thought*, (CUP 1997); Brian Tierney, *The Idea of Natural Rights. Studies on Natural Rights, Natural Law and Church Law*, (Scholars Press 1997); Michel Villey, *La formation de la pensée juridique modern* (puf 2013); Christoph Menke, *Critique of Rights* (Polity 2020). For a brief survey, see Thomas Mautner, 'How Rights Became "Subjective"' (2013) 26 Ratio Juris 111–123.

[29] Christopher Stone, 'Should Trees Have Standing? Toward Legal Rights for Natural bjects' (1972) 45 SCal LRev 450, 450–456.

but structurally the expansion is possible. And strategically, the expansion became interesting in order to make sure that environmental concerns could be litigated and enforced. In 1972, there was very little environmental law altogether. Arguing for rights of nature thus played out nicely within a tradition in which legal and social change had traditionally been achieved through private plaintiffs, and granting rights was a way for the law to recognize justifiable interests. Calling these approaches rights-based should therefore not suggest that it did not matter at all who the recipient of rights was – proponents of rights of nature are all in support of nature and of environmental protection. It does mean, however, that the argument starts with granting rights as a strategic step, not as recognition of a preexisting subjecthood.

The proposal was thus radical in one way but conventional in another. The trajectory drawn by Stone suggests a *method* of moving the borders existing between subjects and objects of law. That method was not new – this had been done before and could be done again. The idea was not to grant rights to everyone and everything. In fact, Stone, like others in the rights-based tradition, struggles to define where the boundary between subjects and objects should now lie – suggesting, in other words, that an abolition of the distinction was not intended. Instead, rights of nature in this tradition closely resemble rights of other nonnatural entities, like a Hindu statue which the Privy Council recognized as early as 1925.[30]

Despite Stone's article (and some other scholarly writings),[31] the main *model case* for rights-based literature today lies in the same countries as that of the nature-based approaches: Ecuador, Bolivia, Colombia, and New Zealand. This seems to mark a similarity with nature-based approaches, but closer inspection demonstrates an important difference: What counts as a model is, first and foremost, the legal form that RoN have achieved. The model is the Ecuadorian Constitution, not the underlying indigenous ontologies – indeed, when the latter are named, this often happens in a superficial, sometimes romanticizing way.[32]

It cannot come as surprising that the *attitude* of this rights-based approach is not revolutionary but reformist. Its goal is not – like that of the nature-based tradition – to replace the existing paradigm of law with another one. Rather, its aim is to use the existing paradigm as a framework within which interests of nature can better be protected. Often, therefore, the granting of rights to nature is presented not as a sudden break with tradition but, to the contrary, as an almost necessary consequence of increased protection for environment on the one hand and human rights on the other. The *project* is one of reform, not revolution.

---

[30] *Pramatha Nath Mullick v Pradyumna Kumar Mullick*; see William Seton Gordon, 'Legal Rights of an East Indian Idol' (1925) 11 ABA J 431–432; PW Duff, 'The Personality of an Idol' (1927) 3 CLJ 42–48.

[31] See Tănăsescu, 'Theoretical Foundations' (n 2) 103ff., also in terms of differences.

[32] See Dirk Hanschel and Annette Mehlhorn, 'Law and Anthropology. An Interdisciplinary Perspective on Rights of Natur' in this volume, 159–161.

Table 2: Rights of Nature

	Nature-Based RoN	*Rights-Based RoN*
Similarities	Rights to nature Policy interests Latin America as model	
Disciplinary traditions	Indigenous ontologies	Rights theories
Methods	Transcending distinction between man and nature, or subject and object	Shifting the border between man and nature, or subject and object
Model cases	Southern communities	Southern legal texts
Attitudes	Revolutionary	Reformist
Projects	World reform: Overcome the existing paradigm of law	Law reform: Improve the existing paradigm of law

3. GLP and RoN as a Hybrid of Hybrids

Taken together, each field provides a useful lens for viewing the other. The lens of global legal pluralism allows a richer understanding of the global RoN discourse as more than just an amalgamation of localized, indeed nationalized discourses and regulations. A GLP perspective allows one to understand the interactions between these localized discourses as constitutive of the global discourse. On the other hand, it is useful to use RoN as a particularly productive concrete example for a richer understanding of GLP. This is so in large part because RoN represent the still relatively rare case of a Global South invention that influences the Global North.

The interconnection is instructive in other ways, too. Neither GLP nor RoN are internally coherent fields claiming a broad consensus. Instead, each is characterized by an internal friction. Remarkably, the interconnection reveals that the internal friction within each of the two fields – GLP and RoN – is replicated within the other. The friction within the field of GLP exists between one approach that is legalistic and normative and another that is critical and emancipatory. A comparable friction exists within RoN between a more legalistic understanding that emphasizes the 'rights' part of RoN and an emancipatory understanding that focuses more on 'nature' and its subjectivity.

The result is that the interconnection between GLP and RoN is twofold. There are two coherent interconnections – elective affinities – between the legalistic versions of GLP and between the emancipatory versions, respectively. And there are two new frictions – between legalistic GLP and emancipatory RoN on the one hand and between emancipatory GLP and RoN on the other. Bringing together GLP and RoN thus creates a hybrid of hybrids.

a) Incompatible Combinations

The frictions can be explained quickly because they create incompatibilities. It is very difficult to combine the rather legalistic idea of GLP as pluralized global law with the revolutionary potential of nature-based RoN. Pluralized global law's interest in ordering plurality is not reconcilable with the radical rejection of subject-object differences in indigenous ontologies. That tension sits uneasily on both sides. On the one hand, pluralized global law is not, indeed almost cannot be, interested in radical ontologies because these would undermine its very project. What we find instead, frequently, are rather superficial references to 'indigenous ontologies' as a undifferentiated amalgam of general nature-friendly attitudes. On the other hand, nature-based RoN will use pluralized global law, at most, strategically. That is so because pluralized global law emerges, ultimately, from the same kind of legal tradition that creates a threat for nature.

Equally difficult, on the other hand, is a combination of the anthropological idea of globalized legal pluralism and the legalistic concept of rights-based RoN. Anthropologists and legal pluralists are, of course, very interested in RoN.[33] But their interest does not typically lie in their particular form as rights, rules, regulations, or institutions; rather, it targets their expression and their political and symbolic power.[34] They are therefore suspicious of RoN as a governance project. Rights-based RoN, on the other hand, has only superficial interest in legal pluralism. Certainly, the idea of indigenous normativities that recognize RoN is attractive from a legalistic perspective, but more as a type of deferred competence in the sense of ordered legal pluralism than as something more radical.

b) Reform: Pluralized Global Law and Rights-Based RoN

Beyond these frictions, two natural affinities exist. The first can be called reformist and belongs to the more legalistic versions of both themes: pluralized global law on the one hand, rights-based RoN on the other. The legalistic approach to GLP inherent in pluralized global law is quite compatible with the similarly legalistic theories of rights – both emerge, ultimately, from the same Northern tradition of law and of rights. Pluralized global law can account, relatively easily, for new kinds of normativities and new kinds of rights. It takes very little, therefore, for pluralized global law to include these new developments – a little openness to inventions from the South, a little openness for a broadened idea of who can hold rights, but not much more. The border between subject and object has shifted a little, but that is familiar to GLP with its concern about shifting the border between law and non-law. And rights-based RoN, on the other hand, can easily claim and find their place within a plural understanding of law. If rights are claims to recognition, then a plural understanding is congenial to a formulation of interests in the form of rights.

---

[33] Hanschel and Mehlhorn, 'Law and Anthropology' (n 32).

[34] See n 17, above.

The model cases – the constitution of Ecuador, the legislation of Bolivia, the treaties of New Zealand, the jurisprudence of the Colombian Court – may look radical in their substance, indeed they are still often thought of as such, especially from the perspective of the Global North. In their formal structures, however, they are quite familiar to pluralized global law. In this paradigm, the pluralism inherent in the idea of RoN is not a problem because it has already been sanitized by the recognition of these rights by the state. RoN can be accommodated within the ordered legal pluralism that is the goal of pluralized global law, and indeed they can be viewed as welcome instruments towards an improvement of existing law, especially under the threat of the climate catastrophe.

All in all, this combination represents, therefore, a thoroughly reformist agenda. Reform is certainly preferable to stasis. At the same time, it seems likely that critics who do not find pluralized global law to be radical enough will voice their same criticisms against rights-based RoN and vice versa. RoN are not used, in this combination, to challenge the idea of order inherent in pluralized global law. And GLP, by contrast, presents no threat to the legalistic form of rights. Instead, pluralized global law and rights-based RoN mutually support each other.

Table 3: Reformist Combination

	Pluralized Global Law	*Rights-Based RoN*
Disciplinary traditions	Law and rights	
Methods	Normative arguments to shift the border between subject and object	
Model cases	Southern legal texts as parts of a global law	
Attitudes	Legal reform to overcome problems of plurality	
Projects	Improve existing global law by accounting for RoN	

c) Revolution: Globalized Plurality and Nature-Based RoN

A very different picture emerges from a combination of the two nonlegalistic conceptions of GLP and RoN: globalized plurality and nature-based RoN. Here, a very different affinity of traditions emerges, that between anthropology as a field keenly interested in non-Northern ontologies and these ontologies themselves. This affinity leads to a very different methodological approach. From the perspective of anthropology, we find attempts to understand the relation between humans and nature (or their place within nature) as radically different ways of being, thereby requiring a more radical, nonhierarchical version of legal pluralism. From the perspective of these ontologies, on the other hand, the anthropological concept of a deep, nonordered pluralism is certainly more attractive. So is the nonnormative approach: Southern ontologies are united in their refusal to accept the Northern distinction between is and ought, between ontology and normativity. Here, the central position is assumed by local communities coupled with an interest in their practices rather than a reductive attention to the legal texts they produce.

The consequence is that RoN and GLP, combined in these two variants, achieve a revolutionary potential. Using nature's subjectivity and Southern communities as a starting point represents a radical opposition to the universalism of Northern law, including the centrality of humans and their rights within it. Starting from RoN in a Southern context requires acceptance of polycentricity and thus abandonment of the Northern universalism that characterizes the dominant paradigm of law. The result is a radicalized global legal pluralism that accounts not only for non-state normativities as equivalent to state law but even for nonhuman normativities.

Such a result is no longer compatible with the existing global paradigm of law and can no longer be brought into the order desired by Northern law. RoN of this kind can no longer be tamed through governance: their claims are more radical. This presents a threat and a promise – a threat to the established status quo of modern law and a promise of alternative, plural, modes of legality. Notably, RoN, including in a nature-based fashion, are not sufficient to formulate a new paradigm. In fact, it is doubtful whether the formulation as rights can truly bring about a new paradigm rather than getting stuck half way between the old and the new. Whereas RoN were possibly too radical for pluralized global law, they may not be radical enough for globalized pluralism.

This is only one concern about the revolution promised by this combination. Another is its potential infeasibility. The discourse about RoN has, so far, been deliberately more reformist than revolutionary: proponents put more hope in change within the existing system than change of that system itself. Their critics who want more radical solutions often fail to consider whether claims for more radicalism are at all realistic. And indeed, just as anthropologists normally do not translate their sometimes radical analyses into radical politics (and would indeed, more often than not, reject any plea to do that), so the proponents of nature-based RoN have little interest in replacing the current rules of the world. A revolutionary approach thus risks remaining hypothetical only.

Table 4: Revolutionary Combination

	Globalized Legal Pluralism	*Nature-Based RoN*
Disciplinary traditions	Anthropology of indigenous ontologies	
Methods	Thick description of the relation between man and nature	
Model cases	Local communities in the South	
Attitudes	Plurality as revolutionary challenge to Northern law	
Projects	Emancipation from dominance of the existing paradigm of law	

4. GLP and RoN between the Global North and Global South

The point is not to suggest a priority of one or the other combinations; this cannot in any event be done abstractly and without detailed analysis. At this stage, the analysis remains deliberately abstract. It is not useless, however, because it achieves a more modest but nevertheless important goal. The analysis has demonstrated that the combination of GLP and RoN creates two relatively coherent combinations. One is legalistic, the other one is not. One looks at rules, the other at practices and ontologies. One aims at governing RoN, the other welcomes their plurality. One is reformist, the other one is revolutionary.

Does this distinction match the (problematic) distinction between Global North and Global South?[35] Is the Global North reformist, and the Global South revolutionary? So far I have left these questions out. The Global South has appeared as the most important origin of RoN, but it has not been connected with either strand of GLP or RoN. And indeed, a simple parallel – Global South is revolutionary, Global North is reformist – would be too simple. The concept of the Global South/Global North dichotomy itself rejects a simple geographic dichotomy – there is a North in the South and a South in the North, and the Global South is better understood as an idea than as a mere territory. (The South is 'global' for a reason.) Thus, we should expect to find both reformers and revolutionaries. Indeed, as we will see, the reality of RoN in Latin America lies between revolution and reform. In the North, however, not much revolution is to be seen. And in this limited sense, there is then a partial affinity between revolution and the South. If revolution is to come, it is through the South. And if it is to become global, the North must open itself more than it has so far, even where – albeit with much hesitation – it does recognize RoN. Translation from the Global South to the Global North is necessary and not seamless.[36] Its potential, however, is enormous.

II. Comparisons

1. Introduction

Viewing the connection of RoN and GLP as a hybrid of hybrids provides a useful perspective. This becomes clear from a comparison of the country reports received in response to a questionnaire formulated with a view to this project.[37] One

---

[35] See Sina Theres Kloß, 'The Global South as Subversive Practice: Challenges and Potentials of a Heuristic Concept' (2017) 11 The Global South 1–17. My own brief analysis is in Ralf Michaels, 'Internationales Privatrecht und Globaler Süden' (2025) IPRax 377, 378–379.

[36] Jenny García Ruales and Yaku Viteri Gualinga, 'A Well-braided (Knowledge) Braid – Lessons Learned from the Kawsak Sacha and the Forest Beings to Europe' in Jenny García Ruales and others (eds), *Rights of Nature in Europe. Encounters and Visions* (Routledge 2024) 27ff.

[37] See Annex 639ff.

problem of country reports in comparative law is that they replicate, and thereby reiterate, the separation of law in the world into distinct national legal orders – the very kind of structure GLP calls into question. Comparison, however, can provide a perspective that manages to cut through this distinction and reveal connections and frictions that transcend national boundaries.

Essentially, what we find is a global legal discourse on RoN. Its main separation is not between countries that recognize RoN and those that do not. First, that relationship is in flux, as RoN are quickly being adopted in more and more jurisdictions. Second, given that most recognition of RoN happens on a substantive, often even local, level, it is hard to group legal orders this way. What is possible, however, is to group countries as participants in, or bystanders to, that very discourse, as evidenced by the difference of intensity with which RoN are discussed in different legal systems. Among those participating in the discourse, we find, in every country, proponents and opponents of RoN. In some countries, proponents of RoN have won out, in others, they are contesting the status quo. In some countries that recognize RoN, their enforcement is weak, whereas in some countries that refuse official recognition RoN nonetheless play a significant role in debates.

2. *Recognition of Rights of Nature*

The rapid spread and broad adoption of rights of nature in legal systems is astonishing. It has been less than twenty years since their first adoption in national legal orders, notably Ecuador (2008),[38] Bolivia (2009/2010/2012),[39] and New Zealand (2014/2017).[40] Bonilla rightly refers to these countries as poietic since they provided the paradigmatic legal formulations of rights of nature that shaped the subsequent spread and discussion.[41] The Latin American models, in particular, are regularly referred to in discourse and projects all over the world, though frequently stripped of the particular context in which they emerged. Now, there are a good number of countries in which rights of nature have been recognized in some form, this in addition to their increasing recognition in international law.

---

[38] Constitution, art 10(1); see Gabriela Espinoza, 'Ecuador. The Nature of the Rights of Nature' in this volume, 213–214.

[39] The 2009 Constitution referred to Pachamama in its Preamble; the Law on the Rights of Mother Earth (Ley de Derechos de la Madre Tierra 2010 (Law 71)) and the Framework Law on Mother Earth and Integral Development for Living Well (Ley Marco de la Madre Tierra y Desarrollo Integral para Vivir Bien 2012 (Law 300)) firmly establishes rights of nature; see Cletus Gregor Barié, 'Bolivia. The Rights of Mother Earth: Fifteen Years of Solitude' in this volume, 258–264; see also Cletus Gregor Barié, *The Plurinational Challenge: Intercultural state-building and the rights of nature in Bolivia* (PhD Thesis VU Amsterdam 2024), <https://doi.org/10.5463/thesis.596>.

[40] Catherine Iorns, 'Aotearoa New Zealand. Protecting Nature through Upholding Indigenous Rights and Human Responsibilities' in this volume, 292–309.

[41] Bonilla, 'The Globalization of Rights of Nature' (n 5) 8.

Responses to the paradigmatic formulations can be fruitfully distinguished as mimesis and resistance.[42] Mimesis first happened in countries of the Global South, and countries of the Global North have mostly been reluctant. However, beyond that the distribution of rights of nature does not follow a simple North/South pattern. Instead, we find most support in the Americas, meaning both South and North America, though in South America recognition is frequently a matter of national law whereas in North America rights of nature are mostly recognized locally. There is also a considerable level of recognition in the Pacific region, no doubt inspired by the New Zealand example. By contrast, African countries, with the exception of Uganda, and Asian countries, with the exception of India, have not been very open to rights of nature.

From a comparative law perspective, it makes sense to separate these legal systems. From the perspective of GLP, by contrast, it makes sense to see them all as interrelated regimes within the bigger framework of pluralistic global law.[43] That global law is, in the case of RoN, strongly based on discourse. And this suggests viewing the different instances of poiesis, mimesis, and resistance as different parts of the global law of RoN. From this perspective, the main difference is between those who participate in this global discourse and those who do not. Some participants in the discourse are *adopters*, some are *resisters*, with a certain blurring in-between – depending on who ultimately wins the debate about adopting or not adopting RoN. But some countries are not so much resisters as *ignorers* who have no occasion to adopt RoN because they hardly participate in the discourse at all.

a) Adopters

Official adoption of RoN happens in different forms and on different levels. National recognition is surprisingly rare. Ecuador seems to be alone in establishing rights of nature in its national *constitution*, though in Mexico the subnational constitution of the city of Mexico did the same,[44] and Pachamama is even found in the Preamble of the Bolivian Constitution.[45] Similarly, New Zealand appears to stand alone in basing its recognition of RoN on *treaties* – here, granting legal personhood to a river was an ingenious solution for an impasse as to whether pro-

[42] Bonilla, 'The Globalization of Rights of Nature' (n 5) 8.

[43] See Ralf Michaels, 'Globalization and Law: Law Beyond the State' in Reza Banakar and Max Travers (eds), *Law and Social Theory* (2nd edn, Hart 2013) 287–303; Ralf Michaels, 'Transnationalizing Comparative Law' (2016) 23 MJ 352–358.

[44] Constitución Política de la Ciudad de México (2016), art 13, A(3), Gaceta Oficial última reforma GO 8 August 2023; for a proposal for a national constitutional provision, see Héctor Mauricio Villegas Rivera and Angelina Isabel Valenzuela Rendón, 'Exploración sobre el reconocimiento a nivel constitucional en méxico de la naturaleza como sujeto de derecho', (2023) 1 Revista internacional de derecho público 209–226.

[45] Nueva Constitución Política del Estado 2009, Barié, 'Bolivia' (n 39) III.1.

perty should belong to the Maori or the state.[46] Also, national *legislation* is relatively rare. Bolivia is the country with the most important national legislation, establishing rights of nature in two statutes. Uganda adopted a national statute in 2019, apparently the first African country to do so.[47] Panama did as much in 2022.[48]

One reason for the relative dearth of national regulation is that parliamentary initiatives are are rare and rarely successful;[49] Spain is a rare exception. In the Global North, few political parties push for it[50] Nationwide pressure for RoN is hard to maintain, given that their recognition is often a minority issue. For such reasons, more happens on a local sphere. Rights are often given not to nature at large but to specific entities, and these are local. Granted, local entities can also be recognized by national legislation, as was the case for the Mar Menor in Spain.[51] But local matters may be of little interest to a national legislator. That may be a reason why the rights of nature provided for in Uganda and even in Mexico City were never implemented in national legislation.[52] In Ecuador, legislative implementation as provided for in the constitution tends to assimilate rights of nature to environmental regulation, thereby missing what is specific about RoN.[53]

Mostly, therefore, we find *local legislation* establishing such rights. This is the case in federal systems – the United States, Canada, and also Brazil.[54] In Australia, the State of Victoria recognizes the Yarra River as 'one living natural entity',

---

[46] Iorns, 'Aotearoa New Zealand' (n 40) 293–294 (Whanganui River), 295 (Te Urewera), 295–296 (Taranaki Maunga).

[47] National Environment Act No 5 of 2019, s 4(1); Tajudeen Sanni 'Uganda. The First African Country to Recognize Rights of Nature' in this volume, 353ff.

[48] Ley Nº 287, de 2022, ley que reconoce los derechos de la Naturaleza y las obligaciones del Estado relacionadas con estos, available at <http://files.harmonywithnatureun.org/uploads/upload1168.pdf>.

[49] Laura Affolter and Sian Affolter, 'Switzerland. Mapping the Scene' in this volume, 624–626, Parliamentary Initiative 21.436–440 (2021).

[50] Elena Ewering, Janina Reimann and Tore Vetter, 'Germany. Rights of Nature: A Paradigm Change for the German Legal System' in this volume, 481–482; Phillip Paiement, 'Netherlands. Rights of Nature, Globalization, and Legal Pluralis' in this volume, 530–532.

[51] Ley 19/2022, de 30 de septiembre, para el reconocimiento de personalidad jurídica a la laguna del Mar Menor y su cuenca, BOE 237 (3 October 2022) 15131 <https://www.boe.es/eli/es/l/2022/09/30/19>; Marie-Christine Fuchs, 'Rights of Nature Reach Europe', (*Verfassungsblog*, 24 February 2023) <https://verfassungsblog.de/rights-of-nature-reach-europe/>; Ludwig Krämer, 'Rights of Nature in Europe: the Spanish Lagoon Mar Menor Becomes a Legal Person' (2023) 20 JEEPL 5–23.

[52] Espinoza, 'Ecuador' (n 38); Sanni, 'Uganda' (n 47).

[53] Espinoza, 'Ecuador' (n 38) 230–232.

[54] Antonio Carlos Wolkmer and Debora Ferrazzo, 'Brazil. Isolated Legislative and Judicial Inroads in a Dependent Capitalism Landscape' in this volume, 377–379, for rejection on the national level, see ibid. TAN 26; Márcia Rodrigues Bertoldi and Roberta Fortunata Silva, 'Direitos da natureza e acesso à justiça: A Ampliação dos Atores Legitimados em Ações Coletivas Para uma Justiça Socioambienta' (2020) 29 Revista Direito em Debate 118, 123–124.

though it stops short of granting it personhood.[55] Similarly, the first adoption of rights of nature in Europe was by a local community, the Italian Civitá Castella, which declared itself a nature's rights zone in 2018.[56]

And not infrequently, rights of nature are recognized in *indigenous* or other non-state regulations whose recognition by the state is uncertain – this was the case in Québec for the Magpie River.[57]

On the other hand, moving from the subnational to the *supranational level*, adoption is again more frequent – no doubt because statements in international treaties and resolutions are less immediately binding. In 2014, the Global Alliance for the Rights of Nature (GARN) established a Rights of Nature Tribunal.[58] The Tribunal holds meetings, recently in coincidence with meetings of COP, and renders decisions that are non-binding in a legal sense but promise to influence debate like its models the International War Crimes Tribunal and the Permanent Peoples' Tribunal.[59] It remains a matter of debate whether the Tribunal equates RoN to traditional liberal and rights-based institutions or whether it allows for true pluriversality.[60]

Finally, *courts* play a crucial role for the adoption of RoN. Thus, in Colombia, recognition of the River Atrato, the Amazon rainforest, Salamanca Island Parl, and Los Nevados National Natural Park as subjects of rights was handled by the Constitutional Court, beginning in 2016 with the Atrato River Case; lower courts added numerous other entities.[61] The legislator merely approved these decisions

---

[55] Yarra River Protection (Wilip-gin Birrarung murron) Act 2017, pt 1 s 1(a) (Austl.); see David Takacs, 'We Are the River' (2021) U Ill L Rev 545, 588–589.

[56] See <https://youtu.be/2sERUH7gXL4>. On the concept of nature's rights zones, see <https://www.natures-rights.org/natures-rights-zone>.

[57] Stéphanie Roy, Gaële Gidrol-Mistral and Alexandra Popovici, 'Québec (and Canada). Nature, Culture, Rupture: Tools for Environmental Protection in a Liberal Anthropocentric Legal System' in this volume, 574.

[58] Michelle Maloney, 'Building an Alternative Jurisprudence for the Earth: The International Rights of Nature Tribunal' (2016) 41 Vermont Law Review 129. For an interesting comparative assessment of the Tribunal's claims to legitimacy compared to those of the League of Nations, see Tim Lindgren, 'In the Name of Nature: Making the League of Nations, the International Rights of Nature Tribunal and International Law' Leiden JIL, First View, <https://doi.org/10.1017/S0922156525100344>.

[59] On these models, see Andrew Byrnes and Gabrielle Simm (eds), *People's Tribunals and International Law* (CUP 2018); Regina Menachery Paulose (ed), *People's Tribunals, Human Rights and the Law – Searching for Justice* (Routledge 2020).

[60] Amaya Querejazu, 'Are the International Tribunals of Rights of Nature pluriversal?' (2024) 38 International Relations 369–387.

[61] On the *River Atrato Case* (2016), see Tania Luna Blanco and David de la Torre Vargas, 'Colombia. Graffiti on the Colombian Legal Architecture: The Emergence of Non-human Subjects and the Rights of Nature' in this volume, 325–327; Yenny Vega-Cárdenas and Nathalia Parra, 'Nature as a Subject of Rights: A Mechanism to Achieve Environmental Justice in the *Atrato* River Case in Colombia' in Amissi Melchiade Manirabona and Yenny Vega Cárdenas (eds), *Extractive Industries and Human Rights in an Era of Global Justice* (LexisNexis Cana-

ex post.[62] In 2024 a Peruvian court recognized the Rio Marañón as a holder of a very detailed list of specific rights.[63] In India, the Supreme Court invoked, as early as 1995, the need to move from an anthropocentric to an ecocentric approach,[64] and courts have recognized rights of nature for the Ganga and other rivers.[65] In Germany, a lower court recognized rights of nature in two decisions,[66] though the lengthy reasoning of the decisions has been widely criticized as doctrinally unsound and as impermissible judicial activism.[67] And even in Ecuador, where RoN are recognized in the constitution, it fell to the Constitutional Court to elaborate the contours of RoN and thereby make them effective.[68] Altogether, these are very pluralist findings: recognition of RoN happens on many levels and through many different institutions.

b) Resisters

The number of countries which have not adopted RoN remains greater than that of adopters – among our country reports we can count China,[69] Cyprus,[70] Czech

---

da 2019) 127–161 for references to subsequent decisions see Blanco and de la Torre Vargas, 'Colombia' (this note) 332–335.

[62] Blanco and de la Torre Vargas, 'Colombia' (n 61) 334.

[63] STC Exp. N° 00010-2022-0-1901 <https://iuslatin.pe/historica-sentencia-rio-maranon-y-sus-afluentes-son-declarados-sujetos-derechos-expediente-00010-2022-0-1901/>. The rights include 'derecho a fluir, para garantizar un ecosistema saludable, el derecho a brindar un ecosistema sano, el derecho fluir libremente de toda contaminacion; el decercho a alimentar y ser alimentado por sus afluentes, el derecho a la biodiversidad; el derecho a que se la restaure, Derecho a la regenerachion de sus ciclos natrales; Derecho a la proteccion, preservacion y recuperacion.' For commentary, see Diana Camila Aguilar Castañeda, 'Protección legal y jurisprudencial de los derechos de la naturaleza' (2024) 13 Revista Quaestio Iuris 9–24; Mariano Andres Bustamante Jimenez, 'Los Ríos como Titulares de Derechos en el Perú' (2025) 18 YachaQ: Revista De Derecho 75–89. René Kuppe points out that the court refers to the river as a holder of rights, but not as a legal subject: René Kuppe, 'Rechte ohne Rechtssubjektivität?' (*Völkerrechtsblog*, 26 March 2025) <https://voelkerrechtsblog.org/rechte-ohne-rechtssubjektivitat/>.

[64] *TN Godavarman Thirumulpad v Union of India* [2012] 3 SCC 277 para 14; see Satish C Shastri, 'Environmental Ethics Anthropocentric to Eco-centric Approach: A Paradigm Shift' (2013) 55 Journal of the Indian Law Institute 522–530; Poonam Kanwal, 'Ecocentric Governance: Recognising the Rights of Nature' (2023) 69 Indian Journal of Public Administration 440–452.

[65] Manjeri Subin Sunder Raj, 'India. Sowing the Seeds of "Nature Rights"' in this volume, 347–350.

[66] LG Erfurt 2 August 2024, 8 O 1372/21, LG Erfurt 17 October 2024, 8 O 836/22; Ewering, Reimann and Vetter, 'Germany' (n 50) 484–486. Previously, the same judge had made two referrals to the CJEU asking it to recognize rights of nature, but the cases were settled; see Ewering, Reimann and Vetter, 'Germany' (n 50) 483.

[67] Ewering, Reimann and Vetter, 'Germany' (n 50) 484 with references in fns 124, 125.

[68] Espinoza, 'Ecuador' (n 38) 233–238.

[69] Jacques deLisle, 'China. Limited Prospects for the Rights of Nature in an Inhospitable Legal-Political Environment' in this volume, 393.

[70] Rokas, 'Cyprus' (n 23) 429–432.

Republic,[71] Germany,[72] Italy,[73] Japan,[74] Netherlands,[75] Norway,[76] Romania,[77] South Africa,[78] and Switzerland.[79] From a state-centered monist perspectives, these are all simply countries with no RoN. From a pluralist perspective, by contrast, if we include initiatives and discourse, the picture becomes more complex. If we look at the Global North, we find that resistance to RoN on the official level happens within a discursive field in which RoN are discussed intensely and find a good amount of support outside of official law.[80]

Globally, the number of initiatives is, unsurprisingly, greater than that of adoptions – a recent survey maps 495 initiatives in 40 countries and territories.[81] Initiatives tend to focus on concrete entities – the Waddenzee or the Maas River in the Netherlands, for example.[82] Even if these initiatives are not as successful as that for the Mar Menor in Spain,[83] they raise attention. Remarkably, two thirds of all initiatives emerged in the Americas, and among 151 local initiatives, nearly half (72) took place in the United States, four times as many as the next ranked country (Brazil).[84] The USA saw, in fact, the first legal acknowledgement of a right of nature – prior the Ecuadorian Constitution – though with far less immediate impact: in 2006, a local ordinance of the Tamaqua Borough of Pennsylvania declared that 'Borough residents, natural communities, and ecosystems shall be considered to be 'persons' for purposes of the enforcement of the civil rights

---

71 Jakub Harašta, 'Czech Republic. An Anthropocentric Approach to Environmental Protection' in this volume, 446.

72 Ewering, Reimann and Vetter, 'Germany' (n 50) 463, but see LG Erfurt 2 August 2024, 8 O 1372/21, LG Erfurt 17 October 2024, 8 O 836/22.

73 Domenico di Micco and Michele Graziadei, 'Italy. Establishing the Rights of Nature in Italy – Between Tradition and Change' in this volume, 493–514.

74 Keisuke Abe, 'Japan. Bringing the Alternative View to Fruition' in this volume, 515–516.

75 Paiement, 'Netherlands' (n 50).

76 Sören Koch, Esmeralda Colombo and Catalina Vallejo Piedrahíta, 'Norway. Rights of Nature in the Norwegian Legal Culture: To Be or Not to Be?' in this volume, 549.

77 Diana Botău, 'Romania. Unfertile Ground for the Rights of Nature' in this volume, 539–540.

78 Louis J Kotzé 'South Africa. Protecting Nature through an Anthropocentric-Oriented Rights Paradigm' in this volume, 367–374.

79 Affolter and Affolter, 'Switzerland' (n 49) 612–619.

80 Martha-Cecilia Dietrich and others, 'Stories of Systemic Failure? Landscaping the Rights of Nature in Europe' (2024) 8 JLA 46–76; García Ruales and others (n 36).

81 <https://ecojurisprudence.org/>; see Alex Putzer, John Cook and Ben Pollock, 'Putting the Rights of Nature on the Map. A Quantitative Analysis of Rights of Nature Initiatives Across the World – Second Edition' (2025) 21(1) Journal of Maps, <https://doi.org/10.1080/17445647.2024.2440376>.

82 Paiement, 'Netherlands' (n 50) 528–529.

83 Teresa Vicente Giménez and Eduardo Salazar Ortuño, 'An Ecological Citizenship's Triumph: From the Popular Legislative Initiative to the Rights Granted for the Mar Menor' in Jenny García Ruales and others (n 36) 83ff.

84 Putzer and others (n 81) 1–3.

of those residents, natural communities, and ecosystems.'[85] It seems fair to say that the ordinance has garnered more intellectual interest than practical importance as a basis of enforceable rights. Nationally, the USA do not recognize RoN, and many of the initiatives in the USA are municipal regulations granting legal personhood to nature that were later struck down by the courts.[86] A rights of nature movement remains active nonetheless and achieves the symbolic value associated with recognitions of RoN that are not ultimately legally successful.[87]

NGOs push for adoption of RoN in other countries as well, among them Australia[88] and Germany.[89] Moreover, we find a high number of indigenous initiatives for RoN, for example in Australia.[90] The difference between NGOs and indigenous initiatives is artificial: NGOs and indigenous communities frequently work together, and some NGOs are constituted by such communities.[91]

A particularly important avenue are court cases. Remarkably, whether such cases are successful or not is often secondary. In many countries, invocations of RoN are rejected by the courts, sometimes even where plaintiffs are successful on other grounds.[92] In some countries they are named as an aside, such as in a lower court decision from São Paulo.[93] In Germany, on the other hand, a lower court's recognition of RoN[94] was effectively obiter dictum, but the wide and nearly universal doctrinal critique can also be seen as a sign that the topic has entered

---

[85] Tamaqua Borough Ordinance no 612 of 2006 (Pennsylvania, United States), sec 7.6 <https://ecojurisprudence.org/initiatives/tamaqua-borough-sewage-sludge-ordinance/>. For background, see Mari Margil, 'NGOs in the United States. Birth of a Movement' in this volume, 165–175.

[86] Kathleen M Mannard, 'Lake Erie Bill of Rights Struck Down: Why the Rights of Nature Movement Is a Nonviable Legislative Strategy for Municipalities Plagued by Pollution' (2021) 28 Buff Env LJ 39; Alexandra Huneeus, 'The Legal Struggle for Rights of Nature in the United States' (2022) Wis L Rev 133, 135–136; Sam Bookman, 'The Puzzling Persistence of Nature's Rights' (2025) 165 ULR.

[87] Margil, 'NGOs' (n 85) 165–175; Craig M Kauffman and Pamela L Martin, *The Politics of Rights of Nature: Strategies for Building a More Sustainable Futire* (Massachusetts Institute of Technology Press 2021) 163ff; Huneeus (n 86) 133ff; Bookman (n 86).

[88] Australian Center for the Rights of Nature, *Recognising the Rights of the Reef* (2021), <https://perma.cc/2EJ9-8N42>; Takacs (n 55) 593 f.

[89] Ewering, Reimann and Vetter, 'Germany' (n 50) 473–481.

[90] Michelle Lim, Anne Poelina and Donna Bagnall, 'Can the Fitzroy River Declaration Ensure the Realisation of the First Laws of the River and Secure Sustainable and Equitable Futures for the West Kimberly?' (2017) 32 Australian Environment Review 18; Takacs (n 55) 590–592.

[91] See II.4, below, 74–76.

[92] On Belgium, see TFI Bruxelles 17 June 2021, *VZW Klimaatzaak v Kingdom of Belgium and others* <https://climatecasechart.com/non-us-case/vzw-klimaatzaak-v-kingdom-of-belgium-et-al/>; see Antoine de Spiegeleir, 'The Belgian Climate Case: From Federalism Idiosyncrasies to Arboreal Novelties' (2023) 7 Chinese Journal of Environmental Law 261–272.

[93] Wolkmer and Ferrazzo, 'Brazil' (n 54) 385.

[94] Note 66, above.

discourse and must be taken seriously.[95] Having said that, courts in countries with greater positivism and less acceptance of judicial lawmaking are less promising avenues for invoking or even creating rights of nature.[96] But even then resistance can pluralize law through debates and discussions.

c) Ignorers

Countries that resist RoN against powerful initiatives in support of them are quite different from countries where non-recognition is a consequence of non-engagement or ignorance. In East Asia, for example, there seem to be few initiatives. This is not necessarily for lack of attention: in China, a quite vibrant scholarly debate exists, though most participants are critical of RoN;[97] the Japanese discussion, by contrast, seems less vibrant.[98] In both of these countries, nature has high cultural value, but indigenous autonomy does not, and courts are unlikely to introduce legal innovations without the nvolvement of the legislator. The same is true in several European countries: Italy has vibrant academic debates (frequently drawing on Latin American examples) but very few visible grassroots movements;[99] this may change as a consequence of a recent constitutional amendment.[100]

In some other countries there is not even academic debate. Such countries do not reject RoN, they simply ignore them. Such absence of any engagement with the issue is hard to assess, of course. Country reports from such countries would presumably be quite useless, except to demonstrate whether the interests behind rights of nature are covered by other rules, or whether they are not protected at all.

3. Characteristics

Subtle distinctions between different implementations are too numerous to assess here. Two of them merit attention, however, because they go to the heart of what rights of nature are about reflect the extent to which they fundamentally challenge traditional concepts of law.

---

[95] Ewering, Reimann and Vetter, 'Germany' (n 50) 484–486; Ralf Michaels, 'Rechte der Natur – ein erfolgreicher Reverse Legal Transplant?' [2024] ZUR 643–649; Andreas Gutmann, 'Fruchtbare Irritationen' (*Verfassungsblog*, 28 October 2024) <https://verfassungsblog.de/rechte-der-natur-erfurt-2/>.

[96] deLisle, 'China' (n 69) 394, 404–405.

[97] deLisle, 'China' (n 69) 397–402.

[98] Abe, 'Japan' (n 74). But see Ueta Masaki [植田将暉], '21世紀の「自然の権利」と大地の人類学 ——法学の存在論的転回? – Nijūisseiki no "jizen no kenri" to daichi no jinrui-gaku – hōgaku no sonzai-ronteki tenkai?' ['Rights of Nature in the 21st Century and Anthropology of the Earth: Steps Toward a Legal Ontological Turn?'] 文化人類学研究 Bunka jinrui-gaku kenkyū [Studies in Cultural Anthropology] 25 (2024) 47–70.

[99] di Micco and Graziadei, 'Italy' (n 73) 511–514.

[100] di Micco and Graziadei, 'Italy' (n 73) 497–498.

a) The Subject of Rights

The first concerns the subject of rights. In some legal orders, that subject is nature or the earth itself. This is what we find in the concept of Pachamama (Mother Earth) in the Ecuadorian Constitution[101] as well as in the statutory law of Bolivia.[102] In other legal orders, rights are given to specific entities only. This second approach is much more frequent: in New Zealand, rights have been given to a river, to a former national park, and, in 2025, to a mountain range that remains a national park;[103] in Colombia[104] and India to rivers;[105] in Spain to a lagoon.[106] On some occasions, nature is recognized as sacred natural sites whose governance is granted to local communities as guardians, as is the case for Benin's Sacred Forests Law.[107] Sometimes, this does not amount to full subjectivity: in Bangladesh, rivers have a dual nature between rightsholders and public trusts;[108] in Italy parks effectively act as substitutes for rights of nature.[109]

Presented like this, the distinction is somewhat misleading. Where rights of nature matter, they do so in specific contexts, and a general invocation of nature at large as a rightsholder has little practical value. Indeed, the German court decision mentioned above was criticized for invoking rights of nature in general without specification, thereby amounting to little more than a tool to increase en-

---

[101] Espinoza, 'Ecuador' (n 38) 213–214.

[102] Bolivian Unity Pact, Barié, 'Bolivia' (n 39) 255.

[103] Iorns, 'Aotearoa New Zealand' (n 40) 301ff; Te Ture Whakatupua mō Te Kāhui Tupua 2025/Taranaki Maunga Collective Redress Act 2025.

[104] Constitutional Court of Colombia 10 November 2016, T-622/16, *Atrato River Case*; Medellin State Superior Tribunal 17 June 2019, 2019-076, *Cauca River Case* (Colombia); First Criminal Court of Neiva-Huila 24 October 2019, 071/2019, *Andres Felipe Rojas Rodriguez and Another v Ministry of Environment and Sustainable Development and Others – Magdalena River Case* (Colombia); see Blanco and de la Torre Vargas, 'Colombia' (n 61) 333; for criticism, see María Ximena González-Serrano 'Rights of Nature, an Ornamental Legal Framework: Water Extractivism and Backbone Rivers with Rights in Colombia' (2024) 52 The Journal of Peasant Studies 322, 329–332.

[105] Raj, 'India' (n 65) 348. In 2017 the High Court of Uttarakhand made a whole list of individualizable entities when it declared 'the [g]laciers, including Gangotri & Yamunotri, rivers, streams, rivulets, lakes, air, meadows, dates, jungles, forests, wetlands, grasslands, springs and waterfalls, legal entity/legal person/juridical person/moral person/artificial person having the status of a legal person, with all corresponding rights […] of a living person, in order to preserve and conserve them.' (High Court of Uttarakhand, 20 March 2017) Writ Petition (PIL) No 140 of 2015, *Lalit Miglani v State of Uttarakhand*.

[106] I.2.b), above, 55–58

[107] <https://gaiafoundation.org/benins-historic-sacred-forest-law-translated-into-english/>; see, more generally, John Studley, *Indigenous Sacred Natural Sites and Spiritual Governance* (Routledge 2018); Piero Zannini and others, 'Sacred Natural Sites and Biodiversity Conservation: A Systematic Review' (2021) 30 Biodiversity and Conservation 3747–3762.

[108] Ishrat Jahan, 'The Dual Status of Rivers in Bangladesh as Legal Persons and Public Trust Properties: The Likelihood of Legal Complexity' (2024) 27 APJEL 211–229.

[109] di Micco and Graziadei, 'Italy' (n 73) 505–509.

vironmental concern. Comparably, in Ecuador, the abstract protection of nature at large also needs to be translated to concrete entities, a task the courts have done for several entities: mangroves, the Los Cedros forest, a river, a river basin, and a monkey.[110]

In fact, the difference between specific rightsholders and nature itself as a subject of rights represents a fundamental jurisprudential difference.[111] Granting rights to specific natural entities is not dramatically problematic for a legal system. It may be unusual to give rights to entities that are neither human nor human-like,[112] but it is not impossible. The law has a long tradition of granting rights to non-humans, and, in this sense, granting rights to mountains or rivers does not require a paradigmatic change. Indeed, as the Italian report makes clear, parks (*Ente Parco*) can already have legal subjectivity under traditional jurisprudential approaches and have, in this sense, a long tradition going back at least to the early 20th century;[113] the proposal in the Netherlands to constitute the Waddenzee as a public natureship entity goes in a similar direction.[114]

By contrast, granting subjectivity to nature itself creates a fundamental challenge to traditional legal thinking. It represents not so much a shift in the barrier between the subjects and objects of law but rather its breakdown – if nature itself is a subject, then nothing in the world is merely an object anymore. Humans themselves are part of nature, which means that any litigation involving nature has an element of self-reflexivity in it. Subjective rights place the rightsholder in at least partial distance from the whole;[115] granting rights to that whole itself creates an *Aufhebung* of the entire distance. This may explain the difficulties of the Ecuadorian Constitutional Court in explaining the rights of specific entities as not being separate from nature at large.[116] Indeed, the Ecuadorian Constitution describes nature as a rightsholder, but not as a legal person.[117]

Notably, RoN are not always necessarily formulated as rights. In Aotearoa New Zealand, what is protected are the intrinsic value of nature, the legal personhood

[110] Espinoza, 'Ecuador' (n 38) 214.

[111] See I.2.b), above, 55–58.

[112] See I.2.b)(2), above, 56–58.

[113] di Micco and Graziadei, 'Italy' (n 73) 505–509.

[114] Paiement, 'Netherlands' (n 50) 533; Tineke Lambooy, Jan van de Venis and Christine Stokkermans, 'A Case for Granting Legal Personality to the Dutch Part of the Wadden Sea' (2019) 44 Water International 786, 798.

[115] Menke (n 28).

[116] Espinoza, 'Ecuador' (n 38) 235–237.

[117] Espinoza, 'Ecuador' (n 38) 233; see also Brad Coombes, 'Nature's Rights as Indigenous Rights? Mis/recognition through Personhood for Te Urewera' (2020) 1 Espace populations sociétés; Brad Coombes 'Personifying Indigenous Rights in Nature? Treaty Settlement and Co-Management in Te Urewera' in Rani-Henrik Andersson, Boyd Cothran and Saara Kekki (eds), *Bridging Cultural Concepts of Nature: Indigenous People and Protected Spaces of Nature* (Helsinki UP 2021) 29–60.

of nature, and human guardianship for nature.[118] Tănăsescu proposes relations rather than rights as the proper legal instrument.[119]

b) The Representation of Nature

A related issue concerns the question of who can represent nature in exchanges with others, including before courts. The frequent ridicule directed at RoN that mountains cannot speak and rivers cannot come to the courtroom is, of course, gratuitous – the same is true for corporations and municipalities. For these latter entities, however, it is clear who can represent them. For nature, this is less clear.

Some systems entrust specific groups as guardians or custodians.[120] Frequently, these are indigenous communities that are in close vicinity to the respective piece of nature whose right is at stake. This is the case for the Te Awa Tupua in New Zealand and the Magpie River in Canada[121] Sometimes, specific institutions are named or even established, as for example in the Indian cases granting rights to the Ganga and Yamuna Rivers and to Glaciers.[122] In Bangladesh, the Supreme Court appointed the National River Conservation Commission as guardian of all rivers.[123] The Colombian Supreme Court combined both approaches in the Atrato River case, ordering the creation of a body made up of both government representatives and local communities.[124]

Other systems allow everyone to bring suit to enforce rights of nature, such as in Ecuador, Bolivia, and Uganda.[125] In Ecuador this has led to a highly diverse

---

[118] Iorns, 'Aotearoa New Zealand' (n 40) 283.

[119] Mihnea Tănăsescu (n 1) 128ff; Garvit Shrivastava, From Rights to Relation – Rethinking the Rights of Nature Framework (*Völkerrechtsblog*, 9 October 2025) <https://voelkerrechtsblog.org/from-rights-to-relation/>.

[120] Tignino, 'International Law' (n 23) 134–136.

[121] Iorns, 'Aotearoa New Zealand' (n 40) 309ff; Roy, Gidrol-Mistral and Popovici, 'Québec (and Canada)' (n 57) 574.

[122] Tignino, 'International Law' (n 23) 134f; Erin O'Donnell, 'At the Intersection of the Sacred and the Legal: Rights for Nature in Uttarakhand, India' (2018) 30 JELP 135–144.

[123] *Human Rights and Peace for Bangladesh v Government of Bangladesh and Others* [2016] HCD (Writ Petition No 13989 of 2016) <http://www.supremecourt.gov.bd/resources/documents/1048627_W.P.13989of2016.pdf> (in Bengali); Ferdousi Begum, 'Implications of Conferring Legal Entity to the Turag River of Bangladesh: Recognising the Rights of Nature' in S Shanthakumar, Nupur Chowdhury, and Stellina Jolly (eds), *Environment Protection and the Law – Perspectives from South Asia* (Bloomsbury 2024) 101–120; Tignino, 'International Law' (n 23) 135; see also Tanjila Tamanna and Raspiatur Rashpi, 'Legal Personhood of Rivers in Bangladesh: Judicial Responses and Legal Implications' (2025) 2 International Journal of Science and Development 38–49; Imtiaz Ahmed, *Rights, Rivers and the Quest for Water Commons: The Case of Bangladesh* (Springer 2021).

[124] *Atrato River Case* para 9.32; Blanco and de la Torre Vargas, 'Colombia' (n 61) 328.

[125] Bolivia: Constitution, art 135, Barié, 'Bolivia' (n 39) 268; Uganda: National Environment Act 2019, s 4; see Sanni, 'Uganda' (n 47).

array of plaintiffs in the 50 or so court cases that have been brought.[126] Many other countries allow for some kind of public litigation for environmental concerns.[127] Stewardship is an idea that can also apply in the Global North.[128]

Again, this represents a deep jurisprudential divide, though not one as clear cut as with regard to subjectivity. Where specific groups are entrusted with enforcing rights of nature, their role is not immensely different from that of a custodian in a traditional context or even stewardship without a rightsholder, such as for the Fraser River in British Colombia.[129] More importantly, however, in such a model it makes little difference whether nature has rights of its own or is merely an object of special importance. In theory, conflicts of interest could arise between the interests of nature and those of its custodians. In practice it is hard to see how such conflicts could ever be resolved against the custodians given that they are also entrusted with defining the interests of nature.

Some legal orders allow everyone to bring suit (*actio popularis*).[130] In such jurisdictions, RoN do not necessarily represent more of a jurisprudential challenge; public interest litigation is a familiar institution in the law.

4. *Intralegal Relations*

In itself, the recognition of RoN says little about their practical importance. On the one hand, even legal orders without RoN may as a practice matter achieve similar results through environmental protection. On the other hand, even recognized RoN must compete, and possibly lose to, other interests. A comprehensive comparison is impossible here, but the structural issues can be demonstrated.

a) RoN and Environmental Protection

Rights of nature coexist with rules and institutions of environmental law and environmental protection. Indeed, in many legal systems that do not recognize RoN, environmental law is viewed as either a substitute for, or a possible intermediary step towards, rights of nature. It acts as a substitute if it creates, in effect, the same kind of protection as would rights of nature. Thus, it has been claimed that the EU already – effectively – recognizes rights of nature as simply the correlate of the obligations others have towards nature.[131] Such obligations can act as an in-

---

[126] Espinoza, 'Ecuador' (n 38) 221.

[127] Eg Wolkmer and Ferrazzo, 'Brazil' (n 54) 385.

[128] Jéremie Gilbert and others, '"Caring for Nature": Exploring the Concepts of Stewardship in European Philosophoes, Spiritual Traditions, and Laws' in García Ruales and others (n 36) 45ff.

[129] Roy, Gidrol-Mistral and Popovici, 'Québec (and Canada)' (n 57) 576.

[130] Guillermo Schumann Barragan, 'Civil Litigation for Environmental Damages – Are the So-Called Rights of Nature an Alternative?' (2024) 14 IJPL 73, 77ff.

[131] Yaffa Epstein and Hendrik Schoukens, 'A Positivist Approach to Rights of Nature in Europe' (2021) 12 JHRE 205–227.

termediary step if increased environmental protection leads, eventually, to adoption of rights of nature

Such arguments underestimate what is peculiar about subjectivity and rights of nature. Rights in a Hohfeldian sense are mere correlates of obligations, but rights in the sense of property rights as well as legal subjectivity are more: they are reasons for obligations.[132] Similarly, claims for subjectivity and rights of nature are more than claims for higher levels of protection; they are calls for the expansion of rightsholders. For such reasons existing laws on environmental protection may actually be detrimental for the introduction of rights of nature, either because such rights seem unnecessary on top of what is already there, or because such rights, if formally adopted, are in fact equated with traditional environmental protection.

The core difference between environmental protection and rights of nature is that the former is anthropocentric and the latter are not. Rights of nature are, therefore, in conflict with environmental protection. It is a weakening of rights of nature if they are equated, in implementing legislation, to anthropocentric environmental protection.

b) Rights of Nature and Conflicting Interests

On the other hand, RoN conflict with the interests of others. A particularly important conflict exists with interests in resource extraction. That conflict is characteristic of countries interested in development.[133] Rights of nature often conflict with multinational corporations, but the conflicts can also exist with the rights and interests of indigenous communities. Whereas many indigenous communities may favor RoN, the interests of such communities are not necessarily completely congruent with those of nature, and it would be paternalistic to assume otherwise. In some countries economic development stands in the way of acceptance of RoN.[134] Even where rights of nature are treated as mere supplementary considerations, they will tend to be secondary.

The hierarchy between rights of nature and other rights thus becomes crucial. In Ecuador, rights of nature are in principle superior to interests in extraction, and the Constitutional Court has indeed held so. A similar hierarchical superiority was established in Bolivia's Law 71 (2019), which required all human activity to reach a dynamic balance with Mother Earth, but the subsequent Framework Law 300 (2012) reduced the status of RoN from superiority to complementarity with other human rights and thereby greatly limited their impact. Rights of na-

---

[132] Indeed, when Epstein and Schoukens (n 131) 218, argue that 'these rights have been under-utilized by EU courts thus far', they abandon their analytical framework for a normative claim that is not warranted by their Hohfeldian method.

[133] Barié, 'Bolivia' (n 39) 255ff.

[134] deLisle, 'China' (n 69) 396.

tures are not automatically 'trumps', in Dworkin's terminology; and indeed some criticism of the concept comes from that characteristic of rights.[135]

5. Legal and Political Operators

Official acceptance of rights of nature is a matter for state institutions, but they rarely act on their own impetus. Instead, RoN is a field in which movements and NGOs are of particular importance.

a) Local Communities

Especially in the Global South, indigenous communities are among the most important operators. In Bolivia, whose former President Evo Morales is indigenous himself, they have been a main driver of law reform in general; the adoption of Law 71 (2010) closely followed a proposal by an alliance of local indigenous communities;[136] custom was transformed into law.[137] In other countries, indigenous communities were similarly behind the push for recognition;[138] sometimes it was these communities themselves that exercised the recognition. Moreover, communities are crucial for the task of educating decisionmakers and providing expertise.[139]

It is important to realize that such local communities do not exist only in the South, and the demand for rights of nature in the North frequently comes from local communities as well. In Denmark, the planned designation of a natural habitat in Copenhagen as a building zone spurred protestations by locals that the habitat should have rights.[140]

b) Local and International NGOs

Other important actors – having significant overlap with indigenous communities – are local NGOs.[141] The Tamaqua ordinance granting subjectivity to a river was the product of intense work by individuals and NGOs,[142] and the same is true elsewhere. Almost universally, where RoN have been accepted, it has been due

[135] Lael K Weis and Robert Mullins, 'Does Nature Need Rights?' (2025) 20 OJLS 1, 20–24; Mauricio Guim and Michael A Livermore, 'Where Nature's Rights Go Wrong' (2021) 107 Va L Rev 1347, 1369ff.

[136] Barié, 'Bolivia' (n 39) 253ff.

[137] Barié, 'Bolivia' (n 39) 254.

[138] On the USA, see Margil, 'NGOs' (n 85) 172.

[139] See Hanschel and Mehlhorn, 'Law and Anthropology' (n 32) 154 on the program 'Intercultural Community Expertise in Human Rights and Rights of Nature' maintained by the Ecuadorian Foundation for the Defense of Nature and her Rights.

[140] Yvette Lind, 'Denmark. Could Danish Human- and Animal-Rights based Environmental Protection Provide Rights to Nature?' in this volume, 460; see Katarina Hovden, 'Naturens rettigheder og samfundsomstilling' (2022) 2 Samfundsøkonomen 57, 61.

[141] Margil, 'NGOs' (n 85).

[142] Margil, 'NGOs' (n 85) 166–170.

to the activism of NGOs; the same is true where they have not been accepted but are nevertheless a matter of political and judicial debate,.[143] By contrast, where NGOs do not focus specifically on RoN, as for example in Cyprus and Norway, RoN tend to have little chance.[144] This is even more so where NGOs altogether fail to operate due to state restrictions, as is the case in China.[145]

In addition, the rapid transnational expansion of RoN is due to the particularly important work of international NGOs. The Global Alliance of Rights of Nature (GARN),[146] established in 2010, is active in no less than 99 countries. It has spearheaded local movements and facilitated knowledge transfer between communities far apart, bringing them together in an Indigenous Council. Also, it has created visibility, through the (nonformal) International Rights of Nature Tribunal. The Community Environmental Legal Defence Fund (CELDF), founded as early as 1995, has also been as instrumental in working together with local legislation projects.[147] Other transnational NGOs exist as well.[148]

c) *Academics*

The wealth of scholarly writing about rights of nature is also a factor in the adoption of RoN. Although a surprisingly great amount of academic writing confines itself to a mere description of RoN in other countries, most prominently Ecuador, Bolivia, and New Zealand, there are, in many countries, proposals by academic and non-academic authors to adopt RoN.[149] Indeed, the amount of writing on RoN is by now immeasurable.

However, whether academic writing has a concrete impact on the actual recognition of RoN is a different matter, of course. Three factors appear particularly relevant for the impact of such academic writing. First, and most important, is the breadth and depth of academic discussion. Here, we find huge differences between countries featuring a lively debate, such as the United States, and countries that tend to be ignorers in the above-defined way (II. 2. c)) and in which the topic is academically irrelevant, such as Norway. Second, the specific positions taken in academic literature matter. In most countries, those writing about RoN tend to be overwhelmingly supportive, but in some countries there is a great level

---

143 Ewering, Reimann and Vetter, 'Germany' (n 50) 481f. On the role of the Nature Movement Society (Vereniging Natuurmomenten) in the Netherlands, see Anne De Vries-Stotijn, Ilon Van Ham and Kees Bastmeijer, 'Protection Through Property: From Private to River-held Rights' (2019) 44 Water International 736, 742f.

144 Rokas, 'Cyprus' (n 23) 432ff; Koch, Colombo and Vallejo Piedrahíta, 'Norway' (n 76) 555; Botău, 'Romania' (n 77) 599ff.

145 deLisle, 'China' (n 69) 402.

146 For information, see <https://www.garn.org>.

147 <https://celdf.org/rights-of-nature/>.

148 See Bonilla, 'The Globalization of Rights of Nature' (n 5) 27, fn 121.

149 Paiement, 'Netherlands' (n 50) 532f.

of opposition and backlash,[150] and in some the focus of the literature is rather not conducive to adoption of RoN in the way they were conceived.[151] Indeed, more recently it seems that academic discussions have matured in the sense that they are no longer merely aspirational (or simply in favor) or generally dismissive but offer more nuanced assessments. Such maturation is, presumably, more favorable to the recognition of RoN than mere advocacy. Third, one important factor is the general impact academic writing has on official law in a country.

Although much of the discourse is transnational and many of the relevant texts exist in several languages, national scholars are often those having the greatest domestic influence: Erik Kaptein is influential mainly in the Netherlands, Klaus Bosselmann mainly in Germany, and so forth.[152] Local academics matter.[153]

d) Fora

Another important influence comes from fora. The term is used here in two senses – as events and places for people to come together, and as actual tribunals. As to the first, several international fora have been established and proven influential. The most important of those may have been the World People's Conference on Climate Change and the Rights of Mother Earth in Bolivia in 2010, which led to the Universal Declaration of the Rights of Mother Earth.[154] Attended by 30,000 people from 100 countries, it can be marked as an important starting place for the recognition of RoN not only in Bolivia but also beyond.[155] The conference's success has been attributed to its ability to connect with long-standing environmental discourse and to the assertion of an indigenous speaker position.[156] Since then, many fora have been strategically organized to coincide with global political con-

[150] Koch, Colombo and Vallejo Piedrahíta, 'Norway' (n 76) 553f.

[151] For example, in Hungary a main focus appears to be the compatibility of RoN with Christianity; see Ágnes Tahyné Kovács, A természet jogai jelenségről világi és egyházi felfogásban (On the Phenomenon of Rights of Nature: Secular and Ecclesiastical Perspectives) (2024) 14 Pro Futuro 12–20, DOI 10.26521/profuturo/2024/2/15441.

[152] Erik Kaptein, *Rechtsgelijkheid voor de natuur* (ISVW 2021); Klaus Bosselmann, 'Eigene Rechte für die Natur?' (1986) KJ 1–22. See Paiement, 'Netherlands' (n 50) 533; Ewering, Reimann and Vetter, 'Germany' (n 50) 464, respectively.

[153] Wolkmer and Ferrazzo, 'Brazil' (n 54) 387 (on animal rights)

[154] <https://www.garn.org/universal-declaration-for-the-rights-of-mother-earth/>; republished eg in (2011) 14 S Cross U L Rev 203; Bonilla, 'The Globalization of Rights of Nature' (n 5) 31; For background, see the policy paper by Cormac Cullinan, 'The Legal Case for the Universal Declaration of the Rights of Mother Earth' (2010) <https://www.garn.org/wp-content/uploads/2021/09/Legal-Case-for-Universal-Declaration-Cormac-Cullinan.pdf>.

[155] Barié, 'Bolivia' (n 39) 253.

[156] Cristina Espinosa, 'The Advocacy of the Previously Inconceivable: A Discourse Analysis of the Universal Declaration of the Rights of Mother Earth at Rio+20' (2014) 23 JED 391–416.

ferences or other events – the Geneva Forum, for instance, coincides with Earth Day.[157]

As to the second meaning, the International Rights of Nature Tribunal stands out as perhaps the most visible nonofficial forum in which rights of nature are litigated.[158] Since 2014, six international tribunals have been held in Quito (2014), Lima (2014), Paris (2015), Bonn (2017), Glasgow (2021), and New York (2024), supplemented by two regional tribunals in Chile (2019) and Europe (Online 2021) and local tribunals most recently in Tren Mays (2023), Biobio (2024), and Yesah (2024).[159] These tribunals do not render binding decisions. Nonetheless, their holdings have served, occasionally, as reference points especially in domestic litigation.

6. Enforcement

The law in the books is not the law in action, and rights that are not enforced are mere symbols. Rights of nature, in particular, risk being merely programmatic or symbolic, without any real enforcement. Since nature itself cannot assert claims, others have to make such claims on its behalf, and these individuals and entities must have both the incentives and the procedural means to do so. Unsurprisingly, even among countries that have formally recognized RoN, differences in the degree of enforcement are substantial, due to both legal and extralegal reasons.

a) Public enforcement

The administrative state, as the most important enforcer of environmental law, is also in charge of enforcing RoN in many legal orders. Lack of public enforcement can hamper RoN: Bolivia's promising legislation, for example, has very limited impact in no small part because the institutional framework intended for implementation was never established.[160] Similarly, the competence of local and indigenous institutions to enforce the rights of the Magpie River in Canada is limited, at least as long as its status is not recognized by the state.[161]

Indeed, it is doubtful whether public enforcement is best suited for RoN, at least where it is left to general institutions like the public prosecutor. From an administration's perspective, it makes comparably little difference whether what is enforce actual rights or ordinary environmental regulations.[162] Without guardians enjoying a certain level of independence from the government, RoN cannot

[157] Affolter and Affolter, 'Switzerland' (n 49) 627.

[158] Michelle Maloney, 'Building an Alternative Jurisprudence for the Earth: The International Rights of Nature Tribunal' (2016–17) 41 Vt L Rev 129.

[159] See Bonilla, 'The Globalization of Rights of Nature' (n 5) 31.

[160] Barié, 'Bolivia' (n 39) 271.

[161] Roy, Gidrol-Mistral and Popovici, 'Québec (and Canada)' (n 57) 574.

[162] See Wolkmer and Ferrazzo, 'Brazil' (n 54) 386; deLisle, 'China' (n 69) 417.

be effective as such.[163] A right in the Northern tradition provides its holder with some space of autonomy separated from the common interest,[164] whereas the state's task is usually to enforce that common interest. Public participation can matter of course – in Bolivia everyone is entitled or even obliged to report transgressions of the rights of Mother Earth;[165] in China, on the other hand, public participation is slim.[166] In any event, public participation is not dependent on the specific structure of RoN.

Within public enforcement, RoN are more relevant if there is a special entity tasked with their enforcement – an ombudsman for nature, for example, like the envisaged defensoria de la madre tierra in Bolivia,[167] or a public trust like the Queen Elizabeth II National Trust in New Zealand, which negotiates open space or conversions with private landowners.[168] In the Netherlands, the proposal for legal subjectivity for the Waddenzee entails creation of a public 'natureship entity'.[169] Here, too, RoN would not need to be actual rights from the perspective of legal doctrine; enforcement could be no less strong if the relevant considerations were formulated merely as interests. It may, however, strengthen the case rhetorically if such institutions enforce actual rights.

b) Private enforcement

In contrast to public enforcement, private enforcement is less dependent on prosecutorial discretion and may therefore be more effective. It concerns litigation against the state, and perhaps even more important, private litigation against private actors like mining corporations that interfere with RoN.[170]

---

[163] Tignino, 'International Law' (n 23) 137, referring to Erin O'Donnell and Julia Talbot-Jones, 'Creating Legal Rights for Rivers: Lessons from Australia, New Zealand, and India' (2018) 23 Ecology and Society 7, 12 f.

[164] Menke (n 28).

[165] Barié, 'Bolivia' (n 39) 263; Ley Marco de la Madre Tierra y Desarrollo Integral para Vivi Bien 2012, art. 19.

[166] deLisle, 'China' (n 69) 421.

[167] A bill proposing implementing legislation failed in 2021, see Barié, 'Bolivia' (n 39) 266 <https://www.vicepresidencia.gob.bo/IMG/pdf/anteproyecto_de_ley_defensoria_de_la_madre_tierra.pdf>. See also Fundación Solón, 'Què es una defensoría de la Madre Tierra?' (boletín no 2, 2021) <https://fundacionsolon.org/wp-content/uploads/2021/05/boletin_def_naturaleza_final-1.pdf>; Nicole Mansuy and others, 'Hacia una constitución ecológica: acceso a la justicia ambiental' (2022) 14 Justicia Ambiental 223, 258–268.

[168] De Vries-Stotijn (n 143) 743f; Christopher Rodgers and David Grinlinton, 'Covenanting for Nature: A Comparative Study of the Utility and Potential of Conservation Covenants' (2020) 83 MLR 373–405.

[169] Paiement, 'Netherlands' (n 50) 533; Tineke Lambooy, Jan van de Venis and Christine Stokkermans, 'A Case for Granting Legal Personality to the Dutch Part of the Wadden Sea' (2019) 44 Water International 786, 798.

[170] See Symposium: Private Rights of Nature (2022) 11 TEL 463–552.

Normally, the enforcement of rights falls to rightsholders themselves, or those acting on their behalf. This is true for public enforcement (against the State) as well as for private enforcement (against private infringers of the rights). Thus, in countries where guardians or the like are appointed by the law, enforcement of RoN falls to them.[171] Even countries that do not recognize RoN can thereby reach comparable results by allowing local communities to challenge public acts affecting the environment in their geographic areas, as is the case, for example, in Cyprus.[172]

In countries in which the law liberally allows for suits and the political and economic conditions are favorable, litigation plays an important role in enforcement. In states without these requirements, enforcement is bound to fail. The importance of extralegal conditions becomes obvious upon comparing Ecuador and Bolivia, both of which allow everyone to bring suit. This liberal rule has led to some 50 cases having been brought in Ecuador, where the courts have been willing to enforce RoN meaningfully, also against the interests of industry.[173] By contrast, there is very little meaningful litigation in Bolivia, in part because the enforcement procedure for RoN remains undefined and public litigation is heard by lower level judges with little special expertise.[174]

Sometimes, rights of nature are merely a side consideration in suits that have not been brought by a natural entity.[175] This is not actually enforcement of a right. By contrast, even where rights of nature are not recognized as such, they may emerge from interpretations of traditional private law. Roy et al make the interesting proposal that if nature is considered *res communis* (Code Civil of Québec, art 913) – as a thing whose use is common to all[176] – then everyone is affected when that use is impaired and so everyone is allowed to bring claims (Code Civil, art 982),[177] in addition to the State.[178] This would effectively create a popular action to protect nature, though not exactly the same as a claim based on a right of nature. Private property can also serve as a functional equivalent for RoN.[179]

[171] Roy, Gidrol-Mistral and Popovici, 'Québec (and Canada)' (n 57) 583f.

[172] Rokas, 'Cyprus' (n 23) 433, referring to *The Republic v The Local Community of Pyrgos* [1996] 3 CLR 503, 508 (in Greek)

[173] Espinoza, 'Ecuador' (n 38) 230.

[174] Barié, 'Bolivia' (n 39) 267.

[175] See LG Erfurt 2 August 2024, 8 O 1372/21, LG Erfurt 17 October 2024, 8 O 836/22.

[176] Code Civil of Québec, art 913.

[177] Roy, Gidrol-Mistral and Popovici, 'Québec (and Canada)' (n 57) 590.

[178] Yenny Vega Cárdenas and Uapukun Mestokosho, 'Recognizing the Legal Personality of a River. The Magpie River/*Mutehekau Shipu* in Canada' in this volume, 199 with reference to the Water Act (2009).

[179] See De Vries-Stotijn (n 143) 736, 742–743.

7. Theoretical Sources and Foundations

A key factor influencing discussions about both whether and how RoN should be accepted consists in the underlying theoretical approaches.[180] Here, the framework of GLP is particularly helpful. Most such theories are by now transnational, as is the discourse on their relative strengths and weaknesses. Which of these theories prevails in specific legal orders depends on local conditions; the comparative analysis can thus highlight the differences among conditions and their relevance. It is not possible here to trace these actual influences in detail, but it may be useful to differentiate between two original traditions (indigenous ontologies and rights-based approaches)[181] and between two elaborations (environmental concerns and ecocentrism) that attempt, respectively, to escape or to transcend the tensions existing respectively.

a) Indigenous ontologies

One important set of influences consists in indigenous ontologies. We see this in almost all countries that first adopted RoN: Bolivia, New Zealand, Ecuador, Colombia.[182] It may be fair to say that many of these ontologies share certain traits – a recognition of the special role that nature has; a preference for a nonextractive, special relation between humans and their environment; an assumption that humans are part of nature rather than its master; an emphasis on community among human and non-human entities rather than a sharp division between individuals and others.

Nonetheless, these ontologies do not translate directly into RoN.[183] First, what the similarities describe are not so much positive common traits but rather manners of distinction from a European tradition insisting on separating between nature and culture, subjects and objects, humans and non-humans, and individuals and society. In this sense, Europe is an outlier, an exception, and a less Eurocentric perspective would demonstrate this.[184] Beyond that, it is not clear to what extent these ontologies are really the same,[185] and it would be a mistake to treat all such ontologies as the same or to even romanticize them.[186] Frequently cited say-

[180] Tănăsescu, 'Theoretical Foundations' (n 2) passim

[181] See I.2.b) above, 55ff.

[182] Barié, 'Bolivia' (n 39) 259; Iorns, 'Aotearoa New Zealand' (n 40) 285; Espinoza, 'Ecuador' (n 38) 223; Blanco and de la Torre Vargas, 'Colombia' (n 61) 332; on Ecuador, see, more generally, Hanschel and Mehlhorn, 'Law and Anthropology' (n 32) 156–159.

[183] Tănăsescu, 'Theoretical Foundations' (n 2) 106f.

[184] See, generally, Lena Salaymeh and Ralf Michaels, 'Decolonial Comparative Law: A Conceptual Beginning' (2022) 86 RabelsZ 166–188.

[185] See Tănăsescu, 'Theoretical Foundations' (n 2) 107.

[186] Roy, Gidrol-Mistral and Popovici, 'Québec (and Canada)' (n 57) 584; Kirsten Anker, 'Ecological Law and Indigenous Relational Ontologies: Beyond the "Ecological Indian"?' in Kirsten Anker and others (eds), *From Environmental to Ecological Law* (Routledge 2021) 111.

ings like 'bien vivir', 'Ekuanitshit' (Take care of the place where you live),[187] and 'I am the river, the river is me' are only indicative of the complicated ontologies they express and cannot substitute for them. If we understand such ontologies as always being tied to their local surroundings, then influence is possible but adoption is doubtful, and a loose invocation of such ontologies without further analysis is little more than exotic coloration of domestic policy preferences.

Second, although indigenous ontologies can serve as a basis for RoN, the translation is by no means effortless. Southern ontologies are not traditionally based on individual rights; RoN therefore represent a significant alteration.[188] This leads to frictions. On the one hand, ontologies with no space for individual rights at all require a fair amount of adaptation to a RoN model; some of the opposition among indigenous groups to RoN projects can be explained by this. On the other hand, some indigenous communities will base their claims on their own property rights rather than rights of nature. This was the case, interestingly, for the Maori – the introduction of RoN was a compromise between competing property claims of the state and the group – and it is also the case for claims by the Sámi in Norway.[189] In East Asia, a traditional veneration for nature did not translate into RoN.[190]

This suggests, thirdly, that the role of indigenous ontologies is more strategic than genealogical.[191] Although these ontologies provide strong arguments against extraction and in favor of greater protection of nature, they do not, in and of themselves, support formulating RoN as actual rights. It has been suggested that indigenous ontologies are invoked merely to conceal the Northern, and ultimately colonial, role that rights in particular play.[192] This seems true only to some extent. It downplays the role of Southern ideologies and actors in the development. And it downplays the fact that the reverse is true as well: rights are used as a form, strategically, in order to conceal the significant change inherent in ascribing a greater and fundamentally different role to nature, which might not have been possible in other ways, a strategy observable already in the 2010 conference.[193]

---

[187] Vega Cárdenas and Mestokosho, 'Recognizing the Legal Personality of a River' (n 178) 183.

[188] See Ariel Rawson and Becky Mansfield, 'Producing Juridical Knowledge: "Rights of Nature" or the Naturalization of Rights?' (2018) 1 Environment and Planning E: Nature and Space 99–119.

[189] Koch, Colombo and Vallejo Piedrahíta, 'Norway' (n 76) 556.

[190] Abe, 'Japan' (n 74) 515; deLisle, 'China' (n 69) 401.

[191] Mihnea Tănăsescu, 'Rights of Nature, Legal Personality, and Indigenous Philosophies' (2020) 9 TEL 429–453.

[192] Rawson and Mansfield (Fn 188).

[193] Espinosa (n 156) 402–403.

b) Rights-based jurisprudence

Another important – and actually quite different –influence consists in Christopher Stone's seminal 1972 article 'Should Trees Have Standing' and the subsequent dissenting vote by Justice Douglas of the US Supreme Court in the Sierra Club case, which picked this idea up.[194] Stone's interest was spurred, originally, not by nature but by the doctrine of standing and the question, arising in his introductory class on property law, of how broadly it could be extended.[195] He suggested a historical genealogy in which rights were given to subjects previously thought incapable of holding them, such as women and enslaved people. Moral standing implied legal standing. It was no logical problem, therefore, to grant such rights to natural entities, too.[196]

This is true at least in legal systems that already rely strongly on the idea of rights.[197] A legal system in which rights play a significant role already provides the structures within which RoN can potentially fit. On the other hand, states in which individual rights and human rights play a lesser role are unlikely to grant them to nature; this is the case in general for China[198] and specifically for environmental law in Norway.[199]

Some proponents of RoN base them on the recognition of human rights, in different variants.[200] One variant includes in human rights the rights of humans *to* nature; this is not, however, the same as justifying the rights of nature on their own terms. A second variant views human rights less as a system of rights and more as a comprehensive normative order; here, human rights are little more than a reference to desirable values. The most promising is a third variant, namely to extend the protection of human rights from humans to nonhumans, whereby animal rights can sere as an intermediate step.[201] The occasional position arguing that this can be done already in existing law[202] seems questionable, at least for the European Convention on Human Rights given its anthropocentric nature. By contrast, the Inter-American Court of Human Rights has twice held that the human right to a healthy environment (El Salvador Convention, art 11) as encompassed by Article 26 of the American Convention (Progressive Development) protects, 'as an autonomous right', 'components of the environment, such as for-

[194] Stone (n 29) 450; *Sierra Club v Morton* [1972] 405 US 727, 729; Tănăsescu, 'Theoretical Foundations'98f.

[195] See Stone (n 29).

[196] See Tănăsescu, 'Theoretical Foundations' (n 2) 104.

[197] See David R Boyd, *Rights of Nature: A Legal Revolution that Could Save the World* (ECW Press 2017).

[198] deLisle, 'China' (n 69) 394 (rights), 395 (human rights).

[199] Koch, Colombo and Vallejo Piedrahíta, 'Norway' (n 76) 555.

[200] Elena Ewering, Andreas Gutmann and Tore Vetter, 'Finding a Path to Europe for the Rights of Nature' in García Ruales and others (n 36) 180ff.

[201] Wolkmer and Ferrazzo, 'Brazil' (n 54) 384; see also Abe, 'Japan' (n 74) 517.

[202] LG Erfurt 2 August 2024, 8 O 1372/21, LG Erfurt 17 October 2024, 8 O 836/22.

ests, rivers and seas, as legal interests in themselves, even in the absence of the certainty or evidence of a risk to individuals'.

However, the connection between moral and legal standing is less clear than is made out here,[203] and the move from rights to RoN is not inherently smoother than that from indigenous ontologies to RoN. Giving rights to women, slaves, and even animals can rest on their similarity to humans and can be viewed as a mere shifting of the line between subjects and objects of law.[204] Giving rights to nature, by contrast, implies abandoning that line altogether. It requires, in other words, a different ontology. Some criticism of RoN rests precisely on the fact that the reliance on rights makes such a shift harder and not easier.[205]

c) Environmental concerns

Moving from ideology and structure to policy highlights environmental concerns as an important driver. Stone's 1972 article met with a world in which environmental law was not yet widely recognized, except in the United States.[206] Focusing on rights was thus, for him, within the tradition of US law to, in particular, enact social change through private litigation rather than public regulation.[207] In this sense, RoN predated environmental law and helped bring it to existence. Since then, however, environmental law and environmental concerns have grown into an independent field, both in legislation and in scholarship. This even includes nations that were previously opposed, such as China.[208]

This genealogical sequence of events calls into question a natural genealogy from environmental concerns. True, in many jurisdictions, environmental concerns are at the root of the recognition or promotion of RoN.[209] Environmental law is not inherently connected to RoN, however, at least insofar as it relies, today, on public regulation rather than private rights and litigation. The hope that RoN are somehow generally more effective than traditional environmental law is a policy argument in need of evidence, and it reduces the granting of rights to a mere instrument towards certain policy concerns. It also avoidsthe concern that granting RoN may actually be detrimental to environmental protection – by singling out individual interests over collective ones, or by substituting symbolics

---

[203] Tănăsescu, 'Theoretical Foundations' (n 2) 100f.

[204] See the earlier arguments of Stone (n 29) 456–457.

[205] Horatia Muir Watt, *The Law's Ultimate Frontier: Towards an Ecological Jurisprudence* (Hart 2023) 285–292.

[206] di Micco and Graziadei, 'Italy' (n 73) 498ff.

[207] See Ralf Michaels, 'American Law (United States)' in Jan Smits (ed), *Elgar Encyclopedia of Comparative Law* (2nd edn, Edward Elgar 2023) 75, 75–77.

[208] deLisle, 'China' (n 69) 407ff.

[209] See Yaffa Epstein, 'The Rights of Nature and Environmental Law: A Developing Relationship', in Josephine van Zeben and Chris Hilson (eds), *A Research Agenda for Environmental Law* (Edward Elgar 2025) 23, 32–33.

for action.[210] Properly speaking, there is no clear general trajectory from environmental concerns to RoN – all the more so insofar as environmental law remains, by and large, anthropocentric. In Italy, the tradition of environmental protection (problematically introduced under Mussolini) followed a trajectory that led to alternatives to RoN.[211]

Consequently, opponents of RoN only sometimes oppose the goal of protecting the environment. More often they suggest that RoN are not necessary, or even conducive, to such environmental concerns. A state like China, which fears social unrest from environment-based lawsuits, will favor public enforcement of environmental law over the granting of rights.[212]

d) Ecocentrism

Environmental concerns remain anthropocentric. A separate tradition, beginning with Godofredo Stutzin, places RoN in an explicit opposition to anthropocentric approaches to law altogether.[213] Stutzin proposed granting rights to nature not in order to shift the border between man and nature, but by contrast to overcome that distinction as necessarily, and problematically, anthropocentric. Tahanescu suggests that this approach was more influential than Stone's on the Ecuadorian Constitution in 2008;[214] he also draws a line from Godofredo Stutzin's work to Thomas Berry's ecocentric theology of rights[215] and Cormac Cullinan's elaboration of the rights to exist, to have a habitat, and to evolve as part of the Earth Community.[216]

Other scholars questioning the dichotomous separation between nature and culture like Philippe Descola,[217] Michael Serres with his contract with nature and

---

[210] Ludwig Krämer, 'Rights of Nature and Their Implementation' (2020) 17 JEEPL 47, 74–75.

[211] di Micco and Graziadei, 'Italy' (n 73) 505ff.

[212] deLisle, 'China' (n 69) 407.

[213] Godofredo Stutzin, 'Should We Recognize Nature's Claim to Legal Rights? An Essay' (1976) 2 EPL 129; Godofredo Stutzin, 'Die Natur der Rechte und die Rechte der Natur' (1980) 11 Rechtstheorie 344–355; Godofredo Stutzin, 'Un imperativo ecológico: reconocer los derechos de la naturaleza' (1984) 1 Ambiente y Desarollo 79; Tănăsescu, 'Theoretical Foundations' (n 2) 101.

[214] Mihnea Tănăsescu, 'The Rights of Nature in Ecuador: The Making of an Idea' (2013) 70 International Journal of Environmental Studies 846; Tănăsescu, 'Theoretical Foundations' (n 2) 102.

[215] Thomas Berry, *The Great Work: Our Way into the Future* (Bell Tower 1999); Tănăsescu, 'Theoretical Foundations' (n 2) 105–109.

[216] Cormac Cullinan, *Wild Law* (Green Books 2011); Tănăsescu, 'Theoretical Foundations' (n 2) 105. Compare the list of rights granted by the Peruvian STC Exp Nº 00010-2022-0-1901 (n 63) above.

[217] Philippe Descola, *Beyond Nature and Culture* (UCP 2013).

Bruno Latour with his parliament of things,[218] or Donna Haraway are named as influences, too.[219] Some of these texts are more legalistic than others, but all draw to a significant degree on extralegal considerations. Their relation to the indigenous ontologies named above is not always direct; in many cases they serve as something like a Northern substitute for these ontologies, for better or worse. Legal theories with particularly close affinity to rights of nature are Earth Jurisprudence[220] and Earth System Law.[221]

8. *Interlegalities*

Finally, we come to an aspect central to legal pluralism in general and GLP in particular: interlegalities, i.e. relationships between laws.[222] These can come in various forms. They can come as influences (not only as legal transplants) between states or between non-state law and state law or indeed between either of those two and international law. But interlegalities can also come in the very practical situations dealt with in conflict of laws, when rights of nature under one legal system are enforced in another

a) Reverse legal transplants or legal coloniality?

Rights of nature are sometimes described as a 'reverse legal transplant'.[223] 'Legal transplant' conveys a copying from one legal order to another; 'reverse' suggests that the direction of this transplant is not, as usual, from North to South, but the other way round.[224] The implied assumption that transplants 'normally' proceed from North to South is not fully unjustified, though it would be foolish to attribute this to the higher quality of Northern law and to ignore the epistemic domination of colonial systems. In fact, in accordance with global legal pluralism it

---

[218] Michel Serres, 'The Natural Contract' (1992) 19(1) Critical Inquiry 1; Bruno Latour, *We Have Never Been Modern* (Harv UP 1991) 142–145; Bruno Latour, *Politics of Nature* (Harv UP 2004); Bruno Latour, *Facing Gaia* (Polity 2017).

[219] Tănăsescu, 'Theoretical Foundations' (n 2) 104; Paiement, 'Netherlands' (n 50) 534; Ewering, Reimann and Vetter, 'Germany' (n 50) 487.

[220] Natalie Rühs and Aled Jones, 'The Implementation of Earth Jurisprudence through Substantive Constitutional Rights of Nature' (2016) 8 Sustainability 174,; Jeremy J Smith, 'Of Kin and System: Rights of Nature and the UN Search for Earth Jurisprudence' (2022) 47 Transactions of the British Geographers 820–834; Cormac Cullinan, 'Earth Jurisprudence' in Nathanaël Wallenhorst and Christoph Wulf (eds), *Handbook of the Anthropocene: Humans between Heritage and Future* (Springer 2023) 563, 566–567.

[221] Joshua Chad Gellers, 'Earth System Law and the Legal Status of Non-Humans in the Anthropocene' (2020) 7 Earth System Governance 100083; Alice Bleby, 'Rights of Nature as an Expression of Earth System Law' in Timothy Cadman, Margot Hurlbert and Andrea C Simonelli (eds), *Earth System Law: Standing on the Precipice of the Anthropocene* (Routledge 2021).

[222] See I. 1. a) above, 49f.

[223] LG Erfurt 2 August 2024, 8 O 1372/21, LG Erfurt 17 October 2024, 8 O 836/22; see also Michaels (n 95).

[224] Sital Kalantry, 'Reverse Legal Transplants' (2020) 99 North Carolina LRev 49–100.

may be more appropriate to replace the whole idea of legal transplants, which implies a hierarchy between donor and recipient, with something like legal entanglements[225] or contagious lawmaking.[226]

The development of RoN is a case in point. There is some truth in the observation that they represent a ('reverse') transplant from South to North: the first official formulations of RoN in South America and New Zealand remain the most influential sources for RoN elsewhere. And although most adoptions have been in other countries in the South, resistance in the North is dwindling, and RoN are garnering sufficient support there too, following the South American models.[227] However, this reading rests on an antiquated and inadequate conceptualization of legal transplants as communication between legal systems, and therefore on a separation of the legal orders between which these transplants take place. From the perspective of GLP, the processes are much more complicated. Thus, although the Ecuadorian Constitution was the first important legal text to recognize RoN, it was not written on a blank slate, nor were its origins purely domestic: international NGOs added international expertise to the domestic sources, and the Ordinance of Tamaqua in Pennsylvania was one of the influential models.[228] That Ordinance itself, however, rested on the advice from international NGOs and also on Southern knowledge. Moreover, the very name of Tamaqua (which is Lenape and means "the river where the beaver dwells"[229] reveals that the ordinance is not purely one arising from a European-based Northern jurisprudence either: it is a fruit of the European colonial destruction of the lands of which the river is a part. Similarly, the introduction of rights of nature in New Zealand in treaties between the state and the Māori was influenced not only by local ontologies but also by common law precedent as far away as London and as long ago as 1925, namely a Privy Council decision that confirmed the legal personhood of a Hindu idol in India.[230]

Rights of nature are, then, deeply embedded in the history of colonialism and the persistence of coloniality. Colonialism lies at the root of the threats to the survival of our world at large, in treating the world as a mere resource. (*colonus* is

---

[225] Thomas Duve (ed), *Entanglements in Legal History: Conceptual Approaches* (Max Planck Institute for European Legal History 2014).

[226] Natasha Affolder, 'Contagious Environmental Lawmaking' (2019) 31 JEL 187–212.

[227] Bonilla, 'The Globalization of Rights of Nature' (n 5) 44–45.

[228] See, more generally, Bonilla, 'The Globalization of Rights of Nature' (n 5) 39–44.

[229] John Heckewelder and Peter S Du Ponceau, 'Names Which the Lenni Lenape or Delaware Indians, Who Once Inhabited This Country, Had Given to Rivers, Streams, Places, &c. &c. within the Now States of Pennsylvania, New Jersey, Maryland and Virginia' (1834) 4 Transactions of the American Philosophical Society 351, 361; John Heckewelder and William C Reichel, 'Names Which the Lenni Lennape or Delaware Indians Gave to Rivers, Streams and Localities within the States of Pennsylvania, New Jersey, Maryland and Virginia, with Their Significations' (1872) 1 Transactions of the Moravian Historical Society 227, 252; Eugen Wendler, *Friedrich List as a Railway Pioneer in the USA* (Springer 2021) 18.

[230] Note 30, above.

Latin for farmer and colonist alike.). And colonialism and coloniality are at the root of the confrontation between European and other ontologies. RoN become necessary, as a partly European-based structure, only in response to European-based destruction. It is with these caveats that RoN can be analyzed as an instance of legal transplant.

Moreover, RoN are an important example of South-South transfers.[231] The early back and forth between Ecuador and Colombia, the adoption of RoN in other Latin American countries – all of that took and takes place as legal transplants largely free of the North-South dynamic that is often viewed as characteristic for transplants. RoN are a particularly good example for countering such views and also a model case for the necessary South-South comparison in law.

b) Interactions between non-state law and state law

Conceptions of RoN or of legal subjectivity for nature often emerge from indigenous non-state laws. They are frequently the first ones to formally grant such rights or to formulate subjectivity in the language of rights. As non-state law, they do not yet have officiallegal force, unless the lawmaking power of the indigenous community is accepted by the state. Short of such delegated lawmaking, and outside of overt rejection or inadvertent ignorance, states acknowledge non-state laws and the rights they create in two other ways: incorporation (the transformation of non-state law into domestic law) or deference (the transformation of non-state law into facts).[232] Incorporation is of particular importance. In Bolivia, the first act on RoN was strongly influenced by an earlier model turned into a draft proposal by the Unity Pact, an alliance of indigenous communities.[233]

Since indigenous communities cannot all be lumped together into a single presumed unity, it is important to Be cognizant of deliberate attempts among indigenous communities to create alliances. The plurinacionalidad referred to in the constitutions of Ecuador and Bolivia acknowledges as much.[234] What *plurinacionalidad* means in practice can differ significantly, however, not unlike the difference between strong and weak legal pluralism discussed above.[235] In Ecuador, resistance to extractive practices rested on traditions of interaction, which in turn

---

[231] Bonilla, 'The Globalization of Rights of Nature' (n 5) 41–44.

[232] Ralf Michaels, 'The Re-*State*-Ment of Non-State Law: The State, Choice of Law, and the Challenge From Global Legal Pluralism' (2005) 51 Wayne LRev 1209–1259 (2005); 24

[233] Barié, 'Bolivia' (n 39) 254–255.

[234] Article 1(1) of the Ecuadorian Constitution explicitly refers to the state as 'plurinacional'; see also, eg, John Antón Sánchez, 'El modelo de estado plurinacional en Ecuador: Ideas y reflexiones' (2014) 14 Antropología Experimental 91; Daniela Maag, 'One State Under Multiple Nations? The Challenge of Building a Plurinational State in Ecuador and Bolivia (2007–2018)' (2025) 53 Politics and Policy, <https://doi.org/10.1111/polp.70017>.

[235] Chris Hesketh, 'What's at Stake in the Plurinational State Debate? The Case of Bolivia' (2025) 10 Radical Americas, <https://doi.org/10.14324/111.444.ra.2025.v10.1.001>.

enabled a rather strong *plurinacionalidad*.[236] In Bolivia, by contrast, where an alliance among indigenous communities was harder to build, it appears that the state has managed to take over much of the governance of RoN. Two countries do not allow for generalization, but here at least we see a correlation between the strength of plurinationality and the effectiveness of RoN.

c) Influences from paradigmatic cases

Ecuador and Bolivia, and sometimes also Colombia, are regularly referred to as models for rights of nature.[237] The Ecuadorian Constitution is even named as an influence for solutions short of RoN, such as a less anthropocentric approach to environmental protection seen in a court decision in Rio de Janeiro.[238]

This prominent position is a mixed blessing. On the one hand, it is impressive to see the speedy and wide influence of a recent introduction, all the more so since it comes from the Global South, which is usually not viewed as an originator of legal transplants. On the other hand, generalized references often paper over both differences between these respective conceptions and the complex ways in which they operate, their limitations, and their political contestations. Frequently, early adopters of RoN are named collectively, despite the important differences existing between them. They then become general inspirations rather than concrete models, and what we see frequently, especially in the Global North, is a combination of general fascination with the idea of RoN, on the one hand, and a reluctance to grapple with the concrete implications involved, on the other. The ideological influence is certainly strong; the precise impact is often weaker.

To some extent, whether RoN are received is influenced by more general attitudes. Countries that generally resist reception of legal solutions from abroad are, unsurprisingly, also less likely to receive RoN; this is the case, for example, for China.[239] The same is true for those countries unwilling to receive solutions from the Global South based on an implied inferiority or incompatibility – some of the rather hefty reactions to a German court's willingness to receive RoN smack of this. More surprisingly perhaps, to the extent RoN are viewed as Western, they are unattractive to states that reject Western transplants – again China serves as an example.[240]

Sometimes the influence is quite clear. Uganda's law seems clearly influenced by Ecuador's given the relatively high level of similarity.[241] Sometimes, the de-

---

236 Maria Akchurin, 'Constructing the Rights of Nature: Constitutional Reform, Mobilization, and Environmental Protection in Ecuador' (2015) 40 Law and Social Inquiry 937, 950–951.

237 Wolkmer and Ferrazzo, 'Brazil' (n 54) 384.

238 Wolkmer and Ferrazzo, 'Brazil' (n 54) 385.

239 deLisle, 'China' (n 69) 401.

240 deLisle, 'China' (n 69) 397.

241 See Matthew W Hopewell, *The Rights of Nature in Uganda: Exploring the Emergence, Power and Transformative Quality of a "New Wave" of Environmentalism* (PhD thesis, SOAS University of London 2019) 36–37.

cision on whether or not to take inspiration from sources abroad is deliberate: the Ecuadorian Constitutional Court may have had good reasons not to refer to the River Atrato case in Colombia in view of significant differences in the role of RoN as between Ecuador and Colombia.[242] In other instances, references are more general because similarities are less pronounced and the formulation of RoN is more homebuilt. This can mean that local conditions are taken into account better and a formulation of RoN is more tailored to local needs. It can also mean, however, that the significant transformative potential of RoN is not used and the irritation[243] from RoN leads to a domesticated and softened version of RoN. In Spain, for example, recognition of the Mar Menor as a legal person has, it appears, so far not greatly enhanced its protection.

d) Interactions with international law

One interesting phenomenon concerns the interaction between domestic and international law. Legal transplant literature was relatively late to realize that transplants could also happen vertically, between domestic and international law.[244] This dynamic plays a particular role for RoN. Rights of nature emerged prominently from the 2010 World People's Conference on Climate Change and the Rights of Mother Earth, and arguments were made from the beginning for their establishment in international law. This was done in a 2017 Advisory opinion of the Inter-American Court of Human Rights,[245] and that decision was, in turn, used to influence domestic law, eg in Bolivia.[246] There are even decisions by the International Rights of Nature Tribunal, although they are not binding.[247] On the other hand, RoN have become important instruments of international relations for Bolivia and Ecuador and have influenced international law.[248]

e) Cross-border recognition of foreign RoN

A so far underappreciated area of transboundary impact of RoN lies in cross-border recognition. Rivers that cross national borders sometimes create transnational

---

[242] Espinoza, 'Ecuador' (n 38) 235.

[243] After Gunther Teubner, 'Legal Irritants: Good Faith in British Law or How Unifying Law Ends Up in New Differences' (1998) 61 MLR 11–32.

[244] Jonathan Wiener, 'Something Borrowed for Something Blue: Legal Transplants and the Evolution of Global Environmental Law' (2001) 27 Ecology Law Quarterly 1295–1372.

[245] Inter-American Court of Human Rights, Advisory Opinion OC-23/17 (15 November 2017) <https://www.corteidh.or.cr/docs/opiniones/seriea_23_ing.pdf>; Bonilla, 'The Globalization of Rights of Nature' (n 5) 34, 35; Tignino, 'International Law' (n 23) 138. See also, now, the confirmation in Inter-American Court of Human Rights, Advisory Opinion OC-32/25 (3 July 2025) nos 7, 279–286 <https://www.corteidh.or.cr/docs/casos/articulos/seriec_559_esp.pdf>.

[246] Barié, 'Bolivia' (n 39) 270.

[247] Barié, 'Bolivia' (n 39) 270, 271.

[248] Bonilla, 'The Globalization of Rights of Nature' (n 5) 32–34.

initiatives, such as the Appel du Rhône between Switzerland and France.[249] Where that is not the case, the issue arises that RoN are recognized in one state but not the other. If, for example, the Spanish Mar Menor is recognized as a legal subject by foreign legal orders, this has two important implications. First, it means that a RoN is recognized and enforced, effectively, even in legal orders that do not grant such legal subjectivity in their domestic law. Secondly, such cross-border recognition can serve as a powerful impetus for the introduction of a similar status also in domestic law, something we have seen earlier with other institutions like same-sex marriage.

Cross-border recognition can happen in public law and in private law. In public law, the Interamerican Human Rights Court emphasized the possibility of cross-border state liability, though without explicit doctrinal discussion regarding the role of RoN recognized in one state but not the other.[250] Moreover, the Mar Menor has applied, successfully, for recognition as a subject with standing in administrative proceedings in Switzerland and Germany.[251] Somewhat ironically, therefore, the Spanish Mar Menor has a stronger legal position in Germany than does the German North Sea.[252]

In private international law, legal personality is usually determined by the law granting legal personality, though the additional and so far underappreciated problem of whether this applies also to the capability of being a subject needs further attention. Literature is slim,[253] case law is nonexistent, further research is needed.

III. Consequences

What can we learn in general from this comparison of RoN through the lens of GLP? I want to suggest a number of tentative conclusions.

---

[249] <www.id-eau.org>; Affolter and Affolter, 'Switzerland' (n 49) 621–623.

[250] IACHR, Advisory Opinion OC-23/17 para 103; Tignino, 'International Law' (n 23) 138.

[251] See <https://ecojurisprudence.org/initiatives/switzerland-recognizes-the-mar-menor-lagoon-as-a-subject-of-rights-and-foreign-legal-entity/>; Nina Kerstensteiner and Felix Aiwanger, 'Personhood Across Borders: The Mar Menor Before Swiss Authorities' (*Verfassungsblog*, 5 September 2025) <https://verfassungsblog.de/mar-menor-switzerland/>.

[252] On the rejection of a suit brought by the seals of the North Sea, see VG Hamburg 22 September 1988, 7 VG 2499/88, [1988] NVwZ 1058 and Michael W Schröter and Klaus Bosselmann, 'Die Robbenklage im Lichte der Nachhaltigkeit' (2018) ZUR 195; Andreas Gutmann, *Hybride Rechtssubjektivität* (Nomos 2021) 87.

[253] But see Stefano Dominelli, 'A New Legal Status for the Environment and Animals, and Private International Law: Tertium Genus Non Datur? Some Thoughts on (the Need for) Eco-Centric Approaches in Conflict of Laws' (2024) 60 Rivista di diritto internazionale privato e processuale 1165; Lukas Rademacher, 'Schimpansen, Flusspferde und Flüsse vor deutschen Gerichten? Partei- und Rechtsfähigkeit von natürlichen Ressourcen und Tieren im anthropozentrischen Kollisionsrecht' in Stephan Hobe, Heike Krieger and Karsten Thorn (eds), *Gemeinwohlinteressen in einer sich wandelnden Welt* (2025, forthcoming) 53 Berichte der deutschen Gesellschaft für Internationales Recht.

First, the success of RoN is ambivalent. We see an extremely high interest in scholarship: RoN are a popular subject. We also see adoptions of RoN spreading relatively quickly, though it remains to be seen whether this trend continues. By contrast, their legal effectivity is, so far, quite limited. In Ecuador, RoN appear to be established: there is ample litigation and not infrequent success. In many other jurisdictions, however, RoN are recognized but rarely effective. This is sometimes overlooked in comparative law literature that addresses only those legal orders that recognize RoN and focuses only on those court cases in which they are invoked in litigation. Comparatists must also focus on jurisdictions in which rights of nature have not been successful. That is more differ. Yet as we see here, it is instructive: the conditions for an effective adoption of RoN depend on local contexts and are often not favorable.

Second, even where RoN are recognized, this often happens only in a sanitized, domesticated form. Both scholarship and existing legislation, especially in Latin America, suggests that RoN can have a truly revolutionary potential and that the form of rights, though inspired by the Global North, is not incompatible with legal theories that replace anthropocentrism with ecocentrism. The reality is, however, that many countries especially in the Global North adopt a softened version of RoN that amounts, in the end, to little more than greater protection for some entities or the environment as such – but not for nature. And even in the Global South, the revolutionary adoption of RoN sometimes translates into a much less revolutionary role in practice.

Third, RoN prove to be an ideal ground for a GLP analysis, just as GLP is a good lens for RoN. RoN are closely linked to legal pluralism insofar as they often are based in, or defer to, non-state orders, in particular those of indigenous communities. At the same time, RoN have a strong affinity with globalization: the spread of the concept is global, as are the discourses around RoN, and of course nature is global or even planetary. Taking both together, we find an almost paradigmatic situation of GLP.

What do these insights bode for the future? Will the growing climate catastrophe increase the pressure not only for heightened environmental protection but also for increased recognition of RoN, perhaps even in their more paradigm shifting form? Or will RoN, as a hybrid, be unattractive, because they are too radical for some and not radical enough for others? Such predictions are hard to make, but one piece of analysis may be helpful. The 'puzzling persistence of nature's rights'[254] goes together with two weaknesses. The first is the often metaphorical or symbolic nature of RoN: they are invoked but not actually enforced. The second is their domestication: RoN are, in practice, often less revolutionary than in their conception. But both weaknesses may turn into strengths. As to the first, the symbolic power of RoN may prove to be a strength on its own, even regardless of direct enforcement in courts. And as to the second, RoN may yet prove to have more revolutionary potential than is visible precisely because of their ability to

---

[254] Sam Bookman, 'The Puzzling Persistence of Nature's Rights' (2025) Utah LRev 165.

fit, seemingly, into traditional (Northern) jurisprudence – before unleashing their more revolutionary potential. If, to paraphrase Boyd, RoN 'could save the world', then nothing short of a 'legal revolution' will be necessary.[255] We are not yet there by far.

[255] Boyd (n 197). See I.3.c) above, 60.

II. Special Reports

Theoretical Foundations of the Rights of Nature

Origins and Conceptual Characteristics

Mihnea Tănăsescu[*]

This article is concerned with the theoretical origins and underpinnings of the rights of nature.[1] It will start with the various intellectual genealogies of this idea and practice in diverse contexts, before analyzing the conceptual issues that are at stake. Finally, it will show how these issues are key to understanding the breadth of present and future cases. It will use practical examples of rights of nature selectively, in order to illustrate conceptual elements. It is therefore not to be read as an exhaustive analysis of the practical implementation of the rights of nature, but as an elaboration of the key conceptual elements that are present, to a greater or lesser extent, in many cases.

The organizing thesis of this contribution is that the rights of nature, in these early stages of their development, have promising yet often conflicting conceptual components. It would help, therefore, to examine them more critically in light of future implementation. This does not mean that the arguments presented here foreclose future conceptual analysis. Quite on the contrary, this is but a step towards the necessary and continuous interrogation of this new legal and politi-

[*] The manuscript was finalized in January 2023. The assessment of literature and case law, including internet materials, reflects the legal situation at the time.

[1] Parts of section I, detailing historical threads of the rights of nature, have previously appeared in Mihnea Tănăsescu, *Understanding the Rights of Nature: A Critical Introduction* (transcript 2022).

cal practice. The kind of constructive skepticism that I will follow is meant to push the boundaries of theoretical analyses in order to create new problems, new situations that can make one think again.

I. Theoretical Foundations

1. Christopher Stone and Legal Standing

There are several different claimed origins for the idea of the rights of nature, each coming with its own conceptual commitments. The most dominant account of their intellectual genealogy traces them to the work of American legal scholar Christopher Stone. In his 1972 article titled Should Trees Have Standing? – Toward Legal Rights for Natural Objects, Stone explicitly argued that the environment could enjoy legal rights.[2] He developed this line of thinking further in his 2010 book *Should Trees Have Standing? Law, Morality and the Environment*. Stone's arguments are still extremely influential, so it makes sense to start there.

What occasioned Stone's thinking was a lawsuit, brought by the Sierra Club.[3] In *Sierra Club v Morton*,

> 'the U.S. Forest Service had granted a permit to Walt Disney Enterprises, Inc. to "develop" Mineral King Valley, a wilderness area in California's Sierra Nevada Mountains, by the construction of a $35 million complex of motels, restaurants, and recreational facilities. The Sierra Club, maintaining that the project would adversely affect the area's aesthetic and ecological balance, brought suit for an injunction'.[4]

However, the Ninth Circuit ruled that the Sierra Club did not have legal standing to bring the suit. An appeal arrived in front of the Supreme Court, which ended up agreeing with the Ninth Circuit, though Justice Douglas penned a now famous dissent based on Stone's legal argumentation.

Stone's basic argument was that Sierra Club did not sue on behalf of Mineral King Valley because they were interested in protecting their own aesthetic inter-

[2] The 'earliest' version of something – regarding here, for example, the earliest argument for rights of nature – is a claim that should be treated with caution. Usually, upon closer investigation, one finds predecessors that vary by degree from the claimed origin of an idea. For example, Roderick Frazier Nash, *The Rights of Nature: A History of Environmental Ethics* (Univ of Wisconsin Press 1989) 127, quotes a 1964 essay by Clarence Morris that dealt with 'nature's legal rights'. Surely, there were predecessors for that as well! The point is that nothing can be settled by finding the earliest version; historical and intellectual threads are living and always respond to present tugging and wrangling. See Nash (ibidem); Christopher D Stone, 'Should Trees Have Standing. Toward Legal Rights for Natural Objects' (1972) 45 SCal LRev 450; Christopher D Stone, *Should Trees Have Standing? Law, Morality, and the Environment* (OUP 2010).

[3] One of the most influential environmental organizations in the Unites States of America; see <www.sierraclub.org>. See *Sierra Club v Morton* [1972] 405 US 727.

[4] Stone, *Law, Morality, and the Environment* (n 2) xiii.

est; they were interested in protecting the integrity of the place itself. However, the US doctrine of legal standing did not allow them to sue because they could not show that they would be directly impacted by the proposed construction. There was no place in US law for suing on behalf of an environment *itself*, irrespective of damage to the person suing. Stone therefore argued that it would make sense to allow standing to apply directly to the natural entities that the Sierra Club was trying to protect.[5]

Stone shows convincingly that organizations like the Sierra Club have had to resort to subterfuge in order to gain legal standing (doing things like claiming 'aesthetic injury'). He argued that it would be much simpler if the law legitimized their motives to begin with, by granting standing to the natural entities themselves. The dissent that Justice Douglas wrote was based on Stone's paper and argued that 'public concern for protecting nature's ecological equilibrium should lead to the conferral of standing upon environmental objects to sue for their own preservation'.[6] In his work, Stone stressed that this is much less radical than it first appears. In fact, there are many non-human and even non-animate entities that do enjoy legal standing, for example ships and corporations. So, the particular history of rights of nature rooted in Stone's work starts with a concern for achieving legal standing. This concern makes sense for the legal system of the United States but is obviated by public interest standing for environmental matters as recognized in other jurisdictions around the world.

Even though standing may not be an issue in other jurisdictions, the concern over it is fundamental for the rights of nature. The argument is that, even though standing as such may be achieved, a nature without rights does not enjoy the moral standing that it deserves, and therefore its interests are inherently construed from anthropocentric points of view. The idea that legal/moral standing can be achieved through granting nature rights is therefore meant to push the law towards ecocentric practices (more on this later). Given this, we need to look closer at the conundrums raised by the relationship between moral and legal standing.

2. *Legal and Moral Personality*

There is no problem with conferring legal standing on anything at all. The only limiting factor, as it were, is what people empowered to confer such standing consider deserving of it, for pragmatic reasons. In order to make this case, it helps to show that having legal standing comes with the creation of a legal personality: whoever or whatever has legal standing becomes, because of that, a 'person' in

---

[5] The doctrine of standing is not the same everywhere. Stone was specifically reacting to the US version of standing, whereas in other jurisdictions – for example, Finland or New Zealand – individuals or groups not directly affected can still sue on behalf of an environment, claiming that they are defending the common good or interest. This is an important contextual element in understanding the genesis of the standing argument in rights of nature advocacy. See Visa AJ Kurki, *A Theory of Legal Personhood* (OUP 2019).

[6] Stone, *Law, Morality, and the Environment* (n 2) xiv.

front of the law. Legal personality and legal standing are a package; you cannot have one without the other.

According to some legal scholars,[7] legal personality is granted by the law in a highly fluid and malleable fashion. This means that a legal person is that entity that the law declares to be a legal person. The interesting question is *why* certain entities are deemed, by the law, to enjoy legal personality and others are not. And that is precisely the terrain on which the rights of nature develop. Both in terms of advocacy and theory, rights of nature advocates argue that there are no valid a priori reasons to use legal personality for some entities but not for nature.

For Stone, as well as for many of his followers, the question of legal standing for nature is intrinsically tied to its moral standing: nature should have legal standing because it is morally worthy of such.[8] In Stone's language,

> 'there is something of a seamless web involved: there will be resistance to giving the thing "rights" until it can be seen and valued for itself; yet, it is hard to see it and value it for itself until we can bring ourselves to give it "rights" – which is almost inevitably going to sound inconceivable to a large group of people'.[9]

This makes it seem as if moral and legal personality are related in a vectored fashion: if something has moral standing, then it is apt for getting legal standing; conversely, granting legal standing should soften the moral imagination. Yet this thought is not entirely supported by legal practice, nor by the way in which moral considerations tend to work.[10] This is not to say that some entities that are morally considerable do not receive legal status on that account. Nor is it to say that the law has no bearing on how morality develops. But it is to say that there is no automatic relationship between the two.

The easiest way to see this is to think about entities that enjoy legal standing without also enjoying, on that account, moral standing. Ships and corporations, for example, do not enjoy moral standing just because they have the legal kind. Conversely, many cultures extend moral standing to ancestors and spirits, but without this translating into legal status akin to the concept of a 'legal person'. The point is that, though the two kinds of standing are entangled within the rights of nature throughout their history, this entanglement itself can be questioned rath-

---

[7] Ngaire Naffine, 'Who are Law's Persons? From Cheshire Cats to Responsible Subjects' (2003) 66(3) MLR 346; Ngaire Naffine, *Law's Meaning of Life: Philosophy, Religion, Darwin and the Legal Person* (Bloomsbury 2009); Ngaire Naffine, 'Liberating the Legal Person' (2011) 26(1) Can J L Soc 193; Ngaire Naffine, 'Legal Persons as Abstractions: The Extrapolation of Persons from the Male Case' in Visa AJ Kurki and Tomasz Pietrzykowski (eds), *Legal Personhood: Animals, Artificial Intelligence and the Unborn* (Springer 2017); Anna Grear, 'Law's Entities: Complexity, Plasticity and Justice' (2013) 4(1) Jurisprudence 76.

[8] This borrows from animal rights discourse. For the classic treatment, see Peter Singer, *Animal Liberation. A New Ethics for Our Treatment of Animals* (HarperCollins 1975).

[9] Stone, 'Legal Rights for Natural Objects' (n 2) 456.

[10] For a non-judgment driven account of moral thought, see Alice Crary, *Beyond Moral Judgment* (Harv UP 2007).

er than assumed.[11] My view is that it has to be analyzed on a case-by-case basis rather than seen as an automatic feature of the rights of nature as such.

Around the same time that Stone was writing his famous legal article, Godofredo Stutzin was developing the bases of environmental advocacy in Chile. As early as 1973,[12] he penned articles calling for the rights of nature. His arguments were similar to Stone's but also contained a different emphasis that continues to influence rights of nature theory and practice today.

Stutzin's main concern and theoretical contribution is seen through the argument that the rights of nature represent an *ecocentric* turn in the history of law. Stutzin saw them as responding to what he called an 'ecological imperative'. He argued that granting nature rights 'implies overcoming the anthropocentric bias of law'.[13] This apparently simple formulation has had far-reaching consequences for the way in which rights of nature are understood, and therefore also for the way in which legal provisions are written.

If, as Stutzin argues, the rights of nature are necessarily an ecocentric move, then they potentially place the environment before human interests and consideration, inasmuch as there is a conflict between valuing nature for its own sake or for instrumental purposes. This implies that it is through granting nature rights that environmental problems can be fixed because the law would be forced to consider the environment irrespective of human interests. This line of argumentation is visible in most rights of nature theories today. The assumption that rights of nature are ecocentric is one of the most present ones in the relevant scholarship today.[14]

---

[11] The relationship between moral and legal conceptions of nature is not unique to the genealogical lines that I am drawing here. Many different cultures and indeed legal traditions posit complex relationships between morality and the law, sometimes not making a distinction between them at all (or, rather, only contextually). I therefore do not mean to say that the relationship between legal and moral rights in the history and practice of the rights of nature is uniquely derived from the sources I explore here. As we will see, there are many other legal traditions that come to have a degree of influence on the rights of nature, not least on the relationship between its moral and legal rights.

[12] See Godofredo Stutzin, 'Un imperativo ecológico: reconocer los derechos de la naturaleza' (1984) 1(1) Ambiente y Desarrollo 97, recounting the history of his own argumentation; Also Farith Simon, 'La naturaleza como sujeto de derechos en la Constitución ecuatoriana: la construcción de una categoría de interculturalidad (Nature as a Subject of Rights in the Ecuadorian Constitution: The Construction of a Category of Interculturality)' in Estupiñán Achury and others (eds), *La naturaleza como sujeto de Derechos en el Constitucionalismo Democrático* (Universidad Libre 2019).

[13] Estupiñán Achury and others (n 12).

[14] See, inter alia, Alberto Acosta and others, 'Dejar el crudo en tierra o la búsqueda del paraíso perdido. Elementos para una propuesta política y económica para la Iniciativa de no explotación del crudo del ITT' (2009) 23 Polis. Revista Latinoamericana 429; Alberto Acosta, 'Toward the Universal Declaration of Rights of Nature: Thoughts for Action' (2010) 24(1) AFESE Journal; Alberto Acosta, *El Buen Vivir: Sumak Kawsay, una oportunidad para imaginar otros mundos* (Icaria 2013).

'Every day it becomes more obvious', Stutzin wrote, 'that if we want sustainable and long-lasting solutions to the ecological problems we have created, we cannot continue ignoring the existence of a nature with its own interests'.[15] This means that nature's rights are formulated as *recognized*, not created or granted by humans. The role of the human is not of fashioning a legal mechanism, but of using legal mechanisms to translate what is already the case. So, the moral standing of nature obviously demands legal standing, the two being inseparable. Furthermore, once the law catches up with the supposedly obvious fact of nature's moral standing, ecological problems can be solved *because of* this alignment of the law with moral sensibility.

This general outline of advocacy and theory has been very durable and potent. I would argue that Stutzin's influence on the rights of nature, though much less acknowledged than Stone's, has so far been more pronounced. It has, at the least, had a great influence on one of the first codifications of these rights, in Ecuador's 2008 constitution.[16] In the specific intellectual history of the rights of nature in Ecuador, these are partly rooted in Indigenous activism around territorial rights and for the adoption of Sunak Kawsay ('buen vivir' or 'good living') as a model of development. In the Ecuadorian civil society associated with this movement, the idea of ecocentrism that Stutzin advocated found a potent and influential expression.[17]

Whereas we can reasonably interpret Stone's work as being equally rooted in the pragmatism of legal standing and the morality of legal personhood, for Stutzin it was about righting a wrong. The concept of right approaches here the older idea of natural rights, that is to say the correct form of something, and its correct treatment, as dictated by nature itself. For Stutzin it was the literal personality of nature that demanded we recognize its rights.

The legal person, in strictly legal terms, is a fiction that can be granted to many kinds of entities inasmuch as the law deems it necessary.[18] But the very terms of legal *person* or legal *personality* already point towards the moral traces that are

---

[15] Stutzin (n 12) 97. Author's own translation.

[16] See Mihnea Tănăsescu, 'The Rights of Nature in Ecuador: The Making of an Idea' (2013) 70(6) International Journal of Environmental Studies 846.

[17] See, inter alia, Alberto Acosta, *Buen vivir Sumak Kawsay: una oportunidad para imaginar otros mundos* (Abya-Yala 2012); Alberto Acosta and Esperenza Martínez (eds), *El buen vivir: una vía para el desarrollo* (Abya-Yala 2009); Adriana Norma Martínez and Adriana Margarita Porcelli, 'Relevancia de los Derechos de la Naturaleza en las comunidades indígenas a la luz de la Corte Interamericana de Derechos Humanos' (2007) Pólemos, Portal Jurídico Interdisciplinario; Chuji Gualinga and others, *Derechos colectivos de los pueblos y nacionalidades* (IWGIA 2009); Antonio Luis Hidalgo-Capitán, Alejandro Guillén García and Nancy Deleg Guazha (eds), *Antología del pensamiento indigenista ecuatoriano sobre Sumak Kawsay* (Universidad de Huelva CIM Ediciones 2014).

[18] See Erin O'Donnell, 'Rivers as Living Beings: Rights in Law, but No Rights to Water?' (2020) 29(4) Griffith LRev 643; also see n 7.

etched within this legal concept.[19] Stutzin doesn't speak of the possibility of formatting nature as a legal *entity*, but rather of the personal qualities of nature that demand a *recognition* of its rights.[20]

3. Theology of Rights

Stutzin's argumentation found another expression through a spiritualist tradition that theologizes the recognition of nature's inherent value through the concept of rights. The most influential early proponent of an *ecotheological* version of nature's potential rights was Thomas Berry, himself building on a long tradition that theologized the idea of rights, as rooted in the concept of natural rights. In turn, Berry decisively influenced the work of Cormac Cullinan, which became – through his book *Wild Law* – an important foundation for rights of nature scholarship and practice.[21]

The story that Berry advanced is best exemplified by the title of his last book, *The Great Work*.[22] This, the culmination of his activities, reunited ideas that he had presented throughout a series of earlier publications as well as in his teaching and public engagements. For the present purposes, several elements of Berry's account of the Universe are relevant.

Berry's story is a grand narrative of *the Universe*. This follows directly from theology, which is interested in the greatest level of explanation for observable phenomena. Wishing to reconcile Christian theology with modern science, especially cosmology and ecology, he focused on a narrative that explained the way in which the Universe – the greatest possible unit – came into being and evolved. Berry also relied on scientific theories, like evolution, and worked theology around them (just like his great influence, Teilhard de Chardin). So, instead of a theological universe that arranged things according to God's plan, Berry argued for an evolutionary universe created by God precisely so as to be self-generating.[23]

Berry considered the universe to be a unity. This principle of unity is a way of reconciling the theological unity of creation (one creator and one creation) with the interrelatedness that ecology had been uncovering since its 19th century beginnings. The fact of interrelatedness is seen to fit within the theological idea of unity by interpreting the universe as a community: given that everything is related

---

[19] Note 7.

[20] In legal theory, there is another salient distinction between a legal *subject* and a legal *person*, with the legal subject encompassing, potentially, a more agnostic view of the entity thus created. For the purposes of this contribution, I will use legal person, as I think it reflects better its use in rights of nature so far, and contrast it with legal entity, which is also supported by some extant cases. See Mihnea Tănăsescu, 'Rights of Nature, Legal Personality, and Indigenous Philosophies' (2020) 9(3) TEL 429.

[21] Cormac Cullinan, *Wild Law* (Siber Ink 2011).

[22] Thomas Berry, *The Great Work: Our Way into the Future* (Crown 2011).

[23] See also PJ Robinson, 'Integrity of Creation and Christian mission' (1991) 19(2) Missionalia: Southern African Journal of Mission Studies 144.

to everything else, everything must be a participant in the great community of being (in the Great Work).

Because of humanity's privileged role within creation, it is only through human consciousness that this great community can know itself. Because of this, humans are in a responsible position, becoming *guardians* of the great mystery.[24] The juxtaposition of the ideas of unity and community, together with the privileging of human consciousness, leads towards a picture of the world that is both hierarchical (in the theological tradition of the Great Chain of Being)[25] and, by its own account, ecocentric. But the theological underpinnings of Berry's argument question the ecocentric–anthropocentric distinction because it is ultimately the responsibility of humans to uphold the order of creation by refashioning their law to fit with the interrelatedness of a universal community.

Berry's idea of community seems inspired by the idea of an ecosystem in early 20th century ecology. It is an interpretation of the fact of interrelation that selects ecological concepts that cohere with theological commitments. In *Wild Law*, Cullinan extracts from Berry's theology several different nature rights that he argues are the most fundamental ones: the rights to exist, to have a habitat, and to evolve as part of the Earth Community. Cullinan's rights, as inherited from Berry's theology, are predicated at the level of the totality and presuppose the existence of an Earth Community. This thinking has had a profound influence on several cases of rights for nature so far.

The history of the rights of nature that goes from Stutzin to Berry to Cullinan is quite different from Stone's work. Stone was arguably a pragmatist, not really interested in developing the concept of nature as a totality (as opposed to a locality). For Stone, standing pragmatically applies to anything, and if people want to speak on behalf of environments, the law can accommodate their wish.[26] However, the ecotheological history that I have briefly sketched is instead interested in advancing a framework that subsumes any given situation under the Great Work, the totality that imposes a series of 'fundamental rights' that have to be *recognized* (as opposed to granted). Cullinan tellingly uses the expression Great Jurisprudence to describe his framework, in a direct reference to Berry.[27]

In another early work on the concept of rights of nature, Roderick Nash[28] discussed how this idea emerged in the English-speaking world out of earlier conceptions of natural rights, successively modified through human rights revolutions (abolitionism and women's rights first and foremost), theories of animal rights,

---

[24] For a discussion of guardianship, see Tănăsescu (n 20); Tănăsescu (n 1).

[25] See Phillipe Descola, *Beyond Nature and Culture* (UCP 2013); Phillipe Descola, 'Modes of Being and Forms of Predication' (2014) 4(1) Hau: Journal of Ethnographic Theory 271.

[26] In many jurisdictions it already does, without appealing to legal personality or rights at all. See Kurki (n 5).

[27] It may be worth pointing out that theological thinking does not present itself as a framework but rather as a revelation of the truth (which leads to recognized, not granted, rights).

[28] Note 2.

and eventually the rights of nature themselves. Though himself subscribing to an 'expanding circle of moral concern' view, he nonetheless revealed the fine webbing that holds together apparently disparate thinkers and traditions around the idea that rights are a *recognition* of something that is already there, and that this recognition can be expanded without limits (to eventually encompass everything).

Nash further showed how theology was never far away from the earliest environmental thinking, nor from the influential debate about the inherent value of nature. The passage from the 19th to the 20th century was a fruitful period for the merging of ecology with value theories and theology. John Muir, for example, the famed father of US national parks, explicitly derived the values that he saw inhering in the natural world from the 'fact' of creation. Muir was a founding member of the Sierra Club, whose activities would animate Stone's thinking.

The intellectual lines sketched so far are but the most dominant and well-documented ones. In my view, what it means for an intellectual history to be dominant is that it silently influences the ways in which multiple rights of nature develop. Many examples of rights of nature so far share one or another of the basic characteristics foretold by the different genealogies described above.

Yet these histories do not *exhaust* the potential elements that have had a major influence on rights of nature. Oral traditions and ad hoc practices are just as important, as we will see later on. So are the legal norms dominant in different jurisdictions that toy with the idea of granting nature rights. Judicial activism, for example, may or may not play a role in developing rights of nature, largely depending on whether such activism is from the outset a possibility within a given legal system. Similarly, organized civil society has an ongoing influence where relevant groups advocate for the rights of nature, therefore etching their own cultural histories within this practice. So, if written twenty years from now, the history of the rights of nature will surely look very different.

4. Legal Pluralism

An especially salient theoretical and practical foundation (see below) of the rights of nature has been their connection with indigenous philosophical traditions and activism. There are several claims and conceptual elements that deserve close attention. The dominant view is that the relationship between indigeneity and the rights of nature is simultaneously one of activism and one of philosophical conception. More widely, the potential mixing of Western and Indigenous law has become increasingly important, giving rise to a scholarly and increasingly practical tradition of legal pluralism. Here, I want to tackle these important issues.

I want to discuss legal pluralism as it pertains to the rights of nature specifically, but it is important to realize that this is but a sub-concern of a much wider movement for the recognition of diverse legal traditions. The basic idea is that Western law, dominant in practically all Indigenous lands, can and should be seen as one among many different systems of law. The task, then, is to pursue hybrids that will allow more space to formerly excluded people. The creation of space within the law is crucial because the law has itself been a primary instrument of exclusion.

So, one of the most salient questions for our discussion is to what extent the claimed affinity of Indigenous thought with the rights of nature serves to advance the agenda of legal pluralism. How much can these rights achieve in terms of making space for different traditions of indigenous law to substantially change legal theory and practice?

The idea that rights of nature either translate or directly come out of indigenous thought is almost universally accepted. I have argued against the dominance of this view,[29] so here I want to summarize the main points that I think complicate this relationship more than has heretofore been appreciated. I want to focus on the idea of a legal person, indispensable to the rights of nature so far, as well as the concept of rights itself.

As Joanna Bourke[30] and Costas Douzinas[31] have repeatedly shown, the legal person has been modelled on an idealized version of a human with characteristics that have gone hand in hand with those of dominant social classes. Naffine, Grear, and Davies[32] further support the view that legal personality and its automatic implication in rights theory and practice originally derives from criteria pertaining to a 'normal human individual'. Arstein-Kerslake[33] has shown how the model legal person has routinely excluded people with disabilities or with identities that differ from the moral standard inherent in the concept. What I want to point out is that the idea of conformity to a standard, in the history of liberalism, understands the person as an *individual*. The rights of the legal person are therefore primarily the rights of an individual qua individual.

The individual is a moral category, as much as a legal one. The individual, in liberal theory, is the ultimate unit of concern; Western political and legal traditions are committed to upholding its primacy.[34] The legal person cannot but participate in this history, drawing its intellectual credentials from it. Perhaps this is clearest in the doctrine of human rights. Its various merits notwithstanding (it is not the merits of rights that are under discussion here, but their conceptual elements), human rights are the rights of individuals and are applied differentially depending on which particular individuals are of concern.

---

[29] Tănăsescu (n 20); Tănăsescu (n 1).

[30] Joanna Bourke, *What it Means to Be Human: Historical Reflections from the 1800s to the Present* (Catapult 2011).

[31] Costas Douzinas, *The End of Human Rights: Critical Thought at the Turn of the Century* (Bloomsbury 2000).

[32] Note 7; Margaret Davies, 'Persons, Property, and Community' (2012) 2(2) feminists@law <https://journals.kent.ac.uk/index.php/feministsatlaw/article/view/37>.

[33] Anna Arstein-Kerslake, *Restoring Voice to People with Cognitive Disabilities* (CUP 2017).

[34] Daniel Bonilla Maldonado, *Legal Barbarians: Identity, Modern Comparative Law and the Global South* (CUP 2021).

The poor, for example, are systematically excluded from the human rights of the rich. As I have argued before,[35] 'ethnic minorities, colonized people, women at the intersection of disadvantageous social and economic factors, all have the same rights in theory but different ones in practice'. The idea of a legal person that goes hand in hand with the individualization of rights is, therefore, beyond its undeniable evolution, one that maintains a certain connection to its moralist history.

The legal personality of nature arises out of this history. And it has traditionally accompanied the concept of rights; the two are only separable analytically. But let's take it the other way: Where in the particular Indigenous thought thus far influential in rights of nature may we find the idea of nature as a moral person in a form deserving legal recognition of the kind now afforded? There are, to be sure, Indigenous traditions that Western thought has translated as recognizing the personality of nature. Andean ones, for example, important in the cases of Ecuador and Bolivia, refer to features of the natural world in ways that may be analogous to personhood. Pachamama, for example, features in the Ecuadorian constitution as synonymous with Nature and Mother Earth.

There are many different Indigenous traditions, and one should not amalgamate them all into one kind of indigeneity. Some of these traditions have historically not been particularly friendly to nature, a point that rights of nature often brush over. However that may be, in the cases surveyed so far there has obviously been indigenous involvement, which elsewhere I have called 'strategic'.[36] And in these particular groups, one of the common threads has been a certain rootedness in specific territories that are part of a more-than-human family of relations.[37]

But Nature and Mother Earth are general concepts above specific instantiations, resembling the totality discussed earlier. Marisol de la Cadena,[38] for ex-

---

[35] Tănăsescu (n 1).

[36] Mihnea Tănăsescu, 'Nature Advocacy and the Indigenous Symbol' (2015) 24(1) Environmental Values 105.

[37] Daniel Bonilla Maldonado quite rightly points out that today many indigenous communities are thoroughly urban. The assumption behind much of the literature stressing inherent ties to ancestral territories is that indigenous people are primarily rural, which is not empirically the case. This point is extremely important, but in the cases involved here there is a strong self-identification of indigenous activism for the rights of nature with rural territories, despite the fact that individuals within these movements may reside in urban areas. Furthermore, as can be seen through Māori advocacy, the idea of ties with specific territories exists in urban settings as well. Many urban areas in New Zealand are claimed as meaningful ancestral territories by urban Māori groups (for example, Orakei in Auckland; see IH Kawharu, 'Biculturalism and Inclusion in New Zealand: The Case of Orakei' (2008) 50(1) Anthropologica 49. So, even though we should always leave space open for multiple indigenous (and other) identification, it is not a stretch to underline, in the particular context of this argument, the importance of particular territories (whether as actual abodes or as ancestral memories fundamental for currently recuperating traditions). See Daniel Bonilla Maldonado, 'Indígenas urbanos y derechos culturales: los límites del multiculturalismo liberal' (2011) 7(2) Revista Direito GV 569.

[38] Marisol de la Cadena, *Earth Beings* (Duke UP 2015).

ample, develops the relationship between places and particular communities, pointing out consistently that these relationships are not analogous to a Humanity–Nature relationship, nor are they reliant on the idea of personhood at all.

> 'What she calls Earth Beings are not approximations of Mother Earth, but kinds of creatures that act in *their* specific way and which enter into very precise relationships with surrounding communities (which, themselves, are not mere collections of individuals)'.[39]

Her work shows that our idea of a person may not be commensurate with some Andean philosophies that do not personify the environment as much as dehumanize the human. Human beings appear as mutable and multiple, their relationships to places are not those of an individual towards another individual feature, but rather of beings that are mutable into each other (and therefore violate what we think of as a person in Western history), relationships that are always *more* than what they appear to be. There is an indescribable excess to any being, the person included, that becomes visible when engaged with a specific territory and world.

Taking legal pluralism seriously, we could say that one potential of the kind of thought de la Cadena describes is to propose a different notion of the *human* person, congruent with the great variety of the surrounding world. This kind of proposal would not necessarily be of rights, because rights must attach to bounded entities. One of the theorists that has done a lot to make the case for the existence of qualitatively different worlds is anthropologist Eduardo Viveiros de Castro.[40] But much of the anthropological corpus in general can be read as a record of misunderstanding the ontological differences between different groups of people. De Castro speaks about *equivocation*, which de la Cadena explains as 'not […] a simple failure to understand'. Rather it is 'a failure to understand that understandings are *necessarily* not the same, and that they *are not related* to imaginary ways of "seeing the world" *but to the real worlds that are being seen*.' She continues: 'as a mode of communication, equivocations emerge when different perspectival positions – views *from* different worlds, rather than perspectives about the same world – use the same word to refer to things that are not the same'.[41]

The equivalence between Nature and Pachamama can be seen as this type of equivocation, a presumed equivalence of perspective about a similar world, when in fact they may convey radically different worlds. The Indigenous ontologies that Viveiros de Castro draws upon are not just descriptions of one same world but rather of completely different ones. In their interaction with colonial moder-

---

[39] Tănăsescu (n 1) 44 (emphasis in original).

[40] Eduardo Viverios de Castro, 'Images of Nature and Society in Amazonian Ethnology' (1996) 25(1) Annual Review of Anthropology 179; Eduardo Viverios de Castro, *Cannibal Metaphysics* (Univocal Publishing 2014); Eduardo Viverios de Castro, 'Perspectivisme et multinaturalisme en Amérique indigène' (2014) 138–139 Journal des anthropologues 161; Eduardo Viverios de Castro, 'Exchanging Perspectives: The Transformation of Objects into Subjects in Amerindian Ontologies' (2019) 25 Common Knowledge 21.

[41] De la Cadena (n 38) 110 (emphasis in original).

nity, Indigenous worlds have had to 'translate' themselves, so as to adapt and survive in a world that was not their own.

Today, we are living through a moment that offers new possibilities for legal pluralism and hybridization, and it matters that this moment does not end up reproducing the power inequalities that have characterized Indigenous–state relations everywhere. This is why it becomes important to question the way in which legal personhood and rights are wedded, and to ask what other conceptions, stemming from other traditions, may be valuable going forward.

If we switch focus from Andean thought to Māori traditions, we can further reflect on legal pluralism. The characteristics of the dynamic between the settler state and Indigenous groups remains largely the same. But in terms of legal pluralism, Māori *tikanga* (laws, ways, customs) has had a decisive influence on the claims settlements that led to the Te Urewera and Whanganui Acts. In the negotiations that led up to Te Urewera Act, legal pluralism is seen in the resistance to adopting a rights language, the use of "legal entity" instead of "person" (though not consistently), and the shunning of guardianship models (another notion of Western origin, with its implication of *directing* the environing world; see also the discussion of Berry, above).

Te Urewera Act therefore avoids some concepts that are usually seen as connecting Indigenous thought with the rights of nature. And yet the law, being still largely in the framework of a Western legal tradition, mandates management plans that are very much in line with managerial thinking that is decidedly not part of *tikanga* (law or custom). However, the management plan that, for example, the board of Te Urewera drafted – Te Kawa o Te Urewera – sidesteps those kinds of managerial schemes as well and grounds itself fully in *tikanga*. Very significantly, one way in which this is done is by recognizing the need for Tūhoe themselves to relearn Tūhoe ways. This contrasts with the usual Western view that idealizes Indigenous people by denying them history and by supposing that they are inherently in harmony with their environment.

The legal pluralism that is present in the Ecuadorian and New Zealand cases is, albeit it in different ways, very significant as examples of hybridization in action. The organized Indigenous communities in Ecuador (CONAIE especially), for example, have advocated successfully for the kinds of constitutional reforms that have led to the rights of nature. But this does not mean that the resulting hybrid faithfully translates what, in other contexts, these communities themselves may describe as their legal traditions.

In the intellectual genealogy of the rights of nature in Ecuador, Indigenous thought on Indigenous rights and Sumak Kawsay has been important, and increasingly customary legal systems around the world are themselves claiming affinity with the rights of nature.[42] Yet we need to recognize the complexity and

---

[42] See, for example, Craig M Kauffman and Pamela L Martin, *The Politics of Rights of Nature: Strategies for Building a More Sustainable Future* (Massachusetts Institute of Technology Press 2021).

fragility of experiments in legal pluralism, something that some Indigenous scholars have recognized for a long time. As Gualinga and others suggest, commenting on the 2008 Ecuadorian Constitution,

> 'this article [on the rights of nature] can be detrimental to nationalities and peoples because they depend directly on nature for their livelihoods, and this can be distorted and claimed by any individual, therefore causing grave conflicts and constitutional contradictions'[43] (but see discussion of Constitutional Court cases below).

Similarly, in the management of Te Urewera, Te Kawa is still bound by the law within which it functions; thus, for example, permits for different activities have to be issued by the board, a practice that modifies *tikanga*. But Māori law is not timeless; it has been modified by centuries of interaction with the settler reality that has been and continues to be New Zealand. Many Māori groups, with different traditions, are *reinventing* traditional practices, so in a sense their own traditions are hybrids across time. The idea of legal pluralism, significantly advanced by rights of nature practice so far, has a lot of potential to generate further surprising arrangements. But, following the very ethical commitments of this kind of legal practice, we should remain vigilant so as not to smother Indigenous contributions through a too-fast assimilation of traditional Western concepts.

II. Conceptual Characteristics and Some Cases

The partial theoretical foundations I have sketched cannot by themselves tell us much about the constituent concepts that have come to play an important role in both theory and practice. Importantly, the differences between cases[44] have started to show theoretical and conceptual possibilities that were not necessarily visible by only accounting for the genealogy of the rights of nature. Local contexts and traditions have modified these decisively, and it is largely within those interactions between theory and practice that the most relevant extant conceptual elements are to be found.

Most obviously, the notions of 'rights' and 'nature' are key elements to understand, but I do not want to treat them at length here because that has been done elsewhere.[45] Instead, I want to focus on conceptual conundrums that mostly arise from the implementation of rights of nature and that are important for understanding their practical application.

---

[43] Gualinga and others (n 17) 151. Author's own translation. It must also be reiterated that Indigenous people are greatly diverse, both among and within groups, and that we can therefore not speak about 'Indigenous opposition' or 'Indigenous support' as such.

[44] I do not use this term in the technical legal sense but rather in the more colloquial sense of an instance of something.

[45] See Mihnea Tănăsescu, *Environment, Political Representation and the Challenge of Rights: Speaking for Nature* (Springer 2016).

1. Level of the Law

Rights of nature cases so far have appeared at various levels: municipal, regional state, and national state, through either laws or constitutional provisions, and in international declarations. The mechanisms through which the various laws and ordinances have appeared also vary, from constitutional assemblies to settlement claims processes to judicial activism. These variations are in part due to differences between the legal systems where rights of nature have appeared. But, perhaps more importantly, they are also due to the political opportunities present in each particular context. These to a large extent define what is and isn't possible in any given historical moment for the rights of nature. And through the particular political context of each case, we also get theoretical innovations that inspire further cases.

I will pick four different cases that each illustrate a different level of the law and a different way of getting there. I will present them chronologically, for two reasons. First, each case partly builds on preceding ones, and therefore the order of appearance is genealogically significant. Second, conceptual innovations appear in each subsequent case, sometimes in ways that are made more intelligible by knowing previous instances of rights of nature. I will not include here international declarations, because they have not yet been successful, though this may of course change in the future.

a) United States Municipal Ordinances

The first rights of nature in practice appeared in the United States (2006). Tamaqua Borough, Pennsylvania, adopted a municipal ordinance granting rights to nature, understood as the municipal area.[46] Sections 7.6, 7.7, and 12.2 of the ordinance seem very much influenced by the work of Christopher Stone, foregrounding standing as vitally important.[47] However, equally important is the background leading to this historic ordinance. Since 2006, similar municipal ordinances have been adopted in dozens of municipalities across the US.[48]

Tamaqua Ordinance no 612 was designed to oppose specific actions by corporations within the municipal area.[49] The area of Pennsylvania where Tamaqua is located has traditionally been connected with resource exploitation, especially mining. But, around the turn of the century, Tamaqua faced a new threat: the disposal of toxic sludge. If corporate actors filed the right paperwork, sludge disposal could

---

[46] The Tamaqua Borough Ordinance no 612 of 2006 (Pennsylvania, United States).

[47] It is perhaps not surprising that the North American strand of rights of nature, exemplified by Stone's work, would have the greater influence in that country. They also draw on the ecotheology of Berry and Cullinan, as seen through the idea of Nature as community.

[48] For a continuously updated list, see: <https://celdf.org/>.

[49] It is worth noting that in Ecuador, long before the constitutional rewriting in 2008, indigenous and wider civil society activism was also focused on opposing corporate resource extraction and pollution. It is also this experience of demanding territorial and prior consultation rights that needs to be considered as important background in that case.

not be stopped. The rights of nature were therefore conceived as a tool to oppose, by more successful legal means, industrial activities within the municipal area.

The Tamaqua ordinance specifically makes it unlawful for 'any corporation or its directors, officers, owners, or managers to interfere with the existence and flourishing of natural communities and ecosystems' (s 7.6). The ordinance establishes standing for the municipal area (in itself), and for residents of the Borough to act as representative of the area's rights. The ordinance declares that 'Borough residents, natural communities, and ecosystems shall be considered to be "persons" for purposes of the enforcement of the civil rights of those residents, natural communities, and ecosystems'.

Legal personality, rights, and standing are clearly linked. The ordinance also grants residents of the Borough 'a fundamental and inalienable right to a healthy environment'. This kind of third generation human right[50] often accompanies rights of nature. The argument is that they are mutually reinforcing: if nature has rights, the right of people to a healthy environment (however that may be defined) has a better chance of being respected. Yet rights of nature and *to* nature may also be in tension, especially if it is unclear what specific rights nature may have in a given case, as well as which human groups have authority and power to determine the content of nature's rights and of human rights to nature.

There are many questions that this formulation of rights for nature raises. I find two of them particularly important. First, the ordinance wants to establish nature's right to restoration. This is a practical innovation that was taken up in different cases of rights of nature worldwide. It is closely connected with the genesis of these ordinances in opposing corporate industrial activities. A right to restoration would impose on corporations, if found guilty of violating the rights of municipal nature, an obligation to remedy the harm done.

This right shows the tensions inherent in the concept of nature as an eco*system*, a *community* that is in *balance*. This idea of Nature imposes a baseline, i.e., a fixed standard given by human observation of the environment at a particular time. In the case of the Tamaqua ordinance under discussion, restoration should be done as to benefit the 'natural community' by reverting it to a pre-disturbance state. What the previous natural state may be is always open to debate and to largely political processes that have to determine it.

Two issues complicate the idea of baseline restoration. Because of climate change, reverting to a baseline is likely impossible. Additionally, there is no unproblematic way of choosing a baseline. As a fundamentally historical concept, there are no pre-determined criteria for preferring one moment in history over another.[51] Imagine that an abandoned old coal pit is currently an oasis for local

[50] Karel Vasak, 'Pour une troisième generation des droits de l'homme' in Christophe Swinarski (ed), *Studies and Essays on International Humanitarian Law and Red Cross Principles in Honour of Jean Pictet* (Martinus Nijhoff Publishers 1984).

[51] However, proposals for overcoming these difficulties do exist. See, for example, Linnea Luuppala, 'Rights-Based Restoration' in Daniel P Corrigan and Markku Oksanen (eds), *Rights*

birds. Would this place, if affected by corporate activities presently, have to be reverted to an abandoned coal pit, or to another pre-mining state? If the latter, then which one? Before or after the colonial enclosure that created the current Borough residents?

The second question raised by this ordinance relates to the relationship between local people and local nature. There seems to be an assumption of locals being friendly to the environment, which is one way of accounting for granting standing to residents. But what if a corporate shareholder invested in toxic sludge becomes a Borough resident? The corporation, as a legal person, could also become a resident, complicating the situation further.

I point towards these issues to give an idea of different questions raised by the conceptual apparatus we have started exploring. So far, corporate actors have not become Borough residents in order to manipulate these kinds of laws. The level of the law – municipal – makes these ordinances all but unenforceable when in conflict with higher-level laws. Macpherson (2021a) shows that these laws have been consistently opposed, and often struck down, in court.[52] Courts have taken the view that these municipal ordinances are unconstitutional, on various grounds. Some scholars[53] argue that the goal of these ordinances is to contest the terrain of legal personality, showing that if corporations can be legal persons, so can nature. This much has obviously been achieved, but in terms of implementation, municipal ordinances remain limited by their definitionally narrow application.

b) Ecuadorian Constitution

In contrast to municipal ordinances, the rights of nature in Ecuador entered the stage at the highest level of the law: the constitution of the state. The writing of the new constitution, in 2008, was done by establishing a Constitutional Assembly tasked with drafting it through a series of widely participative consultations. The rights of nature, for the first time in constitutional history, were included in a constitution. This event opened up the possibility of rights of nature everywhere and is therefore very significant for their history and conceptual development. It is widely recognized as one of the most important rights of nature cases to date.

As in the case of Tamaqua Borough, rights are formulated in similar terms to Berry's fundamental right to existence and in a way that addresses the issue of standing central to Stone. Standing is codified in the largest sense. Whereas in the municipal cases standing applied to any resident, here it applies to any per-

---

of Nature: A Re-examination (Routledge 2021). Here, the concept of ecological integrity is advanced as a possible solution.

[52] Elizabeth Macpherson, 'The (Human) Rights of Nature: A Comparative Study of Emerging Legal Rights for Rivers and Lakes in the United States of America and Mexico' (2020) 31 *Duke Environmental Law & Policy Forum* 327.

[53] Erin Fitz-Henry, 'Challenging Corporate "Personhood": Energy Companies and the "Rights" of Non-Humans' (2018) 41(51) PoLAR 85.

son whatsoever, even regardless of nationality.[54] Besides this issue, the duality of rights *for* (art 72) and *to* (art 74) nature is also present. Finally, the issue of restoration appears as a fundamental right.

The rights of nature in the Ecuadorian constitution have been analyzed and commented upon extensively.[55] Here I want to point out several conceptual elements that are relevant for this case, besides the level of the law and the wide and, to some extent, indeterminate standing. This interpretation of standing opens up towards the concept of representation, which asks the question: Who has the moral and political right, over and beyond the legal one, to bring claims on behalf of nature? Any particular legal claim can be heard regardless of who brings it. Each case can therefore only be judged on the merits of the claims themselves.

This, in principle, does not appear to pose particular problems. If it is nature as such that has received rights, it seems logical to allow anyone to legally represent it. But, as I have argued elsewhere, the issue of moral representation, or what I have called political representation, becomes crucial and is always thorny. The rights of nature cannot be just about legal representation because any time a claim is made on behalf of nature it is also made in order to transform a given balance of power on the ground. And seeing that there is no a priori justification for who has a right to speak for nature, it means that these rights can, in theory, be used to modify relations of power in favor of the already powerful. This remains at the discretion of judges because the constitution – through the wide doctrine of standing it adopts and the vague rights it grants – cannot preclude certain outcomes.

In practice, cases of rights of nature in Ecuador have been won through legal/political representation carried out by a variety of actors. The state has also used them successfully, which may be problematic given its historical and contemporary commitment to extractive industries.[56]

This of course does not mean that state actors in connection with extractive industries will always use the rights of nature selectively or will always succeed

---

[54] In Tănăsescu (n 45) I developed at much greater length the discussion on the universality of standing, as well as documenting in detail the political process within the Constitutional Assembly. The reader interested in the Ecuadorian case specifically should also consult that work.

[55] See, inter alia, Craig M Kauffman and Pamela L Martin, 'Can Rights of Nature Make Development More Sustainable? Why Some Ecuadorian Lawsuits Succeed and Others Fail' (2017) 92 World Development 130; Craig M Kauffman and Pamela L Martin, 'Comparing Rights of Nature Laws in the US, Ecuador, and New Zealand: Evolving Strategies in the Battle Between Environmental Protection and "Development"' (International Studies Association Annual Conference, Baltimore, 23 February 2017) <http://files.harmonywithnatureun.org/uploads/upload472.pdf>; Rickard Lalander, 'Rights of Nature and the Indigenous Peoples in Bolivia and Ecuador: A Straitjacket for Progressive Development Politics?' (2014) 3(2) Iberoamerican Journal of Development Studies 148.

[56] See Erin Daly, 'The Ecuadorian Exemplar: The First Ever Vindications of Constitutional Rights of Nature' (2012) 21(1) RECIEL 63; Kauffman and Martin, 'Can Rights of Nature Make Development More Sustainable?' (n 55).

in doing so. It simply means that they can, and that they have. Conversely, indigenous activists, environmentalists, or others interested in protecting nature for its own sake can obtain – and have obtained – judgments in favor of the rights of nature.

For example, some recent cases show the potential of constitutional provisions to work for environmental protection and against the agenda of an extractivist state. In November 2021, the Ecuadorian Constitutional Court reached a decision on the Cedros Protected Forest,[57] where two mining corporations had been authorized to carry out exploration for future extractive activities. The Court ruled that the exploration permits awarded by the state did not take nature's rights into account and that the corporations were therefore to leave the forest immediately and that no future extractive activities were to be allowed.

Similarly, the Court also reached a decision on the flow of Aquepi River.[58] The water from the river was diverted to such an extent that its natural flow could not be guaranteed during the dry season. This, according to the Court's interpretation of the constitution, violated the river's right to its own evolutionary processes. This decision was reached against the state, which was ordered to undertake more stringent studies on the amount of water that can be safely diverted.

These cases are extremely important, not least because they have *erga omnes* effect. They also fan out the possibilities inherent in the Ecuadorian formulation of rights of nature and show how consequential the level of the law can be. The Court has been very active in selecting cases centered around rights of nature and has so far consistently favored these rights over the right to development. It has also consistently struck down current environmental regulations as insufficient and has mandated a complete overhaul of these regulations such that they become compatible with the rights of nature. It is much too early to properly analyze these decisions, but certain elements have to also be followed in future research. In my view, a vital area of research on the Constitutional Court will be the legitimacy of the specifically *political* implications of these judgments. The shifting of power relations between communities and between different rights of the constitution will have to be carefully followed.

Besides the issues already discussed, Ecuador has opened up another important discussion in rights of nature theory: the relationship between these rights and Indigenous philosophical traditions and practices. As we have seen, rights of nature theory and activism have long claimed that these rights are closely connected to 'indigenous cosmovisions' and that they may emanate from their views. The Ecuadorian constitution uses nature synonymously with Pachamama, the Andean Goddess of life's unifying spirit. This reinforces the idea that rights of nature and indigenous thinking[59] are closely related.

---

[57] Corte Constitucional de Ecuador 10 November 2021, 1149-19-JP/20.

[58] Corte Constitucional de Ecuador 15 December 2021, 1185-20-JP/21.

[59] Indigenous thought is usually referred to as beliefs, cosmovisions, spirituality, and so on. In my view, this is an often-unconscious continuation of colonial thinking that has always

In the Ecuadorian Constitutional Assembly, the organized Indigenous communities of Ecuador, key among which was CONAIE, were extremely active. In political terms, they had a lot to do not only with the rights of nature but also with territorial rights, the inclusion of the principle of Sumak Kawsay or *buen vivir*, and the idea of plurinationality that now structures the Ecuadorian state. I have already argued that the alliance between indigenous communities and the rights of nature in the Constitution was a strategic one, coming out of the very well-founded desire to gather as many tools on the part of these communities as possible. But to say that, philosophically, the rights of nature have indigenous origins, or translate indigenous philosophies, is not historically accurate.

I will now turn to other cases that continue our discussion of the level of the law.

c) New Zealand Claims Settlements

There are so far two cases of rights of nature in New Zealand: Te Urewera Act 2014 and Te Awa Tupua/Whanganui River Act 2017. Both cases modify rights of nature theory considerably because they introduce elements that were simply not present before. I have already argued that local contextual conditions make a big difference to the theoretical and practical development of rights of nature, and this is no different here. The innovations in the Aotearoa New Zealand cases are directly related to the process that led to them, so it makes sense to start there.

As with Ecuador before, I cannot describe the history of Māori–Crown relations in Aotearoa New Zealand in the detail that it deserves, but this work is available elsewhere.[60] It is, however, important to place the rights of nature here in the context of a process without which they would not have been possible in this particular form. This is the claims settlement process that has come to dominate Māori–Crown relations for the last three decades.

The first significant contact between Europeans and Māori dates to 1769,[61]

> 'when the *Endeavour*, captained by James Cook, landed on the eastern shores of the North Island. Seventy years and many missionaries and settlers later, the British Crown and many (but not all) Māori chiefs signed the 1840 Treaty of Waitangi, the most important document in New

---

relegated indigenous thought to the realm of belief. It is not, no more than the doctrine of human rights is a matter of belief. Of course, every people must 'believe' its own theories, but that does not make such belief unfounded or simply spiritual. Indigenous thought is as rigorous as Western thought is, so the same terms should be used.

[60] See, inter alia, Vincent O'Malley, 'Tūhoe-Crown Settlement-historical background' (October 2014) Māori Law Review <https://maorilawreview.co.nz/2014/10/>; Katherine Sanders, '"Beyond human ownership"? Property, Power and Legal Personality for Nature in Aotearoa New Zealand' (2018) 30(1) JEL 207; Carwyn Jones, *New Treaty, New Tradition: Reconciling New Zealand and Māori Law* (UBC Press 2016).

[61] Technically, the first known contact with Europeans was 13 December 1642, when Abel Tasman sailed past New Zealand. However, this encounter did not lead to landing or settlement, which did not occur until Cook's arrival.

Zealand's history. After the signing in Waitangi, the Treaty was taken across the island for additional signatures'.[62]

Tūhoe, the inhabitants of Te Urewera, the land that received legal entity status in 2014, have never signed.[63]

The Treaty of Waitangi was signed in two language versions, a Māori and an English one. The history of the difference between these two is extremely important and has been amply debated. One of the most contentious concepts for the purposes of the present discussion is that of *tino rangatiratanga*. Jones explains that the term varies in meaning from 'self-government' to 'sovereignty' or 'full authority'.[64] The Waitangi Tribunal has argued that 'no one single English concept effectively captures the full meaning of the term', in part because, unlike sovereignty in English, it has spiritual connotations as well as implications of dominion over particular territories.[65] In the Māori version, Article two of the Treaty of Waitangi guarantees the chiefs *tino rangatiratanga*. This term opens up towards the rich Māori *tikanga* that was forced into the molds of the setter-state law.

Recent scholarship on the Treaty as well as recent judicial decisions have more or less settled on the opinion that, at the time of signing, the chiefs did not cede their sovereign ability to direct the life of the community or Ownership of their lands.[66] In the English version of Article One of the Treaty, 'sovereignty' was ceded to the Crown, while in the Māori version it was *kawanatanga*, or 'governorship'.[67] English colonists and their successive governments increasingly acted as if the Treaty of Waitangi had transferred sovereignty of Aotearoa to the Crown, while Māori chiefs operated under the understanding that they had retained *tino rangatiratanga*. Tūhoe have been remarkably consistent throughout this history in affirming *mana motuhake*, a term very close in meaning to *tino rangatiratanga*. As Higgins explains, 'distinctions between *mana motuhake* and *tino rangatiratanga* are contextual rather than categorical, but while they have much in common, *mana motuhake* more strongly emphasizes independence from state and Crown and implies a measure of defiance'.[68] This is not surprising given the especially conflictual history between Tūhoe and the Crown.

Similarly, Whanganui *iwi*, the Māori tribe that has lived with the Wanganui River since the Polynesian settlement of Aotearoa, have also been consistent in

---

[62] Tănăsescu (n 20).

[63] Ngai Tūhoe Deed of Settlement Summary (4 June 2013). See also Judith Binney, *Encircled Lands: Te Urewera, 1820–1921* (Bridget Williams Books 2009).

[64] Jones (n 60) 54.

[65] Jones (n 60) 56.

[66] Sanders (n 60); Moana Jackson, 'The Treaty and the Word: The Colonization of Māori Philosophy' in Graham Oddie and Roy W Perrett (eds), *Justice, Ethics and Nez Zealand Society* (OUP 1992).

[67] Andrew Erueti, 'Conceptualising indigenous rights in Aotearoa New Zealand' (2017) 27 New Zealand Universities LRev 715, 717.

[68] Rawinia Higgins, 'Ko te mana tuatoru, ko te mana motuhake' in Mark Hickford and Carwyn Jones (eds), *Indigenous Peoples and the State* (Routledge 2018) 130.

challenging the Crown's right to modify the river, including but not limited to resource extraction and hydroelectric power generation. But neither Tūhoe nor Whanganui *iwi* had had much success in challenging the government, partly because the Treaty of Waitangi was conveniently ignored. This changed drastically towards the end of the 20th century.

The Treaty of Waitangi Act of 1975 inaugurated the Waitangi Tribunal, 'a standing commission of inquiry established to inquire into Māori claims that laws, policies, acts or omissions of the Crown are or were inconsistent with the principles of the Treaty of Waitangi'.[69] The Tribunal only has powers of recommendation, though this has not rendered it powerless. Indeed, 'the tribunal began to have an influence on public policy, despite its lack of powers to compel the government to take notice of its recommendations'. As Belgrave continues,[70] 'it was partly in recognition of this success that in 1985 the fourth Labour Government extended the tribunal's jurisdiction back to 1840, with far-reaching consequences that were only dimly understood at the time'. This set in motion the contemporary era of negotiations between the government and Māori iwi and *hapū* for breaches of the Treaty.[71]

The grant of legal personality to landscapes in Aotearoa New Zealand has to be understood in the post-1985 context of Treaty settlements. It is this period that elevates the Treaty of Waitangi to the most significant document in Māori–Crown relations. And it is this process of settlement of historical grievances that is so far unique in the history of the rights of nature. It also opens up possibilities that could now have been thought outside it. Most significantly, it introduces the issue of sovereignty into the rights of nature.

Tūhoe claims to Te Urewera, like Whanganui *iwi* claims to the Whanganui River, can be interpreted as complex negotiations about who owns the land, or more precisely about who has ultimate authority in governing the lands. These Acts are about the relationship between different Māori groups and specific lands that they have historical claims to. But in the context of a settler state, issues of authority cannot be presented as ones of sovereignty over a territory, even though Tūhoe claims throughout the negotiations would have more accurately been translated into those terms. Instead, the issue of sovereignty was couched as one of ownership, authority, and the legitimacy to govern.

Sanders[72] therefore argues that the treaty settlement negotiations in these two cases were really centered around the issue of ownership: Who owns the land? The New Zealand government refused to pass ownership to Māori groups, as this would have been politically difficult. Instead, the idea of legally crafting the ter-

---

[69] Sanders (n 60) 208.

[70] Michael Belgrave, *Historical Frictions: Māori Claims and Reinvented Histories* (Auckland UP 2013).

[71] Names for Indigenous Māori descent groups. *Iwi* denotes a larger group than *hapū*.

[72] Note 60.

ritories in question as legal entities arose as a way around the issue. Under this arrangement, the territories would technically own themselves.

The rights of nature as a way towards self-owning land is a striking innovation. It arose as a compromise between two parties that had always failed to reach compromises before. But it should be noted here that Te Urewera and the Whanganui River *received only legal entity/person status*. There are no positive rights in the two Acts.

Several conceptual issues are important here. First, Te Urewera Act speaks mostly about Te Urewera as a legal *entity*, whereas Whanganui River Act speaks about the Whanganui River as a legal *person*. This difference may or may not have practical consequences, but theoretically it is very interesting. The idea of personhood, as I have argued, is tied to ideal models of what constitutes a person. A legal entity, on the other hand, may be interpreted as morally agnostic, or at least as opening up possibilities that can move away from personification altogether.

This brings me to the second important issue. Māori *tikanga* plays an important role in these two Acts, coming close to the ethical thrust of legal pluralism (see section 3). But this pluralism is not perfectly balanced: treaty negotiations take place against the backdrop of state-imposed frameworks for discussion. Higgins argues that treaty negotiations force Māori to band in ways that are not based on Māori tradition. She argues that 'the process that is placed upon iwi to create 'mandated large natural groupings' by the Office of Treaty Settlements' is itself an imposed framework. She continues: '[…] the settlement systems are not determined by Māori and often contravene *tikanga Māori*, or any "customary system of authority"'.[73] This may create tensions in Māori communities, as *tikanga* systems of membership might or might not align with official requirements for entering negotiations.

Jones further points out that Māori terms and concepts are mostly present in the preambles and symbolic sections of the Acts, whereas they are mostly absent in the hardnosed provisions making up the bulk of the legislation.[74] But both laws do make a radical move that allows *tikanga* more space, namely the specific granting of legal and political representation rights for the territories to Tūhoe and Wanganui *iwi* respectively.

This brings me to the third important issue. In the case of Te Urewera, the Act forms a board that is the de facto governor of Te Urewera. This institution is to have a majority Tūhoe membership (the first three years it had half Tūhoe, half Crown membership), and it has the crucial task of drafting management plans for the territory. These plans have to be congruent with the law, but they can also incorporate *tikanga* in ways that the Act did not. As I have argued elsewhere,[75] the first Tūhoe management plan relies almost exclusively on Tūhoe concepts

---

[73] Higgins (n 68) 132.

[74] Jones (n 60).

[75] Tănăsescu (n 1).

and largely ignores the formal requirements of management plans as settler governments had heretofore conceived of them.

The Whanganui River Act does something similar in setting up a board with half *iwi*, half Crown membership that is equally tasked with managing the legal person's affairs. The arrangements for both boards are *extremely* detailed, going into every apparently minor rule of functioning. This seems to be in order to enshrine a commitment to democratic deliberation within the management of the legal entities. The Act also designates the board as the *human face* of the river. This has widely been interpreted as setting the board up as a guardian, but there is no legal proof of this. I will tackle the particular issue of guardianship later, when I discuss at greater length the relationship between Indigenous peoples and rights of nature.

This kind of institutionalization is, to date, absent in most other cases of rights of nature. In part, that is surely because of the specific New Zealand context. But it may also be because the idea of representation developed here was not really thought of in connection with the rights of nature to begin with. The issue of standing, for example, so central to rights of nature elsewhere, is here made irrelevant by the detailed way in which governance arrangements are set up. Yes, the respective boards have standing to represent the two newly created entities, but the innovation here is in spelling out exactly what that means, under which circumstances, to what end, and through which precise procedures.

The fourth important issue raised by these cases is the character of rights themselves – and of nature itself. There are no positive rights in the Act, so the legal persons receive only the automatic rights granted by their coming under the authority of the Bill of Rights. More philosophically, the idea of rights is not prominent in either Act but is rather displaced by the idea of responsibility towards specific territories. So instead of nature as understood in the historical foundations explored earlier, we are confronted with territories having specific histories and ties. In terms of rights then, these arrangements could be called 'minimalist', while in terms of nature they are highly specific.[76]

These two cases have contributed significantly to the conceptual richness of the rights of nature. Yet this does not mean that they 'work' better than rights of nature elsewhere! How they will fare in the future is impossible to tell, and they are not a model to be followed elsewhere, precisely because they are rooted in a particular cultural and legal context. But they do give a lot of food for thought, and their originality deserves to be recognized in order to endow other specific contexts in which liberties can be taken with the concept of rights of nature and experimentation made towards yet unknown possibilities. I now turn to two other, yet again very different, experimentations that deserve sustained reflection.

---

[76] See Elizabeth Macpherson, 'Ecosystem Rights and the Anthropocene in Australia and Aotearoa New Zealand' in Demenico Amirante and Silvia Bagni (eds), *Environmental Constitutionalism in the Anthropocene: Values, Principles, Actions* (Routledge 2021).

d) Colombian Judicial Pronouncement

Atrato River, Department of Chocó, Colombia, became a legal person in 2016.[77] This was not achieved through a municipal ordinance, or a constitutional rewriting or amendment, nor through a national law. It was also not the result of claim settlement processes. Instead, Atrato became a legal person through judicial pronouncement. This implies that Atrato's legal status was not subject to a previous political process that had decided, for whatever reasons, to create it.

The Colombian Supreme Court recognized the Colombian Amazon rainforest as a legal subject in 2018. Both of these cases demonstrate, as Calzadilla argues,[78] that 'rights of nature/ecosystems can be recognized by both legislative and judicial channels'. In rights of nature theory, this possibility has not been explicitly formulated, but it does align with the theoretical view of legal personality as that which can be simply declared by a competent authority. For the purposes of illustration, I will only focus on the *Atrato* case here.

Even though this case did not come out of explicitly political processes, this does not imply that it cannot have important political consequence. Quite to the contrary, following the argument so far it is bound to have far-reaching political implications. In fact, these are already starting to be seen, but to appreciate them we need to understand some of the contextual conditions of this case.

In the *Atrato* case,[79] the Constitutional Court proclaimed it a legal person in response to a *tutela* action brought by an NGO – *Tierra Digna* – on behalf of residents affected by illegal mining in the river. A tutela action is a constitutional mechanism allowing any person to 'request any judge in the country to protect his/her fundamental constitutional rights when they are being violated by a state agent or an individual'.[80] The action was brought before Cundinamarca Administrative Court, which denied the action because it decided that it did not bear on fundamental rights.[81]

The Constitutional Court selected this case for revision and concluded that it is in fact a case that clearly bears on fundamental rights. The judgment of the

[77] The decision was made public only in May 2017. See Elizabeth Macpherson and Felipe Clavijo Ospina, 'The Pluralism of River Rights in Aotearoa, New Zealand and Colombia' (2019) 25(6) Journal of Water Law 283. 'Person' here is the translation for the term 'sujeto', which can also be translated as 'subject'. For the purposes of the present argument, I treat both as synonymous.

[78] Paola Villavicencio Calzadilla, 'A Paradigm Shift in Courts' View on Nature: The Atrato River and Amazon Basin Cases in Colombia' (2019) 15(1) LEAD Journal 49, 51.

[79] For this case, I did not carry out independent field work and have instead relied on published literature and documents. For a very detailed account on the wider context of the case, see Sandrine Revet, 'Le fleuve et ses gardiens' (*Terrain*, 3 March 2022) <https://journals.openedition.org/terrain/22695>; Sandrine Revet, 'Les droits du fleuve. Polyphonie autour du fleuve Atrato en Colombie et de ses Gardiens' (2020) 52 Sociétés Politiques Comparées <https://sciencespo.hal.science/view/index/identifiant/hal-03384888>.

[80] Calzadilla (n 78) 52.

[81] Corte Constitucional de Colombia 10 November 2016, T-622/16, *The Atrato River Case*.

Court is binding on all lower courts. Tierra Digna used the tutela mechanism to compel the state to take protective action safeguarding the wellbeing of local residents and of the river. This effort was validated by the highest Court in Colombia.

There are several immediately noteworthy elements. First, this case began as a violation of the rights *to* nature (as well as a host of other human rights) and became a case of rights *for* nature. As Macpherson and Clavijo Ospina argue, this kind of case raises doubts 'about the usefulness of the ecocentric/anthropocentric divide'.[82] The particular framing of the rights of nature as ecocentric is ubiquitous, but the often-contradictory ways in which these appear in practice undermines the claim.

Chocó is the poorest region of Colombia, with an overwhelmingly Indigenous and Afro-descendent population. It goes without saying that the relationship between ethnicity and socio-economic deprivation is not a mere coincidence, but rather the mark of settler societies everywhere. Though the Court's decision is geared towards the protection of Chocó residents from mining activities, the judicial pronouncement arose without extensive prior consultation with local communities. This is in part a result of the mechanism that led to legal personality. Political processes, inasmuch as they are legitimate, are consultative in nature, whereas judicial pronouncements need not be, because their legitimacy comes from procedures internal to the law that are sealed from wider consultative processes.

Earlier, I made a distinction between cases (like Ecuador) that rely on the ecotheology of rights and which therefore frame the recipient of rights in terms of a featureless nature, or a natural community without boundaries, and place-based cases (like New Zealand) that target a particular territory and a particular relationship with it. The *Atrato* case seems to be of the latter kind because a particular river is named and a particular relationship with it (of the local communities) seems to be prioritized.

Yet this can be questioned, for several reasons. First, the Atrato River is not described in any particular detail, and it therefore does not have salient features that would, in some sense, individualize it. In fact, the relevant part of the decision is framed in terms of 'the planet' and 'humanity', and so it seems that the restriction to Atrato River was only a practical constraint, driven by the tutela action and the opportunity to select it, and did not reflect an internal logic of the case. In other words, the Court could arrive at the judgment under the legal circumstances given, but the theoretical framework used is within the ecotheological tradition explored in the first section, though it may surely have intellectual predecessors different from the ones explored in this contribution.

The same is visible if we examine the framing of the local community, which is invoked as a largely homogenous population defined ethnically. This means

82 Macpherson and Clavijo Ospina (n 77).

that there is no recognition of a special right to have standing or to speak for the river in ways that only *this* community could.

This case has been covered internationally as being very similar to the Whanganui case already encountered. But the differences are theoretically more interesting than the similarities that, in my view, remain on the surface of the discussion. One purported similarity is the granting of guardians to the river, yet in the two cases this means very different things. As we have seen, the New Zealand boards are *not* guardians; they are rather institutions that safeguard the special relationship between Māori and their territories, relationships that are not expressed through guardianship, with its paternalistic legal history (guardians are first and foremost those that take care of those that cannot take care of themselves). And the choice of guardians for the Atrato River is not similar to the Whanganui: the Colombian river received two guardians, namely the national government itself and 'ethnic communities living in the Atrato River basin'.[83]

There are several important conceptual issues here. First, the national government, through its duty to uphold the constitution, already had at its disposal ample mechanisms to oppose mining in the Atrato basin. There is therefore no reason to believe that this extra mechanism would have different results. Or, at least, such a claim would need to be proven. Second, the local guardian is primarily identified through ethnic considerations, which is problematic in several ways. The ethnicity of the guardian is taken to be a unitary phenomenon, though there are many different communities living along the river. Furthermore, the assumed closeness between ethnicized people and natural features risks harking back to a colonial trope.

However, the space opened up by the judicial pronouncement has had a local effect in that some communities living along the river have gathered in diverse ways to explore how this judgment could be used in their favor, and in the river's.[84] And they have gone beyond the legal text itself. For example, they have rejected the idea of a sole guardian in favor of a guardianship model formed by 14 people, two from each of the relevant seven communities, represented by one woman and one man.[85]

Cagüeñas et al document how communities have invented a forum of dialogue that will slowly parse through the meaning of representing the river. Instead of the one 'Atrato', the river is emerging as having different characters in different areas of its flow. Local deliberation has started to question, for example, the environmental impacts of traditional mining pre-mechanization and to ask to what extent those can be tolerated. It has also taken an approach that rebukes the claim

---

[83] Calzadilla (n 78) 55.

[84] Note 79.

[85] Diego Cagüeñas, María Isabel Galindo Orrego and Sabina Rasmussen, 'El Atrato y sus guardianes: imaginación ecopolítica para hilar nuevos derechos' (2020) 56(2) Revista Colombiana de Antropología 169; Macpherson and Ospina (n 77).

that locals are ecocentric, inasmuch as important parts of their relationship with the environment are expressed through use.

It is very interesting that in this case there wasn't necessarily a pre-existing ontological and philosophical system that rendered Atrato in a particular kind of way.[86] What comes out of this new experiment in river representation can be extremely significant for the future of Atrato. Conceptually, this case has already shown characteristics that differ from previous cases, and it will probably continue to innovate as it develops. Throughout the variety of cases so far, the aspect that remains constant is that the rights of nature are always related to political processes, whether before or after their pronouncement, or as a consequence of the pronouncement itself.

2. *Implementation*

I have implicitly argued that the conceptual commitments written within rights of nature provisions have an important effect on their practical implementation. This is not a deterministic argument, as if by knowing exactly what concepts are involved (itself an impossibility) we could predict with any degree of confidence what the results may be. Consequences are themselves mutable and often surprising, further informing theory. What can nonetheless be done is to explore the already visible relationships between some salient concepts and practical results. This kind of exercise allows for more sustained reflection on the precise conceptual formulation that may be decided upon in future cases. It also gives important tools of analysis that would otherwise be absent.

I cannot take stock of all implementations so far. Just like with the overview of cases, the point is not to be exhaustive but rather to highlight theoretical strains and problems. Here I want to take stock of two different aspects: the variety of implementation so far that is directly connected to conceptual commitments and the problems of implementation that may arise.

So far, the most tested rights of nature provisions have been in Ecuador, which is not surprising given that it has the longest history, besides some early municipal ordinances in the United States. And given the kind of rights granted in this case, and the broad provisions on standing, results have varied quite a bit. In particular, they have shown that the rights of nature formulated can thus be used by the state as much as by individuals or organizations concerned with environmental protection, and that these rights need not have any privileged relationship with Indigenous communities inasmuch as this is not specified in the law itself.

The first case in the country was the protection of the rights of the Vilcabamba River.

> 'The plaintiffs sued the municipal government for having modified the course of the river through a road construction project. The municipal government undertook the expansion project without having the necessary environmental impact assessments. The resultant material from the

---

[86] Cagüeñas, Galindo Orrego and Rasmussen (n 85); Macpherson and Ospina (n 77).

construction project was dumped in the river, and this caused a modification of its course, which then resulted in flooding the downstream property of riparian landowners'.[87]

They sued the municipal government using the constitutional rights of nature adopted in 2008.[88]

The local government was ordered to remedy the harm caused to the river and issue a public apology. The river has to be restored to the state that the riparian owners preferred. However, if restoration is to be understood strictly, then the rights of nature could also have been used against riparian owners, for the return of the river to an earlier state still. The potential tension between the rights of humans and rights of nature was resolved in this case in favor of nature. The judge explicitly stated that the right to a healthy environment is more important than the right to a better road, even though this is weighing two different kinds of *human* rights against each other.

This kind of weighing is inevitable in the framework of a constitution that grants many rights, of which many will come into conflict with each other. But the most interesting part of this judgment, to my mind, is that it was justified by the judge in terms of the missing environmental documentation that the construction company should have had. In other words, the rights of nature in this case had not yet modified relevant environmental provisions that already tell different actors what they can and cannot do with respect to the environment. This can be fruitfully compared to the *Aquepi* and *Cedros* Constitutional cases that instead ended up ruling for the rights of nature despite environmental permits having been obtained (see II. 1. b) above).

There is no necessity for judgments to go either way. That is part of the point, that the formulation in the Ecuadorian constitution leaves much to judicial discretion. The wide interpretive space is further exemplified in another case, namely one regarding the Mirador mining project (2013). An alliance of Indigenous and environmental groups sued a mining conglomerate and the state for violating nature's rights in a planned mining operation in the Cordillera del Cóndor, a hyper biodiverse region and home to Indigenous populations. The plaintiffs sued because, they argued, the already approved mining concessions would violate several constitutional rights, including the rights of nature. They further claimed that the precautionary principle enshrined in the constitution would be violated by the planned mining activities.[89]

'Unlike in the Vilcabamba case (which was a small-scale project of little national importance), the company planning to do the mining – Ecuacorriente S.A. – had undertaken the necessary environmental impact studies, which were approved by the resource and environmental ministries.'[90]

---

[87] Tănăsescu (n 1).

[88] See Daly (n 56).

[89] For discusión of this case, see Adolfo Maldonado and Esperanza Martinez, *Una Década con Derechos de la Naturaleza* (Abya-Yala 2019).

[90] Tănăsescu (n 1).

The judge ruled that, as all necessary documentation had been filed, the plaintiffs had no basis to assume rights would be violated. Furthermore, the alleged violation would happen in the future, and therefore the plaintiffs could not claim that it would happen with certainty.

Environmental assessment permits played an important role in both cases, albeit in opposite directions. But most importantly, the judges in the *Mirador* case ruled in favor of the mining operation by arguing that the State has a right to pursue strategic industries for the overall development of the country. This is significant because it shows another potential result of the tension between human rights and the rights of nature.

One of the latest and in many ways most significant case in Ecuador is the recent (November 2021) Supreme Court decision to uphold the rights of nature in the case of the Los Cedros Forest (protected since 1995).[91] In 2017, the government gave authorizations to two mining corporations to conduct exploration in the area. The Court revoked the authorizations and prohibited all extractive activities in the forest. The Court explicitly frames the judgment as one aimed at protecting nature itself, and not as being about the effects exploration may have on human beings.[92] Very interestingly, The Court argued that the risk of species extinction that results from mining activities is akin to genocide. This has already been documented as a potential new crime in international law, going by the name of ecocide.[93] It is very significant to see it formalized in this decision.

The *Los Cedros* decision centers primarily around the violation of art 73 of the Ecuadorian Constitution, which reads: 'The State shall apply preventive and restrictive measures on activities that might lead to the extinction of species, the destruction of ecosystems and the permanent alteration of natural cycles'. It is through the precautionary principle that the Court tied the rights of nature to the duty of the state to protect them. The principle, as defined by the Court, has three parts, namely the risk of irreversible damage, the existence of scientific uncertainty, and the preventive measures adopted by the State. This means that art 73 requires the adoption of the strictest environmental regulatory framework, especially in cases where risk and uncertainty are high, as they are in biodiverse and protected areas.

Indeed, art 131 argues that environmental registries must also include the precautionary principle, and art 134 interprets the need for high regulatory standards in relation to the biodiversity and special nature of Los Cedros. Finally, art 345g demands that pertinent authorities, the Environmental Ministry in particular, amend existing regulation in order to better align with the rights of nature.

This case is extremely interesting, for several reasons. First, it uses the constitutional status of the rights of nature in Ecuador to great effect. This is one of the few cases so far to be reviewed by the Constitutional Court, but once this kind

---

91 Maldonado (n 34) and Tănăsescu (n 1).

92 Corte Constitucional de Ecuador 10 November 2021 (n 57) para 65.

93 Ibidem, para 68.

of judgment is reached, it is bound to have deep repercussions, not least because the judgment applies to all other similar cases. The *Los Cedros* verdict therefore speaks in favor of the constitutionalization of the rights of nature. Second, this case reveals that the doctrine of standing has much less purchase than one may have thought, also because of the constitutional status of the rights of nature. Third, it may have a strong influence on future regulatory frameworks.

In this case, the permits were deemed insufficient, but also because the case was about *exploration* and not exploitation. As the exploratory phase is meant to assess the feasibility of the mining project, the permit requirements are not as strict. This opened up the possibility for the Court to use the laxness of the regulatory framework for exploration in reaching its judgment, a kind of judicial activism in its own right. This being said, the decision seems to go in the exact opposite direction of the reasoning employed by the *Mirador* case judges, who argued that future activities cannot be assumed to violate rights inasmuch as the involved parties had the correct permits.

It seems, then, that under the Ecuadorian framework, a lot of the battle for future implementation will revolve precisely around the issue of the aptness of regulations. The Court in this case framed the rights of nature as ecocentric and argued that there is no contradiction between these and the rights of people to a safe and healthy environment. This is in some sense true. But it is also true that the many rights of the constitution may come into conflict with both the rights of and to nature, as in the *Mirador* case. However that may be, it is through regulatory frameworks that judges are likely to assess whether the right to economic development may trump the rights of nature, because if pursued in a certain regulated fashion, it would not compromise the right to a healthy environment.[94]

This is speculative, but it is also the kind of speculation that is inevitably opened up by these extremely interesting decisions. The *Los Cedros* case adds yet another layer of complexity to the rights of nature, not only in Ecuador but everywhere. It is too early to say what the repercussions, beyond those for the Los Cedros Forest, will be.[95] But, on the strength of the evidence so far, I would expect rights of nature litigation to continue varying widely, given the Ecuadorian framework.

For now, the Court has been consistent in striking down certain kinds of development and/or the modification of rivers, but it is far from clear where inherently extractive economic activities (like farming) may, under this framework,

[94] The latest Ecuadorian Constitutional Court decisions are extremely important in the process of starting to define the relationship between regulatory frameworks and the rights of nature. So far, the Court has struck down environmental regulations because it judged them as contravening the rights of nature.

[95] There are other cases as well – 2167-21-EP (19 January 2022), 1185-20-JP/21 (15 December 2021), 1149-19-JP/20 (10 November 2021), 22-18-IN/21 (8 September 2021), 68-16-IN/21 (25 August 2021), 32-17-IN/21 (9 June 2021) – attesting to a current judicial activity that may prove extremely important in the future of the rights of nature.

happen. Lastly, there are many competing interests at play, and court rulings inevitably cuts through these, and not necessarily only when corporate interests are up against 'community' interests. Inevitably, different communities will also have different claims to nature, and courts, whether constitutional or otherwise, will have the very political task of adjudicating between them.

Other frameworks have been less tested, notably the New Zealand one. In those cases, though, the slowness of implementation is tied to their specific frameworks. Unlike cases (most notably India and Colombia) that mandated very fast implementation of their judgments (mere months), the situations of both the Te Urewera and Whanganui Acts have a longer-term approach. The issue of time is explicitly tackled and modelled on the time horizons of the natural entities in question. But, as I have argued, these cases are explicitly political arrangements of governance, and what it therefore means for them to be 'tested' need not take the form of court judgments. Instead, those results can only become visible in time, as the effects of different, Māori-inspired governance arrangements, will become more palpable.

III. Conclusions

At the beginning of this contribution, I set out to extract and analyze some conceptual threads that make the rights of nature both very promising and potentially problematic. I have therefore discussed some of the complexities and tensions within the theory and practice of rights of nature. An overview of this kind cannot be exhaustive, but instructive at best. In concluding, I wish to draw attention to some of the most salient points touched upon, and especially to their implications for further thought. For the sake of brevity and clarity, I will structure the conclusions in separate points.

– The different ways in which the rights of nature have been thought out and implemented raise notable *difficulties of classification*. Kauffman and Martin have classified existing rights of nature provisions in terms of their strength and scope, a method that may indeed encompass a significant number of cases.[96] I have further suggested that, given the variety of ends to which the rights of nature have been applied, purpose is another important category of classification.[97] But other typologies could be made using minimalist versus maximalist conceptions (for example, legal personality as such as opposed to positive rights), types of rights, types of institutionalization. None of these will capture everything, but this kind of exercise is important in order to remain faithful to the practical experimentation that is often ahead of theory.

[96] Kauffman and Martin, 'Comparing Rights' (n 55).

[97] Mihnea Tănăsescu, 'The Rights of Nature as Politics' in Daniel P Corrigan and Markku Oksanen (eds), *Rights of Nature: A Re-examination* (Routledge 2021).

- Rights can be given to *nature as such*, or else to *specific territories*. We have seen the salient distinctions and the very different formulations that the focus of rights may lead to. The consideration of the precise subject of rights depends on the underlying conceptions of nature that can be productively analyzed in each case. These conceptions are also a useful way into the plural legal traditions that may or may not be adequately represented.
- The *level of the law* is an important consideration when proposing or evaluating rights of nature. It is a crucial element both in what can be achieved within a given political process (the feasibility of constitutional amendments or the possibility of judicial activism, for example) and in what results may be spurred by the power of the law at the level at which it is conceived.
- The issue of *legal standing* has been important for the rights of nature from their partial theoretical beginnings in the work of Christopher Stone. Whether standing as such is needed largely depends on the legal system in question, but the connection between moral and legal standing should not be overlooked no matter what legal system we are considering.
- *Legal pluralism* has been extremely important for many rights of nature cases so far, and it is certainly important in their theory. However, the issue of pluralism was not initially dominant in the theoretical foundations of the rights of nature; rather, it emerged as a hard practical problem. The lens of legal pluralism problematizes the relationship between rights in general and as regards Indigenous-state relations, as well as between the rights of nature and different Indigenous ontologies. Indigenous activism has been instrumental in advancing the rights of nature towards legal pluralism.
- Some cases have been built through *top-down* pronouncements, while others reflect *bottom-up* processes. Rights of nature advocacy usually portrays the rights of nature as always participative in their creation, but this is not necessarily so. The Ecuadorian Constitution and the New Zealand claims settlements are good examples of participatory processes. In other cases, like ones initiated through judicial pronouncement, very little participation was possible.
- An implicit issue running through the argument has been the relationship between *recognizing* and *awarding/granting* rights to nature. The view of rights as recognized is related to theories regarding natural rights and explicitly ties the rights of nature to a moral view. The claim is that the recognition of rights is at once a moral and legal move. On the other hand, thinking of rights as granted sees the rights of nature as conventional rights that may be created with or without an explicit relationship to moral rights. My view is that moral arguments are secondary to the power relations that legal personality inevitably modifies. Moral arguments can therefore be understood as political representative claims.[98] Advocates of rights of nature may use moral argumentation, but

[98] Mihnea Tănăsescu, 'Rethinking Representation: The Challenge of Non-humans' (2014) 49(1) Australian Journal of Political Science 40.

this does not mean that they are unproblematically perceiving moral qualities of nature that someone else may fail to see.

- The above point also stresses that the rights of nature *are always political*, and I have argued, here and elsewhere, that they are fruitfully understood through political and not just legal lenses. The processes that lead to their adoption as well as to their implementation always have political motivations and consequences. This is not a side-effect of the rights of nature; it is instead their very point: to modify relations of power. It therefore matters how rights of nature are codified, but there cannot be a strict unidirectional and largely deterministic relationship between conceptual make-up and political effects. Rather, the careful formulation and adoption of rights of nature can merely try to avoid the possibility of their being used cynically by the powers that they are supposed to oppose.

International Law

Rights of Nature and Rights to Nature: Emerging Trends

Mara Tignino[*]

I. Introduction

Since the 2000s, there has been a growing recognition of the rights of nature at the domestic level. This increasing practice has influenced the emergence of a new approach in international law, one which considers that nature has to be protected in its own right. There is a call to shift in international law from an anthropocentric perspective to an eco-centric one on environmental protection. 'Rights of nature' and 'rights to nature' are two different but complementary notions. The first notion promotes the view that nature has a legal personality, guardians to protect it, and legal standing before judicial bodies. The second notion,

[*] This paper was developed in the context of the International Academy of Comparative Law project. The research has been funded by the Geneva Water Hub and was conducted between 2021–2024. The author wishes to express her gratitude to Prof. Daniel Eduardo Bonilla Maldonado and to the three anonymous peer-reviewers who provided feedback on previous versions of the paper. The manuscript was finalized in March 2024. The assessment of literature and case law, including internet materials, reflects the legal situation at that time.

i.e. rights to nature, underlines the obligation of States to protect nature as well as the human right to a healthy environment, as has been recognized at the global level by the Human Rights Council.[1] The complementarity between rights of nature and rights to nature is reflected in the case law of the Inter-American Court of Human Rights[2] as well as in international documents such as the 'World Declaration on the Environmental Rule of Law' of the International Union for Conservation of Nature (IUCN), adopted at the World Environmental Law Congress in 2016.

The recognition of nature as a holder of rights raises the question of whether legislators and judges need to rethink the obligations of States in relation to nature. The anthropocentric and eco-centric perspectives on the protection of nature should not be viewed antagonistically but, on the contrary, as mutually reinforcing.

The paper will firstly examine the main features and limits of the rights of nature in certain domestic jurisdictions. The second part of the paper will analyze the synergies and possible tensions between the rights of nature and the right to nature, namely the human right to a healthy environment.

II. Domestic Practice on the Rights of Nature: A Comparative Law Approach

The transnational movement on the rights of nature is gaining momentum at both domestic and international levels. Already in the 1970s, Christopher Stone argued that trees, rivers, and other natural elements should be given legal personhood and standing to defend their rights.[3] In the ensuing thirty years, the concept of the rights of nature was followed by some scholars.[4] In the first decade of the 2000s, given the unprecedented environmental crisis, scholars in international environmental law increasingly questioned the ability of humans to respond to the continuing deterioration of the Earth's ecosystems.[5] Some scholars have ar-

[1] UNHRC Resolution 48 (2021), UN Doc A/HRC/RES/48/13, *The Human Right to a Clean, Healthy and Sustainable Environment*.

[2] IACHR 15 November 2017, Advisory Opinion OC-23/17, requested by the Republic of Colombia (*The Environment and Human Rights*); IACHR 6 February 2020, *Comunidades Indígenas Miembros de la Asociación Lhaka Honhat (Nuestra Tierra) v. Argentina*; IACHR 27 November 2023, *Habitantes de la Oroya v. Peru*.

[3] Christopher D Stone, 'Should Trees Have Standing? Toward Legal Rights for Natural Objects' (1972) 45 SCal LRev 450.

[4] Roderick Frazier Nash, *The Rights of Nature: A History of Environmental Ethics* (UW Press 1989); Thomas Berry, 'The Origin, Differentiation and Role of Rights' (2001) Earth Jurisprudence Conference 21; Cormac Cullinan, *Wild Law: A Manifesto for Earth Justice* (7th edn, Green Books 2003).

[5] Louis J Kotzé and Rakhyun E Kim, 'Earth System Law: The Juridical Dimensions of Earth System Governance' (2019) 1 Earth System Governance [100003].

gued that the current anthropocentric approach of environmental law 'has failed to keep humanity from crossing critical planetary boundaries', and they affirmed that new legal approaches are necessary to address the current ecological situation.[6] Despite the progressive developments in 'modern' environmental law[7], recognizing the existence of rights in natural entities such as rivers, glaciers, mountains, or forests represents a new approach to addressing the current ecological crisis.[8]

Legislation and decisions of courts in countries such as Ecuador, Colombia, Bangladesh, India, Uganda, and New Zealand recognize rights in elements of nature. The analysis of national practice shows three dimensions as regards the rights of nature: (1) the attribution of legal personality to elements of nature, including rivers and glaciers; (2) the establishment of 'guardians' to safeguard nature and; (3) the standing of nature before national courts. These three features are common in legislation and in the decisions of courts addressing the rights of nature. However, the rights of nature raise complex challenges in terms of their implementation. There are also tensions between the interests of States and the attribution of legal rights to nature.

1. The Legal Personality of Nature

As already mentioned, one of the main features of the discourse on the rights of nature is the attribution of legal personality to elements of nature such as rivers and glaciers as well as ecosystems. For instance, in New Zealand, the Te Awa Tupua (Whanganui River Claims Settlement) Act of 2017 (2017 Act) recognizes the legal personality of the Whanganui River and its ecosystem. According to the Act, the 'Te Awa Tupua is an indivisible and living whole, comprising the Whanganui River from the mountains to the sea, incorporating all its physical and metaphysical elements' (section 12).

Domestic rulings have also attributed legal personality to critical natural ecosystems such as the Amazon rainforest. Colombian case law represents an important contribution in this regard. Even if Colombia's domestic laws do not recognize the legal personality of nature, in 2018 the Supreme Court recognized the Amazon rainforest as a 'sujeto de derechos',[9] an entity in its own right. This ecosystem is the holder of rights, particularly regarding its protection, conservation, maintenance, and restoration by the State and the territorial entities that comprise it.

---

[6] Kotzé and Kim (n 5) 4.

[7] Julien Bétaille, 'Rights of Nature: Why It Might Not Save the Entire World' (2019) 16 JEEPL 35.

[8] Susana Borràs, 'New Transitions from Human Rights to the Environment to the Rights of Nature' (2016) 5 TEL 113; David Richard Boyd, *The Rights of Nature: A Legal Revolution That Could Save the World* (ECW Press 2017); Louis J Kotzé, 'A Global Environmental Constitution for the Anthropocene?' (2019) 8(1) TEL 11.

[9] Supreme Court of Colombia 5 April 2018, STC4360-2018, Radicación n° 11001-22-03-000-2018-00319-01 (the *Rainforest* case).

The Constitutional Court of Colombia also declared the Atrato River as a legal entity with its own rights (para 5.9).[10] The Court recognized the shift from an anthropocentric approach to an eco-centric one (para 5.9). The rights of nature are viewed as the most effective legal expression of an eco-centric approach. Such an approach places nature on an equal footing with humans (para 5.10). The Court found support for its findings in the biocultural rights which recognize the profound and intrinsic connection that exists between nature and local communities (introduction). Moreover, the legal personality of rivers also finds a basis in international law, particularly in human rights law and international environmental law. Examples include the 1989 ILO Convention n°169 on Indigenous and Tribunal Peoples, the Convention on Biodiversity, the 2007 UN Declaration on the Rights of Indigenous Peoples, the 2016 American Declaration on the Rights of Indigenous Peoples, and the 2003 UNESCO Convention for the Safeguarding of the Intangible Cultural Heritage (para 5.19). Moreover, several countries have developed 'biocultural practices' to ensure the protection of rights of indigenous communities and nature. In 2024, the Tribunal of Nauta in Peru has recognized the Marañón river, one of the most important tributaries of the Amazon basin, as an entity with its own rights. Thus, the Peruvian court has followed a similar approach to that of the Colombian domestic tribunals (Ruling, 00010-2022-0-1901-JM-CI-01, 8 March 2024).[11]

2. *The Custodians of the Rights of Nature*

The attribution of legal personality to elements of nature entails the appointment of custodians. Local communities often act as custodians of rivers given their close and sustainable relationship with freshwater.[12] For example, among the Native American tribal jurisdictions which have recognized rights of nature, the Nez Perce General Council has established a legal guardianship body to represent the rights and interests of the Snake River.

National courts have also appointed governmental agencies as custodians of natural elements. In India, the decision *Mohd. Salim v State of Uttarakhand & others*[13] (the *Ganges and Yamuna* case) and the decision *Lalit Miglani v State of*

---

[10] Constitutional Court of Colombia 10 November 2016, T-622/16 (the *Atrato River* case).

[11] Tribunal of Nauta 8 March 2024, Resolution no 14, 00010-2022-0-1901-JM-CI-01.

[12] International Rivers and others. *Rights of Rivers. A Global Survey of the Rapidly Developing Rights of Nature Jurisprudence Pertaining to Rivers*. (2020) <www.internationalrivers.org/wp-content/uploads/sites/86/2020/09/Right-of-Rivers-Report-V3-Digital-compressed.pdf>; Cristy Clark and others, 'Can You Hear the Rivers Sing? Legal Personhood, Ontology, and the Nitty-Gritty of Governance' (2018) 45 Ecology LQ 787; Gabriel Eckstein and others, 'Conferring Legal Personality on the World's Rivers: A Brief Intellectual Assessment' (2019) 44 (6–7) Water International 804; Joshua C Gellers, 'Earth System Law and the Legal Status of Non-Humans in the Anthropocene' (2021) 7 Earth System Governance [100083].

[13] High Court of Uttarakhand 2017, Writ Petition (PIL) No 126 of 2014, *Mohd. Salim v State of Uttarakhand & others*.

Uttarakhand & others[14] (the *Glaciers* case) have identified the institutions representing the Ganges and Yamuna rivers. The guardians of the two rivers were identified as the Director of the 'NAMAMI Gange' conservation program, the Chief Secretary of the State of Uttarakhand, and the Advocate General of the State of Uttarakhand. These institutions were declared as '*persons in loco parentis*'.[15] They were given the task of protecting, conserving, and preserving the Ganges and Yamuna rivers and their tributaries. These officers were tasked with promoting the health and well-being of these rivers. However, the judgment did not clarify which measures should be taken by the officers to achieve this objective. The government, however, appealed the decision of the High Court to the Supreme Court, which agreed and overturned the earlier *Ganges and Yamuna* ruling.[16]

Similar to the *Ganges and Yamuna* case, the High Court of *Uttarakhand*, in the *Glaciers* case, appointed several guardians, including the Chief Secretary of the State of Uttarakhand and a range of legal advisors, academics, and judges to act as 'the *persons in loco parentis*' to conserve and preserve the Glaciers as well as the Gangotri and Yamunotri rivers in the State of Uttarakhand.[17]

In a neighboring country of India, the Supreme Court of Bangladesh has similarly appointed the Bangladesh's National River Conservation Commission as the guardian of all the rivers of the country.[18] The decision in Bangladesh has therefore taken an approach similar to that of India. Unlike in India, however, the Supreme Appellate Division of Bangladesh (Bangladesh's highest court) upheld the 2019 decision in February 2020.[19]

In some countries, indigenous communities act as the guardians of rivers. For example, in Canada, the guardians of the Magpie river (or Mutehekau Shipu river) are appointed by the municipality of Magpie and the Council of the Innu of Ekuanitshit, an indigenous community.[20] Moreover, in the case of New Zealand, the 2017 Act establishes that Te Awa Tupua is to be represented by a guardian Te Pou Tupua, which 'is to be the human face of Te Awa Tupua' and acts in the name of the river and its ecosystem (section 18). The Te Pou Tupua consists of two persons, one appointed by the iwi, the Maori tribe living in the region of the Whanganui River, and one appointed by the government. This office is responsible for the care and well-being of the river and for maintaining relationships with all interested people.

---

[14] High Court of Uttarakhand 2017, Writ Petition (PIL) No 140 of 2015, *Lalit Miglani v State of Uttarakhand & others*.

[15] *Mohd. Salim v State of Uttarakhand* (n 13) [19].

[16] 'India's Ganges and Yamuna Rivers are "not living entities"' *BBC News* (7 July 2017) <www.bbc.com/news/world-asia-india-40537701>.

[17] *Lalit Miglani v State of Uttarakhand & others* (n 14) 65.

[18] Mohammad Sohidul Islam and Erin O'Donnell, 'Legal Rights for the Turag: Rivers as Living Entities in Bangladesh' (2020) 23 APJEL 160.

[19] Islam and O'Donnell (n 18) 161.

[20] Province de Québec, Municipalité régionale de comté de Minganie, Resolution n°025-21 16 February 2021, *Reconnaissance de la personalité juridique et des droits de la rivière Magpie – Mutehekau Shipu*.

Moreover, in the *Atrato River* case, the Supreme Court of Colombia ordered the creation of a body to exercise legal guardianship over the river, including representatives from the government and the local communities.[21] The Court charged this body with implementing the ruling and developing an appropriate institutional response to the court's findings, and it ordered that the guardianship body should be advised by civil society and a panel of experts.[22] The Tribunal of Nauta in Peru also appointed national authorities and indigenous communities as guardians of the Marañón river.[23]

3. The Standing of Nature before National Courts: The Case of Colombia

A central element of the concept of rights of nature is the legal standing of nature before judicial bodies.[24] Relying on its tradition of civil or common law, nature has only limited legal standing in Europe.[25] Contrary to the European practice, some countries of South America, including Colombia, Ecuador, and Peru, have expressly recognized the legal standing of nature before national courts. The practice of Colombia is the most emblematic in respect of the legal standing of natural elements before judicial bodies. Two cases, the *Atrato River* case decided by the Constitutional Court in 2017 and the *Rainforest* case decided by the Supreme Court in 2018, have – without the support of domestic laws – established the legal framework of rights of nature in this country. The Colombian decisions have been accompanied by strong remedial measures adopted by the courts, including the creation of guardianship entities overseen by courts and requirements to report regularly to the judicial bodies.

4. Potential Tensions between the Rights of Nature and the Interests of States

One of the most important challenges in the implementation of the rights of nature is the potential tension between rights of nature, property rights, and the sovereign rights of States. For instance, both Ecuador and Bolivia recognize property rights and the notion of State sovereignty over natural resources. It could be argued that the notions of property and sovereignty may hide economic interests related to the trading and exploitation of natural resources. The rights of nature can collide with national economic development priorities imposing new duties on property owners and limiting sovereign rights in the exploitation of natural resources. These new obligations will sometimes aim to limit the uses of natural resources so that ecological integrity is not compromised. In this context,

[21] *The Atrato River case* (n 10) para 9.32.

[22] *The Atrato River case* (n 10) para 10.21.

[23] Tribunal of Nauta 8 March 2024, Resolution no 14, 00010-2022-0-1901-JM-CI-01.

[24] Lidia Cano Pecharromán, 'Rights of Nature: Rivers That Can Stand in Court' (2018) 7(1) Resources 13.

[25] In 2023, Spain was the first European country to recognize an ecosystem as a legal person. Ludwig Krämer, 'Rights of Nature in Europe: the Spanish Lagoon Mar Menor Becomes a Legal Person' (2023) 20 (1) JEEPL 5.

the rights of nature would need to be considered as part of the principle of sustainable development. Under this perspective, it would be necessary to find a balance between the rights of nature as regards non-human species and rights to nature as held by individuals and communities. The recognition of the rights of nature cannot be unconditional, and it is necessary to examine the possible contradictions between rights of nature, the economic interests of a State, and the socio-economic needs of local communities. There are no easy answers that can be offered by international environmental law. With the aim of avoiding possible conflicts, this paper suggests that the principles of international environmental law, such as the principles of sustainable development, prevention, and precaution, should take into account initiatives taken at the domestic and local level, including those from civil society and citizens.

There are additional challenges in the recognition of the rights of nature. First, identifying the guardians or stewards that could represent nature can be a challenge. Second, there might be a lack of appropriate funding to enforce the relevant domestic legislation or a judgment granting legal personhood to elements of nature. Often, the local actors responsible for implementing the judgments or the legislation may lack technical or financial capacity. Third, a form of independence between the government and the guardians of nature may be necessary to ensure a full enforcement of the involved legal rights.[26]

The analysis of the rights of nature also raises the question of possible tension between the interests of nature and the interests of society. In this context, it is worthwhile to note that Bolivian Law 71 of 2010[27] states that 'the interests of society, within the framework of the rights of Mother Earth, prevail in all human activities and over any acquired right' (art 2). There is a potential tension between this principle and the other eco-centric provisions of this instrument. A general approach based on international environmental law, having reference, in particular, to the principle of sustainable development – inclusive of its economic, social, and environmental dimensions – may help to resolve these tensions. To put Mother Earth at the center of international environmental law might help to ensure a more balanced approach to sustainable development based on a respect for nature and protection of the rights of future generations.

III. The Relationship between the Rights of Nature and the Right to Nature

The affirmation of the rights of nature is based on the court judgments of various countries, including Colombia, Ecuador, Peru, and India, as well as national Constitutions. In parallel to the transnational movement of recognition of the rights

[26] Erin L. O'Donnell and Julia Talbot-Jones, 'Creating Legal Rights for Rivers: Lessons from Australia, New Zealand, and India' (2018) 23(1) Ecology and Society 7.

[27] Ley de Derechos de la Madre Tierra 2010 (Law 71).

of nature, the support for a global recognition of the human right to a healthy environment has grown over the years.

Some authors, like Valérie Cabanes or the UN Special Rapporteur on human rights and the environment, David Boyd, have argued for both approaches, namely the need to recognize the rights of nature and the global recognition of the human right to a healthy environment.[28] Although the human right to a healthy environment is recognized in several countries and international treaties, its global recognition has taken a number of years. It was only in October 2021 that the Human Rights Council recognized this right.[29]

1. The Role of the Inter-American Court of Human Rights in the Recognition of the Rights of Nature and the Right to Nature

The Inter-American Court of Human Rights has contributed to the recognition of the right to nature. At the request of Colombia, the Court issued an Advisory Opinion in 2017 affirming the human right to a healthy environment and examined the obligations of States when they have caused or may cause significant environmental harm, including in a transboundary context.[30]

The Advisory Opinion includes both an anthropocentric and eco-centric approach. The Inter-American Court affirms that nature has its own rights and should be protected not only because of the benefits it provides to individuals or local communities or the effects that its degradation may have on other human rights, such as health, life or personal integrity, but because of its importance to other living organisms (para 62).

Subsequent to the Advisory Opinion, the Inter-American Court of Human Rights recognized the human right to a healthy environment in *Comunidades Indígenas Miembros de la Asociación Lhaka Honhat (Nuestra Tierra) v. Argentina* and in *Habitantes de la Oroya v. Peru*, wherein it found a concrete violation of this right.[31]

It should be noted that the 2017 Advisory Opinion represents the first instance in which an international judicial body points out the linkages between the human right to a healthy environment and the need to protect nature for its own sake dissociated from the benefits provided to humans. In particular, the Court states that:

> '[A]s an autonomous right, the right to a healthy environment, unlike other rights, protects the components of the environment, such as forests, rivers and seas, as legal interests in themselves, even in the absence of the certainty or evidence of a risk to individuals. This means that it protects nature and the environment, not only because of the benefits they provide to humanity or

---

[28] Valérie Cabanes, *Homo Natura. En harmonie avec le vivant* (Buchet|/|Chastel 2017); David Richard Boyd, *The Rights of Nature: A Legal Revolution That Could Save the World* (ECW Press 2017).

[29] UNHRC Resolution 48 (2021) (n 1).

[30] IACHR 15 November 2017 (n 2).

[31] IACHR 6 February 2020 (n 2), paras 202–203; IACHR 27 November 2023 (n°2), paras 115–129.

the effects that their degradation may have on other human rights, such as health, life or personal integrity, but because of their importance to the other living organisms with which we share the planet that also merit protection in their own right. In this regard, the court notes a tendency, not only in court judgments, but also in Constitutions, to recognize legal personality and, consequently, rights to nature.' (para 62)

Therefore, the Inter-American Court recognizes that the right to a healthy environment, as an autonomous right, also protects elements of nature as legal interests in themselves even without damage or risk of harm to individuals. Nature must be protected not only for the benefit of humans but also in respect of its own value. In this way, the Court points out that the anthropocentric and eco-centric approaches to environmental protection can go hand in hand and are not necessarily in conflict. Both approaches should be mutually reinforcing.

An element that would deserve attention is related to the rights of nature in a transboundary context. As the cases of India and Bangladesh show, the rights of nature may involve transboundary water resources flowing in various countries. This raises an additional challenge in the enforcement of the rights of nature by various jurisdictions. The activities carried out by a country may affect the rights of nature in another State. Natural elements do not recognize political borders. The Inter-American Court of Human Rights considers in its 2017 Advisory Opinion that a State 'may be held responsible for any significant damage caused to persons outside their borders by activities originating in their territory or under their effective control or authority' (para 103). It also adds that 'this obligation does not depend on the lawful or unlawful nature of the conduct that generates the damage, because States must provide prompt, adequate and effective redress to the persons and States that are victims of transboundary harm resulting from activities carried out in their territory or under their jurisdiction, even if the action which caused this damage is not prohibited by international law' (para 103).

The jurisprudence of the Inter-American Court of Human Rights represents a precedent for developing the rights of nature and the right to nature. International rulings can inform the development of international law and are often followed by other international courts and tribunals.

2. The 2021 Human Rights Council's Resolution on the Human Right to a Healthy Environment

In October 2021, the Human Rights Council recognized the human right to a healthy environment at the global level.[32] The resolution recognizes the right to a clean, healthy, and sustainable environment as a human right and affirms its importance for the enjoyment of human rights. The right to a clean, healthy and sustainable environment also requires the full implementation of existing multilateral environmental agreements and the principles of international environmental law.

---

[32] UNHRC Resolution 48 (2021) (n 1).

The resolution does not mention the rights of nature as such. However, as the case law of the Inter-American Court of Human Rights shows, the rights of nature can go hand in hand with the recognition of the human right to a healthy environment. In particular, the call for the implementation of multilateral environmental agreements may contribute to the protection of nature in itself and for mankind. Moreover, by referring to the outcome document authored at the 2012 United Nations Conference on Sustainable Development ('The future we want'),[33] which incorporates an anthropocentric approach to international environmental law, the 2021 resolution may indirectly support the rights of nature.[34] As in the case of the right to a healthy environment, the rights of nature have been recognized by an increasing number of countries.

3. International Initiatives on the Rights of Nature

The progress made at the domestic level on the rights of nature has an impact at the international level. There are various international initiatives which focus on the rights of nature. While there is not a binding international text recognizing the rights of nature, the UN General Assembly and other global institutions have started to work on this topic.

A specific UN initiative exists focusing on 'Harmony with Nature', which strongly supports the rights of nature.[35] Since 2011, the UN General Assembly has engaged in thirteen interactive dialogues with UN member States and other relevant stakeholders to commemorate International Mother Earth Day on 22 April.[36] The last interactive dialogue held in 2023 discussed the opportunity to convene an 'Earth Assembly' at the UN in the coming years. The 2023 dialogue also focused on the protection of biodiversity and the role of indigenous communities as custodians of it. One of the aims of these dialogues is the transformation from a human-centered to an Earth-centered paradigm in analyzing the loss of biodiversity at the global level.

Beyond the UN, other organizations, such as the IUCN, have made calls to develop a Universal Declaration of the Rights of Nature that contributes to a new philosophy of human well-being.[37] Principle 2 of the 2016 World Declaration on the Environmental Rule of Law adopted at the IUCN's World Environmental Law Congress states that '[e]ach human and other living being has a right to the conservation, protection, and restoration of the health and integrity of ecosystems. Nature has the inherent right to exist, thrive, and evolve.'

---

[33] UNGA Resolution A/RES/66/288, 11 September 2012, para 3.

[34] UNHRC Resolution 48 (2021) (n°1).

[35] UN, Resolution adopted by the General Assembly on 21 December 2009, *Harmony with Nature*, UN Doc A/RES/64/196.

[36] UN, Resolution adopted by the General Assembly on 22 April 2009, *International Mother Earth Day*, UN Doc A/RES/63/279.

[37] WCC, *Incorporation of the Rights of Nature as the organisational focal point in IUCN's decision-making*, WCC-2012-Res-100-EN [2012].

While it is difficult to argue that the rights of nature have a specific status in international law, regional, domestic, and local practices relying on national or municipal regulations as well as court rulings constitute promising avenues for the recognition of the rights of nature at the international level. The Inter-American Court of Human Rights cases represent a significant step in creating a close relationship between the right to nature and the rights of nature.

Another promising approach at the international level is related to the analysis of the linkages between sustainable development and the rights of nature. Such rights may support the environmental dimension of sustainable development. In particular, the UN General Assembly has noted that a number of countries recognize the rights of nature in the context of the promotion of sustainable development. The UN General Assembly has also pointed out that, to achieve a just balance among the economic, social, and environmental needs of present and future generations, it is necessary to promote harmony with nature based on the recognition of the rights of nature.[38]

4. *Possible Tensions between the Rights of Nature and the Right to Nature*

Several countries have adopted domestic legislation on the rights of nature and the right to nature. Moreover, jurisdictions have also developed both typologies of rights. Among the national practices analyzed in this paper, the case of Bangladesh shows some possible tensions between the rights of nature and right to nature. As already noted, the Supreme Court of this country has granted legal rights to all the rivers of the country. This decision has drawn criticism for its impact on the rights of fragile communities, many of which live in insecure housing on river floodplains.[39] The Court adopted high penalties for violations of its ruling, including eviction from river land. While these penalties have been imposed in an attempt to strengthen the rule of law, they have heavily affected poor people living in the surrounding area of rivers. These penalties could end up exacerbating existing inequality in the Bangladesh society. This situation is similar to the Indian cases, but it is starkly different from the approaches adopted in Colombia and Peru. In the Colombian case, the rights of nature in particular rivers have been recognized as a mechanism that gives effect to the right to nature of communities who depend on the river, relating especially to the human right to a healthy environment but also to other fundamental human rights, such as the human right to water. In 2024, the Peruvian tribunal of Nauta similarly interpreted the rights of the Marañón river as including a call for the establishment of a river basin organization and as supporting the environmental evaluation of economic activities.

---

[38] UN, Resolution adopted by the General Assembly on 21 December 2020, *Harmony with Nature*, UN Doc A/RES/75/220.

[39] Rina Chandran, 'Fears of Evictions as Bangladesh gives Rivers Legal Rights' *Reuters* (4 July 2019) <www.reuters.com/article/us-bangladesh-landrights-rivers-idUSKCN1TZ1ZR>.

Attention must be paid to the fact that the rights of nature do not replicate social structures based on inequality. The approach of the rights of nature should contribute to building or strengthening a set of common values that support the protection of nature as well as processes that actively engage and empower local communities.

Unlike the examples from Colombia and New Zealand, the Bangladesh case does not appear to be explicitly grounded in the values of local communities. It seems to protect the interests of nature in opposition to the interests of people who live along the river and depend upon it. This adversarial relationship risks exacerbating existing social inequalities and leaving primarily the poor vulnerable to the new and severe penalties. It is worth noting that the adoption of institutional frameworks capable of delivering sustainable and lasting change for both people and nature is likely to be beyond the functions and powers of a judicial body. Institutional arrangements serving to implement the rights of nature and the right to nature require action and support from both the executive branch and legislative bodies as well as civil society.

IV. Conclusions

Recognizing the rights of nature is not a panacea for all the challenges associated with the protection of nature and the flaws of existing environmental laws. Recognition of the rights of nature is based on a philosophical paradigm change in respect of the relationship between humans and nature. It requires rethinking the understanding of international environmental law, which puts human beings at the center of its regime of protection. A question arises as to whether multilateral environmental agreements and principles of international environmental law are sufficient to protect the rights of nature or if it is necessary to reconsider these frameworks. The case law of Colombia, Peru, and the Inter-American Court of Human Rights recognizes that the principles at the base of international environmental law are complementary to the rights of nature. The same complementary approach is adopted in other international documents, such as IUCN declarations and the UN initiative on 'Harmony with Nature'.

The granting of legal rights to natural elements at the domestic level may influence the recognition of the rights of nature at the international level and could be supported by the increasing recognition of a right to a healthy environment. For example, the pollution of a river may threaten this right as well as the right of a river in its own regard. Human rights law focused on protecting the fundamental rights of local and indigenous communities may also ensure that nature is protected in term of its values and not only regarding its importance for human beings. At the same time, principles of international environmental law, such as prevention and precaution, continue to have their relevance even in the context of an eco-centric approach.

The complementarity between the rights of nature and the right to nature does not however exclude possible tensions between these rights, especially in a transboundary context. For example, in the case of the Ganges, which flows through India and Bangladesh, the 1996 Farakka treaty does not provide for the rights of nature and a right to nature. The domestic rulings adopted in the two countries may raise challenges as to their implementation. Transboundary cooperation could help to implement the decisions of domestic courts. As supported by scientific studies, advocacy from civil society groups, and political will, the rights of nature and the right to nature need to be enforced on both sides of the border.[40]

[40] Gauri Noolkar-Oak, 'Transboundary cooperation key to enforcing rivers' legal rights in Bangladesh.' *The Daily Star* (11 August 2019) <www.thedailystar.net/opinion/environment/news/transboundary-cooperation-key-enforcing-rivers-legal-rights-bangladesh-1784641>.

Law and Anthropology

An Interdisciplinary Perspective on Rights of Nature

*Dirk Hanschel and Annette Mehlhorn**

I. Introduction

Debates on rights of nature have become increasingly popular both in academia as well as amongst legal practitioners, environmental activists, and local communities. In this field, combining law and anthropology provides perspectives that are both appreciative and critical and that, if taken seriously, have the merit of combining doctrinal legal scholarship with extended ethnographic fieldwork. It challenges the perspective of engrained lawyers by inviting them to be more conscious of the complexity of human living conditions, just as much as it challenges empirical anthropologists by inviting them to help reconstruct what they have deconstructed through their critical analysis. Thus, this article explores what both law and anthropology can contribute to rights of nature debates and, con-

* The manuscript was finalized in April 2025. The assessment of literature and case law, including internet materials, reflects the situation at that time. Unless otherwise indicated, all internet links were last accessed on 6 May 2025.

versely, what this reveals about the possibilities and pitfalls of this particular interdisciplinary convergence.

It becomes evident that, instead of being too enthusiastic about rights of nature concepts and their seeming ability to change the world, it is crucial to ground any argument in empirical analyses of concrete circumstances and demands, of different power-laden processes of translation, and of the different effects that the use of this legal construct can have in specific contexts. Anthropology can play an important part in demystifying rights of nature's foundation and effects. Within an interdisciplinary law and anthropology approach, this is not only to highlight the complexities of law-in-action and different human-nature relationships, but – crucially – to obtain a realistic picture as to the potential of rights of nature.

After briefly mapping out rights of nature debates from law and anthropology perspectives generally, this paper will zoom in on a number of cases and arguments which are meant to illustrate the bandwidth and vigor of possible anthropological contributions. Importantly, our aim is to display the vitality and importance of the interdisciplinary endeavor, to entice curiosity, and not to give an account of anthropological approaches with any claim of completeness or representativeness. While the chosen vignettes and ethnographic analyses refer to specific contexts, the questions they raise are relevant for rights of nature beyond these, going to (often unquestioned) core assumptions. To put it quite bluntly: One case that shatters overly simple or essentialist assumptions regarding rights of nature might be sufficient to generate new conclusions.

II. Overview of the Emerging Global Rights of Nature Debate

Rights of nature debates are mushrooming across academia, NGOs, and local communities. The debate has been intensely promoted by scholars such as Christopher Stone or Klaus Bosselmann as early as in the 1970s and more recently by David Boyd and others.[1] The idea is to grant nature a better say in legal decision-making by putting it on equal footing with human beings. According to advocates of the idea, rights of nature is nothing short of a 'legal revolution', as Boyd titled his influential monograph on the matter.[2] Rights of nature, according to Boyd, can upend the anthropocentric character of law and of dominant understandings of the human-Earth relationship. It is such a perception of rights of nature, which

[1] Christopher Stone, 'Should Trees Have Standing? Toward Legal Rights for Natural Objects' (1972) 45 SCal LRev 450; Klaus Bosselmann, 'Eigene Rechte für die Natur? Ansätze einer ökologischen Rechtsauffassung' (1986) 19(1) Kritische Justiz 1, 22; David Boyd, *The Rights of Nature. A Legal Revolution that Could Change the World* (ECW Press 2017) 174; see also Dirk Hanschel and Elizabeth Steyn, 'Environmental Justice' in *Oxford Handbook of Law and Anthropology* (OUP 2022) 561.

[2] Boyd (n 1).

is inherently moral and political, that has most powerfully influenced the growing debate and movement and is significantly distinct from more pragmatist and legalist accounts.[3]

More recently, rights of nature find increasing recognition in the law worldwide. They are increasingly being discussed at the international and national levels. The most far-reaching realization of rights of nature has occurred in Ecuador, which enshrined the principle of Pachamama in its constitution.[4] The subsequent efforts of the Ecuadorian Constitutional Court put a spotlight on core questions, such as how to operationalize these rights in concrete cases and how to distinguish them, for instance, from the right to live in a healthy environment, which is also expressed in the constitution.[5] Sometimes rights of nature are established directly by legislation, such as in Bolivia, sometimes by way of interpretation of existing broader provisions by courts (as with the rights of the Amazon established by the American Court on Human Rights or the rights of the Atrato River as guaranteed by the Columbian Supreme Court).[6] Rights of rivers are, furthermore, discussed in other places, such as in India (Yamuna, Ganges, etc.) or New Zealand (Wanganui).[7]

The terminology and concrete stipulation varies greatly and at the same time remains rather abstract. Apart from the generic term of 'rights of nature', other expressions are in use, such as nature as a legal person (without naming its specific rights), recognition of legal rights to nature, etc. Sometimes, more specifically, river rights, rights of forests, etc. are coined.[8] In some cases, the guarantees concern ecological integrity as such.[9]

Rights of nature debates have intersected and entered into a productive dialogue with a number of broader fields, such as various streams of critical legal thought, post-humanism, and new realism, to name just a few. Various questions have been widely debated in the literature. For example, how nature and its interests are to

[3] For an overview see, eg, Mihnea Tănăsescu, *Understanding the Rights of Nature: A Critical Introduction* (transcript 2022).

[4] Mihnea Tănăsescu, 'Rights of Nature, Legal Personality, and Indigenous Philosophies' (2020) 9 TEL 429, 430; Seth Epstein and others, 'Liberalism and Rights of Nature: A Comparative Legal and Historical Perspective' (2022) 18 LCH 1, 7.

[5] Boyd (n 1) 174; Nathalie Rühs and Aled Jones, 'The Implementation of Earth Jurisprudence through Substantive Constitutional Rights of Nature' (2016) 8 Sustainability 174, 176; see furthermore Gustavo Prieto, 'The Los Cedros Forest Has Rights, The Ecuadorian Constitutional Court Affirms the Rights of Nature' (*Verfassungsblog*, 10 December 2021) <https://verfassungsblog.de/the-los-cedros-forest-has-rights>.

[6] Tănăsescu (n 4) 430; Corte Constitucional, Sala Sexta de Revision, Colombia 10 November 2016, *Centro de Estudios para la Justicia Social 'Tierra Digna' v President of the Republic*, NoT-5.016.242.

[7] Boyd (n 1) 134.

[8] Peter Burdon, 'The Rights of Nature: Reconsidered' (2010) 49 Australian Humanities Review 69, 79.

[9] David Boyd, 'Recognizing the Rights of Nature' (2018) 32 Natural Resources & Environment 13, 14.

be defined is crucial and has been subject to debate and has met with a number of important points of critique. A number of analyses have pointed to the tendencies and effects of the often-inherent anthropocentrism and anthropomorphism.[10] What nature is and what it wants is by no means clear, and different interpretations can have significantly different outcomes and political implications.[11]

Another decisive and difficult question is who will make claims on behalf of nature against whom. Who will go to court and have legal standing? This question already motivated Christopher Stone in his famous article 'Should trees have standing?',[12] which has been discussed so many times. Questions of representation are difficult to tackle. Parallels are sometimes drawn to companies as legal persons or the attachment of rights to minors. Representation and the need for proxy-constellations is certainly not a new or unique element for rights of nature, but the former has been shown to surge with particular urgency and complexity in this context.[13] In legal theory, rights of nature have given impetus to important debates and innovations in this regard. Empirically, it has been pointed out that the details as to how questions of representation are resolved has far-reaching consequences.[14]

In this vibrant field of debate, it has become evident that some answers can only be given empirically. The details regarding the elaboration, diffusion, implementation, and contestation of the rights of nature can be studied ethnographically, which might lead to unexpected insights, new questions, and certainly complex pictures in the context of different histories, fields of knowledge and institutionality, political structures and contestations, social struggles, and ontologies.

III. Law and Anthropology Perspectives

Combining law and anthropology has the potential to advance the debates regarding rights of nature in a number of productive ways and on a number of levels, ranging from theory to application. This does not come without challenges, but it is also capable of generating very useful and new results.

---

[10] Alexis Alvarez-Nakagawa, 'Who is the Subject of (Non) Human Rights?' in Alexis Alvarez-Nakagawa und Costas Douzinas (eds), *Non-human Rights: Critical Perspectives* (Edward Elgar Publishing 2024); Anna Grear, 'Law's Entities: Complexity, Plasticity and Justice' (2013) 4 Jurisprudence 76; Tănăsescu (n 4) 429–453.

[11] Eden Kinkaid, '"Rights of Nature" in Translation: Assemblage Geographies, Boundary Objects, and Translocal Social Movements' (2019) 44 Transactions of the Institute of British Geographers 555; Rutgerd Boelens and others, 'Riverhood: Political Ecologies of Socionature Commoning and Translocal Struggles for Water Justice' (2023) 50 The Journal of Peasant Studies 1125.

[12] Stone (n 1).

[13] Andreas Fischer-Lescano, 'Nature as a Legal Person: Proxy Constellations in Law', (2020) 32 Law and Literature 237.

[14] Erin Fitz-Henry, 'Distribution without Representation? Beyond the Rights of Nature in the Southern Ecuadorian Highlands' (2021) 12 JHRE 5; Mihnea Tănăsescu (n 3).

1. Natural Tensions and Conversions

How to relate the disciplines of law and anthropology is not a new question, in fact we can look back at cross-disciplinary work for at least a century.[15] The way the disciplines and their possible relationship have been understood has taken different forms and has been met with different degrees of enthusiasm and skepticism.

For many, law and anthropology has been a particularly challenging combination. In his 1983 essay 'Local Knowledge: Fact and Law in Comparative Perspective', Clifford Geertz famously argued that law and anthropology are divided by 'two distinctive manners of imagining the real'.[16] The notion of a fundamental epistemic divide or even incommensurability has not rarely been present in debates about interdisciplinarity.

Law is often understood as universal and normative and perceives anthropology as descriptive and relativistic. In this widespread view, law is trying to generalize, to produce abstract norms, to govern, sometimes developing under-complex notions of concrete circumstances, whereas anthropology is deconstructing, zooming in, and only sometimes prepared to reconstruct.

While including 'external' knowledge from disciplines such as economics or psychology has been widely successful in legal academia, anthropological knowledge often seems to be met with greater resistance. As Jeffrey S. Kahn summarizes:

> '[T]here is a deep anxiety that ethnographic research is not sufficiently rigorous. It is seen as too subjective, too humanistic, too soft, and, in its more recent incarnations, too committed to a postmodern abdication of realist truth-finding and predictive modeling. Even when it is sufficiently realist, it is still too "swampy", in the sense of being overburdened by a specificity that makes predictive modeling challenging'.[17]

On the other hand, anthropology has been perceived as being ideally positioned to offer law what it lacks. Annelise Riles developed this argument. She shows that much thought on interdisciplinarity builds on the image of disciplines as incomplete and interdisciplinarity as a move towards greater completion. Law is incomplete in a specific way: a gap supposedly separates law from the real world. Anthropology is then presented as the science that is concerned with exactly that: the real world. Anthropology then might be seen as a 'strange gift from afar'.[18]

All this certainly is to be questioned on a number of grounds. To start with, anthropology has long also turned to the 'inside' of law. Furthermore, and more

---

[15] Annelise Riles, 'Representing In-Between: Law, Anthropology, and the Rhetoric of Interdisciplinary' (1994) U Ill L Rev 597.

[16] Clifford Geertz, 'Local Knowledge: Fact and Law in Comparative Perspective' in Clifford Geertz, *Local Knowledge: Further Essays in Interpretive Anthropology* (Basic Books 1983).

[17] Jeffrey S Khan, 'Anthropology, Law, and the Problem of Incommensurability' (2022) 73 Ala L Rev 783, 793.

[18] Riles (n 15) 606.

generally, both the skeptical as well as supportive accounts rely on a number of assumptions which deserve scrutiny. They rely on an essentialization of the disciplines, which does not account for their internal diversity and often builds on outdated and prejudiced understandings. Instead, a range of authors have argued that the notion of a divide is exaggerated.[19]

2. Balancing and Questioning the Disciplines

A number of different schools of thought have taken on the interdisciplinary challenge in different ways: legal anthropology, the law-and-society movement, or critical legal studies, to name just a few. Yet what the interdisciplinary endeavor entails exactly has found many different answers, expressed with enthusiasm or frustration. Interdisciplinarity can work and enrich knowledge production in a number of ways.

Interdisciplinary enterprises might be categorized according to the degree to which one discipline is dominant.[20] An important part of legal anthropology is not necessarily interdisciplinary research, as it focuses on law from a purely social-scientific perspective. Other scholars look at anthropology from the perspective of the law.[21] Often starting from the premise of legal pluralism, they examine how anthropology can help to provide empirical evidence of law in action and of various legal systems that interact with each other in various ways. Between and beyond these polar extremes there are a myriad of ways to connect the disciplines and create new spaces. Different degrees and ways of incorporating the other discipline can have robust justifications. Yet sometimes the former are determined not (only) by explicit rationales but by a more hidden (power) dynamic which might lead to frustration among involved researchers.

Often the promises of interdisciplinarity are considered substantial, but so are the challenges. These challenges can occur on different levels, not only on that of the (already disputed) epistemic differences. Those observing and practicing interdisciplinary research drawing on law and anthropology have noted hurdles, such as the need to acquire knowledge of a vast body of literature, which might be at odds with the demands of a fast-moving academia. The latter also complicates the role of a cornerstone of anthropological research, namely the extended fieldwork which might not be viable or encouraged for a scholar in a law faculty. A similar point can be made about concerns regarding publishing and the relevant audiences.

---

[19] Riles (n 15); Jonas Bens, 'Anthropology and the Law: Historicising the Epistemological Divide' (2016) 12 Int JL Con 235; Matthew S Erie, 'The Normative Anthropologist' (2022) 73 Ala L Rev 803.

[20] Mathias M Siems, 'The Taxonomy of Interdisciplinary Legal Research: Finding the Way Out of the Desert' (2009) 7 Journal of Commonwealth Law and Legal Education 5.

[21] Austin Sarat, 'Cultural Expertise and Socio-Legal Studies:' (Special Issue, 778 Studies in Law, Politics, and Society 2019).

Resistance might also arise specifically when the dominant vision of the academic division of labor is questioned. For example, from the perspective of law, 'anthropologists are called forth to perform the role of surveyors of the "savage slot", a role to which they are no longer committed but which they cannot seem to shake.'[22] Anthropological research into, for instance, formal law might be harder to defend in certain contexts of disciplinary prejudice. At the same time, interdisciplinary encounters might create an impetus to question disciplinary images, also within one's own discipline. As just one example, anthropologists might think about the normative aspect of their work.[23]

So, it might be true: 'Getting these two disciplines to speak to each other may invite hand-wringing; efforts often result in a kind of mimed dance, like Sisyphus on roller skates.'[24] We do not doubt, though, that the efforts can be absolutely worth it.

3. Environmental Law and Environmental Anthropology

Rights of nature create a particularly broad space for inter – (sub) disciplinary debates, also articulating environmental law and environmental anthropology in conjunction.

Environmental law has developed rapidly with the rise of problems ranging from pesticides to long-range transboundary air pollution, ozone depletion, climate change, loss of biodiversity, etc. Cornerstones in the early debate have been provided by reports of scientists and novelists, such as Rachel Carson with her book 'Silent Spring',[25] and international conferences starting with the famous Stockholm conference in 1972.[26] Environmental law is entangled with environmental justice and hence with human rights debates that have been mushrooming in the 1970s and 1980s in the United States, focusing on unequal exposure to pollution and health risks that amounted to racial discrimination. Since then, the debate has moved on to include various facets of environmental and climate justice and litigation, sometimes combining the discussion on the commons with individual human rights, such as health, life, property, food, water, or a healthy environment.[27]

In parallel, anthropological studies on human rights have the potential to cross-fertilize the field of environmental anthropology and its impact on environmental human rights.[28] The field of environmental anthropology has grown immensely

---

[22] Kahn (n 17) 793.

[23] Erie (n 19).

[24] Erie (n 19) 803, 804.

[25] Rachel Carson, *Silent Spring* (Houghton Mifflin Harcourt 1962).

[26] Stockholm Declaration of the United Nations Conference on the Human Environment (16 June 1972) 11 ILM 1416.

[27] Hanschel and Steyn (n 1).

[28] Mark Goodale and Sally Engle Merry, *The Practice of Human Rights* (CUP 2007).

in the past decades.[29] Currently, a wide range of approaches and foci are present in the subfield. The ontological turn, which is highly influential in anthropology, has also significantly shaped the field of environmental anthropology. Within this approach, differences are understood not as differences in perceiving and understanding the world, but as an actual multiplicity of worlds. Such an outlook has, for example, led to a scrutinization of the nature-culture divide and an emphasis on different modes of relationality with non-human entities. Such insights are also highly influential in the rights of nature literature. At the same time, researchers in environmental anthropology also make significant contributions to the broader field of political-ecology, focusing on the political and economic power structures that underlie climate change and environmental destruction.[30]

IV. Environmental Rights in Cultural Context

Against this background, the perspective of the research project on Environmental Rights in Cultural Context (ERCC) is to embark on a constant learning process on how rights to a sound or healthy environment in their respective circumstances can be analyzed by using legal analysis on the one hand and ethnographic fieldwork on the other.[31] The idea to combine both stems from the following insight: If substantive environmental rights are inserted into the constitution, they are deemed to represent fundamental values.[32] Yet, to what extent such values actually correspond to local experiences of injustice remains to be examined. In the case of environmental human rights, such experiences would be articulated by the people subjected to severe environmental degradation. Conversely, rights of nature as a different way to recalibrate human – nature relationships should entail responses to violations of fundamental interests of nature or of parts of it. Since most species and other elements of nature cannot speak for themselves (or human beings find it hard to understand their language), we are instead looking mostly at human cosmovisions and life views that embrace nature and consider human existence as part of it. In order to learn more about such processes of interaction and interpretation, we need to move from the abstract legal provisions to concrete case studies and back.[33]

---

[29] For an overview, see eg Michael R Dove and Carol Carpenter, *Environmental Anthropology: A Historical Reader* (Blackwell 2007).

[30] For an overview, see Bengt G Karlsson, 'Political Ecology: Anthropological Perspectives' in Neil Smelser and James Wight (eds), *International Encyclopedia of the Social & Behavioral Sciences* (Elsevier 2015), <https://doi.org/10.1016/B978-0-08-097086-8.12215-9>.

[31] Dirk Hanschel, 'Environmental Rights in Cultural Context' (Max Planck Institute for Social Anthropology 2019) <www.eth.mpg.de/ercc>.

[32] Dirk Hanschel and others, 'Environmental Rights Between Constitutional Law and Local Context: Reflections on a Moving Target' (2022) 23 GLJ 1012.

[33] Hanschel (n 31).

We think that combining law and anthropology is a particularly rewarding and innovative approach to understand these rights in their respective local (cultural) context. This approach can build on the rich literature of environmental anthropology and plural cosmovisions and ontologies whilst taking law and legal doctrine seriously. In the increasingly vast research on rights of nature, this endeavor is still rather novel and particularly promising, since it is based not only on deep ethnographic insight but also on more detached legal reasoning. Concepts travel from communities to the State and back. The hands-on perspective allows us to establish the extent to which rights of nature have traction within communities and whether they correspond to individual or collective experiences of injustice.[34] What their added value is cannot be established merely by philosophy or legal theory and doctrine but must instead be ascertained in concrete circumstances. At the same time, through its focus on environmental rights, the project builds on the tradition of environmental justice research more generally.[35]

V. Anthropological Insights into a Legal Concept – A Critical Assessment of Rights of Nature

Employing the methods illustrated above, we will now take a critical look at rights of nature on the basis of relevant fieldwork and anthropological arguments. Since relevant anthropological research is extensive, any attempt to be exhaustive in this article would be impossible. In what follows, a number of ethnographic vignettes and insights are presented, based on original fieldwork and other relevant studies. We use them to illustrate different ways in which law and anthropology can be combined in order to help generate a deeper and more nuanced understanding of rights of nature. The presented examples are not intended to make a conclusive argument regarding the questions raised, as this is beyond the scope of this article. They are, more than anything, intended to show the way in which ethnographic research can contribute to rights of nature debates, demystifying their (possible) foundations and their diverse coming into being and effects.

1. Applied Knowledge: Anthropological Expertise in Law

In July 2020, the Fundación Pro Defensa de la Naturaleza y sus Derechos (Foundation for the Defense of Nature and Her Rights) and the renowned Universidad Andina in Ecuador started a fascinating program called 'Intercultural Community Expertise in Human Rights and Rights of Nature'. The stated aim of the course is

[34] Tănăsescu (n 4) 429, 439.

[35] Dirk Hanschel and others (n 32).

'to train leaders in community expertise on the rights of nature to intervene in processes of defending territories with these new rights that are in the process of being recognized. […] [The course's] objective is to confront the advances of agribusiness, and oil and mining extraction'.[36]

Rights of nature cases and debates seem to have been accompanied by an innovative openness of the law (or of concrete judges) to extra-legal knowledge. Interpretation and application of rights of nature does not necessarily imply the involvement of multiple forms of knowledge as produced by ethnographic work, and judges often do not seem to feel the need to engage with such a complex endeavor to a noteworthy degree. However, the unique challenges posed by the concept of rights of nature definitely makes it worthwhile to do so. Prominent cases, such as the *Los Cedros* case in Ecuador or the *Atrato* case in Colombia, have an impressive record of including numerous different experts and decisive on-site visits of members of the respective Constitutional Courts.

Anthropology is thus one of the disciplines which assume a crucial role in court. This leads to one specific and much discussed manner in which law and anthropology collaborate: as expert knowledge in concrete legal settings.

Anthropological expertise within legal settings has been addressed and interrogated in different contexts.[37] It has been a central concern in debates regarding a remaking of law that is responsive to plural realities and 'the reflection, regulation and institutionalization of a new way of considering and weighing the cultural difference' in the legal realm.[38] Emic perspectives adopted by anthropologists have produced a critical body of literature, discussing the possibilities but also the limits and pitfalls of presenting others' culture in court. Authors have highlighted tensions between 'objectivist rhetoric evoked in court and seemingly more relativist and subjectivist stances formulated within the confines of the academy',[39] the benefits and drawbacks of strategic essentialism, and the dangers of reproducing orientalist stereotypes.

In the light of far-reaching proposals for democratizing and pluralizing law, the inclusion of anthropological expertise that tends to be on (behalf of) non-dominant groups or individuals has been challenged as often re-placing direct dialogue

---

[36] Fundación Pro Defensa de la Naturaleza y sus Derechos, 'Peritajes comunitarios interculturales en Derechos Humanos y de la Naturaleza' (*Naturaleza con Derechos* 2020) <https://naturalezaconderechos.org/2020/06/24/peritajes-comunitarios-interculturales-en-derechos-humanos-y-de-la-naturaleza/>.

[37] Marie-Claire Foblets, 'Prefatory Comments: Anthropological Expertise and Legal Practice: About False Dichotomies, the Difficulties of Handling Objectivity and Unique Opportunities for the Future of a Discipline' (2016) 12 Int JL Con 231; Armando Guevara Gil, Aaron Verona and Roxana Vergara (eds), *El Peritaje Antropológico – Entre la reflexión y la práctica* (CICAJ 2015).

[38] Armando Guevara Gil and others, 'Introducción' in Armando Guevara Gil, Aaron Verona and Roxana Vergara (eds), *El Peritaje Antropológico – Entre la reflexión y la práctica* (CICAJ 2015) 13.

[39] Olaf Zenker, 'Anthropology on Trial: Exploring the Laws of Anthropological Expertise' (2016) 12 Int JL Con 293, 295.

with, for instance, authorities in affected communities. According to Clavero this reproduces a colonial 'supremacist prejudice'.[40] Even a benevolent approach might not be able to escape what Foucault described as creating 'a whole set of knowledges that have been disqualified as inadequate to their task or insufficiently elaborated: naive knowledges, located low down on the hierarchy, beneath the required level of cognition or scientificity.' We are thus in the midst of debates regarding representation, agency, subjectivity, and subjectivation. Post-colonial studies have taught us that '[b]eing spoken for is probably the key criterion that defines subalternity'.[41]

At the same time, strong cases have been made for the way anthropological testimony can enrich the judiciary. Some of the pioneering contexts in the definition and application of rights of nature have a history of this innovative openness. For example, the Columbian Constitutional Court can be considered a pioneer in this area, or in the Bolivian context the inclusion of anthropological expertise has often been identified as an important element in shifting the interpretative practice away from strict legalism.[42]

A broader debate around applied knowledge within anthropology is that regarding activist research (or its other denomination, militant or engaged research, which are all profoundly polyphonous). Activist research, in which one takes active part in a struggle and carries out research in some form of dialogue with and commitment to that struggle, is not a new idea.

In the field of anthropology regarding legal and rights struggles, activist research has come to constitute an important methodological concern, marked by concrete challenges and contradictions researchers have encountered during their engagement with movements or collectivities. Scholars have felt tensions between their deconstructive cultural critique and the need for different kinds of knowledge in legal struggles. Put very bluntly: Critique, deconstruction, etc. might not seem adequate answers in the face of urgencies experienced by the people the anthropologist aligns herself with. Speed argues that 'activist research allows us to merge cultural critique with political action to produce knowledge that is empirically grounded, theoretically valuable, and ethically viable'.[43] In a similar vein, Hale has advocated for – sometimes – accepting the need for an 'analytical stop', ie not pushing for complexity when this undermines the concrete goals of struggling people.

---

[40] Bartolomé Clavero, 'Tribunal Constitucional en Estado Plurinacional: El Reto Constituyente de Bolivia' (2012) 94 Revista Española de Derecho Constitucional 29, 48.

[41] Viatcheslav Morozov and Elena Pavlova, 'Indigeneity and Subaltern Subjectivity in Decolonial Discourses: A Comparative Study of Bolivia and Russia' (2018) 21 Journal of International Relations and Development 689, 695.

[42] See eg Marina Correa de Almeida, *Lo nuevo en el constitucionalismo Latinoamericano del siglo XXI: aportes de un analisis critico de la ideología jurídica en Bolivia* (UNAM 2017).

[43] Shannon Speed, 'At the Crossroads of Human Rights and Anthropology: Toward a Critically Engaged Activist Research' (2008) 108 American Anthropologist 66.

The rights of nature again brings those political, ethical, methodological, and epistemological questions to the forefront. Insights from activist anthropology can help to conceptualize the tension between wanting to achieve a political goal and the importance of continued cultural critique and deconstruction. Being accountable to different groups and movements (eg the academic community, an indigenous community, and the climate movement) can complicate but also enrich research and help balance deconstruction and necessary 'analytical stops'.[44]

2. How to Protect Nature When There Is No Nature? The Challenges of Ontological Diversity

How 'nature' is understood in rights of nature accounts is highly consequential. It might be one of the more 'obvious' tasks of anthropology to problematize universal and unitary notions of nature. A broad body of literature has shown that nature is a culturally specific concept.[45] What might this mean concretely for rights of nature? In this section snippets into possible anthropological contributions regarding differing human-environment relationships and their relation to rights of nature are presented.

In some of the communities where fieldwork was conducted by one of the authors, the mountains, the rivers, and the land are animated entities which play a crucial role in their daily life and in political organization.[46] Activist and amawt'a[47] Beatriz Bautista, the founder of a small activist organization supporting communities in their (legal) struggles, explains that 'Pachamama is to be understood as a whole, a great family. We humans have a relationship of mutual nurturing with Pachamama.' This might be an example that resonates perfectly with the idea of rights of nature in its most ambitious formulation: unsettling the modernist ontological division between humans and nature and understanding nature as more than a resource. Examples of (indigenous) ontologies and philosophies which question these key tenets of modernity are abundant: the Kwakwaka'wakw people's worldview, where all living things, including plants and landforms, possess

---

[44] See CR Hale, 'Activist Research v. Cultural Critique: Indigenous Land Rights and the Contradictions of Politically Engaged Anthropology' (2006) 21 Cultural Anthropology 96.

[45] Tănăsescu (n 3); Tănăsescu (n 4); Philippe Descola, *Jenseits von Natur und Kultur* (Suhrkamp 2013); Eduardo De Castro, 'Images of Nature and Society in Amazonian Ethnology' (1996) 25 Annual Review of Anthropology 179; Eduardo De Castro and Débora Danowski, *Há mundo por vir? Ensaio sobre os medos e os fins* (Cultura e Barbárie 2014); Eduardo De Castro, *Kannibalische Metaphysiken* (Merve 2019); Peter Skafish, Pierre Charbonnier and Gildas Salmon (eds), *Comparative Metaphysics: Ontology After Anthropology* (Rowman and Littlefield 2016); Maria Puig de la Bellacasa, *Matters of Care: Speculative Ethics in More than Human Worlds* (University of Minnesota Press 2017).

[46] See Anette Mehlhorn, *With, Against and Beyond the Law: Radical Politics, Legal Activism and New Constitutionalims in Plurinational Bolivia* (Martin-Luther-Universität Halle-Wittenberg 2022).

[47] Spiritual guide.

identities and spiritual lives;[48] the reciprocal relationship between Maya Q'eqchi' and the land they live on;[49] the mountains which play a crucial role in Andean communities. Are these all examples of philosophies and ontologies which translate seamlessly into rights of nature? It should come as little surprise that with attention to specific contexts, soon a complex and often contradictory image emerges. Beyond the obvious fact that indigenous people are diverse – 'indigenous' being a complex and dynamic signifier that denotes many different modes of identification and ascription and that is heavily debated among the very people it supposedly describes[50] – ethnographic examples show that relational worlds including humans and other-than-humans are difficult to translate into a rights of nature framework or can even be entirely incompatible with the latter.

Ethnographic studies have shown that in many cases indigenous communities do not have a concept or do not attach importance to 'nature' as an abstract and universal concept; instead, they focus on the concrete territory and the concrete other-than-human-beings they are connected to.[51] The fact that the relationship between a people and its (other-than-human) surroundings might have no space for any concept as 'nature' can have surprising effects, going against the core of rights of nature. A short vignette from personal fieldwork of one of the authors elaborates this point. During a meeting with representatives of an Aymara community in the Bolivian highlands and activists (for indigenous rights), the topic of the role of mountains in the community came up. A lively depiction offered by the community members introduced everyone present to a reality where the mountains are animate, having a specific role in the political organization and everyday life of the community. The activist enthusiastically chimed in, arguing that this was exactly why there should never be mining in the community. The local authorities paused. No, they answered, they would welcome mining, however under the strict rules which they established.[52] The activist tried to argue against this stance, recurring to drastic images. Trying to capture their ontology, he said that for indigenous people it made no difference if the mountain or if one of the present persons were to be tortured. However, he could not convince the members of the local authorities.

---

[48] Douglas Deur, Kim Recalma-Clutesi and Adam Dick, 'When God Put Daylight on Earth We Had One Voice': Kwakwaka'wakw Perspectives on Sustainability and the Rights of Nature' in Cameron La Follette and Chris Maser (eds), *Sustainability and the Rights of Nature in Practise* (CRC Press 2019).

[49] Lieselotte Viaene, 'Indigenous Water Ontologies, Hydro-Development and the Human/More-Than-Human Right to Water: A Call for Critical Engagement with Plurilegal Water Realities' (2021) 13 Water 1, 10.

[50] See eg Andrew Canessa, *Intimate Indigeneities: Race, Sex and Histories in the Small Spaces of Andean Life* (Duke UP 2012).

[51] See, among many, Philippe Descola, *Beyond Nature and Culture* (UCP 2013); De Castro (n 45).

[52] As emphasized further below, it is crucial to remember that this should not minimize the importance of struggles against extractivism led by indigenous peoples.

Other ethnographic accounts further explore this point. Canessa recounts how the people in an Aymara speaking community in which he lived found it highly bizarre that the anthropologist would go for walks for pleasure, enjoying the scenery and fresh air, ie doing what many people would understand as 'valuing nature'. Similarly, he describes how there was no understanding for his pained facial expressions when people frequently threw trash into rivers with no concern for 'the environment'. He explains that while people in the community 'are intimately connected to the mountains and streams and the land that sustains them[,] they are not, however, in the least bit close to "nature" since, for them, "nature" and the "natural" is not beyond culture.'[53] In that sense the reverence many indigenous peoples have for mountains and streams and the earth might be an understanding opposing that of an oil executive exploiting natural resources. It is, however, also opposed to the understanding of nature of the ecologist fighting that exploitation: it is not a conception of nature as something which is beyond human intervention and cultural production.[54]

'Which part of nature do we protect?' asks Beatriz Bautista when debating rights of nature. 'Because we, indigenous peoples, are also part of nature, but we are being sacrificed.' Similar to what has been pointed out also for other indigenous philosophies, this is a relational approach, which should not be equated with ecocentrism as it is often done in the context of rights of nature. Nature is not existing independently or prior to a concrete relationship established with other-than-human entities, nor is there an inherent value in 'nature' independent of these relationships and the use humans make of it.[55] The quote does not only show that holistic or relational ontologies do not necessarily translate into the kind of preservation and environmental protection many (often non-indigenous) proponents of rights of nature envision. It also highlights a crucial point which will be discussed at greater length below: Rights of nature stand in an ambivalent relationship to indigenous struggles for autonomy and can go against collective rights.

What has been said so far is certainly not to question the relevance of many ethnographic examples where the animated character of (other-than-human) entities and their role in communities translates into demands for environmental protection. For example, Viaene describes how according to Maya Q'eqchi' research, participants found that a planned 'hydroelectric project would deeply transgress and violate the sacred value and dignity of the river and the other interconnected elements, such as the lands, side rivers, and animals living in the rivers'. She quotes one woman: 'The river will feel pain and suffer.'[56] Indigenous peoples have often fought against large-scale projects leading to environmental degradation at the risk of their very lives. Among the communities our paper looks at, there are examples of long-standing struggles against mining projects, (also) in order to end the

---

[53] Canessa (n 50) 161–162.

[54] Canessa (n 50) 161–162.

[55] Tănăsescu (n 4).

[56] Viaene (n 49) 14.

massive pollution of the communities' rivers and soil (one such community neighbored on the above-depicted community whose authorities' views on (hypothetical) mining surprised the activist so much). Bautista stresses that indigenous peoples have practiced what rights of nature wants to achieve through 'respect and care expressed in rituals, ceremonies, in this idea of nurture and being nurtured.' What to do with such diverging realities and how (or if) the rights of nature framework can accommodate them is the type of question anthropology can raise.

When agendas seem to converge and communities' interests align with notions of environmental protection as envisioned by many rights of nature proponents, crucial questions remain about the modes and (im)possibilities of translation. As argued above, one great appeal of rights of nature lies in its claim to inclusion of different ways of being with nature. However, can the latter be translated into law? A telling example comes from de la Cadena, who describes how when her indigenous interlocutor rallied to prevent mining on Mount Ausangate, he had to reframe his demands as an 'ecological' defense, while in fact it was the defense of a distinct relationship with the latter. 'This is a translation which moves earth-being to a realm where they are not (in inherent relations) with Runakuna [the people in question], [it] ultimately cancels the reality-making capacity of the practices that such connection enables.'[57] The rights of nature literature often (implicitly or explicitly) references the ontological turn in anthropology, ie the notion that there are not different views on reality but, in fact, different realities. However – can law address such radical diversity? One key insight of the ontological turn is what has been discussed under the concept of equivocations, which does not mean a simple failure to understand. Rather it is 'a failure to understand that understandings are necessarily not the same.'[58] Can law accommodate such insight? And what would be the consequences for rights of nature if it did?

3. *Indigenizing the Law and the Law's Legal Anthropology*

A closer look at what is included as indigenous cosmovision in rights of nature, at the manner in which this is done, and at its effects, further points to important challenges and limitations. On a different level, delving into this topic highlights that anthropological contributions should not be misunderstood as being limited to presenting knowledge about what is outside of law, so as to include that outside more properly. Rather, these contributions also take the official law along with its workings, actors, and institutions as their subject. In that way, anthropological accounts can, for example, criticize what has been termed the law's legal anthropology.[59] The problem addressed in this section finally also points to anthropology's possible roles in normative theory and critical legal projects.

---

[57] Marisol de la Cadena, *Earth Beings: Ecologies of Practice across Andean Worlds* (Duke UP 2015) 99.

[58] De la Cadena (n 57) 110.

[59] Ronald Niezen, 'The Law's Legal Anthropology' in Mark Goodale (ed), *Human Rights at the Crossroads* (OUP 2014).

As mentioned throughout this chapter, references to indigenous knowledge and ontologies are fundamental in most rights of nature arguments (even though in certainly differing ways). Rights of nature are thus closely related to vibrant fields of debate regarding the decolonization and pluralization of law. The way indigenous knowledge and practices are included in rights of nature has, however, been criticized and contested. Anthropology can offer a long-standing body of critical arguments which point to the complex modes and effects of juridifying indigenous (and other) identities, claims, or knowledge.

Anthropological (and other) studies have shown that law is particularly prone to essentializing and exoticizing indigenous peoples.[60] A crucial insight in this varied literature is not 'merely' that legal identities distort or do not fully match complex 'real' ones. Rather it has been pointed out that this reformulation of identities, intentionally or not, relates in specific ways to bigger social and political dynamics, practices, and institutions.[61]

Rights of nature, too, have been analyzed as not being immune from evoking a purist, static, and non-contextualized notion of indigenous peoples. Not infrequently we are 'supposed to accept, as a given, the existence of community cultural forms in harmony with nature, bearers of other knowledges […], which surprisingly have been spared the contamination of the westernising hegemonic discourses'; in other words, the discursive construction of a 'Rousseauian ideal native'.[62] It is a discourse of duality which can see the 'indigenous' only as the incommensurable 'other' to the Western world. This either invisibilizes actual indigenous people and their demands, or it forces them to appear before the law as 'the impossible object of an authentic self-identity',[63] as the 'hyper-real Indian'.[64] In this vein, Petel argues that understanding indigenous communities as nature's evident guardians 'could put pressure on these communities to essentialize their relationship with nature, reinforcing the "ecological noble savage" narrative'.[65]

---

[60] Glen S Coulthard, 'Subjects of Empire: Indigenous Peoples and the "Politics of Recognition" in Canada' (2007) 6 Contemporary Political Theory 437; Frank Hirtz, 'It Takes Modern Means to Be Traditional: On Recognizing Indigenous Cultural Communities in the Philippines' (2003) 34 Development and Change 887; Charles Hale, 'Rethinking Indigenous Politics in the Era of the "Indio Permitido"' (2004) 38 NACLA Report on the Americas 16; Kathleen Birrell, *Indigeneity: Before and Beyond the Law* (Routledge 2016).

[61] We are loosely echoing what Povinelli argues regarding liberal forms of recognition in Australia: Elizabeth A Povinelli, *The Cunning of Recognition: Indigenous Alterities and the Making of Australian Multiculturalism* (Duke UP 2002).

[62] Victor Bretón Solo de Zaldívar, 'Three Divergent Understandings of Buen Vivir in the Ecuador of the Citizens' Revolution' (2017) 12 LACES 188, 195.

[63] Povinelli (n 61) 6.

[64] Alcida Rita Ramos, 'The Hyperreal Indian' (1994) 14 Critique of Anthropology 153.

[65] Matthias Petel, 'The Illusion of Harmony: Power, Politics, and Distributive Implications of Rights of Nature' (2024) 13 TEL 12.

Niezen understands such tendencies in the context of what he provocatively calls anthropology of law's rival, alter ego, or doppelganger: the law's legal anthropology. It is 'a parallel conceptual geography of humanity, with epistemological foundations in legal activism. Legal anthropology's doppelganger achieves the global currency of its ideas – a kind of vernacularization writ large – by plugging into transnational networks of activist organizations without having to initially struggle for recognition and legitimacy.'[66]

Legal anthropology has a complex and multifarious relationship with this law's legal anthropology. As described above, activist anthropologists can have good reasons to be a part of the implied strategic essentialism in their quest to support a cause or a people. Legal anthropology can, however, also analyze this law's legal anthropology as part of the vibrant scholarship that has established the anthropological study of international organizations, formal law, and the State.[67]

Here rights of nature remain to offer a number of important avenues for further studies. The way they are created and disseminated in transnational networks, courts, or as part of State legislation remains understudied, and ethnographic studies promise to provide important insights. These insights can provide analytical clarity and be part of a project of critique as well as reconstruction.[68]

4. What Is the (Added) Value of the Concept of Rights of Nature?

What can a law and anthropology approach say about the added value of rights of nature? For many proponents, the added value of rights of nature is immense. This point is argued for in a number of ways, sometimes internal to legal theory, sometimes deploying rather abstract concepts such as anthropocentrism, sometimes going into the details of a specific legal system.[69] Skeptical voices have cast doubt on the necessity or the significant added value of incorporating rights of nature, arguing, for example, that existing environmental law is not gaining anything.[70]

In the Los Cedros case,[71] it is striking to see how the Ecuadorian Constitutional Court has tried to carve out the substance of rights of nature in the Ecuadorian Constitution from the lawyers' perspective. It has included a substantial array

[66] Niezen (n 59) 185.

[67] See eg Jonas Bens and Larissa Vetters, 'Ethnographic Legal Studies: Reconnecting Anthropological and Sociological Traditions' (2018) 50 J Legal Plur 239; Ronald Niezen and Maria Sapignoli (eds), *Palaces of Hope: The Anthropology of Global Organizations* (CUP 2017).

[68] Annelise Riles, 'Introduction to the Symposium on The Anthropology of International Law' (2021) 115 AJIL 268.

[69] For a good overview, see eg Frank Adloff und Tanja Busse, 'Welche Rechte braucht die Natur? Eine Einleitung und ein Vorschlag' in Frank Adloff und Tanja Busse (eds), *Welche Rechte braucht die Natur? Wege aus dem Artensterben* (Campus 2021).

[70] Julien Bétaille, 'Rights of Nature: Why It Might not Save the Entire World?' (2019) 16 JEEPL 35.

[71] Corte Constitucional de Ecuador 10 November 2021, CASO no 1149-19-JP/20.

of scientific data in order to argue under what circumstances the ecosystem of the cloud forest would be in danger. Whilst this is truly impressive and a landmark case for rights of nature, the reasoning might have further benefitted from a deeper ethnographic analysis as to the consequences of the decision on affected (non-indigenous) communities, which appear to be quite divided over the matter. At the same time, this case shows the limits of the law and of the exercise of judicial decision-making, which is characterized by severe procedural and time constraints. Obviously, courts can only interpret norms that have been inscribed into the legal code (in this case the constitution). Such norms may be incomplete and affected by translation problems when indigenous notions are transposed into the State law. These are limits of the law. But they may also inspire and enrich the legal order and create a counter-weight against careless use of the environment. This case hence illustrates that concrete effects – and thus what may constitute the added value of rights of nature – can be determined only to a certain degree in theory. Ethnographic studies can show and analyze the multiple and sometimes contradictory ways in which rights of nature play out in specific legal, political, and cultural contexts.

Petel argues that rights of nature are 'compatible with various ideological outlooks and political options. As a result, those initiatives may translate into extremely diverse institutional implementations with contrasted outcomes in terms of power distribution.'[72] This is well illustrated by Tănăsescu's analysis of the case of Te Urewera, land of the Tūhoe people in Aoteora (New Zealand). Here, the details on the formulation of the area as a legal entity provided a provisional solution to the question of who has ultimate authority to govern the land. While this question could not be solved in negotiations between Tūhoe representatives and the Crown, constructing Te Urewera as a legal entity has been a way out of this deadlock. This is not to say that there will be no positive effects in regard to environmental protection or that this was not an important issue in the process. Some have however argued that this question was secondary for the Tūhoe' representatives involved in the process.[73]

Ethnographic studies of universals (ie ideas, values, or logics that are presented as if they are globally applicable) have insightfully shown that they react differently in certain socio-historical contexts and create frictions that lead to new situations.[74] In the context of rights of nature, Eden Kinkaid argues that it is crucial 'attending to the manner in which "rights of nature" becomes articulated through specific practices, cultural meanings, material networks, human–environment relations, legal regimes, governance structures, and political projects in cul-

[72] Petel (n 65).

[73] Tănăsescu (n 3) ch 4; Brad Coombes, 'Nature's Rights as Indigenous Rights? Mis/recognition Through Personhood for Te Urewera' (*Espace populations sociétés*, 2020/1–2) <http://journals.openedition.org/eps/9857>.

[74] Anna Lowenhaupt Tsing, *Friction: An Ethnography of Global Connection* (PUP 2005).

turally and geographically specific ways with often counterintuitive outcomes'.[75] Rutgerd Boelens et al. also emphasize the importance of paying attention to concretely embedded practices and policies, because personification processes (in this case, of rivers) 'are mediated by unequal powers and struggles, for instance over the installation of guardianship and regarding how they are subject to legal, political, economic and moral-normalizing interests in everyday battlefields.'[76]

In this same way we can understand that – and how – rights of nature can even become part of political projects which explicitly go against environmental protection. The reconciliation of rights of nature and extractivism, as seen for some time in Ecuador, has arguably been advanced, to give one example of this. Rights of nature have been shown to play an integral part in clearing the way for large-scale extractivist projects, as they have been used to sue artisanal miners, only to be followed by a subsequent granting of mining concessions to major national and multinational actors.[77]

Rights of nature – as they are necessarily enacted in pre-existing circuits of meaning-making and governance – can become politically productive beyond their mandate to protect nature or distinct natural entities.[78] For example, analyzing the establishment of legal personhood for the Ganga (Ganges) and Yamuna rivers, Kinkaid shows how in this case rights of nature could become part of a repressive agenda against minorities in the context of Hindu nationalism.[79]

VI. Conclusions

In sum, our analysis from the perspective of law and anthropology has shown that rights of nature are, like environmental rights, oscillating and a moving target, and they mean very different things in different contexts. What is more, the reasons for conceptualizing and invoking them are manifold. Against this backdrop, some of the rights of nature debates currently conducted in law or in philosophy run the risk of essentializing, simplifying, or romanticizing real-life situations, notions of environmental justice in communities, and the workings and effects of legal concepts. The purpose of this endeavor will usually be very benevolent, but it is often dependent on reducing complexity in a way that allows one to 'handle' the concept and to insert it in exciting academic or governance debates

[75] Eden Kinkaid, '"Rights of Nature" in Translation: Assemblage Geographies, Boundary Objects, and Translocal Social Movements' (2019) 44 Trans Inst Br Geogr. 555.

[76] Boelens and others (n 11) 1141.

[77] Tănăsescu (n 3) ch 5. This is not to argue that there are not different and promising examples from the Ecuadorian context.

[78] Kinkaid (n 75).

[79] Kinkaid (n 75); Arpitha Kodiveri, 'If Nature Has Rights, Who Legitimately Defends Them?' (2019) <www.openglobalrights.org/if-nature-has-rights-who-legitimately-defends-them/>.

that are too often self-referential and sometimes even characterized by a euro-centric quality. Rights of nature might not be what those who understand it as a 'Nirvana concept' hope for: 'a highly attractive and promising new legal tool that embodies the ideal image that it can save humanity from the Anthropocene catastrophe.'[80] Anthropological studies have yielded a much more complex and often more skeptical image. This does not, however, imply a rejection of the idea and its usefulness as such. The possibility to study with ethnographic depth details regarding the creation, application, and reception of rights of nature can help to identify the criteria for their successful use and also critically re-construct and enrich them. Certainly, however, the interdisciplinary combination of law and anthropology is not limited to one particular format, but can instead differ according to epistemological, theoretical, and practical considerations. We have aimed at giving a glimpse into the diversity and richness that can be brought to the debate about rights of nature by studies which combine law and anthropology in different ways. Instead of having made one big argument about rights of nature or about law and anthropology, we hope to have shown a diverse field that is full of potential.

[80] Viaene (n 49).

NGOs in the United States

Birth of a Movement

Mari Margil[*]

In the beginning, it was just an idea, that nature – long treated as an object of commerce or property, of value to humans but without inherent value of its own – could be something more.

It was an idea that had been applied in the past, to transform how certain peoples were treated as objects of commerce and property, and then as now, the idea was subject to ridicule and disbelief.

Fifty years ago, American law professor Christopher Stone posited the idea that nature could be considered as something more – a legal entity with legal rights. Professor Stone published his idea and arguments in a law review article in 1972, titled 'Should Trees Have Standing?'.[1]

It took over thirty additional years for the idea to move from theory into practice. In 2006, a law was approved – by a local government in the state of Pennsylvania in the United States – the first to treat nature as something other than a 'thing,' but rather as a legal entity with rights.

---

[*] This manuscript was submitted in 2022. The assessment of literature and case law, including internet materials, reflects the legal situation at that time.

[1] Christopher D Stone, 'Should Trees Have Standing? Toward Legal Rights for Natural Objects' (1972) 45 SCal LRev 450.

I. Moving From Theory into Practice, the First Rights of Nature Law

Thomas Linzey was the lawyer and activist who worked on that first rights of nature law, adopted by Tamaqua Borough, Pennsylvania, as well as dozens of other laws that were later adopted by local communities across the US. He also assisted with the drafting of Ecuador's 2008 constitution, the first in the world to enshrine the rights of nature, or *Pacha Mama*, within a national constitutional framework.

If Professor Stone is the grandfather of the rights of nature movement – laying out the vision – then Linzey is the father, taking that vision and making it a reality.

'The first rights of nature law was born from necessity,'[2] Linzey explains. 'In community after community, in country after country, environmental laws and legal systems authorize harm of the environment. And so communities, such as Tamaqua, found that looking to the law to protect them, and protect nature, was impossible.'[3]

Like others that came before, Tamaqua Borough and other communities found that the legal system that they looked to for protection was instead authorizing the very harm that they sought to stop.

Environmental laws, within the US and in place in countries around the world, are based on the idea that nature exists to serve humankind. Environmental laws, rules, and regulations are based on this premise, legalizing certain uses of nature – which knowingly cause environmental harm – in order to serve human interests.

For instance, environmental laws authorize the use of millions of gallons of fresh water at every frack well, which causes toxic contamination of the water. Mining laws authorize explosions and extraction of coal and other minerals and ores which pollute the atmosphere and destroy waterways. Energy companies are authorized to burn fossil fuels which pollute the atmosphere and accelerate global warming. Timber companies receive environmental permits to clear-cut forests, which destroys ecosystems, waterways, and habitat for species. All of these industrial activities, and the many, many others authorized by environmental laws, are legalizing the use and harm of nature.

The consequences of this are growing clearer every day.

The 2019 IPBES *Global Assessment Report on Biodiversity and Ecosystem Services*, for example, explained that global species extinction rates are accelerating due to human activity, such that '1 million species already face extinction, many within decades.'[4]

---

[2] Interview with Thomas Linzey (Spokane, Washington, 13 March 2022).

[3] Linzey (n 2).

[4] Sara Díaz and others (eds), *Summary for Policymakers of the Global Assessment Report on Biodiversity and Ecosystem Services of the Intergovernmental Science-Policy Platform on Biodiversity and Ecosystem Services* (IPBES 2019) 12, <https://doi.org/10.5281/zenodo.3553579>.

Rising temperatures from human-caused climate change are bringing more extreme weather events, causing the melting of ice and snow in the Arctic, warming ocean waters, accelerating species extinction, killing off coral reefs, and bringing many more impacts around the world. In January 2022, the European Union's Copernicus Climate Change Service announced that the past seven years were the hottest in recorded human history, 'underscoring the urgency of climate action months after the COP26 summit fell short in its commitments to tackling the crisis.'[5]

This is occurring despite the innumerable environmental laws, rules, regulations, agencies, ministries, treaties, and policies in place around the world.

For Tamaqua, and other communities, countries, and First Nations around the globe – faced with legal systems that legalize environmental harm, and under which the environment is considered to be nothing more than an 'it' – overcoming this requires reimagining how law works, and for whom. It involves a rethinking of humankind's very relationship with the natural world.

This is not the first time people have looked to government and law for solutions, or for justice, and found none. Consider enslaved peoples, women, indigenous peoples, and others who lived under legal systems in which their very humanity was ignored.

As nature is today in so much of the world, people have long been treated as property or commerce under the law, owned and controlled by others. Faced with such legal systems, movements arose to challenge and change them, including the Abolitionists, the Suffragists, and the American Indian Movement.

As Linzey explains, those movements, just like the rights of nature movement, were born of necessity. And, like those past movements, today's rights of nature movement must change the underlying structure of law itself.

As Linzey teaches in legal trainings and community workshops, enslaved peoples were considered property, without legal rights. Slave codes were put into place, not unlike environmental laws today, which established how slave owners could use, sell, and even destroy their 'property'. Women, similarly, have been treated as property. Under coverture laws, a woman's 'legal identity and rights were covered by her husband's, and her own legal identity was nonexistent.'[6]

Broad-based people's movements had to be built to transform how enslaved people, women, indigenous peoples, children, and others were treated by legal systems, constitutions, and laws. These movements were forced to overcome not only legal barriers but societal, cultural, and economic barriers as well – to have

---

[5] Téa Kvetenadze, 'Last Seven Years Were the Hottest on Record, EU Climate Data Show' *Forbes* (New York, 10 January 2022), <www.forbes.com/sites/teakvetenadze/2022/01/10/last-seven-years-were-the-hottest-on-record-eu-climate-data-show/>.

[6] Keri L Engel, 'The Law OF Coverture: Why Call a Woman BY Her Husbands Name?' <https://amazingwomeninhistory.com/law-of-coverture-why-call-a-woman-by-her-husbands-name/>.

those once considered the property of others, to at last be recognized as possessing legal rights protected by law.

Linzey co-founded the Community Environmental Legal Defense Fund in 1995, in which he pioneered the first rights of nature laws. Today, as co-founder of the Center for Democratic and Environmental Rights, he continues to advance this movement.

He recalls the earliest days of the movement, and how today, like then, the people, communities, civil society groups, and governments that take action are primarily those that are facing severe environmental threats and acknowledge that our current relationship with nature is unsustainable. 'It takes a willingness to say that the existing system isn't working, and that we need to replace it with something new,'[7] he says.

Environmental advocacy groups are well-experienced in navigating existing, conventional nature-as-property environmental laws and regulations, mobilizing to try to protect nature under these frameworks. The rights of nature begins from a different starting point with respect to how nature is treated under the law, and how we govern ourselves toward nature. As Linzey explains, it moves activists out from under those frameworks, to a system that begins from a place where the protection of nature is central, as opposed to most environmental laws that are focused on regulating the use of nature.

II. The First Rights of Nature Laws in the World

The development of the rights of nature movement in the US began with communities in Pennsylvania, facing industrial factory farming and the dumping of sewage sludge on farmland.

Sewage sludge – what the waste management industry has re-branded as 'biosolids' – is the solid that remains after sewage treatment plants remove liquid from waste. The waste comes from industrial operations, hospitals, and homes – anything that is put down a sink or toilet – which ends up at a sewage treatment plant for processing. This can include 'pharmaceuticals, hormones, pathogens, bacteria, viruses, protozoa and parasitic worms, as well as heavy metals like lead, cadmium, arsenic or mercury' along with 'PCBs, PFAS, dioxins, BPAs and dozens of other harmful substances […]'.[8]

In states across the US, sludge has been approved for use as a fertilizer to be spread on cropland and for other uses. This is just one of many examples of environmental laws authorizing environmental harm, including contamination of water and air. Linzey assisted communities in Pennsylvania seeking to block the

[7] Linzey (n 2).

[8] Tom Perkins, 'Biosolids: Mix Human Waste with Toxic Chemicals, then Spread on Crops' *The Guardian* (5 October 2019), <www.theguardian.com/environment/2019/oct/05/biosolids-toxic-chemicals-pollution>.

dumping of sewage sludge on farmland and in abandoned mines, by passing prohibitions on sludge dumping and recognizing the rights of nature. Facing the prospect of sludge being used to fill the massive pits left from decades of coal mining, Tamaqua Borough was the first to enact a rights of nature law prohibiting sludge dumping. Other Pennsylvania communities followed.

Communities in other states, facing different environmental threats but the same legal system which authorized the use of nature over the protection of nature, began to take note. North of Pennsylvania, in New Hampshire, local towns were facing the corporatization of water from bottled water companies. The bottled water industry – often called 'corporate water miners'[9] – siphons off millions of gallons of fresh water from local water sources to bottle and sell elsewhere. The environmental consequences from such operations on water ecosystems – including streams, wetlands, aquifers, and water tables – can be significant. Also, communities dependent on well water can find their wells dry up.

With companies such as Nestle growing their bottled water operations, New Hampshire communities sought to protect their water from this industrialization. However, as with Pennsylvania communities that found their state government authorizing the very industrial activities that they sought to protect against – such as the dumping of sewage sludge on farmland – New Hampshire communities learned that their state legislature authorized large groundwater withdrawals, including bottled water operations, under the state's Groundwater Protection Act.[10]

The Act treats water as an item of commerce to be bought and sold. Thus, in looking to their state Department of Environmental Services for help to protect their water, communities found that the agency was instead authorizing the taking of their water by the bottled water industry.

In 2005, Linzey met with community members in the Town of Barnstead, New Hampshire, as well as with members of the town's elected Select Board. He described how environmental laws operate, regulating how nature is used, and explaining why the community – rather than being empowered by the state to protect its water – found that the state was instead authorizing the taking of the water over community opposition.

Following Linzey's presentation, the Select Board requested his assistance in drafting an ordinance to protect their water. The ordinance was shared with community members, and in March 2006, the draft law was put to a vote of the community at their annual town meeting. It passed overwhelmingly. Barnstead became the first community in the US to ban the corporatization of water by banning large-scale water withdrawals.

---

[9] Doug Pibel, 'Communities Take Power: The Citizens of Barnstead, New Hampshire, Used Local Law to Keep Corporate Giants out of Their Water' *Yes! Magazine* (30 July 2007) <www.yesmagazine.org/issue/stand-up-corporate-power/2007/07/30/communities-take-power>.

[10] NH Rev Stat § 485-C:21 (2024), title L – Water Management and Protection.

It was two years later, in 2008, working with Linzey, that the Select Board amended the ordinance to include the rights of nature. The amended ordinance states:

> 'We the People of the Town of Barnstead declare that we have the duty to safeguard the water both on and beneath the Earth's surface, and in the process, safeguard the rights of people within the community of Barnstead, and the rights of the ecosystems of which Barnstead is a part [...].
>
> All residents of the Town of Barnstead possess a fundamental and inalienable right to access, use, consume, and preserve water drawn from the sustainable natural water cycles that provide water necessary to sustain life within the Town. Natural communities and ecosystems possess inalienable and fundamental rights to exist and flourish within the Town of Barnstead. Ecosystems shall include, but not be limited to, wetlands, streams, rivers, aquifers, and other water systems.'[11]

Other New Hampshire communities followed in Barnstead's footsteps, becoming the first outside of Pennsylvania to protect the legal rights of nature. Several communities in Maine – the towns of Shapleigh and Newfield – facing proposed bottled water operations from Nestle, enacted similar laws.

The engagement of communities facing other environmental threats, such as mining, climate change, and fracking, has led to a growing number of communities across the US passing rights of nature laws, with more than three dozen laws now enacted.

In 2010, the city of Pittsburgh, Pennsylvania, became the largest city in the country to protect the rights of nature. Pittsburgh's city council realized that hydraulic fracturing, or fracking, for shale gas was being planned within the city limits, including under cemeteries. Long known as the 'steel capital'[12] of the US, today the city is home to 300,000 people.

Fracking for both oil and shale gas has become a dominant form of fossil fuel development around the world. It involves drilling miles underground, conducting explosions in the underlying rock, and using millions of gallons of fresh water at every frack well. The impacts from fracking include air and water contamination, and global warming pollution, with consequences for both human health and the health of the natural environment.

Seeking to protect against fracking, Pittsburgh worked with Linzey and his organization to develop Pittsburgh's Community Protection from Natural Gas Extraction Ordinance, which secures the rights of natural communities and ecosystems 'to exist and flourish'[13] within the city. Under the law, residents have legal standing to enforce these rights on behalf of nature.[14]

---

[11] Barnstead Water Rights and Local Self-Government Ordinance, as amended, Town of Barnstead, New Hampshire, October 2008.

[12] William S Dietrich II, 'A Very Short History of Pittsburgh' *Pittsburgh Quarterly* (25 August 2008) <https://pittsburghquarterly.com/articles/a-very-brief-history-of-pittsburgh/>.

[13] Pittsburgh's Community Protection from Natural Gas Extraction Ordinance s 4.2, title Six, art 1, ch 618 of the Code of Ordinances, Pittsburgh.

[14] Pittsburgh Code (n 13).

Other communities in other states have followed suit, including in New York, Virginia, New Mexico, California, Ohio, Oregon, Colorado, and most recently Florida.

III. Protecting the Rights of Nature in the Happiest Place on Earth

Florida, a state in the southeastern US, is known for its water – from beaches lining the Atlantic coast and the Gulf of Mexico, to the string of islands called the Florida Keys, as well as natural springs that dot the landscape. Florida is also home to the Everglades, often called the 'river of grass', a more than two million acre wetland ecosystem unlike any other on earth that has seen massive habitat loss from human development. Across Florida, human activity, including building and development, as well as industrial agriculture, have brought water pollution and contamination, massive algae blooms, as well as red tides. Manatees are facing a major die-off, beaches are unusable, and waters are unsafe to swim.

In the face of so many threats to the states' waters, and seeing little meaningful action at the state and federal level, the people of Orange County, Florida, in the central part of the state, where Walt Disney World is located, took action. Working with Linzey and his current organization, the Center for Democratic and Environmental Rights (CDER), they drafted and proposed an amendment to the county Home Rule Charter to recognize the rights of waterways. In 2019 and 2020 the government-appointed Charter Review Commission debated and held hearings on the proposed measure, ultimately approving its placement on the county ballot for a vote of residents.

In November 2020, the people of Orange County approved the measure with 89 percent of the vote, becoming the largest place in the US – at nearly 1.5 million people – to protect the rights of nature. With that vote, a new section was added to the Orange County Charter, establishing rights of both the natural environment and the people of Orange County:

> 'A(1) The Wekiva River and Econlockhatchee River, portions of which are within the boundaries of Orange County, and all other Waters within the boundaries of Orange County, have a right to exist, Flow, to be protected against Pollution and to maintain a healthy ecosystem.
>
> A(2) All Citizens of Orange County have a right to clean water by having the Waters of Orange County protected against Pollution.'[15]

The Orange County charter amendment, as with many rights of nature laws, provides that citizens of the county can file enforcement actions on behalf of waterways to protect their rights. Following the proposal for a 1900-acre commercial and housing development that would destroy acres of wetlands and damage streams and other waterways in the county, in 2021 a lawsuit was filed in the 9th

[15] Code of Ordinances, Orange County, Florida Part I, Article VII, s 704.1(A).

Judicial Circuit Court of Florida to stop the development. The case was brought by several waterways – Wilde Cypress Branch, Boggy Branch, Crosby Island Marsh, Lake Hart, and Lake Mary Jane – as the lead plaintiffs against the developer, Beachline South Residential, LLC, and the Florida Department of Environmental Protection. The plaintiffs are arguing that their right to exist is at stake from the development. This the first rights of nature enforcement case brought in the US, and it is ongoing at the time of this writing.

IV. Tribal Nations Advance the Rights of Nature

Tribal nations within the US are also advancing rights of nature frameworks in both law and policy. This includes the Menominee Tribe of Wisconsin, which adopted a policy recognizing the rights of the Menominee River. The Ponca Tribe of Oklahoma adopted a rights of nature law. The Yurok Tribe in California adopted a resolution recognizing rights of the Klamath River, which flows through the tribe's reservation lands. Additionally, the Nez Perce Tribe in Idaho adopted a resolution recognizing rights of the Snake River.

The White Earth Band of Ojibwe, of the Chippewa Nation, adopted a tribal ordinance in 2018 securing the rights of manoomin, or wild rice. Linzey and the present author assisted in the development of the law, the first to secure rights of a particular plant species. It states:

> 'Manoomin, or wild rice, within the White Earth Reservation possesses inherent rights to exist, flourish, regenerate, and evolve, as well as inherent rights to restoration, recovery, and preservation. These rights include, but are not limited to, the right to pure water and freshwater habitat; the right to a healthy climate system and a natural environment free from human-caused global warming impacts and emissions; the right to be free from patenting; as well as rights to be free from infection, infestation, or drift by any means from genetically engineered organisms, trans-genetic risk seed, or other seeds that have been developed using methods other than traditional plant breeding.'[16]

The law further secures the rights of tribal members to 'harvest manoomin, and protect and save manoomin seeds, within the White Earth Reservation.'[17]

Much like the waterways in Orange County, facing constant and emerging threats, wild rice, which depends on freshwater habitat, is at risk from climate change, development, pollution, and habitat loss, among other threats. A major threat came with the Enbridge corporation's construction of a tar sands oil pipeline through Minnesota, very close to the White Earth Reservation.

In 2021, the state government approved a permit for Enbridge to use five billion gallons of fresh water in the construction and operation of the pipeline. This is already having impacts on wild rice. In August 2021, a case was brought in the

[16] White Earth Reservation Business Committee, White Earth Band of Chippewa Indians, Resolution 001-19-009 (adopted 31 December 2018).

[17] Resolution 001-19-009 (n 16).

White Earth Tribal Court to block the permit. Wild rice is the lead plaintiff in the case – *Manoomin, et al v Minnesota Department of Environmental Protection, et al* – as joined by tribal members seeking to enforce their legal rights. This the first rights of nature enforcement case to be brought in a tribal court.

The White Earth law and court case reflect how integrally the rights of people to protect nature are tied to the rights of nature itself to be protected. With First Nations, such as the White Earth, this is bound up in their constant struggle for tribal sovereignty and self-determination.

The White Earth Tribal Court spoke to this issue in its response to the State of Minnesota's request to have the case dismissed, writing:

> 'The activity at issue here impacts the ecosystem of Manoomin in that it allows Defendants to control the water quantity and quality on which the plant depends. Manoomin is…an important piece of the diets of native peoples. It is also central to the culture and history of the Anishinaabe people and is an integral part of wetland ecosystems and natural communities. The possible impact of Defendant's activities has a "direct effect on the political integrity, political security or the health and welfare of the Tribe" […] the activity threatens the cultural welfare and continuity of the Band […].
>
> In formally adopting […] laws to protect Manoomin […] the Band is exercising its inherent authority to protect a necessary and vital resource. This authority predates the U.S. Constitution and is reflected in the numerous treaties made between the United States and the Anishinaabeg peoples. These treaties are profound expressions of Tribal sovereignty […].'[18]

In January 2022, a second rights of nature case was brought in a tribal court. Filed by the Sauk-Suiattle Indian Tribe, whose reservation is located in the western part of the State of Washington, the case seeks to have the court recognize that salmon – whom tribal members consider a relative – possess legal rights. The Tribe is arguing that the City of Seattle – against who the case was brought – must remedy the harm done to salmon from the dams the city has on the Skagit River. Salmon are greatly at risk from the dams, which do not allow passage upstream to historic spawning grounds.

More tribal nations are taking steps to advance the rights of nature. In March 2022, the Midwest Alliance of Sovereign Tribes – with thirty-six member tribes – adopted a resolution expressing its support for the rights of nature. And in June 2022, the National Congress of American Indians adopted a policy in support of the rights of nature as well. CDER and its partner, Menīkānaehkem, an indigenous grassroots organization, assisted both organizations in drafting the measures.

Within the US, the rights of nature movement continues to expand. Proposals to bring the rights of nature into state law have advanced, particularly in Florida, where the first local law – in Orange County – passed in 2020. The Florida Rights of Nature Network proposed a citizen-sponsored initiative in 2021, and again in 2022, to secure rights of the natural world. These efforts are moving forward.

---

[18] White Earth Band of Ojibwe Tribal Court18 August 2021, Court file no GC21-0428 – *Ma-noomin, The White Earth Band of Ojibwe, et al v Minnesota Department of Natural Resources, et al*, Order Denying Defendant's Motion to Dismiss.

Further, a proposal to incorporate rights of nature protection into regulations implementing the federal Clean Water Act has been shared with the Biden Administration. The National Council for the Implementation of the Rights of Nature – founded by CDER and made up of legal scholars, activists, and members of tribal nations – developed the proposal which is focused on treaty rights.

As the US sought to expand its lands and control over the continent following the Revolutionary War and well into the 1800s, many tribes were forced off their lands. Under pressure, many tribes signed treaties with the US under which they were required to cede their traditional lands to the national government. The right to hunt, fish, and gather was a concession the US government provided some tribes through treaties, yet those rights have often been ignored.

The proposal that was shared with the Biden Administration would require – for any tribe with a rights of nature law in place – that the federal government gain the consent of the tribe for any government-permitted activity that would impact ecosystems or species on lands and waters guaranteed for use by the tribe to hunt, fish, and gather.

In effect, the proposed regulation would provide tribes the ability to protect the rights of nature on those treaty lands.

V. Final Observations

In the US, despite a growing understanding that climate change and other environmental crises are real and mounting threats – experienced by millions of Americans in the form of more intense hurricanes, rainfall, and flooding, as well as extreme heat, drought, and wildfires – there remains fierce industry and corporate opposition to curbing global warming emissions and protecting nature. The advancement of the rights of nature as a new and higher level of environmental protection has gained the attention and hostility of big business.

This has taken the form of lawsuits brought by corporations to overturn local laws which secure the rights of nature while prohibiting certain industrial activities that would violate those rights. In the states of Ohio and Florida, this resistance has also included actions by state legislatures, on behalf of industry, to pass new laws which prohibit communities from passing local rights of nature laws.

Thus, the state legislatures, controlled by corporate interests, have acted to prevent their communities from putting in place stronger protections for the environment than are in place at the state level.

Yet it is precisely for this reason that the people of Orange County, Florida, felt the need to protect the rights of waterways, because of the failure of their state government to protect wetlands, streams, coastal waters, and other water bodies across the state. It is one thing to fail to do an adequate job of protecting the environment, but in states such as Florida, not only is the state failing to do an adequate job, legislators are instead focusing their time on making sure that local communities can't do the job either.

In fact, in Florida, the legislature amended the state Environmental Protection Act so as to prevent communities from exercising such authority before the people of Orange County even voted on their rights of waterways law. The amendment states:

'A local government regulation, ordinance, code, rule, comprehensive plan, charter, or any other provision of law may not recognize or grant any legal rights to a plant, an animal, a body of water, or any other part of the natural environment that is not a person or political subdivision [...] or grant such person or political subdivision any specific rights relating to the natural environment not otherwise authorized in general law or specifically granted in the State Constitution.'[19]

As explained by Steven Meyers, an attorney representing the waterways in the Florida court case seeking to enforce the rights of the waterways:

'For decades Floridians have despaired seeing our beautiful coastal waters, lakes, rivers, springs, and wetlands ruined by seemingly unstoppable development and sprawl. Florida's regulatory system of granting permits to use, consume, and ultimately destroy these waterways has failed epically. The Florida Legislature has abandoned any real effort to protect these precious waters, and the judiciary has proven unable to do so under the current regulatory scheme [...].'[20]

Chuck O'Neal, the human plaintiff in the case (having joined the plaintiff waterways) adds: 'For too long our legislators have told the public we need balance between commerce and Nature, and then [they] folded to pressure from commerce to permit egregious exploitation.'[21]

***

The fight to protect the rights of the natural world continues in Florida and across the US, as well as around the world. It is defining a new relationship between humans and nature, one that for many is in fact a return to an ancient relationship in which nature is understood as a relative, to be respected and protected.

Success requires gaining an understanding that we are part of and dependent on nature, a significant departure from the dominant cultures and economies of today. And it requires us to make necessary shifts – in law and within society – as past people's movements have had to do, in order to achieve a necessary and fundamental change in how we govern ourselves toward nature and how we treat nature under the law.

---

[19] Environmental Protection Act, s (9)(a) of The Florida Statutes, title XXIX, ch 403, part I, 403.412.

[20] 'First U.S. "Rights of Nature" Enforcement Case Filed' (*Center for Democratic and Environmental Rights*, 27 April 2021) <www.centerforenvironmentalrights.org/news/first-us-rights-of-nature-enforcement-case-filed>.

[21] Center for Democratic and Environmental Rights (n 20).

Recognizing the Legal Personality of a River

The Magpie River/*Mutehekau Shipu* in Canada

Yenny Vega Cárdenas and Uapukun Mestokosho[*]

I. Introduction

'Ekuanitshit means: Take care of the place where you live.
Nutshimit means: where my blood comes from.

Uapukun Mestokosho

In February 2021, the *Mutehekau Shipu* (Innu)/Magpie River (English) was declared a living entity with rights by the Innu Council of Ekuanitshit and the Regional County of Minganie in Côte-Nord, Quebec. The first of its kind in Canada, the declaration protects the nearly 300-kilometer-long river, including the connected 7,650 km^2 watershed. A message of hope in difficult times, in which pandemic and climate change have grabbed our attention, the Magpie River changes the story of protecting nature in Canada.

---

[*] The authors thank Alexandra Baer, Jason Stocker, and Emilie Doyon for the English language revision of different parts of the text. They also thank Rita Mestokosho, Innu writer and poet, for the special quotes shared for this paper. The authors are profoundly grateful for the pertinent comments and editions proposed by Lindsay Keegitah Borrows, specialist in Indigenous legal traditions, professor at Queen's University Faculty of Law.

The manuscript was finalized in November 2023. The assessment of literature and case law, including internet materials, reflects the legal situation at that time. The discussion in section IV was supplemented in April 2025 to reflect more recent developments.

Figure 1: The Mutehekau Shipu/Magpie River: Credit Uapukun Mestokosho

The *Mutehekau Shipu* is known as an ancestor and relative for the community of Ekuanitshit, and it is an integral element of their *Nitassinan* (ancestral territory). Affirming Innu biocultural rights is an important objective in recognizing the river's legal identity as a living being. Archaeological research reveals an Indigenous presence in *Nitassinan* dating back 8,000 years[1] and a continuous, regular, and exclusive occupation[2] of the territory by Innu ancestors spanning 3,500 years.[3]

[1] Jean-Christophe Ouellet and Pierre Richard, 'Un archaïque ancien (8500–8000 ans AA) en Moyenne-Côte-Nord: l'apport des sites EbCx-65 et EbCx-66 en Minganie' in Adrian Burke and Claude Chapdelaine (eds), *L'Archaïque au Québec: six millénaires d'histoire amérindienne* (Recherches autochtones au Québec 2017); Michel Plourde, *8000 ans de Paléohistoire. Synthèse des recherches archéologiques menées dans l'aire de coordination du Parc Marin du Saguenay – Saint-Laurent* (Parcs Canada 2003).

[2] Note: the term 'exclusive' does not always mean an occupation to the exclusion of all: exclusivity can be 'shared'.

[3] Donald Holly, *History in the Making: The Archaeology of the Eastern Subarctic* (Altamira Press 2013); Daniel Chevrier, 'Les premières populations humaines: 8500 à 2000 ans avant aujourd'hui' in Pierre Frenette (ed), *Histoire de la Côte-Nord* (Institut de recherche sur la culture, Presses de l'Université Laval 1996).

The river nourishes an important region where Innu families have gathered during summer for thousands of years[4] to practice *Innu Aitun* (the practice of customary and traditional, cultural, social, economic, and subsistence activities linked to the land, such as hunting, fishing, trapping, and gathering).[5] Knowing about the close relationship between the Innu and the *Mutehekau Shipu*, it is not surprising that they affirmed the ancestral river as a person with rights. *Mutehekau Shipu* and *Nitassinan*, the land and the water, the forest and the sky are all integral to Innu culture, identity, and legal traditions.

Many members of the Innu community who occupy the *Nitassinan* uphold the declaration on the grounds of intergenerational principles.[6] They cite an imperative to care for the land and to preserve its state for next generations.[7] Developments like hydroelectric projects that pose serious threats for the *Mutehekau Shipu*[8] simultaneously threaten the rights of the Innu people, their culture, and their way of life, for the past and present as well as the posterity of the lineage.

Importantly, the Innu conception of territory/land entails a fiduciary relationship that aligns with eco-centric philosophies, including kincentric ones.[9] The Innu of Ekuanitshit speak, sometimes in French, of a feeling of belonging to the land. Not only do they see themselves as inseparable from the land, but they are its ancestral guardians since time immemorial.[10]

Noting the congruence between the Innu legal system and traditional beliefs, which treat the river as sacred and as an entity with agency, the Innu considered it fundamental to recognize the *Mutehekau Shipu* as a legal person. The ultimate

---

[4] Conseil des Innus de Ekuanitshit, *Mémoire sur le projet d'aménagement hydroélectrique du site du barrage Magpie sur la rivière Magpie* (BAPE 2004) 4; Ministère de l'Environnement du Québeck, *Rapport d'analyse environnementale: Projet d'aménagement hydroélectrique du site du barrage Magpie sur la rivière* (Rapport d'analyse environnementale, Gouvernement du Québec 2005) 47–48.

[5] Ministère de l'Environnement du Québec (n 4) 47.

[6] Hydro-Québec, *Complexe de la Romaine: Étude de l'impact sur l'environnement* (Étude d'impact, vol 6, ch 39, Gouvernement du Québec, 2007) 39–47; Jean-Paul Lacasse, *Les Innus et le territoire* (Les éditions du Septentrion 2004).

[7] Ministère de l'Environnement du Québec (n 4) 47–48.

[8] Québec: Hydro-Québec, *Plan stratégique 2009–2013* (2009) 22–23; also found on Ministère de l'Environnement du Québec (n 4) and the Romain Project, see Hydro-Québec (n 6); Sylvie Vincent, 'Le projet de la rivière Romaine vu et rapporté par la presse écrite' (2008) 38 Recherches amérindiennes au Québec 148; Hugo Asselin, 'Plan Nord: les Autochtones laissés en plan' (2011) 41 Recherches amérindiennes au Québec 47.

[9] Enrique Salmon, 'Kincentric Ecology: Indigenous Perceptions of the Human-Nature Relationship' (2000) 10 Ecological Applications 1327: 'Indigenous people view both themselves and nature as part of an extended ecological family that shares ancestry and origins. It is an awareness that life in any environment is viable only when humans view the life surrounding them as kin. The kin, or relatives, include all the natural elements of an ecosystem. Indigenous people are affected by and, in turn, affect the life around them. The interactions that result from this "kincentric ecology" enhance and preserve the ecosystem.' (Abstract).

[10] Hydro-Québec (n 6) 39–47; Lacasse (n 6).

goal is to ensure that the Innu and the collectivity thrive in harmony with a healthy river for future generations to come. As they inherited 'her' from their ancestors, they wished to do the same for those who are yet to come/to be born.

Members of the Minganie municipality also agreed on the importance of changing the anthropocentric paradigm to promote different paths of economic and human development in the Cote-Nord region. They quickly realized the importance of protecting this internationally renowned river, which has exceptional potential for hosting whitewater expeditions.[11] Protecting the Magpie simultaneously protects ecological tourism activities such as whitewater rafting and canoeing.[12] To secure this economic potential, the Regional County Municipality (RCM) of Minganie had proposed several local resolutions aimed at protecting the *Mutehekau Shipu* since 2015, despite falling short of full protection from Quebec's hydropower plans.[13]

In response to the threat of potential hydroelectric projects, Innu youth and the inhabitants of the municipality ardently fought to protect the *Mutehekau Shipu* and cemented it as a symbol of their pride, culture, and of the reclaiming of the land.[14] Hand in hand, Indigenous and non-Indigenous people began to manifest a *two-eyed seeing*[15] process as a way to protect the river, similar to the process undertaken between the Māori and the New Zealand government to protect the Whanganui River.[16]

In 2019 the Innu of Ekuanitshit, in alliance with non-Indigenous peoples from the Regional County of Minganie and NGOs such as the Canadian Parks and Wilderness Society (CPAWS) Quebec Chapter, invited the International Observatory on the Rights of Nature[17] to explore the goal of recognizing the river as a living, legal subject. In February 2021, a press conference release announced the first legal acknowledgment of personhood and rights for a Canadian river. By the means of two mirror resolutions, one from the Innu Council of Ekuanitshit and the other

---

[11] International Rafting Federation, *Top 10 Most Improved Rivers* <www.internationalrafting.com/sustainability/top-10-most-improved-rivers/>.

[12] Lorie Ouellet, *La Rivière Magpie. Une rivière de classe mondiale à protéger* (Chicoutimi: Laboratoire d'expertise et de recherche en plein air – Université du Québec à Chicoutimi, 2013) 2.

[13] In 2015, *The North Coast Regional Conference of Deputies* proposed a Study Zone (ZE-E006) of 2,630 km^2 looking forward to the protection of the Magpie River (Resolution CA-P20150129-12). This proposition has been confirmed by other official resolutions (Resolutions 207-14 and 229-18), adopted in 2016 and 2018 by the Regional County Municipality of Minganie.

[14] Innu Council of Ekuanitshit, Resolution 671-082 about the Magpie River (12 January 2016).

[15] Mi'kmaw Elder Albert Marshall brought forward the concept of two-eyed seeing. It means a perspective that weaves together Indigenous ways of knowing and Western knowledge systems.

[16] Brettel Dawson, 'New Zealand/Aotearoa' in Yenny Vega Cárdenas and Daniel Turp (eds), *Une personnalité juridique pour le Fleuve Saint-Laurent et les Fleuves du monde* (JFD 2021).

[17] To learn more about the International Observatory on the Rights of Nature, visit: <www.observatoirenature.org>.

from the Regional County of Minganie, the *Mutehekau Shipu* was recognized as a person with nine rights, including to exist, to flow freely, to its natural biodiversity, to its natural cycles, to restauration, and to take legal action (among others).[18]

In a context of juridical pluralism, this breakthrough places the Innu epistemology at the heart of the evolution of Canadian environmental law. At the same time, the Magpie River declaration ensures the respect for both the right to self-determination and the biocultural rights of the Innu community of Ekuanitshit as rooted in their longstanding relationship with the territory, waterways, and ecosystems they have protected since time immemorial.

This paper explores how the declaration of the personhood and rights of the Magpie River encompasses a decolonial process that advances reconciliation between Indigenous and non-Indigenous peoples and legal traditions in Canada.[19] First, the paper explores the roots of this important realization grounded in the cosmology of the Innu community of Ekuanitshit. We analyze the confirmation of the river's legal personality as a step in decolonizing the Canadian legal system. Second, the paper explains the complex interactions of the diverse juridical orders that co-exist in the Canadian legal system related to water (including the common law, civil law, and Indigenous legal traditions). Third, we analyze how the juridical structure of Canada is conducive to a dynamic approach between different legal orders that can give life to new models of preserving nature for generations to come. Even if rooted in different epistemologies, the various legal orders are in constant dialogue (sometimes one of competition, other times of co-operation), and these multi-juridical interactions shape the Canadian legal landscape. It is time for Indigenous perspectives to shine.

---

[18] Yenny Vega Cárdenas, 'The recognition of the Magpie/Muteshekau Shipu River as a non-human Person' (*Observatoire International des droits de la nature*, 9 March 2021) <https://observatoirenature.org/observatorio/2022/06/11/the-recognition-of-the-magpie-muteshekau-shipu-river-as-a-non-human-person/>; the entire text of the resolutions can be accessed via the links found in the blog.

[19] David Boyd, *The Rights of Nature: A Legal Revolution that Could Save the World* (ECW Press 2017).

II. The *Mutehekau Shipu* as a Living Entity with Rights: The Resurgence of the Innu of Ekuanitshit Cosmology as a Step in Decolonizing the Canadian Legal System

> 'Nous devons reconnaître la force de notre Mère la Terre et la respecter pour qu'elle continue à nous guérir, nourrir et nous apprendre notre place dans l'humanité.'
>
> *Rita Mestokosho*, Innu poet and writer

Ekuanitshit means: 'Take care of the place where you live'. This word represents well the way in which the culture of the Innu people of Ekuanitshit honors the territory and rivers. Innu means 'human', and Innu people consider themselves guardians of the elements of Nature that comprise their *Nitassinan* (Land).[20]

In Innu cosmology, humans are not just part of the land; unity with the earth is sacred. In the words of Innu people, to be Innu, is to be human, is to be sensitive to nature that surrounds the community, is to understand the environment in which we humans evolve, and is to see the beauty and the good in all things, small or large. For the community, land is a school without limits. It is also a pharmacy, a food pantry, and more. Being Innu is a way of life, a culture, a language, a legal tradition, a legacy to be perpetuated for the survival of the community.

As Innu do not conceive of themselves in division between land, forest and water, their vision teaches us that we are all part of the same cosmos. The link that unites the community to the rivers is part of their Indigenous knowledge and way of being. Rivers are considered the veins of the territory, providing water for the survival of humans, as well as other living beings on earth. For Innu communities, rivers are more than just waterways or resources, they are living beings with their own spirit and agency, and they deserve respect. Rivers are considered ancestors, and the community has the responsibility to take care of them, as they support life.

This Indigenous knowledge has been built up through generations of people living in close contact with nature. It is built on the history and lived experiences of Innu people. This knowledge is cumulative and dynamic as it adapts to social, environmental, spiritual, and political relationships.[21]

Long before the present generation of Innu were born, their ancestors travelled the immense territory known as *Nitassinan*. With the rhythm of the seasons, the Innu, in their nomadic way of life, moved from one place to another. Nomadic

---

[20] Uapukun Mestokosho (Performer), *Nutshimiu Aimun- La langue du territoire* (2021). – In this paper, the authors have chosen to use upper-casing when referring to Nature, Land and Nitassinan. The intention is to gently suggest to the reader that, from our perspective, Nature is considered as a subject rather than a mere object.

[21] Uapukun Mestokosho and Shanice Mollen-Picard (Performer), *Shipu* (2016).

lifestyle, was in fact considered a way to allow Mother Earth to have the time to regenerate. Innu were following the cycles of the seasons in order also to preserve Nature. They were in fact, aware of the crucial impact on earth of staying in the same place for too long. *Ekuanitshit* was a summer gathering place. The great-grandparents of the Innu people today arrived from the far north, guided by the rivers that were the highways of the time.[22]

Forced by colonial policies to become sedentary, the Innu of Ekuanitshit settled in Minganie, a peninsula full of natural treasures. Since colonization, outside powers have sought to transform the territory and harness the 'resources' for their own control and benefit. Yet these natural wonders remain and are part of the living memory and identity of the Innu people. Colonization brought an imposition of laws and regulations in order to appropriate what they called 'natural resources', yet within the traditional Innu territory, aiming among other things to divide the community to better rule.[23]

The Innu have been bombarded for a long time with projects leading to the destruction of land, including dams on rivers that brought enrichment to non-Indigenous governments and companies. Faced with this reality, young people are increasingly revolted by this mentality, which is contradictory to Indigenous knowledge, which for its part advocates respect and living in harmony with Mother Earth.[24]

The Ekuanitshit community believe that it is time to stand up and be proud of the heritage bequeathed by Innu ancestors. Innu youth are no longer afraid to say loud and clear:

> We are here to stay in order to protect the territory, our traditions, and especially our identity.[25] I am Innu and I grew up respecting the traditions learned from my grandparents who walked from Labrador to the shores of the St. Lawrence.[26]

The Innu of Ekuanitshit have been the traditional guardians of the *Mutehekau Shipu*/Magpie River for thousands of years. In their cosmology, rivers do not belong to anyone; rather, it is the *Innu* (humans) who have the obligation to protect them. Indeed, no one can own a river, for humans are not detached from nature, but rather humans are part of Nature. Therefore, communities should all be committed to their conservation, so that future generations can also follow the flow.[27] In the words of the Innu writer and poet Rita Mestokosho:

> 'Our traditional knowledge teaches us that for living well, we need water *nipi*, air *neneun*, and land *nitassinan*. Taking care of Mother Earth is also a way of taking care of ourselves, for our living well. Otherwise, we are going to provoke our own extinction. Maybe […] indigenous

[22] Mestokosho and Mollen-Picard (n 21).

[23] Mestokosho and Mollen-Picard (n 21).

[24] Mestokosho and Mollen-Picard (n 21).

[25] Mestokosho and Mollen-Picard (n 21).

[26] Uapukun Mestokosho-McKenzie and Rita Mestokosho, 'Puamu Innu Ishkueu: Dream of an Innu Woman' in *Indigenous Women's Turn to Take the Talking Stick* (FNQLSDI 2018) 261.

[27] Mestokosho and Mollen-Picard (n 21).

traditional knowledge could save the humanity. We consider ourselves as part of the territory, and if we want to take care of ourselves as species, we should take care of the land. We are convinced that if the recognition of the *Mutehekau Shipu* as an entity with rights, could make evolve the Canadian juridical system, we are taking an important step in the good direction for future generations to come.'[28]

The *Mutehekau Shipu* represents a form of healing. Spending time by the river is a therapy for Innu people, who can reclaim their ancestral practices that were silenced because of colonial violence. Innu people suffer greatly from intergenerational trauma from the past. Being recognized today as traditional guardians brings them closer to their ancestors who have always protected the Magpie, known as the *Mutehekau Shipu*. The recognition of the river's rights will protect her for future generations.

Moreover, as it is time for reconciliation between Indigenous and non-Indigenous people, it is time for healing. Healing the land and preserving it is a way to heal the Innu community from the violence it has suffered. The recognition by the government of the declaration of the legal personality of the *Mutehekau Shipu* is a matter of recognizing the importance of the survival of the Innu culture and traditions. Thus, this is not a matter of money, as the government has acted in the case of residential schools, in which it has tried to repair the damages caused to survivors with amounts of money[29] (for some indigenous individuals, residential schools are considered as an act of cultural genocide[30]). For Innu people, it is the land and the waters that give life and heal us all. As many elders of the community say: 'No money can buy what an Innu is, their roots and culture: anyhow, the money would come from the destruction of the lands we have lived on for generations'. Preserving the *Mutehekau Shipu* and other rivers is a question of survival and self-determination. The Innu people reclaim the freedom to relearn from ancient practices and to spread out their knowledge and cosmology. Recognizing rivers as persons, will finally help improve the dominant legal system that has been an important threat for many forms of life in earth.

Many Innu express this spiritual and biocultural link through poetry, including Rita Mestokosho. She writes:

'The river of life. Even before I was born on this earth, the river guided my dream. It flowed straight from the sky. I seek the river of life. The one that flows to soothe my spirit. The one that dances, dizzying and deep. The one that cleanses my destiny.'[31]

---

[28] Conversation with Mrs Rita Mestokosho, Innu writer and poet (4 January 2022).

[29] Government of Canada, *Indian Residential Schools Settlement Agreement* (9 June 2021) <www.rcaanc-cirnac.gc.ca/eng/1100100015576/1571581687074>: 'The Government of Canada will fund up to $27 million over 3 years to support Indigenous partners and communities in a range of activities, including school-specific research and knowledge-gathering on the children who died at residential schools and their burial places.'

[30] Antonio Voce, Leyland Cecco and Chris Michael, 'Cultural Genocide: the Shameful History of Canada's Residential Schools – Mapped' *The Guardian* (8 September 2021).

[31] Rita Mestokosho (Innu writer and poet), *Atik Utei. Le Coeur du Caribou* (Mémoire d'encrier, 2022).

Many Indigenous knowledge systems promote a vision that opposes the colonial understanding of human superiority over other living beings. Colonial conceptions promote a hierarchical relationship between humans and nature, with humans at the top of the pyramid and the rest of nature at the bottom.[32] In contrast, many Indigenous teachings suggest that humans are not only part of nature, but that all species, no matter how small, deserve equal respect. During ancestral activities such as hunting or fishing, traditional customs teach to thank each living being who give their lives for our survival.[33] These practices, shared throughout the generations, show the importance of balancing and respecting the cycle of life.[34]

The recognition of the legal personality of the *Mutehekau Shipu* is a tool that fits well with Innu cosmology. Innu youth were profoundly inspired by the Māori negotiations with the government of New Zealand, which spanned decades and ultimately gifted us with the fruit of legal personality for the Whanganui River.[35] Links were created between Indigenous young leaders around the world, particularly at the second *Mni Ki Wakan*: World Indigenous Peoples Decade of Water Summit.[36] The report adopted at this gathering emphasized the need to put traditional Indigenous ecological knowledge forward for the benefit of the planet.[37] Indigenous youth across the globe, convinced that it is time to act, have taken a stand in this critical moment for the future of Mother Earth. For many, this means restoring and protecting the veins of the earth: the waterways.[38]

In this sense, recognizing the *Mutehekau Shipu* as a person with rights, like the Whanganui River in New Zealand, is a way of translating into Western law what water means in the eyes of an Indigenous cosmovision.[39] As we have seen, for cultures intrinsically attached to the territory, as the Innu people are, rivers

---

[32] Robin Wall Kimmerer, *Braiding Sweetgrass, Indigenous Wisdom, Scientific Knowledge and the Teaching of Plants* (Wilkweed Editions 2013); Boyd (n 19).

[33] Dominique Rankin and Marie-Josée Tardif, *They Called Us Savages: A Hereditary Chief's Quest for Truth and Harmony* (Vidacom 2021).

[34] Kimmerer (n 32).

[35] Dawson (n 16).

[36] World Indigenous Peoples Decade of Water Summit, '2nd Mni Ki Wakan: 2018 Global Report' (International Report 2018).

[37] World Indigenous Peoples Decade of Water Summit (n 36) 3.

[38] World Indigenous Peoples Decade of Water Summit (n 36) 3.

[39] Geneviève Motard, 'Regards croisés entre le droit innu et le droit québécois: territorialités en conflit' (2020) 65 McGill LJ 421, 456:

'La reconnaissance de la personnalité juridique de certains éléments dits « naturels », comme les rivières, est une solution que retiennent de plus en plus les États et qui fait écho à la doctrine qui s'est penchée sur des propositions de même nature. Dns ces cas, il est possible de designer un gardien ayant le mandat d'assurer le respect des droits de cet être animé. […] L'incorporation de principes fondamentaux du droit autochtone à même le droit de l'État, contribuant ainsi à la formation de la mixité du droit étatique et autochtone, est aussi une technique à laquelle on pourrait avoir recours, comme c'est le cas à Hawai'i dans le domaine de la gestion de l'Eau. […] Le principe du fiduciaire public est cardinal à toute la réglementation sur l'Eau.'

are considered to be ancestors. All members of the community have the responsibility to take care of them. If rivers are considered subjects, this means we need to decolonize the thought that rivers are mere objects to be exploited or polluted. The legal personality of natural entities is a legal tool to respond to the escalating environmental degradation and destruction.[40]

Young Innu leaders, Uapukun Mestokosho and Shanice Mollen-Picard, have been international water ambassadors for several years. They are committed to staying connected to water, speaking up for rivers, and reconnecting people with rivers, both Indigenous and non-Indigenous. This commitment has been highlighted in the World Report on Water and Indigenous Youth, in which they have actively participated: 'Indigenous Youth should share traditional knowledge and connections to the environment and show how other cultures value the environment.'[41]

Together with youth of many Indigenous nations, they were deeply involved in the 2018 Water Declaration that put forward that it is important to change the practices imposed by colonization and that youth have a mission to declare that 'water is alive', and that 'we are all drops of water who come together to form an unstoppable force and flow that can break down all barriers that stand in our way'.[42]

The Innu youth put forward the importance of declaring the *Mutehekau Shipu* as a rights-holding entity before their council. Now the river exists in the eyes of the world as a legal entity with rights. It is now a story to tell across Canada and many countries. Narratives are a powerful way to change Canadian law, as Innu ancestors do to preserve culture, traditions, and their legal tradition. The power of narratives will also make non-Indigenous people understand the importance of recognizing the sacred cycles of life, the sacredness of water, and how we are all intimately interrelated. Today, this is a new story to be told, one that allows Indigenous people to go back to their history, to go back deep in their roots and to help those who were uprooted by the trauma of residential schools and other colonial policies. Until the 1960s, the Innu continued to follow a nomadic lifestyle. But the creation of reserves and residential schools impacted the ability to practice their traditional way of life. Today the youth realize that:

---

[40] Prof. Iorns Magallanes explains very well the importance of the collaboration between indigenous perspectives and Western law in these terms:

'These examples from New Zealand illustrate ways in which the law can be used to implement and incorporate indigenous cosmologies with a Western society and legal system and better protect the natural environment in the process. I suggest that it may even alter the mainstream constructions of nature, through normalizing the indigenous constructions. Thus, the protection of indigenous rights to culture and religion could better protect a healthy environment for everyone.' Catherine Iorns Magallanes, 'Maori Cultural Rights in Aotearoa New Zealand: Protecting the Cosmology that Protects the Environment' (2015) 21 Widener L Rev 273, 275.

[41] World Indigenous Peoples Decade of Water Summit (n 36) 6.

[42] Uapukun Mestokosho, quoted in World Indigenous Peoples Decade of Water Summit (n 36) 7.

'The link to the land is still very strong, even today; we have not lost it, [...] and more and more, we want to go back to our roots, so that our future generations can benefit. [...] We have now a very important job, to be able to protect the river.'[43]

Legal personality serves the purpose of letting the world know that the river is alive, that it exists, that it has a history, and that the Innu are part of the river.[44] This language will be used for the conservation of the *Mutehekau Shipu*, but also for other rivers and streams locally, and at the same time globally. Like other Indigenous youth around the world, Innu youth, hand-in-hand with local non-Indigenous people, will be able to ask themselves the following questions: What are we doing to protect water, rivers, nature? How do our everyday choices impact water?[45] More and more new generations of Innu understand that preserving ancestral culture is a question of survival for the Innu Nation, and at the same time, it is a way to promote the survival of the different expressions of life. Climate change, such as the destruction of ecosystems; has been pointed out as the most devastating environmental threat, even to the human species. It has become clear that many of the causes are anthropogenic through choices that pollute and exploit nature in an unsustainable manner.[46] That is why it is important not only for Innu youth but also to the world to listen to indigenous knowledge and learn from the ecosystemic approach.[47]

Innu elders have always taught the importance of being responsible for the land and waterways where they live. Innu youth can be proud of their traditions and must pass them on to new generations. This knowledge, which was almost lost with the uprooting of colonization, becomes a moral obligation of intergenerational teaching that continues to help new generations understand 'the importance of being stewards of the land'.[48]

Today, recognizing that the *Mutehekau Shipu* has a legal personality is a very innovative way to return to traditional Innu values and to make peace with Mother

---

[43] Shanice Mollen-Picard quoted in Joel Balsam, 'This Wild River in Quebec is Now Considered a Person. How Will It Help with Conservation?' *The Globe and Mail* (19 December 2021).

[44] Motard (n 39).

[45] World Indigenous Peoples Decade of Water Summit (n 36) 8.

[46] Intergovernmental Panel on Climate Change, 'Climate Change 2021. The Physical Science Basis' Working Group, UNEP (2021).

[47] André Francisco Pilon, 'Building a New World: An Ecosystemic Approach for Global Change & Development Design' (Munich Personal RePec Archive MPRA Paper No 62124, 21 March 2013): 'To build a world where people are self-resilient, live dignified lives, ensure healthy and prosperous future, and co-existence for all living beings in harmony, justice, and peace requires more than one actor, that actors be aware of each other's actions, and that they dynamically adjust their own behaviour (and possibly their own goals), taking the behaviour of others into account. New paradigms of power, growth, work, freedom and wealth should be associated with new ways of being-in-the-world, in the benefit of environment, health, culture, education and citizenship.' (Abstract, Academia).

[48] Marie-Hélène Beaudry and Sabryna Godbout, *Teach Me Water. A Tale of Seven Generations* (FNQLSDI 2020) 81.

Earth. The acknowledgement of the river as a living entity is a good way to return to traditional Indigenous value systems. This new model allows Innu to continue with their ancestral vision, because it reminds new generations that rivers are like the veins of the territory and that we all must have a holistic vision of them. The *Mutehekau Shipu* will therefore be protected in its entire watershed, as the resolutions have clearly expressed. The recognition of personhood and rights of the river put forward the Innu cosmology that considers that humans are in equal relationship with animals, plants, and ecosystems.

Since Indigenous legal systems have been suppressed by the dominant system, decolonizing law will continue to evolve in a context of necessary reconciliation with Indigenous peoples. As professor John Borrows writes:

> 'If we, as human beings, reconcile ourselves with the earth, it will be much easier for Indigenous and non-Indigenous people to identify the earth as the source to help correlate how we relate together. […] You can think of yourself as an ally with the fish and the birds and the trees. We are citizens together because we are part of a community that is mutually participatory with the impulse to ensure that it isn't being hierarchal and neglecting voices.'[49]

Moreover, the declaration of the *Mutehekau Shipu* as a legal person sheds light on a dynamic process of decolonizing the Canadian legal system. Canada's legal pluralism has been marked by the colonial domination of the common law and civil law systems over the Indigenous legal traditions, but this imbalance is now under transformation. The influence and dialogue between these systems is undeniable and increasingly present.

The recognition of the rights of the *Mutehekau Shipu* also sheds light on the differences, similarities, and influence that can exist between legal traditions in Canada. The common law, civil law, and Indigenous legal traditions have distinct epistemologies regarding Nature and watercourses. We will analyze in the following section the interactions of those legal traditions in respect of the recognition of legal personality for natural entities. It is important to highlight the complexity of the relationships between several actors and jurisdictions in Canada (federal, provincial, municipal governments, Aboriginal band councils, Indigenous governing bodies, civil society, etc.). The legal system in Canada must be understood in terms of other inherent superstructures, such as history, education, economy, language and ethnicity.

---

[49] John Borrows, 'Reconcile Yourself with the Earth' (University of Calgary, 26 November 2017) <https://ucalgary.ca/news/reconcile-earth> accessed 1 June 2022.

III. The Dynamics Inside the Canadian Legal System with Regard to the Rights of Nature Paradigm

'Si vous voulez comprendre un système, essayez de le changer.'
Kurt Lewin, Psychologist, Scientist (1890–1947)

Although the Constitution Act of 1867[50] establishes Canada as a constitutional monarchy, in reality it operates as a pluralistic legal system. This system encompasses the English Common Law system, the French Civil Law system within the province of Québec, and Indigenous legal traditions throughout the ten provinces and three territories.

Even though the recognition of the rights of Indigenous peoples is set out in the Constitutional Act of 1982 under s 35(1) as follows: 'The existing aboriginal and treaty rights of the aboriginal peoples of Canada are hereby recognized and affirmed',[51] the traditional rights of Indigenous peoples have only gradually been taken into account after legal battles before the courts, creating a body of law related to the recognition of Indigenous rights.

Despite these legal challenges, a process of reconciliation has begun between the Canadian government and Indigenous governments, whereby Indigenous law is recognized as an existing legal system which coexists alongside the common law and civil law traditions. It is in this context that the United Nations Declaration on the Rights of Indigenous Peoples Act received Royal assent in June 2021 and thereafter came into force. By the self-determination principle that is enshrined in this Act, Indigenous legal traditions have even more weight in Canadian law.[52]

Under the division of powers set by the Constitution Act of 1867, several governmental powers have jurisdiction over water and the environment. These powers allow us to identify determining actors in the recognition of the rights of nature. Canadian law immediately divides the components of the environment and, in the same way, divides the jurisdictions of the levels of government. Thus, the Constitution of Canada, which distinguishes property and jurisdiction, confirms the fixed state of borders and proceeds to a division of legislative powers between federated and federal entities in this area.[53]

Since the *Mutehekau Shipu* declaration of personhood and the recognition of associated rights have occurred in the Quebec provincial context in respect of the traditional land of Innu people, it is important to identify the relevant legis-

---

[50] Constitution Act of 1867, Statutes of Canada 1867, 30 and 31 Vict, c 3 <https://laws-lois.justice.gc.ca/fra/Const/page-1.html>.

[51] Constitutional Act of 1982, Statutes of Canada 1982, annex B c 11.

[52] United Nations Declaration on the Rights of Indigenous Peoples Act (assented 21 June 2021) SC 2021, c 14, <https://laws-lois.justice.gc.ca/eng/acts/U-2.2/>.

[53] Motard (n 39) 430; Constitution Act 1867, 30 and 31 Vict, c 3, part VI.

lation related to water and the environment in the context of ancestral rights involved.

IV. Legal Basis for the Recognition of Personhood and the Rights of Rivers

The declaration of the *Mutehekau Shipu*/Magpie River as a legal person having rights aligns with an eco-centric framework. It values the intrinsic dignity of the river as a subject of law. The mirror resolutions by the RCM of Minganie and the Innu Council of Ekuanitshit assert the following nine rights:

1. The right to live, exist, and flow;
2. the right to maintain its natural cycles;
3. the right to evolve naturally, to be preserved, and to be protected;
4. the right to maintain its natural biodiversity;
5. the right to maintain its integrity;
6. the right to perform essential functions within its ecosystem;
7. the right to be protected from pollution;
8. the right to regeneration and restoration;
9. the right to take legal action.

The International Observatory of the Rights of Nature (OIDN) was intensely involved in the drafting of the resolutions that were adopted in the case of the Magpie River in Canada. The OIDN acknowledges that the rights of the River as recognized in the resolutions were inspired by the Universal Declaration of the Rights of Rivers written by the Earth Law Center,[54] the *Te Awa Tupua* (Whanganui River Claims Settlement Act)[55] that was adopted in New Zealand recognizing the rights of the Whanganui River, the Colombian Constitutional Court decision of T-622/16 in which the *Atrato* river was recognized as a legal entity with rights,[56] and some Indigenous resolutions adopted in the United States to grant rights to rivers, such as the Snake and Klamath River among others.[57]

---

[54] Petition, Universal Declaration of the Rights of Rivers <www.rightsofrivers.org/> accessed 14 March 2022.

[55] The bill can be consulted at <www.legislation.govt.nz/act/public/2017/0007/latest/whole.html>.

[56] Constitutional Court of Colombia 10 November 2016, T-622/16, *The Atrato River Case*, <www.corteconstitucional.gov.co/relatoria/2016/t-622-16.html>.

[57] White Earth, Resolution 001-19-009 and Resolution 001-19-010 (adopted on 31 December 2018); Yurok Tribal Council, Resolution 19-40, *Resolution Establishing the Rights of the Klamath River* (adopted on 9 May 2019); Joint Tribal Council of the Passamaquoddy Tribe, *St. Croix River and Alewife Resolution* (adopted in 2012); Nez Perce Tribe, Resolution SPGC20-02, *Resolution recognizes rights of Snake river* (adopted on 18 June 2020); Menominee Indian Tribe of Wisconsin, Resolution 19-52, *Recognition of the Rights of the Menominee Rice* (adopted on 16 January 2020).

As a living entity possessing fundamental rights, the *Mutehekau Shipu* will be represented by guardians appointed in 2022 by the Innu First Nation of Ekuanitshit and the Minganie RCM, with their having the duty to act on behalf of the rights and interests of the river and to ensure the protection of its fundamental rights. guardians will be empowered to take legal action on behalf of the river, to claim compensation for damages suffered by the river, and to receive compensation for its benefit. Guardians will also have the responsibility to participate in all the decisions that could affect the Magpie River and should perform their duties in collaboration with the Planning and Sustainable Development Department of the Minganie RCM, as well as with the Innu Council of Ekuanitshit. The nomination of guardians will ensure participation by Ekuanitshit youth and elders who will work in collaboration with the guardians of the Land/*Nitassinan*.

Professor Lacasse points out that for Innu people, the concept of guardianship of the Land is a deeply held obligation.[58] Thus, the nomination will embrace Innu legal concepts, making sure to include Innu language and legal process such as ceremonies symbolizing that water connects us all. The detailed functions of the Magpie River guardians are detailed in both resolutions but can be understood as protecting the river's rights on a fiduciary basis.[59] The main legal bases upon which these resolutions rely are Quebec civil law, the municipal law framework, and Innu law.

Under the National Water Policy (Quebec, 2002) and the *Act to Affirm the Collective Nature of Water Resources and to Promote Better Governance of Water and Associated Environments* (the 'Water Act'),[60] water in Quebec is part of a shared heritage, which is considered common among those who reside within the province. It is stated that it is important to preserve it and improve its management to meet the needs of present and future generations. Furthermore, it is the State, as guardian of the nation's interests, that must be invested with the power to ensure its protection and management.

As stipulated in s 1, this protective status does not imply an appropriation of water, even by the Quebec government. The State is not the owner of these resources, but rather, the steward or the guardian:

> '1. Being of vital interest, surface water and groundwater, in their natural state, are resources that are part of the common heritage of the Quebec nation. As stated in article 913 of the Civil Code, their use is common to all and they cannot be appropriated, except under the conditions defined by this article.'[61]

---

[58] Jean-Paul Lacasse, 'Le territoire dans l'univers innu aujourd'hui' (1996) 40 Cahiers Géographie 185, 200.

[59] The resolutions can be accessed at the end of this article: Cárdenas (n 18); they are available at <https://observatoirenature.org/observatorio/2022/06/11/the-recognition-of-the-magpie-muteshekau-shipu-river-as-a-non-human-person/>.

[60] Act to Affirm the Collective Nature of Water Resources and to Promote Better Governance of Water and Associated Environments, RLRQ ch C-6.2 (Water Act) <http://legisquebec.gouv.qc.ca/fr/showdoc/cs/c-6.2>.

[61] Water Act (n 60) s 1.

Regarding the use of these resources, the Water Act specifies in s 3 that '[t]he protection, restoration, enhancement, and management of water resources are of general interest and contribute to the objective of sustainable development.'[62]

It is important to highlight that in 2009, the province of Quebec amended its legislation to include the principle that a body of water can suffer a harm. This opened a door to recognize an important aspect of water protection also affirmed by the rights of Nature movement. One of the main objectives of this paradigm shift is to acknowledge that Nature can suffer an autonomous prejudice. On behalf of Quebec, the State can bring an action in damages caused to water. Thus, s 8 of the Water Act introduces the concept of strict liability when damage is caused to watercourses and confers a determining role on to the Attorney General of Quebec:

> '8. If damage to water resources, such as the impairment of their physical, chemical or biological properties, ecological functions, or quantitative status, is caused by a person or through a person's fault or illegal act, the Attorney General may institute an action against that person, in the name of the State as custodian of the interests of the nation in water resources. with a view to obtaining one or more of the following:
>
> 1. Restoration to the original state or to a similar state;
> 2. reparation by compensatory measures;
> 3. reparation by the payment of compensation, lump sum or other.
>
> For the purposes of this article, the initial state refers to the state of water resources and their ecological functions that would have existed without the occurrence of such damage, assessed using the best available information.
>
> The obligation to make reparation for damage to water resources or their ecological functions caused through the fault or illegal act of two or more persons is solidary.'[63]

Although the Attorney General can institute such an action on behalf of Quebec, it is not always able to do so because, in certain cases, Quebec itself grants the authorization to pollute. In such circumstances, Quebec cannot readily claim compensation for contamination it has authorized. It is important to note that this particular article has not been used to date, even though it has existed for over twelve years.

It should also be noted that the Act outlines governance measures for associated communities. In this regard, it specifies that water management must be 'based on an integrated, concerted strategy'. Section 13 further outlines that this integrated and concerted management must take principles of sustainable development into account, including those stated in section 6 of the *Sustainable Development Act*.[64]

In addition, the *Quebec Charter of Human Rights and Freedoms* provides every person with a right to live in a healthy environment in which biodiversity is pre-

---

[62] Water Act (n 60) s 3.

[63] Water Act (n 60) s 8.

[64] Sustainable Development Act, RLRQ ch D-8.1.1.

served;[65] and the *Environment Quality Act* provides that '[e]very person has a right to a healthy environment and to its protection, and to the protection of the living species inhabiting it.'[66]

However, despite the legal framework affirming environmental protection, the interests of water as a living being necessary for all living species to thrive does not seem to be sufficiently represented within watershed organizations, regional consultations, or studies concerned with the authorization of projects contrary to the conservation of Nature. Thus, despite advancements, the water laws and legal framework of Quebec are in many ways ineffective and do not provide sufficient protection for Nature as a living entity. The *General report on Water resources and aquatic Ecosystems in the province of Quebec* states the precarity of biodiversity that depends on water to live. The main threat source identified is human activity that causes significant destruction of habitats and degradation of water as a living environment.[67] From this perspective, the attribution of legal personality to rivers is relevant as another legal tool to fill in the gaps left by legislation.

In general, municipalities play a central role in the protection of waterways and the environment. Under s 103 of the Municipal Powers Act, 'continuously or intermittently flowing watercourses' are under the jurisdiction of the Regional County Municipalities, unless the government decides otherwise by order-in-council.[68] In addition, under s 99 of the *Municipal Powers Act*, a Municipality has the power to regulate 'on any regional matter relating to its citizens that is not otherwise regulated.'[69]

Canadian jurisprudence also recognizes the important role municipal governments play in environmental matters. For example, the Quebec Court of Appeal stated in *Wallot*[70] that 'a reading of the [Municipal Powers Act] shows how the Quebec legislator intended to empower municipalities to exercise the full extent of their powers' (para 30). The Supreme Court of Canada recognized in *Hydro-Québec*[71] that environmental problems 'require action by governments at all levels' (para 127). In *Gastem Inc. v. Municipalité de Ristigouche-Partie-Sud-Est*,[72] the Quebec Superior Court stated that municipalities 'must assume their responsibilities for the protection of the environment on their territory in accordance

---

[65] Charter of Human Rights and Freedoms, RLRQ ch C-12 s 46.1: 'Every person has a right to live in a healthful environment in which biodiversity is preserved, to the extent and according to the standards provided by law.'

[66] Environment Quality Act, RLRQ c Q-2 preamble.

[67] Ministère de l'Environnement et de la lutte contre les changements climatiques, *Rapport sur l'État des ressources en eau et des écosystèemes aquatiques du Québec 2020* (Gouvernement Quebec 2020); The report can be accessed at <www.environnement.gouv.qc.ca/eau/rapport-eau/rapport-eau-2020.pdf>.

[68] Municipal Powers Act, RLRQ ch C-47.1 s 103.

[69] Municipal Powers Act (n 68) s 99.

[70] *Wallot v Québec*, 2011 QCCA 1165.

[71] *R. v Hydro-Québec*, [1997] 3 SCR 213.

[72] *Gastem inc. v Municipalité de Ristigouche-Partie-Sud-Est*, 2018 QCCS 779.

with the subsidiarity principle' (para 50). The subsidiarity principle was recognized by the Supreme Court of Canada in *Spraytech* (para 3), by which decision-making and responsibility must rest with the lowest administrative or political level able to act effectively.[73]

Under the *Municipal Powers Act*, municipalities may establish an environmental protection agency or assign this responsibility to a non-profit corporation or legal person.[74] That is why these entities could more efficiently assign legal personhood and rights to waterways within their boundaries for the purpose of ensuring their protection. This is especially true since they are responsible for implementing regional planning and development policies under the Act Respecting Land Use and Planning and Development.[75] Therefore, they have the power to move towards establishing an existence in harmony with Nature.

With regards to the Innu Declaration concerning the *Mutehekau Shipu*,[76] it is important to understand that the Innu are in the midst of a comprehensive land claims treaty process, and the *Mutehekau Shipu* is part of their claimed territory.[77] Canadian courts have already recognized that the Innu of Ekuanitshit have strong prima facie evidence of land use rights in the *Nitassinan*[78] where the *Mutehekau Shipu* is located. Moreover, the Innu of Ekuanitshit have the inherent right to self-determination and can thus freely ensure their economic, social, and cultural development. To do that, the Innu can maintain and strengthen their distinct political, legal, economic, social, and cultural institutions, as enshrined in international law by arts 3 and 5 of the United Nations Declaration on the Rights of Indigenous Peoples. In the same line, arts 25, 29(1), 32(1) of the same Declaration recognize the right of the Innu of Ekuanitshit to preserve, protect, and define the use of the territory, including the *Mutehekau Shipu*, its ecosystem and the species that live there. They have the right to maintain and strengthen their special spiritual ties

---

[73] 114957 *Canada Ltée (Spraytech, Société d'arrosage) v Hudson*, 2001 SCC 40, [2001] 2 SCR 241.

[74] Municipal Powers Act (n 68) ss 101, 93, 94. Section 101 allows RCMs to exercise the powers set out in ss 93 and 94. Section 93 allows an RCM to establish a regulating body to ensure the protection of the environment. Section 94 allows an RCM to delegate these same responsibilities to a non-profit corporation or legal person.

[75] Act Respecting Land Use Planning and Development, RLRQ ch A-19.1 ss 3, 5, 6, and 7 on the preparation of land use and development plans.

[76] The text of the Innu resolution granting rights and personhood to the Mutehekau Shipu River can be accessed at <http://files.harmonywithnatureun.org/uploads/upload1072.pdf>.

[77] Conseil des Innus de Ekuanitshit, Projet Hydroélectrique du complexe de la Romaine (BAPE 2008) s II.

[78] See *Innu Council of Ekuanitshit v Canada (Attorney General)*, 2014 FCA 189, where the Federal Court of Appeal notes that the Innu of Ekuanitshit have strong prima facie evidence of land use rights in their traditional territory (para 90). The Court also recognizes that the ancestral land rights of the Innu of Ekuanitshit are vulnerable and liable to be violated, and it notes that commercial activities constitute a serious risk to the enjoyment of the Innu of Ekuanitshit's land rights (para 90). See also *Innu Council of Ekuanitshit v Canada (Attorney General)*, 2013 FC 418, [2013] 3 CNLR 145, paras 103–104.

to the river and land and to assume the responsibilities that they owe in this regard to future generations.

Moreover, the federal government of Canada has passed a bill integrating the whole text of the UN Declaration on the Rights of Indigenous Peoples.[79]

As we have highlighted in the first part of this paper, rivers are considered by the Innu of Ekuanitshit as a relative. As an ancestor, rivers have agency and spirit. The recognition of the personhood and rights of the *Mutehekau Shipu* is a reflection of their ontologies as have been translated into Western law.

This declaration should also be understood as one of many expressions of Indigenous water law, written and oral, historic and contemporary, that have been adopted by Indigenous peoples across Canada over the years. An example of a contemporary intertribal written declaration includes the 'National Water Declaration' from the General Assembly of First Nations, which recognizes the sacredness of water, the interconnectedness of all life and the importance of protecting water from pollution, drought, and waste.[80] The Syilx Nation in British Columbia passed a water declaration in 2014 which states:

> 'The Okanagan Nation has accepted the unique responsibility bestowed upon us by the Creator to serve for all time as protectors of the lands and waters in our territories, so that all living things return to us regenerated. When we take care of the land and water, the land and water take care of us. This is our law.'[81]

For a variety of Indigenous peoples in Canada, as in many countries around the world, rivers and land are linked to their culture and identity. The Assembly of Indigenous Peoples acknowledges that 'Water is the blood of Mother Earth'.[82] They have expressed their deep relationship with this essential element in a joint declaration concerning water in these terms:

> 'The sacred water element teaches us that we can have great strength to transform even the tallest mountain while being soft, pliable, and flexible. Water gives us the spiritual teaching that we too flow into the Great Ocean at the end of our life journey. Water shapes the land and gives us the great gifts of the rivers, lakes, ice, and oceans. Water is the home of many living things that contribute to the health and well-being of everything not in the water.'[83]

This declaration also recognizes the importance of guardians in taking care of Mother Earth:

> 'Indigenous Knowledge Systems First Nations have a direct relationship with all waters that exist on Mother Earth, and we care, protect and honor them through our traditional ways. Our own Indigenous knowledge systems are the foundation of our Nations. Our knowledge systems inform our relationship with water as an element, a spiritual entity, a resource and a source of

[79] United Nations Declaration on the Rights of Indigenous Peoples Act (n 52).

[80] Assembly of First Nations, *Honouring Water*, <https://www.afn.ca/honoring-water/>.

[81] Declaration adopted by the Syilx People (31 July 2014) <https://syilx.org/about-us/syilx-nation/water-declaration/>.

[82] *Honouring Water* (n 80).

[83] *Honouring Water* (n 80).

life. We care, protect and honor those relationships through our customs, traditions and practices. Many First Nations seek to restore our traditional ways of protecting the health of water and to share these ways with the world.'[84]

For the Haida Nation, the cedar tree is at the core of their worldview. As professor Boyd explains, a 'cedar tree is known as "every woman's sister" providing for and sustaining our existence'.[85] In the words of Terri-Lynn:

'A worldview where the Cedar Tree is a Sister dramatically changes how we interact with the forests. This worldview has guided millennia of sustainable Haida management of the use of forests of Haida Gwaii—beginning with the nurturing of the first trees to Haida Gwaii and continuing into the present day. In fact, each time we take the bark from a cedar tree, Haida law requires that we speak first to the tree to learn how much of the cedar bark we may take. And each time we would test the cedar to see whether it was suitable for construction of canoes or totem poles, we completed a ceremony which we called T'suuaay 'Kiinga, which translates literally as "feeling into the heart of the Cedar"'.[86]

The Supreme Court of Canada has reaffirmed the overarching links between biocultural rights, ancestral territory, and the vitality of Indigenous peoples' cultures and traditions repeatedly in its jurisprudence. For example, in *Tsilhqot'in Nation v British Columbia*,[87] the Supreme Court recognized the link between the land and the survival of Indigenous cultures.[88] In the same decision, the Court states that the Crown has a fiduciary duty to justify an infringement on Indigenous titles to land by considering future generations:

'[T]he government must act in a way that respects the fact that Aboriginal title is a group interest that inheres in present and future generations. […] incursions on Aboriginal title cannot be justified if they would substantially deprive future generations of the benefit of the land.'[89]

In *Delgamuukw v. British Columbia*,[90] the Supreme Court highlights another aspect of Indigenous groups' rights to land, that is, a link between perspectives, including Indigenous legal systems and the need to maintain a community's relationship with its land.[91] In the Court's view, the preservation of Indigenous cultures could be jeopardized if their lands are used in a way that does not respect the unique relationship Indigenous groups hold with them.[92] Judge Lamer, in the same deci-

[84] Assembly of First Nations, *National Water Declaration*, <www.afn.ca/uploads/files/water/national_water_declaration.pdf>.

[85] Boyd (n 19); see also Condesa Strain, 'Warrior Lawyer Profile: Terri-Lynn Williams-Davidson' (*Peter A. Allard School of Law Blog*, 5 April 2021) <https://allard.ubc.ca/about-us/blog/2021/warrior-lawyer-profile-terri-lynn-williams-davidson>.

[86] Michael Jackson, 'On the Front Cover Terri-Lynn Williams-Davidson' *The Advocate* (3 May 2012) 335, 337.

[87] *Tsilhqotin Nation v British Columbia*, 2014 SCC 44, [2014] 2 SCR 257.

[88] *Tsilhqotin Nation v British Columbia* (n 87) paras 74, 84, 86.

[89] *Tsilhqotin Nation v British Columbia* (n 87) para 86.

[90] *Delgamuukw v British Columbia*, [1997] 3 SCR 1010.

[91] *Delgamuukw v British Columbia* (n 90) para 127.

[92] *Delgamuukw v British Columbia* (n 90) para 127.

sion, notes that if the occupation necessary to establish title has been proven, 'there will exist a special bond between the group and the land in question such that the land will be part of the definition of the group's distinctive culture'.[93]

As we have pointed out, Canadian courts already recognized that the Innu of Ekuanitshit have a strong prima facie case in support of rights to their land in *Nitassinan* in *Conseil des Innus de Ekuanitshit v Canada (Attorney General).*[94] In *Nitassinan*, the Federal Court of Appeal notes that the Ekuanitshit Innu have land use rights in their traditional territory.[95] The Court also recognizes that the Aboriginal land rights of the Ekuanitshit Innu are vulnerable and susceptible to infringement, and it notes that commercial activities pose an especially serious threat.[96]

Indigenous nations in Canada hold an inherent right to self-determination and are thus free to pursue their economic, social, and cultural development, as enshrined in international law by s 3 of the United Nations Declaration on the Rights of Indigenous Peoples. They also have the inherent right to preserve, protect, and define the use of their territory as outlined in ss 29(1) and 32(1) of the United Nations Declaration on the Rights of Indigenous Peoples. Moreover, they have an inherent right to maintain and strengthen their distinct political, legal, economic, social, and cultural institutions as outlined by s 5 of the United Nations Declaration on the Rights of Indigenous Peoples. First Nations also have the inherent right to promote, develop, and maintain their legal systems or customs, as laid out by s 34. Within the context of Canada's commitment to harmonize its internal law with the United Nations Declaration on the Rights of Indigenous Peoples, the manifestation of Indigenous peoples' legal traditions through the rights of Nature approach could be afforded legal protection in Canada.[97]

It is interesting to note that the Tsilhqot'in Nation in western Canada have also documented rights and responsibilities as traditional guardians of the *Sturgeon River*, commonly known as the Fraser River, under their own laws. The Tsilhqot'in National Government passed a resolution stating that humans, animals, fish, plants, earth, and water have legal rights.[98]

---

[93] *Delgamuukw v British Columbia* (n 90) para 128.

[94] *Nitassinan in Conseil des Innus de Ekuanitshit v Canada (Attorney General)*, 2014 FCA 189.

[95] *Nitassinan in Conseil des Innus de Ekuanitshit v Canada (Attorney General)* (n 94) para 90.

[96] *Nitassinan in Conseil des Innus de Ekuanitshit v Canada (Attorney General)* (n 94).

[97] Ministère de la Justice Canada, 'Déclaration commune des ministres Lametti et Bennett à la suite de l'adoption par le Sénat du projet de loi C-15, Loi concernant la Déclaration des Nations Unies sur les droits des peuples autochtones' (16 June 2021), <www.canada.ca/fr/ministere-justice/nouvelles/2021/06/declaration-commune-des-ministres-lametti-et-bennett-a-la-suite-de-ladoption-par-le-senat-du-projet-de-loi-c-15-loi-concernant-la-declaration-des-n.html>. See also: United Nations Declaration on the Rights of Indigenous Peoples Act (S.C. 2021, c 14), <https://laws-lois.justice.gc.ca/fra/lois/u-2.2/>.

[98] Tŝilhqot'in Nation, *ʔEsdilaghSturgeon River Law* (28 May 2020).

The recognition of the Magpie River has inspired other initiatives across the country to embrace the rights of Nature movement. Among the initiatives we highlight is the St. Lawrence River project led by the International Observatory of Nature's Rights, which seeks to recognize the personhood and rights of the river. The Observatory has started a conversation with political parties in Quebec and in Canada at federal level, also with many Indigenous territories and municipalities that the St. Lawrence flows through. At this moment, the project has obtained important support from more than four political parties at the provincial and federal level. Indigenous peoples, municipalities, research centers, and NGOs have joined the project in order to present a bill to recognize the personhood and rights of the St. Lawrence River.[99] A collaboratively written book has been launched concerning the personhood and rights of the St. Lawrence River, including a draft bill proposal.[100] Moreover, in 2023, the Assembly of First Nations Quebec-Labrador recognized the legal personhood of the Saint-Lawrence river by the resolution No 04/2023.[101]

The Saint-Pierre River initiative started in conversations with the NGO *Les Amis du Parc Meadowbrook* and the OIDN. The affected community has made a citizens' declaration of the legal personhood of the historical river in Montreal that is threatened by sewage and drainage systems. The NGO *Les Amis du Parc Meadowbrook* is trying to protect the river and is asking the City of Montreal to recognize its right to exist, flourish, and be free from pollution.[102]

Recently the the Alliance Tenàgadino, which has adopted since 2023 a Gatineau River Rights Declaration, has the intention to work on the adoption of official resolutions that will recognize the Gatineau River as a legal person.[103]

The International Observatory of Nature's Rights was also promoting and discussing with the federal government an integration of the recognition of the rights

---

[99] To follow the project, you can consult <https://observatoirenature.org/observatorio/alliance-fleuve-st-laurent/> and also the network page of the Observatory <www.facebook.com/ObservatoireinternationaldesdroitsdelaNature>.

[100] The collective book is an important source of information about the project: Yenny Vega Cárdenas and Daniel Turp (eds), *Une personnalité juridique pour le Fleuve Saint-Laurent et les Fleuves du monde* (JFD 2021).

The book *A Legal Personality for the St. Lawrence River and other Rivers of the World* is under the direction of the president of the OIDN, Yenny Vega Cárdenas, and Daniel Turp, a research associate with the Observatory (JFD 2021). The book has several chapters written by rights of nature experts from around the world.

[101] Assembly of First Nations Quebec-Labrador, Resolution No 04-2023, 'Alliance for the Defence of the St. Lawrence River', <https://observatoirenature.org/observatorio/wp-content/uploads/2023/05/04-2023-Defence-on-St-Lawrence-RIver-Declaration-APNQL.pdf>.

[102] Several newspapers have published articles concerning protection of the Saint-Pierre River: Alexandre Sauro, 'Une personnalité juridique pour la rivière St-Pierre?' *Métro* (May 2021); Allison Hanes, 'Could Recognizing Rights for Nature Help Save It?' *Montreal Gazette* (April 2021); Stephane Giroux, 'Environmental Advocates Turn Attention to What's Left of a River that Predates the City of Montreal' *CTV News* (April 2021).

[103] To follow the process, you can visit <https://www.tenagadinoalliance.ca/>.

of Nature in a federal bill related to the human right to a healthy environment. The Observatory's work has also been international in scope. For example, they have worked by invitation with organizations in Peru to discuss the protection of the Marañon River in the Amazonian region. A legal action has been presented in that country, supported by Canadian research centers, with the goal of recognizing personhood and rights for this sacred river of the Kukama Indigenous community.[104] In 2024, the river was formally acknowledged as a holder of rights by two judicial decisions.[105]

V. Rights of Nature: An Issue of Convergence in the Canadian Legal System

Professor Emeritus Rodolfo Sacco affirmed that no legal systems are static or perfect. On the contrary, they all have intrinsic contradictions that must be deciphered by comparativists.[106] Consequently, legal systems may contain several responses to the same phenomenon, which may lead to contradictions between the sources of domestic law (statutes, jurisprudence, doctrine, legal orders, or customs). In other words, the law is not univocal; rather, it can be ambiguous, pluralistic, and in constant evolution.

For civil law, the general structure of the Civil Code of Quebec (CCQ),[107] like other Romano-Germanic Codes, includes a division between persons, objects, and actions. Nature is considered as an object, over which the holders of rights have the powers of *usus*, *fructus*, and *abusus*. This last characteristic allows, certainly, the destruction of an object, in all legality, as long as one does not cause a damage for another person. Persons, at the top of the pyramid, are holders of rights, and some persons have obligations. Among the categories of persons in Quebec civil law, we find natural persons (men, women, and children) and moral persons, such as corporations, non-profit organizations, trusts, ships, and churches. We also find legal persons created by public law, which are those created by law or by regulation, such as municipalities and, as was the case, the Magpie River, which has been a legal person since February 2021. Animals, too, are no longer considered objects; instead they are sentient beings endowed with sensibilities and biological imperatives according to art 898.1 of the CCQ.

---

104 *Kukama Women's Federation v PETROPERU, Ministry of the Environment and others*, 2° Juzgado Civil – Sede Central, 00784-2021-0-1903-JR-CI-02 (8 September 2021).

105 Yenny Vega Cárdenas, Patricia Urteaga and Alberto Salazar, 'Marañon River rights recognized by a landmark decision in Peru' (*Blog OIDN*, March 2024, updated November 2024) <https://observatoirenature.org/observatorio/en/2024/03/21/maranon-river-rights-recognized-by-a-landmark-decision-in-peru/>.

106 Rodolfo Sacco, 'Legal Formants: A Dynamic Approach to Comparative Law (Installement I of II)' (1991) 39 AJCL 1.

107 Code civil du Québec, RLRQ ch CCQ-1991.

In art 913 of the CCQ, water is a common good. It is also considered a collective resource, being part of the common heritage of Quebec as per the first article of the Water Act.[108] Having expressly excluded the appropriation of water in Quebec, the Water Act (2009) included a guardian or steward role for the Quebec State, which must take care of it for future generations. It clarifies that the government is not the owner of water because its legal status excludes its appropriation.[109]

On the other hand, the epistemology of the common law suggests that water is the property of the Crown. This legal tradition requires things to have owners, hence the importance of giving the Crown the ownership of common goods. Relevant scholars affirmed that even if water is vested in the Crown, this is not equal to a proprietary right to the water, but that it should be more understood as a public trust to protect the resource and control its use by individuals in the public interest.[110]

As we know, the common law system has a different epistemology than the civil law system. It is not a pyramidal analysis, where you can find the classical separation of property, persons, and actions. It is more of a circular reasoning (rights/remedies). This means that the existence of a right in the common law requires an action that makes the right itself operative.[111]

For some Indigenous legal traditions, the relationship between persons and objects differs in an important way.[112] The human is not at the top of the pyramid, nor is there a pyramid system between the various species or elements of Nature. It is rather a question of a relationship of reciprocity between human beings and their environment. Human beings bear the duty to preserve their territory to pass it on to future generations. They take only what they need, thanking Nature and paying her in some traditions for the favors and fruits received.[113] Some Indigenous cosmologies see Nature as an equal, an ancestor, a relative[114] a sacred ele-

---

[108] Act to Affirm the Collective Nature of Water Resources and to Promote Better Governance of Water and Associated Environments, <http://legisquebec.gouv.qc.ca/fr/showdoc/cs/c-6.2>.

[109] Yenny Vega Cárdenas, Inès Benadda, and Andrew Galliano, 'L'évolution du cadre juridique de la gouvernance de l'eau au Québec' in Yenny Vega Cárdenas and Daniel Turp (eds), *Une personnalité juridique pour le Fleuve Saint-Laurent et les Fleuves du monde* (JFD 2021).

[110] Madeleine Cantin-Cumyn, Michelle Cumyn and Claire Skrinda 'L'eau, Chose Commune: Un statut juridique à confirmer' (2000) 79 Revue du Barreau canadien 398.

[111] Geoffrey Samuel, *An Introduction to Comparative Law Theory and Method* (Hart 2014) 100. For the common law, rights give obligations, and rights without correlative obligations are seen as privileges.

[112] Basil H Johnston, *Honour Earth Mother* (Kegedonce Press 2003).

[113] Rankin and Tardif (n 33).

[114] Robert Yelkátte Clifford, 'Listening to Law' (2016) 33 Windsor YB Access Just 47, 52: 'A long time ago, when the Creator, XÁLS, walked the Earth, there were no islands in the WSÁNEĆ territory. The islands that are there today were human beings (our ancestors). At this time, XÁLS walked among the WSÁNEĆ People, showing them the proper way to live. In doing this, he took a bunch of the WSÁNEĆ People and threw them out into the ocean. Each of the persons thrown into the ocean became the islands there today. Each of those islands were given a particular name that reflects the manner in which they landed, their characteristics or

ment possessing a spirit, an agency.[115] This is the closest thing to legal personality in Western law. In this way, the elements of Nature do not belong to anyone because according to this epistemology, the rivers do not belong to man, rather it is man who belongs to Nature.[116] Consequently, the rights recognized in natural entities become, rather, natural obligations and moral human responsibilities for the well-being of the living world. Cree Nation law and Innu law, among others, are grounded within the relationship with the non-human world. As a result, indigenous law, implies duties to land, water, flora, and fauna.[117]

Thus, in thinking about the recognition of legal personhood for rivers and other natural entities, we ask: Are the different legal systems in Canada ready to embrace the rights of Nature movement? We could advance an affirmative answer, given that it is conceivable that the reasoning of a civil lawyer may align, in view of the fact that Quebec civil law already recognises that water is not appropriable. Furthermore, the stewardship of water bodies by the State has been formally acknowledged within the legal system since 2009.

The Code also already recognizes autonomous prejudice being suffered by water, and it allows the creation of legal persons by public law. This legislation could be understood in favor of recognizing legal persons created by municipalities, such as autonomous public bodies, State enterprises, and in the case in point, well-defined natural elements, as was the case for the Magpie River. The Quebec Civil Code also already recognizes that animals are no longer property, which represents a step away from anthropocentrism and toward biocentrism. Eco-centrism is at this point only on the verge of becoming the new paradigm in Quebec, a development which would be more than welcome in the context of the degradation of rivers, the loss of biodiversity, and the effects of climate change that are worsening the environmental situation in the province and globally.

---

appearance, or the significance they have to the WSÁNEĆ People. "James Island" was named ŁEL,TOS, meaning "Splashed on the Face." ŁEL,TOS reflects the way the island landed in the ocean. The southeast face of ŁEL,TOS is worn by the wind and the tide. After throwing the WSÁNEĆ People into the ocean, XÁLS turned to speak to the islands and said: "Look after your relatives, the WSÁNEĆ People." XÁLS then turned to the WSÁNEĆ People and said: "You will also look after your 'Relatives of the Deep.'" This is what XÁLS asked of us in return for the care our "Relatives of the Deep" provide for us.'

[115] Garma International Indigenous Water Declaration, <https://cdn.prod.website-files.com/64370c192e1dad088fb1d81d/64c6e7ea8a42b2304ee24ea9_41.%20Declaration%20-%20Garma%20International%20Indigenous%20Water%20Declaration.pdf>: 'We the Indigenous Peoples of the World DECLARE that: Water is not a commodity. Water is a spirit that has a right to be treated as an ecological entity, with its own inherent right to exist. […] We assert that water has a right to be recognized as an ecological entity, a being with a spirit and must be treated accordingly. For the Indigenous Peoples water is essential to creation; Ancestral beings are created by and dwell within water. We do not believe that water should solely be treated as a resource or a commodity.'

[116] Supra section II (Innu Cosmology).

[117] Darcy Lindberg and Nêhiyaw Âskiy Wiyasiwêwina, *Plains Cree Earth Law and Constitutional/Ecological Reconciliation* (University of Victoria 2020) 14.

For a common law lawyer, it could be argued that the recognition of rights is effective only if there is also recognition to take legal action. Therefore, the recognition of rights for the Magpie River would have no effect if the public entities had not recognized the right to sue and defend those rights. The circular rationality is also found in the resolutions duly adopted by the local authorities.[118] Moreover, in common law systems, the obligations of good stewardship and respect for ecological processes are correlative obligations of those who have property rights.[119] The stewardship legal concept is easily found and understood in common law systems, more than in civil law systems.[120]

For Indigenous legal traditions, the natural world is often understood as living, and therefore humans have particular relationships and responsibilities towards their ecological kin/relatives.[121] While the term of legal personality is transsystemic, these legal systems could see this concept as a legal tool that gives life to their legal understanding that Nature is an equal, a type of person (ancestor, sister, etc.), or an entity with agency. Nature as a holder of rights means recognizing her as an equal, instead of relegating her to the bottom of the pyramid, as is the case in Western legal systems.[122]

As the rights of Nature model tends to nominate guardians of the land, Indigenous legal traditions are well suited to give meaning to what this could actually look like in practice.[123] They see themselves already as guardians of their territories, not owners. It is important to understand that when they become guardians recognized by the dominant legal system, they are taking back their relationship with land, water, and natural surroundings.[124]

---

[118] The resolutions can be accessed at the end of this article: Vega Cárdenas (n 18).

[119] David Grinlinton, 'The Continuing Relevance of Common Law Property Rights and Remedies in Addressing Environmental Challenges' (2017) 62 McGill LJ 633, para 68.

[120] Grinlinton (n 119).

[121] Richard Atleo, *Principles of Tsawalk. An Indigenous Approach to Global Crisis* (University of British Columbia Press 2012).

In Nuu-chah-nulth, the word *tsawalk* means 'one'. It expresses the view that all living things – humans, plants, and animals – form part of an integrated whole brought into harmony through constant negotiation and mutual respect for the other. Contemporary environmental and political crises, however, reflect a world out of balance, a world in which Western approaches for sustainable living are not working.

[122] Kimmerer (n 32); Boyd (n 19).

[123] Indigenous Leadership Initiative, *A National Indigenous Guardians Network* <https://www.ilinationhood.ca/publications/backgrounder-a-national-indigenous-guardians-network> accessed 3 August 2022.

'Indigenous Guardians: A modern take on an ancient tradition of caring for the land. Across Canada, approximately 30 teams of Indigenous Guardians are working to conserve and manage their lands. They monitor wildlife, patrol protected areas and reduce the impacts of climate change. In the process, they honor their cultural traditions and train the next generation of leaders. This model is proven to deliver major social and environmental benefits. But far more is possible.'

[124] Motard (n 39); Magallanes (n 40).

Recognizing rights for rivers, as was the case for the Magpie, does not mean anthropomorphizing natural entities. Rights are linked to the entity's intrinsic nature, so rivers have the right to flow, to exist, to have their native biodiversity, to be restored, and to be preserved, among others.

In summary, we present below the main differences and convergences between Western law, indigenous epistemologies, and the rights of Nature:

Rights of Nature, indigenous epistemologies and Western law: some differences and convergences

Issue	Western Law	Indigenous (Innu) Epistemologies	Rights of Nature
Nature's status	Nature as an object; exploitation and destruction admitted, and limits for compensation.	Nature as a living being with agency, an equal, water has a spirit, rivers and nature are considered as ancestors.	Nature as an entity, subject of rights with legal personality (as a person).
Vision	Superiority of humanity over Nature.	Humans as part of Nature.	Humans as part of Nature.
Rights and responsibilities	At the beginning, only European men had rights. Personhood is a question of importance, power and sacredness. At the beginning of the discovery of the Americas, Indigenous peoples, black people, and women were not considered as persons. Now, not only all humans have rights, but also corporations, churches, ships, and trusts. Nature is still mostly considered as a mere resource, an object.	Indigenous laws take into account responsibilities rather than rights. Indigenous law looks to Nature and land to provide principles of law and order (John Borrows). Many rules to protect Nature: when to fish, when to hunt, when to use water. Ask for permission before going to a lake. Ceremonies to pay forward what the earth provides (food, agriculture, animals, fish, etc.).	Humans have the responsibility to preserve Nature, provide clean water, respect nature cycles, respect other species that live on the earth/in water; they have the obligation to restore and prevent degradation.
Property	Ownership of land and territories. Concept of private property.	No private property; rather, collective territories. Communities are stewards or guardians of the land, not the owners.	Nature has the right to live, prosper, and be preserved. Beyond private property. Rivers cannot be commercialized. Guardians act in best interest of Nature (rivers); they do not act as owners.

Issue	Western Law	Indigenous (Innu) Epistemologies	Rights of Nature
Powers	Owners have different powers over Nature, the possibility to take advantage of the land in exclusivity of others ('usus'), to take the fruits ('fructus'), and to destroy the land ('abusus').	Some Indigenous peoples conceive themselves as those who take care of the Earth. Power to protect the land.	Guardians will take care of Natural entities. They will seek the best interest to preserve the ecosystems. (As humans are part of Nature, a healthy Earth will be to the advantage of collectivity.)
Water	Some law frameworks consider water as a commodity (can be sold and bought from the market). Other law frameworks consider water as a common good. Western rationality separates water from the ecosystem. Water is a resource. Resource means something that we use to achieve an objective.	Holistic approach; everything is interconnected. Water is part of the river. Rivers have agency; they have a spirit. Water is sacred; it is a source of life.	Water is part of the ecosystem. Rivers are subject of rights. They are not considered as objects. They are considered as living environments. Rivers and ecosystems should be respected in a holistic perspective.
Structural Rationality	Humans and corporations are at the top of the pyramid. Nature is at the bottom. Humans can use Nature to their own benefit.	Humans are part of Nature. Interdependence of Nature and human. Relationship of reciprocity. Biocultural rights.	Humans are part of Nature, not at the center of the universe. Interdependence between Nature and humans.
Goals	Law framework is protecting the economy and Humans.	Protection of all kinds of living beings and ecosystems. Protecting the cycle of life. Humans should not take more than they need, thinking of future generations.	Protection of the ecosystems as sources of life. The economy should be in harmony with Nature.
Relationship with land	Anthropocentric point of view. Occupation of territory without considering other species and ecosystems.	Symbiotic relationship with the territory. Well-living approach. Biocultural rights.	Ecocentric paradigm. Looks to change the way people interact with Nature.

As we can observe, there is a manifest convergence between indigenous epistemologies and the rights of Nature. However, it is worth noting that in a context of significant degradation of nature, Western legal systems are called upon to recognize the rights of nature. Professor Boyd explains the importance of recognizing the rights of Nature in Western law in the following terms:

> 'Rights are symbolically and politically powerful, as the history of the civil rights, women's rights, indigenous rights, and gay rights movements demonstrates. They are not a magic wand that can be waved to solve problems instantly, but they are a proven means of securing progress in the way society embraces previously mistreated communities.'[125]

Sacco explains that the legal formants within a system do not express themselves in a unequivocal way. On the contrary, in most cases there are contradictions between them. We cannot neglect this reality since it allows us to understand the interaction of the sources of law in order to make the norm evolve. What the International Observatory of Nature's Rights has done in their legal support offered to the Innu community and the Regional County Municipality of Minganie is to go beyond the first solution offered in the legal system. We have been able to verify alternative solutions proposed by other legal formants, in this case, in the Innu legal tradition. The Observatory has, following an in-depth study, apprehended the components of the Canadian system as well as its legal forms and the interaction between the components, looking for convergence.

Studies relating to the application or non-application of colonial State law in Indigenous territories merit in-depth study with the help of legal anthropology. Indeed, through this discipline, we can better understand the need to respect and apply Indigenous peoples' own right to their territories as an essential element of their self-determination, their right to autonomy, and their own governance. It is evident that the case of the *Mutehekau Shipu* encompasses the utilization of indigenous epistemologies in conjunction with non-indigenous individuals that also support that perspective. This demonstrates that the process of decolonization is underway. First Nations have always asserted their right to self-determination, and their legal traditions continue to influence the Canadian legal system, including through the recognition of the rights of Nature, as was the case in New Zealand.[126]

In short, the law-in-context approach proposed by Roberto Sacco, also known as the structural approach, makes it possible to reject the idea of the uniqueness of norms, especially in a pluralistic system such as Canada, where normative interactions are rich. To support the postulate of uniqueness would lead Canada and other countries to erroneous results by suggesting that there is only one legal solution. Therefore, Sacco's approach is helpful in analyzing the evolution of the law, as his posture opposes static legal dogmatism and insists on a dynamic analysis of the law to capture the complexity of the law in action.

---

[125] Boyd (n 19).

[126] Jacinta Ruru, 'Listening to Papatūānuku: a Call to Reform Water Law' (2018) 48 Journal of the Royal Society of New Zealand 215.

VI. Conclusion

It is time for Quebec and Canada to transform the structures and systems that are at the root of climate change and environmental degradation. The recognition of rivers as living environments is already enshrined in the new Quebec legislation. Embracing the rights of Nature movement in the province and across Canada is a step in the right direction so as to preserve rivers and natural entities for future generations.

There are various indications that we are moving towards a harmonization of environmental law with eco-centered decision-making approaches. It is globally known that the objectivization of Nature is at the origin of climate change, the destruction of Nature, and environmental and ecological injustices. The last report of the Intergovernmental Panel on Climate Change IPCC states clearly in Chapter Three that '[i]t is unequivocal that human influence has warmed the global climate system since pre-industrial times.'[127] It also states that climate change has driven detectable changes in the global water cycle since the mid-20th century,[128] provoking heavy precipitation and flooding, but also aridity and drought.

The attribution of a legal personality to Nature and, more specifically, to rivers sets the bar very high. The challenges are important, because the environmental norms are reduced to one very important norm: respecting the rights of the river. This way of protecting Nature through the recognition of legal rights seems to be easier to verify, to claim, and to monitor – more than a law framework centered on prohibition. In fact, infringers will be more easily detected. There will no longer be a right to pollute, but an obligation to clean up pollution before discharging it into waterways. The various users of water will have to adjust quickly to a new reality, since water will no longer be considered as a resource exclusively available to humans, but as a living environment, a living being with personality.

This does not mean that humans can no longer use water, as we are also a species that is part of the ecosystem. The paradigm shift will necessarily and imperatively lead to sustainable practices. There will be no more room for closed eyes, plugged ears, or dead ends, for the guardians will be the eyes, the voice, and the light to protect the rights of the river born in law with the watchful eyes of the world.

Defending the rights of Nature is a joint struggle to protect the human rights to water, to a healthy environment, to live, to health, to self-determination, and to ensure bio-cultural rights that aim to protect this intrinsic link between humans and the land that surrounds them. Even if the anthropocentric perspective is still dominant in the Canadian legal system, we cannot ignore that a paradigm shift

[127] Working Group 2021 (n 46) 282: 'Human activity has warmed the world by about 1° C since pre-industrial times, and the impacts of this warming have already been felt in many parts of the world.'

[128] Working Group 2021 (n 46), chs 8 and 15 ff.

is transforming our environmental law systems across the country and around the word. Professor David Boyd,[129] a law professor at the School of Public and Global Affairs at the University of British Columbia, remarked that this 'fascinating' development may also foster more 'innovative' ways to integrate Indigenous law into the Canadian legal system, which has failed to acknowledge that Canada has three founding legal systems and 'not two as many are still taught in law school'.[130]

Climate change and environmental degradation are facilitators of zoonotic spillovers. Science has proven that those factors have an important effect on transmissions chains.[131] Recognizing personhood in natural entities could change human behavior and the way we relate with nature. We could try to prevent pandemics as healthy rivers will lead to healthy communities.

As many affirm, why explore possibilities of life on other planets, while on our own planet, we cannot respect different expressions of life? We need to hear the voices of our ancestors, and Indigenous knowledges to protect life on earth. As Innu elders say: Water unites us all wherever we are on Earth. We are all interdependent as we share the same planet. Water travels everywhere, it goes up in the sky, changes into clouds and then comes down again to wash our worries away and nourish all our being. The memories of our ancestors live in that same water that travels through all of us. That same water from where we all come from. From the womb of our mothers.

---

[129] Boyd (n 19).

[130] David R Boyd, Special Rapporteur on human rights and the environment, quoted in Vega Cárdenas (n 18).

[131] Xavier Rodó, Adrià San Jose and Karin Kirchgatter, 'Changing Climate and the COVID-19 Pandemic: More than Just Heads or Tails' (2021) 27 Nat Med 576.

III. Poietic Legal Orders

Ecuador

The Nature of the Rights of Nature

Gabriela Espinoza[*]

In Ecuador, beginning with the constitutional recognition that nature is a subject with rights, at least three stages of development can be identified regarding this recognition's content, its relationship with other legal concepts, and the scope of its application and enforceability. These stages are presented chronologically and reflect an evolution in discussions on doctrine, law, and jurisprudence that have been progressively moving, though slowly and not without difficulties, towards an understanding of the nature of the rights of nature.

First, in the immediate aftermath of the adoption of the Constitution, academic circles and social movements focused on discussing the rationale behind this constitutional recognition. This was a moment of great constitutional enthusiasm cen-

[*] The manuscript was finalized in May 2022. The assessment of literature and case law, including internet materials, reflects the legal situation at that time. The discussion on legal subjectivity in section III. was supplemented in May 2024, to reflect more on recent developments. Section II. 1. b) was added in February 2025.

tered primarily on foundational debates. The discussion focused on matters such as who was pushing for the recognition of nature; what the relationship was between nature and the worldview of indigenous peoples; what it meant to have the concepts of 'Pacha Mama' and '*sumak kawsay*' as constitutional categories; and what the main reasons were for the recognition of nature as a subject of rights. I will call this stage 'Early Constitutional Enthusiasm'.

Second was a stage linked to the legislative development of the rights of nature and a rather rhetorical judicial development of their content. This stage was marked by a conceptual hybridization between constitutional environmental rights (which focus on human interests) and the autonomous rights of nature. The content of the rights of nature was not yet developed in this stage, but the idea that another subject, not constituted by humans, has constitutional rights began to make its way into Ecuadorian legal culture. I will call this stage 'Intermediate Hesitant Exploration'.

Finally, the third stage marked the beginning of an important development in jurisprudence with two main levels of focus. The first involved explaining the criteria used to define the content recognized as rights of nature, and the second consisted in identifying which part or parts of nature have rights, or whether the rights apply to nature as a whole.

Continuities exist between these stages. Characteristics of one stage that continue into the next sometimes produce tensions. For example, the relationship between the right to a healthy environment and the rights of nature, very characteristic of the second stage, is addressed in the third stage as a relationship yet to be disentangled. At other times, the continuities are either more subtle or more acute. For example, the analysis of why nature has rights (substantiation) is still present in the second stage in a superficial way, and it then deepens and regains interest in the third stage. I will call this stage 'Singularization and Consolidation of Nature as a Subject of Rights'.

Finally, there are also ruptures between stages. For example, the doctrinal discussion in the first stage on how indigenous communities' knowledge could guide the interpretation of the rights of nature is a debate neglected in the following stages. In the third stage, the language of biology has been the key to disentangling the contents of the rights of nature. A return to the indigenous worldview as a tool for interpreting these rights could be the pending task of a fourth stage.

This report is divided into three parts, as follows. In the first part, I will briefly explain the main substantive and adjective characteristics of the rights of nature, referring to constitutional jurisprudence to better illustrate the issues raised. In the second part, I will pause to explain each of the stages, which I refer to as substantiation, exploration, and concretization. Finally, I will explain, in comparative terms, what it means that in Ecuador nature is a 'subject of rights' rather than a 'legal person'.

I. Basic Concepts

The Constitution of Ecuador, approved in 2008, recognizes that nature is a subject of rights and identifies what these rights are. Furthermore, it relates this recognition to the indigenous worldview known as *sumak kawsay* and stipulates that the principles of interpretation for constitutional (human) rights apply equally to the rights of nature. It also determines who can represent nature and how to claim its rights.

1. Recognition and Content of the Rights of Nature

The Constitution frames this recognition in its preamble, with a relational approach between humanity and nature, stating,

> 'We, the sovereign people of Ecuador […]. Celebrating nature, Pacha Mama, of which we are a part and which is vital for our existence […]. Decide to build a new form of citizen coexistence, in diversity and harmony with nature, to achieve good living, sumak kawsay […]'.[1]

The scope of the normative value of the Preamblecan be seen in the express declaration of the recognition of nature as a subject of rights. Article 10 of the Constitution states: 'Nature shall be a subject of those rights recognised by the Constitution'. Art 71 identifies what these rights are and also who can claim them:

> 'Nature or Pacha Mama, where life is reproduced and realised, has the right to full respect for its existence and the maintenance and regeneration of its vital cycles, structure, functions and evolutionary processes. Any person, community, people or nationality may demand from the public authority the fulfilment of the rights of nature. In the application and interpretation of these rights, the principles set out in the relevant provisions shall be observed. The State shall encourage natural and legal persons and collectives to protect nature, and shall promote respect for all the elements that make up an ecosystem.'

Several characteristics of the Constitution's vision of nature and the content of its rights can be identified in this constitutional provision. Firstly, by naming nature as 'Pacha Mama', the Constitution draws a clear relationship with indigenous knowledge and communities; it is they who recognize it as *Mother Earth* and who attribute an intrinsic value to it. Moreover, in connection with the preamble, respect for nature and harmonious coexistence with it constitute the path towards *sumak kawsay* – a symbolic category that means *life in harmony* in the worldview of several indigenous peoples. I will discuss the role played by indigenous communities in this recognition in the second part of this text.

Secondly, as in the preamble, the articles recognize that it is there, in nature, that life is reproduced and realized, and they go on to recognize two rights. The first is a general recognition of respect: integral respect for nature's existence. The second is a specific recognition of rights expressed in biological terms: re-

[1] Translation by the author.

spect for the maintenance and regeneration of a) nature's vital cycles, b) structure, c) functions, and d) evolutionary processes.

Although the recognition of rights is to *a whole*, to all of nature, without specifying particular ecosystems or components, the constitutional text also alludes to respect for 'all the elements that make up an ecosystem'. This is one of the most relevant discussions in current constitutional jurisprudence, as the Constitutional Court has gradually identified ecosystems or elements of nature in order to protect them as subjects of rights. For example, the Court began by declaring mangroves to be subject of rights,[2] then a forest called Los Cedros,[3] then a river called Aquepi,[4] then a micro-basin of an urban river known as Monjas,[5] and finally a woolly monkey called Estrellita.[6] The process of concretization of the protection of nature, its ecosystems, and its elements will be discussed in more detail in the second part of this text.

2. Relationship Between the Constitutional Rights of Nature and Constitutional Principles and Rights

The Constitution also states that 'nature has a right to restoration' and that this is 'independent of the obligation of the State and natural or legal persons to compensate individuals and collectives that depend on the affected natural systems'.[7] The purpose of this express distinction is to separate the contents of the rights of nature, with respect to their reparation, from the individual, collective, or general content of the right to a healthy environment. The relationship between rights to a healthy environment and rights of nature establishes the fact that although they are two different domains, they can also be complementary.[8] I will return to this discussion of complementarity and tension in the second part of this text.

The rights of nature also have connections with other constitutional rights and principles. In particular, the Constitution establishes a causal and mediated relationship between the rights of nature and the Andean worldview of *sumak kawsay*, generally translated as 'good living'. The Constitution of Ecuador establishes that good living is the aim of the political and social economic development regime in the country. In order to achieve good living, people, communities, and nation-

---

[2] Corte Constitucional del Ecuador 8 September 2021, 22-18-IN/21.

[3] Corte Constitucional del Ecuador 10 November 2021, 1149-19-JP/21.

[4] Corte Constitucional del Ecuador 15 December 2021, 1185-20-JP/21.

[5] Corte Constitucional del Ecuador 19 January 2022, 2167-21-EP/22.

[6] Corte Constitucional del Ecuador 27 January 2022, 253-20-JH/22.

[7] Constitución del Ecuador, art 72.

[8] The relationship, and perhaps interdependence, between the right to a healthy environment and the rights of nature is also well formulated in art 66(27) of the Constitution, which recognizes 'the right to live in a healthy, ecologically balanced environment, free of contamination and in harmony with nature'. In the case of Los Cedros forest, this relationship is also further developed. See: Corte Constitucional del Ecuador 10 November 2021, 1149-19-JP/21, paras 239–252.

alities must enjoy their rights, but exercise them responsibly within the framework of interculturality, respect for diversity, and in *harmonious coexistence with nature*.[9] Both the means and the effect of good living are harmonious coexistence and complementarity between humans and nature 'insofar as they make up common life systems'.[10]

The recognition of the rights of nature in Ecuador must therefore be understood within the framework of two important concepts: interculturality and *sumak kawsay*. The idea that nature is its own subject, 'Pacha Mama', is a conception that stems mainly from indigenous worldviews, as does the idea that a harmonious relationship with nature is the path to *sumak kawsay*.

The constitutional recognition of ancestral Andean knowledge (on the purpose of life: *sumak kawsay*; and on its means: respect for nature) in the Western vocabulary of rights translates into an intercultural dialogue. It is recognized that nature is a *subject* that has *constitutional rights*, and that respecting them implies not only recognizing a new subject that has enforceable rights on an equal level with human rights, but also the need for a new way of relating to nature, understanding it now as a subject to be respected and not only as an object of protection.

Furthermore, establishing that the aim of Ecuador's economic, social, political, and cultural regime is to achieve good living implies a central recognition of respect for and conservation of nature. If a harmonious relationship with nature, which has the right to respect for its existence and the maintenance and regeneration of its vital cycles, structure, functions, and evolutionary processes, is indispensable to reach this aim, then guaranteeing that aim implies a significant limit to the exploitation of natural resources and, in general, a guiding principle for development policies.[11]

---

[9] This is established in art 275 of the Constitution. The rights of nature, good living, interculturality, and sustainable development are the criteria used by the Constitution to define what is understood by 'development regime'. It states: 'The development regime is the organised, sustainable and dynamic set of economic, political, socio-cultural and environmental systems that guarantee the realisation of good living, of sumak kawsay'. This development regime, defined by the Constitution, has among its objectives pursuant to art 276(4) '[t]o recover and conserve nature and maintain a healthy and sustainable environment that guarantees people and communities equitable, permanent and quality access to water, air and soil, and to the benefits of subsoil resources and natural heritage'.

[10] Corte Constitucional del Ecuador 10 November 2021, 1149-19-JP/21, para 50.

[11] Defining good living in a concept that encompasses the unwritten philosophy of various indigenous worldviews is not a simple undertaking. The philosophy has political, economic and social implications and uses. For approaches from the authors' own experience of indigenous worldviews, see: Carlos Viteri Gualinga, 'Visión indígena del desarrollo en la Amazonía' (2002) 3 Polis <https://journals.openedition.org/polis/7678>; Raúl Llasag Fernández, 'El *sumak* kawsay y sus restricciones constitucionales' (2009) 12 Revista Foro 113 <https://repositorio.uasb.edu.ec/bitstream/10644/2306/1/07-TC-Llasag.pdf>; Nina Pacari, 'El Sumak Kawsay tiene que ver con un sistema de vida comunitario' in Francisco Hidalgo Flore (ed), *Proceso Constituyente y Buen Vivir* (Universidad Central del Ecuador 2021) <www.cetri.be/IMG/pdf/proceso_constituyente_

Thus, for example, the Constitution determines that economic policy, public debt, trade policy, management of non-renewable natural resources, and forms of production must consider biophysical limits, conservation, and the impacts that such activities may have on nature.[12] In order to frame these obligations, the Constitution is clear in stating that in cases of doubt, in environmental matters, the interpretation most favorable to nature is to apply (*pro natura principle*).[13]

In addition to the pivotal relationship between *sumak kawsay* and the rights of nature, it is possible to identify a web of interconnections with other constitutional rights in the Constitution and in the jurisprudence of the Constitutional Court. One such example is the relationship of the rights of nature with water. The Constitution recognizes the right of people to water[14] and also mandates the conservation of nature to preserve the sources and quality of water.[15] At the same time, it recognizes that water constitutes a vital element both for nature and for

y_buen_vivir_definitivo_/ene_022.pdf>. It is interesting to note that the translation of sumak kawsay is not uncontroversial; for the Amazonian indigenous community of Sarayaku, for example, the concept translates into 'life in plenitude'. For an analysis of its meaning for state-building and its relation to the rights of nature, see: Alberto Acosta, *Buen Vivir y Sumak Kawsay* (Abya Yala 2012). Dania Quirola Suárez, 'Sumak Kawsay. Hacia un nuevo pacto social en armonía con la naturaleza', in Alberto Acosta (ed), *El buen vivir. Una oportunidad para imaginar otros mundos* (Abya Yala 2012). For an analysis that seeks to systematize its multiple meanings, see: Philipp Altmann, 'Buen Vivir como propuesta política integral: Dimensiones del Sumak Kawsay' (2016) 3(1) Mundos Plurales, Revista Latinoamericana de Políticas y Acción Pública 55 <https://revistas.flacsoandes.edu.ec/mundosplurales//article/view/2318/1475>.

[12] Constitución del Ecuador, art 284 'Economic policy shall have the following objectives [...] 4. To promote the incorporation of added value with maximum efficiency, within the biophysical limits of nature and respect for life and cultures'. Art 290 'Public debt shall be subject to the following regulations [...] 2. It shall be ensured that public debt does not affect sovereignty, rights, good living and the preservation of nature'. Art 306 '[...] The state shall promote imports necessary for development objectives and shall discourage imports that negatively affect national production, the population and nature.' Art 317 'Non-renewable natural resources are part of the inalienable and imprescriptible heritage and assets of the State. In managing them, the State shall prioritise intergenerational responsibility, the conservation of nature [...].' Art 319 '[...] The State shall promote forms of production that ensure the good living of the population and shall discourage those that threaten their rights or the rights of nature [...]'.

[13] Constitución del Ecuador, art 395 'The Constitution recognises the following environmental principles: 1. The State shall guarantee a sustainable model of development, environmentally balanced and respectful of cultural diversity, that conserves biodiversity and the capacity for natural regeneration of ecosystems and ensures the satisfaction of the needs of present and future generations [...] 4. In case of doubt about the scope of legal provisions on environmental matters, these shall be applied in the sense most favourable to the protection of nature'.

[14] Constitución del Ecuador, art 298 '[...] Latifundia and land concentration are prohibited, as well as the monopolisation or privatisation of water and its sources'.

[15] Constitución del Ecuador, art 276(4) 'The development regime shall have the following objectives: '4. To recover and conserve nature and maintain a healthy and sustainable environment that guarantees people and communities equitable, permanent and quality access to water, air and soil, and to the benefits of subsoil resources and natural heritage and assets'.

the existence of human beings.[16] As it is a vital element for nature, the care and conservation of river basins and ecological flows, as well as, in general, the use of water must be sustainable in such a way that it does not threaten the vital cycles of nature.[17]

The case of the River Monjas is paradigmatic because it reveals the multiple connections of the rights of nature with other rights. This is an urban river polluted with sewage and grey water discharge. The pollution and, in turn, the discharge of sewage and rainwater is so massive that it has rendered the river dysfunctional: it has destroyed its valleys and vegetation, eroded its floor, and widened its banks. The quality of the water is so poor and the quantity so massive that the ecological flow of the river has been completely altered.

The Constitutional Court declared the violation of the rights of the River Monjas. But it also recognized that the contamination of the body of water and the lack of water sanitation constitutes a violation of the constitutional right to water of the people living around the river basin. It also indicated that the contamination of the river and the massive discharge of water was due to deficient territorial organization that did not take into account the hydrography of the city of Quito (high rainfall).[18]

Consequently, it declared the violation of the inhabitants of Quito's constitutional right to the city (in Ecuador, the city is a constitutional right). Finally, the aggressive erosion of the riverbed has caused one of the gorges through which the river flows to become very deep and produce a high risk of mass displacement. The houses, which are getting closer and closer to the gorge, could fall in. Therefore, the Court declared the violation of the right to a safe habitat of the owners of a real estate property (which is also an asset) that is close to collapse.

3. *Tensions Between the Constitutional Rights of People and of Nature*

Rights of nature can also conflict with other rights. The Constitution states that the rights it has recognized (around 90) are all of equal status, including the rights of nature. And as with all rights, no one right is absolute; all rights allow for lim-

---

[16] Constitución del Ecuador, art 318 'Water is a strategic national asset of public use, an inalienable and imprescriptible possession of the State, and constitutes a vital element for nature and for the existence of human beings. All forms of privatisation of water are prohibited'.

[17] Constitución del Ecuador, art 411 'The State shall guarantee the conservation, recovery and integral management of water resources, river basins and ecological flows associated with the water cycle […]. The sustainability of ecosystems and human consumption shall be a priority in the use and exploitation of water'.

[18] In general, bodies of water may receive rainwater. The problem in Quito, as in other parts of the world, is that the city's ground has lost its capacity to absorb rainfall because the ground has become 'sealed'. This is due to the processes of construction and cementing over of ground: the fewer green areas there are, the less capacity the soil has to absorb rain. In places like Quito, where it rains a lot, the rain that has nowhere to percolate goes to the sewers and from there to the River Monjas.

its and regulations.[19] The rights of nature may have a greater potential for conflict with classical civil rights such as the rights to freedom of enterprise or legal certainty.[20]

For example, there was much debate surrounding the Constitutional Court's recognition of mangrove ecosystems as a subject of rights. One regulation, issued by the President of the Republic, sought to allow all types of 'productive activities or public infrastructure' to be carried out in mangroves, subject to environmental authorization. The Court considered that this generic and indeterminate norm allowed the margin of exploitation of the mangrove to be extended to other productive activities that could affect its ecosystem, and that the inclusion of 'new productive activities' should be introduced by law, as it could mean a limit to the rights of nature, which are reserved by law.

The activities of shrimp companies in mangroves, and in general, the possibility of carrying out extractive activities without considering the elements of nature, is a subject of great debate. For instance, one of the dissenting opinions in this case stated:

> '[t]he Constitution protects the rights of nature, guaranteeing its conservation and restoration, which is not incompatible with the exploitation of natural resources, since they can be used for the benefit of society, in a rational, sustainable and sound manner'.[21]

In another case, the Constitutional Court recognized that a forest called Los Cedros was a subject of rights. In this forest, several permits had been issued for mining prospecting. But the forest, located in the north-west of Ecuador, has a high degree of diversity of fauna and flora, an ecosystem with endemic species, and a

[19] Such limits and regulations must be proportional and reasonable, and in the case of Ecuador, they must be established by 'organic' laws, ie laws that require reinforced majorities. Exceptions to this statement should also be made for rights that cannot be limited, such as torture, arbitrary deprivation of life, and certain minimum procedural guarantees. See, in general: United Nations, General Comments No 24 and 29, Human Rights Committee.

[20] This is a problem that reflects a significant contrast with the other countries that recognize some kind of rights of nature. In Ecuador, the rights of nature are recognized at the constitutional level and this text establishes, in turn, that all rights have equal hierarchy. Therefore, *prima facie*, the constitutional rights of nature have the same weight and importance as the rights of humans. The balance between them has to be established on a case-by-case basis. In the cases of New Zealand and Bolivia, for example, the rights have 'legal' status. They are not constitutional rights, so their protection is less reinforced and does not have the same weight as rights recognized (entrenched) in constitutions. The case of municipal ordinances in the United States that recognize nature's right 'to flourish' have an even weaker legal standing. Above them are the state constitutions. For a detailed comparison, see: Craig M Kauffman and Pamela L Martin, 'Constructing Rights of Nature Norms in the US, Ecuador, and New Zealand' (2018) 18(4) Global Environmental Politics 43. The article demonstrates how the contexts of each country have led to different contents and levels of enforceability of the rights of nature. However, due to the timing of its publication, its analysis of the rights of nature in Ecuador does not consider the third phase as explored in this present chapter.

[21] Corte Constitucional del Ecuador 8 September 2021, 22-18-IN/21, Dissenting Opinion of Constitutional Judge Carmen Corral Ponce.

high number of endangered, unique, and rare species. Los Cedros is considered a biodiversity corridor. In this context, the Constitutional Court found that the authorization for the mining activity violated the rights of the forest and prohibited any type of extractive activity. It concluded that, in light of the precautionary principle, the public authority should not have issued environmental permits for the mining activity. It therefore upheld the decision to annul such permits for mining concessions.

The case generated a great deal of controversy.[22] Could permits for mining concessions be annulled in this way? Wouldn't this decision violate the legal security of the companies that, trusting in the decision of the administrative authorities, had invested and started operations? In the dissenting opinions, the judges, for example, pointed out that 'extractive activity alone does not generate destruction of nature (species, ecosystems) nor does it violate the rights of nature [...]'.[23] These discussions on the balance between the effective enjoyment of some rights and the possible limits to others will continue to be part of the debate, but it is perhaps in cases involving extractive activities where the tension can be seen most clearly between the rights to freedom of enterprise and legal security and the rights of nature (especially when it is understood that the aim of economic development should be good living: harmonious coexistence between people and nature).

4. Enforceability and Obligations of the State and of Natural and Legal Persons with Regard to the Rights of Nature

The constitutional text also establishes several obligations for the State that derive from the rights that it recognizes for nature. It establishes obligations of respect and promotion,[24] prevention,[25] and reparation.[26] It also establishes as duties and responsibilities of Ecuadorians 'to respect the rights of nature, preserve a healthy environment and use natural resources in a rational, sound and sustainable manner'.[27]

---

[22] The decision was a majority ruling with five judges voting in favor of annulling the permits and four against.

[23] Corte Constitucional del Ecuador 10 November 2021, 1149-19-JP/21, Dissenting Opinion of Constitutional Judge Teresa Nuques Martínez.

[24] Constitución del Ecuador, art 71 'The State shall encourage natural and legal persons and collectives to protect nature, and shall promote respect for all the elements that form an ecosystem'.

[25] Constitución del Ecuador, art 73 'The State shall apply precautionary and remedial measures for activities that may lead to the extinction of species, the destruction of ecosystems or the permanent alteration of natural cycles.'

[26] Constitución del Ecuador, art 72 'In cases of serious or permanent environmental impact, including those caused by the exploitation of non-renewable natural resources, the State shall establish the most effective mechanisms to achieve restoration [...].'

[27] Constitución del Ecuador, art 83(6).

The Constitution of Ecuador establishes that all the rights it recognizes, including those of nature, are enforceable and, therefore, fully justiciable; that they are directly and immediately applicable by and before any public or judicial official; and that the lack of a legal norm cannot be invoked to justify their violation or disregard.[28] As the rights of nature are constitutional rights, they are fully enforceable through constitutional guarantees.

The Constitution essentially recognizes two types of guarantees: normative and jurisdictional guarantees. The former has to do with the guarantee that all non- statutory legal norms are developed subject to the reservation of law and in accordance with the Constitution. For this reason, the rights of nature can and have been used to declare unconstitutional laws or regulations that violate their contents.[29] The latter guarantees have to do with activating legal mechanisms to request judges to protect the rights of nature on a case-by-case basis. The Constitution recognizes several mechanisms, such as acción de protección [writ of protection or 'amparo']. Finally, the Constitutional Court can review cases of rights violations to correct the criteria used by lower judges,[30] or to establish general precedents in cases that it has selected for their relevance.[31]

To date, the Constitutional Court, through concrete and abstract judicial review, has recognized, as discussed above, the violation of the rights of mangroves, forests, rivers, and wild animals. The analysis of the contents of the rights of nature that have been developed at the legislative and jurisprudential level will be presented in the second part of this text. For now, suffice to say that in Ecuador the rights of nature are fully justiciable.

Reparation for cases has included: the declaration of unconstitutionality of norms that violate the rights of nature, the annulment of permits for mining activities, the decontamination of rivers, the recognition that wild animals can file habeas corpus, but most importantly, the express recognition that the reparation of the right depends on the type of ecosystem or element of nature in question.

---

[28] Art 11 of the Constitution states that 'The exercise of rights shall be governed by the following principles: 1. Rights may be exercised, promoted and demanded individually or collectively before the competent authorities; these authorities shall guarantee their enforcement [...] 3. The rights and guarantees established in the Constitution and in international human rights instruments shall be directly and immediately applicable by and before any public, administrative or judicial servant, ex officio or at the request of a party. There shall be no conditions or requirements for the exercise of constitutional rights and guarantees that are not established in the Constitution or the law. Rights shall be fully justiciable. The lack of a legal norm may not be invoked to justify their violation or disregard, to dismiss the action on those facts or to deny their recognition.'

[29] Corte Constitucional del Ecuador 8 September 2021, 22-18-IN/21; Corte Constitucional del Ecuador 9 June 2021 32-17-IN/21. The way to monitor compliance with this guarantee is to check the constitutionality of all legislative instruments.

[30] Corte Constitucional del Ecuador, 15 December 2021, 1185-20-JP/21; Corte Constitucional del Ecuador, 19 January 2022, 2167-21-EP/22.

[31] Corte Constitucional del Ecuador 10 November 2021, 1149-19-JP/21; Corte Constitucional del Ecuador, 253-20-JH/22, 27 January 2022.

This has been one of the reasons why, even though the Constitution recognizes the rights of nature as a whole, the Court has also decided to expressly recognize mangroves, rivers, forests, and wild animals as rights holders.

Finally, the Constitution provides that any person, community, people, or nationality can demand the enforcement of the rights of nature (*actio popularis*). The cases that have reached the Constitutional Court on the violation of the rights of nature have been brought by natural persons;[32] by animal welfare, ecological, and environmental organizations;[33] by public authorities;[34] by municipalities;[35] and by nine girls from the Amazon.[36] Additionally, the Ombudsman's office is legally authorized to promote the protection of the rights of nature and to bring legal actions on its behalf.[37]

II. The Ongoing Development of the Rights of Nature

Since the recognition of the rights of nature in the text of the Constitution, three stages of their development can be identified. The first stage was dominated by academic exploration of the struggles and actors that had led to the recognition of these rights in the Constitution. This phase focused on substantiation.

The second stage, was characterized by the drafting of laws and the first rulings that timidly sought to clarify the concept of these rights. This period represents an intermediate hesitant exploration. In the third stage, the Constitutional Court started to develop and solidify the autonomous content of the rights of nature, thereby affirming and consolidating nature as a distinct subject of rights. This was a stage of singularization and consolidation.

1. First Stage: Early Constitutional Enthusiasm

Once the rights of nature had been approved and recognized in the Constitution, the task of explaining them initially fell to the academic world. At this time, texts

---

[32] Corte Constitucional del Ecuador 19 January 2022, 2167-21-EP/22.

[33] Corte Constitucional del Ecuador 8 September 2021, 22-18-IN/21.

[34] Corte Constitucional del Ecuador 10 November 2021, 1149-19-JP/21.

[35] Corte Constitucional del Ecuador15 December 2021, 1185-20-JP/21.

[36] Sala Multicompetente de la Corte Provincial de Justicia de Sucumbíos, 21201-2020-00170.

[37] Article 3 of the Ley Orgánica de la Defensoría del Pueblo of 2019 states (translation by the author): 'Duties. – The duties of the Ombudsman's Office, in accordance with the provisions of the Constitution of the Republic of Ecuador and international human rights instruments, are: […] b) To prevent violations of human rights and nature; c) To promote public awareness, education, advice, advocacy, and monitoring of human rights and nature; and, d) To protect and safeguard human rights and nature.'

were produced by Ecuadorian[38] and foreign[39] authors, initiating what could be considered the canon of the rights of nature in Ecuador. The texts that were produced focused, in particular, on the substantiation and history of the recognition of the rights of nature. I will begin by summarizing these historical descriptions, focusing on the role of social and environmental movements. I will then briefly review the discussion on substantiation.

a) Histories of recognition

In Ecuador, the recognition of the rights of nature has been based mainly 'on historical and ecological components, in consideration of a combination of two ele-

[38] The most representative authors include Alberto Acosta, Ramiro Ávila, Agustín Grijalva, Raul Llasag, Esperanza Martínez, and Julio Prieto. See: Alberto Acosta and Esperanza Martínez (eds), *Derechos de la Naturaleza. El futuro es ahora* (Abya Yala 2009); Alberto Acosta and Esperanza Martínez (eds), *La Naturaleza con derechos. De la filosofía a la política* (Abya Yala 2011); Ramiro Ávila, 'El derecho de la naturaleza: fundamentos', in Carlos Espinosa and Camilo Perez (eds), *Los Derechos de la Naturaleza y la Naturaleza de sus Derechos* (Ministero de Justicia y Derechos Humanos 2011) <https://biblio.flacsoandes.edu.ec/libros/digital/57146.pdf>; Agustín Grijalva, *Constitucionalismo en el Ecuador* (CEDEC 2012) <http://agustingrijalva.com/wp-content/uploads/2016/06/constitucionalismo_en_ecuador.pdf>; Raul Llasag 'Derechos de la naturaleza: una mirada desde la filosofía indígena y la Constitución', in Carlos Espinosa and Camilo Perez (eds), *Los Derechos de la Naturaleza y la Naturaleza de sus Derechos* (Ministero de Justicia y Derechos Humanos 2011) <https://biblio.flacsoandes.edu.ec/libros/digital/57146.pdf>; Esperanza Martínez 'Los Derechos de la Naturaleza en los países amazónicos', in Alberto Acosta and Esperanza Martínez (eds), *Derechos de la Naturaleza. El futuro es ahora* (Abya Yala 2009); Julio Prieto, *Derechos de la naturaleza. Fundamento, contenido y exigibilidad jurisdiccional* (CEDEC 2013) <www.garn.org/wp-content/uploads/2021/09/Prieto_DDN_2013.pdf>. An in-depth and recent analysis of the trajectory of the first decade of the rights of nature can be found in: Adolfo Maldonado and Esperanza Martínez (eds), *Una década con derechos de la naturaleza* (Abya Yala 2019) <https://www.naturalezaconderechos.org/wp-content/uploads/2020/06/Una-década-con-Derechos-de-la-Naturaleza-03-25-2019.pdf>.

[39] Eduardo Gudynas, Boaventura de Sousa Santos (who even gave talks to the Constituent Assembly members), and Eugenio Zafaroni were the most representative foreign authors. See Eduardo Gudynas, *El mandato ecológico. Derechos de la naturaleza y políticas ambientales en la nueva Constitución* (Abya Yala 2009); Eduardo Gudynas, *Derechos de la Naturaleza. Ética bicéntrica y políticas ambientales* (CLAES 2014) <http://gudynas.com/wp-content/uploads/GudynasDerechosNaturalezaLima14r.pdf>; Eugenio Zafaroni, 'La naturaleza como persona: de la Pachamama a la Gaia', in Carlos Espinosa y Camilo Perez (eds), *Los Derechos de la Naturaleza y la Naturaleza de sus Derechos* (Ministero de Justicia y Derechos Humanos 2011) <https://biblio.flacsoandes.edu.ec/libros/digital/57146.pdf>; Boaventura de Sousa Santos, *Derecho y Emancipación* (CEDEC 2012) <www.boaventuradesousasantos.pt/media/Derecho%20y%20Emancipación.pdf>. In a collaboration between the Instituto de Estudios Ecologistas del Tercer Mundo in Ecuador and the Universidad del Bosque in Colombia, a book was dedicated exclusively to the rights of nature: Diana Murcia, *La naturaleza con derechos* (Chasqui Editores 2012) <http://biblioteca.clacso.edu.ar/Ecuador/ieetm/20170626043529/pdf_1395.pdf>.

ments: on alludes to a return/recognition of ancestral knowledge, and the other to the awakening of ecological awareness'.[40]

The ancestral component refers to the generally respectful and harmonious *relationship* that indigenous peoples in Ecuador have with nature – even though different groups may conceive of nature in distinct ways – and to a recognition of their traditional *knowledge* of its life cycles.

Several indigenous organizations went to the Constituent Assembly to request the recognition of the rights of nature. But there was also a sizeable representation of indigenous people as Constituent Assembly members. This cultural component was very important in order to justify the recognition of these rights, as well as the related concepts of *sumak kawsay*, interculturality, and plurinationality.[41] For example, the minutes of one of the Constituent Assembly's discussions state that 'the concept of "subject of rights" shows that this new legal conception of human relations with nature was assimilated by Andean cultures long ago'.[42]

Indigenous leader Nina Pacari sees the constituent process as a point of opening for the introduction of indigenous peoples' concepts – a moment that is also the product of many struggles of the indigenous movement in Ecuador. Thus, she explains,

> 'the indigenous movement sees this as the moment to introduce the themes of plurinationality and interculturality, as well as sumak kawsay. Although we must not forget that there was resistance to these proposals in the Assembly itself, but at the time, those who resisted were isolated'.[43]

The relationship and constitutional recognition of *sumak kawsay* and the rights of nature imply that in order to understand their connection with and application to the construction of law, it is necessary to look at the knowledge and experiences of indigenous peoples. This constitutes a recognition of the various normative systems that make up the Constitution and reinforces the idea of legal pluralism.

For example, in the case of Los Cedros forest, the Constitutional Court recognized that

> 'The rights of nature require that in order to harmonise their relationship with nature, it is human beings who must adapt themselves appropriately to natural processes and systems, hence the importance of having scientific knowledge and community knowledge, especially indigenous knowledge due to their relationship with nature, about such processes and systems'.[44]

---

[40] Julio Prieto, *Derechos de la naturaleza. Fundamento, contenido y exigibilidad jurisdiccional* (CEDEC 2013) 30.

[41] For an analysis of this relationship, see 'Pluralismo jurídico y derechos de la Naturaleza' (2020) 34 Revista de Derecho Foro <https://revistas.uasb.edu.ec/index.php/foro/issue/view/109>.

[42] Asamblea Constituyente ecuatoriana 2008, Acta 58, 6 June 2008, p 67.

[43] Nina Pacari, 'El Sumak Kawsay tiene que ver con un sistema de vida comunitario', in Francisco Hidalgo Flor (ed), *Proceso Constituyente y Buen Vivir 2007–2022* (Universidad Central del Ecuador 2021) 20.

[44] Corte Constitucional del Ecuador 10 November 2021, 1149-19-JP/21, para 52.

However, despite the express recognition that indigenous knowledge should function as an interpretative tool for understanding what the rights of nature are and how they apply, in the subsequent stages of development, this source – which is now a source of law – has not received much attention. This is one of the major ruptures between this stage of development of the rights of nature, where the role of the indigenous worldview in understanding the natural world is highlighted, and the following stages, where this knowledge has not been used to illuminate these rights' content and scope.

The other component on which the recognition of the rights of nature is based is the ecological component. This component stems not only from transnational networks and environmental movements that pushed for its recognition,[45] but also from the concern of the Assembly members themselves for the state of nature. In the report that was drafted for the approval of the texts recognizing the rights of nature, the Assembly members expressed that it is 'relations between humanity and nature based on subordination and utilitarianism'[46] that could lead to the disappearance of forests and the extinction of various species. They stressed

> 'the need to change the development paradigm, the legal approach and the relationship that human beings have with their environment, in order to avoid or at least mitigate the unpredictable consequences that will occur if we maintain the anthropocentric system of exploitation'.[47]

These concerns about the system of production and consumption, about the idea of economic development, and about the current state of nature even led the Constituent Assembly members to question the capacity of environmental legislation and to warn that 'Kyoto failed because it considered nature as an object, and not as a subject'.[48] Ecological awareness, pushed by environmental groups, permeated the discussion in the Constituent Assembly, which looked at the possibility of recognizing the rights of nature as a mechanism to address the ecological crisis. This component is very present in the third stage of development, where, through jurisprudence, the need to protect nature in order to avoid climate catastrophes, species extinction, and water pollution, among other things, is echoed.

---

[45] The idea that nature should have rights and some kind of legal standing does not originate in the Ecuadorian Constitution, nor is its source solely indigenous worldviews. Indeed, there are several transnational actors who, prior to the enactment of the Ecuadorian Constitution, were pushing the debate on the need to reconsider human relations with nature by including legal mechanisms that autonomously protect nature. For example, the *Community Environmental Legal Defense Fund* (CELDF) has been a relevant actor in pushing for the recognition of the rights of nature in municipal legislation in the United States. With the CELDF's help, in 2006, Tamaqua Borough, Pennsylvania, enacted Ordinance 612 which, for the first time, recognized that 'Borough residents, natural communities, and ecosystems shall be considered to be "persons" for purposes of the enforcement of the civil rights of those residents, natural communities, and ecosystems.'

[46] Acta 58 (n 42) p 67.

[47] Acta 58 (n 42) p 13.

[48] Acta 58 (n 42) p 17.

b) Operators

As noted, indigenous organizations played a crucial role in advancing the debate and securing the recognition of the rights of nature in Ecuador. Their influence on the constitutional text is most evident in the adoption of the term 'Pacha Mama' to refer to nature. For indigenous communities, nature is an entity with intrinsic value, and this worldview was incorporated into Ecuadorian constitutionalism in two key ways: (1) through the language of rights, recognizing nature as a subject of constitutional rights, and (2) through the substantive content of these rights, which draws on biological and ecological concepts – guaranteeing the integral respect for nature's existence, as well as the maintenance and regeneration of its vital cycles, structure, functions, and evolutionary processes.

During the constitutional process, the translation of indigenous respect for Pacha Mama into Western legal language aligned with efforts by civil society and academic organizations advocating for the rights of nature. A particularly prominent actor was the Community Environmental Legal Defense Fund (CELDF), founded in 1995, which actively participated in the Constituent Assembly debates and provided advisory support on environmental protection matters. CELDF has long promoted the recognition of 'natural communities' and 'ecosystems' as rights-bearing entities, a perspective that likely shaped Ecuador's constitutional provisions on the rights of nature.[49]

The technical vocabulary used in defining these rights also reflects a significant influence from the deep ecology movement. Alberto Acosta, President of the Constituent Assembly and a key figure in the recognition of nature's rights, played a central role in embedding this ecological philosophy into the constitutional framework. An economist, politician, and environmentalist, Acosta championed the integration of deep ecological principles in the Assembly's deliberations.

The intellectual foundations of deep ecology can be traced back to Aldo Leopold's 1949 work, 'A Sand County Almanac', in which he introduced the 'Land Ethic'. Leopold argued for extending ethical considerations to nature, proposing that humans are part of an interdependent ecological community that includes soil, water, plants, and animals.[50] This perspective laid the groundwork for environmental ethics.

In 1989, American scholar and environmental historian Roderick Nash further developed these ideas in 'The Rights of Nature: A History of Environmental Ethics', where he argued for expanding ethical concern beyond human interests to

---

[49] For example, CELDF has supported the recognition of these rights in several municipalities across the United States. In 2006, the Borough of Tamaqua in Schuylkill County, Pennsylvania, adopted an ordinance recognizing 'the residents of the Borough, natural communities, and ecosystems as persons for the purpose of enforcing the civil rights of residents, natural communities, and ecosystems.' Other municipalities have since followed this example.

[50] Aldo Leopold, *A Sand County Almanac. And Sketches Here and There* (OUP 1940).

encompass nature as a whole.[51] Nash contended that nature possesses intrinsic value, warranting moral and legal recognition.

Subsequent thinkers have continued this intellectual trajectory. Since 1992, theologian and environmentalist Thomas Berry has advocated reimagining humanity's place within what he calls 'the Earth community'. According to Berry, recognizing that 'the Earth is primary and humans are derivative' requires a paradigm shift in law, ethics, and governance. He argues that legal systems, historically complicit in environmental destruction, must be restructured to acknowledge nature's right to exist.[52]

Berry's ideas significantly influenced South African environmental lawyer Cormac Cullinan, who expanded on them in his book Wild Law.[53] Cullinan played a leading role in drafting the Universal Declaration of the Rights of Mother Earth, and his work forms the foundation of the emerging field of Earth Jurisprudence, which has informed advocacy efforts by various transnational organizations.

In Ecuador, Wild Law was translated and published by Universidad Andina Simón Bolívar,[54] with a preface by Ramiro Ávila Santamaría, a professor at the university and former judge of the Constitutional Court. Ávila was the lead judge in the landmark case that first interpreted the meaning of the rights of nature, recognizing mangroves as rights-bearing entities. In this ruling, the Constitutional Court placed the 'systemic principle' at the core of its reasoning – an idea also present in the works of Berry and Cullinan – which asserts that legal recognition should extend to ecosystems based on their essential role within the complex web of life. By applying this principle, Ecuador's Constitutional Court has affirmed the legal personhood of natural elements and ecosystems, reinforcing their protection within the country's legal framework.[55]

The recognition of the rights of nature in Ecuador was the result of a convergence of indigenous worldviews, environmental advocacy, and academic contributions, both domestic and international. Indigenous organizations provided the foundational ethical and philosophical justification, while legal scholars and environmental movements helped translate these values into constitutional language. At the same time, political actors like Alberto Acosta facilitated the integration of these ideas into the constitutional framework. The role of the judiciary, particularly the Constitutional Court, has been instrumental in interpreting and enforcing these rights, further solidifying their legal standing. However, the im-

[51] Roderick Nash, *The Rights of Nature: A History of Environmental Ethics* (The University of Wisconsin Press 1989).

[52] Thomas Berry, *The Great Work* (The River Press 1999).

[53] Cormac Cullinan, *Wild Law. A Manifesto for Eearth Justice* (Siber Ink & Green Books 2002).

[54] Cormac Cullinan, *El Derecho Salvaje. Un manifiesto por la justicia de la Tierra* (UABS 2019).

[55] Corte Constitucional del Ecuador 8 September 2021, 22-18-IN/21, para 37.

plementation of these rights continues to face resistance, particularly from economic and political actors with vested interests in resource extraction.

c) Substantiation

The discussion on the rationale for recognizing the rights of nature presupposes different views of nature. For some, it is necessary to recognize the rights of nature because nature itself has intrinsic value (similar to recognizing that people have rights simply because they are people). For others, recognizing these rights is imperative because it is the best way of combating the ecological crisis and preserving humanity (as if to say that if we protect nature, we protect humanity). There are others yet who support taking political positions on the need for fairer and more equitable economic development, and an ethical imperative to care for nature. These are what I will refer to as essentialist, utilitarian, and political substantiations, respectively.

The utilitarian substantiation is an instrumental-style position that justifies the recognition of the rights of nature either because they can be a tool to protect the human environment or because they can be effective in combating the ecological crisis or climate change. The protection of nature, recognizing its quality as a subject of rights, is justified to the extent that it would impose enforceable duties and obligations that would help to prevent and mitigate the possible risks that its deterioration may bring in the future and that may negatively affect people's lives.[56]

The analysis of the substantiation of rights is characteristic of this first stage of development but is also present, with specific applications, in the third stage of development. There, for example, in recognizing that mangroves are subjects of rights, the Constitutional Court emphasized that these fragile systems require care because their trees and soils have the capacity to absorb ten times more carbon than a terrestrial ecosystem, which would also help 'to mitigate climate change'.[57] Although this type of substantiation remains a view that privileges the preservation of human life, it also recognizes the importance of caring for ecosystems. Moreover, in the jurisprudence of the Court, this reasoning does not seem to exclude the parallel recognition of the intrinsic value of nature.

The essentialist criterion recognizes intrinsic values in nature that are inherent to it by the simple fact of its existence. This substantiation, in the theoretical map of the recognition of the rights of nature, considers nature as an entity in itself and not as an object, which makes it possible to make a conceptual leap from 'protection of the environment' to 'respect for nature'. But it also implies a transformation in the way we see the world. Understanding nature as another subject

---

[56] The Constitution even recognizes the need to preserve nature in order to tackle climate change. Art 413 'The State shall adopt adequate and cross-sectoral measures to mitigate climate change by limiting greenhouse gas emissions, deforestation and atmospheric pollution; it shall take measures to conserve forests and vegetation, and shall protect the population at risk'.

[57] Corte Constitucional del Ecuador 8 September 2021, 22-18-IN/21, para 13.

that has equal value and that, like us, requires equal respect and consideration means moving away from an anthropocentric model – where nature is a secondary or functional element in people's lives – and towards what has been called *biocentrism*.

The biocentric turn emphasizes the non-instrumental value of nature, of intrinsic virtues that do not depend on the valuations of others or on relational characteristics with other processes or objects.[58] The idea of ascribing an intrinsic value to nature, the heart of biocentrism, is driven by the worldviews of indigenous peoples, although it also has philosophical and political roots.[59] One of the practical effects of the biocentric perspective is that it sets limits to environmental management and economics that relies on the economic valuation of natural resources; in other words, it opposes the commodification of nature.

This kind of theoretical analysis of the substantiation based on the intrinsic value of nature is also present in the third phase of development. Here, the Constitutional Court has alluded to the intrinsic value of nature in at least two cases. In the case of the recognition of rights of Los Cedros forest, the Court emphasized that the State's obligation to take precautions and restrict activities that threaten the extinction of species and ecosystems

> 'is not a power or a conditional option, but a constitutional obligation derived from the intrinsic valuation that the Constitution makes of the existence of species and ecosystems, through the rights of nature. Indeed, the risk in this case does not necessarily relate to the effects on human beings, although they may be included, but to the extinction of species, the destruction of ecosystems or the permanent alteration of natural cycles or other types of serious or irreversible damage to nature, independently of such effects.'[60]

Similarly, in the framework of the rights of nature, the Court recognized the rights of a woolly monkey named Estrellita. Before explaining the reasons for this recognition, it elaborated on the logic and justification for the recognition of the rights of nature. There, with a nod to Kantian ethics, it pointed out that

> '[n]ature, therefore, is observed as a subject of rights with an intrinsic value, which implies that it is an end in itself and not merely a means to the attainment of the ends of others. This means that nature can only be analysed as a means if and only if its status as an end in and of itself is not ignored.'[61]

Finally, the political substantiation understands the need to protect nature as a commitment to transform society: from a society that subordinates nature, treats it as an object, and exploits it within limits only determined by the human right

---

[58] For an analysis of the biocentric turn, as part of the so-called 'deep ecology' current, see: Eduardo Gudynas, 'La ecología política del giro biocéntrico en la nueva Constitución de Ecuador' (2009) 32 Revista de Estudios Sociales 34 <https://journals.openedition.org/revestudsoc/16260#quotation>.

[59] See Stephen A Laubach, *Living a Land Ethic* (The University of Wisconsin Press 2014) and the literature cited in notes 50–54.

[60] Corte Constitucional del Ecuador 10 November 2021, 1149-19-JP/21, para 50.

[61] Corte Constitucional del Ecuador 27 January 2022, 253-20-JH/22, para 57.

to a healthy environment, to a society that lives in harmony with nature, treating it as a subject and taking from it what it needs, in a sound and sustainable way. This proposal goes hand in hand with political ecology, which seeks to position the idea that the struggle for equality and justice must be accompanied by debates about nature and the environment.[62]

The analysis of this substantiation, which is strongly developed in the first stage of development, is also present in the third one. There, the Constitutional Court has maintained positions of this type when, for example, it stated in the *Los Cedros* case that

> 'This vision of nature as a simple source of resources to be exploited at will has been profoundly questioned by various aspects of the natural sciences and humanities. The rights of nature represent this challenge in the world of law.'[63]

Nonetheless, the recognition of the rights of nature has not been without its criticisms. These can be summarized as falling into three arguments. The first is the idea that if the aim is to protect nature so that we do not become extinct as a species, then the better method is to strengthen environmental legislation. Secondly, some argue that the recognition of a new subject may make human rights irrelevant.[64] Thirdly, others posit that it is inappropriate to give rights to those who have not been assigned responsibilities and who are not capable of acting.[65] As with any expansion of legal categories, the recognition of new subjects as rights holders has always been controversial, as was the case with women, slaves, and even children.[66] This is one of the challenges of taking the rights of nature seriously and making them effective; I will briefly discuss this in another section.

This debate on whether nature should have rights, which was equally characteristic of the first stage of development, was transformed in the following stages into understanding what these rights are and what part of nature should have such rights. The first question concerns what kind of rights nature has: do these rights refer only to rights to restoration and regeneration, or could they also include other rights such as freedom, for example, in the case of wild animals? The second question is whether nature has rights as a whole, or whether individual ecosys-

---

[62] Several international instruments, as one of their central features, promote the idea of a sustainable relationship between environmental resources and human well-being, such as the *World Charter for Nature*, adopted by the United Nations Assembly in 1982, which recognizes that the human species is part of nature and that life depends on the uninterrupted functioning of natural systems. Another is the *Brutland Report (Our Common Future)*, which developed the term sustainable development for the first time.

[63] Corte Constitucional del Ecuador 10 November 2021, 1149-19-JP/21, para 49.

[64] Farith S Compaña, 'Derechos de la naturaleza: ¿innovación transcendental, retórica jurídica o proyecto politico?' (2013) 13(15) Iuris Dictio 9.

[65] Ricardo Crespo Plaza, 'La naturaleza como sujeto de derechos: ¿símbolo o realidad jurídica?' (2009) 8(12) Iuris Dictio 31.

[66] For a detailed analysis of the continued expansion of standing, see: Christopher Stone, 'Should Trees Have Standing. Toward Legal Rights for Natural Objects' (1972) 45 SCal LRev 450.

tems and elements of nature also have rights. These are questions whose answers are analyzed more acutely in the third stage of development.

2. Second Stage: Intermediate Hesitant Exploration

Since the constitutional recognition of the rights of nature in 2008, around 50 cases claiming violation of these constitutional rights had been brought to court by May 2021.[67] The claims are varied: the violation of rivers due to pollution or the construction of large hydroelectric dams, the violation of ancestral territories – connected to the right to prior consultation – due to authorizations for the exploitation of natural resources, the violation of protected reserves or protective forests due to the issuing of mining permits or other productive activities, the violation of the rights of endangered wild animals (condors and jaguars) due to hunting, and the violation of mangrove forests due to illegal logging. Of these cases, up to the date indicated, few reached the Constitutional Court, and most ended with an appeal court ruling.

What is characteristic of these cases – during this stage – is that, despite the complaint on which they were based (ie the rights of nature), the judges, in order to accept or deny the cases, analyzed the complaints in the light of the right to a healthy environment. Thus, for example, one of the first cases to come before the Constitutional Court sought to declare unconstitutional the special regime that limits the rights of mobility and residence on the Galapagos Islands. The Court noted that such restrictions are constitutional because they are intended to protect 'the rights of nature and of people to live in a healthy environment', adding, 'all the more so if we consider that the current Constitution has the highest standards of environmental protection (rights of nature) with regard to comparative law […]'.[68]

In another case, the Constitutional Court heard a complaint about illegal exploitation of mining. Although the claim was formulated using the rights of nature, the Constitutional Court indicated that acting without an environmental permit for mining exploitation was a violation of the rights of nature. The analysis of these rights became entangled in the labyrinth of permits and special types of mining licenses. This same pattern is also characteristic of the cases that did not reach the Constitutional Court. For example, in a case claiming that the planting of pine trees affected a páramo (alpine tundra ecosystem) called Tanabana, the appeal judge established that environmental permits existed for such planting, that technical reports indicated that environmental regulations had been complied with, and that the pine plantation was located on private territory, thus concluding that the rights of nature had not been violated.[69]

---

[67] The number has been determined from the database provided by the Legal Observatory of the Rights of Nature, <https://www.derechosdelanaturaleza.org.ec/casos/>.

[68] Corte Constitucional del Ecuador para el período de transición 26 April 2012, 017-12-SIN-CC.

[69] Sala Especializada de la Corte Provincial de Justicia de Chimborazo, 06334-2014-1546.

In another case, perhaps one of the most paradigmatic at this stage, the judges heard about the contamination of the River Vilcabamba. The State authorities were polluting the river, dumping soil, and removing water, claiming that these actions were part of road construction. The first-instance judge rejected the claim that the rights of nature had been violated. The court of appeals upheld this determination.

Although the ruling contains important passages on how harm to nature can constitute 'generational damage', the analysis of the violation of nature again relied on the legal standards associated with the right to a healthy environment. The judges argued that whoever carries out a potentially harmful activity is obligated to prove that 'there is no environmental damage'. They concluded that refusing to protect nature due to a lack of evidence would be unacceptable. In a case of doubt 'or where it may be presumed that environmental damage has been produced by contamination, it is the absence thereof that must be proved'.[70]

The understanding of the rights of nature through the right to a healthy environment is found in many other decisions. Cases are resolved on the basis of the criteria of environmental legislation (which protect the human environment) and not on the basis of the contents of the rights of nature (which protect the integrity of nature's structure and life cycles). This is somewhat foreseeable because environmental regulations are designed and enforced with the well-being of the environment for human beings in mind. This means that the environment must maintain certain levels of water, air, and soil quality, not for the sake of the species that live there, but to ensure human well-being. Therefore, addressing violations of an environmental norm may be inadequate or insufficient to address the integrity and impacts of the rights of nature. Furthermore, the way to demonstrate that the right to a healthy environment has not been violated requires a lower standard of protection than the rights of nature. While in the first case, protection is exhausted with the demonstration of the environmental impact study and/or the respective license, in the second case, these studies are not sufficient to demonstrate that the life cycles, structure, functions, and evolutionary processes of nature have not been violated, because nature's legal interest is different.

At the legislative level, the most important norms that have been developed since the recognition of the rights of nature also confuse them with, superimpose them on to, or assimilate them to the purposes of environmental regulation. For example, although the Organic Environmental Code refers to the rights of nature independently of the right to a healthy environment, it considers the protection of the former to be fulfilled as long as environmental norms have been observed. It recognizes that its purpose is both to guarantee the right of people to live in a healthy and ecologically balanced environment as well as to protect the rights of nature. In fact, this norm indicates that '[t]he National Environmental Authority shall define the territorial environmental criteria and shall develop the technical guidelines on the life cycles, structure, functions and evolutionary processes of

[70] Sala Especializada de la Corte Provincial de Justicia de Loja, 11121-2011-0010.

nature',[71] something that has not happened through the actions of said authority, but rather through the jurisprudence mentioned in this text. As far as this Code goes, it is determined that risky activities must ensure that their impacts 'do not affect the dynamics of populations and the regeneration of their life cycles, structure, functions and evolutionary processes, or impede their restoration' (art 190), yet the scope of these criteria is never defined.

Similarly, the Integrated Criminal Code devotes 19 articles in a chapter titled 'Crimes against the environment and nature or Pacha Mama', but it never mentions life cycles or any of the constitutional criteria on the rights of nature. This chapter categorizes crimes against water, air, and soil,[72] and in each case, the same condition applies: the criminal act must be committed in violation of existing regulations. This means that an offence occurs only when environmental regulations are breached, without any consideration of the life cycles, evolutionary processes, or structure of nature.

Although this overlapping relationship between the rights of nature and the environment influenced the stage of judicial and legal exploration and obscured the independent content of the rights of nature, this does not mean that the two rights do not share any common interests. The relationship between the two needs to be clarified and analyzed, as it is ultimately a matter of the relationship between the rights of two subjects. Although nature has specific rights, these could be respected without excluding human beings when, for example, humans coexist harmoniously with nature. In the third stage of development, this relationship between environment and nature is still present and there are still critical knots to be unraveled.

It should be noted that this stage did not end when the next stage began. Although, as I will analyze below, in the third stage the Constitutional Court began to elaborate on the autonomous and detailed contents of the rights of nature, the rather superficial and exhausted analysis of environmental law persists to this day in the decisions of first and second-level judges. This can be explained in two ways.

The first relates to the time that has elapsed between the recognition of these rights in the 2008 Constitution and their explicit and detailed development in the jurisprudence of the Constitutional Court, which begins in August 2021. During this time, constitutional justice, entrusted to trial judges, has become accustomed to approaching the rights of nature through a superficial analysis of the foundations of the recognition of these rights (linked to the first stage) and deeming their contents exhausted by those of the right to a healthy environment (characteristic of the second stage). The second explanation is that the Constitutional Court's jurisprudence, being so recent, has not yet permeated the reasoning of all first and second-level constitutional judges. This is a process which, as has been characteristic for the evolution of the analysis and application of these rights in Ecuador, is likely to progress but to do so slowly.

[71] Código Orgánico del Ambiente, art 6.

[72] Código Orgánico Penal Integral, arts 251–253.

3. *Stage Three: Singularization and Consolidation*

Having passed the stage of judicial and legal exploration of the rights of nature, the Constitutional Court then faced the challenge of analytically unraveling these rights. After more than a decade, there were still no cases that applied and explained what life cycles, functions, structures, and evolutionary processes mean as criteria for the protection of nature. The Constitutional Court began to make up for this gap in August 2021.

As mentioned earlier, this marked the beginning of a significant jurisprudential development that can be studied on two main levels: firstly, by analyzing the criteria that define the content recognized as the rights of nature (the object of the right); and, secondly, by determining whether these rights pertain to a specific component of nature, to nature as a whole, or to both (the subject of the right).

a) The criteria of biology: or what nature is

One of the most significant problems with the rights of nature is that their specific content is framed in the language of biology: the right to integral existence and to the maintenance and regeneration of its life cycles, structure, functions, and evolutionary processes. These concepts cannot be understood and addressed using our classic legal categories. The challenge of interdisciplinarity and the attempt to understand what these concepts mean began with a case on the possibility of diverting riverbeds.

A regulation allowed the diversion or any modification of the natural course of rivers if required for a mining project and with prior authorization from the competent State authority. Although the case does not specifically elaborate on the rights of nature, the regulation was declared unconstitutional because the Court considered that the modification of the natural course of rivers could affect their ecological flow, ie the quantity and quality of water that the river needs to sustain ecosystems. Affecting the ecological flow, according to the Court, can in turn affect the functions and morphology of the river and affect other ecosystems that depend on it. It concluded that the diversion of the natural course of a body of water could have an impact on constitutional rights, and therefore its limitation required a legal reservation.[73] This case opened the door for the Court to analyze what a violation of the contents of nature itself would look like.

Therefore, when the Court heard the case that allowed, by regulation, productive activities of any kind in mangroves, it had an opportunity to elucidate concepts and explain what the object of the rights recognized as belonging to nature is. It explained that the alteration of the elements of any of the stages of the cycle (evaporation, condensation, precipitation, etc.) implies the violation of the vital water cycle, for example, due to pollution or global warming.[74] Referring to the precedent, it emphasized how an infrastructure project that affects the ecological

[73] Corte Constitucional del Ecuador 9 June 2021, 32-17-IN/21, para 59.

[74] Corte Constitucional del Ecuador 8 September 2021, 22-18-IN/21, para 30.

flow can break the structure and alter the functions of the river because it could affect the ecosystemic processes that take place there. Regarding evolutionary processes, it added that disrupting the elements that sustain long-term processes of permanent change that allow for adaptation constitutes a violation of the rights of nature.

With this framework, the Court began the specific application of the recognized rights of nature to concrete cases in which their violation was alleged. Thus, in a case involving the diversion of the flow of the River Aquepi, which affected a community's access to water because the water was being used for irrigation projects, the Court indicated that 'the functions of the river that allow and sustain the life of the human species as well as other species and vegetation' were being altered.[75] What is most interesting about this case is the two levels of analysis of the criteria for the content of the rights of nature: firstly, the idea of 'function' and, secondly, the idea of the 'integrity' of the criteria.

In relation to the first criterion, the function of the river considers both the quantity of water necessary for the existence of abiotic ecosystems, as well as for biotic communities, which expressly includes the human species, for whom the river fulfils the function of providing water for human consumption. Secondly, as for the integrity of the criteria defining the criteria of nature, the Court observed that the diversion of the river changes its structure and morphology, which impedes the fulfilment of its functions and therefore also disrupts its evolutionary process.

On the evolutionary processes, the Court delved into the constitutionality of the granting of permits in the protected Los Cedros forest, on the basis of the ecological principle of tolerance and natural cycles. It explained that, based on this principle, natural systems 'can only function adaptively within an environment whose basic characteristics have not been altered beyond what is optimal for that system'.[76]

Thus, if an environment is profoundly altered, the existence and reproduction of cycles is compromised, adaptive behavior becomes difficult, the functions of nature are limited, and plant and animal species are threatened with extinction. For the Court, the extinction of a species, considering the relationships it establishes with other species in the ecosystem, can lead to a chain of extinction and thus violates the right to evolutionary processes. The picture described above illustrates well the idea of the integrity of nature's right to have its existence respected in its entirety.

More specifically, in the case of the monkey Estrellita, being a member of an endangered species, it was established that her captivity and death violated the evolutionary processes of nature. The Court pointed out that the domestication and humanization of wild animals are phenomena that affect ecosystems and the balance of nature, as they produce the progressive decline of animal populations, which is aggravated in the case of endangered species. The Court associated the

[75] Corte Constitucional del Ecuador 15 December 2021, 1185-20-JP/21, para 47.

[76] Corte Constitucional del Ecuador 10 November 2021, 1149-19-JP/21, para 44.

evolutionary process with life cycles: such phenomena limit the reproductive processes of wild animals and their population dynamics, which can lead to the extinction of a species.[77]

The Court's analysis thus far reveals a position on nature and its rights, and a methodology for understanding them. On the one hand, the Court has taken the position of applying the rights of nature, based on protection, to all levels of ecological organization, where it has identified specific rights holders, as will be seen below. On the other hand, the Court has embarked on a journey into nature, and in order to understand its existence and functioning, it has had to employ other tools of analysis that are not part of the legal culture.

On this journey, the Court is challenged to equip itself with an arsenal of concepts, vocabulary, and literature that is foreign to the training of lawyers, but which is indispensable to move forward in understanding and protecting nature through the rights that have been recognized for it.

b) Towards concretization: the systemic principle

One of the most novel features of the Constitutional Court's recent jurisprudence is the recognition of increasingly specific ecosystems or elements of nature as rights holders. First, it recognized all mangrove ecosystems as rights holders. Then, it recognized a specific forest, Los Cedros, as a holder of rights to the existence and maintenance of cycles, functions, structure, and evolutionary processes. This was followed by two rivers: the Aquepi and the Monjas.[78] And finally, a wild animal: Estrellita.

If the Constitution has recognized that nature as a *whole* has rights, why recognize mangroves, rivers, forests, and wild animals as rights holders? It would be enough to indicate that, as they are part of nature, their rights have been violated. However, the Court has initiated a process of 'concretization' of nature, associated with the systemic principle, the reasons for which I will explain below.

---

[77] Corte Constitucional del Ecuador 27 January 2022, 253-20-JH/22, paras 116–117.

[78] It is noteworthy that in these cases the Constitutional Court of Ecuador does not refer to the famous River Atrato case in Colombia. This is particularly relevant because the Ecuadorian Constitutional Court often closely examines and draws inspiration from the jurisprudence of the Colombian Constitutional Court. Several factors may explain this omission. First, in the *River Atrato* case, the recognition of its rights derives from the 'ecological' and 'ecocentric' interpretation of various provisions of Colombia's Constitution. By contrast, in Ecuador, the recognition of river rights follows directly from the Constitution's explicit recognition of the rights of nature. Second, the River Atrato case is more complex due to its focus on the relationships between the river and the indigenous communities that depend on it – a dimension that Ecuador's Constitucional Court has yet to develop in its river jurisprudence. Similar to the New Zealand cases, the *River Atrato* ruling recognizes indigenous communities as the river's 'guardians'. In Ecuador, however, nature is recognized as a subject of rights, and anyone has legal standing to represent it in court; as such, 'legal guardianship' is unnecessary. Instead, it is possible that the Ombudsman's Office will be tasked with overseeing the implementation of rulings in favor of nature. As a result, both the formulations of rights and the mechanisms for redress concerning rivers have developed differently in each jurisdiction.

Nature can be synonymous with *earth*, with the entire biosphere. Thus, locating or delimiting protection or, in the Court's words, 'protecting its elements and ecosystems', is a process of concretization – or even individualization – of nature that fulfils at least one of the following four functions.

Firstly, it helps to delimit what is to be protected. Nature is a complex being; the Court has noted that it 'is made up of an interdependent and indivisible set of biotic and abiotic elements', so specifying protection is not only reasonable but even desirable. The Court has indicated that this identification and the 'jurisdictional declaration of the subject of rights allows for the determination of its particular characteristics, such as the identification of its name, location, history, concrete specification of its life cycle, structure, functions and evolutionary processes and of the harm that may occur'.[79] This precise identification makes it possible to understand what function any given element or ecosystem fulfils in order to understand the impact of its being affected.

Secondly, the identification of the rights holder within nature helps to delimit what is to be provided with reparation. Therefore, this has a practical side that makes it possible, according to the Court, 'to establish the specific obligations of the State with regard to these elements, as only in this way can the most appropriate reparation measures be established from a systemic perspective'.[80]

Thirdly, the declaration of subjects of rights, in nature, is also a symbolic expression: one that extols that all parts of nature, even those that already have rights by being part of nature, are also individually considered to be elements or ecosystems with intrinsic value. Of course, insofar as these are part of nature, this intrinsic value must take into account the systemic nature of their protection; that is, they must be part of a network whose structure, cycle, and functioning depend on that ecosystem or element.

And finally, the declaration has the sense of emphasizing the interdependence and interrelationship between ecosystems, specifically identified as subjects of rights, and the communities that are settled in or depend on them. Thus, the relationship between the community living in the Aquepi river basin is established between two subjects of rights. Similarly, the relationship with the coastal communities that depend on and live with the mangroves is established between two subjects of rights. The Court has pointed out that the specific recognition of rights holders 'makes sense in order to guarantee the ultimate aim of the constitutional recognition of the rights of nature, which is to protect nature and build a new form of coexistence of citizens, in diversity and harmony with nature'.[81]

Having said that, the declaration and precise identification of ecosystems or elements of nature as rights holders, as part of nature, is a declaration that is part of the systemic principle. The declaration, therefore, of their entitlement or infringement must take into account the systemic relationships maintained by that

[79] Corte Constitucional del Ecuador 15 December 2021, 1185-20-JP/21, para 54.
[80] Corte Constitucional del Ecuador 15 December 2021, 1185-20-JP/21, para 54.
[81] Corte Constitucional del Ecuador 15 December 2021, 1185-20-JP/21, para 53.

ecosystem or element. This is a significant understanding of nature in which the Court has framed its jurisprudence, and it has therefore indicated that 'nature is a complex subject that must be understood from a systemic perspective because nature is made up of an interrelated, interdependent and indivisible set of biotic and abiotic elements (ecosystems). When one element is affected, the functioning of the system is altered. When the system changes, this also affects each of its elements'.[82] Thus, to the extent that nature has rights, so do its systemic components.

The Court, however, has warned that in a general and abstract manner, the recognition of nature as a subject of rights in the Constitution does not require specific recognitions to promote and protect nature and each of its constituent elements.[83] This, in theory, is the general rule. The express recognition of subjects of rights would be exceptional and would serve to note that the elements of nature (considered individually) play a role in ecosystems, from which they derive their integral and individual value, so that their protection makes sense because of their 'systemic importance'.[84] Although the reasons for this – systemic – concretization make sense for the above-mentioned considerations, their application does not seem to be the exception but rather the general rule in the Court's current jurisprudential development.

Moreover, the recognition of specific rights holders within nature has not been a peaceful discussion. This discussion has depended, to a large extent, on what judges have understood by ecosystems and elements of nature, and it also reveals the difficulty of analyzing biological concepts from the perspective of law.

In the mangroves case, the Court took on the challenge of expressing its understanding of nature, ecosystems, and their elements. Here the idea was inaugurated that 'concrete', individual protection is related to the nature of nature:

> 'Nature is made up of an interrelated, interdependent and indivisible set of biotic and abiotic elements (ecosystems). Nature is a community of life. All its constituent elements, including the human species, are linked and have a function or role. The properties of each element arise from interrelationships with the rest of the elements and function as a network. When one element is affected, the functioning of the system is altered. When the system changes, this also affects each of its elements.'[85]

Therefore, the key to individualized protection lies in the interrelationships of the elements and ecosystems that are woven into nature's web. In the case of mangroves, the Court recognized that they are a fragile ecosystem and that unchecked intervention could disrupt several cycles (especially the carbon cycle), potentially preventing mangroves from fulfilling essential functions, including serving as a natural barrier against tsunamis.

---

[82] Corte Constitucional del Ecuador 8 September 2021, 22-18-IN/21, para 27; Corte Constitucional del Ecuador 15 December 2021, 1185-20-JP/21, para 44.

[83] Corte Constitucional del Ecuador 8 September 2021, 22-18-IN/21, para 42; Corte Constitucional del Ecuador 10 November 2021, 1149-19-JP/21, para 43; Corte Constitucional del Ecuador 19 January 2022, 2167-21-EP/22, para 123.

[84] Corte Constitucional del Ecuador 8 September 2021, 22-18-IN/21, para 37.

[85] Corte Constitucional del Ecuador 8 September 2021, 22-18-IN/21, para 27.

Yet the declaration of Los Cedros forest as a subject of rights revealed the difficulty of understanding the relationships and systems of nature. Certainly, the protection of Los Cedros by recognizing its status as a rights holder raised several questions. Is Los Cedros forest an ecosystem, like the mangrove ecosystem in general? What is the important role it plays with the rest of the ecosystems that merits its protection due to these systemic relationships? Would the Court then begin to protect all elements of nature by recognizing them as rights holders? What is the limit?

In a concurring opinion in this case, the constitutional judges noted that the recognition of Los Cedros forest as a subject of rights did not respond to a systemic approach and indicated that it is important to separate 'the content of the right from its ownership and to detail the reasons why it considers the protection of specific elements of nature as part of the sphere of protection of the rights of nature to be indispensable'. In other words, unlike the case about mangroves, where the mangrove ecosystem is recognized in a general way, the particular identification of the titular subject of Los Cedros forest would not be justified because it is a delimited forest and, therefore, an *element* of nature that does not require specific recognition but rather protection through the systemic principle proposed by the Court.

The recognition of specific elements of nature as rights holders will not cease to be a matter of debate. There is no doubt that this recognition must be accompanied by reasoning as to why such recognition is required. However, the limit of this recognition should be based not on the type of element or ecosystem whose rights are violated, but rather on their systemic importance. Take, for example, the carbon cycle. Carbon is just another element of nature, yet altering its cycle could impact the conditions of the biosphere that enable life as we know it. Anthropogenic climate change is a case in point. Or take bees, which, while not exclusive to any ecosystem, are a fundamental element in the ecological balance of the planet due to their pollination function. This function may well be worthy of specific protection under the rights of nature.

In the Los Cedros case, the Court did not declare *the forests* to be violated, as it did the *mangroves*. Perhaps the difficulty in this case was to understand the abstraction of the multiple interrelationships that occur in the biosphere. In the end, the delimitation and reasons for the concretization of nature is one of the most novel debates in Ecuador, and the challenge is to understand all these concepts from a biological perspective and to address the impact of their implementation in the legal system.

III. Legal Person vs Subject of Rights

Understanding the concept of legal personhood, or who and what constitutes a legal person, is crucial for determining who matters in the eyes of the law, who can act, why some have more agency than others, and who is disregarded. Despite its significance, the question remains dispersed within legal subjectivity theories, particularly emerging from legal literature on the legal personality of corporations.[86]

As early as 2001, a Harvard Law Review note anticipated that 'the law of the person is full of profound ambiguities and significant tensions',[87] emphasizing that this issue would be exacerbated by technological advances and economic progress. Animal and environmental rights precisely aim at addressing this problem.[88]

Conceptually, within legal subjectivity theory, the idea of a legal person has two major interpretations. The first, more popular one, identifies the legal person with the human being. This human is conceptualized with intrinsic characteristics such as dignity, autonomy, and consciousness. The paradigmatic model of this legal person is an adult, endowed with legal rights. From this perspective, legal rights 'operate as a kind of overlay on an underlying set of characteristics that make such rights both relevant and intelligible.'[89] Thus, in law, the human being, and the metaphysical characteristics assigned to it, form the foundational concept of the legal person; it is upon these characteristics that define humanity that the concept of legal personhood is also built. This position is identified by Grear as 'naturalistic'.

On the other hand, a positivist orientation understands the concept of legal personhood in purely formalistic terms: it makes no reference to any kind of content or underlying person with intrinsic characteristics. The legal person is what the law designates as a bearer of legal status. It is a highly abstract view, but it still regards the category as a central mechanism of law.

I will explain why neither of the two adequately explains or is suitable for understanding the rights of nature.

---

[86] See John Dewey, 'The Historic Background of Corporate Responsibility' (1926) 35(6) Yale LJ 655; Nicholas James, 'Separate Legal Personality. Legal Reality and Metaphor' (1993) 5(2) Bond LRev 217.

[87] Dave Fagundes, 'Note: What We Talk About When We Talk About persons: The Language of a Legal Fiction' (2011) 114(6) Harv LRev 1746 <https://papers.ssrn.com/sol3/papers.cfm?abstract_id=921133>.

[88] See: Visa Kurki and others, *Legal Personhood: Animals, Artificial Intelligencia and the Unborn* (Springer 2017); Jon Garthoff, 'Decomposing Legal Personhood' (2019) 154(4) Journal of Business Ethics 967 <www.jstor.org/stable/45022919>; Charles Foster and Jonathan Herring, *Identity, Personhood, and the Law* (Springer 2017); Visa Kurki, 'Legal Personhood and Animal Rights' (2021) 11(1) Journal of Animal Ethics 47; Laura Spitz and Eduardo M Penalver, 'Nature's Personhood and Property's Virtues' (2021) 45(1) Harv Envt'l L Rev 67; Andreas Fischer-Lescano, 'Nature as a Legal Person: Proxy Constellations in Law' (2020) 32(2) Law and Literature 237.

[89] Anna Grear, 'Law's Entities: Complexity, Plasticity and Justice' (2013) 4(1) Jurisprudence 81.

1. The Legal Person as Natural Person

This approach conceives the legal person as the natural person, as long as they are endowed with reason, consciousness, will, autonomy, and dignity. Thus, the reasons for granting legal personality (or legal status) to the person, that is, to recognize them as someone before the law and, consequently, to recognize rights and obligations, are based on the reasons for recognizing that being with the characteristics of what is understood as human. What is human and what constitutes humanity has also been the subject of constant critique and progression.

The idea of respecting the rights of individuals is fundamental to the traditional grammar of constitutionalism, which is based on the notion that constitutional States are designed to guarantee their protection. With the birth of the modern State, the individual effectively became the unit of law. The very idea of individual characteristics as reasons for recognizing rights is also the premise upon which the entire international human rights system is built.

However, the characteristics of humanity as conditions for also being a legal person have been a contentious issue throughout history. For example, in Ecuador, from the first Constitution of 1830 to the Constitution of 1878, only white (or mestizo) men, of legal age, who could read and write, were considered persons with rights or who could demand things through the law. The characteristic of humanity for centuries materialized in the recognized capacities of men: reason and autonomy.

From this perspective, humanity itself is a restrictive concept, and for many years, what the law considered as a legal person was similarly restricted. To qualify as a person and to have legal personality, legal theory and practice conceived humans as rational persons fully competent to initiate actions in court, sue, or be sued, and act on their own behalf for their own rights. The subject that best represents this model is the capable contracting man.

The Inter-American Court of Human Rights (IACHR) has indicated that the term 'person' refers to human beings.[90] Here, the correspondence is exact: the legal persons of international human rights law, for the purposes of protection under the ACHR, are individuals of the human species. Thus, for the IACHR, the legal person to which its protection is directed is a human who, consequently, is a rights holder and, therefore or for that reason, has legal personality. This Court has stated that the right to recognition of legal personality implies 'the capacity to be a rights holder (capacity to enjoy) and to have duties; the violation of that recognition implies an absolute disregard for the possibility of being a holder of those rights and duties.'[91]

This approach also fits within the classic civil recognition of legal persons. The Civil Code recognizes two: 'Persons are natural or juridical [...] All individu-

[90] IACHR 26 February 2016, *Opinión Consultiva OC-22/16*, para 38.

[91] IACHR 19 June 1998, *Benavides Cevallos vs Ecuador*, para 179.

als of the human species, regardless of their age, sex, or condition, are persons. They are divided into Ecuadorian and foreigners.'[92]

The juridical person, on the other hand, is a 'fictional person' to whom the Civil Code also grants legal personality through legal capacity or capacity to act. That is, the possibility of having rights and obligations refers to the capacity to exercise civil rights and assume civil obligations, and to be represented judicially and extrajudicially. This legal capacity implies 'being able to bind oneself, without the ministry or authorization of another.'[93]

The civil recognition of natural and legal persons' capacity to bind themselves has perhaps been the source of the greatest confusion and debate about just who can be the subject of rights. Again, the possibility of having rights presumes that the subject who holds them has the capacity to exercise rights, assume obligations, and bind themselves: they are a subject with reason, will, and conscience.

The identification of a subjective right through the characteristics of its subject brings several problems. It emphasizes the right holder as an agent and as an elector. Grounding the conception of rights in autonomy could potentially be incompatible with attributing rights to beings that cannot exercise powers or renounce them, such as babies, animals, or nature.

The discussion once again focuses on legal capacity – the ability to exercise and claim a right – as a prerequisite for being a legal person. Capacity presumes the existence of an individual with reason and will. The issue with defining a subject in this way is that (i) it conditions the validity of rights – and legal personality – on the capacity to demand their fulfillment, and (ii) it limits the scope of rights to the ability to assume obligations. However, (i) requiring the capacity to assert rights as a condition for legal personality would deny dignity to individuals as the basis for possessing rights. Thus, it would deny legal personality to those who, unable to claim their rights themselves, currently employ other mechanisms for the purpose, eg children through their parents or legal representatives; or legal persons through their managers or legal representatives.

Therefore, it does not seem plausible that the capacity to demand the fulfillment of rights and the capacity to obligate oneself are necessary requirements to be a person of law (or legal person) and to have legal personality. Under this framework it is also difficult to conceive nature as a 'legal person'.

2. The Legal Person as an Abstraction

The second viewpoint on who constitutes a legal person is indifferent to whom the law designates as a 'legal person,' as long as they have the capacity to hold rights and participate in legal relationships. Naffine understands capacity here not as the competence to claim rights but as the ability for a person to make their

[92] Ecuador, Civil Code, prov 40 and 41.

[93] Ecuador, Civil Code, prov 1461.

own decisions when the law recognizes their capacity to establish a legal relationship or not.[94]

The legal person understood as an abstraction devoid of any moral or substantive dimension emphasizes the entity's capacity to function adequately for the law to also function. From this viewpoint, the division between natural persons and legal persons is unnecessary; 'all that is needed for something to be a person is for the law to grant it a formal capacity to bear a right or an obligation.'[95]

Kelsen clearly represents this formal and abstract idea of what a person means. For him, a person 'designates a bundle of obligations, responsibilities, and subjective rights; a set, therefore, of norms.'[96] The person would be 'the support of the duties, responsibilities, and subjective rights that result from those norms'.[97]

For him, this way of conceiving persons has the advantage of discarding the idea that a person possesses a substance or essence that carries duties, responsibilities, and subjective rights. Rather, viewing the person as a personification of a set of legal norms 'avoids the error of considering it as something distinct from the set of duties, responsibilities, and rights constituted by those norms'.[98]

In order to understand the rights of nature and open the field of legal subjectivity, the concept of 'legal person' understood in purely abstract terms might initially seem convenient. One can argue that, to the extent the law recognizes something as a legal person, this legal recognition should suffice to ensure respect for its rights. However, there are two main issues with this approach. First, the term 'person' remains in the concept, even if only in an abstract sense. It is difficult to reconcile an extralegal understanding of what it means to be a 'person' with the notion of something that does not fulfill the conditions typically associated with being a person or a human being.

Secondly, even if the first problem could be overcome, in the case of nature, it does not seem desirable to strip the concept of 'nature' of its meaning. The subject of nature, as a holder of rights, has particularities that prevent an entirely abstract and formal analysis of its status as a subject of the law.

The moral considerations of why nature has characteristics that morally deserve recognition and respect are important in Ecuadorian constitutionalism because these same reasons also guide the application of its rights. Hence, the Ecuadorian case requires understanding nature not as a 'legal person' or even a 'legal entity' in any of the variations mentioned above, but as a *rights-bearing subject*.

---

[94] Ngaire Naffine, 'Who are Law's Persons? From Cheshire Cats to Responsible Subjects' (2003) 66(3) MLR 346, 350.

[95] D P Derham, 'Theories of Legal Personality', in Leicester Webb (ed), *Legal Personality and Political Pluralism* (Melbourne UP 1958), cited by Naffine (n 94) 352.

[96] Hans Kelsen, *Teoría Pura del Derecho* (Eudeba 2000) 102.

[97] Kelsen (n 96) 102.

[98] Kelsen (n 96) 104.

The recognition of nature as a new subject of law challenges classical notions of subjectivity, expressed through the idea of the 'person' and the model of legal personality, to include new claims and claimants in the law.

Hence, the Constitution in Ecuador recognizes nature as a 'rights holder'. It is not a 'legal person' with 'legal personality'. This designation is not irrelevant, nor is it just a play on words. The term's use has important consequences for the doctrine of fundamental rights. The rational for being a *subject of rights*, and not a legal person, is not tied to notions of rationality or dignity to which the gaze inevitably turns when asking for the foundation of a new subject to whom rights are granted.

These notions are crucial for understanding the nature of the new subject and the nature of its rights. Therefore, it is also not desirable to perceive nature as a 'legal entity' in the formal and abstract manner suggested by positivist positions.

This new *subject* of rights (not just a subject of law) has no obligations and cannot acquire them. In this sense, it cannot bear – as in the paradigm of legal personality – rights as long as it has obligations. This subject only imposes duties on others. Where there is a duty towards nature, there is a corresponding right for nature. The determination of its legal subjectivity does not depend on the existence of obligations on its part; nor on its capacity to assert its rights. In the latter case, the Constitution provides for *actio popularis* so that any person can claim its rights.

In conclusion, this way of conceptualizing nature as a 'subject of rights' implies recognizing other models of legal subjectivity that do not refer to the attributes of individuals, their capacity to exercise or demand rights, or their capacity to bear obligations.

3. *Rights of Nature: a New Model of Legal Subjectivity*

Recent literature on the rights of nature addresses the concept of 'legal personhood' or 'legal personality' as implicit concepts in the recognition of nature's rights. Thus, it is automatically assumed that recognizing these rights implies that nature is a 'legal person'.[99]

The Ecuadorian case not only 'opens up' the doctrine of rights by including a new subject of rights but also creates a new doctrine to understand the legal subjectivity of nature. In contrast, most cases recognizing rights of nature con-

---

[99] See Mihnea Tănăsescu, 'Rights of Nature, Legal Personality and Indigenous Philosophies' (2020) 9(3) TEL 429; Brad Coombes, 'Nature's Rights as Indigenous Rights? Mis/recognition Through Personhood for Te Urewera' (2020) 1(2) Espace populations sociétés. In particular, Tănăsescu points out that 'the Ecuadorian Constitution pushes to conceive nature as a legal person, a conception that runs the risk of blending moral and legal notions in ways that are ultimately futile because the idea of personhood is always shaped by a particular type of being (usually the human person, with a set of characteristics).' In my view, this is an incorrect conception of how nature is perceived in Ecuador, where it is referred to as a 'subject' and is recognized with rights inherent to its own nature.

tinue to analyze legal subjectivity through the 'legal person' and 'legal personality' framework. I will briefly analyze the cases of New Zealand and India to better understand the differences with the Ecuadorian case.

In New Zealand, Te Urewera Park became 'a legal person [who] has all the rights, powers, duties, and responsibilities of a legal person.'[100] Additionally, the court ruling established the 'Te Urewera' Board, which is responsible for managing the park, with two-thirds of its composition being Maori. The court ruling regarding the Whanganui River and its national park also applies a similar model. This model is one of 'legal personality' which, consequently, applies concepts of that model, including the idea of 'legal personhood' that aligns more with the concept of a legal entity.

Indeed, parks and rivers are personified, shedding light on several issues. The model allows for the recognition not only of rights (which is positive) but even obligations (duties or responsibilities). But how do we understand the obligations that nature would have? One possible answer could be that nature provides 'environmental services' and that is its obligation. However, what happens when, for example, due to heavy rains, a river overflows and causes deaths. How do we take nature to court to hold it accountable?

The answer could go through the route of representation. For example, the 'Te Urewera' Board would be responsible. But that would lose sight of the reason for granting rights to nature: to protect it. The idea of recognizing parks and rivers as 'legal persons', with duties and responsibilities, seems to aim more towards giving nature the ability to act: to exercise its 'will and autonomy' and, through the Board – as an example – enter into contracts or take out a loan at the bank. Here, then, the idea of the legal personality model and the powers of administration of the Board make sense.

From the perspective of understanding nature as an autonomous subject of rights, represented only for litigation, the idea that a Board represents it and carries out any act that implies obliging it seems to divert attention from the purpose of granting rights: to give it autonomy to claim its rights and demand compensation for damages it suffers. Additionally, in the case of Te Urewera, not only the Tuhoe are part of the Board, but also public officials of the Crown. Therefore, if the idea was to strengthen the quality of Tuhoe guardians, the composition of the Board weakens it as well as their autonomy.[101] But it also weakens the potency of the discourse regarding rights for non-human persons. It ends up adopting a model applied to the conception and enforceability of human rights, reducing its effectiveness for nature.

Another case that also adopts the model of legal subjectivity through the categories of legal personhood and legal personality is the case of the Ganges and Yamuna rivers in India. In this case, the Uttarakhand High Court recognized these rivers and their

[100] New Zealand, Act Te Urewera, 2014, sec 11.1.

[101] For an in-depth analysis, see Coombes (n 99).

'tributaries, streams, and all natural courses derived from their waters as a legal person/legal entity/living entity with the status of a legal person with all the rights, duties, and responsibilities [rights, duties and liabilities] of a living person to preserve and conserve the Ganga and Yamuna rivers.'[102]

Additionally, it declared the Secretary of State of Uttrakhand and the Attorney General of the State of Uttrakhand as persons in loco parentis so that they are responsible for protecting, conserving, and preserving the rivers and their tributaries. The in loco parentis doctrine is applied to parents as legal representatives of their children.

The decision to frame the rights of nature within the classical concepts of legal subjectivity through the legal person model was not without criticism. This way of understanding the rights of nature implies assigning reciprocal duties to the rivers and therefore accepting that the rivers must fulfill duties. Furthermore, it implies a state representation system that views nature as a minor. Several lawyers have pointed out these issues: 'When you extend duties and responsibilities to natural entities, that opens up a new set of problems [...]. To some extent, it's counterproductive. Why get into those problems?'[103] The Supreme Court revoked this judgment precisely because of the difficulty of understanding rivers as legal persons. Echoing the arguments of the State of Uttrakhand against the decision of the Uttarakhand High Court and pointing out that recognizing rivers as legal persons would imply being able to sue them for floods, the Supreme Court established that the decision of the Uttarakhand High Court was unworkable and that rivers could not have the same legal status as human beings.

From this comparative perspective, the conception of nature as a subject of rights in the Ecuadorian case is better understood. By not relying on the traditional model of legal subjectivity, it recognizes nature as an autonomous subject of rights, with its own reasons to be recognized as such and with its own rights according to its nature. What seems inevitable is the mechanism of representation to claim its rights, but even there, no specific guardian or representative was assigned; rather, all people are allowed to request the direct redress of its rights.

4. Conclusions

The development of the rights of nature in Ecuador has sought to resolve questions that are foreign to the world of law: what is nature, how are its rights understood and applied, what happens when these rights collide with those of people, and how does their enforceability affect the legal system? The answers to these

[102] India, High Court of Uttarakhand 2017, Writ Petition (PIL) No 126 of 2014, *Mohd. Salim v State of Uttarakhand & others*.

[103] 'La Corte Superior de Madrás – India dictamina que la naturaleza tiene un estatus similar al de los humanos, y que los humanos están obligados a protegerla' (*Naturaleza con Derechos*, 10 May 2022) <www.naturalezaconderechos.org/2022/05/10/la-corte-cuperior-de-madras-india-dictamina-que-la-naturaleza-tiene-un-estatus-legal-similar-al-de-los-humanos-y-que-los-humanos-estan-obligados-a-protegerla/>.

questions can be approached and studied by looking at three stages of development that I have referred to as substantiation, exploration, and concretization.

These stages allow us to see how the development of a concept, with components from both science and ancestral knowledge, has progressed, though not necessarily in a linear fashion. Nor has it been a process without multiple ruptures, instead being one that has been slow and fragmented, with weak continuities and with tensions. The stages demonstrate the phenomenon of astonishment, understanding, disinterest, and enthusiasm for any new concept that is foreign to law. Thus, there was a first stage characterized by excitement, enthusiasm, and a desire to understand what this new concept was all about and where it came from. Then came an intermediate stage of jurisprudential and legal development, which, lacking clear guidance, hesitantly attempted to distinguish this new content from the old (the environment), eventually falling into a period of lethargy marked by a certain disinterest. Finally, there emerged a stage of renewed enthusiasm, fueled by years of unfinished business, which sought to understand what nature is and how it is to be protected.

In this final stage, the Constitutional Court's jurisprudence on the content and identification of the subjects of rights of nature is very promising. The Court has demonstrated that it takes the concepts that make up rights of nature seriously, and that is no easy task. The rights of nature push courts, lawyers, and the legal forum to question dominant legal concepts, such as the concept of legal subjectivity, while also requiring the incorporation of biological principles, which are not intuitive to us as lawyers, as well as ancestral knowledge, which is not typically taught in law school.

Perhaps this last factor is one of the gaps still pending in jurisprudence and law: whereas courts may have the ability to invoke the scientific concepts of natural cycles, structure, functions, and processes, the ability to understand them from the point of view of indigenous peoples remains elusive. This is a dialogue of knowledge, a necessary meeting of cultures to which recognition of the rights of nature invites us, and it is a path of legal pluralism that remains to be travelled.

The rights of nature must still be explored not only as a valuable tool to shed light on how to integrate ancestral knowledge into the sources of law in a legal system but also, in a more practical way, in order to think about how their contents can be used to combat the global ecological crisis and to face challenges – such as climate change – through the recognition of other forms of development.

Finally, this expansion of the subject of constitutional rights highlights how Ecuadorian constitutionalism constructs and understands fundamental rights, particularly the concept of rights holder. The inclusion of nature as a subject of constitutional rights challenges classical notions of legal subjectivity and entails an expansion of moral consideration based on theories of environmental ethics and the ancestral knowledge of indigenous peoples.

Bolivia

The Rights of Mother Earth: Fifteen Years of Solitude

Cletus Gregor Barié[*]

I. Introduction

In terms of the rights of nature, Bolivia is a contradictory case. On the one hand, it appears to be the vanguard in the recognition of the rights of Mother Earth, as it was one of the first countries in the world with specific legislation on the matter, along with the United States in 2006, Ecuador in 2008, and New Zealand in 2012.[1]

[*] This article was written as part of a doctoral research project on the challenges of plurinational state-building, carried out at the Vrije Universiteit Amsterdam, under the supervision of Ton Salman and Michiel Baud, and defended in May 2025. The manuscript was completed in August 2023. The evaluation of literature and case law, including internet material, reflects the legal situation at that time. A preliminary English version of this text as part of the full thesis is available: Cletus Gregor Barié, *The Plurinational Challenge: Intercultural state-building and the rights of nature in Bolivia* (Vrije Universiteit Amsterdam 2024), <https://doi.org/10.5463/thesis.596>.

[1] Craig M Kauffman, 'Mapping Transnational Rights of Nature Networks & Laws: New Global Governance Structures for More Sustainable Development' (International Studies Association Annual Conference, Toronto, 29 March 2020) <http://files.harmonywithnatureun.org/uploads/upload924.pdf>.

Being the 'first cases of explicit recognition of nature as a legal entity',[2] these legal innovations have attracted the interest of numerous researchers and are a regular reference in debates on the new Latin American constitutionalism and alternative approaches to development.[3]

On the other hand, under the successive administrations of President Evo Morales (2006–2019), the Bolivian government relied early on the intensification of the massive exploitation and export of its raw materials to take advantage of the boom in international prices in the approximate period between 2003 and 2014.[4] In this way, broad Bolivian social and productive sectors joined, tacitly or explicitly, a 'commodity consensus',[5] which implied a marked neo-extractivist orientation of the economy.[6] As an additional factor, one which certainly increases the complexity of the issue, since 2010 Bolivia managed to position itself internationally as one of the most determined countries in denouncing the predatory effects of globalization and in promoting the recognition of the rights of Mother Earth in multiple international forums.

Fifteen years after the incorporation of the rights of Mother Earth as a cross-cutting issue in public policies through a Short Law (Law 71 of 2010)[7], it is worth asking about the practical effects of these legal innovations in such a contradictory context. What does pro-Mother Earth legislation in Bolivia currently consist of, and how is it integrated into relevant sectoral policies, eg regarding the environment, land, and natural resources? What judicial bodies guarantee these rights and how can they be asserted? What has the implementation process been like and what has been the scope and practical impact of these instruments so far?

The objective here is precisely to explore these issues, focusing on the tensions between the original aspiration of guaranteeing the permanence of Mother Earth's life cycles and the implementation paths ultimately adopted. We thus attempt to fill a gap in research, as studies on the impact of Mother Earth rights in Bolivia are incipient and often limited to an analysis of legal provisions and gen-

---

[2] María Valeria Berros, 'Challenges for the Implementation of the Rights of Nature: Ecuador and Bolivia as the First Instances of an Expanding Movement' (2021) 48 Latin American Perspectives 192, 195.

[3] Antonio Carlos Wolkmer, Maria de Fátima S Wolkmer and Debora Ferrazzo, 'Derechos de la Naturaleza: para un paradigma político y constitucional desde la América Latina' in L Liliana Estupiñán Achury and others (eds), *La naturaleza como sujeto de derechos en el constitucionalismo democrático* (Universidad Libre de Colombia 2019) 71.

[4] José Antonio Ocampo, 'Commodity-Led Development in Latin America' (2017) 9 Revue internationale de politique de développement 51.

[5] Maristella Svampa, '"Consenso de los Commodities" y lenguajes de valoración en América Latina' (2013) 244 Nueva Sociedad 30. All translations from Spanish are by the author.

[6] Hans-Jürgen Burchardt and others (eds), *Nada dura para siempre: Neo-extractivismo tras el boom de las materias primas* (Universidad Andina Simón Bolívar and International Center for Development and Decent Work (ICDD) 2016); Eduardo Gudynas, *Extractivismos. Ecología, economía y política de un modo de entender el desarrollo y la naturaleza* (CEDIB 2015).

[7] Ley de Derechos de la Madre Tierra 2010 (Law 71).

eral assessments.[8] This is partly due to the difficulty of establishing criteria to measure the effectiveness of this new type of legislation,[9] as the analysis requires a rethinking of the foundations of traditional environmental law. This inadequate conceptual elaboration is compounded by practical limitations in accessing up-to-date information on both legal aspects (eg through agro-environmental courts) and current environmental policy. The official information available proves to be rather limited.[10]

We draw on some significant contributions to the discussion, including the analysis of Kotzé and Villavicencio Calzadilla,[11] who focus on constitutional tensions and introduce a broad perspective on the legal-political context. Lalander's work[12] also opens up a series of reflections on the contradictions between policies aimed at resource extraction and a safeguarding of the rights of Mother Earth.

In addition, Kauffman and Martin's[13] recent analysis of the politics of nature rights and land jurisprudence raises a number of criteria for comparing progress. The matrix we will use here with guiding questions to assess legal texts (see *Table 1*) is based precisely on the set of indicators developed by these authors and their ideal models of pro-Mother Earth norms.[14] In this vein we differentiate between two dimensions: scope of rights refers to how nature is defined and what catalogue of rights it is granted; legal strength refers to the mechanisms and instruments to guarantee their enforcement and their degree of mainstreaming and articulation with other legislation, eg regarding the environment, natural resources, or forests.

---

[8] Marco Aparicio Wilhelmi, 'Nuevo Constitucionalismo, derechos y medio ambiente en las constituciones del Ecuador y Bolivia' (2011) 9 Revista general de derecho público comparado 1; Cletus Gregor Barié, 'Nuevas narrativas constitucionales en Bolivia y Ecuador: el buen vivir y los derechos de la naturaleza' (2014) 59 Latinoamérica. Revista de Estudios Latinoamericanos 9; Andrea Carolina Herrera-Martínez, *La madre tierra como sujeto de derechos en las constituciones de Ecuador (2008) y Bolivia (2009)* (Universidad Católica de Colombia 2020); Pinto Calaça and others, 'La naturaleza como sujeto de derechos: análisis bioético de las Constituciones de Ecuador y Bolivia' (2018) 18(1) Revista Latinoamericana de Bioética 155.

[9] Michele Carducci and others, *Towards an EU Charter of the Fundamental Rights of Nature* (EESC 2020) 23.

[10] Interviews with Schwarz (2021) and Requena (2021).

[11] Louis J Kotzé and Paola Villavicencio Calzadilla, 'Somewhere between Rhetoric and Reality: Environmental Constitutionalism and the Rights of Nature in Ecuador' [2017] TEL 401.

[12] Rickard Lalander, 'Rights of Nature and the Indigenous Peoples in Bolivia and Ecuador: A Straitjacket for Progressive development Politics?' (2014) 3(2) Revista Iberoamericana de Estudios de Desarrollo 148; Rickard Lalander, 'Entre el ecocentrismo y el pragmatismo ambiental. Consideraciones inductivas sobre desarrollo, extractivismo y los derechos de la naturaleza en Bolivia y Ecuador' (2015) 6(1) Revista Chilena de Derecho y Ciencia Política 109; Rickard Lalander, 'Indigeneidad, descolonización y la paradoja del desarrollismo extractivista en el Estado Plurinacional de Bolivia' (2017) 8(1) Revista Chilena de Derecho y Ciencia Política 49.

[13] Craig M Kauffman and Pamela L Martin, *The Politics of Rights of Nature. Strategies for Building a More Sustainable Future* (Massachusetts Institute of Technology Press 2021) 61.

[14] Kauffman and Martin (n 13) 15, 61.

Our view of norms is inspired by legal anthropology,[15] especially by a series of works that have since the 1990s investigated the role of multi- and plurinational legal reforms in Latin America and their contribution to democratization.[16] In particular, we are interested in how social actors, as acting subjects, resort to legal discourse and approaches (*rights-based demands* in the words of Altmann)[17] to expand their possibilities to act for themselves (in the sense of *agency*, according to Ortner)[18] and to guide their collective action. Indigenous agency is understood here as 'the capacity of ethnic actors to act on themselves and their environment'.[19] Agency is always negotiated in concrete situations, especially in the context of indigenous peoples' relationship with the state, in which there is an 'interplay between parties, where the limits and concrete forms for this relationship are disputed, reworked, created and updated'.[20] In this sense, the norms discussed on Mother Earth reflect the negotiation capacity and balance of power between different actors at a given moment. The legal texts we will discuss can thus be understood as a temporary product of a play of forces, as 'the result of pact[s] between different actors, which freeze a certain balance of power, subordinating some and privileging others'.[21] With the enquiry developed here on the advancement of the rights of Mother Earth in Bolivia, we seek to contribute to a broader reflection on the possibilities of social agency of indigenous organizations in the framework of a new plurinational state.

The research is mainly based on a set of data obtained in twenty-one semi-structured interviews with indigenous leaders and representatives, environmental experts, and state officials in 2020 and 2021 (see selection of interviewees, below). A first cycle addressed the different notions of the plurinational character of the state, the role of the government as 'official custodian' of this new ideology, and

---

[15] Esteban Krotz (ed), *Antropología jurídica. Perspectivas socioculturales en el estudio de derecho* (Anthropos Editorial 2002).

[16] Willem Assies, Gemma van der Haar and André Hoekema (eds), *El Reto de la diversidad. Pueblos indígenas y reforma del estado en América Latina* (El Colegio de Michocán 1999).

[17] Philipp Altmann, '"The Right to Self-determination": Right and Laws Between Means of Oppression and Means of Liberation in the Discourse of the Indigenous Movement of Ecuador' (2016) 29 International Journal for the Semiotics of Law – Revue internationale de Sémiotique juridique 121.

[18] Sherry B Ortner, *Anthropology and Social Theory. Culture, Power, and the Acting Subject* (Duke UP 2006).

[19] Christian Martínez Neira, Patricia Rodríguez and Miguel Leone Jouaunny, 'La vía política indígena: Agencia y nuevas formas de politización en América Latina' (2019) 52 Polis 5, <https://doi.org/10.32735/S0718-6568/2019-N52-1378>. All translations of quotations from Spanish to English are made by the author.

[20] Samanta Guiñazú, Florencia Trentini and Nadia Ameghino, 'Agencia(s) indígena(s) en políticas públicas participativas en Norpatagonia: políticas de comanejo y relevamiento territorial' (2019) 52 Polis 45, <https://doi.org/10.32735/S0718-6568/2019-N52-1366>.

[21] Ruben Martinez Dalmau, 'El proceso constituyente: la activación de la Soberanía', in n Íñigo Errejón and Alfredo (eds), *¡Ahora es cuándo, carajo! Del asalto a la transformación del Estado en Bolivia* (Intervención Cultural 2011) 37.

perspectives on the practical progress made and the remaining challenges. One surprising finding was that the construct of plurinationality has begun to become part of the social imaginary of the state and has established itself as a new frame of reference against which government action is assessed.[22]

In a second series of interviews, the focus was specifically on the issue of Mother Earth rights, their genesis, their conflictive transformation into legislation, and the incipient steps of implementation. According to most interviewees, respect for Mother Earth is the philosophical and spiritual basis for many of the state reform proposals made by indigenous groups that – through a post-constitutional process and after a struggle over the interpretation of the scope of reform – have been transformed into an explicit juridical-legal approach. The interviews identified novel aspects that contradict some of the findings of the literature mentioned above. One example is the alleged lack of jurisprudence on the subject, an aspect that we suggest should be analyzed in a more nuanced way.[23] In addition to the interviews, we reviewed proposals and statements by indigenous organizations in order to understand the scope of the initial proposals and to contrast them with the legislation that was finally approved.

The article is organized in four parts. In the first section we situate the emergence of pro-Mother Earth approaches in the context of the search for a refoundation of the state, in which the impetus and initiative of civil society was gradually absorbed and replaced by a new government under the Movement Towards Socialism (MAS). In this conflictual phase, most indigenous organizations, temporarily brought together in a coalition known as Unity Pact, merged their different implicit ideas of respect for Mother Earth in a legislative proposal with strong transformative potential. In the second part, we will analyze the implications of a new legal framework that now refers to the rights of Mother Earth. Specifically, we will evaluate the constitution and two specific laws based on the aforementioned dimensions of scope and legal strength.

In the third part we explore the legal effectiveness of these innovations, reviewing three aspects: the functioning of the institutions created, the jurisprudence, and the adequacy of the remaining legislation for the new conception of life cycle care. We also identify – as a novel phenomenon – an increasing connection and networking between environmental lawyers and local communities, and we describe the growing incorporation of pro-Mother Earth approaches in actions in defense of territories. In the conclusion we summarize the advances and pending challenges resulting from these legal innovations and discuss their potential for strengthening indigenous agency. Undoubtedly, the Bolivian experience offers lessons for other regions as to the challenges involved in rethinking the relationship with nature from a legal perspective.

---

[22] Cletus Gregor Barié and Moira Zuazo, 'Plurinational state and ecological conscience in Bolivia: The case of the 2019 forest fires' (2022) 114 ERLACS 89, <https://erlacs.org/articles/10.32992/erlacs.10862>.

[23] Kauffman and Martin (n 13) 118.

II. Emergence of the Rights of Mother Earth

The rights of Mother Earth emerged in a context of a struggle between different social actors regarding the construction of a new state model since the beginning of the century. In a first phase, which lasted until approximately 2009/2010, the protagonism and transforming impulse came from civil society, especially from peasant and indigenous organizations. It was in this context of social effervescence and intense intercultural dialogue that the proposal for a plurinational state was developed and solidified as an alternative to the current republican state. Plurinationality as an agenda and political slogan began to consolidate as a kind of 'meta-concept' that aggregated several others, including the decolonization of society, the direct political representation of indigenous peoples, the establishment of their territorial autonomy, and the search for good living in harmony with nature.[24]

In a second phase, which began in 2010, this transformative impulse was taken up and channeled by the Movement Towards Socialism government, which placed the plurinational state and indigenous and peasant peoples and their approaches to good living and the protection of Mother Earth at the center of the official discourse, while in parallel attempting to reduce the agency of the indigenous organizations. These tensions over the scope of state reforms also implied different visions of its role in addressing cultural diversity: on the one hand there is a conception of community-based reconstruction of the state which seeks 'to respect cultural difference in its multiple forms and protection of marginal peoples [= indigenous minorities] from the state'. On the other hand, there is the discourse of a strong central state, based on an 'ecumenical indigeneity for a majority of Bolivia's citizens'.[25]

The debate over the rights of Mother Earth took place precisely during the transition between an indigenous-peasant protagonism and a concentration of power from a government that defined itself as popular. In this context, two actors with sometimes overlapping and sometimes conflicting interests were competing for the authority to define the scope of these new rights: the Unity Pact, a temporary alliance of indigenous peoples from all over Bolivia, and the Movement Towards Socialism, which, as the governing party under the presidential leadership of, first, Evo Morales (2006-2019) and, later, Héctor Arce (from 2020), took the agenda of social change under its banner.

---

[24] Daniel Bonilla Maldonado, 'El constitucionalismo radical ambiental y la diversidad cultural en América Latina. Los derechos de la naturaleza y el buen vivir en Ecuador y Bolivia' (2019) 42 Revista Derecho del Estado 3; Jorge Resina de la Fuente, *La plurinacionalidad en disputa. El pulso entre Correa y la CONAIE; dinámica y marcos discursivos de la contienda por lo plurinacional* (Abya-Yala 2012); Salvador Schavelzon, *Plurinacionalidad y vivir bien/ buen vivir. Dos conceptos leídos desde Bolivia y Ecuador post-constituyentes* (Abya-Yala and CLACSO 2015).

[25] Andrew Canessa, *Conflict, Claim and Contradiction in the New Indigenous State of Bolivia* (desiguALdades.net 2012) 33 <www.desigualdades.net/Working_Papers/Search-Working-Papers/Working-Paper-22-_Conflict_-Claim-and-Contradiction-in-the-New-Indigenous-State-of-Bolivia_/index.html>.

1. The Unity Pact Proposal

Founded in September 2004 in Santa Cruz de la Sierra, the Unity Pact represents the culmination of several decades of coordination and articulation between different subaltern rural actors. Faced with the growing crisis of a governance model based on short-term political pacts, the Unity Pact managed to bring together the most important peasant and indigenous organizations and articulated demands for reform to be carried out through the Constituent Assembly.[26] Initially, the Unity Pact was a close ally of President Evo Morales and, during the constituent deliberations (2006–2007), it became one of the most influential actors among Bolivia's progressive sectors, 'a kind of collective organic intellectual'.[27]

In contrast to Ecuador, where several of the existing indigenous organizations are national in scope, the best known being CONAIE (Confederation of Indigenous Nationalities of Ecuador), in Bolivia they are historically fragmented by region. There are marked ideological differences between the lowland organizations, represented by CIDOB (Confederación de Pueblos Indígenas del Oriente Boliviano), and the highland ones, such as CONAMAQ (Consejo Nacional de Ayllus y Markas del Qullasuyu) and CSUTCB (Confederación Sindical Única de Trabajadores Campesinos de Bolivia). These include the significance and political implications of ethnic identity, conceptions of territory and land ownership, and the manner in which they interact with government bodies.[28] These profound discrepancies also affected the dynamics within the Unity Pact, contributing to its fragile state as an alliance.[29]

The impetus to deepen the issue was given at the World People's Conference on Climate Change and the Rights of Mother Earth, held in Tiquipaya (Cochabamba) in April 2010 at the initiative of the Bolivian government. There, delegations from 129 countries adopted a Universal Declaration of the Rights of Mother Earth and proposed the creation of an international tribunal against environmental crimes.[30] The meeting represents a major milestone in the formation of an in-

---

[26] Fernando Garcés (ed), *El Pacto de Unidad y el proceso de construcción de una propuesta de constitución política del estado* (Centro Cooperativo Sueco 2010); María del Pilar Valencia García and Iván Égido Zurita, *Los pueblos indígenas de tierras bajas en el proceso constituyente boliviano* (CEJIS 2010).

[27] Luis Tapia, 'La fuerza de octubre', in Helen Alvarez and others (eds), *De regreso a octubre. La 'guerra del gas'; 10 años después* (Página Siete and El Cuervo 2013) 127, 131.

[28] Claudia Arce, 'Las organizaciones indígenas y sus relaciones con el gobierno (2010-2014)' in Maria Teresa Zegada (ed), *Las naciones indígena originario campesinas en el horizonte plurinacional* (Centro Cuarto Intermedio 2014) 97; Ton Salman, 'Movimientos sociales en Bolivia en tiempos del MAS' (2011) 14(29) Tinkazos 21.

[29] Interview with Zuazo (2014).

[30] Estado Plurinacional de Bolivia, 'Conferencia Mundial de los Pueblos sobre el Cambio Climático y los Derechos de la Madre Tierra. Discursos y Documentos Seleccionados' (Conferencia Mundial de los Pueblos sobre el Cambio Climático y los Derechos de la Madre Tierra, Tiquipaya, Cochambamba, 20 to 22 April 2010). <www.cancilleria.gob.bo/webmre/sites/default/files/libros/Cmpcc%20discursos%20y%20documentos%20seleccionados.pdf>.

ternational network of exchange and support in the promotion of the rights of nature, as a contribution to an emerging 'informal system of global governance'.[31] Inspired by international debates and featuring the participation of intellectuals such as theologian Leonardo Boff, environmental lawyer Cormac Cullinan, and Ecuadorian economist Alberto Acosta,[32] the indigenous organizations of the Unity Pact met for nine months in order to reach a consensus on their own proposal for a Draft Bill on the Rights of Mother Earth. In accord with Wilson Rocha,[33] we understand this process of translating and codifying the millenary custom of respect for Pachamama as a case of custom transformed into law.

From the outset, this proposal rejected the notion of natural resources as an expression of 'the relationship of dispossession of nature by the capitalist world-system', and describes them instead as 'beings, substances, elements, components, bounties of Mother Earth'.[34] Mother Earth is defined as 'a dynamic living system formed by the indivisible community of all living beings'[35] and for the purposes of its protection it is considered a collective subject of public interest.[36] It has the right to life, to diversity, to the functionality of water cycles, to clean air, to equilibrium, and to restoration.[37] In cases of doubt regarding the protection and defense of rights, the authorities must favor Mother Earth (*in dubio pro natura* principle). An Ombudsman's Office oversees the enforcement, promotion, dissemination, and fulfilment of these rights.[38] Free and informed prior consultation is also guaranteed in the case of extractive projects and legislative measures affecting indigenous peoples.[39] The draft also sought to achieve a cross-cutting scope, as it was intended to become 'the basis at all levels and sectors of the Plurinational State for the participatory formulation and implementation of policies for the use and exploitation of the components of Mother Earth'.[40]

The proposal of the Unity Pact thus adopts an approach in which the whole of nature, its life systems, and its components are rights holders (*Nature's rights model*). This broad approach contrasts with the legal personality model of a specific ecosystem (*legal personhood model*), as applied, for example, in Colombia

[31] Kauffman (n 1) 27.

[32] Gabriel Mario Santos Villarreal, 'Conferencia Mundial de los Pueblos sobre Cambio Climático y Derechos de la Madre Tierra' (Conferencia Mundial de los Pueblos sobre el Cambio Climático y los Derechos de la Madre Tierra, Tiquipaya, Cochambamba, 20 to 22 April 2010) <www.diputados.gob.mx/sedia/sia/spe/SPE-CI-A-08-10.pdf>.

[33] Interview (2021).

[34] Pacto de Unidad, *Anteproyecto de Ley de Madre Tierra. Primer borrador de consenso de la Asamblea Legislativa plurinacional con el Pacto de Unidad el 24.11.10, Complementado el 22.02.11 en consenso on el Pacto de Unidad* (2011) 7.

[35] Pacto de Unidad (n 34) art 4.

[36] Pacto de Unidad (n 34) art 5.

[37] Pacto de Unidad (n 34) art 7.

[38] Pacto de Unidad (n 34) art 37.

[39] Pacto de Unidad (n 34) art 42.

[40] Pacto de Unidad (n 34) art 2(2).

(Magdalena River), India, or New Zealand.[41] It also introduces a principle of maintenance of integrity, which not only implies restoration in the event of damage, but also guarantees that the functionality of the different ecosystems will be safeguarded continuously (right of equilibrium).[42] A strength of this model is the breadth of situations that can be considered as violating the rights of nature. The state and society are also legally obliged to act decisively and proactively, ie they cannot turn a blind eye to the destruction or overexploitation of nature. The project combines institutional mechanisms, eg the oversight of the Defensoría de la Madre Tierra (Ombudsman's Office), with flexible instruments, eg allowing a complaint to be filed by any citizen who sees these rights violated (see *Table 1*).

Despite its high degree of conceptual elaboration, the text was only partially taken up in the subsequent legislative process. Following approval of the new Constitution in 2009, an event to which it had made major contributions, the Unity Pact gradually lost its capacity for dialogue with the government, until it disintegrated at the end of 2011. Since then, other civil society organizations have managed to position themselves as legitimate spokespersons of organized civil society. A prime example is CONALCAM (Coordinadora Nacional por el Cambio), an entity which has been described as 'a form of unconditional articulation of social organizations with the government'.[43]

2. *The Multiple Ideological Sources of the MAS*

Since its foundation in 1997, the Movement Towards Socialism (MAS) basically brought together three groups: coca growers and colonizers, indigenous and peasant organizations, and, to a lesser extent, some urban circles sympathetic to Marxism and indigenism. The founding members conceived MAS as a 'political instrument' of different social organizations and not as a conventional party, which is also reflected in the second part of its acronym, IPSP (Political Instrument for the Sovereignty of the People). MAS discourse revolves around the defense of sovereignty and natural resources, such as the coca leaf and hydrocarbons, in the face of external intervention. These nationalist and anti-imperialist components are often complemented by the demand for recognition of the rights of indigenous peoples and a socialist rhetoric, introduced mainly by urban intellectual groups.[44]

Evo Morales and his vice president, Álvaro García Linera, were elected three times in presidential elections: 2005 (53.7%), 2009 (64.2%), and 2014 (61.4%). National voting in 2019 was surrounded by suspicions of irregularities and fraud,

---

[41] Kauffman and Martin (n 13) 15.

[42] Pacto de Unidad (n 34) art 7(5).

[43] Maria Teresa Zegada and others, *La democracia desde los márgenes: Transformaciones en el campo político boliviano* (CLASCO and Muela del Diablo 2011) 257.

[44] María Reneé Barrientos Garrido, *De las calles a las urnas: Discurso político y estrategias identitarias del movimiento cocalero y su 'instrumento político': MAS-IPSP* (Master thesis, Universidad de Salamanca 2010).

which led to the resignation of the president. Jeanine Áñez, second vice-president of the Senate, then assumed the presidency amid a debate over the constitutional nature of her presidential succession.[45] Finally, in Bolivia's October 2020 general elections, MAS won again, and its candidate, Luis Arce, was elected with 55.1% of the vote.

The MAS government's economic policies are often characterized as 'developmentalist' because they focus on economic growth, expansion of the domestic market, and the fight against extreme poverty. However, this characterization is inaccurate, since one of the objectives of the development theories promoted in the middle of the last century by the Economic Commission for Latin America (ECLAC) was to facilitate a late industrialization, which did not occur. In fact, Bolivia experienced between 2011 and 2014 a 'reprimarization of its economy', with a decline in its manufactured products.[46]

Although the different MAS governments prefer to use the concept of 'social extractivism' instead of neo-extractivism, they have widely defended this policy as a transition strategy from dependence to a more autonomous development.[47] The importance of maintaining and even expanding the extractive industry as a 'strategic sector' also comes from a constitutional mandate and is explicitly reflected in successive public policy plans and programs, among others, the 2006–2010 Development Plan and the 2012 Patriotic Agenda.[48] It was also accompanied by a wide range of long-term public policies, including the promotion of new hydrocarbon and oil concessions with strong state participation, the consolidation of a new intermediary mining sector (the so-called *cooperativistas*), and permission for the use of transgenic crops and policies for the consolidation of large agricultural properties. This orientation towards a commodity export model was then installed not only as a 'transitory' approach, as often pointed out by the MAS government, but as a structural transversal axis.[49]

---

[45] BBC, 'Bolivia crisis: Jeanine Áñez declares herself interim president', *BBC News* (London, 13 November 2019), <www.bbc.com/news/world-latin-america-50399640>.

[46] Manuel De La Fuente, 'Extractivismo y conflictos en Bolivia', in Manuel De La Fuente, Tania Ricaldi Arévalo, Angel Saldomando (eds), *Lógias de Desarrollo, Extractivismo y Cambio Climático* (UMSS-CESU and Reinventerra 2017) 233, 236; Solange Sardan Matijašević, *Extractivismo en Bolivia en el periodo 2000–2014: ¿Qué tan agudo es el caso?* (Friedich-Ebert-Stiftung Bolivia 2017) 5.

[47] Eduardo Gudynas, *Derechos de la naturaleza y políticas ambientales* (PDTG and others 2014); Gisela V Rodriguez Fernandez, 'Neo-extractivism, the Bolivian State, and Indigenous Peasant Women's Struggles for Water in the Altiplano' (2020) 13(1) Human Geography 27.

[48] Ministerio de Planificación del Desarrollo, Agenda patriótica 2025. 13 pilares de la Bolivia digna y soberana (2012), <https://siteal.iiep.unesco.org/bdnp/3612/agenda-patriotica-2025>; Ministerio de Planificación del Desarrollo, Plan nacional de desarrollo. 'Bolivia digna, soberana, productiva y democrática para vivir bien': lineamientos estratégicos (2007), <https://siteal.iiep.unesco.org/bdnp/613/plan-nacional-desarrollo-bolivia-digna-soberana-productiva-democratica-vivir-bien-2006-2011>; Rodriguez Fernandez (n 47).

[49] Fernanda Wanderley, *¿Qué pasó con el proceso de cambio en Bolivia? Ideales acertados, medios equivocados, resultados trastrocados* (CIDES-UMSA 2013).

As a reflection of their ideological diversity, and in parallel to the consolidation of an extractivist economic policy, the successive MAS governments incorporated the new indigenous approaches almost systematically in state symbols and official discourse. The MAS government's protagonism in the defense of Mother Earth and the appropriation of the different indigenous conceptions began in its first administration (2006-2009). Subsequently, with the departure of several intellectuals and activists involved in the defense of indigenous rights and the environment, these references lost force and became more rhetorical. As an example of the symbolic role that the topic has assumed, under the leadership of the Minister of Foreign Affairs and later Vice-Chancellor, David Choquehuanca, the government to this day regularly convenes national events related to the subject.[50] Mother Earth has thus been integrated as a substantial ingredient of the national symbolism 'in which the indigenous peoples and decolonization [...] are part of a new state narrative'.[51]

The fervent evocation of the Pachamama is also addressed to international public opinion in order to position Bolivia within a group of countries critical of the terms of operation of the global economy. In this context, Bolivian diplomacy achieved a long series of international milestones. In 2009, the United Nations General Assembly declared 22 April as International Mother Earth Day. That same year, Miguel d'Escoto named Evo Morales 'World Hero of Mother Earth'. Since then, a series of resolutions and reports have been presented at the United Nations on the subject, including a feasibility study on the recognition of Mother Earth's rights.[52] Recently, at the Glasgow Climate Summit 2021, the Bolivian delegation achieved mention of Mother Earth in the final document.[53] Bolivia continues to be perceived by many international activists as a vanguard country on the issue of Mother Earth rights.[54]

---

[50] *Re-encuentro con la Pachamama – en vivo* (2021) <www.facebook.com/share/18rjwWnhGy/?mibextid=wwXIfr>.

[51] Yuri F Tórrez and Claudia Arce, 'La estética del Estado Plurinacional: ¿Racialización o visibilización del indígena?' (*Hurgando el avisperso. Blog de Yuri F. Tórrez*, 14 June 2017) <https://hurgandoelavispero1.wordpress.com/2017/07/14/la-estetica-del-estado-plurinacional-racializacion-o-visibilizacion-del-indigena-por-claudia-arce-y-yuri-f-torrez/>.

[52] Bartolomé Clavero and Carlos Mamami, *Study on the need to recognize and respect the rights of Mother Earth* (United Nations 2010) <https://digitallibrary.un.org/record/676394?ln=en&v=pdf>. Bolivia also chaired in 2014 the Group of 77, which now integrates 134 countries, and it convened a summit in Santa Cruz de la Sierra together with China, which concluded with a call for a 'new order for living well.' Several international declarations incorporated Bolivian approaches to Mother Earth, including the 2012 United Nations Conference on Sustainable Development in Rio de Janeiro (known as Rio+20), the 2012 Alternative World Water Forum in Marseille, France, and the 2012 World Conservation Congress.

[53] Washington Post Staff, 'The Glasgow climate pact, annotated' *The Washington Post* (Washington D.C., 13 November 2021) <www.washingtonpost.com/climate-environment/interactive/2021/glasgow-climate-pact-full-text-cop26/>.

[54] Interview with Requena (2021).

III. Pro-Mother Earth Legislation

Although Mother Earth is mentioned in several Bolivian laws in a secondary way, to understand the spirit and scope of her rights it is sufficient to review the current Constitution and Laws 71 and 300, as well as the associated regulations, In light of the ambitious proposal made at the Tiquipaya meeting to build 'new models to re-establish harmony with nature', this set of norms seems rather modest.[55]

1. Ephemeral Appearance of Pachamama in the Constitution of 2009

During the Constituent Assembly, the issue of the rights of Mother Earth was raised in the Commission of Renewable Natural Resources, Land, Territory and Environment[56] and then introduced in a preliminary version of the constitutional text through the novel category of 'crimes against the environment'.[57] However, the finally approved Constitution[58] maintains the human being as the main holder of rights, either as an individual or as part of a cultural collectivity, and the development and welfare of human beings is held as the highest objective. Therefore, it is individual persons who have the right 'to a healthy, protected and balanced environment'[59] and who can act in defense of the right to the environment.[60] Likewise, indigenous nations and peoples have the right to 'live in a healthy environment, with adequate management and use of ecosystems'.[61] In any case, Bolivians are called upon to 'protect and defend an environment adequate for the development of living beings'.[62]

The Constitution also provides for two novel mechanisms to enforce environmental rights, which have had some relevance in practice. First, a new Agro-environmental Tribunal, as the highest instance,[63] which, among other things, resolves questions regarding the use and exploitation of renewable natural resources, water, forests, and biodiversity, and regarding practices that endanger the ecological system and the conservation of species or animals.[64] Another novel legal instrument is the popular action against any act or omission of the authorities or individuals that violates environmental rights.[65] This action can be exercised by any

---

[55] Estado Plurinacional de Bolivia (n 30) 89.

[56] Juan Carlos Pinto Quintanilla (ed), *Enciclopedia histórica documental del proceso constituyente boliviano* (Estado Plurinacional de Bolivia 2009) 1865–2022.

[57] Magna Asamblea Constituyente, Constitución Política del Estado. Aprobada en grande sobre los informes de mayoría (Sucre 2007).

[58] Nueva Constitución Política del Estado 2009.

[59] Nueva Constitución Política del Estado 2009, art 33.

[60] Nueva Constitución Política del Estado 2009, art 34.

[61] Nueva Constitución Política del Estado 2009, art 30(2)(10).

[62] Nueva Constitución Política del Estado 2009, art 108(16).

[63] Nueva Constitución Política del Estado 2009, art 186.

[64] Nueva Constitución Política del Estado 2009, art 189(1).

[65] Nueva Constitución Política del Estado 2009, art 135.

person, individually or on behalf of a collective, even where only the threat of a violation exists,[66] a preventive aspect that, according to Aparicio Wilhelmi,[67] is 'of enormous importance in the environmental field'.

Alongside a conventional environmentalist conception, which places human beings and communities and their economic development at the center of concerns, the first signs of a different conception also appear. As an innovation, Pachamama, the indigenous deity of fertility, enters the scene in the preamble: 'Fulfilling the mandate of our peoples, with the strength of our Pachamama and thanks to God, we re-found Bolivia'. She is thus the guarantor and spiritual guide of social change.[68] Another concept taken from the indigenous cosmovision is 'living well', in the Aymara language 'suma qamaña', which is assumed as the structural basis and ethical-moral principle of society[69] (preamble and art 8(I) and (II)) and which even guides a new plural economic model.[70] The idea of *living well*, which exists under different notions and expressions for various indigenous peoples of the region, refers to the aspiration for a full life in harmony with oneself, one's community, and the natural environment. A still-evolving concept, it combines the idea of material well-being with peaceful social coexistence and solidarity.[71] In the Bolivian Constitution, living well is not clearly defined and appears alongside the concept of development, which continues to function as the dominant paradigm.[72]

The constitutional text is also influenced by a holistic ecosystem approach, contained in the Convention on Biological Diversity (1992) and widely disseminated by multilateral organizations, which seeks to safeguard the structure, functioning, and key processes of ecosystems, rather than just focusing on one species or human activity.[73] The Constitution, for example, assigns the various organs of the state (national and subnational) with the obligation to 'preserve, conserve and contribute to the protection of the environment and wildlife by maintaining the ecological balance and the control of environmental pollution'.[74] By elevating ecological balance to a position of higher purpose, comparable to the fulfilment of civil and political rights, the text seems close to adopting a theoretical position of ecological constitutionalism.[75]

---

66 Nueva Constitución Política del Estado 2009, art 136.

67 Wilhelmi (n 8) 15.

68 Joel Colón-Ríos, 'Constituent Power, the rights of nature, and universal jurisdiction' (2014) 60 McGill LJ 127, 150.

69 Nueva Constitución Política del Estado 2009, Preamble and art 8(1), (2).

70 Nueva Constitución Política del Estado 2009, art 306(1), (3), art 313.

71 Ivonne Farah and Véronica Tejerina, *Vivir Bien. Infancia, género y economía, entre la teoría y la práctica* (CIDES-UMSA 2013).

72 Barié (n 8).

73 UNEP-WCMC, *Ecosystem approach* (2019) <www.biodiversitya-z.org/content/ecosystem-approach>.

74 Nueva Constitución Política del Estado 2009, arts 299(2)(1), 342, 380(2).

75 Rudolf Steinberg, *Der ökologische Verfassungsstaat* (Suhrkamp 1998).

To conclude, the Bolivian Constitution reflects ideological tensions and oscillates between conventional developmentalism, the defense of Pachamama,[76] and an ecosystemic approach. Ultimately, it is up to the state to reconcile these contradictory mandates. On the one hand, it is assigned the function of 'guaranteeing the well-being, development, security, protection and equal dignity of individuals, nations, peoples and communities',[77] giving priority to the industrialization of renewable and non-renewable natural resources to overcome dependence on the export of raw materials.[78] On the other hand, it must ensure the conservation and protection of the environment within the framework of sustainable development and in harmony with nature.[79]

2. *The Law on the Rights of Mother Earth (Law 71)*

In December 2010, in view of President Evo Morales' need to present concrete actions at the World Summit on Climate Change in Cancun, the ruling party's bench accelerated the approval by 'urgent vote' of the Law on the Rights of Mother Earth (Law 71) in the Plurinational Assembly. Given the time pressure, they opted for a Short Law, a compact format of ten articles that requires a shorter legislative process. Law 71 is mainly based on the proposal of the Unity Pact mentioned above, and in part it even reproduces it verbatim.

Mother Earth is defined as a 'dynamic living system made up of the indivisible community of all living systems and living beings,' including humans.[80] The introduced concept of 'living systems' is used as a synonym for ecosystems, describing them as 'complex and dynamic communities of plants, animals, microorganisms and other beings and their environment, where human communities and the rest of nature interact as a functional unit'.[81] For the purposes of protecting and promoting its rights, Mother Earth is considered a 'collective subject of public interest'.[82] She has, among others, the right to maintain the integrity of her life systems, the right to preserve her biodiversity (without being genetically manipulated), and a right to the functionality of her hydrological cycles.[83] It also guarantees the preservation of the quality and composition of the air, the balance between its components, its timely and effective restoration, and the prevention of contamination and toxic and radioactive waste.[84]

Given this broad set of rights, it is especially incumbent upon the state to ensure their fulfilment. The obligations in this regard are very broad and range from

---

[76] Eduardo Gudynas, '¿Por qué Bolivia no reconoce los derechos de la Naturaleza?' *Nature Rights Watch* (8 June 2018) <https://naturerightswatch.com/nosotros-about/>.

[77] Nueva Constitución Política del Estado 2009, art 9(2).

[78] Nueva Constitución Política del Estado 2009, arts 9(6), 311(II)(3), 316(6), 319.

[79] Nueva Constitución Política del Estado 2009, arts 9(6), 299(II)(1), 342.

[80] Ley de Derechos de la Madre Tierra 2010, art 3.

[81] Ley de Derechos de la Madre Tierra 2010, art 4.

[82] Ley de Derechos de la Madre Tierra 2010, art 5.

[83] Ley de Derechos de la Madre Tierra 2010, art 7(I)(1)–(3).

[84] Ley de Derechos de la Madre Tierra 2010, art 7(I)(4)–(7).

the development of public policies for prevention, protection, precaution, and defense of Mother Earth to the incorporation of clean alternative sources in the energy matrix and promotion of the issue in multilateral and regional relations.[85] In addition, it establishes an Ombudsman for Mother Earth as a body for the promotion, dissemination, and enforcement of her rights.[86]

Law 71 has an almost constitutional ambition: the plurinational state must guarantee these new rights 'at all levels and territorial spheres and through all its authorities and institutions'.[87] But this ambition remains in generalities, without indicating responsible institutions, procedures, and deadlines. Looking again at the two criteria of scope and legal strength, we find that Law 71 defines a scope of application and guarantees very similar to those of the aforementioned proposal of the indigenous organizations, but with a legal force that is not very explicit (see *Table 1*).

Despite these limitations, the Law of the Rights of Mother Earth as 'the world's first statutory law granting rights to nature' that aims to 'overcome the Cartesian dualism that disassociates people from Mother Earth'[88] has defined a new horizon of what is legally conceivable. In fact, it transitions from a still ambiguous and hesitant vision of the Constitution regarding the care of nature to a position of defending the integrity of ecosystems. Connecting with current debates in legal philosophy,[89] we can then understand it as a first milestone in a possible route of transition from an anthropocentric to an ecocentric legal tradition. In the first tradition there is a marked hierarchical differentiation between humans and non-humans, in which the value of the latter is determined primarily by their utility. In contrast, ecocentrism recognizes 'an intrinsic value of nature, totally independent of its instrumental value for humanity'.[90] The shift in perspective implied by ecocentrism is drastic, comparable to the 'Copernican revolution' that replaced medieval geocentrism with Renaissance heliocentrism.

3. Framework Law on Mother Earth and Integral Development for Living Well (Law 300)

In Bolivia, the expression is often used that someone erases with their elbow what they write with their hand. Precisely this seems to have happened in October 2012, when the Plurinational Assembly ratified a Framework Law on Mother Earth and Integral Development for Living Well (Law 300), done with the purpose of specifying the scope of Law 71. Once again, it was the government's in-

---

[85] Ley de Derechos de la Madre Tierra 2010, art 8(1)–(7).

[86] Ley de Derechos de la Madre Tierra 2010, art 10.

[87] Ley de Derechos de la Madre Tierra 2010, art 8.

[88] Kotzé and Calzadilla (n 11) 407–409.

[89] Peter Burdon (ed), *Exploring Wild Law. The Philosophy of Earth Jurisprudence* (Wakefield Press 2011).

[90] Paola Villavicencio Calzadilla and Louis Jacobus Kotzé, 'Living in Harmony with Nature? A Critical Appraisal of the Rights of Mother Earth in Bolivia' (2018) 7 TEL 397, 398.

ternational agenda that set the accelerated pace of the legislation's elaboration, after almost two years of legislative inactivity on the issue. Law 300 was designed primarily to position Bolivia for the UN Conference on Sustainable Development Rio+20 in June 2012, and it introduces a different set of issues. It is no longer a matter of specifying and operationalizing the rights of Mother Earth, but rather one of defining the objectives of integral development, setting the conditions for access to the components of Mother Earth (ie natural resources), and outlining the corresponding institutional framework.[91] The topics covered are therefore very broad, ranging from environmental education issues[92] to climate change compensation mechanisms.[93] In theory, its status as a framework law means that it has a higher rank than other laws and that it is 'of preferential application for the development of specific laws, policies, norms, strategies, plans, programs and projects',[94] such as those on mining, hydrocarbons, and water.

In Law 300, Mother Earth maintains its status as a collective subject of public interest.[95] This law mentions the catalogue of rights enumerated in Law 71,[96] but it modifies their legal hierarchy and relativizes them with respect to other rights. For example, the right to the maintenance of the integrity of living systems in Law 71[97] now appears as one of several 'principal aspects' to be considered in conservation.[98] Similarly, while Law 71 categorically established the right of Mother Earth to preserve her diversity without genetic modification,[99] reference is now made only to the progressive elimination of genetically modified crops.[100] The right to the 'preservation of the functionality of water cycles' in Law 71[101] is transformed into 'guaranteeing the right to water' and the imperative to consider both the conservation of living systems and the satisfaction of the domestic needs of people and productive processes.[102] Finally, the right to the restoration of living systems, guaranteed by Law 71,[103] is now translated into a 'principle' (not a right), according to which individuals and the state and individuals must 'carry out an integral and effective restoration or rehabilitation of the functionality of the same'.[104]

---

[91] Ley Marco de la Madre Tierra y Desarrollo Integral para Vivir Bien 2012 (Law 300), arts 1, 3(1), 4.

[92] Ley Marco de la Madre Tierra y Desarrollo Integral para Vivir Bien 2012, art 13.

[93] Ley Marco de la Madre Tierra y Desarrollo Integral para Vivir Bien 2012, ch IV.

[94] Ley Marco de la Madre Tierra y Desarrollo Integral para Vivir Bien 2012, art 2.

[95] Ley Marco de la Madre Tierra y Desarrollo Integral para Vivir Bien 2012, art 4(1)(a).

[96] Ley Marco de la Madre Tierra y Desarrollo Integral para Vivir Bien 2012, art 7.

[97] Ley de Derechos de la Madre Tierra 2010, art 7(1)(1).

[98] Ley Marco de la Madre Tierra y Desarrollo Integral para Vivir Bien 2012, art 16(9).

[99] Ley de Derechos de la Madre Tierra 2010, art 7(1)(2).

[100] Ley Marco de la Madre Tierra y Desarrollo Integral para Vivir Bien 2012, art 24(8), (9).

[101] Ley de Derechos de la Madre Tierra 2010, art 7(3).

[102] Ley Marco de la Madre Tierra y Desarrollo Integral para Vivir Bien 2012, art 27(1).

[103] Ley de Derechos de la Madre Tierra 2010, art 7(1)–(6).

[104] Ley Marco de la Madre Tierra y Desarrollo Integral para Vivir Bien 2012, art 4(5).

Law 300 also defines competent authorities and specifies a whole series of protection mechanisms. Following a long tradition in Bolivian law, this law distinguishes between administrative protection through regulations and sanctioning bodies as opposed to judicial protection.[105] All public authorities, the Public Ministry, the Defensoría de la Madre Tierra,[106] and the Agro-environmental Tribunal are competent instances, which are even obliged to act in cases of violation. In addition, any individual or collective person has the duty to report any transgression of the rights of Mother Earth.[107] On the other hand, the right to free, prior, and informed consultation[108] has been reduced to the level of 'participation in audits' for extractive projects.[109]

The described strategic reorientation of Law 300 towards development issues occurred, according to some testimonies, in early 2012 on the direct instructions of Vice President Álvaro García Linera, who also oversaw the drafting process and celebrated its conclusion. One of the main reasons for this political directioning seems to have been the intention to moderate the expectations created by the debates around Law 71 on the potential scope of the rights of Mother Earth and to avoid at all costs posing any legal obstacles to the economic project already underway: 'The existence of the rights of Mother Earth, as they were conceived in Law 71, was going to delay industrial exploitation projects in mining or hydrocarbons to the point of making them unviable'.[110]

Law 300, therefore, abandons as a general objective that all human activities reach a dynamic balance with the cycles of Mother Earth.[111] The rights of Mother Earth are now formally placed on a hierarchical level of equality with other human rights. This 'principle of complementarity' creates a situation of competition with other rights, such as the right of the population to live free from material poverty.[112] The rights of Mother Earth are thus weakened in the face of the new higher objective of achieving 'integral development'.[113] Concepts such as development, natural resources, exploitation, and growth, already surpassed in the Short Law (Law 71), reappear now, although the government team that participated in the elaboration stressed on several occasions that the concept of development 'is not a development synonymous with progress or Western-style development', but 'an intermediate phase to achieve Living Well'.[114]

---

[105] Ley Marco de la Madre Tierra y Desarrollo Integral para Vivir Bien 2012, arts 34–38.

[106] Without specifying its functions beyond art 10 of Law 71.

[107] Ley Marco de la Madre Tierra y Desarrollo Integral para Vivir Bien 2012, art 39.

[108] Constitutionally guaranteed in art 30(2)(15).

[109] Ley Marco de la Madre Tierra y Desarrollo Integral para Vivir Bien 2012, art 26(2).

[110] Interview with Higueras (2021).

[111] As established in art 2(1) of Ley de Derechos de la Madre Tierra 2010 (Law 71).

[112] Ley Marco de la Madre Tierra y Desarrollo Integral para Vivir Bien 2012, art 4(d).

[113] Ley Marco de la Madre Tierra y Desarrollo Integral para Vivir Bien 2012, art 5(3).

[114] RO Halkyer and DP Balanza, 'Ley Marco de la Madre Tierra y Desarrollo Integral para Vivir Bien' (8 April 2013) <http://reneorellanahalkyer.blogspot.com/2013/04/ley-marco-de-la-madre-tierra-y.html>.

Returning to the two dimensions mentioned above to evaluate pro-nature legislation, we see that the Framework Law drastically reduces the scope of Mother Earth's rights, reorienting it towards an anthropocentric approach. It extends the tasks of the state to economic promotion and multiplies the competent institutions, without specifying mandates, instruments, or deadlines, which ultimately weakens its legal force. Rather than the result of a 'pragmatic' attitude, as euphemistically qualified by Lalander,[115] in which the government would be faced with a dilemma between environmental protection and urgent social inclusion through rapid income, the substantial rights of Mother Earth, which are not even expressly developed, are finally blurred in philosophical disquisitions, when it is indicated, for example, that to know how to listen is to 'read the wrinkles of grandparents'[116] or that '[t]o live well is to dance in gratitude of Mother Earth'.[117] Researcher Fernanda Wanderley[118] argues that the deepening of the anthropocentric approach could correspond to an economic project clearly delimited from its beginnings. According to this hypothesis, 'the MAS sought to discursively manage both development models, though incompatible with each other [...] But all this on the margins of the economic strategy and without comprehensive and consistent actions'.

The eventful approval procedure of 'one of the most controversial laws of the Morales government'[119] and its developmentalist content provoked a strong debate and ended up deepening the rift between the indigenous organizations and the government. Critics of the Law also included the recently deceased Bartolomé Clavero,[120] a prominent Spanish constitutionalist, Eduardo Gudynas[121] (2018), a specialist in extractivism, and Raúl Prada,[122] a Bolivian intellectual and the former Minister of Planning. The latter placed the Law in a broader context, describing it as a 'legislative operation to supplant the Constitution of the Plurinational State' and lamenting that 'the rights of the Pachamama now serve to cancel them from the indigenous peoples and Mother Earth herself'.[123]

---

[115] Lalander (n 12).

[116] Ley Marco de la Madre Tierra y Desarrollo Integral para Vivir Bien 2012, art 6(7), an expression originally coined by Foreign Minister David Choquehuanca.

[117] Ley Marco de la Madre Tierra y Desarrollo Integral para Vivir Bien 2012, art 6(3).

[118] Fernanda Wanderley, 'La pregunta del millón' *El País* (Tarija, 26 August 2016) <https://elpais.bo/opinion/20160826_227232-la-pregunta-del-millon.html>.

[119] HS Lohman, *'Se han adueñado del proceso de lucha'. Horizontes comunitario-populares en tensión y la reconstitución de la dominación en la Bolivia del MAS* (Autodeterminación and SOCEE 2015) 214.

[120] Bartolomé Clavero, 'Bolivia: Ley de la Madre Tierra contra derechos de pueblos indígenas o Legislación contra Constitución' (*Servicio de Noticias Ambientales (sena)*, 11 June 2012) <https://senaforo.net/2012/06/11/fob1307/>.

[121] Gudynas (n 76).

[122] Raul Prada, 'Bolivia: Ley de la Madre Tierra contra el derecho de los Pueblos Indígenas' (*Oikologías*, 12 June 2012) <https://pradaraul.wordpress.com/2012/06/12/bolivia-ley-de-la-madre-tierra-contra-el-derecho-de-los-pueblos-indigenas/>.

[123] Prada (n 122).

IV. Challenges in Implementation

So far we have reviewed the main characteristics and internal consistency of pro-Mother Earth legislation and identified a number of challenges, considering the dimensions of scope and legal strength. However, these turn out to be only two of a number of factors in thinking about the applicability and efficiency of such legislation. As we have tried to illustrate in the case of direct representation,[124] the translation of legal innovations inspired by indigenous cosmovisions into tangible legal practices is more like an exhausting obstacle course than a simple stimulus-response mechanism. The long road to implementation begins with the construction, discussion, and condensation of proposals; it proceeds to their public debate; and it then continues forward with political negotiation (eg, in the legislature or in a constitutional assembly). After constitutionalization and the creation of the corresponding laws and regulations, and based on the new principles and postulates set forth therein, the elaboration of public policies and programs begins. In this process, a series of not readily seen contextual factors also play a role, such as the general level of compliance with the laws (rule of law), the political will to apply the new laws, the institutional culture (especially in the judicial branch), and the symbolic impact and broader social acceptance of these innovations. Such a complexity of interdependencies makes analysis difficult and even calls into question the very notion of effectiveness:

> 'The problem with the meaning of term effectiveness in relation to environmental law relates to the fact that environmental law has usually been […] placing attention on the formal compliance with a norm […]. The capacity of a given prescription to effectively address and solve the problem at stake has almost never been taken into account. This explains why in many circumstances it has proven quite difficult to assess the true effectiveness of a given environmental law norm.'[125]

As a contribution to this discussion and in an exploratory way, we propose the review of three aspects to give an idea of the practical impact and effectiveness of pro-Mother Earth legislation beyond the technical quality of the norm itself: the established institutional framework, the resulting jurisprudence, and the coherent articulation of these rights with existing or future legislation.

1. Institutional Framework

Law 300[126] creates the Plurinational Authority of Mother Earth. However, contrary to its name, this institution is not focused on the protection of Mother Earth's rights, but rather on the accompanying public policies related to climate change. Its main task is to manage funds, especially of international origin, and to coor-

---

[124] Cletus Gregor Barié, 'Representation of indigenous peoples in times of progressive governments: lessons learned from Bolivia' [2020] LACES 1.

[125] Carducci and others (n 9) 23.

[126] Ley Marco de la Madre Tierra y Desarrollo Integral para Vivir Bien 2012, art 53(1), (2).

dinate and facilitate the implementation of climate change mitigation and adaptation mechanisms.[127] The novelty, moreover, occurs mainly in the name and is not one of structure or mandate, since this Plurinational Authority is formally the successor to the National Climate Change Program (PNCC) of the Ministry of Environment and Water.[128]

Law 300 also provides for the creation of two other new institutions: first, a Plurinational Council for Good Living in Harmony and Balance with Mother Earth (Consejo Plurinacional para Vivir Bien, en Armonía y Equilibrio con la Madre Tierra,), for 'monitoring, consultation and participation in the elaboration of policies, plans, programs and projects',[129] headed by the president. Secondly, the Law refers to the Ombudsman of Mother Earth as one of the bodies 'obliged to activate the administrative and/or jurisdictional instances, with the purpose of demanding the protection and guarantee of the rights of Mother Earth'.[130] No specific deadlines were given for the creation of the two aforementioned institutions through special laws or regulations, which so far have not been created. In April 2021, President Luis Arce announced that the legislature was about to pass the Law of the Ombudsman of Mother Earth,[131] but its approval was ultimately suspended due to the opposition of interest groups, mainly mining cooperativists, who described it as 'an attack on the mining, metallurgical and manufacturing sector'.[132]

The 2009 Constitution also created a new instance, the Agro-environmental Tribunal, as the 'highest specialized Tribunal of the agro-environmental jurisdiction',[133] with the explicit task, among others, of resolving 'lawsuits on actions that threaten fauna, flora, water and the environment; and lawsuits on practices that endanger the ecological system and the conservation of species or animals'.[134] However, despite the fact that the Law of the Judiciary (Law 25 of 2010) establishes a period of two years to adapt the codes, so far no Agro-environmental Procedural Code has been approved, which makes it difficult to address and resolve

[127] Autoridad Plurinacional de la Madre Tierra, *Audiencia pública de rendición de cuentas final gestión 2020* (La Paz 2020).

[128] Ley Marco de la Madre Tierra y Desarrollo Integral para Vivir Bien 2012, Transitional art 2.

[129] Ley Marco de la Madre Tierra y Desarrollo Integral para Vivir Bien 2012, art 52(1).

[130] Ley Marco de la Madre Tierra y Desarrollo Integral para Vivir Bien 2012, art 39(1); see also arts 36 and 39(I)(4).

[131] 'Bolivia prepara ley para creación de Defensoría de la Madre Tierra' (*Agencia Boliviana de Información*, 4 April 2021) <https://abi.bo/index.php/noticias/sociedad/5115-Bolivia-prepara-ley-para-creaci%C3%B3n-de-Defensor%C3%ADa-de-la-Madre-Tierra-(amplia)>.

[132] 'Atentado contra sector minero metalúrgico y manufacturero: "Ley de Defensoría de la Madre Tierra" conculcará derechos de los productores' (*La Patria*, 23 April 2021) <https://impresa.lapatria.bo/noticia/1034793/ley-de-defensoria-de-la-madre-tierra-conculcara-derechos-de-productores>.

[133] Nueva Constitución Política del Estado 2009, art 186.

[134] Nueva Constitución Política del Estado 2009, art 189(1). See also Law 300, art 36 and art 39(1)(4).

any related conflicts. Currently, the Agro-environmental Tribunal processes claims based on internal guidelines, without the legal certainty of a statutory procedure.[135] In 2021, the magistrates of the Agro-environmental Tribunal presented Vice President David Choquehuanca with a proposal for this Code which, however, generated resistance from some assembly members, so it was not approved. According to magistrate Elva Terceros,[136] 'there is a collective belief that, if procedural instruments are created to protect the rights of Mother Earth, it will not be possible to develop natural resource exploitation activities that affect nature'.

There are other agencies with different functions in agrarian and environmental matters.[137] Budget allocations to institutions specialized in environmental protection, especially SERNAP and MMAYA, have remained at very low levels in the last decade, which even seriously limits their basic functioning. When observing this institutional framework as a whole, it is clear that so far the rights of Mother Earth have not been incorporated into their institutional guidelines, and there have been no reported cases in which the institutions have acted ex officio in defense of these rights, despite the fact that their mandate would allow them to do so.[138]

2. *Incipient Jurisprudence*

Substantial jurisprudence related to the rights of Mother Earth has not yet been developed in Bolivia.[139] According to environmental lawyer Paola Cortés,[140] enforcement has not only not taken off but has even regressed in some aspects. This situation contrasts with that of Ecuador, where there are more than twenty rulings related to topics such as river pollution, the use of transgenics, mining, and even shark management.[141]

---

[135] Tribunal Agroambiental, Guía de proceso en materia ambiental. Acuerdo de Sala Plana SP. TA. N° 015/2020 (Tribunal Agroambiental 2020) <www.tribunalagroambiental.bo/2020/11/25/guia-de-procesos-en-materia-ambiental/>.

[136] Interview (2021).

[137] Worthy of mention are: the Authority for Social Control and Inspection of Forests and Lands (ABT), which prosecutes, controls, supervises, and regulates the forestry and agrarian sectors; the National Agrarian Reform Institute (INRA), in charge of planning, executing and consolidating the agrarian reform process in the country; the National Protected Areas Service (SERNAP), in charge of safeguarding the country's protected areas; the National Forestry Development Fund (FONABOSQUE), a decentralized institution of the Ministry of Environment and Water (MMAYA) that finances programs and projects aimed at sustainable forest management; the Jurisdictional Administrative Mining Authority (AJAM) under the Ministry of Mining and Metallurgy, responsible for the direction, superior administration, registration, control, and supervision of mining activity.

[138] Interview with Higueras (2021).

[139] 'Roles distintos en relación a la Madre Tierra' (*Fundación Solon*, 6 May 2021) <https://fundacionsolon.org/2021/05/06/roles-distintos-en-relacion-a-la-madre-tierra/>.

[140] Interview (2021).

[141] UN, *Rights of Nature Law and Policy* (2022) <www.harmonywithnatureun.org/rightsOfNature/>.

In Bolivia, in theory any person or organization that has knowledge of a violation of the rights of Mother Earth can and even must report it.[142] Accordingly, there is both administrative protection (through public authorities) and jurisdictional protection.[143] For the latter, three competent jurisdictions have been established: the ordinary jurisdiction, exercised by the Supreme Court of Justice, other courts, and more than a thousand lower-level judges; the Indigenous Peasant Jurisdiction, administered by its own authorities; and the agro-environmental jurisdiction, exercised by the Agro-environmental Tribunal together with, currently, thirty-six environmental courts throughout the country.[144]

However, the procedure for claiming the rights of Mother Earth is still not defined, and it is often unclear which jurisdiction to follow, as Cortés[145] points out: 'There is no critical path to file a complaint. It is very discouraging for those who want to take legal action.' To date, there is also no definition of environmental crimes. There are several cases in which the Plurinational Supreme Court declined jurisdiction and explicitly referred the cases to the Agro-environmental Tribunal, which, in any case, has so far received a very small number of complaints. One of the reasons for the low use of the agro-environmental jurisdiction, compared to the courts of ordinary jurisdiction which are overloaded with cases, is also the lack of knowledge about this new instance, even among judges and lawyers.

Many complaints related to environmental issues are brought by means of a 'popular action', a new type of legal proceeding foreseen in article 135 of the Constitution, which is filed when collective rights are violated. One of the problems with the popular action process is that it is generally resolved by ordinary courts, meaning that the case will be heard by a judge who does not have any specialized knowledge of environmental matters. According to Cortés,[146] 'ordinary law is of little use for environmental issues, this is also reflected in the experiences of other countries'. In general, there is a very limited group of experts and jurists working in the area of environmental law in Bolivia.

In analyzing the evolution of jurisprudence, it is useful to distinguish three levels of incorporation of the rights of Mother Earth: the complementary use of these rights, their use within conventional environmental law, and, finally, their substantive approach.

At the first level we find a whole series of supreme orders, sentences, and resolutions, especially from the Plurinational Supreme Court, in which Mother Earth is mentioned superficially or is added in some secondary aspect accompanying the main argument. An example of this secondary argument mechanism can be seen in a popular action that was filed by an agrarian union in the department of Cochabamba in 2016 regarding the community use and supply of drinking wa-

---

[142] Ley Marco de la Madre Tierra y Desarrollo Integral para Vivir Bien 2012, art 39(3).
[143] Ley Marco de la Madre Tierra y Desarrollo Integral para Vivir Bien 2012, arts 35, 36.
[144] Nueva Constitución Política del Estado 2009, art 179(1).
[145] Interview (2021).
[146] Interview (2021).

ter. In that case, to emphasize the importance of the water resource, the plaintiffs resorted to metaphorical language, stating that water 'for the peasant and indigenous sector is like the blood of Mother Earth'[147] and describing it as 'the sap of Mother Earth that allows the subsistence of her children'. The Court considered this intercultural aspect in its judgment and recognized that 'within a biocentric vision, where living beings, the earth and whatever is attached to it complement each other', the right to water is an 'autonomous right intimately related to the environment'.[148] However, the main argument in favor of the complainants was that access to drinking water and health were fundamental rights.

In a second type of jurisprudence, Mother Earth is no longer a secondary component of the argument but the main focus, albeit circumscribed by an anthropocentric vision. In Uncía, a mining town in the Altiplano, one of the significant cases in this category took place. Guided by agro-environmental judge Maribel Ruiz Molina, representatives of an indigenous community (Ayllu Sikuya) initiated in 2018 a conciliation proceeding with several mining cooperatives in regards to massive water pollution affecting their crops. In a process of training and dialogue, the implications of the new rights of Mother Earth were discussed for the first time, and these considerations were a decisive factor in convincing the mining companies to reverse the decades-long situation of contamination. According to Judge Ruiz Molina,[149] the concept of Mother Earth rights was not immediately understood, not even by the indigenous population, as the first extractive activities in this area dated back some forty years. The case went through different administrative channels, with the involvement of the competent municipal authorities, and resulted in a substantial improvement of environmental conditions, through a package of measures that included decontamination of the riverbed, reforestation, and the construction of a tailing dam.[150] In this case, the focus of the dispute was the impact on the communities and not the recovery of the ecological systems, which in this legal area would imply a radical rethinking of mining activities.

In a third line of jurisprudence, the environment is a subject of rights. In this approach, the protected legal good is not an individual right but a diffuse collective right. The application of this approach in Bolivia is recent:

- In 2011, when Law 71 had just been passed, a group of citizens from Tarija succeeded in transferring wild animals from a zoo which was in precarious condition to specialized centers outside the city. Interestingly, instead of following a conventional line of argument around animal protection, the Constitutional Court broadly interpreted the constitutional right to a 'healthy, protected

[147] Tribunal Constitucional Plurinacional 3 March 2016, *Acción de amparo constitucional por derecho de acceso al agua*, 0273/2016-S1.

[148] Tribunal Constitucional Plurinacional 3 March 2016 (n 147) 15.

[149] Interview (2021).

[150] Tribunal agroambiental, 'Tribunal agroambiental culmina ciclo de conversatorios sobre litigio ambiental estratégico' (29 May 2020) <www.tribunalagroambiental.bo/2020/05/29/tribunal-agroambiental-culmina-ciclo-de-conversatorios-sobre-litigio-ambiental-estrategico/>.

and balanced environment'.[151] As a basis for granting the requested protection, it stated that there is a 'diffuse interest whose ownership rests on all members of the community and its affectation involves not only and essentially man, but on the contrary, on equal terms, all the elements of which the environment and its various ecosystems are composed' and that, therefore, 'when a component of the biological chain is neglected, mistreated or eliminated, it has a negative impact on the entire natural system, of which human beings are part'.[152]

– In a more recent ruling on the expropriation of public land in the municipality of Quillacollo in Cochabamba, the Agro-environmental Tribunal resorted to Law 300 and applied the principles of precaution and prevention of possible damage to Mother Earth so as to ensure the water recharge of the site in question.[153]

Given the relative novelty of the institutionof guardianship of Mother Earth, Bolivian jurists also frequently refer to international legislation and jurisprudence. A significant case that has inspired jurisprudence in this context was an advisory opinion of the Inter-American Court of Human Rights, a regional court based in San José, Costa Rica. In the ruling issued by the Court, several fundamental rights 'require, as a necessary precondition for their exercise, a minimum environmental quality, and are profoundly affected by the degradation of natural resources'.[154] In other words, the legal good of a balanced environment precedes human rights to a certain extent since 'environmental degradation and the adverse effects of climate change affect the effective enjoyment of human rights'.[155]

On another level, the jurisprudence of the International Tribunal for the Rights of Nature contains quite elaborate ecocentric arguments. In Bolivia, there are two rulings on the matter. In the case of the construction of the TIPNIS road, to which we will refer below, this Tribunal established in 2019 that

> 'the Plurinational State of Bolivia has violated the Rights of Nature and [the rights of] the indigenous peoples of TIPNIS as defenders of Mother Earth and failed to comply with its obligation to respect, protect and guarantee the Rights of Mother Earth as established in the national legal framework and related international regulations'.[156]

The International Tribunal for the Rights of Nature also investigated a series of forest fires that occurred between July and October 2019 in the Chiquitanía region, an extensive plain located between the Gran Chaco and the Amazon in south-eastern Bolivia. The event caused the loss of approximately five million hectares, making it one of the most serious ecological disasters of the last decade.

---

[151] Tribunal Constitutional Plurinacional 7 December 2011, *Tutela por traslado de zoológico*, 2010-22309-45-AP.

[152] Tribunal Constitutional Plurinacional 7 December 2011 (n 151) III 4.

[153] Sentencia Agroambiental Plurinacional 17 April 2019, *Demanda contencioso administrativa por adjudicación de predio*, S1ª Nº 22/2019.

[154] Corte Interamericana de Derechos Humanos 15 November 2017, *Obligaciones estatales en relación con el medio ambiente*, Opinión consultiva OC-23/17.

[155] Corte Interamericana de Derechos Humanos 15 November 2017 (n 154) 47.

[156] International Tribunal for the Rights of Nature 15 May 2019, *Case of The Isiboro Sécure National Park And Indigenous Territory (TIPNIS)*.

The impact was especially strong in indigenous territories and protected areas.[157] The Tribunal ruled on behalf of 'those animals and plants that have no voice, those environmental refugees whose rights have been violated by these fires',[158] that the state, in partnership with agribusiness, committed ecocide,[159] understood as 'massive damage or destruction of the "ecological system", that is, of biodiversity and ecosystems produced by humans'.[160] In short, it considered that the responsibility of the authorities was twofold, since they not only failed to take adequate measures to extinguish the fire but also created the legal conditions to make it possible. Although these decisions are not binding, its verdicts and legal analysis have a certain impact at the international level. In Bolivia, the ruling was widely commented on in the media.

3. Coherence and Articulation of Mother Earth's Rights

Apart from Laws 71 and 300 (and their accompanying regulations), there are very few legal instruments with ecocentric approaches in Bolivian law to which defenders of Mother Earth can turn. Recently, in September 2021, a group of municipal authorities in the department of La Paz resolved to declare the Sacred Lake Titicaca (in Quechua Quta Mama) as a subject of law.[161] In 2020, legislators also debated a bill that would classify ecocide as a crime against humanity,[162] but the process was suspended.

The international treaties and conventions that Bolivia has ratified also apply as part of a set of norms of higher constitutional hierarchy (the so-called constitutional block).[163] These include the Convention on Biological Diversity (1992), the Convention for the Protection of the Flora, Fauna and Natural Scenic Beauty of the Americas (OAS, 1940), and the Convention on Wetlands (1971). Regarding the special relationship of indigenous peoples with their territory and habitat, ILO Convention 169 (1989), the United Nations Declaration on the Rights of Indigenous Peoples (2007), and several other regional and international instruments and jurisprudence also apply. The Escazú[164] ratified by Bolivia in 2019, has

---

[157] Fátima Monasterio, *Deforestación e incendios forestales en Bolivia y Derechos Humanos y de los Pueblos Indígenas* (Cejis 2019) 2.

[158] International Tribunal for the Rights of Nature 20 August 2020, *Chiquitanía, Chaco and Amazonía v. the Plurinational State of Bolivia.*

[159] International Tribunal for the Rights of Nature 20 August 2020 (n 158) arts 128; 129.

[160] International Tribunal for the Rights of Nature 20 August 2020 (n 158) art 78.

[161] RTP Bolivia 22 October 2021, Senado aprueba reconocimiento de la Quta Mama como sujeto de derechos, <www.facebook.com/rtpbolivia/posts/senado-aprueba-reconocimiento-de-la-quta-mama-como-sujeto-de-derechosel-pleno-de/2377136519087199/>.

[162] Proyecto de Ley de 'Tipificación del Ecocidio como Delito de Lesa Humanidad'. PL no 275/2020-2021.

[163] Nueva Constitución Política del Estado 2009, art 410(1).

[164] Comisión Económica para América Latina y el Caribe, Acuerdo Regional sobre el Acceso a la Información, la Participación Pública y el Acceso a la Justicia en Asuntos Ambientales en América Latina y el Caribe (2018).

the potential to strengthen access to environmental information and adequate public participation in resource extraction projects.

In addition to the absence of a coherent and integrated legal nucleus that protects the rights of nature in its subjective (content) and adjective (process) dimensions, two additional challenges can be identified. On the one hand, legislation existing prior to the new Constitution of 2009 is still in force and retains a markedly anthropocentric approach, thereby requiring urgent updating as foreseen in Law 300,[165] among others, this problem exists with regards to the Environmental Law (Law 1333 of 1992) and the National Agrarian Reform Service Law (Law 1715 of 1996).

A second and even more serious aspect is the additional introduction of new legislation having a high potential to violate the rights of Mother Earth. In technical terms, these are 'counterweights' and 'neutralizing elements'[166] that may have an impact on a growing 'implementation gap'.[167] In the Bolivian case, some authors even speak of 'deconstitutionalization'[168] and 'domestication'[169] of constitutional postulates given the additional fact that the control mechanisms provided by the Constitution[170] have not been effective in these cases.

There are several public policy areas with a high risk of 'neutralizing' or 'eroding' the ecocentric principles contained in an incipient form in Bolivian legislation, including the mining sector, hydrocarbons, land and resource use, land use planning, and as relating to indigenous peoples. One such example is the Mining and Metallurgy Law (Law 535 of 2014), which restricts the constitutional principles of offering prior, free, and informed consultation to indigenous peoples. Also illustrative in this regard was the development of legislative measures that promote the expansion of the agricultural frontier in lowlands, popularly also called the 'incendiary package' because of its incentivizing forest burning:

> 'The fires that hit Santa Cruz [in 2019] have been a direct consequence of the oversized expectations aroused by recent government policies of de-protection of forests, cancellation of sanctions and fines for illegal logging and burning, and settlement authorizations'.[171]

---

[165] Ley Marco de la Madre Tierra y Desarrollo Integral para Vivir Bien 2012, art 3(3).

[166] Cletus Gregor Barié, *Pueblos indígenas y derechos constitucionales en América Latina. Un panorama* (2nd edn, Abya-Yala and others 2003) 553.

[167] Concept developed by Rodolfo Stavenhagen, see Wilhelmi (n 8) 19.

[168] Boaventura de Sousa Santos and José Luis Exeni Rodríguez (eds), *Justicia indígena, plurinacionalidad e interculturalidad en Bolivia* (2nd edn, Abya-Yala and Fundación Rosa Luxemburg 2013) 723.

[169] Fernando Garcés, 'The Domestication of Indigenous Autonomy in Bolivia. From the Pact of Unity to the New Constitution' in Nicole Fabricant and Bret Gustafson (eds), *Remapping Bolivia: Resources, Territory, and Indigeneity in a Plurinational State* (SAR Press 2011) 47.

[170] Nueva Constitución Política del Estado 2009, art 196-204.

[171] Fundación TIERRA, *Fuego en Santa Cruz. Balance de los incendios forestales 2019 y su relación con la tenencia de la tierra* (2019) 79–80.

4. Growing Presence of Mother Earth in Social Conflicts

Faced with insufficient mechanisms to protect the rights of Mother Earth, and even basic conventional environmental rights, many local socio-environmental conflicts tend to escalate and proceed beyond the institutionally established channels. The Atlas of Environmental Justice,[172] edited by the Autonomous University of Barcelona, currently identifies 42 cases of communities involved in large-scale environmental conflicts in Bolivia. Current examples are the protests over oil exploration in the Takovo Mora Guaraní indigenous territory, the debate over the construction of a hydroelectric dam in the Madidi National Park, and the conflict over hydrocarbon exploration and exploitation in the Tariquía National Reserve (Department of Tarija) by the Brazilian company Petrobras.[173]

Some of the socio-environmental conflicts transcended the local level and had an impact at the national level. In 2011, for example, several lowland indigenous organizations mobilized against the construction of a highway through the Isiboro Sécure Indigenous Territory and National Park (TIPNIS), which we referred to earlier. In the course of this conflict, which stirred Bolivian opinion for months and left several people injured, various indigenous organizations demanded the right to free, prior, and informed consultation and claimed the intangibility of the territory.[174] Apart from referring to the rights of indigenous peoples, which are more established in Bolivia than those of nature, the indigenous organizations' statements and declarations during their marches frequently referred to the protection of the 'Loma Santa' as a variation of the more generic concept of Mother Earth, understood as a safe haven that protects against all kinds of threats.[175]

On another level, in October 2019, as a reaction to the government's inaction in the face of forest fires in the Chiquitanía, for the first time in Bolivia there was a national mobilization explicitly in favor of greater environmental protection.[176]

---

[172] EJAtlas – Global Atlas of Environmental Justice <www.ejatlas.org>.

[173] Saúl Marcelo Chinche Callizaya, 'Tariquía, nuevo atentado a la Madre Tierra' *Pagina Siete* (La Paz, 13 April 2019) <https://lilaandrea.github.io/Crisis/contenido/DEFENSORES%20CHACO/20190413_939586-1_Tariquia_nuevo_atentado_a_la_Madre.html>.

[174] Centro Cuarto Intermedio, *Los trasfondos del TIPNIS. Conversatorio dirigido por el Area de Reflexión del Centro Cuarto Intermedio* (Estudios&Ensayos 2012); Fundación UNIR Bolivia, *Informe sobre conflictividad en Bolivia. (Junio a Diciembre de 2010)* (Bolivia 2011); Anna F Laing, *Territory, resistance and struggles for the plurinational state. The spatial politics of the TIPNIS Conflict* (PhD thesis, University of Glasgow 2015).

[175] Pavlína Springerová and Barbora Vališková, 'Territoriality in the Development Policy of Evo Morales' Government and its Impacts on the Rights of Indigenous People: The Case of TIPNIS' [2016] CJLACS 147, 165.

[176] Gustavo Rodríguez Cáceres and Rosemary Amils (eds), *Los incendios en la Chiquitanía el 2019: Políticas devastadoras, acciones irresponsables y negligencia gubernamental* (CEDIB and LaLibre 2020); CEJIS, *Indígenas de la Chiquitanía presentan proyectos para abrogar tres leyes y dos decretos que permitieron los incendios del 2019* (CEJIS, 27 February 2020) <www.cejis.sinnersite.com/noticias/indigenas-de-la-chiquitania-presentan-proyectos-para-abrogar-tres-leyes-y-dos-decretos-que-permitieron-los-incendios-del-2019/>.

CIDOB called for the 'Tenth Indigenous March for Life and in Defence of the Forest', demanding, among other things, the declaration of a national disaster to expedite international aid, the repeal of laws that allow deforestation, and the cessation of new settlements in the region.[177] In their petitions, references to the legal framework related to indigenous and Mother Earth rights were widespread. A great diversity of social actors joined the strike, including students, environmentalists, feminists, and researchers as well as a large segment of Bolivia's non-governmental organizations.[178] These examples illustrate the growing diffusion and entrenchment of ecocentric ideas among different social actors and their incorporation into the repertoire of collective actions, which challenges the thesis of Kauffman and Martin,[179] who argue that 'Bolivian social movements have not invoked the rights of nature'.

In the face of growing internal divisions among traditionally indigenous and peasant organizations – some of which defend their internal autonomy (known as "organic" sectors), while others adhere firmly to the government – new independent collectives have emerged in recent years. An outstanding example is the National Coordinator for the Defence of Indigenous Peasant Territories and Protected Areas of Bolivia (CONTIOCAP), which brings together community organizations affected by extractive projects and has become a source of information and early warning on environmental risks in the country. Bolivia historically has a wide network of non-governmental organizations related to environmental, indigenous, and territorial issues, and these organizations remain active despite government attempts to restrict their space in civil society (*shrinking space of civil society*).[180]

The exchange between environmental advocates and specialists at the Latin American level is increasingly intensifying, as was indicated by several interviewees, especially in Peru, Argentina, and Chile, and as is often promoted by regional organizations such as AIDA,[181] the international environmental news platform Mongabay,[182] and multilateral agencies such as UNDP (United Nations

[177] L Escobar, 'Indígenas organizan la X Marcha por la Vida en defensa del bosque' *Pagina Siete* (11 September 2019) <www.cabildeodigital.com/2019/09/indigenas-organizan-la-x-marcha-por-la.html>.

[178] 'Nació coordinadora en defensa de la naturaleza y naciones' *El Dario* (26 September 2019) <www.pub.eldiario.net/movil/index.php?n=9&a=2019&m=09&d=26#google_vignette>.

[179] Kauffman and Martin (n 13) 135.

[180] Helmut K Anheier, Markus Lang and Stefan Toepler, 'Civil Society in Times of Change: Shrinking, Changing and Expanding Spaces and the Need for New Regulatory Approaches' (2019) 13 Economics <http://dx.doi.org/10.5018/economics-ejournal.ja.2019-8>. These organizations include CEBEM (Centro Boliviano de Estudios Multidisciplinarios), CEDLA (Centro de Estudios para el Desarrollo Laboral y Agrario), CEJIS (Centro de Estudios Jurídicos e Investigación Legal), CIPCA (Centro de Investigación y Promoción del Campesinado), FAN (Fundación Amigos de la Naturaleza), Fundación Tierra, and UNITAS (Unión Nacional de Instituciones para el Trabajo de Acción Social).

[181] <https://aida-americas.org/en>.

[182] <www.mongabay.com>.

Development Program). Lui Laura Zapana,[183] an activist lawyer, for example, shared lessons learned in Bolivia with members of the Chilean Constituent Assembly in 2021. The growing identification of citizens (especially women) involved with the rights of nature has even been noted in one of the first surveys on the subject (although it focused only on the metropolitan area) conducted by the NGO Ciudadanía: '9 out of 10 people, believe that plants and animals have as much right to exist as human beings'.[184]

However, given the wide variety of sectors that favor and support public policies of extraction and the intensive use of natural resources, many protests tend to be limited in scope. Over the last fifteen years, a new dominant bloc of traditional and emerging groups in the economy and politics has consolidated in Bolivia, supported by a broad regulatory regime. Without going into a detailed analysis, it is generally possible to differentiate between, on the one hand, the groups traditionally linked to the export and agro-industrial sector,[185] and, on the other hand, the existence of a broad sector of previously semi-formal and subordinate actors, such as cooperativists, coca growers, or colonizers, who under the government of Evo Morales consolidated a legal status and benefited extensively from his policies.

V. Final Reflections

Since the 1990s, indigenous peoples in Latin America have been one of the main drivers of state reforms.[186] However, as Gutiérrez Chong et al.[187] observe – taking up on a reflection by Valladares on the legalization of indigenous demands[188] – despite their hard work of social mobilization and political advocacy, the institutional and legal changes achieved have often yielded a 'bitter harvest', ie very limited practical results in relation to the effort involved. This is also true for the recognition of the rights of Mother Earth in Bolivia, understood as a demand that seeks social transformation based on rights (rights-based demands).

183 Interview (2021).

184 Comunidad de Estudios Sociales y Acción Pública, 'Ciudadanía ambiental en las regiones metropolitanas de Bolivia' (2018) 5 Ciudadatos 18.

185 Among them are the CAO (Cámara Agropecuaria del Oriente), CAINCO (Cámara de Industria, Comercio, Servicios y Turismo de Santa Cruz), ANAPO (Asociación de Productores de Oleaginosas y Trigo), and FEGA-SACRUZ (Federación Departamental de Ganaderos de Santa Cruz), all with strong political ties.

186 Assies, van der Haar and Hoekema (n 16).

187 NG Chong, JM Reséndiz and FS Espinosa, *Cultura política indígena. Bolivia, Ecuador, Chile, México* (Universidad Nacional Autónoma de México 2015) 42.

188 Laura Valladares, 'El movimiento indígena en México: constructores y excluidos de la nueva democracia', in Scott Robinson Studebaker, Héctor Tejera Gaona and Laura La Valladares de Cruz (eds), *Política, etnicidad e inclusión digital en los albores del milenio* (Miguel Angel Porrua 2007) 169, 196.

It was the Unity Pact which in 2010 translated the millenary custom of respect for Pachamama into a coherent and elaborated legal language. In this way it outlined a possible roadmap for moving towards an ecocentric model. However, with the MAS government taking the lead on the issue, the original proposals lost their scope and the social agency of the indigenous promoters was reduced. Laws 71 and 300 were drafted at the rapid pace of international conferences in which Bolivia strived for a certain protagonism. While Law 71 is a partial synthesis of the proposal of the Unity Pact, Law 300 proposes a competition between the rights of nature and social rights, such as the right to live without poverty. This created an erroneous conceptual collision between the rights of nature and social rights. Finally, the scope of rights of Bolivian pro-Mother Earth legislation was reduced by an anthropocentric turn, and its legal strength was diluted in fictitious institutions.

Looking at the efficiency and impact of this legislation, we find that its contribution to the strengthening and maintenance of specific ecosystems has been minimal. None of the laws unfurled their transformative potential: the institutional framework foreseen for the protection of nature, which has been entrusted to the Defensoría de la Madre Tierra (Ombudsman's Office) and the Consejo Plurinacional de la Madre Tierra (Plurinational Council of Mother Earth) as consultative bodies, has never been established. The new Agro-environmental Tribunal, despite having a wide network of offices in the country, is underutilized, since most environmental cases go through the ordinary justice system. As for jurisprudence, it is still incipient, and cases in which the environment is protected as a legal good are an exception. In the face of the extensive development of extractivist legislation, the rights of Mother Earth in Bolivia have been isolated and relegated to 'solitude'. In spite of all these inconsistencies, the two laws in question nevertheless constitute the only current legal sources defending the rights of Mother Earth in the midst of an almost completely anthropocentric weave of legislation. These laws serve as a key source at the regional level and are a mandatory point of reference.

However, the recognition of the rights of Mother Earth has had a certain symbolic effect: Bolivian society is now more aware of environmental issues, as evidenced by the mobilizations during the fires in the Chiquitanía region in 2019. There is a new generation of social organizations defending their territories and beginning to engage in exchange at the regional level, such as the National Coordinator in Defence of Native Indigenous Territories, Peasants and Protected Areas (CONTIOCAP). There are also broad social sectors, some with strong local and community roots and others of more urban and youthful origin, that are taking up the cause of environmental protection as their own and that are frequently making reference to the rights of Mother Earth, although they reject its instrumental and rhetorical use.

The Bolivian case offers several lessons. It shows the importance of placing the maintenance of ecological cycles as a higher objective of development (or good living), preferably at the constitutional level. As long as society aspires to development with strong anthropocentric characteristics, the rights of Mother Earth will easily remain in the background. It also leaves evidence of the importance of institutions that defend these rights ex officio and of specialized courts. Finally, the rights of Mother Earth must be articulated in a coherent manner with relevant legislation, regarding, for example, mining, forests, indigenous peoples' rights, and natural resources. The legal preconditions and the accumulated practical experience are in place, such that Bolivia – beyond grandiloquence – could well take more consistent steps towards an ecocentric development model.

Table 1

Criteria for measuring the extent of pro-nature norms. Elaboration by author based on Kauffman and Martin.[189] Points highlighted in italics indicate aspects that put the rights of nature at risk due to their anthropocentric orientation.

<table>
<tr><td colspan="2">How is nature defined? (scope)</td></tr>
<tr><td>Unity Pact Proposal[190]</td><td rowspan="3">– Dynamic living system made up of the indivisible community of living organisms
– It consists of living systems as functional units and contains components for their utilization</td></tr>
<tr><td>Law 71 of 2010[191]</td></tr>
<tr><td>Law 300 of 2012[192]</td></tr>
<tr><td colspan="2">Who owns the right? (scope)</td></tr>
<tr><td>Unity Pact Proposal</td><td rowspan="3">Mother Earth is a collective subject of public interest</td></tr>
<tr><td>Law 71 of 2010</td></tr>
<tr><td>Law 300 of 2012</td></tr>
<tr><td colspan="2">What rights does nature have? (scope)</td></tr>
<tr><td>Unity Pact Proposal</td><td>– Maintenance of integrity and diversity, without being genetically altered
– Preservation of the functionality of water cycles
– Preservation of air quality and air composition
– Rights to the balance and continuation of her cycles and to the reproduction of life processes
– Timely and effective restoration of livelihood systems
– Principle of *in dubio pro natura*</td></tr>
<tr><td>Law 71 of 2010</td><td>– Maintenance of the integrity and diversity, without being genetically altered Preservation of the functionality of water cycles
– Preservation of air quality and air composition
– Rights to the balance and continuation of her cycles and to the reproduction of life processes
– Timely and effective restoration of livelihood systems
– Pollution-free living</td></tr>
</table>

189 Kauffman and Martin (n 13) 15 and 61.

190 Pacto de Unidad (n 34).

191 Ley de Derechos de la Madre Tierra 2010.

192 Ley Marco de la Madre Tierra y Desarrollo Integral para Vivir Bien 2012.

Law 300 of 2012	– Not specified, reference to Law 71
What type of standard is it and what is its hierarchy? (scope)	
Unity Pact Proposal	Preliminary draft law
Law 71 of 2010	Short Law, requiring further legislative development
Law 300 of 2012	Framework law
How do the rights of nature relate to other rights? (strength)	
Unity Pact Proposal	– Bolivians exercise the rights of Mother Earth in a way that is compatible with their individual and collective rights
Law 71 of 2010	– Bolivians exercise the rights of Mother Earth in a way that is compatible with their individual and collective rights – Individual rights are limited by the exercise of the collective rights of Mother Earth – Conflicts between rights must be resolved in a way that does not irreversibly affect the functionality of livelihood systems – Society's interest prevails over any rights
Law 300 of 2012	– Compatibility and complementarity of Mother Earth rights, with indigenous rights, civil, political and social rights, rights to live free of poverty
How are they specifically linked to indigenous peoples' rights? (strength)	
Unity Pact Proposal	Free, prior, informed, and compulsory consultation for the use and exploitation of the components of Mother Earth in indigenous territories
Law 71 of 2010	–
Law 300 of 2012	Compatibility and complementarity with collective and individual rights of indigenous and aboriginal peasant nations and peoples, intercultural, and Afro-Bolivian communities
What are the responsibilities of the state? (strength)	
Unity Pact Proposal	– Respect, protect, and guarantee the rights of Mother Earth for the well-being of present and future generations. Develop public policies, systems, and measures for prevention, warning, protection, precaution, or restriction – Develop public policies for the restoration of the integrity of Mother Earth's vital cycles, processes, and balances – Promote and support practices compatible with the rights of Mother Earth – Establishing a comprehensive state and participatory planning system – Ensuring food sovereignty – Prohibit the entry, import, introduction, and marketing of toxic materials, including genetically modified organisms – Developing policies to adapt to the effects of climate change – Define policies to ensure energy sovereignty – Developing education policies

	– Demand global recognition of the differentiated environmental debt in foreign policy – Ensuring an institutional design that allows for protection – Establish policies that stimulate the development and application of ever better technologies and practices – Incorporate the quantification and integral qualification of economic processes in interrelation with Mother Earth
Law 71 of 2010	– Respect, protect, and guarantee the rights of Mother Earth for the well-being of present and future generations – Ensuring the conditions for Mother Earth's diverse life systems to absorb damage, adapt to disturbances, and regenerate themselves – Develop public policies and systematic actions for prevention, early warning, protection, and precaution – Develop balanced forms of production and consumption patterns, considering the regenerative capacities and the integrity of Mother Earth's vital cycles, processes, and balances – Develop policies to defend Mother Earth at the plurinational and international level from the overexploitation of its components, from commodification and from the causes of global climate change and its effects – Developing policies to ensure energy sovereignty – Demand international recognition of environmental debt through financing and technology transfer – Promote peace and the elimination of all nuclear, chemical, and biological weapons and weapons of mass destruction – Promote the recognition and defense of the rights of Mother Earth at the multilateral, regional, and bilateral levels of international relations
Law 300 of 2012	– Ensure the continuity of the regenerative capacity of Mother Earth's components, zones, and life-systems. Carry out restoration or rehabilitation of the functionality of life-systems – Conserving components, zones, and life systems – Drawing up specific rules and providing for technical-administrative bodies to sanction acts or omissions – Promoting sustainable consumption habits – Establish clean production processes – Registering Mother Earth's components with high strategic value – Crimes related to Mother Earth are not subject to statute of limitations – Promoting intra- and intercultural education and dialogue regarding knowledge and knowhow – Moving towards *phasing* out pollution of Mother Earth – Promoting the industrialization of Mother Earth's components – Ensuring the sustainability of the state itself in order to achieve a good standard of living – Mainstreaming integrated development in policies, standards, strategies, plans, programs, and projects – Facilitating equitable access to Mother Earth's components

	– Democratizing access to the means and factors of production – Promote sources of decent employment within the framework of integral development
What are the responsibilities of individuals and communities? (strength)	
Unity Pact Proposal	– Defending and respecting the rights of Mother Earth – Participate in the generation of proposals – Report any act that violates the rights of Mother Earth – Assume new production and consumption habits and practices – Practice sustainable and balanced use and exploitation of the components – Carry out a comprehensive and effective restoration of the damage caused – Individuals and collectives are obliged to respect, protect, and guarantee the rights of individuals and groups – Supporting actions aimed at conservation and protection – Take out insurance cover before engaging in risky activities
Law 71 of 2010	– Defending and respecting the rights of Mother Earth – Participate actively in the generation of proposals aimed at respecting and defending – Report any act that violates the rights of Mother Earth, its life systems, and/or its components – Adopting new production practices and consumption habits – Ensuring the sustainable use and exploitation of the components – Promote harmony on Mother Earth in all areas of its relationship with the rest of the human communities – To call for actions oriented to the conservation and/or protection of Mother Earth
Law 300 of 2012	– Restoration in cases of damage caused – Participate in the prioritization of her needs within the framework of participation and social control – Promote sustained and permanent processes of decommodification – Duty to report rights violations – In the case of high-risk use, commitments can be made through economic instruments
Who can exercise the rights? (strength)	
Unity Pact Proposal	All Bolivian women and men
Law 71 of 2010	
Law 300 of 2012	
What are the competent judicial instances? (strength)	
Unity Pact Proposal	Constitutionally recognized jurisdictions
Law 71 of 2010	–
Law 300 of 2012	Ordinary jurisdiction, agro-environmental jurisdiction, and indigenous jurisdiction

Who ensures compliance? (strength)	
Unity Pact Proposal	– Defensoría de la Madre Tierra (Ombudsman's Office) – Administrative sanctioning bodies at all levels – Public Prosecutor's Office
Law 71 of 2010	– Defensoría de la Madre Tierra
Law 300 of 2012	– Defensoría de la Madre Tierra – The public authorities, at any level of the state, within the framework of their competences, Public Prosecutor's Office, Agro-environmental Tribunal
What institutional framework is established? (strength)	
Unity Pact Proposal	– Defensoría de la Madre Tierra ensures enforcement, promotion, dissemination, and compliance – Ministry of Mother Earth – Specialized Agroenvironmental Directorates to strengthen criminal prosecution – Consejo Plurinacional de la Madre Tierra (Plurinational Council of Mother Earth) as a consultative body
Law 71 of 2010	– Defensoría de la Madre Tierra ensures enforcement, promotion, dissemination, and compliance
Law 300 of 2012	– Defensoría de la Madre Tierra – Consejo Plurinacional para Vivir Bien, en Armonía y Equilibrio con la Madre Tierra, (Plurinational Council for Living Well in Harmony and Balance with Mother Earth) as an instance for monitoring, consultation, and participation in the elaboration of policies, plans, programs, etc. – Mother Earth's component registry system – Autoridad Plurinacional de la Madre Tierra, (Plurinational Mother Earth Authority) and various other climate change-related instruments
What is the degree of mainstreaming of rights? (strength)	
Unity Pact Proposal	Foundation at all levels and sectors for the formulation and implementation of policies for the use and exploitation of Mother Earth's components
Law 71 of 2010	The state must ensure implementation at all levels and in all territorial areas and through all its authorities and institutions
Law 300 of 2012	– Scope extends to all sectors of the central level of the Plurinational state of Bolivia and the autonomous territorial entities – Preferential application for the development of specific laws, policies, regulations, strategies, plans, programs, and projects – Gradual implementation

Table 2: Persons interviewed[193]

Ruth Alipaz	Leader of the National Coordinating Committee for the Defence of Native Indigenous Peasant Territories and Protected Areas of Bolivia (CONTIOCAP)	7 December 2021
Agustín Cahuana Choque	Indigenous authority, National Council of Ayllus and Markas of the Qullasuyu (CONAMAQ)	16 December 2020
Paola Cortés Martínez	Lawyer specializing in environmental issues	15 October 2021
Enrique Cunai	Deputy for the special indigenous constituency of Beni in the Plurinational Assembly; secretary of the Committee of Amazon Region, Land and Territory	22 December 2020
Marcelino Higueras	Natural resources engineer; president of the Chuquisaca branch of the Committee for the Defence of National Heritage (CODEPANAL).	28 December 2021
Fátima Monasterio	Lawyer and environmental activist	13 August 2021
Cecilia Moyoviri Moye	Senator; Chairwoman of the Committee on Land and Territory, Natural Resources and Environment	17 December 2020
Maribel Ruiz Molina	Agro-environmental judge of Uncía	23 December 2021
Diego Pacheco	Researcher and civil servant in various positions	11 August 2014
Cecilia Requena	Senator for Comunidad Ciudadana	24 November 2021
Wilson Rocha Vera	Climate change and indigenous peoples specialist	21 September 2021
Vivian Schwarz	Director of Citizenship – Community for Social Studies and Public Action	22 June 2021
Elva Terceros	Magistrate and former President of the Agro-Environmental Tribunal	21 and 29 January 2022
Miguel Angel Uche	Leader of the Central de los Pueblos Étnicos Mojeños de Beni (CPEM-B), Indigenous Multi-ethnic Territory	1 December 2022
Lui Laura Zapana	Lawyer and environmental activist	20 November 2021
Moira Zuazo	Political scientist; former staff member of ILIDIS Foundation (Friedrich-Ebert-Stiftung)	19 March 2014

[193] I sincerely thank the interviewees for their time and generosity in sharing their knowledge and experiences. I am also grateful to Raquel Melina Aguilar Jiménez, political scientist, for facilitating the interviews in times of pandemic and for many inspiring conversations.

Aotearoa New Zealand

Protecting Nature through Upholding Indigenous Rights and Human Responsibilities

Catherine Iorns[*]

I. Introduction

Aotearoa[1] New Zealand has adopted three key tools for environmental protection that are increasingly being referred to as rights of nature: intrinsic value of na-

[*] Many thanks are due to my research assistant Jonathon Sylvester LLB for being so helpful, patient, and willing to undertake a range of disparate tasks on different topics, not all the results of which made it into this paper. The manuscript was first written in 2023, updated in April 2024 and May 2025 due to developments in relation to the Taranaki Maunga, and finalized after editing in October 2025. The assessment of literature and case law, including internet materials, reflects the legal situation at that time. All internet sources cited were last accessed on 8 October 2025. All New Zealand legislation is available at <www.legislation.govt.nz>.

[1] Aotearoa is the Māori name for New Zealand; both names are often used together to denote respect for both.

ture, legal personhood for nature, and human guardianship of nature. New Zealand has been recognized internationally for its innovative recognition of legal personhood for nature. It is less well-known for the intrinsic value afforded to nature in law, its recognition of animal sentience, or for the frameworks for guardianship and responsibility for nature that have been adopted in law.

Interestingly, within New Zealand, the adoption of these tools is not referred to as adopting or implementing rights of nature. Indeed, while New Zealand is a party to most international environmental agreements and prides itself in being a strong protector of the environment, that pride is not reflected in the national state of the environment.[2] In contrast, respecting and defending *human* rights garners strong political support, and it is through New Zealand's human rights framework that some eco-centric legal tools have been adopted.

The human rights most relevant to the adoption of legal personhood and guardianship for nature have been indigenous rights. This focus on indigenous rights has arisen because Aotearoa New Zealand is a relatively young state that arose as a result of colonization after the 1840 Treaty of Waitangi between the British Queen and the indigenous Māori. The effects of that colonization were the loss of Māori lands, territories, resources, authority, culture, and the relationships with nature that they previously enjoyed.[3] These losses have been the source of grievances that, since the late-1980s, have been the subject of reparations settlements between the New Zealand government and Māori tribes. It is these reparations agreements that have contained most of the co-management and co-governance guardianship arrangements and, most notably, three situations where nature has been made a legal person. The reason that legal personhood was chosen and guardianship bodies were created is because these mechanisms reflect the indigenous view that aspects of nature are their ancestors. New Zealand has merely put that view of nature as kin into law and identified its legal guardians.

---

[2] For example, New Zealand has been ranked 18th out of 179 in the world for its proportional environmental impact (where the 1st position corresponds to the greatest environmental impact proportional to total resource availability – ie, New Zealand is one of the worst 20 countries of the world). Corey JA Bradshaw, Xingli Giam and Navjot S Sodhi, 'Evaluating the Relative Environmental Impact of Countries' (2010) 5(5) PLoS ONE e10440 (2010), <https://doi.org/10.1371/journal.pone.0010440>. New Zealand still produces the most landfill waste per capita of all developed countries, nearly all reptiles and most native birds are threatened with extinction, and freshwater quality continues to decline. Ministry for the Environment, *Our Environment 2025* (8 April 2025) <https://environment.govt.nz/publications/our-environment-2025/>.

[3] This is only a summary list; what has been lost is much more extensive, especially when the flow-on effects are included. As a result of settlement over the last 200 years, Māori now make up only approximately 17 percent of the population of approximately 5.3 million; see, eg, Statistics New Zealand on population numbers and ethnicity at 30 June 2024 <www.stats.govt.nz>.

II. Theoretical Foundations for Rights of Nature

The theoretical frameworks that underpin the use of the mechanisms adopted in Aotearoa New Zealand vary, depending on the mechanism. The legal personhood and guardianship arrangements are based on the Māori philosophy or worldview of kincentrism and human guardianship (*kaitiakitanga*), along with the recognition of indigenous rights and reparations for historical breaches of them. In contrast, the recognition of intrinsic value for nature has been based on the need for respect for and protection of the environment.

1. Māori Cosmology: Kincentrism and Guardianship

The key theoretical foundation for the legal personhood and guardianship examples is Māori cosmology, which upholds a kincentric view of the environment and human relationships with it.[4] Indigenous histories typically tell of how people today descended from, are part of, and are genealogically related to the natural world and are dependent on it.[5]

Importantly, 'the spiritual and physical realms are not closed off from each other, as they tend to be in the European context'.[6] Typically, the animals, plants and land are imbued with spirits, while these spirits also inhabit the people and look after them. Thus the people in return treat animals, plants, or other natural features as kin, and look after them respectfully. This construction of the relationship between people and their environment is one that enables humans to continue to live as peoples in and as part of the natural world.

In terms of a hierarchy, the indigenous construction of nature effectively reverses the Western hierarchy: humans are not seen as having any right or even ability to completely dominate nature and are instead seen as its guardians.[7] Some

---

[4] For an explanation of a kincentric approach, see, eg, Enrique Salmon, 'Kincentric Ecology: Indigenous Perceptions of the Human-Nature Relationship' (2000) 10(5) Ecological Applications 1327.

[5] For a North American view, see Richard Erdoes and Alfonso Ortiz, *American Indian Myths and Legends* (University of New York Press 1984). This book contains 166 such stories, covering creation, the sun, moon and stars, heroes, warriors, love, the trickster, 'Animals and Other People', ghosts and the spirit world, and 'Visions of the End'.

[6] *Ngati Rangi Trust v Manawatu-Wanganui Reg'l Council* [2004] NZEnvC 172 [26].

[7] Waitangi Tribunal, *Ko Aotearoa Tēnei: A Report into Claims Concerning New Zealand Law and Policy affecting Maori Culture and Identity, Te Taumata Tuarua* (Wai 262, 2011) 269. *Kaitiakitanga*, or guardianship or stewardship, is one of the fundamental principles underlying Māori society, beliefs and practices, ibid at 237. Waitangi Tribunal reports abound with claimant evidence of their status and traditional practices as *kaitiaki* (guardians) in the past and continuing today; ibid 252. For a summary of 'Kaitiakitanga today', see ibid at 244–248. The Waitangi Tribunal's reports are available by their Wai numbers at <https://www.waitangitribunal.govt.nz/en/publications/tribunal-reports/by-wai-number>.

indigenous stories refer to animals or natural features as guardians of the people[8] or even as their masters.[9] One related belief is that if people do not behave in the right way with respect to nature, then nature will take its revenge – for example, the animals might not offer themselves up for food or it might not rain when needed.[10] Unsurprisingly, such a relationship does not envisage humans owning nature in the Western, liberal sense.[11] As has been described in respect of Māori:

> The *kaitiaki* relationship with the environment is not the transactional or proprietary kind of the Western market, and it does not rest on 'ownership'. Rather, like a family relationship, it is permanent and mandatory, binding both individuals and communities over generations and enduring as long as the community endures.[12]

This indigenous cosmology and construction of nature is so different from the liberal (Enlightenment) construction that it has frequently been thought of as unable to coexist with and within a Western, liberal, democratic society. Yet, despite such difficulties, the New Zealand examples of personhood, guardianship, and intrinsic value for nature show that it is in fact possible.[13] Indeed, adopting the indigenous approaches implemented in the legal personhood examples will likely entail an alteration of these underpinnings. For example, the theoretical foundation for nature as an ancestor is not that of a conservationist ethic, whereby a national park is seen as separate from the people who inhabit it. Instead, the key

---

[8] See generally Waitangi Tribunal, *Te Kahui Maunga: The National Park District Inquiry Report* (Wai 1130, 2013) 95–98. The Waitangi Tribunal describes a *kaitiaki* [guardian] spirit who appeared in the form of a seagull, ibid at 97.

[9] See, eg, Winona LaDuke, *All Our Relations: Native Struggles for Land and Life* (1999), at 51. Daniel Ashini, chief land rights negotiator for the Innu, describes the Innu religion as being: 'based on a belief in animal masters and other forest spirits. When we hunt we must show respect for animal masters […]. We do not overhunt or overtrap areas where animals are scarce. If we do not show respect in this way, the animal masters get angry and punish us by not giving us any animals at a later date.'

[10] Waitangi Tribunal, *The Whanganui River Report 1997* (Wai 167, 1999) 108. For example, the Whanganui River is described as 'an ancestral treasure handed down as a living being related to the people of the place […]. It governed their lives, and like a tupuna [ancestor], it served both to chastise and protect.' at 46.

[11] See, eg, evidence of Turama Hawira of the Māori tribe Ngati Rangi: 'we are defined by our ancestral mountain, our ancestral rivers and our ancestral land. They are the source of our wellbeing – spiritually, intellectually and physically. We do not separate our wellbeing from the wellbeing of our taonga tupuna [natural treasures handed down from our ancestors]. Nor can we possess them. They do not belong to us – we belong to them.' *Ngati Rangi Trust v Manawatu-Wanganui Reg'l Council* [2004] NZEnvC 172 [29]. For a North American example: 'The four leggeds came before the two leggeds. They are our older brothers, we come from them […]. We cannot say that we own the buffalo, because he owns us.' LaDuke (n 9) 135 (quoting Birgil Kills Straight, Oglala Lakota).

[12] Waitangi Tribunal, *Ko Aotearoa Tēnei* (n 7) 269.

[13] For more information on and examples of Māori cosmology in New Zealand law, see Catherine J Iorns Magallanes, 'Maori Cultural Rights in Aotearoa New Zealand: Protecting the Cosmology that Protects the Environment' (2015) 21 Widener L Rev 273.

goal is to maintain and strengthen the connection between the people and the place by enabling the people to exercise guardianship responsibilities.

2. *Intrinsic Value(s) of Nature*

A quite different theoretical foundation exists for the recognition of the intrinsic worth or value of nature. This concept emerged as a counter narrative in the 1970s to dominant, anthropocentric concepts of ecosystem services.[14] The ecosystem services approach is that nature is valuable only to the extent that it provides benefits to humans;[15] it is the dominant approach in the New Zealand legal system and its laws.

A simple definition of intrinsic value is 'the right of other living things to exist for their own sake, regardless of the interest of human beings'.[16] A similar definition is the 'value that exists independently of humans' and by virtue of what a natural entity is, 'not because of human attitudes *toward* it',[17] nor what it can do for people.[18] All of these definitions share the element that value is independent of human interests. Putting this concept in terms of the rights of nature approach, intrinsic value represents the fundamental right to life or existence, and thus likely also the right to fulfil evolutionary processes.

While this concept is fairly simple, it does not address precisely what subjects possess intrinsic value, nor what properties qualify a natural entity to have intrinsic value. Different philosophers have taken different approaches. Some believe in a broad approach whereby all teleological centers of life,[19] or all experiencing subjects of life, possess intrinsic value.[20] Others specify that intrinsic value is attributable to entities which possess rationality,[21] or that it is derived by

[14] Chelsea Batavia and Michael Paul Nelson, 'For Goodness Sake! What Is Intrinsic Value and Why Should We Care?' (2017) 209 Biological Conservation 366, 367.

[15] Hua-Xia Sheng and others, 'Ecosystem intrinsic value and its application in decision-making for sustainable development' (2019) 49 Journal for Nature Conservation 27, 27.

[16] Karen Cronin, 'Ecological Principles for Resource Management: A Summary' in Ministry for the Environment, *Fundamental Issues in Natural Resource Management* (Working Paper no 1, 1988) 46. Cronin elaborates: 'the intrinsic value of ecosystems is about the value of existence and integrity of life on earth, beyond human existence or expression of human need'. Karen Cronin, 'A Note on the Concept of "Intrinsic Values" and its Interpretation for Resource Management Law Reform' in Ministry for the Environment, *Resource Values* (Working Paper no 10, 1988) 6.

[17] Jennifer Caldwell, 'Intrinsic Value: Thoughts from a Legal Perspective' in Ministry for the Environment, *Implementing the Sustainability Objective in Resource Management Law* (Working Paper no 25, 1988) 5. Emphasis included in original text.

[18] John A Vucetich, Jeremy T Bruskotter and Michael Paul Nelson, 'Evaluating Whether Nature's Intrinsic Value Is an Axiom of or Anathema to Conservation' (2013) 29 Conservation Biology 321, 323.

[19] Paul Taylor, 'The Ethics of Respect for Nature' (1981) 3 Environmental Ethics 197.

[20] Tom Regan, *The Case for Animal Rights* (University of California Press 1983).

[21] Marc D Davidson, 'On the Relation between Ecosystem Services, Intrinsic Value, Existence Value and Economic Valuation' (2013) 95 Ecological Economics 123, explaining Kant's position.

virtue of having interests.[22] One argument is that living organisms embody intrinsic value simply by fulfilling their individual and evolutionary processes.[23] For example, plants develop and defend a biological identity.

Across the differing perceptions of what properties are the source of intrinsic value, it is commonly agreed that individual organisms each have intrinsic value. However, views differ over whether collective entities such as ecosystems can possess intrinsic value. Some argue that collective entities cannot possess intrinsic value, as species do not have interests distinct from those of the individual organisms that comprise them.[24] Others, however, argue that ecosystems do possess intrinsic value, which is derived from their role in preserving the integrity, stability, and beauty of the biotic community.[25] While these debates may seem more theoretical than practical, any reference to intrinsic value in law requires choices to be made about which definition and subject to adopt, and what obligations they might give rise to on the part of humans.

III. Recognition of the Intrinsic Value of Nature in Law

Aotearoa New Zealand has a legal system whereby Parliament is sovereign and is not legally bound by a higher-order written constitution. It inherited this from the United Kingdom upon colonization, along with the common law system for the courts' role. Judicial decisions are thus based on a mix of statutory interpretation and common law rules and fundamental values developed by courts. As legislation has primacy, it is the primary focus.

1. Legislation

Several significant pieces of New Zealand's environmental legislation contain recognition of the intrinsic worth or value of nature. Key examples are the Reserves Act 1977, the National Parks Act 1980, the Conservation Act 1987, the Resource Management Act 1991, the Hazardous Substances and New Organisms Act 1996, and the Marine and Coastal Area (Takutai Moana) Act 2011. The theoretical choices made in the statutes are addressed below. In addition, several other statutes specific to particular natural areas refer to intrinsic values of nature.[26]

---

[22] Ronald Sandler, 'The Value of Species and the Ethical Foundations of Assisted Colonisation' (2010) 24 Conservation Biology 424, 425.

[23] Holmes Rolston, 'What Do We Mean by the Intrinsic Value and Integrity of Plants and Animals?' in William T Noonan and Harold Glasser (eds), Conservation and Moral Reasoning (CUP 1992). Rolston also implies a hierarchy whereby the degree of intrinsic value varies according to adaptive fit and natural selection. He uses this hierarchy to justify the destruction of intrinsic value through processes such as predation.

[24] See, eg, Sandler (n 22) 426.

[25] Aldo Leopold, *A Sand County Almanac and Sketches Here and There* (OUP 1949).

[26] See, eg, Water Conservation Orders in respect of the Kawerau and Rangitata Rivers, the two statutes conferring legal personality on Te Awa Tupua and Te Urewera, which are discussed

Because New Zealand does not have an overarching legal code, separate statutes on different topics are adopted at different times. As such, each statute can adopt different definitions and concepts, even on similar topics. In addition, judicial interpretations of a term in one statute will not necessarily apply to the same term in another statute.

The Reserves Act 1977,[27] the National Parks Act 1980,[28] and the Marine and Coastal Area (Takutai Moana) Act 2011[29] all have very similar purposes and references to the intrinsic worth of nature.[30] None of the Acts define the term 'intrinsic worth' further, so it is left to take its ordinary meaning.[31] However, interestingly, nature's intrinsic worth, although ecocentric in abstraction, has still been seen as rooted in human enjoyment of nature. For example, the courts have determined that the view from a scenic reserve is part of its intrinsic worth;[32] this has justified the Minister of Conservation acting under the Reserves Act to grant a lease to enable construction of a gondola on a scenic reserve to take advantage of the view. In determining when and where 'protection and preservation are desirable', the decision-making is based on what is 'desirable in the [human] public interest'. Thus, the measure of nature's intrinsic worth has been established anthropocentrically.

The Conservation Act 1987 refers to 'intrinsic value' as part of the definition of 'conservation', of 'nature conservation', and of 'preservation'. While also undefined, the drafting makes it clear that intrinsic worth is a key factor in decisions about what to conserve, preserve, and protect. The New Zealand Supreme Court has held that the definition of 'conservation' and its inclusion in the purpose of the Act means that the existence of 'intrinsic conservation values' are key to the existence and protection of conservation land; only where such values no longer exist can the protected status of conservation land be revoked.[33] Thus, intrinsic

---

later in this chapter, some forest and heritage area legislation, and the Antarctica (Environmental Protection) Act 1994.

[27] See Reserves Act 1977, s 19.

[28] National Parks Act 1980, s 4.

[29] See the purpose of this Act, s 4.

[30] For example, the National Parks Act 1980 provides in s 4(1):

'It is hereby declared that the provisions of this Act shall have effect for the purpose of preserving in perpetuity as national parks, for their intrinsic worth and for the benefit, use, and enjoyment of the public, areas of New Zealand that contain scenery of such distinctive quality, ecological systems, or natural features so beautiful, unique, or scientifically important that their preservation is in the national interest.'

[31] The ordinary meaning of the text is interpreted in light of the purpose of the legislation: Interpretation Act 1999, s 5(1).

[32] *Summit Road Society (Inc) v Minister of Conservation* [1990] 14 NZTPA 216.

[33] *Hawke's Bay Regional Investment Co Ltd v Royal Forest and Bird Protection Society of New Zealand Inc* [2017] 1 NZLR 1041. In this judicial review, the conservation status of land was unable to be revoked for a dam, as it still had inherent conservation value. The Minister's (highly politically charged) decision to revoke that conservation status was quashed.

values, even if vaguely defined, have enabled the strong protection sought by the Act.

The Resource Management Act 1991 and the Hazardous Substances and New Organisms Act 1996 take a different approach. They both refer to 'intrinsic values of ecosystems' in a list of decision-making principles that must be taken into account when exercising functions under each Act.[34] Importantly and helpfully, the concept of intrinsic values of ecosystems is defined in the statute; these values include:

'(a) their biological and genetic diversity; and
(b) the essential characteristics that determine an ecosystem's integrity, form, functioning, and resilience.'

2. Judicial Interpretations of Intrinsic Value of Nature

The 'intrinsic values of ecosystems' in section 7(d) in the Resource Management Act has had the most judicial consideration and application of any of the legislation on this topic. The Act governs the granting of permits for many activities that might affect the environment, and most developments that come before the courts will be affecting ecosystems to an extent. As decision-makers under the Act need only 'have particular regard to' the intrinsic values of ecosystems, intrinsic values are not as fundamental to the definition of what is to be protected nor to decision-making as they are in the context of the Conservation Act. Despite the weak instruction to decision-makers, courts have been able to utilize the provision and clarify its application.

Just because intrinsic value is recognized in law doesn't mean it will be protected; it can be outweighed by other factors. For example, approval was given for an open cast coal mine on a plateau containing rare plant species and a unique and endangered ecosystem.[35] In that decision, the Environment Court considered the application of section 7(d), with a focus on the ecosystem's integrity:[36]

'Section 7(d) requires particular regard for the intrinsic value of ecosystems. We have no doubt that if the mine is consented on the conditions proposed there will be a reduction in the value of ecosystems on the mine site. Connectivity and the current assemblages will be disrupted or lost by changes such as the loss of some plant species [that] are likely to be permanently lost to the site, and some of those have considerable rarity value.'

In addition, the Environment Court found that 'there is clearly an adverse effect on the proposal on significant habitats of indigenous fauna,' with it being 'a century or more before similar habitats to those presently existing would be available on the mine site'.[37] However, despite these factors, the Court granted the min-

---

[34] See Resource Management Act 1991, s 7(d) and Hazardous Substances and New Organisms Act 1996, s 6(b).

[35] *West Coast Environmental Network Inc v West Coast Regional Council and Buller District Council* [2013] NZEnvC 47.

[36] Ibid [319].

[37] Ibid [324].

ing consent and agreed to the applicant's suggested biodiversity offsets elsewhere in order to compensate for the clear biodiversity losses that would occur from the mining.

Despite such examples of loss of nature with significant value, there are also examples where intrinsic value of nature has been upheld by decision-makers. One notable case denied an application for permits to operate marine farms within the habitat of a threatened seabird.[38] The Court stated that this decision was made having particular regard to the intrinsic values of King Shag, a threatened species which would be adversely affected by a decision to grant the consent, due to disturbance from human activity.[39]

When considering the content of intrinsic biodiversity values, courts have clarified that 'not all ecosystems are equal', and in the context of the RMA, indigenous biodiversity possesses a higher degree of intrinsic value than introduced species.[40] It has also been stressed that intrinsic values are those which 'can be identified objectively by scientists if given enough time and money to study them' and are 'not anthropocentric values'.[41]

Overall, the different New Zealand statutes have adopted different approaches to the implementation and protection of the intrinsic values of nature. In resource management permitting, intrinsic values have been taken as one factor to be balanced when deciding whether or not to grant an activity permit.[42] Where there is conflict, it has been viewed as a factual question about the extent of effects and the overall balance. Whereas in relation to the Conservation Act, where conflict between human values was not built into the legislation, conflict arose in practice through competing interests behind decision-making.[43] What legislators and governments have not done is provide any guidance on how to resolve such conflicts, such as through providing a list of priority principles. For example, philosophers have suggested fundamental priority principles of self-defense, proportionality, minimum wrong, distributive justice, and restitutive justice;[44] these have not been taken up by legislators or courts.

Despite this lack of guidance, the inclusion of intrinsic value in New Zealand legislation has helped to signify the importance of protecting the environment in the face of constant human desires for its use. Perhaps as we recognize more rights

---

[38] *Clearwater Mussels Ltd v Marlborough District Council* [2018] NZEnvC 88 [117].

[39] Ibid [117].

[40] *Lindis Catchment Group Inc v Otago Regional Council* [2019] NZEnvC 166 [172].

[41] *Oceana Gold (New Zealand) Ltd v Otago Regional Council* [2019] NZEnvC 41 [74].

[42] See *Auckland Volcanic Cones Society Inc v Transit New Zealand* 31 March 2003, High Court Auckland, AP 123-SW02; *Oceana Gold (New Zealand) Ltd v Otago Regional Council* [2019] NZEnvC 41; *Creswick Valley Residents Association Inc v Wellington City Council* [2015] NZENVC 149. Note that the legislation labels such activity permits as 'resource consents'.

[43] *Royal Forest and Bird Protection Society of New Zealand Inc v Minister of Conservation* [2016] NZHC 220.

[44] Paul W Taylor, 'Priority Principles' in Paul W Taylor, *Respect for Nature: A Theory for Environmental Ethics* (Princeton UP 1986) 199.

of nature, including its legal personhood, courts and other decision-makers will get more used to the concept of balancing rights according to priority principles, just as is already practiced in a human rights context.

IV. Legal Personhood for Nature

New Zealand law has now accorded legal personhood to nature three times: for a river, for a very large area that was previously a national park, and for a mountain area that remains a national park. This section of the Report summarizes the background factors, actors, and features of each personhood example.

1. Background Factors and Actors

As with other settled countries around the world with surviving indigenous peoples, Māori have made legal and political claims to those things that they have lost through colonization. Beginning in the 1970s, New Zealand has established inquiry and compensation laws, procedures, and practices in order to resolve such claims.[45] Each tribal settlement with the government is given effect through bespoke legislation.

The settlement of indigenous tribal grievances has led to the creation of a wide range of measures designed not just to compensate for losses but also to better recognize and reestablish indigenous knowledge and authority systems and to revitalize indigenous tribal governance, especially over their traditional territories and resources. As well as the return of lands, territories, and resources, this has led to significant numbers of co-management and co-governance agreements with Māori tribes. A key aspect of this is recognizing the worldviews of the tribes in question and their cultural exercise of responsibility for and guardianship over the natural world. It is through this recognition and system for resolving indigenous grievances that the legislature has adopted the innovative use of legal personhood for natural features.

Legal personhood has been established only by Parliamentary statute. While it has widespread Parliamentary support, this legislation only came to Parliament via the governing Executive. The reason for the government putting this legislation before Parliament is that it was the result of a negotiated process between the government and the relevant Māori tribes in settlement of historical grievances over colonization.[46] Legal personhood was a tool chosen to reconcile the different positions of the government and the tribes over the land and resources

---

[45] For more information, see, eg, Catherine Iorns, 'Reparations for Maori Grievances in Aotearoa New Zealand' in Federico Lenzerini (ed), Reparations for Indigenous Peoples: International and Comparative Perspectives (OUP 2008).

[46] For more information about the grievance settlement and reparations process, see, eg, Iorns (n 45).

in question. Thus the key legal and political operators in the establishment of legal personhood for nature in Aotearoa New Zealand have been Parliament, the government (ie, Executive), and Māori tribes.

Within the government, the Ministers in charge of the different government departments relevant to the land in question play an extremely important role in deciding what can be negotiated, both politically and legally. Such grievance settlement negotiations are undertaken by a separate government office; this was the Office of Treaty Settlements within the New Zealand Ministry of Justice for the legal personhood examples discussed in this chapter, but it is now Te Arawhiti: The Office for Māori Crown Relations.[47]

The first such example to be agreed upon in negotiation between the government and tribes was made with Whanganui Iwi in relation to the Whanganui River. This was first agreed to in 2012,[48] finalized as an agreement in 2014,[49] and enacted in statute in 2017.[50] The length of time taken between general agreement in 2012 to passage as a statute in 2017 illustrates the detail that needed to be negotiated. The settlement legislation – especially one establishing a new legal personality for the river – necessarily involved a wide range of stakeholders. In addition to the 12 tribes along the river (and other tribal groups that exercised customary rights and responsibilities in respect of the Whanganui River), other stakeholders included the different landowners up and down the river, as well as central government departments, local government authorities, and statutory, commercial, and other community bodies with interests in the river and its management. Establishment of the governance framework and body for the new legal person required a significant amount of time for negotiation. All of these stakeholders can at least be considered as part of the political operators involved in the establishment of the new legal person, and some of them with legal, statutory roles (such as the various central and local government bodies). This is demonstrated by the composition of the strategy group, Te Kopuka na Te Awa Tupua, which includes up to 17 stakeholder representatives: from the Whanganui Iwi, from the local and central government, from the conservation, recreation, and tourism sectors, and from the operator of the electricity generator on the river.

Perhaps the most interesting aspect for the purposes of transnational comparison is in respect of how the suggestion for personhood arose. As mentioned above,

---

47 See, eg, <https://www.tearawhiti.govt.nz/>. The change in name is partly intended to represent a change in focus, moving away from a focus exclusively on the settlement of historical grievances so as to include positive future relations and upholding the government's indigenous rights obligations.

48 Tutohu Whakatupua – Whanganui River Settlement Agreement (2012) (NZ).

49 Office of Treaty Settlements, Whanganui Iwi (Whanganui River) Deed of Settlement (2014); Ruruku Whakatupua: Whanganui River Deed of Settlement – Te Mana o Te Iwi o Whanganui, cl.1.4 (2014), <https://www.govt.nz/assets/Documents/OTS/Whanganui-Iwi/Whanganui-River-Deed-of-Settlement-Ruruku-Whakatupua-Te-Mana-o-Te-Iwi-o-Whanganui-5-Aug-2014.pdf> [hereinafter 'Te Mana o Te Iwi o Whanganui'].

50 Te Awa Tupua (Whanganui River Claims Settlement) Act 2017.

the ownership and control of the river was in Māori tribal hands before colonization. The tribes thus wanted both ownership and control to be returned to them, both of the riverbed and its water. The government considered this to be politically unacceptable, given the large number of different uses and stakeholders that existed today; settlement negotiations were at an impasse for more than a decade. The Office of Treaty Settlements proposed a novel solution whereby no one owned the riverbed and it would be given legal personality, allowing it to own its own riverbed. This was supported through legal precedent: a case of the Privy Council (from when it was New Zealand's highest court) affirmed that a Hindu idol or statue had legal personality.[51] The Privy Council stated that courts had long recognized that an idol was 'a juristic entity' and:[52]

> 'has a juridical status with the power of suing and being sued. Its interests are attended to by the person who has the deity in his charge and who in law is its manager with all the powers which would, in such circumstances, on analogy, be given to the manager of the estate of an infant heir. It is unnecessary to quote the authorities; for this doctrine, thus simply stated, is firmly established.'

It was thus clear that such personhood for inanimate objects was legally possible according to New Zealand common law.

Personhood also resonated with the tribal relationship with the river as an ancestor, and the guardianship arrangements better represented their relationship with their ancestor. Even if tribal legal and beneficial ownership would better implement tribal demands, this personhood solution represented and applied tribal cosmology and enabled the impasse in negotiations over ownership to be broken.

Ownership of the water in the river was also sought by the tribes. However, this was refused by the government on the basis that the current legal position in New Zealand was that freshwater or waterways cannot be owned as private property; only rights to use it are accorded by the government.[53] The legal person established in the statute is defined broadly and extensively, comprising the Whanganui River, encompassing all its tributaries and physical and metaphysical elements; most people are thus surprised to find that the river owns only its river bed, not its water.[54] Note that the reasons for leaving out the water are not legal but political: if the government recognized *any* ownership of water, then Māori tribes would likely make a claim for compensation for the taking of water from them pursuant to colonization; not surprisingly, the government wants to avoid this.[55]

---

[51] *Pramatha Nath Mullick v Pradyumna Kumar Mullick* (28 April 1925) [1925] UKPC 33 <https://www.bailii.org/uk/cases/UKPC/1925/1925_33.html>.

[52] Ibid 2–3.

[53] For a summary of this position and the government's arguments in relation to the Whanganui River, see Dame Anne Salmond, 'Fresh water is beyond "entrenchment"' (*Newsroom*, 20 November 2022) <https://newsroom.co.nz/2022/11/29/anne-salmond-three-waters-beyond-entrenchment/>.

[54] Te Awa Tupua is recognized in s 12, with the definition of Whanganui River in s 2. Vesting of title to the Crown-owned parts of the riverbed is contained in s 41.

[55] The Waitangi Tribunal has found that Māori tribes were guaranteed an exclusive right to control access to and use of the water in their territories – ie the equivalent of ownership of such

This history clearly illustrates how the establishment of legal personhood for nature was not done pursuant to any movement for the better environmental protection of the river nor any conceptual framework utilizing rights of or for nature. It was genuinely a political negotiation over how best to recognize indigenous rights of reparation for the taking of the river through colonization. Helpfully, it garnered widespread social and political support partly because of this novel use of personhood for nature.

The situation with Te Urewera is similar. In 2012, the negotiations between Tūhoe and the government were similarly at an impasse over possible return of the ownership of Te Urewera to the Tūhoe tribe. Tūhoe wanted the national park returned to their ownership and control, and the government refused to hand over ownership of a national park, as 'ceding ownership of Urewera National Park to Tūhoe would have set a dangerous precedent for other treaty claims'.[56] After the initial agreement over legal personality for the Whanganui River, Te Awa Tupua, Tūhoe began negotiating with the government over using legal personhood for their reparations settlement, whereby nobody would own the national park but it would own itself. This agreement was finalized quickly – later in 2012[57] – and legislation to implement it was introduced in 2013 and passed in July 2014.[58]

Even though these negotiations began after those regarding the Whanganui River, they were finalized much more quickly. This was due largely to the much smaller numbers of stakeholders that needed to be involved. The lands in question were already a national park, managed by the Department of Conservation. There wasn't the need for local government or other large business interests to be involved, nor other central government departments apart from reportedly the Minister of Finance.[59] In that respect, it was a comparatively simpler negotiation.

These two settlement agreements established a new precedent and now even other tribes who have already reached a settlement with the government in respect of their historical grievances are investigating ways to establish legal per-

---

water: Waitangi Tribunal, The Stage 2 Report on National Freshwater and Geothermal Resources Claim (Wai 2358, 2019). However, the government is not bound to change its water law accordingly and has not done so.

[56] 'Jon Key outlines Te Urewera's precedent' (*Radio New Zealand*, 18 May 2010) <https://www.rnz.co.nz/national/programmes/morningreport/audio/2297402/john-key-outlines-te-urewera's-precedent>. In May 2010, the Prime Minister rejected handing over ownership of such an important national asset to Tūhoe, even though the Office of Treaty Settlements had earlier been negotiating settlement on this basis. See, eg, Derek Cheng, 'Tuhoe hold emergency talks over Key "betrayal"' (*New Zealand Herald*, 13 May 2010) <https://www.nzherald.co.nz/kahu/tuhoe-hold-emergency-talks-over-key-betrayal/VOVDTKTJJQHHYHXDNBNTG43FRM/>.

[57] 'Historic Treaty Settlement for Tuhoe' (*Stuff*, 11 September 2012) <https://www.stuff.co.nz/national/politics/7653374/Historic-Treaty-settlement-for-Tuhoe>.

[58] Tūhoe Claims Settlement Act 2014.

[59] See, eg, Cheng (n 56).

sonhood for such ancestors. The third personhood example for Taranaki Maunga was passed into law on 1 February 2025.[60]

The key lesson to take from this historical background is that the legal personalities in question did not arise from the rights of nature movement nor were they adopted in order to establish rights for nature. Instead, they arose out of an attempt to settle historic indigenous grievances that resulted from colonization. They thus arose from human rights considerations. Yet even while the human rights results are consistent with the UN Declaration on the Rights of Indigenous Peoples, they did not arise from trying to comply with such international obligations. Instead, they were the product of domestic political imperatives to settle these historic grievances, and they are arguably leading the comparative international practice on implementation of the rights of indigenous peoples.[61]

2. *Te Urewera*

Te Urewera National Park was the largest in New Zealand's North Island with a total land area of 212,673 hectares.[62] It was created in the 1950s out of ancestral land confiscated from Tūhoe without their agreement.[63] Key to settlement of the tribal grievances associated with this taking[64] was Tūhoe regaining control of the Te Urewera National Park and the ability to exercise their chieftainship and governance authority (*mana motuhake*) and guardianship (*kaitiakitanga*) over Te Urewera, along with explicit acknowledgement of the metaphysical relationship of Tūhoe with their ancestral land.[65] The method of achieving this was through the creation of legal personhood for the park, whereby Te Urewera would obtain its own separate identity as 'a legal entity' with 'all the rights, powers, duties, and liabilities of a legal person',[66] and have a legal guardian to manage and protect its interests.

a) Te Urewera the legal person

It has been commented that the 'Te Urewera Act makes clear that Te Urewera ceases to be vested in the Crown, ceases to be Crown land and ceases to be a na-

[60] Te Ture Whakatupua mō Te Kāhui Tupua 2025/Taranaki Maunga Collective Redress Act 2025 (hereinafter 'Taranaki Maunga Act').

[61] For more information on the impact of international standards and New Zealand's human rights laws, see, eg, Catherine J Iorns Magallanes, 'International Human Rights and their Impact on Domestic Law on Indigenous Peoples' Rights in Australia, Canada and New Zealand' in P Havemann (ed), Indigenous Peoples' Rights in Australia, Canada and New Zealand (OUP 1999) <http://ssrn.com/abstract=2175388>.

[62] Department of Conservation, *Te Urewera National Park Management Plan* (Wellington: Department of Conservation 2003) 1. See also Ngai Tuhoe, *Te Urewera* (2022), <http://www.ngaituhoe.iwi.nz/te-urewera>.

[63] Department of Conservation (n 62) 1; see also Tuhoe (n 62).

[64] These grievances are summarized in the Tūhoe Claims Settlement Act 2014, s 9.

[65] Te Urewera Act 2014, s 9(36).

[66] Te Urewera Act 2014, s 11(1).

tional park (s 12).'[67] The subject of rights is simply Te Urewera, defined as 'the legal entity created by section 11 or, as the context requires, the place encompassing Te Urewera land'.[68] '*Te Urewera land* means the land held from time to time in the name of Te Urewera and subject to this Act',[69] where 'land' is not further defined. The territory encompasses mostly native forest but also two lakes and their numerous linked waterways, and of course the flora and fauna that live within them. While the National Park was created largely out of two water catchments, in relation to the questions set for this comparative project, it might be better to describe the subject of legal personhood as ecosystems.

Despite this dry and sparse legal definition, the 'Background' to the Act describing Te Urewera reflects both the Tūhoe worldview and Earth laws and rights of nature approaches.[70] It is clear that seeing the forest as a real person – a living, spiritual, or metaphysical being – was more important than making it a legal person. But legal personhood was the best tool to enable Te Urewera to stand apart from human ownership within the current legal system.

The legislation vests a fee simple estate in the Te Urewera lands that establish the entity,[71] and then declares that the land is inalienable.[72] Thus Te Urewera technically holds the title to itself. Apart from that, such 'rights, powers, duties, and liabilities' are not stated and are left to be inferred from law. This can include owning property generally, entering into agreements (including contracts), and enforcing its rights, including via court action should that be necessary. The analogy most often used is a company, which is a legal entity that is created pursuant to a statute, and which can have all those things done in its name.

b) Guardianship body

Just as with a company, Te Urewera does not do any of these things itself, but it pursues its interests through people who are appointed to that role. Indeed, establishing and exercising the guardianship for Te Urewera is more important to Tūhoe than legal personhood itself. As a result, Te Urewera is now no longer managed by the Department of Conservation, nor is it managed by Tūhoe. Instead, the new legal entity is managed by the *Te Urewera* Board, which must 'act on behalf of, and in the name of, Te Urewera' and 'provide governance for Te Urewera'.[73]

The Board currently consists of nine members, six of which are appointed by Tūhoe trustees and three of which are appointed by the Minister of Conserva-

---

[67] Jacinta Ruru, 'Tuhoe-Crown Settlement – Te Urewera Act 2014' (2014) 14(1) Māori LRev 1, 1.

[68] Te Urewera Act 2014, s 2, 'Te Urewera'.

[69] Te Urewera Act 2014, s 2, 'Te Urewera land'.

[70] Te Urewera Act 2014, s 3.

[71] Te Urewera Act 2014, s 12 and Schedule 1.

[72] Te Urewera Act 2014, s 13.

[73] Te Urewera Act 2014, s 12.

tion.[74] This membership was designed to accommodate the development of tribal capacity post-settlement and reflects the primacy of Tūhoe in the arrangement.

The Te Urewera Board gives Te Urewera the ability to make its own decisions about the management of its land.[75] The Act grants the Board extensive powers and obligations, including the power to make by-laws and grant activity permits within Te Urewera.[76] Notably, the Board has the ability to grant permits for 'taking, cutting, or destroying indigenous plants' and for 'trapping, taking, hunting, or killing indigenous animals.'[77] These permits may be granted only where 'the preservation of the species concerned is not adversely affected; the effects on Te Urewera are no more than minor; and the grant of a permit is consistent with the management plan'.[78] However, such permits provide a significant inroad into the protections afforded under the National Parks Act 1980, at least conceptually. The indigenous wildlife in a national park is protected absolutely, reflecting the traditional wilderness-conservation approach, whereby nature is wholly protected from human activity within protected areas.[79] By contrast, the new Te Urewera law upholds the indigenous perspective that nature can be protected in conjunction with human use, if managed properly.[80] For many conservationists, this is seen as granting more rights to humans than to nature.

Eight principles govern the implementation of the Act, which all decision-makers including the Board need to abide by. These principles focus primarily on different aspects of environmental protection, but they also protect indigenous relationships with the park as well as public access for the benefit of all New Zealanders.[81]

The Board, when making decisions, must also 'consider and provide appropriately for the relationship of iwi and hapū and their culture and traditions with Te Urewera'.[82] The legislation expressly states that the purpose of this duty is to 'recognise and reflect' both *Tūhoetanga* (Tūhoe identity and culture) and 'the Crown's responsibility under the Treaty of Waitangi'.[83] In addition to these principles, Tūhoe spirituality is directly provided for in Board decision-making.[84]

---

[74] For the first three years there were eight members, four of which were appointed by Tūhoe trustees and four of which were appointed by the Crown, Te Urewera Act 2014, s 21.

[75] Te Urewera Act 2014, s 18(3).

[76] Te Urewera Act 2014, subpart 4.

[77] Te Urewera Act 2014, Schedule 3(1) and (5).

[78] Te Urewera Act 2014, Schedule 3(2).

[79] Colin Michael Hall, *Wasteland to World Heritage: Preserving Australia's Wilderness* (Melbourne UP 1992) and Juhani Saarinen, 'Wilderness Use, Conservation and Tourism: What Do We Protect and For and From Whom?' (2015) 18(1) Tourism Geographies 1, 3.

[80] For an elaboration of the relevant indigenous concepts, see, eg, Iorns Magallanes (n 13).

[81] Te Urewera Act 2014, s 5.

[82] Te Urewera Act 2014, s 20(1).

[83] Te Urewera Act 2014, s 20(2).

[84] Te Urewera Act 2014, s 18(2).

c) Implementation

The Board released their management plan for Te Urewera, *Te Kawa o Te Urewera*, in 2017 ('Te Kawa').[85] This plan shows a new vision of the rights and responsibilities between humans and the environment. The overriding purpose of Te Kawa is to delineate the dignity and command the respect for nature and to reaffirm the subordinate place of humans within it.[86] As the Introduction notes, 'Te Kawa is about the management of people for the benefit of the land – it is not about land management.' It emphasizes duty and 'responsibility […] not power or rights' throughout the document. It expressly states: '[T]he Board through Te Kawa is disrupting the notion of our false superiority over the natural world […] we are returning to our place in nature, as her child.'[87]

Despite the primary right implemented in the legislation being that of Te Urewera owning its own land, Te Kawa notes that Te Urewera is not defined by property rights. On the contrary:[88]

> 'The use of property rights by the western legal system has hidden from view the concept of nature; rendered her parts as natural resources now capable of rival priorities competing with other household choices. These human granted rights have displaced our devotion for Papatūānuku with ownership now serving individual advantage. These motives suppress the vision of Te Urewera as distinct life. Yet, property rights do not give life nor do they encourage the connectedness of all living things for life. In this altered relationship people struggle to unify to perceive the whole which is nature collectively. Needlessly our fracturing of nature has sponsored our own fragmentation. The use of property rights to regulate human disputes arising from human society is no longer permissible in and of Te Urewera. Te Urewera may never again be owned by people.'

Implementation of the management plan and guardianship is undertaken by Tūhoe guardians on the ground. Traditional Tūhoe elders who understand the forest (or the 'bush' is it is typically referred to) have been training youth to better understand the life-cycle of the forest and its ecosystems – to listen to the voice of Te Urewera. These guardians relate their observations and understandings back to the post-settlement tribal governance body – Te Uru Taumatua – that works between the statutory governors and the traditional guardians; all work together in managing people for the benefit of the land.

Examples have been given of what it means to manage people for the benefit of the land and to make decisions in Te Urewera's interest. The key criterion is to ask whether the decision benefits the land or the people. This has been applied to maintenance of biodiversity, such as through managing the impact of different species, especially what are regarded as pest species. The introduced possum eats birds and their eggs, and eats the new growth on many trees; they have signifi-

---

[85] Te Kawa o Te Urewera: The Te Urewera Management Plan (2017), <http://www.ngaituhoe.iwi.nz/Te-Kawa-o-Te-Urewera>.

[86] Ibid 15.

[87] Ibid.

[88] Ibid 24.

cantly damaged the forest ecosystem and this currently threatens the existence of some species of birds. The Department of Conservation in other national parks, and previously in this one, used to kill possums by dropping '1080' (sodium fluoroacetate) poison pellets over wide areas of countryside because it was cheap and effective. Tūhoe have stopped using the poison and instead hunt individual possums for their meat and fur, reducing possum numbers even if not eliminating them. In a similar vein, they have experimented with sealing gravel roads through the forest with a non-toxic alternative to asphalt.

The key criterion of benefiting the land can also be applied to people and the built environment. For example, some hiking trails and bridges that were washed out in weather events have not been rebuilt because they were said to be in the wrong place for Te Urewera, even if they were benefiting the hikers. The rules for camping have been made more restrictive in order to benefit the land, such as in relation to camping, lighting fires, cutting vegetation, and carrying out any waste. The first Living Building[89] in Aotearoa New Zealand was built for Te Uru Taumatua.[90]

It has not all been smooth operationally, due to internal tensions within Tūhoe over the constitution and operations of Te Uru Taumatua. The tensions have arisen largely around the roles and power of traditional tribal authorities within the statutorily-mandated regime, and thus compared with Te Uru Taumatua. Disputes have ranged from on-the-ground management of Te Urewera, to the selection of the trustees of Te Uru Taumatua.[91] A legal challenge to trustees was taken all the way to the Supreme Court and decided in 2024.[92] This legal dispute has now been resolved, but the tensions between traditional tribal authorities and the new governance bodies established by statute remain.

In summary, the key aspect of this settlement agreement and the Te Urewera Act for Tūhoe has been the recognition of their authority over their traditional lands and territories and their ability to exercise authority through being guardians of their ancestor, Te Urewera. Legal personhood is the tool that has enabled all of this to happen, through removing ownership of Te Urewera from humans. But legal personhood on its own has not been enough; also necessary is the formal establishment of Tūhoe as guardians, both at the Board level and on the ground. But care needs to be taken integrating new governance bodies with traditional tribal authorities and traditions.

---

[89] For more information on the Living Building Challenge see, eg, <https://living-future.org/lbc/>.

[90] See, eg, <https://living-future.org/case-studies/te-kura-whare/>.

[91] See, eg, 'Court favours Tūhoe Collective' (*Eastern Bay*, 9 October 2024), <https://easternbayapp.co.nz/news/articles/67046d28f6a2e500e53489e8>.

[92] *Paki Nikora and Parearau Polly Alice Nikora on behalf of te Kaunihera Kaumātua of Tūhoe v Tamati Kruger on behalf of Tūhoe – Te Uru Taumatua Trust* [2024] NZSC 130 <https://www.courtsofnz.govt.nz/assets/cases/2024/2024-NZSC-130.pdf>.

In terms of rights of nature, the key right of nature here is normally seen as the right of the forest and water body ecosystems to exist and maintain their ecological cycles. However, the approach taken is not of typical, protective conservation; it instead focuses on kincentric guardianship. Perhaps the key right of nature here is the right to have a guardian look after its interests, similar to the right of an elderly relative to have a guardian when they do not have the capacity to act. However, elder care is not typically thought of as a right but as a responsibility of the family members; it is this responsibility that is the priority and main focus of the guardians of Te Urewera. This reinforces why the arrangement is not labelled as a 'rights of nature' approach in Aotearoa New Zealand, but one implementing indigenous rights to be the legal guardians of their nature-ancestor.

3. The Whanganui River – Te Awa Tupua

The Whanganui River, is the longest navigable river in New Zealand, with a length of 290 kilometers, and passes through the traditional territory of a number of Whanganui Māori tribes or *iwi*.[93] The River is central to the local tribal identity, providing 'both physical and spiritual sustenance to Whanganui Iwi from time immemorial'.[94] For Whanganui River tribes the river is their ancestor, and the people are inseparable from the river. The intimate connection to the river underpins their responsibilities of care, protection, management, and use of the Whanganui River.[95] It is this belief that has given rise to the Whanganui proverb, '*Ko au te awa, ko te awa ko au*' (I am the river and the river is me).[96]

As a result of colonization following 1840, the Crown asserted both legal and actual control of the river.[97] The Crown assumed ownership and undertook riverbed works, gravel extraction, and water diversions for a hydroelectric power scheme.[98] Despite this, the Whanganui Iwi claimed that they were still the rightful guardians (*kaitiaki*) of the river and of its life force (*mauri*), and that it should be returned to them, along with the right to control its management.[99] As discussed above, it was the negotiations to resolve these grievances that resulted in the establishment of the river as a legal person.

---

[93] David Young, 'Whanganui Tribes' (*Te Ara – The Encyclopedia of New Zealand*, 22 September 2012, updated 22 March 2017) <https://teara.govt.nz/en/whanganui-tribes>.

[94] Waitangi Tribunal, *Ko Aotearoa Tēnei* (n 7) 269.

[95] Office of Treaty Settlements, *Whanganui Iwi (Whanganui River) Deed of Settlement* (2014); Te Awa Tupua (Whanganui River Claims Settlement) Act 2017 (Te Awa Tupua Act).

[96] See, eg, Te Awa Tupua Act, s 13(c).

[97] Te Awa Tupua Act, s 69(5) and (7).

[98] Te Awa Tupua Act, s 96(9).

[99] Grievances over the river were the subject of petitions to parliament, Royal Commission reports, Waitangi Tribunal claims, and numerous court cases from 1938 until 2010: Te Awa Tupua Act, s 69(5).

a) The river as a legal person

The Te Awa Tupua Act 2017 created a new legal entity for the river in and of itself, Te Awa Tupua.[100] The Act recognizes the indivisible connection between the Whanganui Iwi and the Whanganui River, as well as the river's metaphysical status as a living being.[101] The constitution of the Whanganui River as Te Awa Tupua is broad, effectively defining it by its catchment, and including connected waterbodies and their beds.[102]

The personhood statement is the same as for Te Urewera: 'Te Awa Tupua is a legal person and has all the rights, powers, duties, and liabilities of a legal person.'[103] This endowment of legal rights can include owning property generally, entering into agreements (such as contracts, for example), and enforcing its rights, including via court action should that be necessary. Importantly, decision-makers under 27 other statutes are expressly required to 'recognise and provide for the Te Awa Tupua status' and its values – 'Tupua te Kawa'[104] – when their decision relates to the river or to any activity within the catchment that affects the river.[105]

b) Rights, powers, duties, and liabilities

The first key right of the new legal person is the right to own property and, in particular, to not be owned by anyone else. Therefore, just as with Te Urewera, it owns its own riverbed.[106] As with any other landowner, those rights can be asserted in law. This vesting of the riverbed does not quite fit the traditional Māori cultural conception of the river. One reason it does not fit well is that the riverbed has been separated from the water, and ownership of the water is not included as part of the settlement; the river is thus not legally the indivisible whole that the Whanganui Iwi would have liked.[107] In contrast, a positive aspect is that the whole catchment, including water, is in the definition of Te Awa Tupua, and included for management purposes, even if it does not own it.

A second reason that this vesting does not quite match the traditional Māori cultural conception of the river is that having title to land does not in itself represent traditional Māori concepts of responsibility and control; Māori see themselves as 'users of something controlled and possessed by gods and forebears'.[108] However, such ownership title is said to be as close a fit as can be attained with-

---

[100] Te Awa Tupua Act, s 14.

[101] Te Awa Tupua Act, s 13.

[102] Te Awa Tupua Act, s 7.

[103] Te Awa Tupua Act, s 14.

[104] The intrinsic values of Te Awa Tupua are identified in Te Awa Tupua Act, s 13.

[105] Te Awa Tupua Act, s 15. These statutes are listed in Schedule 2, along with 3 statutes where decision-makers must have particular regard to Te Awa Tupua's status and Tupua te Kawa.

[106] Te Awa Tupua Act, s 41.

[107] Waitangi Tribunal, *The Whanganui River Report* (n 10) 35.

[108] Waitangi Tribunal, *The Whanganui River Report* (n 10) 35.

in the concepts offered by the current legal system. Ownership of the river was 'the heart, the core, and the pith of this [Whanganui iwi] claim',[109] and vesting title to the riverbed in the river itself gives practical effect to the perception of the river as being 'beyond possession'.[110]

The powers to be exercised by Te Awa Tupua are more varied than those exercised by Te Urewera. In comparison with Te Urewera, the legal landscape around Te Awa Tupua is much more complex. This River is not wholly enclosed by (previous) national park boundaries and is instead part of the normal legal landscape, with multiple overlapping laws, decision-makers, and users/stake-holders. It is thus necessary to specify how this legal person can interact with other legislative processes. Interestingly, the statue provides that Te Awa Tupua may be treated as a charitable entity, a public body, and/or a body corporate, depending on the purpose.[111]

The duties and liabilities of Te Awa Tupua are not fully delineated in the legislation, but some liabilities and responsibilities are specifically excluded. The exclusions are for ones that arise in respect of the land and activities that already existed at the time of vesting. Some are specifically identified as remaining with the Crown (government).[112] Another specific exclusion is land tax: Te Awa Tupua riverbed land is exempt from annual land taxes.[113]

c) Guardians

The legislation embraces the Māori principle of guardianship (*kaitiakitanga*) in that legal guardians must be appointed to uphold and protect the interests of the Whanganui River. Two persons 'of high standing', one appointed by the Crown and one appointed collectively by all tribes with interests in the river are to be the river's guardians (*kaitiaki*) and must act in the name of Te Awa Tupua.[114] This office of guardianship, called Te Pou Tupua, must 'promote and protect the health and wellbeing of Te Awa Tupua,' 'act and speak on behalf of Te Awa Tupua', and

---

109 Waitangi Tribunal, *The Whanganui River Report* (n 10) 332.

110 Waitangi Tribunal, *The Whanganui River Report* (n 10) 46.

111 Te Awa Tupua Act 2017, s 17.

112 Te Awa Tupua Act 2017, s 56 and Schedule 5. For example, as per Schedule 5, cl 1: 'The liabilities excluded by section 56 are liabilities for –

(a) remediation of any historical contaminated site on the relevant land if the contamination was caused at any period when the land was held by the Crown:

(b) an existing structure on the bed of the Whanganui River if the structure was established before the date on which the relevant land vested:

(c) an existing activity (whether authorised under a consent or designation, as a permitted activity, or by or under any legislation) if the activity was authorised before the date on which the relevant land vested:

(d) public access to, and use of, the Whanganui River, including for navigation.'

113 Te Awa Tupua Act 2017, Schedule 5, cl 4. Such land taxes are called rates, levied by local authorities in New Zealand.

114 Te Awa Tupua Act 2017, s 18.

act 'in its interests and consistently with Tupua te Kawa [its intrinsic values]'.[115] While the office of Te Pou Tupua embodies a co-governance arrangement, with both tribal and government members, its powers enable stronger legal respect for the environment in line with Māori cosmology.[116]

It is Te Pou Tupua that exercises the powers and undertakes any responsibilities of Te Awa Tupua, such as asserting any rights as landowner and participating in other statutory processes, such as those listed above. They will also liaise with the decision-makers under the 29 other statutes listed in Schedule 2 of the legislation, in order to represent the River and its interests. A key duty of Te Pou is that it:[117]

> 'must develop appropriate mechanisms for engaging with, and reporting to, the iwi [tribes] and hapū [subtribes] with interests in the Whanganui River on matters relating to Te Awa Tupua, as a means of recognising the inalienable connection of those iwi and hapū with Te Awa Tupua.'

The Pou Tupua is supported by an advisory group, Te Karewao,[118] of at least three persons, including one person appointed by the iwi with interests in the Whanganui River (other than Whanganui Iwi) and one person appointed by the relevant local authorities.[119] This group deliberately excludes Whanganui Iwi as they are represented in the Pou who are being advised, although they are expected to be brought in for any matters affecting their part of the river.[120] The Act outlines the powers of Te Karewao, including the power to request and receive 'information or technical advice from any department of State if the information or advice requested relates to that department's functions'.[121] Legal liability of members for their actions as members is expressly limited:[122]

> 'The persons appointed to Te Karewao are not personally liable for any action taken or omission made in their capacity as members of Te Karewao, but only if the action or omission relates to their functions and powers under this Act and they have acted in good faith.'

As well as the small advisory group, there is a large strategy group – Te Kōpuka – that is made up of 'representatives of persons and organizations with interests in the Whanganui River, including iwi, relevant local authorities, departments of State, commercial and recreational users, and environmental groups.'[123] 'The purpose of Te Kōpuka is to act collaboratively to advance the health and well-being of Te Awa Tupua.'[124]

---

[115] Te Awa Tupua Act 2017, s 19.

[116] For more information on the work of Te Pou Tupua, see their website: <https://www.tepoutupua.nz/>.

[117] Te Awa Tupua Act 2017, s 19(2)(b).

[118] Te Awa Tupua Act 2017, s 27.

[119] Te Awa Tupua Act 2017, s 28.

[120] Te Awa Tupua Act 2017, s 28(2).

[121] Te Awa Tupua Act 2017, Schedule 3, cl 7.

[122] Te Awa Tupua Act 2017, Schedule 3, cl 9.

[123] Te Awa Tupua Act 2017, s 29(2).

[124] Te Awa Tupua Act 2017, s 29(3).

In order to implement responsibility and guardianship for the River, the Act provides for the development of *Te Heke Ngahuru*, a 'Whole of River Strategy' that addresses and advances the health and well-being of the Whanganui River.[125] A new strategic vision was launched 15 September 2023.[126] Turama Hawira – one of the Pou guardians – commented:[127]

> 'Te Heke Ngahuru emphasises a community-led approach guided by indigenous values,' says Mr Hawira. 'These indigenous values are known as Tupua Te Kawa, a set of intrinsic values which represent the essence of Te Awa Tupua. These ideas are being closely observed by indigenous and non-indigenous communities worldwide as it paves the way towards a more sustainable future.'

The legislation also establishes *Te Korotete o Te Awa Tupua*, a 'River Fund', 'to support the health and wellbeing of Te Awa Tupua'.[128] It was established with NZD $30 million from the government,[129] but Te Pou can seek and receive other contributions from any source.[130] At earlier stages of the negotiation this was described as a clean-up fund for the river; the Te Awa Tupua Act allows it to be allocated 'on a contestable basis'.[131] This Fund is managed by the Office of the Guardian.

d) Implementation

Since enactment in 2017, Te Pou Tupua have established guiding principles that 'logically guide all interactions in relation to the health and wellbeing of the Whanganui River and its tributaries for the future'.[132] They have undertaken their statutory functions and represented the River in many statutory processes, and with numerous external bodies. Most liaison is with local municipal authorities and with municipal resource planning and permitting processes. One of the first actions of the local City Council was to appoint a staff member within the Council to liaise with the new guardians.

Such involvement was apparently slow to get started. Gerrard Albert – the lead negotiator and the then Chairperson of Nga Tangata Tiaki O Whanganui, the post-settlement governance body for the Whanganui Iwi – commented as follows about the first three years:[133]

---

125 Te Awa Tupua Act 2017, ss 35–38.

126 'Te Pou Tupua Celebrate Launch of Te Heke Ngahuru' (*Te Pou Tupua*, 18 September 2023) <https://www.tepoutupua.nz/post/te-pou-tupua-celebrate-launch-of-te-heke-ngahuru>.

127 Ibid.

128 Te Awa Tupua Act 2017, ss 57–59.

129 Te Awa Tupua Act 2017, s 57(2): 'the Crown contribution'.

130 Te Awa Tupua Act 2017, s 58(3).

131 Te Awa Tupua Act 2017, s 59.

132 See <https://www.tepoutupua.nz/tupuatekawa>.

133 Julia Hollingsworth, 'This River in New Zealand is Legally a Person. Here's How It Happened' (*CNN*, 11 December 2020) <https://edition.cnn.com/2020/12/11/asia/whanganui-river-new-zealand-intl-hnk-dst/index.html>.

> 'The problem with paradigm shifts is that they can take a while to work effectively. And in the three years since the legislation has been in place […] there have been some problems.
>
> For example, the river's guardians weren't consulted about plans for a new bicycle bridge over the river […]. They don't support it as they don't see why the additional development is necessary, but they don't want to kick up a fuss.
>
> […] We don't want to dictate, we want to be part of our community. But we do not want to be downgraded and ignored and used as the kicking board any longer […].'

Despite some such setbacks, Gerrard Albert was also reported as preferring 'diplomacy to threats of legal action – he'd rather foster goodwill with the community'.[134] In the ensuing years, goodwill is exactly what has been fostered, with the primary focus of Te Pou being on education about the paradigm shift required to see the river as a person – as their ancestor – and on relationship-building.[135] Te Pou Tupua see the potential being the recognition both of the river as an indivisible whole, in its own right, and of its descendants and traditional guardians having the authority to act as guardians consistent with their customary laws and practices.[136]

As with Te Urewera, the key right of nature here is the right of the river ecosystems to exist and maintain their ecological cycles. But this legislative settlement recognizes that the people's existence is intertwined with that of the river and that part of their existence is to act as guardians to look after its interests. This is again a focus on responsibilities of the people more than on the rights of the river.

4. Taranaki Maunga

The most recent natural feature to be given legal personhood in Aotearoa New Zealand is the Egmont National Park, containing Mount Taranaki.[137] This is the result of another indigenous reparations settlement agreement for essentially the same types of indigenous grievances as both of the other agreements discussed above.[138]

As with Te Urewera and Te Awa Tupua, the Crown is giving up ownership of the national park and is establishing a new legal entity called Te Kāhui Tupua that will own the land and thus own itself. A guardianship body – Te Tōpuni Kōkōrangi[139] – represents the new legal person. All powers must be exercised in accordance with the intrinsic values of Te Kāhui Tupua, called Nga Pou Whakatu-

---

134 Ibid.

135 See, eg, <https://www.tepoutupua.nz/faqs>.

136 Ibid.

137 Taranaki Maunga Act 2025 (n 60), in force since 2 February 2025.

138 See, eg, the 'Summary of historical account' in Taranaki Maunga Act, s 4. See also the statements made upon introduction of the Bill in Parliament: Hansard, Te Pire Whakatupua mō Te Kāhui Tupua/Taranaki Maunga Collective Redress Bill – First Reading, 9 April 2024 <https://www.parliament.nz/en/pb/hansard-debates/rhr/combined/HansDeb_20240409_20240409_24>.

139 Taranaki Maunga Act, s 27.

pua, which 'are the foundation for the governance, management, and administration' of the new legal person.[140] These values are similar in concept to those of Tupua te Kawa.[141]

Unlike the Te Urewera arrangement, the National Park continues to exist, but with a different official geographic name – Te Papa-Kura-o-Taranaki.[142] Similarly, the mountain that forms most of the park, instead of being called Mount Egmont or Mount Taranaki, will be called by its Māori name, Taranaki Maunga.[143] Further, the National Parks legislation continues to apply, but the body administering it changes to a co-governance arrangement.

One key additional protection is an explicit ban on mining in the park,[144] and certain minerals will be vested in the new legal person, thereby overriding the Crown Minerals Act 1991 in that respect.[145] Additional protections in relation to Crown Minerals are that some decisions that would normally have been made by the Minister of Conservation are now going to be made jointly with the new legal representatives of the park.[146] Another specific protection identified is for the intellectual property rights associated with the name Te Kāhui Tupua.[147]

Te Tōpuni Kōkōrangi will be 'the human face and voice' of the new legal person and will perform the functions in relation to the new national park and under the National Parks Act 1980.[148] Its functions are very similar to those of Te Pou Tupua,[149] but its membership is more like the original Te Urewera board, with eight members, four appointed by the Minister of Conservation, and four appointed by the Iwi trustees.[150] As the manager of a national park, it is a public body, and relevant public statutes apply to Te Tōpuni Kōkōrangi.[151] Yet for the purposes of asset management – if Te Kāhui Tupua comes to own assets – then it must be managed by an asset management company which the guardians are empowered to set up under the Companies Act 1993.[152]

The range of differences and similarities between this Bill and the other two examples of legal personhood illustrate how each arrangement is bespoke and

[140] Taranaki Maunga Act, s 20. These intrinsic values are contained in s 19.

[141] Te Awa Tupua Act, s 13.

[142] Taranaki Maunga Act, s 23(1).

[143] Taranaki Maunga Act, s 23(2).

[144] Taranaki Maunga Act, s 88.

[145] Taranaki Maunga Act, s 61.

[146] Taranaki Maunga Act, s 90.

[147] Taranaki Maunga Act, s 38.

[148] Taranaki Maunga Act, s 28.

[149] Taranaki Maunga Act, s 29.

[150] Taranaki Maunga Act, s 31. These are the trustees of the post-settlement governance entity, Nga Iwi o Taranaki, representing the eight Taranaki Iwi in this agreement. See s 10 for the list of iwi.

[151] For example, the Official Information Act 1982, Ombudsmen Act 1975 and Public Audit Act 2001 apply: Taranaki Maunga Act, s 34.

[152] Taranaki Maunga Act, ss 41–45.

tailored to the wishes of the relevant negotiating iwi, in combination with what the government is politically willing to negotiate at the time.

5. *Influence Elsewhere*

News of New Zealand's legal personhood examples has reverberated around the world and has prompted many calls for it to be used in other countries. The most significant uses have been by judges to help justify their decisions to create legal personhood in their countries. Two examples are those of the Uttarakhand court in relation to the personhood for glaciers and related waters, and the Bangladesh Court in relation to all rivers in Bangladesh. These examples are covered in more detail elsewhere in this book, so I will note them only briefly here.

In March 2017, the Ganges River and the linked Yamuna River were declared by an Indian court to be legal persons.[153] Ten days after this decision, another decision declared related glaciers and other natural features to be legal persons and in need of protection.[154] Both decisions used personhood as a tool for environmental protection, whereby the poor environmental state justified such an extraordinary measure,[155] along with the appointment of guardians to protect these natural features. The second decision directly cited the New Zealand Te Urewera Act 2014.[156] This example was one of several illustrations of what the court termed the 'New Environment Justice Jurisprudence'.[157]

The Colombian Constitutional Court in its 2016 decision on the Rio Atrato[158] did not refer explicitly to the New Zealand examples.[159] However, I am told that they were the subject of much discussion among lawyers, law clerks, and judges as examples of a potential new approach to integrating protection for both nature and indigenous rights. The New Zealand laws did thus not provide a formal, legal influence but served as an informal example of personhood and guardianship to uphold the biocultural rights and responsibilities that were at the heart of that case.

The 2019 Bangladesh decision declaring rivers to have rights as juristic entities and the Water Commission as their guardian did refer explicitly to the New

---

[153] High Court of Uttarakhand 2017, Writ Petition (PIL) No 126 of 2014, *Mohd. Salim v State of Uttarakhand & others*.

[154] High Court of Uttarakhand 2017, Writ Petition (PIL) No 140 of 2015, *Lalit Miglani v State of Uttarakhand & others*.

[155] *Mohd. Salim v State of Uttarakhand* (n 153) [10]: 'The […] Rivers […] are loosing [sic] their very existence. This situation requires extraordinary measures to be taken to preserve and conserve Rivers Ganga and Yamuna.'

[156] *Lalit Miglani v State of Uttarakhand & others* (n 154) 42.

[157] *Lalit Miglani v State of Uttarakhand & others* (n 154) 42.

[158] Constitutional Court of Colombia 10 November 2016, T-622/16, *The Atrato River Case*.

[159] At the time of the *Atrato River* decision, the Te Urewera Act 2014 was law, and the Te Awa Tupua developments had been in the news, whereby a negotiated agreement was in the process of being put into law.

Zealand examples.[160] Indeed, news reports of the decision referred to the examples as though all such decisions were linked.[161]

Many non-judicial commentators have labelled the New Zealand examples as adopting a 'rights of nature' approach.[162] As I have mentioned, this is not the correct label for any of the three examples that New Zealand has adopted to date. These examples have utilized the same tools of personhood and guardianship, but what is most notable is that New Zealand seems to have pioneered the establishment and appointment of guardianship bodies in law. The US approach of stating that nature has rights in order to give standing to people to represent it in court does not establish nor appoint any guardians for nature. It is this US approach that was followed in Ecuador. What is different about the New Zealand – and the Indian, Colombian, and Bangladesh examples – is that guardians are defined and appointed in order to uphold the interests of the new legal entity. This is significant because the focus thus shifts to what we need to do as guardians in order to uphold our responsibilities to protect nature. The establishment of guardians in law therefore doesn't leave the protection of these interests to the chance that a person might want to take a role as guardian on behalf of the river, such as by asserting standing in court. It requires the exercise of this role at all times and allows it to be exercised in other fora, not solely courts. I suggest that we could call this a focus on guardianship responsibilities more than on nature's rights.

V. Guardianship for Nature

New Zealand law provides for guardianship for nature, including for the recognition and/or establishment of guardians as stewards. The focus in this section is on general provisions rather than the specific arrangements for the legal persons discussed above. Of particular note is the recognition in New Zealand law of Māori guardianship over their traditional territories and 'resources' according to Māori traditional laws and customs; the Māori term for this is *kaitiakitanga*. This legal recognition is part of efforts to better uphold Māori rights in legislation and to better incorporate Māori cosmology, especially in relation to the natu-

[160] Bangladesh High Court 3 February 2019, Writ Petition No 13989 of 2016, *Human Rights and Peace for Bangladesh v Government of Bangladesh and other*, translation from Bangla by Mohammad Sohidul Islam, and described in Mohammad Sohidul Islam and Erin O'Donnell, 'Legal Rights for the Turag: Rivers as Living Entities in Bangladesh' (2020) 23 APJEL 160.

[161] See, eg, Danyal Hussain, 'Bangladesh declares its rivers "legal persons" to ensure their waterways are protected, following in the footsteps of Colombia, India and New Zealand' (*Daily Mail Online*, 2 July 2019) <https://www.dailymail.co.uk>.

[162] See, for example, Mari Margil, 'The Standing of Trees: Why Nature Needs Legal Rights' (2017) 34(2) World Policy Journal 8, 9–10; and Craig M Kauffman and Pamela L Martin, *The Politics of Rights of Nature: Strategies for Building a More Sustainable Future* (Massachusetts Institute of Technology Press 2021) 15–17 and 141–161.

ral world and their traditional territories – and in relation to rights of traditional governance over them.[163]

Rights of guardianship are significant for rights of nature approaches because they uphold a relationship between people and nature that does not see nature simply as a resource for human benefit but as holding intrinsic value. Māori customary approaches adopt a kinship relationship whereby each tribe and subtribe have special connections to natural features that are their ancestors and that are thus seen and treated as persons in fact, whether or not they are recognized as such in law. Recognizing indigenous and community guardianship can thus provide practices that are more respectful of nature, in line with those envisaged by advocates of rights of nature approaches.

In addition to Māori customary guardianship, the broader ethic of stewardship has also been recognized in NZ resource management law, where the human responsibilities to the natural world are emphasized. While not based on a kinship relationship with nature, the ethic of stewardship also adopts a more respectful relationship; it can thus be another tool to help implement a rights of nature approach and provide a basis for recognition of more substantive rights for nature in the future. Such community guardianship tools and mechanisms – indigenous and non-indigenous – can be adopted *whether or not* legal personhood for nature has been established.

1. The Resource Management Act 1991

The Resource Management Act was the first legislation to include reference to *kaitiakitanga* in 1991. This Act establishes New Zealand's legislative framework for decision-making and planning regarding the use of the environment.[164] Helpfully, this Act also includes reference to 'the ethic of stewardship' alongside *kaitiakitanga*.[165] The requirement for both of these concepts is that decision-makers

[163] For more information on Māori cosmology and its incorporation in New Zealand law, see, eg, Iorns Magallanes (n 13).

[164] At its core is sustainable management; the Act was the first in the world to implement the 1986 Brundtland concept of sustainable development in environmental management legislation. Brundtland Report, *Our Common Future* (1986). The Act's purpose is 'to promote the sustainable management of natural and physical resources'. Section 5(1). Its definition:

'(2) In this Act, sustainable management means managing the use, development, and protection of natural and physical resources in a way, or at a rate, which enables people and communities to provide for their social, economic, and cultural well-being and for their health and safety while –

(a) sustaining the potential of natural and physical resources (excluding minerals) to meet the reasonably foreseeable needs of future generations; and

(b) safeguarding the life-supporting capacity of air, water, soil, and ecosystems; and

(c) avoiding, remedying, or mitigating any adverse effects of activities on the environment.'

[165] Resource Management Act 1991, s 7; this is the same provision containing reference to the intrinsic values of ecosystems as was discussed above.

have 'particular regard' to *kaitiakitanga*[166] and to the ethic of stewardship[167] when making planning documents and making decisions over permits ('resource consents') for uses of natural resources.

Kaitiakitanga is defined as:

> 'the exercise of guardianship by the tangata whenua[[168]] of an area in accordance with tikanga Maori[[169]] in relation to natural and physical resources; and includes the ethic of stewardship.'

It was expressly envisaged by Parliament that *kaitiakitanga* can be exercised only by traditional Māori groupings and guardians, whereas the ethic of stewardship recognizes a similar responsibility on the part of non-Māori.[170] The Environment Court has commented that:[171]

> 'the provisions about kaitiakitanga, the ethic of stewardship, and the quality of the environment, might be regarded as more or less synonymous – expressing the need for resources to be treated and used with care, and with consciousness of the needs of future generations to have access to them.'

Stewardship thus 'appears to involve a responsible compromise of landowners' dominion over their land'.[172]

This provision has produced positive protection for guardianship interests, both during the planning and development process, and in decisions on resource consents. For example, in relation to *kaitiakitanga*, it was decided early on that, where Māori had an ancestral relationship with a natural resource, before any permit could be issued in respect of that resource, decision-makers must recognize that ancestral link, provide for the maintenance of that relationship, and provide for the tribe's guardianship – in accordance with Māori culture – over that resource in the future.[173] There have, subsequently, been many decisions upholding and providing for a range of aspects of tribal guardianship, even where the land in question is not owned by the tribal guardians.[174]

Application of the (non-indigenous) ethic of stewardship has entailed the creation of Community Liaison Groups as 'local stewards for the area'. This has been

---

166 Resource Management Act 1991, s 7(a).

167 Resource Management Act 1991, s 7(aa).

168 '*[T]angata whenua*, in relation to a particular area, means the iwi [tribe], or hapu [subtribe], that holds mana whenua over that area'; '*mana whenua* means customary authority exercised by an iwi or hapu in an identified area'. Resource Management Act 1991, s 2.

169 '*[T]ikanga Maori* means Maori customary values and practices'. Resource Management Act 1991, s 2.

170 See, eg, *Takamore Trustees v Kapiti Coast District Council* [2003] 3 NZLR 496 [2003] NZRMA 433 [81]–[87].

171 *JARA Family Trust v Hastings District Council* [2015] NZEnvC 208 [38]. The Court said similarly in *Avatar Glen Ltd v New Plymouth District* [2016] NZEnvC 78 [117].

172 *Maniototo Environmental Society Inc v Central Otago District Council* ENC Christchurch, C103/09, 28 October 2009 at [211].

173 See *Haddon v Auckland Reg'l Council* [1994] 3 NZRMA 49 (PT) 52, 57–59.

174 See, eg, Iorns Magallanes (n 13).

done according to the view that 'the ethic of stewardship relates to citizens participating in the careful and responsible management of our natural resources in the interest of long-term sustainability'.[175]

2. *Coastal and Marine Area Guardianship*

There are various ways that customary guardianship (*kaitiakitanga*) rights of Māori over the coastal and marine area are recognized in New Zealand law. The first is the basic right of participation by traditional guardians in defined decision-making processes regarding the coastal and marine area over which they are guardians.[176] Such rights of participation are protected in law in respect of marine reserves, marine mammal sanctuaries, conservation protected areas, concessions, marine mammal standings, and commercial watching permits.[177] The next level is participation in the management of local fishery areas; these can be established where the area is a traditional customary food gathering area of particular spiritual and cultural significance to the Māori group (tribal or subtribal).[178] Management includes the formulation of regulations for the 'conservation and management of the fish, aquatic life or seaweed.'[179]

Fisheries regulations may enable tribal guardians to regulate and enforce customary food gathering according to traditional tribal practices.[180] Such regulations may also establish reserves for Māori customary marine management practices, including food gathering, that exclude commercial fishing.[181] Protected Māori customary rights may also be recognized so that they can be exercised in accordance with traditional guardianship customs.[182] Customary marine title – including title to the foreshore and seabed – can also be declared, so as to recognize the special relationship enjoyed between tribal groups and the marine environment according to their culture and traditions.[183]

3. *Co-governance and Co-management*

Alternative governance and management arrangements have been used successfully to increase the use of practices that better respect our natural world. Indige-

---

[175] See, eg, *Wellington City Council v Wellington Regional Council* ENC Wellington W020/09, 17 March 2009, at 8.3.

[176] These are called 'participation rights'. See Marine and Coastal Area (Takutai Moana) Act 2011, s 47(1)–(2).

[177] Marine and Coastal Area (Takutai Moana) Act 2011, s 47(3).

[178] Fisheries Act 1996, s 174 (NZ). Section 185 of the Fisheries Act provides the power to recommend the making of regulations. See The Ministry for Primary Industries, *Māori customary fishing information and resources* <https://www.mpi.govt.nz/fishing-aquaculture/maori-customary-fishing/maori-customary-fishing-information-and-resources/>.

[179] Fisheries Act 1996, s 185.

[180] Fisheries (Kaimoana Customary Fishing) Regulations 1998, reg 11(1), (2).

[181] Fisheries (Kaimoana Customary Fishing) Regulations 1998, regs 27–29.

[182] See Marine and Coastal Area (Takutai Moana) Act 2011, s 51(e), subpart 2.

[183] Marine and Coastal Area (Takutai Moana) Act 2011, s 58.

nous co-governance and co-management arrangements are increasingly being used in Aotearoa New Zealand to increase Māori participation in decision-making over territories and natural resources and to increase the incorporation of Māori ethics and practices.[184] Thus, for example, even where a government may not be in a position to transfer title to land and resources to indigenous peoples, nor to create a legal person, nor to transfer complete responsibility for its management away from the title holders, the management of those resources can be delegated to and/or shared with indigenous peoples. Conversely, title to land may be transferred back to indigenous peoples, but its management may still be subject to government regulation; its governance and or management can thus be shared between indigenous peoples and government.[185] Such methods and mechanisms can be an effective way to implement rights of nature tools and approaches, whether or not legal personhood for nature has been established.

There is a wide range of types and examples of co-management and co-governance arrangements in Aotearoa New Zealand. Some have been agreed to pursuant to environment- or resource-related legislation and policy; most have been agreed to as a result of indigenous grievance negotiations and settlements.[186] Some have been voluntary agreements between Māori tribal authorities and local government.

The more recent agreements have been able to incorporate Māori cosmology and enable the exercise of *kaitiakitanga* by traditional guardians and the exercise and maintenance of Māori customs (*tikanga*). This exercise has adopted stronger emphases on environmental protection. In one voluntary agreement, this changed emphasis caused some previous practices for public parks to change, including limiting visitors to walk-on only, increasing the number of trails so as to better increase the connection between people and the places, and removing infrastructure especially to better recognize the sacred nature of the places.[187] It provides

---

[184] The term 'natural resources' is used simply by convention; no implications are being made about anthropocentric views of the environment.

[185] For example, in Aotearoa New Zealand there are situations where title to land (including lakebeds) has been returned to Māori as a result of a grievance settlement, but where it is co-managed jointly by Māori and the government and may still be subject to legislative regulation. See, for example, Ngāi Tahu Claims Settlement Act 1998, Ngāi Tahu (Tūtaepatu Lagoon Vesting) Act 1998, Ngaa Rauru Kiitahi Claims Settlement Act 2005, Te Arawa Lakes Settlement Act 2006, Affiliate Te Arawa Iwi and Hapu Claims Settlement Act 2008, Port Nicholson Block (Taranaki Whānui ki Te Upoko o Te Ika) Claims Settlement Act 2009, and Ngatikahu ki Whangaroa Claims Settlement Act 2017.

[186] For examples, see, eg, Catherine Iorns Magallanes, 'Māori Co-governance and/or Co-management of Nature and Environmental Resources' in Roderick Joseph and Richard Benton (eds), *Waking the Taniwha: Māori Governance in the 21st Century* (Thompson Reuters, 2021). This chapter includes as Appendix 2 a table from the Office of Treaty Settlements identifying more than twenty such bodies established pursuant to Treaty Settlements; ibid 338–381.

[187] See, for example, Auckland Council, *Te Mahere Whakakotahi i ngā Whakahaere mō ngā Tūpuna Maunga: Tūpuna Maunga Integrated Management Plan*, available at <www.new.aucklandcouncil.govt.nz>.

a good example of what can be achieved simply through good relations and practices between local government and indigenous guardians.

These guardianship and co-governance examples are not typically thought of as rights of nature examples, but they fit into the broader concepts of Earth Laws that uphold different human relationships with nature. Upholding human responsibilities for nature focuses on the overarching goal of rights arguments and thus should be considered as part of a 'rights of nature' approach.

VI. Conclusion

This chapter summarizes the three rights of nature approaches in Aotearoa New Zealand. The key aspects and conclusions are:

1. New Zealand has adopted three key tools that are part of the rights of nature approach: intrinsic value of nature, legal personhood for nature, and human guardianship of nature.
2. New Zealand has recognized legal personhood three times: for a river, for what was formerly a national park (including forest, rivers, and a lake), and for a mountain.
3. New Zealand law generally has focused on reframing and creating responsibilities for nature, more than recognizing rights of nature. There has been very limited wider recognition of the rights of nature movement, although the sentience of animals has also been recognized in statute.[188]
4. The New Zealand examples have been adopted through Parliamentary legislation, not court decisions. New Zealand's system of parliamentary supremacy creates both an opportunity for and an impediment to adopting rights of nature.
5. While there are no direct rights of nature in a constitutional form, the aspects of rights of nature that have been adopted can be enforced through the New Zealand legal system through a range of mechanisms, such as claims brought by the representatives of legal personalities, judicial review of decision makers, or through resource management and similar statutory processes. They can also be implemented through participation by the representatives in administrative processes.
6. The examples of legal personhood in Aotearoa were developed in order to better uphold Māori rights, particularly to uphold the traditional Māori view that nature is an ancestor. The development of New Zealand's approach has been influenced by decolonization movements and an indigenous Māori worldview.

---

[188] For a discussion of the New Zealand law on animal sentience, see, eg, Catherine J Iorns Magallanes, 'Chapter 22: New Zealand', in Adam Zelle and others (eds), *Earth Law: Emerging Ecocentric Law – A Guide for Practitioners* (Aspen, CO 2021) 597–599.

7. The Māori rights upheld are consistent with the UN Declaration on the Rights of Indigenous Peoples, but the examples did not arise from trying to comply with such international obligations.
8. New Zealand's adoption of legal personhood for nature was not influenced by other legal systems nor by the rights of nature movement. However, it has been informed by the colonial traditions of New Zealand's common law, with a legal precedent for it coming from a UK Privy Council court decision from India.
9. The New Zealand examples of legal personhood for nature have been considered internationally, such as by the initial decision on Ganges and Yamuna River personhood and by the Bangladesh Constitutional Court. To date there has been significant global interest in the examples of legal personhood.
10. It can be said that New Zealand pioneered both the establishment of legal personhood for nature and formal guardianship bodies, where both make important contributions to rights of nature and human responsibilities for it.

I suggest that the above discussion has shown that the debate in Aotearoa New Zealand over the key tools for implementing rights of nature – legal personhood and guardianship – has been based on indigenous rights, not on environmental rights, and has been a result of domestic political imperatives, not international influences. While legal personhood is certainly part of the rights of nature toolbox, rights are not the only way of conceiving it and do not provide the best label for describing the New Zealand developments. Instead, where legal personhood has been adopted in Aotearoa New Zealand, it is better framed as implementing human responsibilities and as a part of *kaitiakitanga* (guardianship) obligations.[189] The rights employed are human rights, particularly the right to be a (legal) guardian of nature.

Yet, the concept of treating the lands as an ancestor, and giving that ancestor legal standing – or, as the Privy Council said, making it a juristic entity – along with establishing a guardianship arrangement in law in order to better look after that ancestor's interests, is exactly the kind of tool that we need to better uphold nature's interests. It establishes in law a different relationship between people and nature, and it suggests a different way of managing that relationship. As Te Kawa o Te Urewera states, it is not about managing the land, but managing the people and how they treat that land.

The legal personhood examples have also entailed a shift in the way the law can allocate property rights by enabling nature to hold inalienable, freehold title effectively to itself.[190] They entail a shift from an external ownership model, which

---

[189] See, eg, Catherine J Iorns Magallanes, 'From Rights to Responsibilities using Legal Personhood and Guardianship for Rivers', in Betsan Martin, Linda Te Aho and Maria Humphries-Kil (eds), *ResponsAbility: Law and Governance for Living Well With the Earth* (Routledge 2019).

[190] Bernardo Filgueira and Ian Mason, *Wild Law: Is there any Evidence of Principles of Earth Jurisprudence in Existing Law and Practice* (UK Environmental Law Association & The Gaia Foundation Research Paper 2009) 5.

prioritizes the uses that humans derive from a resource, to an indigenous kincentric, self-ownership model, which requires no human justification for affording nature protection and guardianship as an ancestor.

This is reinforced by all of the legislative examples protecting the new legal persons for their intrinsic value.[191] This combination of legal personhood and intrinsic worth suggests that they are valued above and beyond any human use to which they might be put. Indeed, the concepts of intrinsic value for nature and guardianship (stewardship) deserve more attention as basic foundations of a more respectful relationship with nature; such respectful relationships could lead to a wider range of substantive rights in the future.

There is significant debate about the best way to protect the environment, and whether people need to be removed from it, or if it is better if respectful connections are maintained, with human uses of nature that are both culturally and ecologically sustainable. The adoption of the legal personhood model in Aotearoa New Zealand marks an acceptance of the perpetual link and spiritual connection between indigenous peoples and their lands, as well as a willingness to uphold, without limitation, the intrinsic values and interests of nature. While the examples discussed in this chapter did not make *all* rivers or national parks in Aotearoa New Zealand legal persons, these developments nonetheless represent significant steps for the protection of both indigenous rights and the environment.

[191] See, eg, Te Urewera Act 2014, s 3(8).

IV. Mimetic Legal Orders

Colombia

Graffiti on the Colombian Legal Architecture: The Emergence of Non-human Subjects and the Rights of Nature

Tania Luna Blanco and David de la Torre Vargas[*]

I. Introduction

In its 21 July 1971 edition, the *New York Times* published the story behind TAKI 183, a 17-year-old who had written his pseudonym[1] in indelible marker across walls, statues, vehicles, and representative objects of New York City, thus giving visibility to the contemporary graffiti movement. His job as a messenger allowed him to etch his mark at some of the city's most iconic locations, such as at John F. Kennedy International Airport, Broadway, some of the most established universities, and, in general, on any surface flat and clean enough to be transformed by his tag. As he went about his deliveries, TAKI gained numerous imitators along the way – Joe 136, BARBARA 62, EEL 159, YANK 135, and LEO 136 – and even replied to his detractors that if the city wouldn't go after those who had

[*] The manuscript was finalized in the middle of 2024. The assessment of literature, jurisprudence, and other source materials reflects the legal and academic context as of September 2024. All translations from Spanish sources are the author's own.

[1] The pseudonym TAKI 183 combines the diminutive of the Greek name Demetrius (Taki) with a reference to 183rd Street, where TAKI lived at the time he started tagging streets, '"Taki 183" Spawns Pen Pals' *The New York Times* (New York, 21 July 1971) 37.

filled subway cars with political advertising, it shouldn't go after a few young people who left their names on the same places.[2]

The graffiti movement transformed cities and introduced a discussion into the public debate on whether these disruptive expressions could be considered art, all while the efforts of those in power to erase the tags that seemed to multiply and flood the city's material and symbolic space became futile. However, graffiti coexisted with more traditional structures and surfaces, some of which even resisted the collective phenomenon and were repainted in immaculate colors repeatedly, to maintain the halo of uniformity that envelops tradition and allows it to sustain itself. In this context, disruptive graffiti tags reminded surfaces of their permanent tension with transformation and heterogeneity in a world that hardly remains static and unalterable.

In Colombia, and in times of the proclaimed revolution of the rights of nature, we are witnessing a contemporary movement with new tags appearing on the most classical buildings of the country's legal architecture,[3] ones that announce the emergence of non-human subjects originating from nature. Though these new subjects face conservation reactions that question their emergence, far from disappearing, they are multiplying. They can be analyzed using the metaphor of the emergence of graffiti in contemporary art to explore a debate that continues to be open and under permanent construction as regards its dialogue with the law.

This article is located in our present, from which some authors have announced a radical and transformative revolution in Colombian law that has triggered the recognition of diverse subjects and modified our political, social, and legal frameworks for understanding nature.[4] We accept the moment of confrontation that the

[2] '"Taki 183" Spawns Pen Pals' (n 1).

[3] 'Architecture is the technique that converts an empty space into a habitable or purpose-oriented space […]. It has always been the desire of jurists to delimit the legal terrain until the (curiously fictitious) elaboration of the modern state was achieved. On the other hand, the idea of "closed" or constructed space provides certainty and security in legal ontology'. José Ramón Narváez Hernández, 'Arquitectura jurídica. Certeza y seguridad en el derecho' (2002) 6 *Revista Telemática de Filosofía del Derecho* 247.

[4] Germán Andrade, '¿El fin de la frontera? Reflexiones desde el caso colombiano para una nueva construcción social de la naturaleza protegida' (2009) 32 Revista de estudios sociales 48; Felipe Clavijo Ospina, 'Una revolución llamada derechos de la naturaleza' in Natalia Castro Niño and Wilfredo Robayo Galvis (eds), *Emergencia climática: prospectiva 2030* (Universidad Externado de Colombia 2020) 647, 651; Juan Felipe García Arboleda and María Cristina Hernández, 'La Corte Constitucional y la transformación del gobierno de los recursos naturales en Colombia: el caso de los mineros tradicionales de Marmato' in Natalia Orduz Salinas (ed), *En La Corte Ambiental. Expresiones ciudadanas sobre los avances constitucionales* (Fundación Heinrich Böll 2018); Javier Alfredo Molina Roa, 'La irrupción del biocentrismo jurídico. Los derechos de la naturaleza en América Latina y sus desafíos' (2016) 6 Ambiente y Sostenibilidad 64; Javier Alfredo Molina Roa, 'Derechos de la naturaleza. Elementos para un debate' in María del Pilar and García Pachón (eds), *Reconocimiento de la naturaleza y de sus componentes como sujetos de derechos* (Universidad Externado de Colombia 2020); Diana Murcia, 'El Sujeto Naturaleza: elementos para su comprensión' in Alberto Acosta and Esperanza Martínez (eds), *La*

law and its different regimes are experiencing through their interaction with classical, modern, and contemporary debates on the subject. However, far from embracing the emancipatory movement of radical social transformation through the law, this paper will describe the intermediate path that the Colombian legal system has taken as a scenario in which fleeting emancipations – events of recognition for non-human subjects from nature – coexist with stabilizing reactions from the most classical structures of the legal system that resist being transformed.

We will then refer to those classical categories on which the main premises of law as a system have been erected,[5] among them the concept of the *persona* and the separation between persons and things in the Romano-Germanic tradition.[6] Based on this construction, only a person can be attributed rights, obligations. or legal relationships, which implies an equivalence between a person and a subject of law that civil codification has broadly expanded with little controversy since the 19th century.[7]

But at the present time, the definition of what is considered a *person*, or a *thing* is very much open to reinterpretation. Countries such as Colombia are witnessing a constant struggle in which diverse actors confront each other in judicial and extrajudicial scenarios to dispute the power of definition around these classic categories, even in these times in which the flow of theories and normativities between the Global North and South feed exchanges that offer possible interpretations to redefine the *human being–nature* relationship, which in turn has different impacts on the grammar of modern constitutionalism[8] and the traditional legal architecture of Colombian law.

In line with this purpose, this article is divided into three main sections. The first presents the reader with an overview of the openings that Colombian law has allowed for the legal recognition of nature since the emergence of Colombian environmental law as a discipline, it has favored early inclusions in its legislation. The second analyzes the fleeting transformations that, in dialogue with international instruments for the protection of human rights and taking advantage of the space opened up by the 1991 Constitution and the constitutional bloc in Colombia, have taken place in the form of an explosion of various legislation and judicial and administrative decisions. This explosion has had an expansive effect,

---

naturaleza con derechos de la filosofía a la política (Abya Yala 2011); Amanda Villavicencio Peña, 'La naturaleza con derechos. De la filosofía a la política' (2012) 34 Crítica Jurídica Nueva Época 205.

[5] Visa AJ Kurki, 'A Short History of the Right-Holding Person' in *A Theory of Legal Personhood* (OUP 2019) 31.

[6] Kurki (n 5); FC de Savigny, *Sistema del derecho romano actual*, vol II (F Góngora y Compañía Editores 1879).

[7] A Guzmán Brito, 'Los orígenes de la noción de sujeto de derecho' (2002) 24 *Revista de Estudios Histórico-Jurídicos* 151.

[8] Daniel Bonilla Maldonado, 'El constitucionalismo radical ambiental y la diversidad cultural en América Latina. Los derechos de la naturaleza y el buen vivir en Ecuador y Bolivia' (2019) 42 Revista Derecho del Estado 3.

mainly for the legal-political recognition of new subjects of law originating from nature, with special emphasis on the *River Atrato* case as a milestone in the discussion that is being raised. Finally, the third section shows that the current scenario should be interpreted with moderate optimism, and it explains that the road ahead remains unstable and uncertain when it comes to the definition or redefinition of the legal relationship between human beings and nature, despite the new normative instruments that are being issued in Colombia. In this way, it will be shown that, beyond the proclaimed revolution of the rights of nature, the more classical legal architecture in Colombia has been graffitied but not radically transformed.

The text concludes by acknowledging the significant revolution surrounding the rights of nature, driven by various judicial decisions in the Colombian context. At the same time, it calls for caution against any totalizing vision that overlooks the nuances, unresolved points of debate, and the ongoing areas of legal and conceptual work that remain open. Thus, the buildings that make up the architecture of our legal system today carry some tags that stand out in the transnational landscape in the style of the first New York graffiti, which reminds us of the relevance of discussing why; despite having witnessed so many regulatory changes in Colombia, we find so few transformations in socio-environmental reality.

II. From the Conservationist Agenda to Green Constitutionalism: Openings on the Colombian Stage

On the Latin American stage, Colombia was early in recognizing its commitment to 'the preservation and restoration of the environment and the conservation, improvement and rational use of renewable natural resources' (Decree Law 2811 of 1974, art 2.1).[9] It did so through the creation of the National Institute of Re-

[9] The background can be traced back to the national forestry legislation passed in 1909 and its subsequent developments initiated during the 1950s, these developments including the recommendation of the Currie Mission (Law 2 of 1959; Law 3 of 1961), which led to the creation of the Natural Resources Division of the Ministry of Agriculture and the first faculties of forestry engineering, biology, and marine sciences in the country. The 1950s were a time when state bodies emerged that would aim to reconcile utility and conservation interests, among them the Corporación Autonoma Regional del Valle del Cauca (Regional Autonomous Corporation of Valle del Cauca, CVC), founded in 1954. Between 1954 and 1968, five other regional autonomous corporations were created in Colombia (Manuel Alberto Rodríguez Becerra, 'El Código Nacional de Recursos Naturales Renovables y de Protección al Medio Ambiente: el conservacionismo utilitarista y el ambientalismo' in Universidad Externado de Colombia, *Evaluación y Perspectivas del Código Nacional de Recursos Naturales de Colombia en sus 30 años de vigencia* (2004) 162). The institutional design that allowed these autonomous corporations to function still exists (Political Constitution, art 150, 7).

newable Natural Resources and the Environment in Inderena in 1968[10] (Decree 2420 of 1968) and a Renewable Natural Resources and Environmental Protection Code, which was issued through the extraordinary powers granted to President Misael Pastrana Borrero by Law 23 of 1973 and which was motivated by the Stockholm Summit of 1972 as a precursor.

While the Stockholm Declaration (see UN, 1973), the creation of the World Commission on Environment and Development in 1983, the World Charter for Nature (1986), and the Brundtland Report (1987), among other efforts, had a resounding impact in environmental matters around the world, Colombia was at the same time initiating a green agenda as an analogue to this international discussion. Prior to this agenda, the Colombian legal system already had mechanisms for the management of some domains – such as the oversight of hydroelectric power plants, irrigation works, and the water quality in urban centers and dams – during the 1960s, which anticipated the Stockholm Declaration, and which were characterized by scattered regulations and administrative procedures.[11]

The space opened by the Renewable Natural Resources and Environmental Protection Code began to generate new normative recognitions for the regulation of the citizen-nature relationship, resulting not only from legal commitments, such as those adopted by the Sanitary Code (Law 9 of 1979) and the National Statute for the Protection of Animals (Law 84 of 1989), but also from regulations, such as those created by Decrees 877 of 1976 (uses of forest resources), 1337 of 1978 (ecological education and environmental preservation), 1415 of 1978 (creation of the Joint Commission on Environmental Affairs), 1608 of 1978 (compensation, restoration, protection, and promotion of wildlife), 1541 of 1978 (regulation of water resources), 1608 of 1978 (management and preservation of wildlife), 1741 of 1978 (special management areas), 1715 of 1978 (landscape protection), 2104 of 1983 (solid waste), and 1594 of 1984 (water uses and solid waste).[12]

---

[10] The acronym Inderena initially stood for Instituto Nacional para el Desarrollo de los Recursos Naturales Renovables (National Institute for the Development of Renewable Natural Resources). As Manuel Rodríguez Becerra recounts, Inderena initially focused on the exploitation of resources, but in 1976, in response to the strong environmental movements that were beginning to develop throughout the world and particularly in Latin America, it changed its name to Instituto Nacional de los Recursos Naturales Renovables y del Medio Ambiente (National Institute for Renewable Natural Resources and the Environment), keeping the same acronym but promoting a broader scheme of environmental protection (Rodríguez Becerra (n 9)). Little information is available on the creation of Inderena in 1968; however, it coincided with a global discussion at the United Nations, condensed in Resolution E/RES/1346 (XLV) of the same year, which focused attention on the so-called problems of 'human environment' and committed to promoting efforts to 'protect and improve the natural surroundings in the interest of man' (p. 8).

[11] Rodríguez Becerra (n 9) 155.

[12] Manuel Alberto Rodríguez Becerra, *Memoria del primer ministro del medio ambiente: 7 de febrero-6 de agosto de 1994, informe al congreso de la República* (República de Colombia, Ministerio del Medio Ambiente 1994).

The green agenda implied a change of focus from the model of resource use that had prevailed – ever since the 19th century – to one that was consolidated in the mid-20th century.[13] In this way, the Colombian state's interest in advancing from a focus on resource exploitation (utilitarian conservationism) to one that prioritized environmental protection was in dialogue with global agendas on economic development and was in line with the opinions and recommendations of the 45th session of the United Nations Economic and Social Council, which incorporated the proposals of the Intergovernmental Conference of Experts on the Scientific Basis for Rational Use and Conservation of the Resources of the Biosphere (Resolution 1346 [XLV], 1968).

Despite the vast legislation on the subject, Colombia encountered a great deal of resistance that emerged from the interaction of the new norms with the existing legal frameworks. Utilitarian conservationism, the vision around which the very idea of the republic was built, attempted to create a degree of nuance with an environmental protection agenda that did not change, in practice, the very vision of ownership and economic exploitation in terms of the use of natural resources.[14]

Nevertheless, generating high tensions during implementation, Colombia engaged with various aspects of environmental public policy at two specific moments of time. The first of these, from 1973 to 1976, opened the way to a commitment to environmental legality, and between 1990 and 1993, the Colombian state underwent a new boom cycle, this time with the entry into force of the 1991 Constitution and the inclusion of around 50 articles which, individually and in systematic relation to others, represent the state's commitment to an ecological constitution.[15]

The Colombian Constitutional Court, the body authorized to interpret the 1991 Constitution (art 241), emphasized from the outset that our constitutional order is composed of a triptych, taking into account that at least three constitutions coexist within the constitutional text: a social, an economic, and an ecological one; the latter posits a vital connection between human beings and nature that translates into an important pillar of the Colombian constitutional order. However, despite the 1991 Constitution's significant recognition of the environment, whereby the environment is even understood as a right and a collective good (arts 79, 80, 95.8, and 333), as well as the institutional framework created by Law 99 of 1993 (National Environmental System [SINA]), the tensions derived from the

[13] Rodríguez Becerra (n 9) 161.

[14] C Ruiz Páez, *El pensamiento ecológico de Simón Bolívar* (Academia Boyacense de Historia 1994); Isaías Tobasura Acuña, 'El legado Ecológico del Libertador' (2011) 32 Revista Luna Azul 135.

[15] Manuel Alberto Rodríguez Becerra, '¿Hacer más verde al Estado colombiano?' (2009) 32 *Revista de Estudios Sociales* 18, 19; Constitutional Court of Colombia 17 June 1992, T-411/1992, 28 June 2001, C-671/2001, 27 July 2010, C-595/2010, 24 August 2011, C-632/2011 and 5 March 2014, C-123/2014, among others.

triptych of society, economy, and nature remain. The constitutional green state, even with an abundance of legislation that is recognized as a guarantor, stands in contrast with the numerous socio-environmental conflicts that the country is experiencing.[16] It is necessary to emphasize that regulation does not operate in a vacuum and that the normative stakes that states develop interact with pre-existing visions. In Colombia, the utilitarian conservationist vision coexists with a vision of environmental protection that, despite the stabilizing reactions to which this text will refer, has been conquering more and more space for nature.

The following section presents some of the most representative emergences of non-human subjects that have occurred in the Colombian legal order so as to question public policy and environmental regulations; they denounce an exceedingly superficial commitment as well as the adoption of an anthropocentric vision, according to which nature must be protected only to the extent that it is useful for human beings. However, it will be shown that new discourses have begun to question this prevailing vision, making use of all kinds of constitutional and legal actions to redefine the person-nature relationship, from slogans that demand giving these subjects a greater role as protagonists to even redesigning the legal system to allow for full recognition.

III. Fleeting Emancipations and Partial Recognitions: New Tags on the Traditional Buildings of Colombian Legal Architecture

One of the first graffiti tags to appear on traditional Colombian legal architecture emerged on 10 November 2016. On that day, the Colombian Constitutional Court ruled 'TO RECOGNISE the River Atrato, its basin and tributaries as an entity subject to rights'.[17] Until that time, the River Atrato was classified as an asset of public use and property of the nation, according to the Renewable Natural Resources and Environmental Protection Code and the Colombian Civil Code (Laws 57 of 1887 and 153 of 1887).

The Constitutional Court's qualification of the River Atrato was unprecedented in its case law. In the terms used by Geels and Schot,[18] this marked a *disruptive change*[19] that generated a transformation with respect to the traditional forms of law on the basis of which Colombia's classical legal institutions were built: (a) the di-

[16] Comptroller General of the Republic, 2020.

[17] Constitutional Court of Colombia 10 November 2016, T-622/16, *The Atrato River Case*, 5.6.

[18] Frank W Geels and Johan Schot, 'Typology of sociotechnical transition pathways' (2007) 36 Research Policy 399.

[19] 'Disruptive change corresponds to changes that occur infrequently, develop gradually, but have a high-intensity effect in one dimension' (Geels and Schot (n 18) 404).

vision between subjects and objects;[20] (b) the anthropocentric approach;[21] and (c) the subjugation of things – including those of public use – to the concept of property.[22]

However, this graffiti on the traditional legal architecture did not emerge spontaneously or as an isolated event. Although no tag had yet appeared that showed signs of the emergence of a new subject, voices were already announcing its arrival. This is why there are references in Colombian jurisprudence, prior to the *River Atrato* case, to the possibility of recognizing the rights of nature.[23]

For example, in 2011, the Constitutional Court's Judge Eduardo Mendoza Martelo stated that 'nature is not only conceived as the environment and surroundings of human beings, but also as a subject with rights of its own, which, as such, must be protected and guaranteed'.[24] Another decision that the *River Atrato* case recognizes as transcendental in opening the way for the recognition of the rights of nature is Ruling T-080[25] (with the ruling of Jorge Iván Palacio, the same judge who delivered the opinion in the *River Atrato* case), which analyzes the case of the Lorsban dumping in the Bay of Cartagena and provides clues on understanding the case's history under the influence of Andean constitutionalism:

> 'Along these lines, constitutional jurisprudence has considered ancestral knowledge and alternative currents of thought and has come to hold that "*nature is not only conceived as the environment and surroundings of human beings, but also as a subject with rights of its own, which, as such, must be protected and guaranteed*".'[26]

For its part, the Council of State used this wording in 2013 in a ruling that – although later overturned via a *tutela* action[27] (a constitutional injunction in Co-

[20] The Civil Code evidences this separation in its first two books, dedicated to persons and property, respectively.

[21] Elizabeth Jane Macpherson, 'Derechos constitucionales, derechos humanos, derechos indígenas: el lado humano de los derechos de la naturaleza' in María del Pilar and García Pachón (eds), *Reconocimiento de la naturaleza y de sus componentes como sujetos de derechos* (Universidad Externado de Colombia 2020).

[22] Ugo Mattei, *Bienes comunes. Un manifiesto* (Editorial Trotta 2013).

[23] Javier Alfredo Molina Roa, *Derechos de la naturaleza. Historia y tendencias actuales* (Universidad Externado de Colombia 2014); Javier Alfredo Molina Roa, *Los derechos de los animales. De la cosificación a la zoopolítica* (Universidad Externado de Colombia 2018).

[24] Constitutional Court of Colombia 24 August 2011, C-632/2011, 8.10.

[25] Constitutional Court of Colombia 20 February 2015, T-080/2015.

[26] T-080/2015 (n 25) 5.2.3, italics in the original. The ruling referred to includes the following footnote: 'Constitutional Court, Ruling C-632 of 2011. Position in harmony with Latin American and aboriginal constitutionalism that has shown a special sensitivity to the "pacha mama": "In Ecuador and Bolivia, the political divorce between the government and the indigenous peoples has its origin in the contradiction between an extractivist economic policy and the sumak kawsay or suma qamaña ("good living") regime recognised in the Constitution, thought of as an alternative development scheme." This, however, does not prevent the official policies of these governments from sometimes running counter to the principles they preach.'

[27] The *tutela* action is a constitutional mechanism in Colombia that allows individuals to request immediate judicial protection of their fundamental rights. It functions as a form of constitutional injunction.

lombian law) – extended the category of subject of rights[28] to animals and even to plant species and water resources.[29]

However, as mentioned, it was not until October 2016 that the category of subject of rights was used in a decision of the Constitutional Court in relation to the River Atrato, which will be explained in more detail below.

IV. The Pioneering *River Atrato* Case: Our TAKI 183?

The trial that led to the *River Atrato* ruling took place within the traditional legal architecture and made use of its tools, but it ended up making a disruptive mark on the most classical buildings of the Colombian legal system.

Communities that have historically suffered from the complicated problem of contamination and degradation of the River Atrato (located in the department of Chocó) filed a tutela action to stop the intensive and large-scale use of various illegal mining and logging methods carried out with heavy machinery (dredgers and backhoes) and highly toxic substances (such as mercury) in the River Atrato

[28] In Colombian constitutional law, the "category of subject of rights" refers to the legal recognition of entities – traditionally human beings, but increasingly extended to animals, ecosystems, or rivers – as holders of fundamental rights. This conceptual development stems from an evolving interpretation of constitutional protections under the 1991 Constitution, influenced by Andean constitutionalism and Latin American legal pluralism.

[29] Administrative Litigation Chamber, Section 3, Subsection C, Judgment of 26 November 2013, Case No. 25000-23-24-000-2011-00227-01(AP), Opinion by Gil Botero, J: 'Having made the above philosophical, ethical and legal journey, it is up to the Chamber to establish its position in relation to the following legal problem: Is it possible to sustain the existence of autonomous and direct rights of animals and plant species in Colombia? In the opinion of this Court, in the terms of Article 4(c) of Law 472 of 1998, there is an express recognition by the legislator of direct and autonomous rights in favour of animals and plant species in our national territory. This purpose is unquestionable if we analyse the new regulations contained in Law 1,638 of 2013, which prohibits the use of wild animals in circuses or circus shows; that is, once again the legislator has assigned rights to living beings other than human beings, specifically, the rights not to be mistreated and not to live in precarious conditions. Thus, for the Colombian legislator, animals and plant species (eg forests, the Amazon, páramos, water sources and resources etc.) are subjects of rights and, therefore, through a popular action any person can request their protection acting as an unofficial agent of these entities, though it is not possible to affirm that this is a collective subjective right belonging to society. On the contrary, this represents the express recognition by the constituent and legislator to attribute value to animals and plant species, for which, in each specific case, the judge must make a ruling of weighting or proportionality between the competing or opposing interests. In other words, it is not necessary for animals or plant species to be considered legal or moral persons for them to be able to appear before the administration of justice to seek protection of their rights recognized by the community itself, as the appropriate constitutional procedural mechanisms exist to guarantee the protection of the respective rights' (3.4).

as well as in its basins, marshes, wetlands, and tributaries.[30] The plaintiffs asked the constitutional judge to issue a tutela to protect the fundamental rights to life, health, water, food security, a healthy environment, culture, and territory of the plaintiff ethnic communities, and, consequently, to issue a series of orders and measures that would allow for the articulation of structural solutions for the serious health, socio-environmental, ecological, and humanitarian crisis in the Atrato river basin along with its tributaries and surrounding territories.[31]

In this case, the Constitutional Court decided to grant the actors protection of their fundamental rights by adding a creative order in the fourth paragraph of the 'resolutions' section: 'TO RECOGNISE *the River Atrato, its basin and tributaries as an entity subject to rights* of protection, conservation, maintenance and restoration by the State and ethnic communities'.[32]

To implement this decision, the national government, together with the ethnic communities living in the Atrato river basin, was ordered to exercise guardianship and legal representation in respect of the river's rights; they were called *guardians of the river*. They were given the task of acting, in turn, within the guardians' commission, together with an advisory team also made up of the Humboldt Institute and WWF Colombia (World Wildlife Fund). They were also able to receive support from other public and private institutions; universities (regional and national); academic and research centers working in natural resources; and environmental (national and international), community, and civil society organizations that wished to be involved in the project to protect the River Atrato and its basin.

From this perspective, the court's ruling recognizes rights of the river and orders both the creation of a group to represent it as well as a panel of experts who are to guide the implementation of the ruling together with the community. Attributing subject-of-rights status to a river implies, in the first place, that a new approach to its protection is required. The recognition of a new subject also gives

[30] 'In the case of Chocó, the history of the settlement of rural areas, mainly during the 19th century, allows us to understand the type of dispersed distribution along the rivers that characterise the region. The settlements make the river a central space in all the economic, domestic and socio-cultural activities of the local inhabitants. The river even constitutes the main factor of cultural identity of this region' (*River Atrato* case (n 17)).

[31] The case aroused the interest of different organizations and entities that presented opinions (twenty-six) at the request of the court or that intervened as citizens (seventeen), including: Universidad Nacional de Colombia, Faculty of Natural Sciences; Asociación Nacional de Empresarios de Colombia (ANDI); Instituto Humboldt; Universidad de los Andes, Department of Anthropology; Pacific Environmental Research Institute; FIAN International; Universidad de Antioquia, Faculty of Anthropology; Intervention of Universidad del Rosario, Faculty of Natural Sciences; Interamerican Association for Environmental Defense; Asociación Colombia de Minería; Universidad Tecnológica del Chocó; Universidad de Antioquia, Instituto de Biología; Instituto Colombiano de Antropología e Historia; Universidad Nacional, Faculty of Anthropology; Fundación Activos Culturales Afro; Amicus Curiae of the Ombudsperson's Office; Centro de Estudios para la Justicia Social (Dejusticia).

[32] *River Atrato* case (n 17), the underlined portion ('as an entity subject to rights') appeared in bold in the original ruling; the additional italics have been added by the authors.

evidence of a change in the legal tradition whereby rights are considered as an exclusive prerogative of individuals, understood from the classical vision promoted by Roman law. This new approach highlights collective action and prompts several questions: How did the Constitutional Court arrive at this determination? What were the reasons for leaving a challenging message in the traditional legal architecture, much in the same way as TAKI 183? To answer these questions, we will analyze the reasoning used in the *Atrato River* case.

V. The Reasoning Behind the Measure: Theoretical Background

The Constitutional Court begins its argument by stressing the importance of the environment for the 1991 political charter and pointing out its character as an overriding interest. To demonstrate its relevance in the constitutional order, the Court refers to the concept of an *ecological constitution* and affirms that a healthy environment has a tripartite constitutional dimension: as a principle, as a fundamental and collective right, and as an obligation falling upon the authorities, society, and individuals.

The *Atrato River* case lays the foundations for the introduction of new theoretical perspectives based on the pluralist approach of the political charter itself and pointing to a dynamic and permanent relationship between the Constitution and the environment. At the same time, it stresses the possibility of 'varying or reformulating the traditional paradigms on which the relationship between people and their environment has been structured', and it points out that, given the environmental deterioration facing the planet, measures must be implemented that it characterizes as 'serious and stricter'.

Similarly, with the support of previous rulings, the ruling points out that it is possible to establish at least three theoretical approaches to protecting the environment: *anthropocentric*,[33] *biocentric*,[34] and *ecocentric*.[35] Justifying the thrust of the latter view, it argues that nature should be considered as a subject of rights and a living entity composed of multiple forms of life.

The reference to the ecocentric approach was the main foundation for declaring the River Atrato as a subject of rights. This approach had been used as a concept

---

[33] The Constitutional Court defines this as an approach which 'conceives of the human being present as the sole raison d'être of the legal system and natural resources as mere objects at the service of the former' (*River Atrato* case (n 17) 5.6).

[34] Defined in the ruling as the vision that 'vindicates more global and supportive conceptions of human responsibility, which advocate [in equal measure] humankind's duties towards nature and future generations' (Constitutional Court of Colombia 7 May 2002, C-339/2002, 5.8).

[35] Such approaches 'conceive of nature as an authentic subject of rights [and categorize nature as] a living entity composed of multiple other forms of life and cultural representations' (*River Atrato* case (n 17) 5.10).

in constitutional jurisprudence prior to the *Atrato River* case,[36] and it has continued to be referred to in the jurisprudence of the Constitutional Court.[37] By means of an argumentative marriage between the ecocentric approach and the recognition of biocultural rights, the ruling provides an opportunity to redefine the relationship Colombians have with nature, leaving behind the classical concepts of property law and the frameworks for understanding of environmental law:

'the greatest challenge facing contemporary constitutionalism in environmental matters consists in safeguarding and effectively protecting nature, the cultures and forms of life associated with it and biodiversity, not simply because of the material, genetic or productive utility that these may represent for human beings, but because, as a living entity composed of other multiple forms of life and cultural representations, they are subjects of individualisable rights, which makes them a new imperative of comprehensive protection and respect on the part of states and societies.'[38]

In terms of biocultural rights – one of the central elements of the ruling – the Court provides a comprehensive definition that is central to the *River Atrato* case:

'So-called biocultural rights, in their simplest definition, refer to the rights that ethnic communities have to administer and exercise tutela autonomously over their territories – in accordance with their own laws, customs – and the natural resources that make up their habitat, where their culture, traditions and way of life are developed based on the special relationship they have with the environment and biodiversity.'[39]

Based on this new understanding proposed by the ruling, environmental law must be nourished by knowledge from outside of the law, so that it can understand nature beyond the possibilities offered by the law itself. However, in order to link the concept to formal sources of law and thus give legitimacy to its argument, the Court cites Ruling T-955/2003,[40] in which Convention 169 of the International Labour Organization (ILO) was interpreted broadly, and it emphasizes that the concept of *biocultural rights* is a legal device that results from the set of collective rights of ethnically differentiated communities in recognition of their bio-

[36] The concept was developed by the Constitutional Court in 24 August 2011, C-632/2011, C-123/2014 (n 15), T-080/2015 (n 25), 21 September 2015, T-606/2015, and 18 November 2015, C-699/2015.

[37] The ecocentric approach has been developed in the Constitutional Court of Colombia cases of 1 February 2017, C-041/2017, 18 October 2017, C-644/2017, 23 May 2018, C-048/2018, and 8 May 2019, C-186/2019. In the latter case, the Constitutional Court held: 'In light of the transcribed consideration, the anthropocentric and ecocentric conceptions differ in that the former is based on the idea that human beings, being superior to the rest of nature, are considered legitimate masters of it and, to that extent, value is attributed to it according to its contribution to the sustainability of human life. In contrast, the ecocentric view assumes that nature has an inherent and intrinsic value, regardless of whether it is useful to human beings, such that it acquires relevance in its own right' (C-186/2019, 3.4.1.1.).

[38] *River Atrato* case (n 17) 5.2. In this regard, the ruling relies on another court decision, also written by Judge Jorge Iván Palacios (Constitutional Court of Colombia 16 July 2015, C-449/2015).

[39] *River Atrato* case (n 17) 5.2.

[40] Constitutional Court of Colombia 17 October 2003.

logical and cultural diversity,[41] derived from the constitutional protection of arts 7, 8, 79, 80, 330, and 55 (transitional provisions) of the 1991 Political Constitution.[42]

The basis for the recognition of the rights of nature in general and of rivers in particular in the decision of the Colombian Constitutional Court is inscribed within different theoretical currents that are grouped within the so-called ecocentric paradigm.[43] The theoretical bases for the recognition of the rights of nature have important precedents in the approach taken by Christopher Stone in 1972 and in Justice Douglas's dissenting opinion of the same year in the United States Supreme Court's decision in *Sierra Club v Morton*.[44] The approaches pioneered in Latin America by Godofredo Stutzin[45] have also been highlighted.

Significant contributions have also been identified for the support of the rights of nature in the *new Latin American constitutionalism*;[46] in the principle of good

---

[41] On this point, the Court uses the body of constitutional law in Colombia to argue about the space that international human rights law has opened up to address the relationship between biological and cultural diversity (*River Atrato* case (n 17)).

[42] For the recognition of biocultural rights, it also turned to the writings of Kabir Sanjay Bavikatte and Tom Bennett Bavikatte, 'Community Stewardship: The Foundation of Biocultural Rights' (2015) 6(1) Journal of Human Rights and Environment 7. The scope and importance of biocultural rights has also been developed by Javier Gonzaga Valencia Hernández, 'El derecho a la participación en las decisiones ambientales: un enfoque desde la justicia ambiental y la justicia constitucional' in Natalia Orduz Salinas (ed), *La Corte Ambiental. Expresiones ciudadanas sobre los avances constitucionales* (Fundación Heinrich Böll 2018) and Jessica Paola Melo Parra, 'Vínculo entre la Constitución Cultural y la Constitución Ecológica' in Natalia Orduz Salinas (ed), *La Corte Ambiental. Expresiones ciudadanas sobre los avances constitucionales* (Fundación Heinrich Böll 2018).

[43] Alcides Antúnez Sánchez and Eduardo Díaz Ocampo, 'El pluralismo jurídico y los derechos a la Pachamama' (2018) 20(27) DiXi Revista, <https://doi.org/10.16925/di.v20i27.2398>; Myriam Ximena Arenas Orbegozo, 'Who Are We After Worlds Collide. Explorations into the Sources of sui generis Legal Persons in Colombia: The Atrato River' (Master Thesis, Tilburg University 2019) <https://arno.uvt.nl/show.cgi?fid=149409>; Stefan Knauß, 'Conceptualizing Human Stewardship in the Anthropocene: The Rights of Nature in Ecuador, New Zealand and India' (2018) 31 Journal of Agricultural and Environmental Ethics 703; Carolina Valladares and Rutgerd Boelens, 'Extractivism and the Rights of Nature: Governmentality, "Convenient Communities" and Epistemic Pacts in Ecuador' (2017) 26(6) Environmental Politics 1015.

[44] Christopher D Stone, *Should Trees Have Standing? Law Morality and the Environment* (OUP 2010); *Sierra Club v Morton*, 405 U.S. 727 (1972).

[45] Godofredo Stutzin, 'Un imperativo ecológico: reconocer los derechos de la naturaleza' (1984) 1(1) Ambiente y Desarrollo 97.

[46] Silvia Bagni and Giorgia Pavani, 'La comparación jurídica como elemento epistemológico para el estudio y la investigación en el ámbito de los derechos de la naturaleza, la interculturalidad y los procesos de paz' in Manual Alberto Restrepo Medina (ed), *Interculturalidad, protección de la naturaleza y construcción de paz* (Universidad del Rosario 2019) 67; Bonilla Maldonado (n 8); Liliana Esupiñán Achury, 'Neoconstitucionalismo ambiental y derechos de la naturaleza en el marco del nuevo constitucionalismo latinoamericano. El caso de Colombia' in Liliana Estupiñán Achury and others (eds), *La naturaleza como sujeto de Derechos en el Constitucionalismo Democrático* (Universidad Libre 2019); Rubén Martínez Dalmau, 'Fundamentos para el reconocimiento de la naturaleza como sujeto de derechos' in Liliana Estupiñán

living[47] and the rights of mother earth;[48] in the Earth Jurisprudence movement;[49] in the aspects that defend the conception of nature as a system (system thinking) within the natural sciences, as in Jamie Murray;[50] and even in justifications regarding the origin and necessity of recognizing rights of nature based on the concept of natural contracts.[51]

The declaration which established the River Atrato as a subject of rights opened a relevant legal discussion about the basis and scope of these rights. However, many Colombian judges, noticing the deficiency of traditional instruments for the defense and management of the environment in practice, took up the approaches of *River Atrato* case and, in search of ecological protection in their different territorial spheres, replicated the main notes of the declaration establishing rivers as subjects of rights.

VI. Graffiti Calling for a Redirection of Approaches and Ways of Relating to the Environment

Characterized by the explicit consecration of an ecocentric approach, as well as by the recognition of multicultural perspectives derived from the constitutional interpretation of multiculturalism and ethnic diversity in Colombia, the acknowledgment of non-human elements of nature as legal subjects has opened the way for new legal frameworks and rights-based environmental approaches – although the debate remains far from settled.[52]

Achury and others (eds), *La naturaleza como sujeto de Derechos en el Constitucionalismo Democrático* (Universidad Libre 2019); Eugenio Raúl Zaffaroni, *La Pachamama y el humano* (Ediciones Madres Plaza de Mayo 2011).

47 Nilda Garay Montañez, 'Los bienes comunes y el buen vivir en el constitucionalismo latinoamericano: de las exclusiones históricas a las inclusiones cuestionadas' (2018) 21(42) Revista Prolegómenos. Derechos y Valores 135.

48 Alberto Acosta, 'Construcción constituyente de los derechos de la naturaleza. Repasando una historia con mucho futuro' in Liliana Estupiñán Achury and others (eds), *La naturaleza como sujeto de Derechos en el Constitucionalismo Democrático* (Universidad Libre 2019); Alberto Acosta, 'Los derechos de la naturaleza: una lectura sobre el derecho a la existencia' in Alberto Acosta and Esperanza Martínez (eds), *La naturaleza con derechos de la filosofía a la política* (Abya Yala 2011).

49 C Cullinan, 'If Nature Had Rights' (2008) 27(1) *Orion Magazine* 26; Oliver A Houck, 'Noah's Second Voyage: The Rights of Nature as Law' (2017) 31(1) Tulane Env LJ 1 <https://journals.tulane.edu/elj/article/view/2394>; Jamie Murray, 'Earth Jurisprudence, Wild Law, Emergent Law: The Emerging Field of Ecology and Law – Part 1' (2014) 35 Liverpool LRev 215; Jamie Murray, 'Earth Jurisprudence, Wild Law, Emergent Law: The Emerging Field of Ecology and Law – Part 2' (2015) 36 Liverpool LRev 105 (2015).

50 Murray, Part 2 (n 49).

51 Michell Serres, *El contrato Natural* (Pre-Textos 2004).

52 We understand non-human subjectivities as being the result of a process of subjectivation through which we make overall advances in the social and cultural construction of a subject which is an alternative to the human person and which can be recognized by law.

Some examples of the cascading reaction generated by the *River Atrato* case come not only from judicial provisions in the High Courts but also from trial judges and administrative authorities at the national, departmental, and municipal levels, individuals who have seen in regulatory power the possibility of recognizing nature as a subject in specific territorial contexts. For example, the Colombian Supreme Court of Justice classified as a subject of rights the Amazon;[53] Salamanca Island Park;[54] and Los Nevados National Natural Park.[55] And its Criminal Cassation Chamber indicated that it would accept the declaration of *subject of rights* made by the Civil Cassation Chamber in relation to the Amazon.[56] Echoing this declaration, courts and tribunals of different categories and specialization have recognized different ecosystems as such, including the Pisba páramo (alpine tundra ecosystem);[57] the Rivers Cauca,[58] Combeima, Cocora, Coello,[59] Otún,[60] La Plata,[61] Magdalena,[62] Pance,[63] Quindío,[64] and Fortalecillas;[65] Lake Tota;[66] and the Las Hermosas Páramo Complex.[67]

For its part, the Special Jurisdiction for Peace (JEP) concluded that the ancestral territory (Katsa Su) should be recognized as a victim of ongoing armed conflict.[68] Similarly, the Chamber for the Recognition of Truth, Responsibility and Determination of Facts and Conduct accredited as victims the territories of the

---

[53] Supreme Court of Justice of Colombia 5 April 2018, STC 4360-2018, *Amazon* case.

[54] Supreme Court of Justice of Colombia 17 June 2020, STC 3872-2020.

[55] Supreme Court of Justice of Colombia 25 November 2020, STL 10716-2020.

[56] Supreme Court of Justice of Colombia 4 December 2019, AP5169-2019, relying on STC 4360-2018 (n 53).

[57] Administrative Court of Boyacá 9 August 2018.

[58] Medellin State Superior Tribunal 17 June 2019, 2019-076, *Cauca River* case: 'Three: TO RECOGNISE the River Cauca, its basin and tributaries as an entity subject to rights to protection, conservation, maintenance and restoration by EPF and by the State'.

[59] Administrative Court of Tolima 30 May 2019, *Tolima Rivers* case: 'FOUR: TO RECOGNISE the Rivers Coello, Combeima and Cocora, their basins and tributaries as entities subject to rights to protection, conservation, maintenance and restoration by the State and the communities'.

[60] Juzgado Cuarto de Ejecución de Penas y Medidas de Seguridad de Pereira, Regional Court 11 September 2019, 036/2019 (tutela).

[61] Single Municipal Civil Court of La Plata 19 March 2019 (tutela).

[62] Juzgado Primero Penal del Circuito con Funciones de Conocimiento Neiva–Huila 24 October 2019, 071/2019 (first-instance tutela). In the second instance, the proceedings were declared invalid.

[63] Juzgado Tercero de Ejecución de Penas Y Medidas de Seguridad 12 July 2019, 31/2019 (tutela).

[64] Administrative Court of Quindío 5 December 2019.

[65] First Criminal Circuit Court of Neiva-Huila, 20 May 2021. In the second instance, the tutela action was declared inadmissible.

[66] First Civil Court of Sogamoso 27 November 2020, ST-0047.

[67] Superior Court of the Judicial District of Ibagué 15 September 2020.

[68] JEP, Chamber for the Recognition of Truth, Responsibility and Determination of Facts and Conduct, Order SRVBIT-079.

Afro-Colombian organizations Recompas and Barbacoas[69] and also the world-territory of the Eperara Siapidaara people, Eperara Euja.[70]

The national government has issued, among others, Decrees 1148 of 2017, and 1597, 1732, and 1289 of 2020, which designate the respective representatives of those natural entities that have been recognized by the judges as subjects of rights. Decree 348 of 15 July 2019, issued by the governor of the department of Nariño, lists the department's strategic ecosystems and protected areas as subjects of rights, and, although it did not materialize, a draft legislative act was passed in Congress to amend the Colombian Political Constitution and recognize that 'nature, as a living entity subject to rights, shall enjoy the protection and respect of the State'.[71]

Focusing on the rulings that address the declaration of elements of nature as subjects of rights, we find that all of them make explicit reference to the *River Atrato* case, which they recognize as the originator of these decisions.

The rulings that recognize rights of nature are not mere replicas of the *River Atrato* case; they adapt the declaration of rivers as subjects of rights to the specific protective needs of each case, thus showing that judges find such a determination useful. That said, it is not possible to speak today of a radical constitutional move to recognize nature as a subject of rights in the Colombian legal system,[72] but only of fleeting emancipations. The creativity of litigants in using the tutela action (Political Constitution 1991, art 86) should also be highlighted, with litigants defending the validity of the tutela action and discarding popular actions (art 88) as the essential protection mechanism. Such an approach recognizes the intimate relationship between the collective rights violated and the fundamental rights of human beings, including those belonging to future generations, even when a reference to the very category of subject of rights is missing in court decisions. For many, evading the debate is not a minor issue, instead translating in practice into a symbolic recognition that affects the material possibilities of compliance with the ruling, as some dissenting opinions have warned:

'It is not enough to make general statements and provide answers of the same nature. This would constitute mere emotion and general knowledge or information leading to the issuing of orders that will surely not be complied with. What is instead required is concrete investigation and issuing concrete decisions supported by reliable and clear evidence, whose compliance can be demanded. I ask myself in this case, how is the tutela judge going to act when the short periods

[69] SRVBIT-018 of 2020.

[70] SRVBIT-094 of 2020.

[71] Bill C279, 2020, 1.

[72] Pierre Brunet has carried out a comparative study of cases related to the recognition of rights of rivers in New Zealand, India, and Colombia. In his study, he highlights the differences in the problems these cases sought to address. He points out that it is precisely these differences which has the consequence that the recognition of rights of nature cannot yet be considered a *transnational model* (Piette Brunet, *Para un análisis del discurso jurídico* (Universidad Externado de Colombia 2019)).

of time granted for the immense orders to be fulfilled have elapsed and, having failed to develop or achieve the objectives, sanctions for non-compliance are requested?'[73]

VII. Stabilizing Reactions and Calls for Change

Just as interesting debates were generated on whether graffiti was or was not art, decisions recognizing rights for non-human elements have generated diverse reactions of both support and defiance. The divergent positions in scholarship clearly reflect two extremes: a critical stance towards the *River Atrato* case and, at times, towards the recognition of the rights of nature; and, on the other hand, a stance that reflects an openness to new conceptions of law – to the breaking of paradigms – and that points to the need to seek new approaches to the protection of nature in Colombian law.

Those who do not share the case's approach and content qualify it as a ruling that contributes little or nothing to protection and see it only as a semantic conquest or even as a smoke-and-mirror product of questionable legal activism.[74] These stabilizing tendencies have been characterized by support for the traditional design of legal architecture, and advocates of this view vehemently argue that the new marks destabilize the established order.

The issue has given rise to a large amount of academic output. Some recent publications lean towards a critique of the new approach and are more inclined towards stabilizing tendencies,[75] although others have positioned themselves within broader visions of nature.[76] For its part, another academic work defends a greater openness to new theoretical currents with an interdisciplinary approach that supports the recognition of the rights of nature.[77] Still others propose accepting the invitation of Andean constitutionalism with more radical moves.[78]

Pimiento Echeverry, who has devoted much of his research output to a property-law based protection of public goods, has pointed out that declaring rivers as subjects of rights does not generate greater protection, and he categorizes the re-

[73] Dissenting opinion, Judge Álvaro Fernando García, *Amazon* case (n 53) 53–54.

[74] Camilo Andrés Garzón, 'Los "derechos a la naturaleza" pueden ser más humo que protección' *La Silla Académica* (4 February 2020) <www.lasillavacia.com/historias/historiaacademica/los-derechos-a-la-naturaleza-pueden-ser-mas-humo-que-proteccion-/>.

[75] María Del Pilar García Pachón and Lisneider Hinestroza Cuesta, 'El reconocimiento de los recursos naturales como sujetos de derechos. Análisis crítico sobre los fundamentos y efectividad de la sentencia del río Atrato' in María del Pilar García Pachón (ed), *En Reconocimiento de la naturaleza y de sus componentes como sujetos de derechos* (Universidad Externado de Colombia 2020).

[76] Castro Niño and Robayo Galvis (n 4).

[77] Everaldo Lamprea Montealegre, *El derecho de la naturaleza: una aproximación interdisciplinaria a los estudios ambientales* (Universidad de los Andes and Siglo del Hombre Editores 2018).

[78] Esupiñán Achury (n 46).

cognition of rights as a 'political declaration [that] contributes little or nothing in the consecration of a better water protection regime and that seems more like a formal ideological response than a real increase in the water protection model'.[79] García Pachón and Hinestroza Cuesta[80] have pointed out that the *River Atrato* case has problems in its reasoning and call for further legal construction and structuring. For their part, Guzmán and Ubajoa consider that

> 'it is inappropriate to grant legal personality to nature in general, and to its elements in particular, because such granting is not capable of generating human change in the way of conceiving the environment. This change can only emanate from the internalisation of an irreproachable truth: environmental damage represents harm to the human beings themselves. [This implies] diminishing the importance of the rights of human beings.'[81]

The voices in favor highlight the changes that the *River Atrato* case, as the most important example of this movement, has generated, and emphasize its contributions to Colombian law and the redefinition of the notion of the *subject*. For example, García Arboleda and Hernández[82] consider the ruling to be a milestone in the history of the social rule of law in Colombia, as does Molina Roa, who has emphasized that this declaration is

> 'an irreversible legal reality, the product of a social and legal movement with decades of trajectory and legal precedents that did not hesitate to catalogue nature or animals as legal goods with intrinsic value and, therefore, recipients of state protection.'[83]

---

[79] Julián Andrés Pimiento Echeverry, 'El dominio público del agua en el derecho colombiano' in María del Pilar García Pachón (ed), *Tratado de derecho de Aguas. Derechos de aguas colombiano para el siglo XXI. Tomo I* (Universidad Externado de Colombia 2018) 82–83.

[80] García Pachón and Hinestroza Cuesta (n 75).

[81] Luis Felipe Guzmán Jiménez and Juan David Ubajoa Osso, 'La personalidad jurídica de la naturaleza y de sus elementos versus el deber constitucional de proteger el medio ambiente' in María del Pilar García Pachón (ed), *Reconocimiento de la naturaleza y de sus componentes como sujetos de derechos* (Universidad Externado de Colombia 2020) 208; in the same vein, see also Carlos Alberto Chinchilla Imbett, 'La equiparación a sujetos de derechos de los animales y los ecosistemas. El uso impropio de la categoría "sujeto de derechos" para establecer nuevos límites a la autonomía individual' in María del Pilar García Pachón (ed), *Reconocimiento de la naturaleza y de sus componentes como sujetos de derechos* (Universidad Externado de Colombia 2020) and Ángela María Amaya Arias and Diana Geraldine Quevedo Niño, 'La declaratoria de la naturaleza como sujeto de derechos. ¿Decisión necesaria para la efectividad de las órdenes judiciales?' in María del Pilar García Pachón (ed), *Reconocimiento de la naturaleza y de sus componentes como sujetos de derechos* (Universidad Externado de Colombia 2020).

[82] García Arboleda and Hernández (n 4).

[83] Molina Roa, 'Derechos de la naturaleza' (n 4) 151. In the same vein, see also Bagni and Pavani (n 46); Esupiñán Achury (n 46); María Daniela de la Rosa Calderón y Diego Felipe Contreras Pantoja, 'Fundamentos constitucionales de los recursos hídricos en el ordenamiento jurídico colombiano' in María del Pilar García Pachón (n 79); Clavijo Ospina (n 4) 651. The *River Atrato* case has also been the subject of statements in foreign literature; we highlight the writings of Silvia Bagni, 'Los derechos de la naturaleza en la jurisprudencia colombiana e indiana' (2018) 7(9) Revista Jurídica Derecho 33; Elizabeth Jane MacPherson, *Indigenous Water*

However, the debate remains open and is articulated around progressive and reactionary voices, so it is important to assume a perspective that allows us to take stock and think about what we have pending.

VIII. Moderate Optimism: What Do We Have Pending?

The rulings that recognize the rights of non-human subjects coincide in that they do not delimit the scope and nature of such rights, meaning that it is difficult to speak of a transformation of the traditional legal frameworks for understanding the classic categories of persons, as well as the relationship between persons and things in Colombian law.

The idea that nature has inherent rights that must be protected is ancient and has its roots in both Western and non-Western systems.[84] What is particularly novel, therefore, is its inclusion in contemporary legal systems, through different litigation strategies of environmental movements that have aimed to push for national[85] and even international legal change to motivate the transformation of classic categories.[86] The consideration of rivers as subjects of rights – and the general recognition of rights for non-human elements of nature – is a *key emerging concept* that marks a path towards new ways of humans relating to the environment in the Colombian context and also new ways of the legal system relating to other social systems. The recent introduction of this concept into the legal system evidences the need for development and delimitation of the concept's contours, and it creates an important research agenda for the law in interaction with other disciplines or areas of knowledge.

Rights in Law and Regulation: Lessons from Comparative Experience (CUP 2019); Randall S Abate, *Climate Change and the Voiceless: Protecting Future Generations, Wildlife, and Natural Resources* (CUP 2020) 120–172; Craig M Kauffman and Linda Sheehan, 'The Rights of Nature: Guiding Our Responsibilities through Standards' in Stephen J Turner and others (eds), *Environmental Rights. The Development of Standards* (CUP 2019); Cristy Clark and others, 'Can You Hear the Rivers Sing: Legal Personhood, Ontology, and the Nitty-Gritty of Governance' (2018) 45(4) Ecology LQ 787; Erin O'Donnell, *Legal Rights for Rivers: Competition, Collaboration and Water Governance* (Routledge 2019).

[84] Kauffman and Sheehan (n 83).

[85] Linda Sheehan, 'Realizing Nature's Rule of Law Through Rights of Waterways' in Christina Voigt (ed), *Rule of Law for Nature: New Dimensions and Ideas in Environmental Law* (CUP 2013).

[86] The Inter-American Court of Human Rights has noted – since the issuance of Advisory Opinion 23 of 2017 and as highlighted in the Judgment of 6 February 2020 in the case *Indigenous Communities of the Lhaka Honhat (Our Land) Association v Argentina* – that elements of nature such as rivers, seas, and ecosystems deserve protection as 'legal interests in themselves, even in the absence of the certainty or evidence of a risk to individuals. This means that nature must be protected', not only because of its 'benefit' or 'effects' with respect to human beings, 'but because of its importance for the other living organisms with which we share the planet'.

From this perspective and as Bonilla Maldonado[87] notes, the legal discipline is presented as a space of struggle, where the construction of new subjectivities, such as recognition of the rights of nature, evidences a process of transformative hybridization that makes room for other visions and generates quality legal products; or, to use the term coined by Esquirol, movements that create *legal capital*.[88]

However, it is erroneous to assert that the jurisprudence of the Constitutional Court in Colombia is consistently moving towards the adoption of ecocentric or biocentric approaches to nature and animals. On the contrary, in the constitutional rulings themselves we witness a persistent debate between those who consider that the Colombian legal system maintains an anthropocentric approach with alterations (marks on the traditional legal architecture) and those who consider it to have shifted completely towards an ecocentric approach or towards intermediate models. This shows how, even though disruptive marks may be recognized, the identifying characteristics of a liberal constitution are maintained.[89]

In a 2019 ruling in which the Constitutional Court declared sport hunting unconstitutional, finding it contrary to the current constitutional order that prohibits hunting animals exclusively for sport, the court reiterated its primary commitment to nature based on an anthropocentric approach.[90] The clarifications and dissenting opinions accompanying this decision exemplify the thesis that this text has sought to defend in relation to the permanent struggle that exists to give content to classic legal notions. For Justice Gloria Stella Ortiz, for example, the Court omitted in the aforementioned ruling the jurisprudential line that has been developed, thus moving away from the anthropocentric approach to law and coming to place itself in an intermediate scenario within the philosophical and legal discussion surrounding the issue.[91] For her part, Justice Diana Fajardo clarified her vote to defend the ecocentric approach struck by the Court to date, and she em-

---

[87] Daniel Bonilla Maldonado, *Los bárbaros jurídicos. Identidad, derecho comparado moderno y el sur global* (Siglo del Hombre Editores and Universidad de los Andes 2020).

[88] Jorge L Esquirol, *Las ficciones del derecho latinoamericano* (Siglo del Hombre Editores and Universidad de los Andes 2014) 63.

[89] Bonilla Maldonado (n 8).

[90] Constitutional Court of Colombia 6 February 2019, C-045/2019: 'Thus, our Ecological Constitution or Green Constitution has defined a political-legal programme of action that includes the environment as a scenario of our Social Rule of Law, where rational beings, in the exercise of their dignity (Political Constitution, art 12), in turn display dignified behavior towards human and non-human beings who share their space, as a correlate of their related duties (Ruling C-666 of 2010)'.

[91] In this regard, Judge Ortiz stated: 'Having described these positions, the Colombian legal system is not entirely anthropocentric, nor does it recognize animals as holders of subjective rights. The approach adopted by our legal system and by this Corporation in several of its pronouncements, is an approach that tends to be more intermediate in the international philosophical discussion, to the extent that it clearly requires avoiding animal suffering, while recognizing that by virtue of their quality as rational beings and in accordance with their dignity, it is people who have specific duties towards animals, to avoid their suffering and pain as sentient beings and to advocate for their protection and care' (C-045/2019 (n 90) 4).

phasized that the profound call to open up the law to other discourses and knowledge in light of contemporary debates implies recognizing non-human subjectivity in the Colombian legal system.

In contrast, in her dissenting opinion on the ruling on the hunting of animals for sport, Justice Cristina Pardo demonstrates a stabilizing reaction. In her view, there can be no recognition of non-human subjects where, for her, what is human – specifically embryos or fetuses – has not yet been fully recognized by the legal system within the classic category of personhood. With a contrasting argument, the judge emphasizes that Colombia prefers to protect animals rather than commit itself to protecting the life of the unborn, and she rejects the decriminalization of abortion.[92] The tension in the Constitutional Court, which is evidenced in its dissenting opinions, is again evident in the recent ruling that declared unconstitutional the legal provision allowing sport fishing.[93]

Taking all of this into account, in the constitutional actions that have come to their attention, the Constitutional Court in Colombia, as well as judges of different instances and levels, have been more daring when it comes to recognizing the existence and rights of non-human elements originating from nature than of non-human animals. However, far from having a solid precedent, we currently have a chaotic line of jurisprudence that remains under construction. Even though the *River Atrato* case is so forceful that it even denounces the economic development model that has resulted in irreparable damage to nature in collusion with the State,[94] constitutional and public policy decisions continue to be issued in which free economic and business initiative (Political Constitution, 1991, arts 333 and 334) redefine the social and ecological function of property that has been built up over time (art 58).

---

[92] In this regard, Judge Pardo stated: 'It prohibits the unnecessary disposal of their life. Regarding animals, it maintains that their life may not be disposed of and is protected, unless action to the contrary is necessary. Animal life cannot be disposed of for trivial reasons. On the other hand, in the case of the unborn, its undisputed sentient condition becomes irrelevant. In cases where it is perfectly feasible to preserve both lives, more weight is given to the option for death, even though abortive procedures may be more painful than shooting an animal. Thus, for example, in Ruling SU-096 of 2018, the Court endorsed the vacuum aspiration of the uterus of a human being at six months gestation, a procedure contraindicated at this stage and one which involves fetal dismemberment. Since I do not participate in this illogical and unscientific form of reasoning, I disagree with the previous decision' (C-045/2019 (n 90)).

[93] Constitutional Court of Colombia 27 April 2022, C-148/2022.

[94] On this point, the Constitutional Court quotes Arturo Escobar: 'sustainable development as a global strategy "[...] is the latest attempt to articulate modernity and capitalism. It implies the resignification of nature as environment, the re-inscription of the earth as capital under the perspective of science, the reinterpretation of poverty as the effect of environmental destruction, and the development of new models of management and planning contracts vested in states that act as arbitrators between nature and the people. [...] This development discourse has been the most effective operator of the politics of representation and identity in much of Asia, Africa and Latin America since the post-war period"' (*River Atrato* case (n 17) 5.15).

In such decisions, arguments are used that are based on the neoliberal model and accommodated in the interpretations given to the principles of precaution, prevention, sustainability, and environmental non-regression. These arguments put environmental human rights in check by calling for legislators (freedom of configuration) to determine their content, mainly in response to the democratic debate and the economic, social, and political reality of the country,[95] thus instrumentalizing democracy.[96]

For this reason, the conflict derived from the recognition of non-human subjects and the rights of nature in the Colombian context remains alive and reflects profound epistemic and ideological tensions. The debate in the Constitutional Court, and among various courts and judges in Colombia, not only demonstrates that marks have been made on the most traditional buildings of the Colombian legal architecture, but is also evidence of how, time and again, some jurists resist change and demand that legal buildings be re-established so as to leave no trace of the reconfigurations driven by the vicissitudes of time. However, these defiant marks are appearing more and more frequently and, even though they cannot proclaim a transformative revolution of the rights of nature, the legal discourse is nourished by fleeting emancipations that today allow us to speak of rivers, ecosystems, animals and, in general, of non-human subjects with rights in some contexts.

The legal tradition that resists change forces us to remember that history is written and rewritten repeatedly, and that the law is shaped by ongoing daily struggles – between forces pushing for transformation and others seeking stabilization. Over time, these tensions leave marks etched into the most classical buildings of the legal architecture, much like the tags of pioneers like TAKI, whose graffiti spread across the length and breadth of New York's buildings in the 1970s.

---

[95] Constitutional Court of Colombia 8 September 2021, C-300/2021.

[96] María Cecilia Roa-García and Diana Milena Murcia-Riaño, 'El Acuerdo de Escazú, la democracia ambiental y la ecologización de los derechos humanos' (2021) (1) Naturaleza y Sociedad. Desafíos Medioambientales 31, 43.

India

Sowing the Seeds of 'Nature Rights'

Manjeri Subin Sunder Raj[*]

Nature rights or rights of nature have been one of the most significant set of rights that the world has come to recognize, albeit a bit too late. Right from its infancy, this 'new' set of rights has been able to revolutionize the way in which traditional environmental law has been conceived, and it has opened the doors facilitating radical change. It stands as a paradigm shift from a conservative environmental legal system to a more inclusive, robust concept – Earth Jurisprudence.

Having gained a strong foothold in various parts of the globe, this emerging line of jurisprudence seems nothing short of a fairy tale – a way in which the relationship between humans and nature has been redefined – and if adopted in the true sense as it is conceived, it is a sure shot path towards achieving a better world. Delving into the vast legal landscape of *rights* and *duties*, this work tries to encapsulate the myriad changes that have been brought about and tries to put forth a strong case for adopting an eco-centric approach – especially in countries like India, where one does notice that this line of jurisprudence has not yet been fully adopted in its truest sense.

[*] The manuscript was finalized in February 2025, and the references cited, including materials quoted from the internet, pertain to and portray the position at that point of time.

I. Affording Legal Protection to the Environment in the Indian Legal System

After gaining independence in 1947 following over two centuries of British rule, India enacted her Constitution in 1949.[1] Having borrowed numerous concepts from across a number of Constitutions, the Indian Constitution, is one of the longest in the world and tries to chart a path towards achieving all those ideals mentioned in its Preamble.

However, conspicuous by their absence are provisions relating to the environment. The original Constitution of India was devoid of any such particular provision which spoke about 'protecting the environment and its constituents'; this is probably because of the fact that during those times, countries all around the globe were more inclined towards development rather than protection of the environment, having been at the receiving end of two devastating world wars.

However, in 1976 the Constitution was amended[2] – and a specific reference was made to the environment. The reason for the same can be traced to the Stockholm Conference of 1972, which was attended by the then Prime Minister of India, Mrs. Indira Gandhi. There, she made her oft quoted remark that *poverty is the greatest polluter*,[3] which opened a can of worms. As a result of this, the 42nd Constitutional Amendment made two specific changes in respect of the environment by inserting arts 48A and 51A(g):

> 'Article 48A – Protection and improvement of environment and safeguarding of forests and wild life.
> The State shall endeavour to protect and improve the environment and to safeguard the forests and wild life of the country.'

> 'Article 51A Fundamental Duties
> It shall be the duty of every citizen of India
> (a) […]
> (g) to protect and improve the natural environment including forests, lakes, rivers and wild life, and to have compassion for living creatures.'

The former was included as a Directive Principle of State Policy (DPSP), which imposed a duty on the state to take care of the environment, and the latter was set out as a fundamental duty, which was cast on the citizens of India.

This act of amendment did open up newer possibilities, and it was able to pique interest in the topic; however, the implementation of these newly inserted arti-

---

[1] Available at <https://legislative.gov.in/constitution-of-india>.

[2] Constitution (Forty-second Amendment) Act 1976, s 10 (with effect from 3 January 1977).

[3] For more about the story behind this remark, see <https://thewire.in/books/indira-gandhi-nature-pollution>.

cles created a bit of a concern as, unlike fundamental rights, fundamental duties[4] and DPSP[5] are not really enforceable by courts.

II. Incorporating Environmental Rights into the Indian Legal System

While the trajectory of environmental rights did pose an initial problem, especially as regards their implementation, the judges were able to circumvent this, thanks to their proactive efforts, wherein they relied on one of the most – if not the most – important articles of the Constitution of India:

'Article 21. Protection of life and personal liberty. – No person shall be deprived of his life or personal liberty except according to procedure established by law.'

By affording a very liberal interpretation to art 21, courts started to treat it as a repository of rights, and they started reading into art 21 all sorts of rights, which included environment rights as well. By resorting to such an expansion of the scope of art 21, the courts in India were able to tread newer paths and chart newer courses towards achieving some of the ideals which had been mentioned in the Preamble.[6] It is important to note at this stage that the impact that international law, especially international environmental law, did have on India was so profound that it was capable to be read into Indian Law as such.[7]

---

[4] See art 32 – Right to constitutional remedies. See also art 226 – Power of High Courts to issue certain writs. The latter is wider in scope as it enforces both fundamental rights as well as legal rights, whereas art 32 deals only with enforcement of fundamental rights.

[5] Constitution of India, art. 37 – Application of the principles contained in this Part: 'The provisions contained in this Part shall not be enforceable by any court, but the principles therein laid down are nevertheless fundamental in the governance of the country and it shall be the duty of the State to apply these principles in making laws.'

[6] Constitution of India, Preamble: '*We, the people of India*, having solemnly resolved to constitute India into a *sovereign socialist secular democratic republic* and to secure to all its citizens:

justice, social, economic and political;
liberty of thought, expression, belief, faith and worship;
equality of status and of opportunity; and to promote among them all
fraternity assuring the dignity of the individual and the unity and integrity of the Nation;
in our constituent assembly this twenty-sixth day of November, 1949, do *hereby adopt, enact and give to ourselves this constitution*.'

[7] Constitution of India, art 253 – Legislation for giving effect to international agreements: 'Notwithstanding anything in the foregoing provisions of this Chapter, Parliament has power to make any law for the whole or any part of the territory of India for implementing any treaty, agreement or convention with any other country or countries or any decision made at any international conference, association or other body.'

One such instance of the Indian court relying on international law while there was no specific domestic law was the *Vishaka* case, AIR 1997 SC 3011, which read into Indian law the pro-

While discussing the mining activities that were taking part in the Doon Valley, the Supreme Court in *The Rural Litigation and Entitlement Kendra Dehradun and Others v State of Uttar Pradesh and Others*,[8] had a chance to delve into the matter of environment protection. It was the first case in India which dealt with such aspects:

'involving issues relating to environment and ecological balance and the questions arising for considerations are of grave moment and significance not only to the people residing in the Mussoorie Hill range forming part of the Himalayas but also in their implications to the welfare of the generality of people living in the country.'[9]

Directing that the lime stone quarries be shut down, the judgment points out that this is the only way even if it causes much hardship.

'This would undoubtedly cause hardship to them but it is a price that has to be paid for protecting and safeguarding the right of the people to live in healthy environment with minimal disturbance of ecological balance and without avoidable hazard to them and to their cattle, homes and agricultural land and undue affectation of air, water and environment.'[10]

Though such instances were there, one notices that it was only when a specific right was read into art 21 that the courts in India vested the environment with such rights – albeit in an anthropocentric manner. In *Subhash Kumar v State of Bihar and Others*[11] the Supreme Court of India held that the

'[r]ight to live is a fundamental right under art 21 of the Constitution and it includes the right of enjoyment of pollution free water and air for full enjoyment of life.'

This opened up a liberal approach of reading into art 21 the right to live in a healthy environment – and soon many judges followed suit. However, one must notice that courts still followed an anthropocentric approach – one where the decisions aimed at protecting the environment for human's sake.

III. Interspersing Religion, Culture, Tradition, and Environmental Protection

The Indian environmental ethos has been in vogue for many a century and has throughout the ages been able to intersperse religion, culture, and tradition and bring out a unique method of environmental protection. It is also interesting to note that courts, too, have had a chance to delve into such practices and traditions, and use them to foster environmental protection. The socio-cultural angle

---

visions of CEDAW and laid down specific guidelines. Later steps led to the enactment of the Sexual Harassment of Women at Workplace (Prevention, Prohibition and Redressal) Act, 2013.

[8] 1985 AIR 652.

[9] From the judgment rendered by Justice PN Bhagwati.

[10] Idem.

[11] 1991 AIR 420.

of such practices, one can notice, was harnessed by courts, time and again – and the real victory lies in the acknowledgment that has been extended to such practices regarding the role that they play in environmental protection.

Harnessing such beliefs, cultures, and traditions has had its fair share of recognition – 'man has to share earth with his fellow beings',[12] and that 'every being possesses a living spirit, did cater to the fact that inanimate objects were worshipped'.[13] There are also resplendent examples of communitarian practices which toed an eco-centric approach. These practices were keen on treating Mother Nature as a living being – their mother – and on taking care of her.

Numerous examples from across India bear testimony to the innate connection that exists between the protection of nature and cultural practices. The Demazong Buddhist eco-cultural landscape in Sikkim Himalayas, the Apatanis of Arunachal Pradesh,[14] Mendha in Maharashtra,[15] traditional dwellers' role in Karnataka,[16] all have emphasized and furthered the inter-connection. One also comes across a number of folklores which are part and parcel of these tribes – and all of them portray the inter-connectedness. The Idu Mishmi folklore states that sparrows taught humans cultivation and as a result they are protected during harvest season. Common traits of fear and pain were also included in the Rengma Naga folklore, which states that plants and animals used to plead for mercy when man wanted to kill an animal or cut down a tree.[17] There have been instances of tribes considering animals as their siblings – as is evident by such a portrayal by the Ramo tribe, who treated tigers as brothers.[18] There were also numerous folksongs prevalent in the North East that were against hunting and killing.[19]

---

[12] Nirmal Kumar Bose, *Tribal Life in India* (National Book Trust of India 2002) 61.

[13] Ramesh Chandra Majumdar, *An Advanced History of India* (MacMillan 1973) 20.

[14] Suresh Chand Rai, 'Traditional Ecological Knowledge and Community-based Natural Resource Management in Northeast India' (2007) 4 Journal of Mountain Science 248.

[15] Vivek Nirmale and others, 'Assessment of Indigenous Knowledge of Coastal Fisherfolk of Greater Mumbai and Sindhudurg Districts of Maharashtra' (2004) 3 Indian Journal of Traditional Knowledge 27).

[16] M Parinitha and others, 'Ethno-botanical Wealth of Bhadra Wildlife Sanctuary in Karnataka' (2004) 3 Indian Journal of Traditional Knowledge 37.

[17] For more, see Verrier Elwin, *Myths of the North East Frontier of India* (North East Frontier Agency 1958).

[18] M Dhasmana, *The Ramos of Arunachal* (Concept Publishing Company 1979).

[19] Abhik Gupta and Kamalesh Guha, 'Tradition and Conservation in Northeastern India: An Ethical Analysis' (2002) 12 Eubios Journal of Asian and International Bioethics 15.

Table 1[20]

Animal(s) Protected through Taboo	Tribe/Ethnic Group	State in N.E. India
Deer, wild boar, and other animals in mating season; pregnant females and young animals; leader of a group of deer/wild boar.	Various tea garden communities, Hrankhawl, Hmar, and Debbarman tribes	Assam, Tripura, Mizoram
Elephants, tigers, monkeys, owls, vultures, house crows, ravens	Various tea garden communities	Assam
Elephants, song birds, certain snakes	Hrankhawl	Assam, Tripura
Monkeys, otters	Pnar	Assam, Meghalaya
Elephants, eagles, parrots, hill maynas	Dimasa	Assam
Hoolock gibbons, tigers, pythons, wild goats, bulbuls (birds), frogs	Rongmai naga	Assam, Manipur
Tortoises, snails, *Channa morulius* (fish), small eels, some catfish, snakes	Ningthouja clan of Meitei	Manipur, Assam
Eggs and meat of all animals, some catfish, small eels, *Channa morulius*, snails	Khuman clan of Meitei	Manipur, Assam
Field rats	Khabanganba clan	Manipur, Assam
Sparrows	Moirang clan	Manipur, Assam
Parrots, owls, elephants, monkeys, jackals	Muslim trapper	Assam
Sparrows, jackals, crows, eagles, vultures	Muslim nomad	Assam
Monkeys, elephants, songbirds	Muslim nomad	Assam
All poisonous and non-poisonous snakes	Worshippers of Goddess ÅeManasa'	Assam
Herons, egrets, and cormorants in heronries during mating/nesting season	Most communities in valley areas	Assam

These examples have helped chart a new course – albeit without legal recognition being afforded – because culture and tradition have always been inclined towards environmental protection.

[20] Source: Gupta and Guha (n 19).

IV. Adopting Eco-centrism: The Indian Court Experience

Courts in India have time and again ensured that such a right – to live in a healthy environment – has been protected. Though anthropocentric in nature, one does realize that the recognition of this right is a significant step towards better environmental protection.

While customs and traditions were used by the courts to refer to best practises, the decision rendered by Justice Chinnappa Reddy in *Sachidananda Pandey v State of West Bengal*[21] reproduced the entire text of the conversation between the wise Indian Chief of Seattle and the Great White Chief of Washington and threw light on a very fundamental question – is there a concept of ownership?

Illegal mining activities that were ongoing in a tiger reserve was the matter to be decided upon in *Tarun Bharat Sangh, Alwar v UOI.*[22] Having the issue brought to its notice, the Supreme Court made it clear that the laws that have been enacted are to be followed, and there is no reason to be overly concerned about balancing economical and ecological interests.

The Supreme Court described the *interdependency* of socio-cultural aspects and environmental protection as a topic to be concentrated more on in *M C Mehta v UOI.*[23]

In *T.N. Godavarman Thirumulpad v Union of India & Others*,[24] the Supreme Court came up with the path-breaking observation that '[e]nvironmental justice could be achieved only if we drift away from the principle of anthropocentric to eco-centric'.[25] This was a welcome change, especially as it put across the imminent need to move towards an eco-centric approach. By harnessing such elements that were part of the myriad customs, traditions, and practices that were there amongst the many communities in India, the court's observation threw light on the importance that needs be placed on them – as well as on the ways in which they can be harnessed in law.

Quite suddenly, one notices a spurt in the number of such instances, and courts have made use of this opportunity to foster better environment protection. A specific example can be the opinion of the court in *Aruna Roy v UOI*,[26] where the court opined that 'Religion is the foundation for value base survival of human beings in a civilized society. The force and sanction behind civilized society depends upon moral values'.

It is against this backdrop that the court had a chance to look into the *Niyamgiri* case,[27] a case where local communities raised their voice against mining ac-

---

[21] [1987] 2 SCC 295.

[22] [1992] Supp (2) SCC 448.

[23] [1992] 1 SCC 358.

[24] Writ Petition (C) No 202 of 1995, <https://indiankanoon.org/doc/187293069/>.

[25] Idem, at para 14, words of Justice KS Radhakrishnan.

[26] [2002] 7 SCC 368.

[27] *Orissa Mining Corporation v MOEF* [2013] 6 SCC 476.

tivities that were taking place on a locally revered mountain – 'niyamraja' – which they alleged was against their customary beliefs and traditions. The opinion of the court, allowing the petition indicating that such traditions and customs need be taken into account, opened new avenues.

Soon, Indian courts started to take due notice of the developments that were ongoing worldwide. One might notice that the shift was slow, but it certainly was present – as was clear from the observation made by the Supreme Court in *Animal Welfare Board of India v A. Nagaraja & others*,[28] wherein the court stressed the fact that such laws should be based on the dual concepts of 'animal welfare' and 'species' best interests'. Though human necessity was laid down as a justifiable exception, a positive duty was cast on humans to take care of animals. One can read between the lines and observe that the court did actually confer legal rights to such animals,[29] though not explicitly. As part of the obiter in the judgment, this observation played an important role in the expansion of this line of jurisprudence.

You are met with either *ridicule or fear* whenever a new entity is given rights, observed Christopher Stone,[30] and one does notice that his prophecy has turned out to be quite true – if one delves into the trajectory of cases in India.

As far as Earth Jurisprudence is concerned, a new dawn broke in the judgment rendered by Justice Rajiv Sharma in *Mohammed Salim v The State of Uttarkhand*.[31] Public interest litigation (PIL) was filed questioning the illegal constructions and encroachments on the banks of the river Ganga. The PIL also questioned the failure of the new state of Uttarakhand, which was carved out from the state of Uttar Pradesh, to form a Ganga Management Board. The court delved into the matter in hand and held that the Ganga along with its tributary, Yamuna, needs to be deemed a juristic person / legal person / living entity having the status of a legal person with all *corresponding rights, duties, and liabilities of a living person*. The court also delved into the guardianship principle and appointed the Director of the NAMAME Gange Project, the Chief Secretary of the State, and the Advocate General as 'persons in loco parentis as the human face to protect, conserve and preserve River Ganga and Yamuna and their tributaries'.

Later, in *Lalit Miglani v State of Uttarakhand and others*,[32] the court had to deal with a PIL filed to secure legal entity status for natural entities. It was heartening to note that the court afforded legal status to glaciers (including Gangotri

[28] [2014] 7 SCC 547.

[29] Para 30 of the judgment.

[30] Christopher D Stone, 'Should Trees Have Standing? – Toward Legal Rights for Natural Objects' (1972) 45 SCal LRev 450, 455.

[31] Writ Petition (PIL) No 126 of 2014, decided on 20 March 2017. See, <www.tribuneindia.com/news/uttarakhand/courts/uttarakhand-hc-accords-human-status-to-ganga-yamuna/379739.html>. In this case, the Ganga and the Yamuna and all their tributaries, streams and natural water flowing continuously or intermittently with them were treated as legal persons.

[32] Writ Petition (PIL) No 140 of 2015, available at <www.livelaw.in/uttarakhand-hc-declares-air-glaciers-forests-springs-waterfalls-etc-legal-persons/>. In this case, legal rights of person-

and Yamunotri), rivers, streams, rivulets, lakes, air, meadows, dales, jungles, forests wetlands, grasslands, springs, and waterfalls. It was also held that '[a]ny harm caused to these entities was to be treated akin to harm caused to humans'.[33]

However, things did not stay rosy for a long time. Following the decision in the *Mohammed Salim case*, the State of Uttarakhand filed an appeal before the Supreme Court and put across the difficulty that it faced in respect of the applicability of the High Court judgment. The contention made was that the rivers passed through various states and the High Court ruling, which appointed certain state officials as guardians of the river, would pose administrative and implementation hurdles. Taking due cognizance of the issue, the Supreme Court stayed the order of the High Court in *The State of Uttarkhand and others v Mohammed Salim and others*.[34]

One might think that this would have been the end of the road, but thankfully later decisions – albeit of High Courts – did provide much need strength and succor. A PIL was filed before the Uttarakhand High Court – *Narayan Dutt Bhatt v UOI and others*[35] – wherein the plea was to restrict the unchecked movement of horse drawn carts between India and Nepal. The plea maintained that there were few to no checks at all and that the animals were ill-treated – with most of them not being vaccinated as well. The suit also made a case for instances of animal cruelty. As a result of such a situation, the petition put forward a prayer that the entire animal kingdom needs to be conferred status as a legal entity/legal person.

The court had a rich case before it, allowing it to delve into the different aspects related to animal protection, and it held that

> 'the entire animal kingdom including avian and aquatic species is to be treated as legal entities having a distinct persona with corresponding rights, duties and liabilities of a living person'

Over and above such a point, the court also ensured that a positive duty was cast on the citizens to play their part and role in securing such rights for animals. The citizens were appointed as loco-parentis for the entire animal kingdom and were responsible for taking care of the welfare protection of the animals.

> 'All the citizens throughout the State of Haryana are hereby declared persons in loco parentis as the human face for the welfare/protection of animals'.

---

hood were afforded to Gangotri and Yamunotri glaciers and the surrounding areas, including meadows, waterfalls, lakes, and forests.

[33] Idem, at 65.

[34] Petition to Special Leave to Appeal No 016879/2017. Order available at <www.livelaw.in/sc-stays-uttarakhand-hcs-order-declaring-ganga-yamuna-rivers-living-legal-entities-read-order/>. Also see, <www.outlookindia.com/website/story/supreme-court-stays-uttarakhand-hc-order-declaring-ganga-yamuna-a-living-entity/299507>.

[35] Writ Petition (PIL) No 43 of 2014, available at <https://indiankanoon.org/doc/157891019/>.

A later decision in *Karnail Singh v State of Haryana*[36] also had to deal with a situation regarding animal rights; namely, as it was highlighted in the petition, cattle that were being transported to the state of Uttar Pradesh from the State of Haryana had to suffer in pitiable conditions. Delving into the matter, the court reiterated what was held in the *Narayan Dutt Bhatt* case.

There was also an instance of the Punjab and Haryana High Court taking *suo-motu* cognizance based on a letter that was sent to it by a public-spirited citizen; in the letter, the sad state of the Sukhna lake was brought to the court's notice. In *Suo motu v Chandigarh Administration*,[37] the High Court held that the Sukhna Lake is

> 'a legal entity/legal person/juristic person/juridical person/moral person/artificial person for its survival, preservation and conservation having distinct persona with corresponding rights, duties and liabilities of a living person, and the citizens of the state were appointed as guardians as well'.[38]

In April 2022, the Madurai Bench of the Madras High Court, in *A Periyakaruppan v The Principal Secretary*,[39] specifically mentioned that the rights of nature should be recognized – and that nature should be afforded legal personhood. However, the saga does not end there.

V. Lessons from International Developments

As one delves into the impact that international law, especially international environmental law, has had on the Indian legal system, one cannot but help notice its enormity.

Following the Stockholm Conference, India had to showcase her steps towards fostering environment protection, and changes were initiated in her legal system, especially in art 21 of the Constitution, which was widened in its scope and ambit by judges in a number of instances. Soon it included, amongst others, the right to live in a healthy environment as well.[40]

This was just the beginning of a lengthy process, wherein one could notice the trajectory of the development of Indian environmental law. Soon, India also enacted a number of rules and regulations, in addition to the legislation, and on the environmental protection front it was doing a pretty decent job.

---

[36] [2019] SCC OnLine P&H 704, available at <www.livelaw.in/pdf_upload/pdf_upload-361239.pdf>.

[37] Writ Petition (C) 18253 of 2009, decided on 2 March 2020. See <http://files.harmonywithnatureun.org/uploads/upload994.pdf>.

[38] Idem.

[39] Writ Petition (MD) No 18636 of 2013, see <www.casemine.com/judgement/in/625ef382b50db9cc5192cd20>.

[40] *Subhash Kumar v State of Bihar and Others* (n 11).

With specific reference to rights of nature, international law did play an important role and it was mentioned in the earliest judgments in India which pertained to nature rights. However, the influence was at a minimal level, given that not many decisions were rendered which promoted giving such a right to nature – or which extended legal personality to nature and natural entities. The affording of legal personhood to the Whanganui River in New Zealand was one of the earliest of instances when international legal developments had an impact on Indian courts. Justice Rajeev Sharma, the first judge in India who read rights of nature into the Indian context, relied on the New Zealand case and tried to stitch an argument for extending legal personality to non-human entities in India as well. However, this was not done in the same manner as was seen in the New Zealand case, given that he read in not just rights but also duties and liabilities – which could be understood as a dampener.

To date, the Indian Supreme Court has not truly appreciated the wide array of international instances and cases which provide legal personhood to non-human as well as non-living entities. However, that has not affected the rights of nature movement, which has had quite a good outing in international law.

Still, what the Indian legal system lacks is a reason to once again embrace the very notions of environmental protection which were strongly and deeply rooted in her culture and tradition – and to afford these notions legal validity.

VI. The Long Trudge Towards Realizing a Long-Cherished Dream

Indian courts have followed a traditional approach of granting rights following a case specific approach. While this does ensure that rights are conferred, one does feel that a more concrete plan, if in place, will be able to shape the rights of nature movement, especially in a country like India.

What does strike one, though, is that Indian courts have always fallen back on the fundamental duty that has been placed on the citizens under Art. 51A(g) and helped assimilate it to better effect.

A PIL was filed before the Supreme Court of India, by an NGO – the People's Charioteer Organization – which sought that the entire animal kingdom be given a legal entity status, and the approach of the Supreme Court was far from what defenders of rights of nature have strived for. The judge orally remarked:

'Do you mean animals can sue or be sued? That is the meaning of legal entity. Is your dog equal to you?'[41] He also mentioned that the petition was very unlikely to be considered.[42]

---

[41] <www.tribuneindia.com/news/nation/sc-agrees-to-hear-plea-seeking-to-declare-animal-kingdom-as-legal-entity-138799>.

[42] <www.tribuneindia.com/news/nation/sc-agrees-to-hear-plea-seeking-to-declare-animal-kingdom-as-legal-entity-138799>.

This takes us back to understanding the depth of what Stone had observed a long time back![43]

The decision rendered by the Court, in which it denied the extension of legal personhood to animals, takes us backward – not just in time, but also in the jurisprudential realm.

One can only fervently hope that the day when nature is given rights is not too far away.

[43] *Orissa Mining Corporation v MOEF* [2013] 6 SCC 476.

Uganda

The First African Country to Recognize Rights of Nature

Tajudeen Sanni[*]

I. Recognition and Implementation of the Rights of Nature in Uganda

In 2019, Uganda became the first African country to recognize nature as having rights under the National Environmental Act No 5 of 2019 (NEA).[1] The acknowledgment of nature as a bearer of rights represents a significant departure from the conventional human-centered perspective that has long defined environmental law and policy in Uganda. By conferring legal standing upon nature itself, the Act allows for legal action to be taken on behalf of ecosystems and natural entities when they suffer harm or their rights are infringed. This legislative innovation thus signifies a movement towards a more earth-centered, eco-centric (or geocentric) paradigm, aligning Uganda with a growing international movement that seeks to recognize and enforce the rights of nature (RoN) as a means of enhancing environmental stewardship. The adoption of this new framework fosters a more practical and enduring method of environmental governance, holding the promise of transforming Uganda's capacity to conserve its ecological heritage. Indeed, the Act not only strengthens mechanisms for environmental protection

[*] The manuscript was finalized in December 2025. The assessment of literature and case law, including internet materials, reflects the legal situation at that time. The author deeply appreciates the inputs from Prof Ralf Michaels and Dharmita Prasad and the additional editing done in the editorial services department at the Max Planck Institute in Hamburg.

[1] The National Environment Act No 5 of 2019, <https://ulii.org/en/akn/ug/act/2019/5/eng@2019-03-07>.

but also opens new avenues for legal advocacy dedicated to upholding RoN within Uganda's judicial landscape.[2]

NEA, section 4(1), titled 'Rights of Nature', recognizes that 'nature has a right to exist, persist, maintain and regenerate its vital cycles, structure, functions, and its processes in evolution.' The language of the section appears to confer a generalized and universal status upon natural entities. However, section 4(4)[3] suggests that these rights apply only for conservation areas identified by the Minister responsible for the environment under section 51 of the NEA.

Section 4(4) requires the Minister responsible for the environment to enact regulations for the purpose of operationalizing the rights under section 4(1). Such regulations have yet to be enacted. The delay can be attributed to institutional challenges, including complexity in defining conservation areas and potential conflicts with existing land use regulations. Developing these regulations requires extensive collaboration between environmental experts, legal practitioners, and policymakers, as well as careful consideration of ecosystems that require legal protection. Additionally, there may be a lack of political urgency to prioritize these regulations due to competing national development goals. This regulatory gap is further complicated by the lack of sufficient capacity to address the scientific and legal complexities involved in recognizing nature's rights.

The NEA provides a legal framework for recognizing community-based conservation. A notable example is the role of customary governance systems,[4] such as those practiced by the Bagungu people near Murchison Falls National Park.[5] These systems emphasize environmental stewardship through cultural norms that restrict unsustainable resource uses. By reinforcing local governance structures and recognizing customary laws, Uganda can enhance biodiversity conservation in a manner that aligns with both community interests and national environmental goals.[6] The integration of Earth Jurisprudence principles into conservation ef-

---

[2] Laura Alessandra Nocera, 'La Natura come soggetto di diritto nell'ordinamento dell'Uganda' (2021) 3(2) Nuovi Autoritarismi E Democrazie: Diritto, Istituzioni, Società, 188.

[3] Section 4(4): 'The Minister shall, by regulations, prescribe the conservation areas for which the rights in subsection (1) apply.

[4] Custom is recognised as a source of law in Uganda under s 15 of the Judicature Act, Cap 16. This has been affirmed by the courts in many cases, eg in *Mifumi (U) Ltd & 12 Others v Attorney General*, Constitutional Appeal No 02 of 2014 (SC). This is in line with the National Objectives and Directive Principles of State Policy of the Constitution of the Republic of Uganda 1995, Cultural Objective XXIV. See also Jamil Ddamulira Mujuzi, 'Reconciling Customary Law and Cultural Practices with Human Rights in Uganda' (2020) 41(2) Obiter 239.

[5] Caitlin Skurky, 'Toward a Decolonized Non-Anthropocentric Legal System' (2021) 34 Geo Envt'l L Rev 351, 368; Mugisa Bob, *Assessment of Bagungu Community Participation in Wildlife Conservation in Murchison Falls National Park* (Special Project Report, Makerere University 2023) <https://dissertations.mak.ac.ug/bitstream/handle/20.500.12281/18252/Mugisha_CAES_BSC.pdf>.

[6] African Biodiversity Network, *The Importance of Community-Led Conservation in Africa's Climate Crisis*, <https://africanbiodiversity.org/the-importance-of-community-led-conservation-in-africas-climate-crisis/>.

forts – such as recognizing the intrinsic rights of ecosystems – demonstrates a progressive approach that mirrors the success of community conservancies elsewhere. Strengthening such legal and institutional mechanisms would further empower Ugandan communities to protect forests, wetlands, and wildlife, thereby ensuring long-term ecological sustainability.

In line with international human rights norms, Uganda's National Environment Act correlates nature's rights with the right to a clean and healthy environment, as enshrined under article 39[7] of the Ugandan Constitution.[8] Moreover, this recognition aligns with Africa's regional commitments, particularly those of the African Commission on Human and Peoples' Rights (ACHPR), which has called for the recognition of sacred natural sites and territories (SNST) and their governance systems.[9] These international frameworks support Uganda's move towards embedding Earth jurisprudence in national law, thus making the protection of nature a broader, global responsibility.

In Uganda, scholars and legal practitioners are navigating this shift, exploring how RoN can complement existing environmental laws. Many view RoN as a means to enhance legal protections for the environment, addressing deficiencies in traditional frameworks that may not fully address ecological degradation.[10] A few express concerns about potential conflicts, particularly regarding the attribution of legal personhood to natural entities and the practicalities of enforcement in modern courts.[11] This evolving discourse reflects a broader global trend, where legal systems are increasingly recognizing nature as a rights-bearing entity, aiming to harmonize human and environmental interests.[12] Therefore, in the Ugandan context, the relationship between the right to a clean and healthy environment and RoN is generally viewed as complementary. Both aim to protect and preserve the environment, though they may differ in focus and application. The challenge lies in integrating these frameworks to ensure that legal protections are robust, enforceable, and truly reflective of both human and ecological needs.

---

[7] Art 39 Right to a clean and healthy environment: 'Every Ugandan has a right to a clean and healthy environment.'

[8] Caiphas Brewsters Soyapi, 'The Courts and the Constitutional Right to a Clean and Healthy Environment in Uganda' (2019) 28 RECIEL 152.

[9] See ACHPR, *Resolution 372 (LX) 2017 on the Protection of Sacred Natural Sites and Territories*, 8–22 May 2017, <https://achpr.au.int/en/adopted-resolutions/372-resolution-protection-sacred-natural-sites-and-territories-achprres>.

[10] See, eg, African Institute for Culture and Ecology, *Earth Jurisprudence and Rights of Nature in Uganda*, <https://africeug.org/earth-jurisprudence-rights-of-nature/>.

[11] This point came up in 2023 in a discussion on rights of nature with Prof Sam Tindifa, a community leader and law professor at Islamic University in Uganda.

[12] Matthew W Hopewell, *The Rights of Nature in Uganda: Exploring the Emergence, Power and Transformative Quality of a 'New Wave' of Environmentalism* (MSc dissertation, University of London 2019), <https://indisproject.org/wp-content/uploads/2020/11/Matthew-Hopewell-Rights-of-Nature-UG-Dissertation.pdf>.

II. Legal and Political Operators

The process of having RoN recognized within Ugandan law took several years and involved a number of groups. It emerged from a 2013–2014 series of workshops in Kampala that examined the shortcomings of Ugandan environmental law.[13] The effort was spearheaded by a group of lawyers, the Advocates for Natural Resources and Development (ANARDE), who worked towards national legislation through lobbying and drafting activities.[14] The National Association of Professional Environmentalists (NAPE) and the African Institute for Culture and Ecology (AFRICE) worked with local communities, collecting relevant knowledge, and also provided training. AFRICE, in particular, worked at the community level through collaboration with the Bagungu of Buliisa, the Banybutumbi of Lake Edward, and the Bassese of the Kalangala Islands in Lake Victoria, all of whom have rich traditional practices for the conservation and protection of forests, wetlands, and water bodies, including all the biodiversity that they contain.[15] The African Biodiversity Network (ABN) and the UK-based Gaia Foundation enabled cross-African interchange. The Open Society Initiative for East Africa (OIEA) served as a donor.[16] Knowledge was shared between all groups.

After the enactment of the NEA in 2019, advocacy has continued at different levels. At the national level, NGOs and environmentalists have continued to push for implementing regulation by preparing a draft legal framework and policy proposals. At a local level, the Buliisa Council, in collaboration with the Indigenous Bagungu People, have passed a resolution and a subsequent ordinance recognizing the customary laws of the Bagungu Custodian Clans (Nganda Zilamansi zya Bagungu).[17] The Bagungu have since activated clan structures (including custodians and women elders) for the enforcement of rights of nature.[18]

---

[13] Hopewell (n 12) 21, 33ff.

[14] Alex Putzer, Tineke Lambooy, Ignace Breemer and Aafje Rietveld, 'The Rights of Nature as a Bridge between Land Ownership Regimes' (2022) 11(3) TEL 501, 518.

[15] 'Uganda Takes the Lead in Recognition of Customary Law' (*AFRICE*, 2022) <https://africeug.org/uganda-takes-the-lead-in-recognition-of-customary-law/>. See also the interview with Dennis Tabaro, 'Indigenous Biocultures and Rights of Nature in Uganda' (*Global Assembly*), <https://globalassembly.medico.de/en/rights-of-nature/case-studies/interview-indigenous-biocultures-and-rights-of-nature-in-uganda>.

[16] Hopewell (n 12) 34.

[17] Resolution on the Customary Laws of Bangungu Custodian Clans, 6 November 2019, <http://files.harmonywithnatureun.org/uploads/upload989.pdf>; The Local Governments (Buliisa District) (Bagungu Customary Laws) Ordinance 2020; see <https://africeug.org/uganda-takes-the-lead-in-recognition-of-customary-law/>; <https://gaiafoundation.org/uganda-recognises-rights-of-nature-customary-laws-sacred-natural-sites/> and <https://ecojurisprudence.org/wp-content/uploads/2022/11/Buliisa-Ordinance.pdf>.

[18] Tabaro (n 15).

III. Enforcement

The NEA provides for public enforcement of rights of nature. Section 4(3) of the NEA states that "[g]overnment shall apply precaution and restriction measures in all activities that can lead to the extinction of species, the destruction of the ecosystems or the permanent alteration of the natural cycles." This section operates in tandem with the national objectives and directive principles of State Policy XIII of the Constitution, which mandates the State to 'protect important natural resources, including land, water, wetlands, minerals, oil, fauna and flora on behalf of the people of Uganda'. Uganda's principal environmental regulatory agency, the National Environment Management Authority (NEMA), enforces laws aimed at conserving natural resources and biodiversity.[19] Some local communities have expressed distrust towards the government,[20] though others view the government enforcement efforts as an important safeguard.

However, with respect to RoN, mechanisms for private enforcement may be even more important. Under section 4(2) of the NEA, 'a person has a right to bring an action before a competent court for any infringement of rights of nature under this Act.' Moreover, section 134(2) NEA empowers any person to apply for an environmental restoration order before the court even if such person is not suffering any harm and has no direct or personal interest in the matter. This, in essence, establishes a robust public interest litigation element to enforce RoN. Section 4(2) NEA does not limit its application based on an anthropocentric conception of environmental damage, instead extending it to an eco-centric perspective in the enforcement of RoN. This public interest litigation mechanism serves as a powerful tool in the hands of citizens, enabling them to act on behalf of nature and ensuring that individuals or corporations cannot escape accountability for ecological damage by simply claiming a lack of personal harm on the part of applicants. These provisions are supplemented by article 50[21] of the Constitution,

---

[19] See NEA, ss 8–10.

[20] Makanga Ronald Kakumba, 'Ugandans want government action on environmental problems affecting their communities', Afrobarometer Dispatch No. 577 (25 November 2022), <https://www.afrobarometer.org/publication/ad577-damage-control-ugandans-want-government-action-on-environmental-problems-affecting-their-communities/>; Iryne Kinsumbya, *Compliance to environmental laws in the use and management of Lutembe wetland in Ddewe village, Wakiso district, Central Uganda* (MA Dissertation, Makerere University, 2021), <https://dissertations.mak.ac.ug/handle/20.500.12281/8852>; George Lubega Matovu, *Challenges in Monitoring and Enforcement of Environmental Laws in Uganda* (Greenwatch report, 2025), <https://greenwatch.or.ug/sites/default/files/documents-uploads/challenges_in_monitoring_and_enforcement_of_environment.pdf>.

[21] Art 50 Enforcement of rights and freedoms by courts:

'(1) Any person who claims that a fundamental or other right or freedom guaranteed under this Constitution has been infringed or threatened, is entitled to apply to a competent court for redress which may include compensation

(2) Any person or organisation may bring an action against the violation of another person's or group's human rights.

which grants any person the right to seek legal remedies for a violation of a basic right, regardless of whether the violation affects them or someone else. These provisions are significant in widening citizens' *locus standi* for redressing environmental wrongs against nature in general and enforcing RoN in particular. Any person is thus granted the power to invoke this right by filing a claim before the courts of Uganda to seek redress for violation of environmental rights.[22]

A special role for enforcement lies with local communities,[23] in line with community-based conservation. Rights of nature are viewed as simultaneously empowering communities. Enforcement by local communities includes the revitalization of ecological knowledge. But it also includes regulation: 'clan structures (including custodians and women elders), for restorative justice where violators of the Rights of Nature are punished and made to restore, atone or compensate for the violation of nature's rights.'[24]

The enforcement of environmental rights in Uganda has encountered considerable challenges, particularly regarding access to justice. Environmental litigation tends to be complicated and often remains out of reach for many citizens.[25] Judges interviewed shortly after the adoption of the NEA seemed skeptical about its judicial enforcement.[26] Nonetheless, individuals and organizations in Uganda have actively taken the government to court to enforce environmental rights. These legal actions often challenge government inaction or policies that harm the environment, with plaintiffs asserting that their constitutional and environmental rights have been violated. For example, *Tsama William and Others v Uganda's Attorney General and Others*[27] saw residents of Bududa district seeking justice for the government's failure to mitigate the risks of landslides, which had devastating environmental and human impacts. In *The Environment Shield Limited & Another v Jinja City Council & Another*,[28] a civic environmental company and journalist successfully challenged the local government's plan to cut

(3) Any person aggrieved by any decision of the court may appeal to the appropriate court.

(4) Parliament shall make laws for the enforcement of the rights and freedoms under this Chapter.'

22 For environmental cases that were instituted in that context, see *Advocates Coalition for Development and Environment (ACODE) & Anor v Attorney General & National Environment Management Authority* HC Misc Cause No 0100 of 2004 (12 July 2005), <https://www.globalhealthrights.org/wp-content/uploads/2013/02/HC-2004-Advocates-Coalition-for-Development-and-Environment-v.-Attorney-General.pdf>; *The Environmental Action Network Ltd (TEAN) v Attorney General & National Environment Management Authority* HC Misc Appl No 39 of 2001, <https://www.tobaccocontrollaws.org/litigation/decisions/the-environmental-action-network-ltd-tean-v-the-attorney-general-national-environment-management-authority-nema-2>.

23 Hopewell (n 12) 29–30.

24 Tabaro (n 15).

25 See, eg, Samantha Atukunda K Mwesigwa and Peter Davis Mutesasira, 'Climate Litigation as a Tool for Enforcing Rights of Nature and Environmental Rights by NGOs: Security for Costs and Costs Limitations in Uganda' (2021) 15(2) Carbon & Climate LRev 139.

26 Putzer and others (n 14) 523.

27 High Court of Uganda at Mbale, <https://www.climatecasechart.com/document/tsama-william-and-others-v-ugandas-attorney-general-and-others_0748>.

28 High Court of Uganda at Jinja, <https://media.ulii.org/media/judgment/108666/source_file/ENVIRONMENT_SHIELD_FINAL_RULING.docx>.

down urban Mvule trees, highlighting the environmental and cultural value of these trees. These cases underscore the vital role of public interest litigation in enforcing environmental rights in Uganda, ensuring that both the government and its agencies are held accountable for actions that compromise the environment.

One notable issue in the current legal framework is the absence of a dedicated environmental court which could simplify legal procedures and ensure that environmental cases are addressed with the necessary expertise and urgency. Such a specialized court would provide focused adjudication tailored to the country's unique environmental challenges and enhance public access to timely and competent environmental justice.[29] Moreover, reinforcing the legal framework for environmental advocacy is crucial in empowering citizens. By improving access to justice and guaranteeing the enforceability of RoN, Uganda can advance toward a stronger and more effective system of environmental governance. This effort involves establishing institutional mechanisms that are readily accessible to the public, such as specialized environmental adjudication. Such institutions would not only protect environmental rights but also play a vital role in promoting the long-term sustainability of Uganda's natural resources.[30]

It is not only the right to a clean and healthy environment that correlates with RoN; there are a number of other rights that also help to realize this framework. For example, for the purpose of environmental advocacy involving rights of nature, freedom of association (article 29 of the Constitution) is critical, especially considering the communal nature of traditional communities that serve as custodians of natural monuments. People can more successfully campaign for RoN by creating and participating in non-governmental organizations and civil society groups. An association that involves indigenous people as keepers and protectors of nature would be vital, and thus the right of association correlates with and complements the rights of nature.[31] Additionally, the right of access to information is critical in allowing the public to effectively advocate for environmental protection: the public must be aware of environmental concerns and the sources of those threats. The public needs to have access to information that allows them to respect and advocate nature's right to flourish. Uganda's constitution features some of the most robust rules on access to information in the world.[32] It allows broad access to information stored by the government, except where the dissemination of the information is likely to jeopardize the state's security or sovereignty or infringe on any other person's right to privacy.[33]

---

[29] Charlotte Kabaseke, 'The Right to Access to Justice and the Need for an Environmental Court in Uganda' (2023) 3 EALS Human Rights & Rule of Law Journal 27.

[30] Kabaseke (n 29).

[31] Kenneth Kakuru and Irene Ssekyana (eds), *Handbook on Environmental Law in Uganda*, vol 1 (2nd edn, 2009) 59–60, <https://www.greenwatch.or.ug/greenwatch-publications/handbook-environmental-law-uganda-vol1>.

[32] Constitution, art 41; Access to Information Act 2005, <https://ulii.org/en/akn/ug/act/2005/6/eng>.

[33] UDHR, art 9, African Charter on Human and Peoples' Rights (done at Banjul, 26 June 1981; entered into force 21 October 1986), OAU Doc CAB/LEG/67/3 Rev 5, reproduced in 21 ILM 59 (1982).

IV. Theoretical Sources and Foundations

As elsewhere, a main theoretical source named by participants in the law-making process are non-Western ontologies.[34] They include the emphasis that man is part of nature and, conversely, that nature has authority over man. The relationship between humans and nature is viewed as one of cooperation, not competition or exploitation. This interconnectedness is derived from concrete African traditions and cultures. Such ontologies and cultures contrast with Western anthropocentrism and techno-scientism as well as Western ideas of exploitation. As a consequence, RoN are viewed as being closely connected to decolonization. However, some see a generational gap in which non-Western ontology and culture resonate more with elders than with younger generations.

A philosophical grounding for RoN may thus be seen in many of the Ugandan people's traditions, spiritual beliefs, ceremonies, customary laws, and arts and crafts, which have existed since time immemorial. Nature conservation has a long history in the traditional heritage and cultural ethos of Africa in general and of Uganda in particular.

A fuller appreciation of how RoN can take root in Uganda also requires recognizing the multiple normative orders that shape environmental governance and land relations in post-colonial societies. These include the formal statutory and political structures inherited from the post-independence state; the chthonic or customary legal traditions that continue to guide communal land tenure, natural sites, and ecological stewardship; the authority exercised by civil society and community-based institutions; the normative influence of international and regional soft-law frameworks that increasingly articulate ecocentric principles; and the significant role of local and multinational corporate actors whose economic activities shape environmental outcomes.[35] These overlapping normative spheres generate a complex legal landscape within which RoN must be interpreted, contested, and institutionalized, making it imperative to examine how they converge, reinforce, or conflict with one another in the Ugandan context.

A look at diverse ancient cultures in Uganda reveals how people have co-existed harmoniously with nature. Among the Bahima and Karamojong pastoralist communities, traditional practices focused on balancing human activity with nature. The Bahima would traditionally shift grazing areas to allow the land to regenerate, preventing over-exploitation.[36] Similarly, the Karamojong moved their herds based on seasonal rains and pastures, ensuring land recovery.[37] Like many

---

[34] Hopewell (n 12) 22ff.

[35] Putzer and others (n 14) 519f.

[36] See W Wurzinger, A M Okeyo, D Semambo and J Sölkner, 'The sedentarisation process of the Bahima in Uganda: An emic view' (2016) 4 African Journal of Tropical Agriculture, <https://www.internationalscholarsjournals.com/articles/the-sedentarisation-process-of-the-bahima-in-uganda-an-emic-view.pdf>.

[37] Greg FG, HU Ndume and FV Okoyoyo, 'Way of Life and Grouping Practices of Bahima Pastoralists' (2019) 7(3) AASTGB 28.

indigenous small holders farmers in parts of East Africa,[38] the Bakiga (the Kiga people) and Banyankore practiced leaving land fallow, allowing the soil to regenerate, enhancing productivity while and maintaining local ecosystems. The Bakiga practiced a unique type of farming known as terracing designed to prevent soil erosion.[39] All these methods are aimed at helping preserve resources and maintain ecological balance.

However, rising population pressures have made it harder to sustain these practices, leading to land scarcity, overuse, and environmental strain. As human settlements and agriculture expand, these once-sustainable traditions are increasingly difficult to maintain. This strain on land resources has led to over-exploitation, environmental degradation, and a loss of biodiversity, making it clear that traditional methods alone can no longer protect nature effectively.[40]

Nonetheless, the influence of traditional Ugandan and African practices cannot be overlooked. The strong cultural connection to nature, embodied in customs and customary practices, plays a key role in the preservation of natural resources. Indigenous practices and local governance systems must be seen as allies in the legal framework for the protection of nature. Uganda's rich tradition of environmental stewardship provides a solid foundation for integrating modern legal principles with time-honored conservation practices. Indigenous communities in East Africa, like the rest of the continent, have long practiced environmental stewardship, and recognizing their legal traditions could strengthen the country's RoN initiatives. Instead of retaining a one-size-fits-all approach derived from Western legal systems, Uganda can develop place-based, culturally relevant, holistic legal frameworks that grant natural entities legal standing, ensuring their protection while preserving indigenous ecological practices beyond a general statement of RoN in the law. This aligns with the global call for international law to effectively move beyond anthropocentric perspectives and embrace ecocentric approaches that recognize the intrinsic value of nature in all aspects of life.

Indeed, the traditional Ugandan communities regard even their occupations and means of livelihood such as farming as a shared responsibility that sustains people and the environment they inhabit.[41] Like in the rest of Africa,[42] their cooperative norms and respect for seasons, soils, water bodies and local ecosystems

---

[38] PKR Nair, 'Alternative and Improved Land Use Systems to Replace Resource-Depleting Shifting Cultivation' (1987), available at <https://www.fao.org/4/ad085e/AD085e09.htm>.

[39] See Robert Tweheyo, Dastan Bamwesigye, Miriam Kiconco and 'Climate Change Mitigation in Southwestern Uganda Using Indigenous Knowledge' <https://www.preprints.org/manuscript/202401.0422>.

[40] 'Uganda – Environmental' (*The World Countries*) <https://theworldcountries.com/geo/environment/facts/Uganda>.

[41] See Margaret A Rugadya, 'Pastoralism as a conservation strategy' (Uganda Country Paper prepared for IUCN study, 2006), available at <https://iucn.org/sites/default/files/import/downloads/uganda_country_study.pdf>.

[42] Admos Chimhowu, 'The "new" African Customary Land Tenure. Characteristic, Features and Policy Implications of a New Paradigm' (2019) 81 Land Use Policy 897–903.

reflect the position that nature has an ethical standing and must be accorded care and not be subjected to degradation.[43]

The same principle applies in the context of play and celebrations as there are a plethora of festivals, songs, customs, and myths in Uganda that celebrate nature's gifts and reveal not only the personality of nature but also the traditional close sense of connectedness and harmony between humans and nature.[44] Examples include the *Agwara* dance among the *Alur* community of Northern Uganda,[45] the *Ekyooto* festival of Banyankore, and rituals by the Acholi to appease a spiritual force called the *Jok*[46] during times of drought or agricultural nee.[47] The country's vast cultural diversity provides a wealth of ecologically and spiritually literate knowledge and practices from which to learn. This is a phenomenon that cuts across the African continent in general, and not just Uganda as a country. In modern times, although many of the practices involving rituals have diminished, the cultural memory remains significant, reflecting the deep-rooted connection between spirituality and environmental stewardship among the people. While many have embraced the Abrahamic religions,[48] the people of Uganda's historical respect and recognition for nature have evolved rather than disappeared.

RoN offers a transformative approach to reconciling diverse land ownership structures in post-colonial societies. Historically, many of these societies operate under fragmented legal frameworks, where colonial land tenure systems coexist with indigenous and customary practices. The recognition of RoN provides an avenue for integrating these systems, fostering a more cohesive legal approach that respects both environmental sustainability and indigenous land rights. By assigning legal personhood to nature, RoN creates a framework where environmental governance is not solely reliant on human-centric property rights but also considers nature as a legal entity with enforceable rights.

---

[43] Doreen Misanya, Valtentina C Tassone, Aad Kessler, Paul Kibwika and Arjen EJ Wals, 'Analysing Farmers' Learning for Socio-Ecological Stewardship in Eastern Uganda: A Transformative Learning Ecology Perspective' (2023) 95(1) NJAS: Impact in Agricultural and Life Sciences 31.

[44] See, eg, Alfdaniels Mabingo, '"African Dances are Valid Knowledge": Dance Teachers' De/construction of Meanings from Cultural Heritage Dances in Uganda' (2019) 20(3) Research in Dance Education 311.

[45] Agwara dance: Helping to preserve the Alur culture for centuries (2017), <https://okello22blog.wordpress.com/2017/11/24/agwara-dance-helping-to-preserve-the-alur-culture-for-centuries/>.

[46] Bethwell A Ogot, 'The Concept of jok' (1961) 20 African Studies 123.

[47] See 'Banyankore Culture of Uganda' (18 May 2024), <https://www.twesesafaris.com/news/the-banyankore-culture-of-uganda>.

[48] According to the 2014 Uganda National Population and Housing Census, Christians constituted about 83–84% of the population, Muslims 13.5–14%, traditional or animist religions 0.1%, other religions 1.4–1.6%, and those with no religion 0.2%; Uganda Bureau of Statistics, *National Population and Housing Census 2014: Main Report* (UBOS, 2016) Table 4.1 <https://unstats.un.org/unsd/demographic/sources/census/wphc/Uganda/UGA-2016-05-23.pdf>. In all groups RoN legislation is considered to be in line with their respective traditions.

V. International and Regional Law Influences

The regional lawmaking landscape has contributed to the adoption of RoN in Uganda. As mentioned earlier, the African Commission on Human and Peoples' Rights adopted resolution ACHPR/Res.372 in 2017, which calls on all states parties to recognize sacred natural sites and territories (SNST), their customary governance systems, and the rights of the custodian communities. Uganda, as a signatory to this resolution, committed to recognizing and protecting sacred natural sites and territories. It is not a coincidence that Uganda's NEA was enacted in 2019, just two years after the adoption of the resolution. By signing the resolution, Uganda demonstrated its intention to align domestic environmental law with regional commitments, thereby reinforcing the protection of ecosystems and biodiversity. SNST are important ecological, cultural, and spiritual sites for indigenous groups, and they are understood to be vital to the Earth's overall health and resilience. They serve as wildlife havens and play an important role in preserving biodiversity and ecosystem integrity. The conservation of SNST thus hinges on the acknowledgment of custodian communities' rights.

A central aspect of this advocacy involved the revitalization and integration of customary ecological governance systems. Communities such as the Bagungu in western Uganda have traditionally protected sacred natural sites, embodying principles that align with the RoN framework. By recognizing and incorporating these indigenous practices into national legislation, Uganda not only honors its age-old cultural heritage but also strengthens the practical implementation of RoN by leveraging longstanding community-based conservation methods and processes.[49]

This trend is mirrored at the national level. Following several years of community conversations, Bagungu villages in Uganda came together in November 2018 to construct eco-cultural maps and document their customary rules. The community recognized RoN and the future generations of all species as part of these Earth-centered rules.[50] The Bagungu have been advocating for the recognition and protection of their ancestral area and rights as custodian communities, employing these community-developed resources.

The recognition of RoN in Uganda reflects a broader global movement towards eco-centric legal frameworks, as seen in countries like Ecuador and New Zealand. Indeed, ANARDE based its legal report for the legislature not only on local knowledge but also on models from Ecuador and New Zealand, both of which had enacted rights of nature legislation before Uganda.[51] At the same time, Uganda's approach, which integrates statutory law with customary governance systems, offers a unique model that could inform and inspire similar initiatives in other jurisdictions seeking to harmonize environmental conservation with cultural practices.

---

[49] Hopewell (n 12).

[50] See above, 356.

[51] ANARDE, *Rights of Nature Gain Ground in Uganda's Legal System* (Gaia Foundation/ANARDE, 2 April 2019), <https://gaiafoundation.org/rights-of-nature-gain-ground-in-ugandas-legal-system/>.

A comparative analysis of Ecuador and Uganda highlights the varying degrees to which RoN can be institutionalized within national legal frameworks.[52] Ecuador's 2008 Constitution explicitly recognizes nature as a rights-bearing entity, embedding RoN within its environmental governance structures. By contrast, the effectiveness of RoN in Uganda depends on their institutionalization within existing legal and governance structures. In post-colonial societies, where land ownership often reflects historical inequities, the adoption of RoN can foster institutional interplay between state law, customary law, and international environmental principles. This interplay enables legal adaptation, where environmental concerns are integrated into land tenure systems without disregarding the cultural and historical dimensions of land use. By recognizing nature as a stakeholder, RoN encourages participatory governance, where local communities, governmental agencies, and environmental organizations collaborate to enforce ecological protections.

VI. Efficacy

Over the last decades, there has been a growing understanding and acknowledgment of the intrinsic qualities of sacred natural locations and their accompanying customary governance systems. Various government entities and conservation agencies, as well as civil society groups and guardians, have called for the adoption and implementation of strict rules for the protection of natural and cultural sites and other vital biodiversity areas. Similarly, civil society organizations such as NAPE have collaborated with international organizations to address the impacts of mining and other extractive activities on these sites, emphasizing their cultural and ecological significance. Their efforts include implementing the relevant environmental laws and international agreements which provide a framework for environmental management and the protection of natural sites.[53]

The UNESCO Universal Declaration on Cultural Diversity (2001), the UNESCO Convention for the Safeguarding of Intangible Cultural Heritage (2003), the Convention on the Protection and Promotion of Diversity of Cultural Expressions (2005), the UN Declaration on the Rights of Indigenous Peoples (2007); IUCN 2008 Resolution 4.038 Recognition and safeguarding of sacred natural places in protected areas[54] and 2012 recommendation 147 Sacred Natural Sites – Support for Custodians and Customary Laws in the Face of Global Risks and Challenges[55] are among some of the relevant international agreements.[56] Uganda has been

[52] Putzer and others (n 14) 519.

[53] NAPE and Gaia Foundation, *Mining and its Impacts on Water, Food Sovereignty and Sacred Natural Sites and Territories: Advocating for Recognition and Protection of Water, Food Sovereignty and Sacred Natural Sites and Territories in Uganda's Oil Region* (July 2014), <https://sacrednaturalsites.org/wp-content/uploads/2014/09/napegaiaminingreport.pdf>.

[54] <https://sacrednaturalsites.org/items/iucn-2008-resolution-4-038-recognition-and-conservation-of-sacred-natural-sites-in-protected-areas/>.

[55] <https://portals.iucn.org/library/node/44114>.

[56] AFRICE, *Customary Governance and Earth Jurisprudence* (2021), <https://africeug.org/customary-governance-and-earth-jurisprudence>.

part of these efforts, and the implementation of RoN is expected to further help in the realization of these international obligations.

Community-led conservation initiatives play a crucial role in addressing Uganda's environmental challenges by integrating indigenous knowledge with sustainable resource management. They empower local communities to take charge of conservation efforts, ensuring ecological resilience and sustainable livelihoods.

Nonetheless, the jury is still out on the impact of the recognition of RoN in Uganda, as the law is relatively new. The practical enforcement of RoN presents challenges, particularly in balancing environmental protection with economic development initiatives, such as oil exploration in ecologically sensitive regions.[57] The effectiveness of RoN will depend on the development of specific regulations and the establishment of mechanisms that ensure environmental considerations are prioritized in decision-making processes. The role of NGOs and local communities remains crucial in monitoring and advocating for the faithful implementation of these rights.[58] It is anticipated that RoN could have a significant impact on the conservation of nature and environmental protection at a time when Uganda is engaged in oil exploration activities in an ecologically sensitive part of the country.

VII. Conclusion: The Importance of RoN

Uganda's environment is facing significant threats that require urgent attention. Key among these are deforestation, land degradation, climate change, and biodiversity loss. Uganda's forests are rapidly shrinking due to logging, agriculture, and other human activities.[59] This loss of forest cover not only threatens the country's biodiversity but also exacerbates soil erosion and water scarcity, undermining the agricultural sector that the majority of Ugandans rely on. Similarly, land degradation is widespread, driven by factors like overgrazing, poor agricultural practices, and deforestation, leading to declining soil fertility and reduced agricultural productivity.[60] This issue is noted in the UNEP (United Nations Environment Programme) report[61], which further points to the loss of wetlands and erosion as major contributors to environmental challenges.

Moreover, climate change is significantly impacting Uganda's physical environment.[62] The World Bank Climate Knowledge Portal reveals that the country is

---

[57] Jack Losh, 'Uganda Joins the Rights-of-Nature Movement but Won't Stop Oil Drilling' (*National Geographic*, 2 June 2021) <https://www.nationalgeographic.com/environ ment/article/uganda-joins-the-rights-of-nature-movement-but-wont-stop-oil-drilling>.

[58] Hopewell (n 12).

[59] Michèle von Kocemba and others, '"The Key Issue is a Land Question." – Five Decades of Forest Cover Change in the Mount Elgon Protected Area System' (2025) 22 JMS 776.

[60] Maia Call and Clark Gray, 'Climate Anomalies, Land Degradation, and Rural Out-migration in Uganda' (2020) 41 Population and Environment 507.

[61] UN Environment Programme, 'Disasters and Conflicts, Uganda', <https://www.unep.org/topics/disasters-and-conflicts/country-presence/uganda>.

[62] Nuwagira Upton and Igga Yasin, 'Review of the Past, Current, and the Future Trend of the Climate Change and its Impact in Uganda' (2022) 5 EAJENR 115.

experiencing more extreme weather events – such as droughts floods – and unpredictable rainfall patterns.[63] These climatic shifts adversely affect agriculture, water supply, and the livelihoods of millions of Ugandans. UNEP also identifies that Uganda's natural resources, particularly its water bodies and ecosystems, are under growing pressure from both human activity and climate-related changes.[64] Given these environmental threats, the introduction of RoN can serve as an effective legal tool to mitigate these problems. By granting legal standing to ecosystems and natural entities, Uganda can provide a legal mechanism for protecting forests, wetlands, and other critical environments. This shift in the legal framework, which acknowledges the intrinsic value of nature, offers the potential to better safeguard the country's natural resources and address the escalating environmental challenges.

The expansion of settlements and agricultural activities can have a significant detrimental impact on biodiversity and water resources, directly conflicting with RoN principles. For example, swamps and wetlands, which are vital for regulating water quality and providing habitats for diverse species, are frequently drained or converted into farmland or urban areas, putting pressure on ecosystems that depend on the natural balance. In Uganda, this tension is visible in regions like the Murchison Bay and Katonga Wetlands, where human settlements and agricultural expansion are increasingly threatening critically vital ecosystem. Although RoN could mitigate such issues by legally protecting these ecosystems, the economic needs of communities and the demands for land for housing and agriculture often take precedence over environmental preservation.[65]

While the RoN movement has gained traction in various jurisdictions, its potential in post-colonial societies remains underexplored. The success of RoN will largely depend on the extent to which legal systems can accommodate non-human entities as rights bearers while addressing land ownership conflicts that are deeply rooted in historical injustices. As more nations explore the adoption of RoN, there is an urgent need for jurisprudential development that clarifies the legal standing of nature and its representation in court. Future research needs to focus on how RoN can be integrated into land reform policies, ensuring that environmental governance is both legally enforceable and culturally adaptable. Uganda's NEA incorporates some elements of RoN but remains in its formative stages. The challenge lies in its practical implementation, as Uganda's legal system still heavily relies on human-centered property rights. However, the incorporation of RoN into statutory law presents an opportunity to develop a legal framework that harmonizes statutory land ownership laws with customary practices, ensuring ecological protection within a localized context.[66]

---

63 <https://climateknowledgeportal.worldbank.org/country/uganda>.

64 UNEP, Measuring Uganda's Climate Vulnerabilities <https://unepccc.org/project/measuring-ugandas-climate-vulnerabilities>.

65 'Uganda – Environmental' (n 40).

66 Putzer and others (n 14) 518.

South Africa

Protecting Nature through an Anthropocentric-Oriented Rights Paradigm

Louis J Kotzé[*]

I. Absence of Rights of Nature Provision in South African Law

Like in the majority of other countries on the African continent,[1] there is no provision – either contained in the Constitution of the Republic of South Africa, 1996 (Constitution); or in the environmental framework law, the National Environmental Management Act 107 of 1998 (NEMA); or in any sectoral environmental law – that explicitly or implicitly provides for rights of nature.[2] The issue of rights of nature has also not arisen explicitly in any court case to date. The closest that South African law comes to recognizing rights of nature is in s 24 of the Constitution, the environmental human right provision:

---

[*] I wish to acknowledge with gratitude the inputs and guidance of my colleagues, Professors Willemien du Plessis and Anél du Plessis, and Dr. Caiphas Brewsters Soyapi, on an earlier draft of this report. The manuscript was finalized in 2023 and reviewed in April 2025 to include relevant updates.

[1] Oluwabusayo Temitope Wuraola, 'The Legal Rights of Natural Entities: African Approaches to the Recognition of Rights of Nature' in Michael Addaney and Ademola Oluborode Jegede (eds), *Human Rights and the Environment under African Union Law* (Palgrave Macmillan 2020).

[2] For a comprehensive overview, see Ivana Custic, 'Reflecting on South Africa's Recent Jurisprudence Relating to Estuaries through the Lens of Rights of Nature: Is it Providing Key Guidance or Evidence of Missed Opportunities?' (LLM Dissertation, University of Cape Town 2017).

'Everyone has the right

a. to an environment that is not harmful to their health or well-being; and
b. to have the environment protected, for the benefit of present and future generations, through reasonable legislative and other measures that
    i. prevent pollution and ecological degradation;
    ii. promote conservation; and
    iii. secure ecologically sustainable development and use of natural resources while promoting justifiable economic and social development.'

The environmental human right only has a whiff of an ecocentrism ethic that would mirror the central tenets of a right of nature through its inclusion of the phrase 'ecologically sustainable development'.[3] But this, in and of itself, is insufficient to construct a rights of nature provision. Ecologically sustainable development is generally considered an expression of so-called 'strong sustainability', which is very different from 'weak sustainable development'. Weak sustainable development

'assumes equal importance of the three social, economic and ecological sectors as three "interlocking circles". In this image, the area of sustainable development is where social, economic and ecological interests overlap. Such [an] approach favours the search for the (lowest) common denominator';[4]

which inevitably results in the prioritization of short-term socio-economic growth at the cost of the environment. Strong sustainability, on the other hand, 'sees the ecology as the overarching system, society as part of it and the economy as part of both systems.'[5]

Even if that could have been sufficient to construct a rights of nature clause, any ecocentric ambition that such a provision could have had in human rights terms is inevitably counter-balanced by s 24's predominant anthropocentric framing, a framing that is reinforced and supported by South Africa's anthropocentric legal system.[6] The wording of the environmental right, while it does provide in some

---

[3] S 24(b)(iii) of the Constitution. See, further Melanie Murcott, 'Transformative Environmental Constitutionalism's Response to the Setting Aside of South Africa's Moratorium on Rhino Horn Trade' (2017) 6(84) Humanities 1.

[4] Klaus Bosselmann, 'Strong and Weak Sustainable Development: Making Differences in the Design of Law' (2006) 13 South African Journal of Environmental Law and Policy 40, 43.

[5] Bosselmann (n 4) 43. See also Louis J Kotzé and Sam Adelman, 'Environmental Law and the Unsustainability of Sustainable Development: A Tale of Disenchantment and of Hope' (2023) 34 Law and Critique 227.

[6] Werner Scholtz, 'The Anthropocentric Approach to Sustainable Development in the National Environmental Management Act and the Constitution of South Africa' (2005) 1 Journal of South African Law 69; Louis J Kotzé, 'The Judiciary, the Environmental Right and the Quest for Sustainability in South Africa: A Critical Reflection' (2007) 16(3) RECIEL 298; Louis J Kotzé, 'Phiri, the Plight of the Poor and the Perils of Climate Change: Time to Rethink Environmental and Socio-Economic Rights in South Africa?' (2010) 2(1) JHRE 135; Louis J Kotzé and Anél Du Plessis, 'Some Brief Observations on Fifteen Years of Environmental Rights Jurisprudence in South Africa' (2010) 3(1) Journal of Court Innovation 157.

detail for environmental protection in general terms, is not aimed at environmental protection for the sake of it; it is aimed at environmental protection so as to ensure that present and future human needs are met (including health and well-being) through the 'use of natural resources' and so as to promote 'justifiable economic and social development'. The right neither refers to non-human health and well-being, nor does it reference present or future non-human generations; it is squarely focused on promoting human interests.

The tensions between the type of anthropocentrism that s 24 pursues, and the need for ecological sustainable development that is also recognized in s 24, is evident from judgments such as *WWF South Africa v Minister of Agriculture, Forestry and Fisheries and Others*,[7] where the High Court stated that:

> 'Conservation and sustainable development, which are placed to the fore by s 24 of the Constitution, ss 2(3) and 2(4)(a) of NEMA, and various paragraphs of s 2 of the MLRA [Marine Living Resources Act], are not only, or even primarily, important because of the pleasure humans derive from healthy and biodiverse ecologies. Many people in the past, the present and the future have depended, do depend or will depend for their economic welbeing on exploiting renewable resources. To enable them to do so, and thus to preserve food security and avoid poverty, one cannot allow the resource of the many to be exhausted for the benefit of the few (I speak relatively of the 'few' current participants in the lobster sector as against all of those who will come after them).'[8]

South Africa has a plural legal system, consisting of a wide range of statutes, constitutional law provisions, customary law provisions and indigenous laws.[9] This is also true of South African environmental law.[10] A substantial number of these laws has been influenced by foreign law, ie, other legal systems, including specifically the Constitution and several environmental laws such as NEMA.[11] The Constitution itself explicitly states the South African legal system's openness to foreign law and legal influences in providing that: 'When interpreting the Bill of Rights, a court, tribunal or forum may consider foreign law'.[12] Yet, despite such openness towards foreign legal cultures and influences, none of the key developments around rights of nature, such as those in Latin America, have yet managed to infiltrate the South African legal system. This is of course not to suggest that this might not still happen in future.

---

[7] *WWF South Africa v Minister of Agriculture, Forestry and Fisheries and Others* (11478/18) [2018] ZAWCHC 127; [2018] 4 All SA 889 (WCC); 2019 (2) SA 403 (WCC) (26 September 2018).

[8] At para 91.

[9] Tshepo Madlingozi and Tracy-Lynn Field (eds), *Introduction to Law and Legal Skills in South Africa* (2nd edn, OUP 2021).

[10] Jan Glazewski, *Environmental Law in South Africa* (Lexis Nexis 2024).

[11] Eg, Francois du Bois and Danie Visser, 'The Influence of Foreign Law in South Africa' (2003) 13 TLCP 593.

[12] South African Constitution, s 39(1)(c).

II. Ensuring Adequate Protection of Nature and the Relationship of Rights of Nature to Socio-Economic Development

Apart from several grassroots level movements and initiatives led by lawyers and activists,[13] there are no formal intentions or processes underway to recognize rights of nature. This does not mean, however, that South African law does not protect nature. Several sectoral environmental acts, such as the National Environmental Management: Biodiversity Act 10 of 2004 (NEM:BA) and the National Environmental Management: Protected Areas Act 57 of 2003 (NEM:PAA), contain elaborate provisions that protect nature. But none of these are framed generally in rights terms or, more specifically, in rights of nature terms.

From a realist perspective, this begs the question: Does one actually require an explicit rights of nature provision in a constitution or law to ensure adequate protection of nature? While there are no straightforward answers to this question, one could argue that sufficient and effective nature protection laws might be sufficient. These laws might not be grounded in an ecocentric/relational ethic of the kind that a rights of nature provision would support, but they nevertheless protect nature. For example, the purpose of the NEM:BA is to provide for

> 'the management and conservation of biological diversity within the Republic and of the components of such biological diversity; the use of indigenous biological resources in a sustainable manner; and the fair and equitable sharing among stakeholders of benefits arising from bioprospecting involving indigenous biological resources'.[14]

NEM:PAA, for its part, aims to provide for the

> 'declaration and management of protected areas; [...] for co-operative governance in the declaration and management of protected areas; [...] to effect a national system of protected areas in South Africa as part of a strategy to manage and conserve its biodiversity; [...] to provide for a diverse and representative network of protected areas on state land, private land, communal land and marine waters; [...] to promote sustainable utilisation of protected areas for the benefit of people, in a manner that would preserve the ecological character of such areas; [and] [...] to promote participation of local communities in the management of protected areas, where appropriate.'[15]

These are all well-intended and comprehensive objectives that create an enabling governance framework that mandates government to protect nature, and depending on how the provisions of these laws are enforced, they can even be highly effective.

---

[13] Steven Robins, '"Rights of Nature": A fundamental challenge to the world as we know it' (*Daily Maverick*, 6 January 2022) <https://www.dailymaverick.co.za/article/2022-01-06-rights-of-nature-a-fundamental-challenge-to-the-world-as-we-know-it/>; Lesai Seema, 'Recognising that Nature Has Legal Rights Is a Key Step to Slowing the Degradation of Our Planet' (*Daily Maverick*, 2 May 2023) <https://www.dailymaverick.co.za/opinionista/2023-05-02-recognising-that-nature-has-legal-rights-is-a-key-step-to-slowing-the-degradation-of-our-planet/>.

[14] NEM:BA, s 2.

[15] NEM:PAA, s 2.

The more pertinent consideration then is perhaps not if and how a rights of nature provision shapes nature conservation laws (South Africa's example suggests that the country has sufficient laws despite the absence of a rights of nature provision), but rather how it might shape and influence the more general socio-economic development trajectory of a country, including its trade, its investment and economic policies, and its policies on population growth, among many others. By way of analogy, in most Latin American countries that have explicit and far-reaching rights of nature provisions, there is still an ongoing debate about the real meaning, impact, and value of these provisions.[16] There is a valid argument that, often, many of these provisions remain merely symbolic despite small pockets of success, and that they have not been able to fundamentally change the course of Latin American countries' extractivist economies. We clearly see, for example, the many challenges posed by the realities of implementing *buen vivir* as playing out in the so-called 'Latin American paradox', in terms of which unabated mining, hydrocarbon activities, and monoculture exports continue to fund social spending and public works.[17]

III. Activity of NGOs and Key Figures in Protecting Nature

While a rights of nature provision does not exist in South African law, numerous NGOs (such as Save the Vaal Environment; the World Wildlife Fund; Friends of the Earth South Africa/groundWork, Birdlife South Africa; the Endangered Wildlife Trust; the Wildlife and Environment Society of South Africa[18]) have been and continue to be active in supporting, enforcing, and relying on the s 24 environmental human right and other environmental laws, notably in their legal aid assistance to indigent communities and class action campaigns against corporations and government. A key example is the Centre for Environmental Rights, which is responsible for litigating most of South Africa's prominent environmental protection cases.[19] But none of these efforts have been explicitly conducted on the back of the rights of nature paradigm.[20]

---

16 Kotzé and Adelman (n 5).

17 Miriam Lang, 'Prologue: Crisis of civilisation and challenges for the left' in Miriam Lang and Dunia Mokrani (eds), *Beyond Development: Alternative Visions from Latin America* (1st transl edn, Fundación Rosa Luxemburg and Transnational Institute 2013) 5.

18 Links to environmental NGO's at <https://www.fosaf.org.za/ngo.php>.

19 <https://cer.org.za/>.

20 An iniative demands the recognition of Table Mountain as a 'living' entity; see <https://www.Tablemountainrights.org/>; Kevin Bloom, 'Towards the Recognition of Table Mountain as a "Living" entity — in Law and Beyond' (*Daily Maverick*, 22 April 2025) <https://www.dailymaverick.co.za/article/2025-04-22-towards-the-recognition-of-table-mountain-as-a-living-entity-in-law-and-beyond/>.

It is probably ironic that while South African law does not provide for rights of nature, one of its most famous and globally influential environmental lawyers, Cormac Cullinan, has been and continues to be a key figure in propagating the rights of nature paradigm based on his key publication, Wild Law: A Manifesto for Earth Justice,[21] where he argues for the recognition of rights for nature. A Wild Law Institute named after the book operates in Cape Town towards the recognition of Rights of Nature worldwide.[22] While mere speculation, one of many reasons why such prominent advocates of rights of nature have been unable to successfully advocate for the inclusion of a rights of nature provision in the South African Constitution (if it was their intention to do so to start with) is possibly because of the complex process of drafting constitutions where broad consensus is sought, instead of privileging any one specific aim of any one specific interest group.[23] Moreover, the Constitution was finalized and adopted in 1996. During that time, rights of nature provisions across the world were only tentatively starting to emerge in some legal systems, and they had not become part of the more mainstream debate and, more importantly, formal legal processes as they are today.

IV. Enforcement of Rights of Nature

South African law provides a very wide *locus standi* to a virtually unlimited range of people to institute proceedings in a court, including claims based on s 24 of the Constitution.[24] But these would not be relevant for present purposes since South African law does not provide for rights of nature. If it might offer such a right in future, the following constitutional provisions would be applicable, and they will offer an equally wide and almost unrestricted access to courts that will allow almost anyone to act on behalf of nature:

Constitutional provisions:

'*34. Access to courts*
Everyone has the right to have any dispute that can be resolved by the application of law decided in a fair public hearing before a court or, where appropriate, another independent and impartial tribunal or forum.'

'*38. Enforcement of rights*
Anyone listed in this section has the right to approach a competent court, alleging that a right in the Bill of Rights has been infringed or threatened, and the court may grant appropriate relief, including a declaration of rights. The persons who may approach a court are –

---

[21] Cormac Cullinan, *Wild Law: A Manifesto for Earth Justice* (Green Books 2003).

[22] <https://www.wildlaw.net/>.

[23] See, for example, Christina Murray, 'A Constitutional Beginning: Making South Africa's Final Constitution' (2001) 23(3) U Ark Little Rock L Rev 809.

[24] Louis J Kotzé, 'Access to Courts and Environmental Dispute Resolution in South Africa: A Survey of Recent Developments in Constitutional State Context' (2010) 2 Onze Omgeving: Tijdschrift van de Vlaamse Vereniging voor Omgevingsrecht 9.

a. anyone acting in their own interest;
b. anyone acting on behalf of another person who cannot act in their own name;
c. anyone acting as a member of, or in the interest of, a group or class of persons;
d. anyone acting in the public interest; and
e. an association acting in the interest of its members.'

Environmental framework law provisions:[25]

'32. (1) Any person or group of persons may seek appropriate relief in respect of any breach or threatened breach of any provision of this Act, including a principle contained in Chapter 1, or any other statutory provision concerned with the protection of the environment or the use of natural resources:
(a) in that person's or group of persons own interest;
(b) in the interest of, or on behalf of, a person who is. for practical reasons, unable to institute such proceedings:
(c) in the interest of or on behalf of a group or class of persons whose interests are affected;
(d) in the public interest; and
(e) in the interest of protecting the environment."

As far as could be established, there has to date not been a case that has been brought before the courts on behalf of nature.

V. International Law

Similar to its stance towards foreign law, South African law warmly embraces and incorporates international law. This is evident from the Constitution that provides: 'Customary international law is law in the Republic unless it is inconsistent with the Constitution or an Act of Parliament'.[26] Moreover, '[w]hen interpreting any legislation, every court must prefer any reasonable interpretation of the legislation that is consistent with international law over any alternative interpretation that is inconsistent with international law.'[27] These provisions are bolstered by the constitutional requirement that '[w]hen interpreting the Bill of Rights, a court, tribunal or forum must consider international law.'[28] Moreover, NEMA sets out elaborate provisions in its Chapter 6 related to international environmental agreements and obligations. None of the latter, however, specifically refer to the incorporation of rights of nature.

Collectively, however, these provisions do create an enabling environment that could potentially facilitate the future incorporation of international laws on rights of nature into South African law, where they exist either as part of a binding agreement that South Africa is a party to, or as part of customary international law that would then apply automatically in the country.

[25] NEMA, s 32.
[26] South African Constitution, s 232.
[27] South African Constitution, s 233.
[28] South African Constitution, s 39(1)(b).

The primary reason that these provisions have not yet been used to incorporate international law related to rights of nature is because such a right does not exist in international law. Rights of nature are at least neither (yet) entrenched in or supported by any binding international agreement, nor do they manifest in customary law provisions. The 2022 United Nations General Assembly Resolution recognizing the right to a clean, healthy, and sustainable environment,[29] is very much formulated in trite anthropocentric *human* rights terms, and does not provide for rights of nature.[30] The United Nations Harmony with Nature Programme is the only UN-level initiative that develops and supports more ecocentric approaches to environmental protection, including rights of nature as a means to achieve this.[31] This is also true for the Universal Declaration of the Rights of Mother Earth that was launched by some states in 2010, but failed to garner the critical universal support it required to become part of the larger international environmental law agenda.[32] In sum, nothing of what is crystalizing through these initiatives has resulted in the formulation and adoption of any meaningful international agreement on rights of nature. Developments in international law with respect to rights of nature remain few and far between and until this changes it is unlikely that South African law will adopt rights of nature as a result of international law.

---

[29] (A/RES/76/300).

[30] Louis J Kotzé, 'The Right to a Healthy Environment and Law's Hidden Subjects' (2023) 117 AJIL Unbound 194, <https://doi.org/10.1017/aju.2023.27>.

[31] <http://www.harmonywithnatureun.org/>.

[32] <https://www.garn.org/universal-declaration-for-the-rights-of-mother-earth/>.

V. Resisting Legal Orders

Brazil

Isolated Legislative and Judicial Inroads in a Dependent Capitalism Landscape

Antonio Carlos Wolkmer and Debora Ferrazzo[*]

I. Existence and Contents of Rights of Nature

1. No Dedicated Federal Provisions Concluded but Local Legislation

The rights of nature are not expressly recognized in the Brazilian legal system. However, constitutional hermeneutics based on a broad interpretation of the prohibition of animal cruelty (Constitution, art 225, § 1, VII, final part)[1] allow for the possibility of recognition, even if based on implicit legal rules.

---

[*] The manuscript was finalized in December 2022. The assessment of literature and case law, including internet materials, reflects the legal situation at that time. Following revision and discussion, the text was supplemented in February 2025 to reflect more recent developments. All translations of Portuguese-language statutes, cases, and literature are author translations except as otherwise noted.

[1] Constitution of the Federative Republic of Brazil 1988 <www.planalto.gov.br/ccivil_03/constituicao/ConstituicaoCompilado.htm>.

There are some state and municipal laws that declare that animals or nature have rights, and while each one deserves to be celebrated, it is possible to foresee conservative resistance arguing that these laws are unconstitutional in regulating 'legal personality'. The argument might go on to claim that the matter should be dealt with in the Constitution, or more precisely in the Civil Code, and that it is the responsibility not of the states or municipalities but of the Union, that is, the National Congress.[2]

One example of successful local legislation is that of the municipality of Bonito, Pernambuco State, which in 2017 stipulated in its organic law[3] that 'The Municipality recognizes the right of nature to exist, thrive and evolve'. Subsequent to this provision, another law was passed, Law 1257,[4] which provides for the creation of the Municipal Policy of Agroecology and Organic Production, with guidelines for the respective Municipal Plan. This law is oriented towards promoting the development and sustainable use of natural resources, which, according to its art 3, includes promoting 'the rights of nature', as recommended by the aforementioned organic law of the municipality. It also addresses issues such as the popular solidarity economy and traditional peoples, who are considered to be those who adopt their own forms of organization and interaction with their territories and natural resources.

Likewise in the state of Pernambuco, the municipality of Paudalho[5] approved legislation for protection of the mineral water source in São Severino, whereby the legislative text includes a discussion about the existence, prosperity, and evolution of nature and justifies public action in the sense of guaranteeing the right to an ecologically healthy and balanced environment.

In 2019, Amendment no 47 to the Organic Law of the municipality of Florianópolis[6] in the state of Santa Catarina was enacted. This amendment gives the municipality the competence to adopt measures that promote diversity and harmony with nature, thus promoting the sustainable management of resources and

---

[2] See the rules on legislative powers in the Constitution of the Federative Republic of Brazil 1988, art 22(I).

[3] Lei Orgânica do Município de Bonito Estado de Pernambuco 2016 (Bonito, Pernambuco, Brazil), <https://transparencia.bonito.pe.leg.br/app/pe/bonito/2/lei-organica-municipal>.

[4] Lei Municipal Nº 1.257, de 7 de julho de 2021 Dispõe sobre a criação da Política Municipal de Agroecologia e Produção Orgânica de Bonito (PMAPO) e estabelece as diretrizes para o Plano Municipal de Agroecologia e Produção Orgânica de Bonito (PLAMAPO) (Bonito, Pernambuco, Brazil), <www.institutoabdalazizdemoura.org.br/wp-content/uploads/2023/11/LEI-N°-1257-2021-SANCIONADO_EXEC.-AGROCOLOGIA.pdf>.

[5] Lei Nº 878 de 20 de dezembro de 2018 Declara como Patrimônio Natural, Ambiental, Cultural, a Fonte de Água Mineral em São Severino do Ramos e dá outras providências (Paudalho, Pernambuco, Brazil), <https://transparencia.paudalho.pe.gov.br/uploads/5310/2/atos-oficiais/2018/leis/1684436308_lei8782018.pdf>.

[6] Emenda à Lei Orgânica Nº 47, de 2019 Altera o art 133 da Lei Orgânica do Município de Florianópolis de 12 de novembro de 2019 (Florianópolis, Santa Catarina, Brazil), <https://leismunicipais.com.br/a/sc/f/florianopolis/emenda-a-lei-organica/2019/4/47/emenda-a-lei-organica-n-47-2019-altera-o-art-133-da-lei-org-nica-do-municipio-de-florianopolis>.

agroecological practices with the objective of achieving 'quality of life of human and non-human populations, respecting the principles of good living and granting nature the right of ownership'.

2. Recent Case Law

The fact is that since we started writing this report, new legal and judicial precedents in addition to those mentioned above have been emerging, and it can be assessed that such a development is occurring with gradually increasing frequency. This indicates a progressive march towards successfully establishing the rights of nature, but it is not a path without setbacks.

In contrast to the legal precedents mentioned above, the Brazilian justice system refused to admit the case regarding legal action in respect of the River Doce, in the state of Minas Gerais, in response to the petition of the lawyer and professor Lafayette Garcia Novaes Sobrinho.[7] These proceedings will be discussed below, but an excerpt from the court decision can be quoted at this point:

> '[O]ur legal system has not conferred attributes of personality to other living beings, such as animals, forests, seas and rivers [...] at least legally, the river cannot claim its rights before the courts [...] With these considerations, I verify the absence of a procedural presupposition of existence, as the legal system does not grant legal personality to the applicant Bacia Hidrográfica do Rio Doce [Doce River Basin] [...]'[8]

With regard to the protection of the rights of nature, the judiciary (Supreme Federal Court, STF) has recognized in its rulings the right of animals not to suffer cruel treatment. This was the case of the Direct Actions of Unconstitutionality (ADI) nos 1856 and 3776,[9] which declared cockfighting unconstitutional. This was also the case of Extraordinary Appeal no 153,531,[10] which declared the *farra do boi* ('festival of the ox') unconstitutional.[11]

---

[7] Lafayette Garcia Novaes Sobrinho, 'Ação judicial a favor da bacia hidrográfica do rio doce' (2018) 13(1) Revista Brasileira de Direito Animal 219. Finally, in 2024, Lei Nº 4.225 was approved, guaranteeing rights to the waves at the mouth of the River Doce, <https://linhares.camarasempapel.com.br/Arquivo/Documents/legislacao/image/LEI42252024.pdf>.

[8] Márcia Rodrigues Bertoldi and Roberta Fortunato Silva, 'Direitos da Natureza e acesso à justiça: a ampliação dos atores legitimados em ações coletivas para uma justiça socioambiental' (2020) 29(54) Revista Direito em Debate 118, 123–124.

[9] Supremo Tribunal Federal, Direito constitucional, ADI 1856, DJE no 106, 3 June 2011; and idem, ADI 3776, DJ no 124, 29 June 2007.

[10] Supremo Tribunal Federal, Direito constitucional, Recurso Extraordinário 153.531-8 Santa Catarina, DJ no 1902-02, 13 March 1998, <http://portal.stf.jus.br/processos/detalhe.asp?incidente=1544862>.

[11] Indeed, these events are a very common practice in the south of the country, particularly in the state of Santa Catarina. Despite terming themselves a 'party', they end up being reduced to the reproduction of cruel practices, with severe mistreatment of animals.

But when the STF found the *vaquejada* unconstitutional in ADI no 4983,[12] it came under criticism and engendered a movement to 'abolish' the jurisdictional protections granted to nature. This was done through an amendment to the Constitution, which became known as the 'PEC[13] da Vaquejada'. As shocking as it may seem, the amendment was not given this name by environmentalists; rather, it emerged from the position of the congresspeople themselves, and even led to the announcement of the formation of a Parliamentary Front in *Defence of Vaquejada*. The effect was that the official documentation to process the proposed amendment in Congress expressly stated that it was a reaction to the declaration on the unconstitutionality of *vaquejada*, as can be seen in the official publications of the Federal Senate.[14] This demonstrates that it was an opportunistic PEC (constitutional amendment proposal) meant to serve the wishes of rural representatives in the face of a particular STF decision that contradicted their interests.

It is also important to mention one of the more recent precedents of the Brazilian Constitutional Court. This is an action referred to as the Argument of Non-Compliance with the Fundamental Precept,[15] which was circulated in 2019 and decided in late 2021. In this case, the Court was asked to declare unconstitutional the interpretation of the regulation that permitted the slaughter of animals captured under conditions of mistreatment or abuse. Brazilian legislation regulates the confiscation of 'products and instruments' under these conditions, stipulating that priority be given to the release of the animals into their habitats and, if this is not possible, to delivery to zoos or qualified institutions or technicians who can assume custody and care. The relevant provisions also stipulate that animals must be kept in conditions that ensure their welfare until they can be relocated. However, the authorities had frequently been obtaining judicial authorization to kill these animals.

Based on the grounds presented in the above decision, the ruling can perhaps be considered the most progressive precedent of the Brazilian Constitutional Court, as it appears to have affirmed in the most explicit and forceful way the recognition of rights for other living beings – beyond human beings. The summa-

[12] Supremo Tribunal Federal, Direito constitucional, ADI 4983, <http://portal.stf.jus.br/processos/detalhe.asp?incidente=4425243>. *Vaquejada* is a sport practiced in northern Brazil where two people on horseback try to bring down an ox by pulling it by the tail between two previously marked tracks.

[13] PEC is the abbreviation for an amendment updating the Constitution (proposed amendment to the Constitution).

[14] Senado Federal, Proposta de Emenda à Constituição Nº 50, de 2016, <www25.senado.leg.br/web/atividade/materias/-/materia/127262>; and Câmara dos Deputados, Proposta de Emenda à Constituição 304/2017, <www.camara.leg.br/proposicoesWeb/fichadetramitacao?idProposicao=2123843>.

[15] Supremo Tribunal Federal, Arguição de Descumprimento de Preceito Fundamental no 640, 20 September 2021, <https://redir.stf.jus.br/paginadorpub/paginador.jsp?docTP=TP&docID=758761538>.

ry of the ruling states that 'art 225, §1, VII, of the Constitution of 1988 stipulates the protection of fauna and prohibits any form of mistreatment of animals, *in order to recognize the inherent value of other forms of non-human life*, protecting them against abuse'[16] (emphasis added). These constitutional resolutions, the minister[17] says, are in line with international standards and the tenets of science that have advocated respect for the life of non-human animals, given their status as sentient beings.

3. Recognition of Specific Rights and Proposed Legislation

Some court decisions are beginning to emerge that recognize specific rights, such as the actions mentioned in the previous section; also seen in a pioneering way in Brazil was the first grant of habeas corpus in respect of a non-human animal, *a horse*, thus establishing the first recognition of the right to physical freedom for non-human animals. This happened not in legislation, but in jurisprudence.

The ruling in question is from the state of São Paulo (2020), and the constitutional protection was granted in an interlocutory appeal ruling[18] despite not having been requested by the parties and not being directly related to the matter discussed in the appeal. This was possible because Brazilian legislation, specifically art 654, § 2 of the Brazilian Code of Criminal Procedure,[19] allows the granting of constitutional protection ex officio by judges and courts when they verify the need to redress a violation or an unlawful threat to the law.

Another relevant precedent was the case involving the parrot Leozinho. This relates to a conflict involving the long-term maintenance of a parrot living with an elderly woman.[20] According to the Brazilian Institute for the Environment and Renewable Natural Resources (IBAMA), the wild animal lived in captivity and its custody should have been transferred to the public authorities, thus preserving legality through judicial guardianship in order to eliminate social expectations of impunity in the face of criminal practices in relation to wild animals. In the first instance court, the judge understood that the complexity of the case could not be resolved through strictly legal means, because after some 15 years of cohabitation, the animal itself had become accustomed to being fed by humans; further, an emotional bond had been established between the parrot and the 75-year-old woman, and they were both integrated into the same family unit. A

---

[16] Note 15.

[17] Title of constitutional judges in Brazil.

[18] Tribunal de Justiça de São Paulo, Agravo de Instrumento no 2139566-66.2019.8.26.0000, 17 June 2020, <www.tjsp.jus.br/>.

[19] Decreto-Lei Nº 3.689, de 3 de outubro de 1941, Código de Processo Penal. Published in Diário Oficial da União, 13 October 1941, <www.planalto.gov.br/ccivil_03/decreto-lei/del3689compilado.htm>.

[20] Superior Tribunal de Justiça, Agravo Interno no Recurso Especial no 1389418/PB (2013/0211324-4), 21 September 2017, <https://processo.stj.jus.br/processo/pesquisa/?aplicacao=processos.ea&tipoPesquisa=tipoPesquisaGenerica&termo=REsp%201389418>.

sudden change in its routine could have caused severe harm to the parrot, as well as to its guardian, who, in addition to her advanced age, was suffering from hypertensive heart disease. The court accepted these arguments and allowed the parrot to remain with its guardian for reasons including the animal's best interests.

In terms of legislation, there are currently some proposals to alter the Constitution and to amend and create laws to advance the recognition of the rights of nature, the rights of animals, and environmental protection. Some examples are discussed below.

In October 2021, discussion began in the Chamber of Deputies regarding an amendment to the Constitution, PEC 37/2021,[21] so as to include, among the fundamental rights set out in its famous art 5, the right 'to an ecologically balanced environment and climate security'. In this case, the exegesis of the article establishes it as a right of human beings, both Brazilians and foreigners, and not as a right of nature, meaning that the Constitution would continue to be anthropocentric.

At the end of the same year, two bills stood out among many others. One was the bill for the creation of a public health system for non-human animals,[22] declaring the right of access to health services as a fundamental right of animals and an obligation of the state, as well as of people and society. The other was a bill establishing shared custody of pets.[23] Finally, there was a third bill, which, after discussions and amendments, arrived in its current form: this bill declares that non-human animals have a *sui generis* legal status as de-personalized subjects of law and that they must receive jurisdictional protection to protect against violations of those rights by the state; moreover, it is prohibited to treat these non-human animals as things.[24]

Evaluating quantitative data, it is clear that this discussion has expanded, and in recent years the surge in legislative proposals has been exponential. As regards non-human animals, in the years 2022 to 2024 more than 1,300 proposals were presented on average each year in Brazil.[25]

The most important thing to note is that these are all proposed laws that, even if approved, will have to overcome great challenges to become effective.

---

[21] Câmara dos Deputados, PEC 37/2021. Altera o art. 5°, <www.camara.leg.br/proposicoesWeb/fichadetramitacao?idProposicao=2304959>.

[22] Câmara dos Deputados, PL 3308/2021, <www.camara.leg.br/proposicoesWeb/fichadetramitacao?idProposicao=2300298>.

[23] Câmara dos Deputados, PL 4375/2021, <www.camara.leg.br/proposicoesWeb/fichadetramitacao?idProposicao=2311683>.

[24] Câmara dos Deputados, PL 6054/2019, <www.camara.leg.br/proposicoesWeb/fichadetramitacao?idProposicao=601739>.

[25] See <www.camara.leg.br/busca-portal?contextoBusca=BuscaProposicoes&pagina=1&order=relevancia&abaEspecifica=true&q=animais>.

4. Article 225 of the Constitution and Other Legal Grounds

Our main basis is the Constitution, art 225, § 1, VII: 'to protect fauna and flora, *prohibiting, in the form of law, practices that jeopardise their ecological function, cause the extinction of species or subject animals to cruelty*' [emphasis added].

But in the case of the habeas corpus ordered in respect of a horse, which is an important precedent, the relevant legislation was the Code of Criminal Procedure, art 654, § 2: 'Judges and courts have the power to issue writs of habeas corpus on their own when, in the course of the proceedings, they find that someone is suffering or is on the verge of suffering unlawful coercion.' In that case, the court granted the horse's freedom without anyone in the proceedings having applied for it, and to do so it made use of this legal provision.

There are other legal sources for judicial protection which are discussed throughout this report, such as Law 4717/65 and Law 7347/85, which regulate, respectively, popular actions and public civil actions. The first provision is an instrument for natural persons and can be used only by citizens; the second is an instrument for institutions, as only institutions such as associations, public defenders, or public prosecutors can initiate a public civil action. Both actions serve to protect the environment.

In Brazil, the possibility of recognizing nature, or specifically non-human animals, as subjects of rights first began to be explored in the judicial field a few years ago. Thus far, there are not many positive precedents in this regard and, among the few attempts that have been made, the refusals of state officials prevail.

In this respect, research and academic studies indicate that the first attempt to obtain recognition of nature as a subject of rights from the Brazilian courts took place in 2011, through a public civil action proposed by the Public Prosecutor's Office of Pará, a Brazilian state. On that occasion, the initial petition sought to prevent the construction of the Belo Monte megaproject, thereby highlighting humanity's march towards the recognition of the rights of nature. But the judge responded that he could not consider Volta do Xingú as a subject of rights, nor did he identify any illegality during the trial, declaring the requests in this action inadmissible.[26]

The Brazilian Constitution states:

> 'Art 225. Every person has the right to an ecologically balanced environment, good for the common use of people and essential for a healthy quality of life, which imposes on the government and the community the duty to defend and preserve it for present and future generations.'

From this text, we note that, the prevailing view in Brazil is still to consider nature as a good to be used by people, meaning that its protection is not linked to its intrinsic dignity but to the utilitarian value it has for society and humanity. For

---

[26] Bertoldi and Silva (n 8) 123.

this reason, in our comparative studies and research, we consider the Brazilian legal system to be anthropocentric.[27]

II. Institutions and Political Actors Promoting Rights of Nature

In terms of institutions or political actors[28] promoting or actually recognizing the rights of nature, it can be observed that there are some court decisions, but these are still not dominant in character. Otherwise, there are few local laws, the most significant of which having already been mentioned.

Among the factors that have contributed to the progress we are experiencing – in short and slow steps – it is probably worth mentioning legislation adopted in Bolivia (2009) and Ecuador (2008)[29] and also the significant increase of interest in animal rights that we observe in our teaching practice. In the Brazilian case, it is not possible to say that both factors play the same role, since the theoretical and conceptual discussion of Latin American legal systems is not necessarily widespread in academia, with the exception of certain counter-hegemonic initiatives.

Even so, animal rights have increasingly become a topic in Brazilian public debates and political agendas, which leads us to believe that academic debates – which are also more inclined towards animal rights – continue to have a greater influence on the subject.

As regards university debates on the rights of nature, our observations in our teaching practice suggest that the primary interest remains mostly fixed on domestic animals, which demonstrates a selective agenda (many students want to investigate animal rights, but they focus more on dogs and cats and not on other species).

III. Execution of Rights of Nature

1. Key Players in Enforcement

In the case of Brazil, the main actors involved in the enforcement of rights of nature are the Public Prosecutor's Office along with NGOs defending the rights of nature or animal rights; and, in some cases, even individuals play a role. But we do not have a specific body to represent nature; when necessary, we have a member of the Public Prosecutor's Office charged with the task of doing this.

---

[27] Antonio Carlos Wolkmer, Maria de Fátima S Wolkmer and Debora Ferrazzo, 'Derechos de la Naturaleza: para un paradigma político y constitucional desde la América Latina' in Liliana Estupiñán Achury and others (eds), *La naturaleza como sujeto de derechos en el constitucionalismo democrático* (Universidad Libre 2019).

[28] Such as non-governmental organizations, universities, and academic movements.

[29] For both, see the pertinent chapters in this volume.

For example, Law 7347 of 1985[30] establishes the public civil action, which is comparable to a class action in the US system, and these actions have been an important instrument for the defense of the environment, although it can only be brought by entities such as NGOs or the Public Prosecutor's Office itself. In any case, the action protects the environment as an *object and not as a subject* of rights. Even so, from the point of view of protection, it has been an important tool.

Law 4717 of 1965,[31] on the other hand, regulates the popular action, which has a more restricted scope than a public civil action in terms of the protected objects, but which allows for a judicialization of the defense of the environment. This action can only be proposed by persons who can prove their citizenship (with a *título de eleitor*, an identification card indicating that the holder is an eligible voter and thus a citizen).

As pointed out in the previous sections, the recognition and defense of the rights of nature does not have an explicit legal basis in Brazil that considers nature to be a subject of rights. As such, this has been a space for a strengthening of critical legal theory, with an alternative use of the law and even claims for an alternative law.[32] This has led to cracks in conservative jurisprudence and an opening up to new understandings, which has occurred not only in the most common forms, such as those mentioned above – public actions or public civil actions – but through everyday efforts that sometimes work, as in the case of obtaining a grant of habeas corpus in favor of a horse.

Considering individual cases, there is a relevant trial that can be considered paradigmatic, as it relates the possibility of defending the rights of nature on the initiative of private individuals, ie any citizen. In 2020, an appeal in the civil liability trial reached the Court of Justice of the State of São Paulo.[33] The case involved the interests of a beekeeper (the petitioner) who worked as a family farmer, keeping bees and extracting honey. After his neighbor sprayed agricultural pesticides from an airplane, the beekeeper noticed high mortality among his bees and took the case to the Brazilian courts, seeking financial compensation but also the defense of the environment and the bees themselves, given their ecological importance. In this trial, as in others, we saw the impact of debates carried out within the legal field using legal concepts, such as 'ecological constitutionalism'. The judges also reflected on the interpretation of art 225 of the Federal Constitution, asking whether it should be anthropocentric or biocentric and concluding

[30] Lei No 7347, de 24 de julho de 1985. Disciplina a ação civil pública de responsabilidade por danos causados ao meio-ambiente, ao consumidor, a bens e direitos de valor artístico, estético, histórico, turístico e paisagístico (VETADO) e dá outras providências. Published in Diário Oficial da União, 25 July 1985, <www.planalto.gov.br/ccivil_03/leis/L7347Compilada.htm>.

[31] Lei No 4717, de 29 de junho de 1965. Regula a ação popular. Published in Diário Oficial da União, 5 July 1965, <www.planalto.gov.br/ccivil_03/leis/l4717.htm>.

[32] Amilton Bueno de Carvalho, *Direito alternativo* (FisicalBook 1998).

[33] Tribunal de Justiça de São Paulo, Apelação cível no 1002508-07.2017.8.26.0615, 5 July 2020, <https://esaj.tjsp.jus.br/cjsg/getArquivo.do?cdAcordao=13719479&cdForo=0>.

that although the Constitution adopted an anthropocentric vision, it enshrined environmental rights in several provisions, which attribute value not only to human life but also to non-human forms of life, which the Court defines as 'ecological legal anthropocentrism'. Influenced by experiences in other countries, such as the Ecuadorian Constitution of 2008, the judges observed a strengthening in recent years of the biocentric vision, according to which economic activity 'must necessarily observe the rights of nature'.

2. *Competence and Procedures for Defending Rights*

In general, nature in Brazil is considered an asset of the Union or the Brazilian states, depending on which forest, river, etc is at issue and where it is located; thus, if it is, for instance, a river that crosses more than one state, it belongs to the Union[34] and must be defended in federal court. Therefore, it will always depend on the environmental damage or threat one is fighting against.

As nature is not recognized as a subject of rights in Brazil, the competence to decide the matter is not defined in terms of nature itself but in terms of who is considered 'its formal owner', which generally means that the competence is in federal institutions; materially speaking, nature is considered the property of the people, by force of the Constitution.

The manner of realizing rights of nature varies greatly – even within the same court – and, in general, the rights are not fulfilled by way of their recognition as such, but because of the human need for survival. Sometimes, not even the latter reason is enough to generate respect for the environment and nature, which shows the vulnerability of our system. The example of the *vaquejada* is emblematic: the STF recognized a right, but Congress did not comply with the declaration of unconstitutionality and simply modified the constitutional text in the 'PEC da Vaquejada'. In one of our studies, we demonstrated such contradictions with reference to some decisions of the Federal Supreme Court.[35] In that work, we included other examples of how weak our system of rights is in relation to nature.

Some legal commentators and court rulings have had an impact on theoretical reflections and even on the system of rights, although this influence in the system of rights is even more timid than in legal research.

From a theoretical perspective, we witness an interdisciplinary field of debate characterized by a convergence of sociological, ontological, ethical, philosophical, legal, economic, and other perspectives. In general, these are critical perspectives, their being marked by resistance or at least a questioning of anthropocen-

---

[34] Cf the Brazilian Constitution of 1988, art 20(III).

[35] Wolkmer, Wolkmer and Ferrazzo (n 27).

trism and speciesism. Authors such as Felipe,[36] Gordilho,[37] Araújo,[38] Regan,[39] and, most of all, Singer[40] have been important voices in this sense, if we limit ourselves to naming a few examples (in so doing not questioning or minimizing the importance of other contributions).

Authors such as Heron Gordilho have also made an important contribution in the more pragmatic field of legal theory. This author, who is a member of the Public Prosecutor's Office, currently Chief Public Prosecutor,[41] coordinates a unit covering animal rights research within postgraduate studies[42] and is editor of the Brazilian Animal Law Journal.[43] He was one of the authors of the world's first petition for habeas corpus – admitted in court – seeking freedom for a non-human inmate, a chimpanzee. Although it did not go to trial, since the animal unfortunately died before the case reached that point, this was an important precedent because its mere admission into court – which had never occurred before – had repercussions throughout the world.[44]

We have tried to reconcile these two dimensions – the theoretical foundations, especially the historical, ethical, political, and legal ones – with an understanding of the impact and development in rights systems. It is for this reason that we have already authored some studies and have others under development, not only on animal rights but rather on nature as a whole. These studies are strongly influenced by the important contributions of Latin American constitutions and Andean constitutions, including their worldview.[45]

IV. Relationship with Other National Legal Systems

As mentioned above, foreign influences inspiring the recognition of rights of nature have included the Bolivian Constitution of 2009 and associated semi-constitutional legislation, especially Law No 71 of 2010, 'Law on the Rights of Mother Earth',[46] and Law No 300 of 2012, 'Framework Law on Mother Earth and Integral Development for Good Living'.[47]

---

36 Sônia T Felipe, *Acertos abolicionistas: a vez dos animais*: crítica à moralidade especista (Ecoânima 2014).

37 Heron José de Santana Gordilho, *Abolicionismo Animal: habeas corpus para grandes primatas* (2nd edn, Edufba 2017).

38 Fernando Araujo, *A hora dos direitos dos animais* (Almedina 2003).

39 Tom Regan, *Jaulas Vazias* (Lugano editora 2006).

40 Peter Singer, *Libertação Animal* (Via Optima editora 2008).

41 See <www.nipeda.direito.ufba.br/>.

42 Cf <http://dgp.cnpq.br/dgp/espelhorh/9247033382457379>.

43 See <https://periodicos.ufba.br/index.php/RBDA/index>.

44 See below, note 50.

45 Wolkmer, Wolkmer and Ferrazzo (n 27).

46 Ley de Derechos de la Madre Tierra 2010 (Law 71).

47 Ley Marco de la Madre Tierra y Desarrollo Integral para Vivir Bien 2012 (Law 300).

Revelations on the possibility of laws of this type within the scope of positive law systems contributed to the emergence, also in Brazil, of a legal theory critical of anthropocentrism.

The Constitution of Ecuador (2008) has also drawn attention to and demonstrated that other rights are possible beyond the construct of anthropocentrism. However, these laws and even foreign case law have had much more impact in academia than in Brazilian judicial processes or legislation. We have already cited the emblematic *River Doce* case, which was an unprecedented action in Brazil;[48] the *River Atrato* case served here as inspiration in support of the initial petition, but the judge who tried the case was not convinced. For this reason, the lawyer who brought the *River Doce* action even expressed his frustration, saying

> '[u]nfortunately, the Federal Court of Minas Gerais refused to recognize the River Doce as a subject of rights. It argued that, in Brazil, the law does not admit this right, *ignoring that this (lack of law) did not prevent the Constitutional Court of Colombia from recognizing the River Atrato as a subject of rights* […] This event revealed something very serious. In Brazil, not even human beings are recognized as subjects of rights, because, if they were, preventive measures would have been taken by the company and the government in the face of the risk of death proven in technical reports prior to the rupture of the dam.'[49] (Emphasis added.)

In turn, Brazilian case law has met with interest abroad. In 2005, the first writ of habeas corpus in respect of an animal in Brazil was successfully filed (but not granted).[50] This case was mentioned and studied in several countries. We believe it may have projected some impact. This precedent was even translated and published by a center for legal and historical studies on animal rights at Michigan State University,[51] along with other judicial precedents and legal instruments from around the world that are considered paradigmatic.

As regards the question of an international application of rights of nature, legal conflicts involving nature, whether or not they reach the courts (judicialized or not), tend to be resolved under property law criteria and thus, very often, by

---

[48] Lafayette Garcia Novaes Sobrinho, 'Ação inédita no país, Rio Doce entra na Justiça contra desastre de Mariana' (*Revista IHU online*, 17 November 2017) <www.ihu.unisinos.br/78-noticias/573741-acao-inedita-no-pais-rio-doce-entra-na-justica-contra-desastre-de-mariana>.

[49] Lafayette Garcia Novaes Sobrinho, 'Ação do Rio Doce' (*Lafayette Inteligência Jurídica*, 2 May 2018) <http://lafayette.adv.br/acao-do-rio-doce/>. – In its original form: 'Infelizmente, a Justiça Federal de Minas Gerais negou o reconhecimento do Rio Doce como sujeito de direito. Argumentou que, no Brasil, a lei não admite este direito, ignorando que isto (falta de lei) não impediu que a Corte Constitucional da Colômbia reconhecesse o Rio Atrato como sujeito de direito […] Esse acontecimento revelou algo muito grave. No Brasil, nem os seres humanos são reconhecidos como sujeitos de direitos, pois, se fossem, teriam sido tomadas medidas preventivas, pela empresa e pelo governo, diante do risco de mortes comprovado em laudos técnicos anteriores ao rompimento da barragem.'

[50] Tribunal de Justiça do Estado da Bahia, 9ª Vara Criminal de Salvador (BA). Habeas Corpus no 833085-3/2005, 28 September 2005, <https://periodicos.ufba.br/index.php/RBDA/article/view/10259/7315>.

[51] See <www.animallaw.info/case/suica-habeas-corpus>.

way of interpretations within a civil-law and commercial logic. That is to say, this is dominantly still the case in Brazil, although it is no longer uniformly true.

Foreign rights cannot be applied with the 'force of law' (coerciveness), but they can be mentioned among the rationale behind a judicial conviction or as the theoretical foundation of decisions. A set of judicial decisions and legal instruments considered important in this respect is presented in the fourth section of our study.[52]

We believe that trends in neighboring countries in recognizing the rights of nature are having some influence in Brazil, as happened with Bonito municipal law.[53] But there is a larger movement among state officials when it comes to the issue of animal rights, and we relate this to the growing academic and social interest in the 'animal cause'. This is not a matter of foreign law, but there is also international law itself, which in Brazil is incorporated by the ratification of treaties, protocols, conventions, and other forms of commitment.

As for the applicability of our rights in foreign systems, we believe that, according to the principle of sovereignty, it is up to each country to define such applicability, although, based on this same principle, the probability that Brazilian rights will be effective abroad is very low.

V. The Role of International Law

Regarding international law, some international commitments touching upon the issue of rights of nature have been incorporated into the rights system in Brazil, such as the Convention on International Trade in Endangered Species of Wild Fauna and Flora;[54] the Vienna Convention for the Protection of the Ozone Layer and the Montreal Protocol on Substances that Deplete the Ozone Layer;[55] and the Convention on Biological Diversity.[56] The latter, in particular, was signed in Brazil in 1992, when the country hosted the Second United Nations Conference on Environment and Development. There are other examples, but the effectiveness of these instruments and of environmental protection in Brazil has not even

[52] Wolkmer, Wolkmer and Ferrazzo (n 27) 97–101.

[53] Lei Orgânica do Município de Bonito Estado de Pernambuco 2016, <http://bonito.pe.leg.br/wp-content/uploads/2023/02/LEI-ORGANICA.pdf>.

[54] Decreto n 3.607 de 21 de setembro de 2000. Dispõe sobre a implementação da Convenção sobre Comércio Internacional das Espécies da Flora e Fauna Selvagens em Perigo de Extinção – CITES, e dá outras providências (Brazil) <www.planalto.gov.br/ccivil_03/decreto/d3607.htm>.

[55] Decreto n 99.280 de 6 de junho de 1990. Promulgação da Convenção de Viena para a Proteção da Camada de Ozônio e do Protocolo de Montreal sobre Substâncias que Destroem a Camada de Ozônio (Brazil) <www.planalto.gov.br/ccivil_03/decreto/1990-1994/d99280.htm>.

[56] Decreto n 2.s519 de 16 de março de 1998. Promulga a Convenção sobre Diversidade Biológica, assinada no Rio de Janeiro, em 5 de junho de 1992 (Brazil) <www.planalto.gov.br/ccivil_03/decreto/d2519.htm>.

risen to the challenge of being effective in enabling human survival, much less in advancing the recognition of the rights of nature.

Our research trajectory has led us to believe that the major international influences on environmental issues in Brazil come from developmental discourse that is conceived within the logic of capitalism and that 'its consequences are the planetary environmental tragedy and the disappearance of peripheral traditions and cultures'.[57] Such an understanding helps in understanding the ineffectiveness of the rights and protection of nature in Brazil.

As to the question of whether there has been a conflict of laws as between international law and rights of nature in Brazil, we cannot presently provide an exhaustive answer, as Brazil is currently confronting serious conflicts arising from environmental disasters, such as the one resulting from the rupture of the dams in Mariana and Brumadinho, in the state of Minas Gerais. The Brazilian state was heavily criticized for its actions – or rather, its lack of action – in relation to these disasters, and even the established powers are being blamed for accelerating the relaxing of legislation when the opposite should be the case, ie when more effective regulatory frameworks for damage prevention and environmental protection should be adopted.[58] In any event, there is an intense search for solutions through judicial processes: almost 85,000 lawsuits have arisen from these disasters.[59] However, there is great difficulty in holding transnational corporations, international conglomerates, or large companies accountable, and this has required intervention by international human rights organizations in order to defend victims of the extractive model and victims of predatory actions which, aiming only for profit, destroy nature and put all forms of life at risk.[60]

The environmental problems arising from mining activities[61] do not lead to a conflict between national laws and international law, since Brazil is a signatory to most international treaties and agreements. The conflict is between ensuring the effectiveness of environmental protection laws and reality, which worsens in certain periods, such as that of the national government of 2019 to 2022, in which

[57] Maria de Fátima S Wolkmer and Debora Ferrazzo, 'O paradoxo do desenvolvimento: direito ambiental e bens comuns no capitalismo' (2018) 15(33) Revista Veredas do Direito 163.

[58] Klemens Augustinus Laschefski, 'Rompimento de barragens em Mariana e Brumadinho (MG): Desastres como meio de acumulação por despossessão' (2020) 2(1) Ambientes: Revista de Geografia e Ecologia Política 98. Inter-American Commission on Human Rights, 'Brasil: Derechos humanos y actividad minera' (*Youtube*, 2016, Video of 55 min 31 sec) <www.youtube.com/watch?v=1RAfL10Jfmk>.

[59] 'Rompimentos de barragens geraram mais de 84 mil ações no TJ-MG' (*Consultor Jurídico*, 11 December 2019) <www.conjur.com.br/2019-dez-11/rompimentos-barragens-geraram-84-mil-acoes-tj-mg>.

[60] Mario Campagnani, 'Organizações denunciam internacionalmente mais de 10 violações da mineração no país' (*Justiça Global*, 7 June 2016) <www.global.org.br/blog/organizacoes-denunciam-internacionalmente-mais-de-10-violacoes-cometidas-pela-industria-da-mineracao-no-pais/>. Inter-American Commission on Human Rights (n 58).

[61] Inter-American Commission on Human Rights (n 58).

there was an astonishing increase in the complaints against environmental damage, while enforcement powers of protective laws were taken away from the authorities responsible for inspection. We recently published a more detailed study including a 'Cartography of Violence under Brazilian Governments',[62] which shows the relationship between authoritarian advances and the dismantling of institutions for the monitoring and protection of nature. Our conclusion: on a planetary scale, the Amazon was the biggest victim, along with all those who defend it. In Brazil, defending nature and the environment is a very dangerous activity, as it is one of the world's countries in which the most environmental activists are killed.[63]

VI. Effectiveness

The main obstacle to the implementation of the rights of nature in Brazil is the capitalist mode of production.[64] We think all the others derive from that.

We mention the *PEC da Vaquejada* episode again; not only because it illustrates the conflict between the economic interests of national and foreign oligarchies with the rights of nature, but also since it illustrates quite clearly that the advances made in jurisprudence have not been effective. More than that, this episode also proves that the powers that be do not attempt to hide that capital and economic interests interfere and prevail in relation to the needs of people and the rights of nature.

In addition to this case, other situations already mentioned – such as the *River Doce* legal action, the accountability of large mining companies, and cattle ranching – indicate the difficulties that Brazilian society still has to overcome in order for the rights of nature to be recognized.

VII. Consequences

We have not yet consolidated the recognition of rights of nature in Brazil. Currently, the mere imminence of this possibility has intensified conflicts and debates. Brazil evidences strong tensions in all its regions when it comes to protecting the environment or the rights of nature. These tensions are more intense in regions having a larger indigenous population, as their way of life confronts

---

62 Antonio Carlos Wolkmer, Felipe de Araújo Chersoni and Debora Ferrazzo, 'Nature and Human Rights of Indigenous Communities in the Amazon: A Cartography of Violence under Brazilian Governments' (2024) 15 Beijing LRev 2518.

63 José Brito, 'Brasil é o 3° em mortes de ativistas ambientais e dos direitos humanos, diz ONG' (*CNN Brasil*, 28 July 2020) <www.cnnbrasil.com.br/nacional/brasil-e-o-3-em-mortes-de-ativistas-ambientais-e-dos-direitos-humanos-diz-ong/>.

64 Wolkmer and Ferrazzo (n 57).

and limits the advance of landowning interests. Tensions even lead to violent conflicts and a high number of deaths, especially among indigenous people and environmental activists.[65]

We consider it difficult to estimate the impact of the recognition of the rights of nature in Brazil because, on the one hand, our country has a long tradition of important and inclusive but ineffective laws and, on the other, there has been a regrettable impunity in the face of abuses of power committed against vulnerable subjects, among which we can include nature, especially as a result of the mercantile logic that is deeply rooted in Brazilian society.

[65] Brito (n 63).

China

Limited Prospects for the Rights of Nature in an Inhospitable Legal-Political Environment

Jacques deLisle[*]

Rights of nature are not recognized in Chinese law. A full-text search of a comprehensive database of national-level Chinese laws, regulations, rules, judicial interpretations, and other legal sources retrieves no references to the Chinese terms that are typically used for 'rights of nature.'[1] Although laws and legal mechanisms to protect the environment have expanded significantly in China, especially during the last decade or two, and environmental issues have been identified as a major policy focus by the regime, rights of nature have not been enacted into law or referenced in laws. Rights of nature have not been a significant part of mainstream debates over environmental law and policy in China.

[*] This chapter reflects developments through May 2024 and source materials up to that date. Final revisions to the text were made in March 2025.

[1] Searches were conducted for 自然权利 (*ziran quanli*) and 自然的权利 (*ziran de quanli*) and 自然体权利 (*ziranti quanli*) and the English terms 'right of nature' and 'rights of nature' in the comprehensive database of Chinese law maintained by Peking University Law School (Chinalawinfo). The same searches yielded the same results in Westlaw China.

I. Legal and Political Contexts: An Inhospitable Environment for the Rights of Nature

Several overlapping features of Chinese law, politics, and policy have been unconducive to acceptance of the rights of nature. First, 'rights' (of any sort) hold a relatively limited – and in some respects recently shrinking – place in Chinese law and law-related discourse. A relatively extensive list of rights is set forth in the constitution of the People's Republic of China – the highest source of law in China. But constitutional provisions are not directly operative in Chinese law. Constitutional rights require legislative implementation. A brief flirtation with making constitutional rights directly enforceable by courts was abandoned in the late 2000s.[2] The rights enumerated in the constitution are rights of human citizens only, and they include no provisions closely related to the environment. The articles in the constitution addressing environmental issues are framed not as rights but as state powers and commitments, including to build or promote the development of an 'ecological civilization' (生态文明/*shengtai wenming*), to 'protect and improve the living environment and the ecological environment' (保护和改善生活环境和生态环境/*baohu he gaishan shenghuo huanjing he shengtai huanjing*), to prevent and control pollution (防治污染/*fangzhi wuran*), and to prohibit 'appropriation of or damage to natural resources' (侵占或者破坏自然资源/*qinzhan huozhe pohuai ziran ziyuan*).[3]

Second, and relatedly, official conceptions of law in China are largely positivist. Rights are seen primarily as creations of the sovereign through positive acts of lawmaking, rather than as matters of natural rights or human rights or rights derivable (by courts or others) from principles outside rights-creating positive law of the state.[4] Ideas of 'rights as trumps' chafe against the notions of relatively unfettered party-state power and discretion that have pervaded contemporary Chinese law. Such notions also are in tension with the regime's strongly instrumentalist conceptions of law.[5] The lack of centrality of rights extends from matters of theory and principle to popular attitudes. In the phrasing of one influential study that captures a widely held view among analysts of Chinese law and

---

[2] Shen Kui, 'Is it the Beginning of the Era of the Rule of the Constitution?' (2003) 12(1) Pacific Rim Law and Policy Journal 199; Thomas E Kellogg, 'The Death of Constitutional Litigation in China?' (2009) 9(7) Jamestown Foundation China Brief <https://jamestown.org/program/the-death-of-constitutional-litigation-in-china/>.

[3] Constitution of the People's Republic of China, preamble and arts 9, 26, 89(6) (2018).

[4] See generally Franz Michael, 'The Role of Law in Traditional, Nationalist, and Communist China' (1962) 9 China Quarterly 124; Ji Wei-dong, 'The Sociology of Law in China: Overview and Trends' (1989) 23(5) Law & Society Review 903, 905; Andrew J Nathan, *Chinese Democracy* (University of California Press 1985) 107–111 (political rights as grants of the state, generally through law, to empower citizens to participate in the state).

[5] See generally Jacques deLisle, 'Law in the China Model 2.0: Legality, Developmentalism and Leninism under Xi Jinping' (2017) 26(103) Journal of Contemporary China, 68; Perry Keller, 'Sources of Order in Chinese Law' (1994) 42(4) AJCL 711.

politics, 'rights consciousness' among the public is relatively weak, compared to 'rules consciousness' – that is, awareness of the rules that the state has created and a sense that they can or should be enforced (including to achieve some of the outcomes that otherwise might be provided by more robust legal rights).[6]

Third, the 'rights of nature' idea that has gained ground in some legal systems in parts of the Global South and the West draws on a long-standing and well-entrenched discourse of human rights that has not been deeply received in Chinese law.[7] Since the early 1990s, China has in principle accepted that human rights are universal.[8] But that acceptance has come with significant relativist caveats.[9] Especially since the early to middle 2010s, the ruling Chinese Communist Party has pushed back against ostensibly universal values (including human rights) as reflecting Western values that are unsuited to China.[10] Although rights related to nature and the environment have not been a principal focus, the challenge posed in this Chinese critique to some of the most long-standing liberal international rights, as well as newer or more uncertain rights (such as a right to democratic governance), betokens skepticism toward the rights of nature as an asserted right of recent, largely foreign origin. Chinese writings about the rights of nature routinely point to their Western roots, sometimes citing this as a basis of their unsuitability for China (as is discussed in the next section of this chapter.)

Fourth, official and quasi-official Chinese lists of human rights include some principles that are in some tension with the rights of nature. At least, or especially, in the context of China's economic policies and laws during the last several

---

[6] See Elizabeth J Perry, 'A New Rights Consciousness?' (2009) 20(3) Journal of Democracy 17; Lianjiang Li, 'Rights Consciousness and Rules Consciousness in Contemporary China' (2010) 64 China Journal 47.

[7] See David R Boyd, 'Recognizing the Rights of Nature' (2018) 32(4) Natural Resources & Environment 13; Viktoria Kahui, Claire W Armstrong and Margarethe Aanesen, 'Comparative Analysis of Rights of Nature (RoN) Case Studies Worldwide' (2024) 221 Ecological Economics 108193; Susana Borràs, 'New Transitions from Human Rights to the Environment to the Rights of Nature' (2016) 5(1) TEL 113.

[8] See State Council Information Office, People's Republic of China, Human Rights in China (1991) <www.china.org.cn/e-white/7/index.htm>.

[9] See generally Yongjin Zhang and Barry Buzan, 'China and the Global Reach of Human Rights' (2020) 241 China Quarterly 169; Jiangyu Wang, 'China and the Universal Human Rights Standards' (2001) 29 Sy J Int L 135; see also Wang Yi, State Councilor and Foreign Minister of the People's Republic of China, Remarks: A People-Centered Approach for Global Human Rights Progress, United Nations Human Rights Council (22 February 2021) <http://geneva.china-mission.gov.cn/eng/dbdt/202102/t20210222_9899531.htm> (calling to uphold 'both [the] universality and particularity' of human rights, including variation based on 'history, culture, social system and level of economic and social development').

[10] See Chinese Communist Party Central Committee, Communiqué on the Current State of the Ideological Sphere, September 2013 ('Document Number 9'), translated at <www.chinafile.com/document-9-chinafile-translation>; Russell Ong, 'Peaceful Evolution, "Regime Change" and China's Political Security' (2007) 16(53) Journal of Contemporary China 717.

decades, a 'right to (economic) development'[11] has been associated with prioritizing rapid economic growth, initially and largely still today in the form of industrialization. This agenda had little regard for environmental costs and consequences, until policy changes since the early to middle 2000s began to ameliorate the one-sided emphasis on near-term growth or 'GDP fundamentalism' in favor of attention to sustainable development goals, 'scientific development', and 'high quality development' that places greater emphasis on the environmental impact and sustainability of economic development.[12]

China's official and orthodox conceptions of human rights also assert that human rights are not above state sovereignty – a position that sometimes verges on a 'right to sovereignty.' This position entails asserting that states have great discretion to make their own choices concerning domestic governance, free from transnational interference and unwanted external influence. Emphasis on this view of the relationship between human rights – as well as international legal norms more generally – and state sovereignty has reflected and supported a restrictive approach to the reception of international legal norms. This is especially so for international norms (and all the more so with still-emerging ones) that restrict the discretion of the state in domestic governance, as robust notions of rights of humans – or nature – would threaten to do.[13]

Finally, Chinese law is not capacious in receiving international legal norms into domestic law. With the partial exception of those embodied in treaties (where views vary among Chinese legal scholars and China's practice has been somewhat inconsistent and evolving), international legal norms and obligations generally do not enter Chinese domestic law without a legislative act (of a general or specific sort).[14] Still-inchoate international legal norms and rules, such as those associated with the rights of nature, are especially unlikely to find a path into Chinese municipal law. In recent years, China has become increasingly vocal (and has echoed earlier, Mao-era positions) in criticizing some aspects of international law as the self-serving rules of Western states, created through a process in which China played no role and with content that often ill-serves China's inter-

[11] See, eg, State Council Information Office, People's Republic of China, 'The Right to Development: China's Philosophy, Practice, and Contribution' (December 2016) <http://www.scio.gov.cn/zfbps/32832/Document/1534707/1534707.htm>.

[12] See the discussion of these issues in section IV.

[13] See, eg, Jacques deLisle, 'From Economic Development to What – and Why?: China's Evolving Legal and Political Engagement with International Human Rights Norms' in Yu Guanghua (ed), *Rethinking Law and Development: The Chinese Experience* (Routledge 2013) 117; 'The Costs of International Advocacy: China's Interference in United Nations Human Rights Mechanisms' (*Human Rights Watch*, 5 September 2017) <www.hrw.org/report/2017/09/05/costs-international-advocacy/chinas-interference-united-nations-human-rights>.

[14] See generally Björn Ahl, *Die Anwendung völkerrechtlicher Verträge in China* (Springer 2009) 355–367; Zou Keyuan, 'International Law in the Chinese Context' (2010) 44(3) Val U L Rev 935; Pitman B Potter, 'China and the International Legal System: Challenges of Participation' (2007) 191 China Quarterly 699.

ests and ill-suits its needs and preferences.[15] International legal norms and rules that, like the rights of nature, could impede the pursuit of other high-priority goals (such as economic development or party-state security) or that might support challenges to party-state authority or agendas are particularly vulnerable to such rejection.

II. The Rights of Nature in Scholarly/Theoretical Discourse: Mixed Views and Limited Prospects

The term 'rights of nature' (typically rendered as 自然权利/*ziran quanli* or, less ambiguously but less frequently, as 自然的权利/*ziran de quanli*) is known in Chinese legal circles. But it has not figured prominently, even in academic and related theoretical discussions. Published Chinese scholarly works on the rights of nature have been notably mixed in their views. They often have been tentative or reluctant in embracing or advocating the rights of nature as something that is or should be recognized in Chinese law or promoted in law reform efforts. They have been relatively pessimistic, or at least guarded, about prospects for the rights of nature in Chinese law.

Chinese scholars are familiar with the discourse that has emerged abroad, primarily in the West. Several commentators frame their discussions of the rights of nature in the context of the development of non-anthropocentric, biocentric, or eco-centric views of the relationship between humans and nature (primarily but not exclusively among foreign analysts), the impact of environmental protection movements and notions of environmental justice in other countries, or the broad arc from natural law to natural rights, human rights, the rights of non-human beings, and on to the rights of nature in ethical and legal thinking, primarily in the West. Some of these authors recount many of the theoretical and practical arguments for and against the rights of nature found in foreign and, much less often, Chinese scholarship. On the 'con' side, they note views that rights belong only to people, that rights require consciousness of rights or a correlative capacity to bear obligations or 'subjective freedom' (主体自由/*zhuti ziyou*) (which nature does not have), and that law is suited only to regulating rights among persons. On the 'pro' side, they note that conceptions of rights – including legal rights – can change and have expanded over time, that traditional approaches which ex-

[15] See generally Robert D Williams, 'International Law with Chinese Characteristics: Beijing and the "Rules-Based" Global Order' (*Global China*, October 2020), <www.brookings.edu/wp-content/uploads/2020/10/FP_20201012_international_law_china_williams.pdf>; Tom Ginsburg, 'Authoritarian International Law' (2020) 114(2) AJIL 221; and Jacques deLisle, 'The Chinese Model of Law, China's Agenda in International Law, and Implications for Democracy in Asia and Beyond' (*ASAN Forum*, 24 October 2018) <https://theasanforum.org/the-chinese-model-of-law-chinas-agenda-in-international-law-and-implications-for-democracy-in-asia-and-beyond/>.

clude rights of nature are incapable of solving current ecological crises, and that the rights of nature can be (as the experience of some foreign jurisdictions shows) incorporated effectively into law.[16]

Evaluations of the possible trajectory of the rights of nature in Chinese law are often far from optimistic. For example, one quite sympathetic assessment of ethical arguments for discerning rights of nature as an evolution that goes beyond human and, in turn, animal rights, concedes that there is no scholarly consensus on whether rights of nature can be established and how, and that robust recognition of the rights of nature would raise difficult and clarification-demanding questions about their relationship to the rights of people.[17] The author does not argue that such rights are ready for reception into Chinese law. Another article that advocates including the rights of nature in Chinese law acknowledges that the proposal – while drawing on established foreign legal ideas – would demand relatively radical (and difficult) reform in China. The authors note that such a change would require significant departures from 'traditional' jurisprudence that – largely due to its anthropocentrism and its reasoning from conventional notions of rights, obligations, and legal relations – has been unreceptive toward the rights of nature.[18]

Some Chinese scholarly commentators argue that rights of nature are at odds with the 'humanist' (人文主义/*renwenzhuyi*) underpinnings of rights generally or Marxist conceptions of law specifically. Still others argue from a more narrowly doctrinal perspective that civil law (at least in China) views only persons as subjects (rather than objects) capable of bearing rights and enforcing them through legal processes.[19]

---

[16] See, eg, 张锋 [Zhang Feng] and 姚昌 [Yao Chang] '自然权利的批判与辩护' [Critique and Defense of Natural Rights] (2006) 16(3) 中国人口·资源与环境 [China Population, Resources and Environment] 34; 郑庭伟 [Zheng Tingwei], '自然的权利：从环境法视角的法理学思考' [Rights of Nature: Jurisprudential Thinking from the Perspective of Environmental Law] (December 2023) 22 中国环境管理 [China Environmental Management] 12; 韩立新 [Han Lixin] and 刘荣华 [Liu Ronghua], '环境伦理学的发展趋势与研究对象' [Development Trends and Objects of Study in Environmental Ethics] (2007) 33(6) 思想战线 [Battlelines of Thinking] 21; see also 张晓琴, [Zhang Xiaoqin], '自然的权利：生成逻辑及其实质' [Rights of Nature: Their Logic and Essence] (2016) 6 南京师大学报（社会科学版） [Journal of Nanjing Normal University (Social Science Edition)] 47; Xia Mingfang, 'A New Revolution in Historical Research for the Twenty-First Century' (2017) 3 RCC Perspectives 3; 徐祥民 [Xu Xiangmin] and 巩固 [Gong Gu] '自然体权利：权利的发展抑或终结' [Rights of Nature: The Development or the End of Rights?] (2008) 4 法制与社会发展 [Legal System and Social Development] 80.

[17] Zhang (n 16).

[18] 王树义 [Wang Shuyi] and 王萌 [Wang Meng], '论自然体权利及其保护' [On the Rights of Nature and Their Protection] (2022) 43(3) 吉首大学学报（社会科学版） [Jishou University Journal (Social Science Edition)] 78.

[19] 隋璐明 [Sui Luming], '论自然体权利的民法困境' [On the Civil Law Dilemma of the Rights of Nature] 2015 2(1) 法制博览 [Legality Vision] 253; Xu and Gong (n 16); 曹明德 [Cao Mingde], '森林资源生态效益补偿制度简论' [A Brief Discussion of the Compensation System for Forest Ecology Resource Benefits] (2005) 23(1) 政法论坛 [Political Science and Law Forum] 133.

Other Chinese legal scholars take skeptical views for pragmatic reasons. One finds that the arguments for grounding the rights of nature in natural rights, natural law, or existing positive law is questionable, in part because such efforts have proved inadequate to the seemingly more manageable task of establishing human rights or rights for people.[20] One assessment points to the limited effect that recognition of the rights of nature has had in other legal systems, and another argues that any embrace of the rights of nature likely would be overshadowed in China by other, more strongly entrenched notions of rights to development and rights to survival.[21]

Yet another analysis by a legal scholar opposes adopting a rights of nature framework, arguing that rights of nature are 'fictional' in character and that using rights-based arguments to address the relevant issues would fail to achieve its goal (of solving the environmental crisis) and would risk undermining the utility of rights theories in more traditional areas.[22] One especially critical pair of scholars argues that embracing the rights of nature in Chinese law would 'open the door wide' to rights of nature claims that would undermine the rights of persons, including in the form of placing ecology-based collective rights above individual rights.[23] Another, more philosophically oriented assessment offers a bleak prognosis for rights of nature, partly on the grounds that the concept is rooted in Western (human) rights-based thinking that has limited traction in Chinese law.[24] In a similar vein, an analysis of the rights of nature in the Ecuadorian constitution argues that the concept is rooted in local indigenous culture and thus likely not suitable for adoption in China.[25]

Some writers argue that the more promising potential solutions in law (and beyond) lie in areas other than rights of nature or their legal recognition. Some see the most promising options as being more special legislation to regulate environmental harm, compensate human victims, or expand opportunities for judicial enforcement through civil litigation.[26] One legal academic argues for relying upon individual and collective rights currently in the constitution and bet-

[20] 何志鹏 [He Zhipeng], '"自然的权利"何以可能' 法制与社会发展' [The Possibility of 'Rights of Nature'] (2008) 1 法制与社会发展 [Legal System and Social Development] 102.

[21] 陈海嵩 [Chen Haisong] '"自然体权利"的宪法确认及其反思' ["Rights of Nature": Constitutional Confirmation and Reflections Thereon] (2012) 14(4) 辽宁行政学院学报 [Journal of Liaoning Administrative Institute] 47.

[22] 刘卫先 [Liu Weixian], '自然体与后代人权利的虚构性' [The Fictional Character of the Rights of Nature and the Rights of Future Generations] (2010) 6 法制与社会发展 [Legal System and Social Development] 19.

[23] Xu and Gong (n 16).

[24] 孔成思 [Kong Chengsi], '自然权利论的困境及反思 – 中西传统道德结构差异的视角' [Dilemmas and Reflections on the Theory of the Rights of Nature – Perspectives from the Different Moral Structures in Chinese and Western Traditions] (2016) 2 华中科技大学学报(社会科学版) [Journal of Huazhong University of Science and Technology (Social Science Edition)] 34.

[25] Chen (n 21).

[26] Sui (n 19); Cao (n 19); Chen (n 21).

ter legislation and implementation of legislation. In the author's view, such an approach is preferable to Chinese constitutional law's adopting a separate 'environment(al) right' (环境权/*huanjingquan*) (meaning a right of people to a good environment) as some have urged because it would enjoy no advantage relative to other specified constitutional rights that are already in place. The author also discusses and rejects the approaches taken in Western countries' framing of environment-related rights as 'human rights' or 'rights of nature,' arguing that the rights of nature concept is controversial, a focus of widespread opposition in countries where it has been adopted, difficult to implement, problematically 'eco-centric,' and pushed by radical environmentalists.[27]

Another scholarly analysis argues that the 'rights of nature' are best addressed through recognition of 'people's rights to/in nature' (人对自然的权利/*rendui ziran de quanli*), partly as a means for establishing people's responsibilities with respect to nature.[28] Others argue, somewhat similarly, for placing more emphasis, including through environmental protection laws and other laws, on people's duties or obligations (义务/*yiwu*) toward nature.[29] Still another assessment suggests that success in avoiding environmental catastrophe may depend less on law (whether or not law recognizes 'rights of nature') than on social change and attention to questions of equity, both domestically and internationally.[30]

Some authors look to Chinese culture and traditions as possible sources of inspiration in finding means to deal with serious environmental challenges. Such analyses generally lean away from embracing the idea of rights of nature in Chinese law. A relatively common point is to stress Confucian and other Chinese traditions that emphasize 'harmony' (和谐/*hexie*) between humans and nature. But this normative frame is not seen as providing a foundation for rights of nature.[31]

---

[27] 彭峰 [Peng Feng] '论我国宪法中环境权的表达及其实施' [On the Expression of Environmental Rights in the Constitution and Their Implementation] (2019) 10 政治与法律 [Politics and Law]. See also Lu Zhongmei, 'Roadmap for the Construction of an Ecological Rule of Law in China' (2013) 34(4) Social Sciences in China 162 (arguing for construction of an environmental legal system that corresponds to China's ecological civilization through improving and rationalizing – especially through attention to sustainability – environmental legal mechanisms, and institutions).

[28] 吕忠梅 [Lu Zhongmei], '环境权入宪的理路与设想' [The Rationale and Assumptions for Incorporating Environmental Rights in the Constitution] (2018) 1 法学杂志 [Legal Science Magazine].

[29] See 白雪 [Bai Xue], '环境伦理视角下人对自然的权利和义务' [Human Rights and Obligations toward Nature from the Perspective of Environmental Ethics] (2018) 28(3) 信阳农林学院学报 [Xinyang Agricultural and Forestry University Journal] 12.

[30] 韩立新 [Han Lixin] and 刘荣华 [Liu Ronghua], '环境伦理学的发展趋势与研究对象' [Development Trends and Objects of Study in Environmental Ethics] (2007) 33(6) 思想战线 [Battlelines of Thinking] 21.

[31] See Kong (n 24). See also Mary E Tucker, 'The Relevance of Chinese Neo-Confucianism for the Reverence of Nature' (1991) 15(2) Environmental History Review, 55 (arguing that China's non-anthropocentric and organic views of man's relationship with nature offer a promising alternative to Western traditions in addressing concerns that are problematically addressed

One scholar has sought evidence for ideas of rights of nature in a minority ethnic group in China, but their conception (as described in the article) does not clearly extend beyond animal rights, and the traditions of a very small population of non-Han ethnicity do not provide a strong cultural basis for China-wide law or legal norms.[32]

A few scholars argue in favor of recognizing rights of nature in Chinese law (even though some are not very sanguine about the chances for success). Although acknowledging that their proposal would entail relatively radical reform and would face daunting obstacles, two authors argue that rights of nature should be legally recognized in China, and that Chinese law should place rights of nature on par with rights of persons. They assert that this approach, although very novel in China, could be based in the Chinese Communist Party's commitment to a 'community of man and nature' (人与自然是生命/*ren yu ziran shi shengming*) – a term closely aligned with the key Xi-era trope of a human community of shared/common destiny (人类命运共同体/*renlei mingyun gongtongti*) – and could draw upon valuable and instructive foreign experience.[33] After undertaking an extensive review of the international debates concerning the normative and practical cases for the rights of nature, another pair of co-authors conclude that the rights of nature should and can be accepted into Chinese law.[34] Yet another author offers a notably people-centered argument for rights of nature, opining that we have moved toward rights of nature 'thinking' (思想/*sixiang*) and endowing nature with rights in large part to protect people's own rights and interests in protecting the environment in which people live and on which they depend for survival.[35] Another zealous advocate for the rights of nature in Chinese law argues that scholarly attitudes have been shifting in favor of recognizing the rights of nature, in part because mounting environmental problems have laid bare the contradictions between the interests of people and nature. On his account, there is now a need to recognize the rights of nature and, further, to balance and accommodate the two sets of rights.[36]

---

by, inter alia, 'rights of nature'). See also Han and Liu (n 30) (views similar to Kong's); 白雪 [Bai Xue], '环境伦理视角下人对自然的权利和义务' [Human Rights and Obligations toward Nature from the Perspective of Environmental Ethics], (2018) 28(3) 信阳农林学院学报 [Xinyang Agricultural and Forestry University Journal] 12 (aligning human rights and obligations toward nature with Zhuangzi's notion of harmony and coexistence between heaven, earth and people); Xia (n 16) (comparing Chinese and Western traditions and arguing for harmony-related notions supporting protection for the environment in Chinese tradition).

[32] 廖国强 [Liao Guoqiang] and 殷国聪 [Yin Guocong], '自然是否享有权利? – 哈尼族"自然的权利"观念探析' [Does Nature Have Rights? Analyzing the Hani People's Concept of 'Rights of Nature'] (2008) 3 云南大学学报(法学版) [Yunnan University Journal (Law Edition)] 154.

[33] Wang and Wang (n 18).

[34] See Zhang and Yao (n 16).

[35] 俞田荣 [Yu Tianrong], '自然法·自然权利·自然的权利' [Natural Law, Natural Rights, and the Rights of Nature] (2005) 1 浙江社会科学 [Zhejiang Social Science] 112.

[36] Zheng (n 16). See also Cao (n 19).

To be sure, restrictions on academic freedom and opportunities to publish in China – especially concerning controversial issues that, like the environment, have a history or prospects of becoming focal points for protests – mean that publicly available works do not represent the full spectrum or a representative sample of thinking on many issues among relevant experts or advocates. But the visible subset is important, and disproportionately so. Unpublished work generally has less effect on law-affecting debates and discourse, and rights of nature is not so sensitive a subject that scholarly work on the topic is likely to be subject to unusually rigid censorship. Indeed, the works cited in the preceding paragraphs in many cases include extensive, and sometimes favorable, discussions of and citations to pro-rights of nature scholarship, primarily but not exclusively from foreign sources, and some authors strongly favor recognizing rights of nature in Chinese law. While Chinese theorists and scholarly advocates do have a range of views, the relatively few who call for recognition of rights of nature focus more on normative issues than on practical and doctrinal impediments or comparisons to alternative approaches. Moreover, as the discussion in other parts of this chapter addresses, pro-rights of nature views have gained no discernible traction in Chinese law, legal practice, and policy discourse.

III. Legal and Political Actors: Weaknesses as Forces for the Rights of Nature

Domestic and international legal, political, and civil society actors in China that might be expected to foster or support recognition or protection of the rights of nature have not brought about the adoption of the rights of nature in Chinese law or significant consideration of such a move.

International and domestic non-governmental organizations of the types that might promote the rights of nature (and have done so in some jurisdictions) have not played that role in China. Space for foreign non-governmental organizations and domestic civil society organizations has consistently been very limited in China. In recent years, international non-governmental organizations, including those focusing on environmental issues, have faced tightening restrictions. The Overseas Non-Governmental Organizations Law (2017) and other legal and policy measures have undercut – or at least unsettled – the ability of such organizations to operate in China and to collaborate with domestic NGOs.[37] Some international NGOs working on the environment have ended operations in China. For

[37] See Overseas Non-Governmental Organizations Law of the People's Republic of China (2017); Heike Hobig and Bertram Lang, 'China's Overseas NGO Law and the Future of International Civil Society' (2022) 52(4) Journal of Contemporary Asia, 574; The China NGO Project, 'List of De-Registered Representative Offices' (*ChinaFile*, 15 June 2022) <www.chinafile.com/ngo/latest/list-of-de-registered-representative-offices> (list including some in environmental areas).

those that remain (and those that have left as well), pressing for recognition of the rights of nature has not been on the agenda of their law reform advocacy work (and such work has been a relatively small part of their portfolios).[38]

Domestic NGOs, too, have faced highly, and in some respects increasingly, restrictive conditions.[39] Although environmental NGOs enjoy comparatively extensive room to operate (relative to NGOs in many other fields), the legal and political constraints on NGOs, and their need to work within the system, make them unlikely vehicles for pressing for relatively radical legal reforms such as recognition of rights of nature. Such groups' legal work has typically focused on aiding or representing pollution victims in litigation, improving public awareness of existing environmental law, and supporting relatively conventional legislative reforms that have not extended to pushing for recognition of rights of nature.[40]

Activist lawyers have confronted similar, and in recent years worsening, limitations. During their heyday in the 2000s, 'rights protection lawyers' (维权律师/*weiquan lushi*) or 'human rights lawyers' (人权律师/*renquan lushi*) sometimes took on environment-related cases, such as representing victims of pollution or allegedly unlawful state-brokered expropriation of land rights for transfer to uses with adverse environmental impacts.[41] But such cases were relatively few, often made limited headway in the judicial process, and did not involve assertion of 'rights of nature.' Moreover, the regime's earlier tolerance and occasional limited support for rights protection lawyering was superseded by a severe crackdown beginning in 2015.[42]

Courts have not been a force for recognizing rights of nature in China. Several factors limit courts' potential roles on this front. The generally positivist cast of Chinese law weighs against court-created rights of nature or broad readings of legislative or regulatory provisions that might be construed to encompass

---

[38] See, eg, 'The Facts on NRDC's Work in China' (*NRDC*, 13 September 2023) <www.nrdc.org/facts-on-china>; 'Natural Resources Defense Council (NRDC)' (*China Development Brief*) <https://chinadevelopmentbrief.org/ngos/natural-resources-defense-council-nrdc/>; 'Natural Resources Defense Council' (*ChinaFile*) <www.chinafile.com/natural-resources-defense-council>; Steven Mufson and Chris Mooney, 'House Republicans Attack Environmental Group over Its Climate Work in China' *Washington Post* (Washington DC, 5 June 2018).

[39] See generally Diana Fu and Emile Dirks, 'The Party's Struggle to Tame Civil Society' in Jacques deLisle and Guobin Yang (eds), *The Party Leads All: The Evolving Role of the Chinese Communist Party* (Brookings Institution Press 2022).

[40] See, eg, 'Center for Legal Assistance to Pollution Victims' (*China Development Brief*) <https://chinadevelopmentbrief.org/ngos/center-for-legal-assistance-to-pollution-victims/>; Mayu Suzuki and Kimberly Go, Center for Legal Assistance to Pollution Victims (CLAPV) Law Cases (*Wilson Center*) <www.wilsoncenter.org/publication/center-for-legal-assistance-to-pollution-victims-clapv-law-cases>.

[41] See generally Rachel E Stern, *Environmental Litigation in China* (CUP 2013) 151–178 (chapter on 'Heroes or Troublemakers? The Lawyers').

[42] Hualing Fu, 'The July 9th (709) Crackdown on Human Rights Lawyers' (2018) 27(112) Journal of Contemporary China 554.

rights of nature.[43] So, too, does the related, highly civilian conception of courts and judging, which leaves relatively little room for common law-style 'judicial lawmaking' including inference of novel 'rights.'[44]

The absence of rights of nature from authoritative, positive law sources is not the only impediment to a judicial role. Courts have had considerable discretion not to accept cases that are, in principle, justiciable. They have been most likely to reject cases that are likely to be controversial or sensitive – a category that includes allegations of large-scale environmental harms and that surely would include bold assertions of rights of nature.[45] Reforms since the middle 2010s have sought, with some success, to reduce this dimension of court discretion,[46] but such change has been too modest to open the courtroom doors to litigants asserting rights of nature, especially in the absence of policy signals, or clear legal authorization, to do so.

A principal judicial pathway for addressing environmental harm has been too narrow to accommodate rights-of-nature claims. Plaintiffs in administrative litigation cases have brought challenges to government decisions that undertake or allow environmental harms, including permitting and licensing decisions. But, before reforms adopted in 2014, administrative lawsuits were limited to 'concrete administrative acts' – essentially, discrete decisions affecting individual natural or legal persons' rights and interests.[47] The 2014 revisions to the Administrative Litigation Law allowed somewhat broader challenges, but the revised law still does not permit challenges to relatively high-level rules or regulations or legislation – the types of suits where 'rights of nature' (if recognized in Chinese law) might provide a basis for invalidating laws that require or authorize environmentally harmful actions.

Chinese courts are subject to a variety of forces that would weigh heavily against rights of nature (and kindred pro-environment agendas). Judges are sub-

---

[43] See the discussion in section I.

[44] See, eg, Shucheng Wang, 'Guiding Cases and Bureacuratization of Judicial Precedents in China' (2019) 14(2), U Pa Asian L Rev 96; Wang Chenguang, 'Law-Making Functions of the Chinese Courts: Judicial Activism in a Country of Rapid Social Changes' (2006) 1(4) Frontiers of Law in China 524 (noting the conventional civil law/common law contrast and that any judicial law-making is controversial in China although it is also inescapable in a rapidly changing society).

[45] See generally Hualing Fu, 'Mass Disputes and China's Legal System' in Teresa Wright (ed), *Handbook of Protests and Resistance in China* (Edward Elgar 2019) (limited role for courts and legal system, given the political emphasis on stability and atomizing mass disputes and a statist approach to permitting articulation of grievances, including concerning environmental issues).

[46] See generally Margaret YK Woo, 'Court Reform with Chinese Characteristics' (2018) 27(1) Wash Int LJ 241; He Haibo, 'How Much Progress Can Legislation Bring? The 2014 Amendment of the Administrative Litigation Law of the PRC' (2018) 13(4) U Pa Asian L Rev 137.

[47] See, eg, Shiling Xiao and Yang Lin, 'Judicial Review of Administrative Rules in China: Incremental Expansion of Judicial Power' (2022) 17(2) JCL 370; Administrative Litigation Law of the People's Republic of China (2014, 2017), art 53.

ject to discipline and sanctions for vaguely defined incorrect decisions, including – under reforms adopted in the mid-2010s – 'lifetime responsibility' for such errors. An innovative judicial decision recognizing rights of nature could well be called a significantly incorrect decision, given the current content of Chinese law, or could lead to less formal adverse career consequences for a judge.[48]

Courts are also subject to multiple channels of influence – both formal and informal – from the Chinese Communist Party and the Chinese state. These include: adjudication committees within courts that are headed by politically vetted court leaders and that have the authority to review and revise decisions made by the judges who hear cases (and that are most likely to exercise that power in novel or controversial cases); party political-legal committees that oversee courts; and officials of party and government entities at the same administrative levels as the court and which have significant power over the courts' funding, appointments process and ability to function effectively and, thus, can achieve alignment of adjudication outcomes with local economic and political interests (a phenomenon known as 'local protectionism').[49] To the extent that such 'political' factors affect courts' behavior, they include much (such as imperatives for economic growth, social stability, and deference to local officials' authority) that is not favorable to rights of nature.[50]

National and local government and party institutions and officials have not advanced the recognition of the rights of nature in China. At the national level, the rights of nature and environmental concerns have been lesser priorities compared to promoting economic development, often at the expense of the environment and the values protected by the rights of nature. Although national priorities in policy and, to a degree, in law have shifted toward greater concern with environmental protection during the last decade and a half or more, they have not included an embrace, or consideration, of the rights of nature.[51]

Within the Chinese bureaucratic state, the organs that bear primary responsibility for environmental issues have been relatively weak actors, compared to agencies focusing on economic development, in shaping laws and policies. Their relatively modest roles as shapers of top-down regulatory measures – including pro-environment undertakings – have not included pursuit of a rights-of-nature approach.[52]

---

[48] Xin He, 'The Party's Control over the Judiciary' in Jacques deLisle and Guobin Yang (eds) (n 39); 'Judge Sows Seeds of Lawmaking Dispute', *People's Daily* (24 November 2003) <http://en.people.cn/200311/24/eng20031124_128871.shtml>; Jim Yardley, 'A Judge Tests China's Courts, Making History' *New York Times* (New York, 28 November 2005) (measures taken against judge who engaged in review of provincial regulation for inconsistency with national law, a move that implied assertion of a significant expansion of powers of judicial review).

[49] See generally Kwai Hang Ng and Xin He, *Embedded Courts: Judicial Decision-Making in China* (CUP 2017).

[50] The anti-rights-of-nature valence of these political factors are discussed in more detail later in this section.

[51] See the discussion in section IV.

[52] These issues are discussed in more detail in section V.

At local levels, the political environment for the rights of nature (and environmental concerns more generally) has been, and to some extent – and despite legal and policy reforms – remains, not hospitable. Although addressing environmental problems has become a more significant element in the laws and policies that local state officers and party cadres are charged with implementing, a key priority has remained local economic performance, typically in the short term.[53] Pursuit of local GDP growth often has come at the expense of the environment. These patterns are especially pronounced in less economically developed areas of China,[54] where pressure to raise growth rates is especially high and where economic comparative advantage often lies in more pollution-intensive industries. In this context, so expansive and state-limiting a principle as the rights of nature unsurprisingly gains little traction.

Societal pressure for recognizing the rights of nature is limited by multiple factors.[55] As the scholarly discourse's framing of the rights of nature as a relatively new idea suggests, the concept has not drawn much public attention or support. The absence of support from the regime also has contributed to a lack of coverage of the issue in China's highly regulated and often state-linked media. To the extent that some commentators' assessments are correct that rights of nature is a relatively poor cultural fit in China, social support for the idea – an idea also not supported by party-state authorities – is unlikely to be robust. More broadly, the relatively limited place of rights discourse and arguably even rights consciousness among the public is not conducive to acceptance of relatively novel rights concepts, such as the rights of nature.

Other factors, noted elsewhere in this chapter, are also inimical to the development of effective societal backing for rights of nature. These include the narrow scope for civil society organizations, which does not encourage sweeping and novel rights claims, and the regime's limited tolerance for rights-articulating modes of addressing environmental issues through legal channels, such as collective lawsuits and public interest litigation. The authorities are wary of extra-judicial collective action, including in the form of environment-focused public protest – which raises the specter of political dissent and disruption of social order, the avoidance of which is a priority for the party-state and a major factor in

---

[53] See, eg, Pierre F Landry, Xiaobo Lu and Haiyan Duan, 'Does Performance Matter? Evaluating Political Selection Along the Chinese Administrative Ladder' (2018) 51(8) Comp Pol Stud 1074.

[54] See Rui Pang and others, 'Pollute First, Control Later? Exploring the Economic Threshold of Effective Environmental Regulation in China's Context' (2019) 248 JEM 109275.

[55] See generally Benjamin van Rooij, Rachel E Stern, and Kathinka Furst, 'The Authoritarian Logic of Regulatory Pluralism: Understanding China's New Environmental Actors' (2016) Regulation and Governance 3 (arguing that the logic of China's authoritarian regime's approach prevents the development of social actors who can bypass the regulatory state and that the regime's approach to new societal actors is an ambivalent mix of need and fear).

the performance evaluations of local-level party and state officials and, to a degree, courts as well.[56]

Across these dimensions, framing agendas in terms of strong 'rights' claims, including – and perhaps especially – novel and foreign-influenced ones such as rights of nature claims, risks provoking more negative responses from officials. Where there has been success in addressing environmental issues in China, it has often come from actors and through mechanisms that do not augur, or seek, recognition of robust claims for rights, including the rights of nature.

IV. A Greener Turn in Law and Policy Pledges – Without 'Rights of Nature'

Pro-environment legal change in China has been framed not in terms of 'rights of nature' but, rather, in other, more established frameworks such as: the regime's long-standing emphases on economic development and social stability; and, arguably, a much older and recently revived cultural tradition that focuses on the interdependence of man and nature, rather than the rights of either.[57] The relatively recent turn in China toward greater emphasis on environmental law (and environmental protection more broadly) reflects a recognition of the challenges of sustainability and the long-term or deferred environmental costs of China's rapid-growth-oriented model – including clean-up costs, inefficient economic growth, adverse public health effects, the loss of agricultural land, and the serious consequences of climate change that China is beginning to face.

The pro-environment shift also responds to concern about threats to social order in the form of environment-based protests, complaints, and lawsuits by citizens who have a rising environmental consciousness or who have been harmed by, or seek to prevent, pollution and other environmental ills. The change is also motivated by other, often newer concerns, such as perceived opportunities for China to reach the next level of economic development by promoting green technology at home and acquiring substantial market shares in global markets for sustainability-related technology (partly as an element of rising global competition with the United States).[58] These diverse elements impelling China's turn to greater concern about environmental issues are not conducive to a rights of nature-centered (or even a rights of nature-encompassing) legal approach.

---

[56] See generally Dali L Yang, 'China's Troubled Quest for Order: Leadership, Organization and the Contradictions of the Stability Maintenance Regime' (2017) 26(103) Journal of Contemporary China 35. See also the discussion in section V.

[57] See, eg, the 'Ideas' and 'Principles' sections of State Council, People's Republic of China, *Integrated Reform Plan for Promoting Ecological Progress* (21 September 2015).

[58] For a relatively optimistic overview on this issue, see Robert V Percival, 'China's "Green Leap Forward"' (2011) 12(3) Vt JEL 633. See also Scott M Moore, *China's Next Act: How Sustainability and Technology Are Reshaping China's Rise and the World's Future* (OUP 2022).

The relatively recent emphasis on environmental issues is a notable departure from a long tradition in policy and politics. Law and policy in the People's Republic of China were long indifferent, or worse, toward the concerns about the environment and sustainability that have impelled efforts to provide strong protections through recognition of the rights of nature in some other jurisdictions. During China's Mao era (from the founding of the People's Republic in 1949 to the middle to late 1970s), the predominant ethos was expressed in the phrase 'Man must conquer/control nature' (人定胜天/*ren ding sheng tian*).[59] Rapid economic development was often a top priority. In its initial years, the PRC broadly followed the environment-disregarding and ecologically damaging Soviet model of industrialization. During much of the PRC's first three decades, economic policy emphasized the development of heavy, often polluting and environment-ravaging industrialization. At times, the leadership pursued radically transformative social and economic 'campaign' policies, including ones with serious negative environmental consequences. One example was the 'Great Leap Forward' drive for rapid, decentralized industrialization (including so-called backyard steel furnaces) and greatly increased crop yields.[60]

Through the early decades of the Reform Era (beginning near the end of the 1970s), economic development – of a more market-oriented and internationally engaged type – took precedence over protecting the environment or addressing the adverse environmental consequences of economic growth. Economic development, increasingly in the form of manufacturing for export and raising citizens' material living standards, outweighed other considerations, including environmental degradation and sustainability. In the radical turn from a planned economy to a more market-oriented one, the creation of state capacity and will to regulate lagged. Regulations – and regulators – for environmental protection were especially late-developing.

China's greatly deepened economic integration with the global economy meant that much manufacturing for global markets – especially in lower-tech and sometimes highly polluting sectors – migrated to China, to serve both foreign export markets and growing Chinese demand. In response to rising international concerns with climate change, the official Chinese position pointed to China's not having been responsible for the creation of the climate crisis (since it was a late-developing nation and had only relatively recently begun to contribute significantly to greenhouse gases) and argued for 'common but differentiated responsibilities' (with rich, earlier-industrializing nations bearing more of the costs of

[59] See, eg, Judith Shapiro, *Mao's War Against Nature: Politics and the Environment in Revolutionary China* (CUP 2001).

[60] See Scott Rozelle, Jikun Huang and Linxiu Zhang, 'Poverty, Population and Environmental Degradation in China' (1997) 22(3) Food Policy 229; Vaclav Smil, 'Environmental Degradation in China' (1980) 20(6) Asian Survey 777.

addressing the climate crisis).[61] Some of the adverse environmental impact of China's industrialization could be credibly attributed to developed countries, whose consumers imported goods from China that were inexpensive in part because they did not reflect their true environmental costs. None of this suggested fertile soil in China for the rights of nature when such ideas began to emerge in legal discourse in the outside world.

In more recent times, and particularly after the Reform Era entered its second quarter-century, broad shifts in policy orientation brought a new focus on environmental issues and a series of laws and legal changes to implement those new priorities, but not openness to the rights of nature. During Hu Jintao's tenure as China's leader, official rhetoric and policy replaced the previously dominant emphasis on 'development' (seen largely in terms of GDP growth) with 'comprehensive' or 'scientific' development (which included attention to environmental issues and sustainable development, and which was raised to a high position of 'theoretical guidance' for the Chinese Communist Party).[62] In Hu's valedictory report to the 18th National Congress of the Chinese Communist Party in 2012, 'ecological construction' reached formal parity with economic, political, cultural, and social construction. This was interpreted – especially after its amplification by Hu's successor Xi Jinping – by environmental law proponents as a mandate for reforming and strengthening environmental protection laws and institutions.[63]

Under Xi, official rhetoric and policy have emphasized the achievement of 'ecological civilization' – a term adopted at the 17th National Congress of the Chinese Communist Party, five years before Xi came to power.[64] Environmental goals, benchmarks, resource commitments, and transformation toward a greener economy have been written into major policy documents, including the most recent Five-Year Plans for the national economy, with 'green development' now declared one of the five basic principles of economic development, and 'high quality development' becoming a central theme. Two of the most significant sets of policy

---

[61] See, eg, 'China Reaffirms the Key Principle of "Common But Differentiated Responsibility"' *China Daily* (1 December 2015); Björn Conrad, 'China in Copenhagen: Reconciling the "Beijing Climate Revolution" and the "Copenhagen Climate Obstinacy"' (2012) 210 China Quarterly 435.

[62] See 'Scientific Outlook on Development Becomes CPC Theoretical Guidance' (8 November 2012) <www.china.org.cn/china/18th_cpc_congress/2012-11/08/content_27041783.htm>; 'Scientific Outlook on Development' *Xinhua* (8 September 2010) <www.chinadaily.com.cn/china/cpc2011/2010-09/08/content_12474310.htm>.

[63] See Hu Jintao, 'Political Report to the 18th National Congress of the Chinese Communist Party' §§ I, II, VIII, translated at <www.mfa.gov.cn/ce/ceus/eng/zt/18th_CPC_National_Congress_Eng/t992917.htm>; Lu (n 27). Notably, in this work Lu does not mention the rights of nature, and in this and other work, Lu rejects the idea of incorporating rights of nature in Chinese law; Lu (n 27) and Lu (n 28).

[64] State Council Information Office, People's Republic of China, *China on Its Way to Promote Ecological Civilization* (11 August 2017) <www.scio.gov.cn/32618/Document/1560769/1560769.htm>. See generally Sam Geall and Adrian Ely, 'Narratives and Pathways towards an Ecological Civilisation in Contemporary China' (2018) 236 China Quarterly 1175.

documents from the party's central committee during Xi's tenure – at the third and fourth plenums of the 18th Central Committee – included calls for accelerating the construction of the 'system of ecological civilization' and protecting the ecological environment through developing 'strict legal structures.'[65] 'Ecological security' was identified as an aspect of comprehensive or overall national security under Hu and, more so, under Xi.[66] Xi – along with Chinese advocates for environmental law and environmental law reform – has even raised the possibility of writing citizens' environmental rights into the state constitution.[67]

Numerous pro-environment laws and regulations reflect and extend the political and policy commitments. Although the constitution includes nothing that points toward the rights of nature, the document's preamble calls for the coordinated development of ecological civilization. Its operative provisions declare that the state protects natural resources (including living ones) and protects the environment and controls pollution and that the State Council (the highest executive organ) has the function and power of directing and managing ecological conservation.[68]

China's Environmental Protection Law, revised in 2014, declares that its purposes include 'protecting and improving the environment, preventing and controlling pollution and other public hazards, safeguarding public health, advancing the construction of ecological civilization and promoting the sustainable development of the economy and society.' It defines the environment broadly to include 'all natural elements and artificially transformed natural elements affecting hu-

---

[65] Decision of the [18th] Central Committee of the Communist Party of China [3rd Plenum], *Decision on Important Issues of Deepening Reform*, <http://politics.people.com.cn/n/2013/1116/c1001-23560979-14.html>; Decision of the [18th] Central Committee of the Communist Party of China [4th Plenum], *Important Issues of Comprehensively Advancing Rule According to Law* <www.chinalawtranslate.com/en/fourth-plenum-decision/>; State Council Information Office, People's Republic of China, *What to Know about China's High-Quality Development* (18 March 2024) <http://english.scio.gov.cn/in-depth/2024-03/18/content_117066836.htm>.

[66] Katja Drinhausen and Helena Legarda, 'Confident Paranoia: Xi's 'Comprehensive National Security' Framework Shapes China's Behavior at Home and Abroad' (15 September 2022) Merics China Monitor, Exhibit 4; 习近平 [Xi Jinping], '坚持总体国家安全观 走中国特色国家安全道路' [Adhere to the Comprehensive National Security Concept and Follow the Path of National Security with Chinese Characteristics], 人民网－人民日报 [*People's Daily Online*] (16 April 2014).

[67] See, eg, 中共中央 国务院关于加快推进生态文明建设的意见 [Opinions of the CPC Central Committee and the State Council on Accelerating the Ecological Civilization Construction] (2015) (setting forth agenda of environmental protection and including emphasis on enforcement measures, green economic development, and strengthening protection of citizens' 'environmental rights'); 'Xi's Thoughts on Ecological Civilization Published' *China Daily* (1 August 2022); Pan Xiang-chao, 'Research on Xi Jinping's Thought of Ecological Civilization and Environment Sustainable Development' (2018) 153(6) IOP Conference Series: Earth and Environmental Science 4 (reference to possible constitutional amendment); Arthur Hanson, 'Ecological Civilization in the People's Republic of China: Values, Action, and Future Needs' ADB East Asia Working Paper no 21, December 2019).

[68] PRC Constitution, preamble, arts 9, 26, 89. See also the discussion in section I.

man existence and development, which includes, among other things, the atmosphere, water, seas, land, minerals, forests, grasslands, wetland, wildlife, natural and human remains, nature reserves, historic sites and scenic spots, and urban and rural areas,' and declares environmental protection to be a 'basic/fundamental policy of the state' (基本国策/*jiben guoce*).[69] The same law also provides that the state's environmental protection-related economic and technology policies are undertaken to promote harmony between humankind and nature (促进人与自然和谐/*cujin ren yu ziran hexie*) and to coordinate environmental protection with economic and social development (使经济社会发展与环境保护相协调/*shi jingji shehui fazhan yu huanjing baohu xiang xietiao*)[70] – a balancing of interests and a human-centered approach that still is far away from 'rights of nature' conceptions that have developed in some other jurisdictions.

The Environmental Protection Law broadly imposes obligations on state and private actors and citizens to protect the environment. It pledges state support in terms of providing funding, supporting research, incorporating environmental protection into government plans and actions, disseminating information and education, and conducting various monitoring, inspection, and supervision of environment-affecting activities, including for enforcing compliance with state standards.[71] Other provisions impose on state institutions a range of obligations to take steps to safeguard the environment and to prevent or sanction environment-harming behavior and provide compensation for harms that occur.[72] Environmental protection duties – and liability for violations – are imposed on individuals and enterprises, as well as state entities.[73]

Especially after the 2014 amendments, the Environmental Protection Law provides for much more detailed regulatory requirements, including concerning emissions targets, licensing of polluting enterprises, formulation of emergency response plans, and disclosure of environmental risks.[74] The Environmental Protection Law,[75] along with the Environmental Impact Assessment Law and in the context of broader 'open government information' reforms, also requires state actors to disclose to the public environment-related information and directs relevant government bodies to conduct – and release – environmental impact assessments of proposed construction projects before approving them.[76]

---

[69] Environmental Protection Law of the People's Republic of China as amended 2014, arts 1–4 (Environmental Protection Law).

[70] Environmental Protection Law, art 4.

[71] Environmental Protection Law, arts 6–27.

[72] Environmental Protection Law, arts 28–39.

[73] Environmental Protection Law, arts 40–43.

[74] Environmental Protection Law, arts 44–52.

[75] Environmental Protection Law, arts 53–56.

[76] See Environmental Impact Assessment Law of the People's Republic of China 2016, 2018; Jamie P Horsley, 'China Promotes Open Government as It Seeks to Reinvent its Governance Model' (22 February 2016) <https://law.yale.edu/sites/default/files/china-law-documents/2016_horsley_china_promotes_open_government.pdf>.

Scores of other laws and regulations have focused on addressing environmental issues, including measures addressing air pollution, damage to marine areas, solid waste disposal, forestry, mining, energy production, wildlife protection, fisheries, construction, environmental emergencies (that is, sudden, large-scale incidents), and more. Some date in their original, rudimentary form to the 1980s or even the end of the 1970s. Almost all have been adopted or revised in the later 2000s and 2010s.

In the early 2020s, a new type of legislation emerged. The Yangtze River Protection Law and the Yellow River Protection Law focused on particular geographic areas, undertook comprehensive regulation for protection of a regional ecosystem, imposed responsibilities across branches of central ministries and local governments, mandated planning for ecosystem protection, and provided for enforcement mechanisms including administrative sanctions and criminal punishments. Another region-focused law was adopted for the Qinghai-Tibet Plateau.[77] Although the laws' focus on named rivers and mountains faintly echoed the rights of nature idea that mountains and rivers could have rights, that idea was not part of the Chinese discourse.

China's Criminal Law prohibits some types of environment-harming actions. Examples include putting falsified information in environmental impact statements and various acts that cause serious pollution or that, in violation of state laws and regulations, damage protected plant and animal species, or harm protected natural areas or import unauthorized waste or invasive species.[78]

Also in the public law realm, administrative law has added environment-specific provisions (supplementing the general mechanisms that administrative law provides for challenging, or seeking reconsideration of, state acts that cause or approve environment-harming acts). The Administrative Licensing Law, for example, directs state authorities to require licenses for activities that directly affect the protection of the ecological environment.[79]

China's civil law adopted a much-noted 'green principle' in the General Provisions of Civil Law (2017) which became (with minor modifications not relevant here) the general part of the Civil Code in 2021. The principal provision states, broadly, that parties to civil legal relations 'shall [应当/*yingdang*] carry out civil activities in a manner that is conducive to saving resources and protecting the ecological environment [保护生态环境/*baohu shengtai huanjing*].'[80] The provision thus imposes environment-related obligations on natural and legal persons but does not speak explicitly to the rights of nature (or to any rights correlative to the stated obligations).

---

[77] Yangtze [Changjiang] River Protection Law (2021), Yellow River Protection Law (2023); Law on Ecological Conservation of the Qinghai-Tibet Plateau (2023).

[78] Criminal Law of the People's Republic of China as amended 2020, arts 229, 338–346.

[79] Administrative Licensing Law of the People's Republic of China as amended 2019, arts 11–12.

[80] Civil Code of the People's Republic of China 2021, art 9 (Civil Code).

The provision was somewhat controversial for bringing into civil law issues of environmental protection that critics argued were properly left to public law. As the text reflects, the 'green principle' has been seen as a relatively vague general imperative with little operative content, being more advisory or in the nature of guidance than a true mandate. It also has drawn criticism for differing from the paradigm of general principles in Chinese civil law because it lacks universal applicability across civil activities and is more thoroughly dependent for its meaning on specific provisions in the Civil Code's later chapters (or other civil law statutes, to the extent that they have not been superseded by the Civil Code).[81]

Obligations to protect the environment (focusing on pollution-prevention, ecological harm compensation, and resource conservation) are found in other sections of the Civil Code. Examples include provisions addressing the use and management of real property[82] and performance of contracts (including property management services contracts).[83] A distinct section focuses on environmental torts and includes both general matters of liability for environmental torts and specific provisions imposing civil liability for actions violating state environmental laws and regulations.[84] Some similar provisions can be found in the laws governing torts, contracts, and property that were superseded by, and largely codified in, the Civil Code.

Beginning around 2017, Chinese environmental law scholars and advocates began to push for an Environmental Code, which was on the National People's Congress legislative agenda by 2023. Codification would gather together previously fragmented environmental laws, raise environmental law, in principle, to equality with the Civil Code (currently China's only code), and thereby signal top-level political support for environmental protection.[85]

Changes in legal procedures and institutions have accompanied the 'greening' of substantive law. Key developments include the establishment of special environmental courts in some Chinese jurisdictions, the authorization of environmental public interest litigation by NGOs and procurators, and a mandate to procu-

---

[81] See Tiantian Zhai and Yen-chiang Chang, 'The Contribution of China's Civil Law to Sustainable Development: Progress and Prospects' (2019) 11 Sustainability 294, 295, 298-300; Mo Zhang, 'Striking the Balance Between Contractual Rights and Obligations' (2021) 5(1) Cardozo International and Comparative Law Review 81, 111–112, 156. See also Jie Ouyang, 'Unleashing the Green Principle in the Chinese Civil Code: Embedding Private Law into the Green Transition' (2023) 12(5) Journal of European Consumer and Market Law 203 (noting, and rejecting, scholars' critiques of the ostensible politicization and 'constitutionalisation and ecologisation' of private law through the Civil Code's green principle).

[82] Civil Code, arts 286, 326, 346.

[83] Civil Code, arts 509, 619, 937–942.

[84] Civil Code, arts 1229–1235.

[85] See Kai Xu and Guangdong Tian, 'Codification and Prospect of China's Codification of Environmental Law from the Perspective of Global Environmental Governance' (2022) 19(16) International Journal of Environmental Research and Public Health 9978; 'China Aims to Codify Law on Environment' *China Daily* (14 November 2023).

rators to oversee administrative authorities charged with environmental protection missions.[86] None of these has moved discernibly toward acceptance of rights of nature in Chinese law.

V. Enforcement Mechanisms: Law's (Limited) Impact

The environment-related provisions in Chinese law – which are not based on rights of nature, but which mandate pursuit of some of the same ends that would be served by acceptance of rights of nature – are enforced through a variety of legal means and other state mechanisms. Overall, China has made progress in recent years in addressing its severe environmental problems, as well as their impact abroad, but progress has been limited and uneven, with ambitious commitments and plans falling well short in practice. Here, too, the rights of nature have not been a factor in these developments.

Environmental divisions have been established in Chinese courts at multiple levels. Such subject matter-specific chambers are exceptional in Chinese courts (with the other principal example being intellectual property courts). The creation of environmental courts reflects, in part, the authorities' recognition of the significance of the issues and the area of law, and the importance of specialized knowledge among judges adjudicating cases in the field. A cadre of environmental lawyers (ranging from quotidian to crusading) has emerged, as have a small group of public interest environmental litigation entities. Major successful court cases on environmental issues have received considerable attention and recognition. For example, the Supreme People's Court in 2018 issued a set of 'Ten Model Cases concerning the Services and Guarantee of the People's Courts for the Development of Ecological Civilization in A New Era,' covering criminal, administrative, and civil examples.[87]

Civil suits brought by injured private parties have been one notable mechanism for enforcing environmental laws and addressing environmental harms.[88] Various laws – especially tort law, now embedded in a book of the Civil Code, and the Environmental Protection Law – provide for civil remedies, including damages, cessation, and repair in cases of environmental harm or ecological damage. Plaintiffs can prevail without proof of fault, or with the burden of proof shifted to defendants, in pollution and ecological damages cases. Punitive damages are available in cases of intentional environmental pollution or ecological damage.[89]

---

[86] These are discussed in the next section.

[87] 最高人民法院发布10起人民法院服务保障新时代生态文明建设典型案例 [Supreme People's Court, Ten Model Cases concerning the Services and Guarantee of the People's Courts for the Development of Ecological Civilization in A New Era] (4 June 2018).

[88] For an overview of relatively early developments, see Alex Wang, 'The Role of Law in Environmental Protection in China' (2017) 8(2) Vt JEL 195.

[89] See Tort Liability Law 2010, arts 65–68; Civil Code, arts 1229–1235; Environmental Protection Law, arts 6, 64.

Under Chinese civil procedure law, such suits can be brought by individual plaintiffs or in collective litigation. Some environmental tort suits have involved numerous plaintiffs (although true class actions are not provided for). Environmental litigation is a rare area where Chinese law allows for public interest litigation by NGOs – ones that are 'qualified environmental protection organizations' (which must specialize in environmental protection and public welfare activities and have done so for at least five years with no record of violations) under the Environmental Protection Law as revised in 2014.[90]

Civil suits by private plaintiffs have had some limited success. Prior to the adoption of the Tort Liability Law and amendments to the Environmental Protection Law in the 2010s, environmental tort cases leading to court judgments were extremely rare.[91] A study of China's relatively new environmental courts finds that they have had low caseloads, been wary of taking on high-impact cases, and treated potentially adverse impacts on economic development and social stability as significant factors in their decision-making.[92]

Individual plaintiffs in environmental cases sometimes face challenges endemic to the Chinese legal system: defendants have better access to legal services, more experience litigating, and support from local party and government actors (which are often intertwined with defendant enterprises through ownership and other ties, or seek to protect such enterprises as sources of local economic growth, jobs, and government revenue, and can influence courts through controls over court budgets and judicial appointment and promotion).[93] Systemic reforms to address the problems of judicial local protectionism and more ad hoc success in generating publicity for some environmental cases (thereby drawing the attention of higher-level authorities or a wider public) have ameliorated some of these impediments, albeit unevenly.

Plaintiffs who have sought to bring suits – sometimes as collective (that is, multi-plaintiff) actions – to address mass torts, including for injuries caused by environmental harms, have faced difficulty in finding courts open to fully adju-

---

[90] See Environmental Protection Law, art 58; Yina Liu, 'Friends of Nature and Public Interest Environmental Litigation' (2019) 3 Chinese Journal of Environmental Law 225 (describing public interest litigation and support for such litigation by an NGO established in China in 1994 and engaged in environmental litigation from 2011 and, especially, following the 2014 reform to the Environmental Protection Law); see also Lei Xie and Lu Xu, 'Environmental Public Interest Litigation in China: Findings from 570 Court Cases Brought by NGOs, Public Prosecutors and Local Government' (2022) 34(1) JEL53.

[91] See Adam Moser and Tseming Yang, 'Environmental Tort Litigation in China' (2011) 41(1) ELR News & Analysis 10895.

[92] Rachel E Stern, 'The Political Logic of China's New Environmental Courts' (2014) 72 China Journal 53.

[93] See generally Xin He, 'The Politics of Courts in China' (2017) 2(2) China Law and Society Review 129; Xin He, 'Do the "Haves" Come Out Ahead in Shanghai Courts?' (2013) 10(1) JELS 120; Randall Peerenboom (ed), *Judicial Independence in China: Lessons for Global Rule of Law Promotion* (CUP 2010).

dicating their claims. Courts often split collective claims into smaller group or individual cases, and sometimes refuse to litigate meritorious claims.[94] Although such litigation has provided limited direct redress, the state has sometimes responded to such lawsuits and the underlying harms by providing structured compensation schemes for victims and sometimes has sanctioned defendants deemed responsible for causing the harms.[95]

Public interest environmental litigation cases have been relatively few (averaging one hundred or so per year during the first several years in which they were allowed) although success rates have been high for the relatively rare cases – some addressing large-scale harms – that have been fully adjudicated.[96] One study finds that prospects for such suits have been limited by lack of resources, poor litigation risk assessment and case selection, and regulatory restrictions on NGOs activities.[97] Another study finds that courts have been politically wary of accepting such suits, especially where they would have significant social or economic consequences.[98] Another analysis attributes failures of environmental public interest litigation to a limited range of permissible defendants, narrow notions of standing, and daunting evidentiary requirements for securing damage awards.[99]

State entities are also empowered to litigate environmental cases. The Administrative Litigation Law authorizes a people's procuracy (which, among other functions, acts as the prosecutor in the Chinese system) to make suggestions and, if those are not followed, bring an enforcement suit in court against a state organ having duties of supervision and administration in 'ecological environment and resource protection' fields where the state organ has exercised its powers unlawfully or failed to perform its duties, resulting in harm to state, public, or social interests.[100] The Civil Procedure Law similarly authorizes the procuracy to bring an enforcement action in court against a state organ for the same types of malfeasance if no other legally authorized entity (such as a properly recognized NGO) brings such a suit, or to support a suit if one is brought by such a legally authorized entity.[101] These powers, too, have beeen used only to a limited extent, re-

[94] See generally Rachel E Stern, 'From Dispute to Decision: Suing Polluters in China' (2011) 206 China Quarterly 294.

[95] See generally Chenglin Liu, 'Socialized Liability in Chinese Tort Law' (2018) 59 Harv Int LJ Online 16 (concerning state pressure on courts not to take mass tort cases and state-arranged plans for widespread but low compensation and victim waiver of further legal claims); Environmental Protection Law, arts 63, 69 (imposing administrative or criminal sanctions in appropriate cases).

[96] Xie and Xu (n 90).

[97] Lei Xie and Lu Xu, 'Environmental Public Interest Litigation in China: A Critical Examination' (2021) 10(3) TEL 441; Liu (n 90).

[98] Stern (n 92) 64–66.

[99] Gao Qi, '"Public Interest Litigation" in China: Panacea or Placebo for Environmental Protection?' (2018) 16(4) China: An International Journal 47.

[100] Administrative Litigation Law, art 25.

[101] Civil Procedure Law, art 58.

sulting in a few thousand procuracy-initiated civil cases per year – often for relatively small-scale cases but with a near-100% win rate.[102]

Other, less environment-specific channels for adjudication of environmental issues are available as well. The Administrative Litigation Law, for example, allows private plaintiffs to bring lawsuits to challenge a range of government actions. Environment-related suits, including those concerning permits for environment-harming activities, are part of the landscape of administrative litigation, but the impact is limited because aggrieved citizens can only bring challenges to relatively specific state actions or (after 2014 reforms) lower-level normative rules, and not to underlying rules, regulations, or statutes, or patterns of state behavior that lead to environmental harms.[103] Although recent reforms have, in principle, limited courts' discretion to turn away potentially meritorious suits, required defendant agency heads (or their delegees) to appear in court, and made courts less dependent on local government entities for funding (and thus presumably less likely to be biased in their adjudication), aggrieved potential plaintiffs have long been deterred from bringing meritorious claims by perceptions that courts are likely to be biased in favor of the defendant local governments or by the threat of retaliation from local officials who do not welcome the prospect of being defendants in such suits.[104] The effectiveness of administrative and civil litigation also has been constrained by the still-limited impact of rules for disclosure to the pubic of environment-related information.[105]

Not surprisingly, given the authoritarian nature of the Chinese party-state, the limited roles and autonomy of courts, and other features of China's legal and political systems, addressing many of the types of environmental issues that are covered by the rights of nature (and much environmental law) in other legal systems is, in China, left to top-down administrative enforcement.[106] Such methods are arguably temperamentally in tension with a 'rights (of nature)'-based approach, and Chinese authorities have shown no affinity for framing an enforcement agenda in terms of rights of nature.

Some environment-related laws and regulations do not provide for enforcement by entities outside the state, and some do not specify enforcement mechanisms at all. The procuracy has the authority to initiate administrative litigation on environmental issues, and under mid-2010s regulations concerning compen-

---

102 See Xie and Xu (n 90).

103 See also the discussion in section III.

104 See generally Kevin J O'Brien and Lianjiang Li, 'Suing the Local State: Administrative Litigation in Rural China' (2004) 51 China Journal 75; Tianhao Chen, Wei Xu, and Xiaohong Yu, 'Administrative Litigation in China: Assessing the Chief Officials' Appearance System' (2024) 259 China Quarterly 744.

105 For a discussion of this legal development and its limited effectiveness, see Alex L Wang, 'Explaining Environmental Information Disclosure in China' (2018) 44 Ecology LQ 865.

106 For an overview of this pattern with a focus on climate change, see Alex L Wang, 'Climate Change Policy and Law in China' in Cinnamon P Carlane, Kevin R Gray and Richard Tarasofsky (eds), *Oxford Handbook of International Climate Change Law* (OUP 2016) 636.

sation for environmental harms, local governments began bringing a small but rising number of environmental enforcement cases. But these powers have been exercised rarely, with only a handful of cases fully adjudicated each year.[107] The Environmental Protection Law[108] and other laws in the field (including the Environmental Impact Assessment Law[109]) provide for administrative fines, orders to cease prohibited activities and repair damage, and, in serious cases, criminal prosecution. Malfeasance or nonfeasance by state authorities with responsibilities in the field are subject to sanctions as well, under the Environmental Protection Law and other environment-focused laws.

The top-level policy and political commitment to addressing environmental issues has had significant real-world effects. Yet several factors still have impeded the effectiveness of the state-led or top-down approaches. The principal state entity in the field has relatively modest clout. In recent years, its bureaucratic stature and enforcement resources have been upgraded. It has evolved from the Environmental Protection Administration, to the Ministry of Environmental Protection, to the Ministry of Ecology and Environment, and it has accrued broadened responsibilities shifted from other entities, including a climate change portfolio previously held by the powerful National Development and Reform Commission. But it is still regarded as a relatively weak bureaucratic actor with limited implementation capacity, particularly in the face of environmental protection-undermining agendas and mandates from other state organs with responsibility for economic development (which has long been associated with significant environmental harm).[110] Also complicating the administrative state's response to environmental challenges is the fragmentation of regulatory and enforcement authority across multiple, often siloed, bureaucracies, including entities that regulate economic sectors that cause economic harm (such as energy, several types of natural resources, and so on) and that are regulated through sector-specific organic laws.[111]

Environmental protection mandates in policy and law also suffer from China's chronic challenges of slippage in implementation. Lower-level officials continue to face effective pressure to emphasize economic performance. Although 'green

[107] See Xie and Xu (n 90).

[108] Environmental Protection Law, arts 57–69.

[109] Environmental Impact Assessment Law, arts 29–34.

[110] See Danny Marks, 'China's Climate Change Policy Process: Improved But Still Weak and Fragmented' (2020) 19(67) Journal of Contemporary China 971; Li Jing, 'China's New Environment Ministry Unveiled' (*Dialogue Earth*, 19 April 2018) <https://dialogue.earth/en/climate/10599-china-s-new-environment-ministry-unveiled-with-huge-staff-boost/>.

[111] See, eg, Bo Zhang, Cong Cao, Junzhan Gu and Ting Liu, 'A New Environmental Protection Law, Many Old Problems? Challenges to Environmental Governance in China' (2016) 28 JEL 325; Lixin Guan, Ge Sun and Shixiong Cao, 'China's Bureaucracy Hinders Environmental Recovery' (2011) 40(1) Ambio 96.

Region-specific law such as the Yangtze and Yellow River Protection Laws and the prospective Environmental Law Code seek to address some of these problems of regulatory fragmentation.

GDP' or environmental performance became a much-emphasized factor in evaluating local-level officials' performance, principal evaluation metrics continue to be economic performance and social stability. Implementation of pro-environment laws and policies is also hampered by the local party-state's often deep entanglement with local polluting and environment-damaging enterprises, which are key providers of economic growth, jobs, revenue, and political support in the relevant area – and, thus, affect matters of vital importance to local officials' success.[112] At best, local officials face competing pressures from multiple directions, contradictory or shifting mandates from above, uncertain consequences for their behavior, and a system that often monitors for the appearance rather than the reality of meeting targets, resulting in a 'muddling through' approach and, thus, middling implementation.[113] Falsified reports and backsliding from initial enforcement are reportedly problems as well.[114]

Some empirical studies of environmental protection law have found it to be in significant part 'symbolic' or 'performative' at national and local levels. According to these analyses, pro-environment laws are adopted and publicized to gain legitimacy with a public that is concerned with such issues even though the laws will not be effectively implemented, and local environmental protection officials work hard to demonstrate sympathy and responsiveness to outraged citizens' environment-related demands even though the officials lack or eschew means to implement pro-environment laws and policies.[115]

Other empirical studies show that, when action is taken to address ongoing environmental harms, it is often haphazard or irrational as a mode of environmental protection. One study finds that when authorities press for implementation of antipollution measures, lower-level officials resort to 'blunt force' implementation, meeting pollution-reduction quotas by shutting down (sometimes only

---

[112] See, eg, Graeme Lang, 'China's "Green GDP" Experiment and the Struggle for Ecological Modernisation' (2010) 40(1) Journal of Contemporary Asia 44; Ran Ran, 'Perverse Incentive Structure and Policy Implementation Gap in China's Local Environmental Politics' (2013) 15(1) Journal of Environmental Policy and Planning 17; Alex L Wang, 'The Search for Sustainable Legitimacy: Environmental Law and Bureaucracy in China' (2013) 37 Harv Envt'l L Rev365; Tucker van Aken and Orion Lewis, 'The Political Economy of Non-Compliance in China: The Case of Industrial Energy Policy' (2015) 24(95) Journal of Contemporary China 798; see also Wei Shen and Dong Jiang, 'Making Authoritarian Environmentalism Accountable? Understanding China's New Reforms on Environmental Governance' (2021) 30(1) Journal of Environment and Development 41 (describing remediation efforts through centralization).

[113] See Denise Sienli van der Kamp, 'Governance by Uncertainty: Changing Patterns in China's Environmental Enforcement' (2023) Studies in Comparative International Development, <https://doi.org/10.1007/s12116-023-09396-x>; Xuegang Zhou, Hong Lian, Leonard Ortolano, and Yinyu Ye, 'A Behavioral Model of "Muddling Through" in the Chinese Bureaucracy: The Case of Environmental Protection' (2013) 70 China Journal 120.

[114] Wang (n 112); Jie Gao, 'Pernicious Manipulation of Performance Measures in China's Cadre Evaluation System' (2015) 223 China Quarterly 618.

[115] Alex L Wang, 'Symbolic Legitimacy and Chinese Environmental Reform' (2018) 48(4) Environmental Law 699; Iza Ding, 'Performative Governance' (2020) 27(4) World Politics 525.

temporarily) enterprises with relatively little regard to whether they are in compliance with relevant laws (and thereby reducing compliance incentives for industry more broadly).[116]

Public participation has a mixed record as well. Citizens with environment-related grievances often resort to petitioning or 'letters and visits' (信访/*xinfang*) and kindred legally informal but institutionalized methods. Through such channels, complainants sometimes (but not always) invoke legal rights and rules in seeking (and sometimes receiving) discretionary redress from officials at various levels in China's multi-tiered party-state.[117]

Citizens' direct collective action has seen uneven results as well, depending on various factors, including the perceived potential impact on social stability and economic interests, and, relatedly, the identity of the public participants. In perhaps the most famous example of successful citizen-initiated environmental action, a largely non-confrontational street presence by Xiamen residents helped stop state permission to locate a polluting chemical plant in their neighborhood. Similar efforts have succeeded in some other cases, but these remain rather rare events, and their success likely reflects the relatively high social and economic status of the complainants and sometimes results in only the relocation of polluting activities to another place.[118] More heated and large-scale activism over environmental issues has more often triggered repression, and the regime's tolerance for public activism on issues that include the environment has shrunk under Xi Jinping.[119]

Societal pressure and preferences also are not uniformly in favor of greater environmental protections. Shutdowns of polluters, as well as other small-scale

---

[116] Denise S van der Kamp, 'Blunt Force Regulation and Bureaucratic Control: Understanding China's War on Pollution' (2020) 34(1) Governance 191.

[117] Jingjing Zeng, Meng Yuan, and Richard Feiock, 'What Drives People to Complain about Environmental Issues? An Analysis Based on Panel Data Crossing Provinces in China' (2019) 11(4) Sustainability 1147; Tseming Yang and Xuehua Zhang, 'Public Participation in Environmental Enforcement… with Chinese Characteristics?' (2011–2012) 24 Geo Int'l Envt'l L Rev 325; see also Xiao Zhu, Taotao Qiu and Dongshu Liu, 'Resisting Public Monitoring in Authoritarian Regimes: Evidence from Local Environmental Litigation in China' (2023) 36(2) Governance 459 (concerning local authorities' avoidance of public monitoring in the face of disclosure-promoting litigation).

[118] Wang Xixin and Zhang Yongle, 'The Rise of Participatory Governance in China' (2018) 13(1) U Pa Asian L Rev 24, 33–39 (the authors also detail other similar, though less well-known incidents); see also H Christoph Steinhardt and Fengshi Wu, 'In the Name of the Public: Environmental Protest and the Changing Landscape of Popular Contention in China' (2015) 75 China Journal 61; Benjamin van Rooij 'The People vs. Pollution: Understanding Citizen Action Against Pollution in China' (2010) 19(63) Journal of Contemporary China 55.

[119] See generally Elizabeth C Economy, 'China's Imperial President: Xi Jinping Tightens His Grip' (2014) 93(6) Foreign Affairs 80; see also Diana Fu and Greg Distelhorst, 'Grassroots Participation and Repression under Hu Jintao and Xi Jinping' (2017) 79 China Journal 100 (finding continuities and differences in the space for contentious participation by individuals and groups under Hu Jintao and Xi Jinping).

enforcement measures, and even the construction of green energy sources – all measures that advance goals that overlap with those addressed by 'rights of nature' – have significant benefits in terms of improving human well-being and equity, but they also (especially in the short term) can impose significant costs on the most vulnerable, who lose jobs, land, and other interests.[120] Such losses raise human rights concerns, primarily about social and economic rights. The means the Chinese state uses to achieve such ends have also drawn human rights-based criticisms for the extension of the surveillance state to the monitoring of environment-affecting activities, the lack of fair process, harsh enforcement measures, and politically charged criminal prosecutions.[121]

VI. International Interactions – Greater Engagement, With Limits and Without Rights of Nature

The lack of recognition of rights of nature in Chinese law has limited rights-of-nature-related interactions between China's legal system, on one hand, and foreign legal systems and international law, on the other. On the 'inbound' side, traits addressed earlier in this chapter underpin this likely durable pattern: China's engagement with international law (including a non-capacious approach to reception of international legal norms or obligations, especially where they are not embodied in a treaty and have questionable claims as rules of customary international law), Chinese domestic law (including a lack of emphasis on robust rights, and a positivist understanding of law), the Chinese legal system (including limited roles for – and significant political and policy constraints on – courts, lawyers, litigants, and the possibility of bringing environment-related claims), China's politics and governance (including tightly restricted space for social and political activism, preference for state-led, top-down approaches to environmental issues, and growing suspicion of legal ideas from abroad), and the still-limited and only-recent emphasis on environmental issues in Chinese policy and lawmaking (as well as perennial implementation difficulties for legal and policy initiatives). To be sure, China's increasingly protective environmental laws and regulations (like many areas of law in Reform Era China) reflect extensive adoption and adaptation of foreign models, but such borrowing or transplantation is not poised to expand to the still-emerging principles of the rights of nature.

[120] See Chunbo Ma, 'Who Bears the Environmental Burden in China? – An Analysis of the Distribution of Industrial Pollution Sources' (2010) 69(9) Ecological Economics 1869; Shihu Zhong, Yunjun Xiong and Guocheng Xiang, 'Environmental Regulation Benefits Whom? Heterogeneous Effects of the Intensity of the Environmental Regulation on Employment in China' (2021) 281 JEM 111877.

[121] See, eg, Yaqiu Wang, 'Beijing's Green Fist' (*Human Rights Watch*, 29 March 2022) <https://www.hrw.org/news/2022/03/29/beijings-green-fist>.

Existing pathways for rights of nature to make their way into Chinese law from the outside are narrow, absent legislative or regulatory action. Chinese courts will apply foreign or international law in some limited circumstances. Chinese law allows for the direct application in civil and administrative litigation of some treaties to which China is a party, including where treaty provisions are inconsistent with domestic law, but the contours of this principle are ambiguous and China is, of course, not a party to any treaty concerning the rights of nature.[122] China has a law on choice of law in cases involving foreign parties and interests that is in many respects consistent with what is found in many jurisdictions, allowing significant latitude to parties to choose the law governing their agreements and, in the absence of such a choice, applying a closest connections test, but prohibiting the application of foreign law where it would be contrary to China's public or social interests.[123] There do not appear to be any such cases that have involved claims under foreign (or domestic) law asserting or addressing the rights of nature.

On the 'outbound' side, China has not promoted recognition of the rights of nature in other jurisdictions or at the international level. China's non-acceptance of the rights of nature into its own laws and policy discourse has given Beijing little leverage, or reason, to press a pro-rights of nature agenda with other states or international institutions. Its rising criticism of international legal norms shaped by others (especially the West) also cuts against Beijing's promoting the rights of nature abroad (as well as at home). China's longstanding (if recently eroding) embrace of sovereigntist norms in international law impedes not only the reception of rights of nature principles from abroad, but also PRC evangelism for international or foreign legal norms that – like the rights of nature – can be significantly constraining of a state's exercises of sovereign regulatory powers in its own jurisdiction (especially in the absence of a treaty to which the state has consented).[124] China's stance might make less likely the recognition of rights of nature in some other legal systems, especially in the Global South, where China's normative and political influence is relatively strong, where China's broader political support can embolden states' governments in resisting international pressure and suasion, and where China's agenda and interests are sometimes at odds with those of weaker states that face more immediate environmental crises.

There are some reasons to believe that China's international engagement on issues related to the environment might move in directions that are more in line with the rights of nature and kindred legal ideas. China is facing significant and

[122] See generally Bing Bing Jia, 'A Synthesis of the Notion of Sovereignty and the Ideal of the Rule of Law: Reflections on the Contemporary Chinese Approach to International Law' (2010) 53 Ger YBIL 11; Jacques deLisle, 'China's Approach to International Law: A Historical Perspective' (2000) 94 Proceedings of the Annual Meeting (American Society of International Law) 267.

[123] Law for Foreign-Related Civil Relationships of the People's Republic of China 2011.

[124] See the discussion in section I.

growing external pressure over the environmentally harmful acts of Chinese entities operating abroad. China received international criticism for its perceived role in stymying progress at the 2009 COP15 Copenhagen meeting on climate change.[125] Chinese companies, especially in extractive industries and infrastructure sectors, draw criticism for environmental harms that might constitute violations of the rights of nature abroad. For example, the self-styled 'International Rights of Nature Tribunal' (an unofficial civil society initiative that convenes alongside the COP climate negotiations) showcased testimony criticizing the environmental impact in Latin America of Chinese-invested infrastructure projects associated with China's signature Belt and Road Initiative (BRI). Host-country political activists have invoked the rights of nature – in court proceedings involving forum jurisdictions' 'rights of nature' provisions, and more broadly – in opposing Chinese mining projects.[126] China also has faced charges that its overseas projects have violated rights of nature-adjacent international legal obligations concerning the human right to live in a healthy and undamaged environment.[127]

China is especially vulnerable to having the behavior of its firms abroad imputed (politically and sometimes legally) to the state because those entities are often state-owned or state-linked enterprises over which the state has considerable control and because China has not imposed more extensive 'green' legal obligations on those companies (with the rules for operations abroad often falling short of the limited requirements that apply to domestic activities).

At the same time, China has been seeking to cultivate a positive image and something of a leadership role in international efforts to address climate change and other environmental challenges. Notably, China's principal retooling of the BRI focused significantly on addressing concerns about environmental impacts of BRI projects abroad (both in host countries and in terms of global impacts such as climate change).[128] China also has begun to conjoin sustainability, ecological

---

[125] See Björn Conrad, 'China in Copenhagen: Reconciling the "Beijing Climate Revolution" and the "Copenhagen Climate Obstinacy"' (2012) 210 China Quarterly 435; Mark Lynas, 'How Do I Know China Wrecked the Copenhagen Deal? I Was in the Room' *The Guardian* (22 December 2009).

[126] See Fermin Koop, 'Environmental Tribunal Questions China-Backed Projects in Latin America' (*Dialogue Earth*, 8 December 2015) <https://dialogochino.net/en/infrastructure/4472-environmental-tribunal-questions-china-backed-projects-in-latin-america/>; Leah Shipton and Peter Dauvergne, 'The Politics of Transnational Advocacy Against Chinese, Indian, and Brazilian Extractive Projects in the Global South' (2021) 30(3) Journal of Environment and Development 240.

[127] See, eg, Efrain Ramos, 'New Report Examines Rights Violations by Chinese Business Activities in Latin America' (*International Federation for Human Rights*, 23 March 2022) <www.fidh.org/en/issues/globalisation-human-rights/business-and-human-rights/violations-chinese-business-latin-america-report>.

[128] See, eg, 'The Green Belt and Road Initiative' (*China Briefing*, 12 November 2021) <www.china-briefing.com/news/the-green-belt-and-road-initiative/>; 'Xi Announces Major Steps to Support High-Quality Belt and Road Cooperation' *Xinhua* (18 October 2023) <https://english.www.gov.cn/news/202310/18/content_WS652f65e6c6d0868f4e8e05bd.html>.

civilization, and the BRI in promoting a 'China model' abroad.[129] China has touted its breakthrough contribution, in collaboration with the United States, to the creation of the Paris Accords, made much of the US's subsequent turn against the agreement during the Trump administration, and undertaken a high-profile renewed commitment to cooperation on climate change with the US announced at the COP26 climate change meeting in Glasgow in 2021.[130]

China's role as host of the COP15 meeting of parties to the UN Convention on Biological Diversity in Kunming in 2021-22 (where rights-based approaches were in the mix) has created additional incentives for – and greater pressure on – China to position itself publicly as a cooperative and, indeed, leading player on environmental issues.[131] The 'Kunming Declaration' – a largely hortatory document adopted by around one hundred members at the first, virtual segment of the meeting and not including the rights of nature – in part reflected China's bid to be seen as taking a leadership role. The quest for stature and influence was perhaps most evident in the declaration's title – 'Ecological Civilization: Building a Shared Future for All Life on Earth'–which included the Xi-era Chinese tropes 'ecological civilization' and 'shared future for all.'[132] So, too, the 'Group of 77 and China' agenda for sustainable development similarly stresses man living in harmony with nature and acknowledges some member states' adoption of a 'rights of nature' approach.[133]

---

[129] See, eg, Michael Faure, 'The Export of Ecological Civilization: Reflections from Law and Economics and Law and Development' (2020) 12 Sustainability 10409.

[130] See generally Gao Xiaosheng, 'China's Evolving Image in International Climate Negotiation: From Copenhagen to Paris' (2018) 8(4) China Quarterly of International and Strategic Studies 213; Kevin Rudd, 'The New Geopolitics of China's Climate Leadership' (*Dialogue Earth*, 11 December 2020) <https://chinadialogue.net/en/climate/the-new-geopolitics-of-chinas-climate-leadership/>; U.S. Department of State, *U.S.-China Joint Glasgow Declaration on Enhancing Climate Action in the 2020s* (10 November 2021) <www.state.gov/u-s-china-joint-glasgow-declaration-on-enhancing-climate-action-in-the-2020s/>.

[131] See, eg, Juliet Lu and Tyler Harlan, 'COP15 in Kunming: A New Role for China in Global Conservation?' (*Wilson Center*, 19 October 2021) <www.wilsoncenter.org/blog-post/cop15-kunming-new-role-china-global-conservation>; Andrea Schapper, 'Procedural Rights for Nature' (2022) 43(5) Third World Quarterly 1197; Michael Taylor, 'Giant Leap for Nature? All Eyes on China to Land New Global Pact' (*Thomson Reuters Foundation*, 13 July 2021) <https://news.trust.org/item/20210712234820-cs0bh/>.

[132] Kunming Declaration – Declaration from the High-Level Segment of the UN Biodiversity Conference 2020 (Part 1) under the Theme: 'Ecological Civilization: Building a Shared Future for All Life on Earth' (October 2021) <www.cbd.int/doc/c/df35/4b94/5e86e1ee09bc8c7d4b35aaf0/kunmingdeclaration-en.pdf>.

[133] See, eg, the declaration 'For a new world order for living well' adopted at the Summit of Heads of State and Government of the Group of 77 held in Santa Cruz de la Sierra, on 14 and 15 June 2014 <www.g77.org/doc/A-68-948(E).pdf>.

Still, all of this stops well short of support, or acceptance, by China of the rights of nature domestically or internationally.[134] It would be a radical change for China to become a global leader in, or proponent of, the rights of nature. Even if China were to accept something more akin to the rights of nature as an international obligation or in domestic law, the realization of such rights or implementation of such obligations would be far from certain. Especially given the many features of Chinese law, institutions, politics, and policy addressed in this chapter, there would still be substantial risk that China would engage in largely 'symbolic' or 'performative' lawmaking and merely 'paper compliance' with rights of nature obligations.[135]

---

[134] Indeed, in 2022, China joined Russia in opposing proposed language on human rights and indigenous rights in a draft biodiversity treaty, favoring the sovereignty-friendly text calling for upholding rights 'in accordance with national legislation,' Chloé Farand, 'Russia, China Oppose "Human Rights" in Nature Talks, Amid Slow Progress to a Deal' (*Climate Home News*, 30 March 2022) <www.climatechangenews.com/2022/03/30/russia-china-oppose-human-rights-in-nature-talks-amid-slow-progress-to-a-deal/> (concerning negotiations in Geneva on a treaty to address biodiversity loss).

[135] See Wang (n 115); Timothy Webster, 'Paper Compliance: How China Implements WTO Decisions' (2014) 35(3) Mich J Int L 535; see also Lori F Damrosch, Maria T Caffi, and Jacques deLisle, 'How International is International Law?' (2017) 111 Proceedings of the Annual Meeting (American Society of International Law) 69, 75–78 (concerning China's portrayal as *infra legem* behavior that some other states and critics view as violations of international legal obligations or corrosive of international legal rules).

Cyprus

Constitutional Adoption of a Right to Nature, but No Formal Recognition of Nature's Rights

Konstantinos A Rokas[*]

I. Introduction

The Republic of Cyprus (hereinafter 'RoC') is somewhat unique, because of its political situation. The Constitution of the RoC, ratified on 16 August 1960, is the result of negotiations conducted in Zurich and London, which led to the independence of Cyprus from Great Britain and created a unitary presidential republic. The northern part of the island has been illegally occupied by Turkey since 1974. Thus, the island of Cyprus has been de facto divided since that year. The southern two-thirds of the island are controlled by the RoC whereas the northern part is illegally occupied and controlled by the Turkish Cypriot community and the Turkish army. A buffer zone exists between the two parts of the island as maintained by UN peacekeeping forces. In 1983, Turkish Cypriots unilaterally declared

[*] The manuscript was finalized in June 2022. The assessment of literature and case law, including internet materials, reflects the legal situation at that time. However, the article includes the 2024 constitutional reform of the Republic of Cyprus (with references updated in April 2025), which incorporated a right to nature under article 7A.

I would like to thank Ms Ioanna Chadjiyianni, lecturer in law at the University of Cyprus, Mr Lefkios Sergidis, former executive director of the NGO Terra Cypria, and Ms Ioanna Samara, attorney-at-law, for their input. I also wish to express my sincere gratitude to the editorial team for their insightful comments and meticulous corrections. Any errors remain, of course, my own.

their independence and proclaimed the 'Turkish Republic of Northern Cyprus' (hereinafter 'TRNC'). The UN Security Council declared that the unilateral declaration of independence was invalid; Turkey is the only country, to date, that has recognized the TRNC as an independent State.

Notwithstanding this particularity, Cyprus faces a number of environmental challenges that affect and will further affect the island and its two communities as a whole: water shortages, forest fires, urbanization, and desertification due to climate change are some of the most pressing ones. And although some efforts to achieve informal coordination between the two communities have been made, there are still many steps to be taken. The existential threat that climate change poses for the island knows no borders, as evidenced especially during the Covid-19 pandemic. For several months, communication and circulation between the two parts of the island was interrupted.

This report focuses on the Republic of Cyprus. The legal system of the country is mixed and deeply influenced by the common law. However, civil law systems, and in particular the Greek legal order, also have a significant influence on the Cypriot legal order, especially in relation to administrative and public law in general.[1] The Supreme Court of Cyprus is occasionally influenced by the case law of the Greek Council of State (ie the Greek Supreme Administrative Court, 'Symvoulio tis Epikrateias'), which in turn is influenced by the French *Conseil d'État*.

It should be pointed out that, despite its unresolved political situation, Cyprus has been a member of the Council of Europe since 24 May 1960 and a full member of the European Union since 2004, although the relevant *acquis* does not apply to the areas not under the control of the RoC. Cyprus has been a member of the United Nations since its independence. The historical context of the adoption of its Constitution helps us to understand the status of rights to nature and difficulties regarding the recognition of such rights. At the time of its adoption, awareness as to the need for environmental protection was minimal. This is why the Constitution of 1960 did not enshrine the right to a clean and healthy environment, contrary to what has been the case in other countries both in Europe and elsewhere.

We will hereafter present some aspects of the Cypriot legal framework on the protection of the environment and more precisely on the situation of the rights to nature and rights of nature in the Republic. We will explain that the protection afforded to the environment remains insufficient and that rights to nature and especially rights of nature are protected to only a limited extent. In practice, the right of environmental NGOs to seek judicial recourse for the protection of nature has proven very difficult to exercise. There is hope that the Convention of Aarhus, as it will be interpreted by the CJEU, along with the explicit incorporation of a right to nature into the Cypriot Constitution in 2024, will mark a shift from this position. A development with the potential to positively influence the

[1] Natalia Charalampidou, 'Cyprus' in Kurt Deketelaere (ed), *IEL Environmental Law* (Kluwer Law International BV 2021) 27, no 2.

rights to nature in Cyprus[2] was also the resolution and recommendation on the adoption of a new protocol to the ECHR on an enforceable right to a 'safe, clean, healthy and sustainable environment'[3] as made by the Council of Europe's Parliamentary Assembly in September 2021.

II. Recognition of a Right to Nature and No General Recognition in the Legal System of Rights of Nature

The Cypriot Constitution does not contain any provision recognizing rights of nature as such. Until 2024, there was neither a provision establishing a duty for the State to protect the environment nor a distinct individual right to the protection of the environment.[4] The Constitution only recognized, under art 23(3), the power of the State to impose restrictions on the right of property in the interest of public health and urban planning.[5] There is neither legislation nor case law recognizing nature as a subject of rights. Under the Cypriot Constitution, only natural and legal persons governed by private law are recognized as holders of rights.[6]

Furthermore, there have been no proposals so far for any recognition of specific rights of nature as a whole, of ecosystems, or of particular components of ecosystems in Cyprus. Thus, for instance, no specific rights are enjoyed by the Larnaka Salt Lakes ecosystem, which is part of the European Network Natura 2000, is covered by the European Habitats Directive (92/43/EEC) as well as the Cypriot Law on the Protection and Management of Nature and Wildlife (153(I)/2003), and which is included in the Ramsar List of Wetlands of International Importance.

A certain degree of protection for rights to nature has been guaranteed through the interpretation of the Constitution since the early 90's. The Supreme Court of Cyprus has interpreted art 7(1) of the Constitution, which guarantees the right to life and personal integrity, as comprising also the right to the environment.[7] It has

---

[2] Ioanna Hadjiyianni and Alexandros Vryonides, 'An Explicit Right to the Environment: The Potential of an Environmental *Lex Specialis* to Further "Green" the ECHR' (*EULaw live*, 11 November 2021) <https://eulawlive.com/op-ed-an-explicit-a-right-to-the-environment-the-potential-of-an-environmental-lex-specialis-to-further-green-the-echr-by-ioanna-hadjiyianni-and-alexandros-vryoni/>.

[3] Ancrer le droit à un environnement sain: la nécessité d'une action renforcée du Conseil de l'Europe, Résolution 2396 (2021) <https://pace.coe.int/fr/files/29499/pdf>.

[4] Iliana Nikolaou and Georgios Tsaousis, *Handbook of the Cyprus Administrative Procedure* (Nomiki Bibliothiki 2021) 53 (in Greek).

[5] Ibid.

[6] Kostas Paraskeva, *Cypriot Constitutional Law, Fundamental Rights and Liberties* (Nomiki Bibliothiki 2015) 20–23 (in Greek).

[7] *The Local Community of Pyrgos v The Republic* [1991] 4 CLR [The Cyprus Law Reports] 503, (in Greek); *Cypriot Republic v Council of Improvement of Geri* [1998] 3 CLR 210 (in Greek); *Thanos Club Hotels v ETEK* [2000] 3 CLR 324 (in Greek); *Local Council of Parekklissia et al. v The Cypriot Republic*, 1628/2005 (in Greek); joint cases 1274/2010, 1275/2010,

been suggested that this case law constitutes a judicial recognition of an actionable right to the environment.[8] However, this view was open to debate, given that the judges appear to believe that the right of associations – and NGOs in particular – to challenge decisions, acts, or omissions of the administration in violation of environmental law, under art 146(2), is dependent on the explicit recognition of a right to nature in the Cypriot Constitution.

Notwithstanding the assessment as to the impact of this case law, there was an ongoing discussion and proposals for the recognition of an explicit right of citizens to a clean, healthy, and sustainable environment and, at the same time, of the State's duty to protect the natural environment.

There have been two proposals by political parties in Cyprus for an amendment of the Constitution so as to include a right to a clean and healthy environment. One of those was formulated in 2022 by 'AKEL', the Progressive Party of Working People, which is one of the major political parties in Cyprus. They have proposed that an additional article be included to the Constitution, namely art 7A, to be worded as follows:[9]

> '1. The protection of the natural environment constitutes an obligation of the State. For its protection, the State has an obligation to take preventive, repressive or remedial measures with respect for the principle of sustainability.
> 2. Each person has a right to a safe, clean, healthy and sustainable environment.
> 3. Each person has the following rights:
>   (a) the right of access to relevant information kept by State authorities without a need to prove a direct and personal interest;
>   (b) the right of access to justice for issues relevant to the environment;
>   (c) the right of effective remedies for the violation of any right arising from this article and related legislation.
> 4. No interference with the exercise of this right is possible unless it is in accordance with the law and necessary in the interest of the security of the Republic or the constitutional order or the territorial integrity or public security or public health or the protection of the rights and freedoms guaranteed to every person under the Constitution.'

The other proposal was made by deputies of the Green Party of Cyprus (*Ecologists' movement and citizens' cooperation*). Their proposal for an amendment to the Constitution would include and protect, as a fundamental right, the right of every person to live in conditions that ensure the protection of their health, the environment and biodiversity.[10]

1276/2010, 1277/2010 and 1278/2010 *Municipality of Ayia Napa et al. v The Cypriot Republic* (in Greek); Charalampidou (n 1) 35, no 17.

[8] Ioanna Hadjiyianni, 'Access to Justice for the Protection of the Environment: the Notion of Legitimate Interest in Cyprus in Light of the Aarhus Convention and its Implementation in the European Union', (2020) 2 Perivallon kai Dikeo 224, 235 (in Greek).

[9] Available at <www.nomoplatform.cy/wp-content/uploads/2022/01/23.02.063.007-2022-2-Tropopoiisi-tou-SINTAGMATOS.pdf> (in Greek). All English-language translations are our own unless indicated otherwise.

[10] Available at <www.nomoplatform.cy/wp-content/uploads/2022/04/23.02.063.049-2022-O-peri-tis-Tetartis-Tropopiisis-tou-Sintagmatos.pdf> (in Greek).

There have not been any discussions as to the introduction of other specific rights.

There are no specific rights recognized in the legal order of Cyprus such as legal personality, life, the ability to enter into contracts, or the ability to inherit. Thus, for instance, the *Agrino* wild sheep of Cyprus, a protected species which is in fact included in the global list of endangered species, does not enjoy legal standing and cannot claim legal personality or any other right that can be granted to a person. The *Agrino* of Cyprus ('Ovis Orientals Ophion', 'κυπριακό αγρινό') is a wild sheep, a descendant of Asian sheep, which has lived in Cyprus for the last 8,000 years and constitutes a national symbol.

As already mentioned, the provision that has constituted the legal basis upon which stakeholders have had judicial recourse against decisions, acts, or omissions of the authorities in violation of environmental law has been art 7(1) of the Cypriot Constitution. It is this provision that has been interpreted as the legal basis for a general right to nature. Case law has also been crucial, especially as regards the interpretation of legal standing. Moreover, EU law rules, as interpreted by the CJEU, and international law – in particular the International Convention of Aarhus and the ECHR – complete the set of rules which are important for understanding how rights of nature have been formulated in Cyprus. Finally, it is worth mentioning that EU law, and in particular art 191(2) TFEU, is the source of the basic principles of environmental law in the Cypriot legal order.[11] Among those principles, the role of the preventive and precautionary principles is especially crucial.[12]

The protection of rights of nature is far from satisfying. Therefore, the main objective was to clearly establish an explicit right to the environment. Such right is meant to provide for general protection of the environment. It would create a clear responsibility of the State to protect the environment and would also ensure that environmental NGOs and associations have the right to bring cases in relation to instances of environmental degradation and to challenge decisions of the administration on environmental matters.

In December 2024, the Parliament of the Republic of Cyprus adopted, by a substantial majority, the 19th constitutional amendment, which explicitly incorporated a right to nature under Article 7A of the Constitution.[13] This article provides:

'1. Everyone has the right to a safe, clean, healthy, and sustainable environment.
2. Everyone has the right:
   (a) to access relevant information held by the competent authorities of the Republic, without being required to prove a direct and personal legal interest, as provided by law;

[11] Charalampidou (n 1) 31, no 11.

[12] Charalampidou (n 1) 31, no 12.

[13] The text of the Constitution is available in Greek at <https://www.cylaw.org/nomoi/enop/non-ind/syntagma/full.html>.

(b) to access justice in matters relating to the environment, in accordance with Article 146; and
(c) to effective remedies for the violation of any right arising from this article and the relevant legislation.

3. The protection of the natural environment constitutes a duty of the State, which is obliged to take preventive, remedial, or restorative measures in accordance with the principle of sustainability.
4. No interference with the exercise of this right is permitted except as provided by law and only when necessary in the interest of the security of the Republic, the constitutional order, territorial integrity, public safety, public order, public health, or the protection of the rights and freedoms guaranteed to every person by the Constitution.'

III. Legal and Political Operators

In Cyprus, there are a number of environmental NGOs that have been active in promoting the recognition of rights of nature and are trying to raise awareness; their work aims at guaranteeing that the rights to nature are afforded effective protection. Some of these organizations have had a long presence and tradition in actively promoting the protection of the environment. Worth mentioning are BirdLife Cyprus, which was created by the merger of two of the most ancient and important organizations ('Cyprus Ornithological Society – *Kypriakos Ornithologikos Syndesmos*', founded in 1957, and 'Cyprus Ornithological Society – *Ptinologikos Syndesmos Kyprou*', founded in 1970), Terra Cypria – The Cyprus Conservation Foundation (founded in 1992), Friends of Akamas (founded in 1986), and Friends of the Earth Cyprus (founded in 1980). All these organizations have been actively promoting the environmental cause and have formed a barrier to uncontrolled development. Apart from the NGOs, Communities (one of the administrative subdivisions in the RoC) also play a role in the protection of the environment as they have standing to seek protection of the environment before the courts. The courts' recognition of the legitimate interest of Communities – such that they can bring proceedings for violations of environmental law (see IV, below) – has led Communities to assume responsibilities in the form of judicial actions and to have an impact on the broader movement for the protection of the environment.[14] Political parties also play a role as regards the promotion of the rights to nature, as has already been highlighted.

The motivations underlying the actions of stakeholders are multiple and most probably vary in nature. It is difficult, in this context, to make a detailed assessment of the exact reasons behind the mobilization of each of these groups and associations. However, the common ground of all their efforts has been the desire to promote more effective environmental protection. Again, these groups, associations, and administrative units have not, to date, had as their main objective

[14] There are 24 associations active in Cyprus in different fields related to environmental protection and the environment in general.

the recognition of specific rights of nature; however, they have contributed, inter alia, to a broader protection of a general right to nature.

IV. Enforcement and Procedures

Access to courts for the exercise of rights regarding the protection of nature, where such protection results from the provisions of the Constitution as interpreted by case law, is very strict. In the case of *The Republic v The Local Community of Pyrgos*,[15] the Supreme Court of Cyprus recognized that a local Community has the right, under art 146 of the Constitution, to challenge the actions of public authorities regarding issues affecting the environment in the proper geographical area of the Community concerned. On account of their mission, Municipalities and Communities have legal standing allowing recourse at first instance before the Administrative Court ('Dioikitiko Dikastirio') for the protection of the rights of nature. The Supreme Court of Cyprus ('Anotato Dikastirio Kyprou') hears appeals against judgments of the Administrative Court.

According to art 146(2) of the Constitution

> '[…] an action may be brought by a person where any existing legitimate interest that such person has, either individually or by virtue of being a member of a Community, is adversely and directly affected by such decision or act or omission.'

Therefore, persons whose existing legitimate interest is adversely and directly affected by a decision, act, or omission of a public authority can bring proceedings in accordance with art 146(2) of the Cypriot Constitution. Not every individual has the right to challenge a decision, act, or omission before the courts; namely, the interested person has to have a legitimate interest in order to have recourse to the courts. This legitimate interest must be present (existing at the time of the filing of the action against the challenged act[16]), direct (related directly to the applicant), and personal (in that it concerns the applicant specifically and personally).[17] 'A legitimate interest is established when there is a reasonable claim that the applicant, in a capacity recognised by law, is likely to incur damage as a result of the contested act';[18] in other words, when there is a specific legal relationship between the applicant and the contested act. And although the notion of 'legitimate interest' must be interpreted as covering not only pecuniary damage

---

[15] *The Republic v The Local Community of Pyrgos* [1996] 3 CLR 503, 508 (in Greek).

[16] *Pitsillos v CBC* [1982] 3 CLR 208; *Charalambous v The Republic* [1996] 3 CLR 73 (in Greek); C Kombos, 'The Supreme Court's of Cyprus and the CJUE's Approach to Standing for Judicial Review and to the Preliminary Reference Procedure' (2010) 16(3) EPL 327, 341.

[17] *Adonis Papantoniou and others v Ministers' Cabinet* (No 1) [1993] 4 CLR 399; Hadjiyianni (n 8) 234.

[18] Hadjiyianni (n 8) 234; Ioanna Hadjiyianni, 'Multi-Level Governance in Action: Access to Justice in National Courts in Light of the Aarhus Convention and its Incorporation in the EU Legal Order' (2020) 26(4) EPL 889.

but also moral prejudice, at the same time it is considerably restrictive in several other ways.

The Supreme Court has been particularly reluctant to recognize a right of environmental NGOs or of the public at large to challenge decisions of the administration in environmental matters.[19] It has repeatedly found that associations lack standing based primarily on the fact that Cypriot law does not provide for an *actio popularis*; to recognize such an action would create an excessive burden for the administration of justice. Nonetheless, Supreme Court Judge Nikolaidis opined in favor of extending the protective scope of the right to life to the protection of the environment; he stressed the importance of NGOs for the protection of the environment and argued in favor of accepting the legal standing of NGOs to challenge decisions of the administration.[20] Unfortunately, Judge Nikolaidis' approach was not accepted by the plenary of the Supreme Court.[21]

Nevertheless, it has been correctly suggested that applicants must rely on, and courts must rule in accordance with, the rules of the International Convention of Aarhus (especially art 9(3) thereof), the Charter of Fundamental Rights of the European Union, and the rules of EU law in general.[22] In addition, recent case law of the CJEU can soundly be advanced as an additional argument in favor of granting legal standing to NGOs focusing on the protection of the environment. The *Protect Natur* judgment of the CJEU[23] states that:

> 'The combined provisions of Article 9(3) of that convention approved by Decision 2005/370, Article 47 of the Charter of Fundamental Rights and Article 14(1) of Directive 2000/60 must be interpreted as precluding national procedural rules that deprive, in situations such as that in question in the main action, environmental organisations of the right to participate, as a party to the procedure, in a permit procedure that is intended to implement Directive 2000/60 and limit the right to bring proceedings contesting decisions resulting from such procedure solely to persons who do have that status.'

It should be noted that the Cypriot Republic has adopted one of the most pro-European interpretations of the principle of supremacy. In the framework of the 2006 amendment of the Cypriot Constitution, the provision of art 1A was added; it reads as follows:

> 'No provision in the Constitution can be deemed as overriding any legislation, acts or measures enacted or taken by the Republic which are obligatory as a Member State of the European Union, nor does it hinder Regulations, Directives or other binding provisions or measures of a legislative nature enacted by the European Union from having legal force in the Republic.'

---

[19] Charalampidou (n 1) 35, no 17.

[20] *Friends of Akamas v The Cypriot Republic* [1998] 4 CLR 767 (in Greek).

[21] *Thanos Club Hotels v ETEK* [2000] 3 CLR 324 (in Greek).

[22] Hadjiyianni (n 8) 238–241; Theognosia Kouspi, 'The Protection of Environment and Access to Justice' (Dikeossyni, 20 February 2020) (in Greek) <https://dikaiosyni.com/katigories/arthra/i-prostasia-tou-perivallontos-kai-i-prosvasi-sti-dikaiosini/>.

[23] CJEU 20 December 2017, C-664/15, *Protect Natur-, Arten- und Landschaftsschutz Umweltorganisation v Bezirkshauptmannschaft Gmünd*, ECLI:EU:C:2017:987.

Therefore, the judiciary could not even rely on the national constitutional identity under art 4(2) TEU – assuming that this would be possible – in order to avoid the impact of EU and international law as regards the standing of associations and NGOs to bring proceedings before the courts.

Under art 146 of the Cypriot Constitution, a person whose existing legitimate interest is adversely and directly affected by a decision, act or omission of a public authority has a right of recourse to the Administrative Court of Cyprus to complain that the relevant decision, act, or omission is contrary to a provision of the Constitution or of any law or is made in excess or in abuse of powers vested in such authority. The relevant action has to be filed within seventy-five days of the date when the decision or act was published or, if not published and in the case of an omission, when it came to the knowledge of the person filing the action.

The Court may confirm, either in whole or in part, the authority's decision, act or omission; or it may declare null and void, in whole or in part, a decision, act, or omission of the authority which is in violation of the provisions of Cypriot law. As a result of Cyprus' following the common law tradition, there is no State mechanism for the enforcement of administrative decisions, such as a mechanism ordering the recovery of expenses in favor of citizens.[24] Because of the lack of a 'law on enforcement of administrative acts, such as the German 'Verwaltungs-Vollstreckungsgesetz' or the Greek 'Kodikas Ispraxis Dimosion Esodon' [*Κώδικας Είσπραξης Δημοσίων Εσόδων*] compliance is ensured through establishing that non-compliance constitutes a penal offence, and recovery of expenses takes place through filing civil lawsuits before civil courts requesting payment of such expenses as civil debts owed to the State'.[25] Thus, civil and criminal courts assume a role in the enforcement of administrative matters and the person aggrieved by a void act or decision of a public authority will also need to institute legal proceedings for the recovery of damages or for another remedy, and for the recovery of just and equitable damages.[26]

In principle, the rights to nature and the protection provided by the judiciary as described above necessitate the assistance of public authorities and of the judiciary. The fact that there are no specific rights of nature, and that no element of nature or animal can be considered as a holder of rights in the Cypriot legal order, also explains why there are, to date, no other means available for the enforcement of rights.

It should, of course, be borne in mind that since Cyprus is a member of the European Union, both the country and its citizens have at their disposal the legal and judicial means available under EU law. At this point, it must also be mentioned that a hesitancy of the Cypriot courts has been observed with regard to

---

[24] Charalampidou (n 1) 28, no 3.

[25] Ibid.

[26] Ibid.

submitting preliminary questions to the CJEU.[27] Nonetheless, NGOs and citizens could inform the Commission of the need to inquire into a Member State's potential violation of EU law in relation to the environment and thereby eventually put in motion the procedure of art 258 TFUE against that state – in this case, Cyprus. Finally, it should be noted that the 2024 reform of the Cypriot Constitution – by which Article 7A explicitly enshrines a right to nature – may enhance the judiciary's recognition of the *locus standi* of legal persons, and particularly of environmental NGOs, to challenge before the courts any decisions, acts or omissions contravening EU and international environmental protection law. It is submitted that, following this significant constitutional reform in Cyprus, actions brought by NGOs can no longer be dismissed on the basis of lack of legal standing.

V. Theoretical Sources and Foundations

With regard to environmental protection no specific theoretical sources and foundations proper to Cyprus can be identified. Rather, the prevailing discourse tends to reflect broader trends observed within the European Union and in Greece. The admission of Cyprus into the European Union has had a tremendous impact on the protection of the environment. The Republic of Cyprus has transposed EU legislation and has, as a result, been equipped with a full and comprehensive legal framework for the protection of the environment.

Other than that, it should be mentioned that Cyprus' first environmental organizations were established in the 60's and the 80's, but many more have been created since the country joined the EU on 1 May 2004. The main motivation behind those initiatives has been a love for nature and, more recently, a sense of urgency in the face of climate change and insufficiently monitored construction activity on the island.

VI. Relationship Between National Legal Systems

As pointed out above, Greek law has been, and continues to be, a source of influence for Cypriot public law. The Supreme Court of Cyprus follows and constantly adopts parts of the case law of the highest administrative court of Greece, ie *Symvoulio tis Epikrateias*. Furthermore, art 24 of the Greek Constitution may have been a source of positive influence in the discussion regarding the recognition of a right of citizens to a clean and healthy environment.

In particular, art 24(1) of the Greek Constitution is worded as follows:

> '1. The protection of the natural and cultural environment constitutes a duty of the State and a right of every person. The State is bound to adopt special preventive or repressive measures

[27] Hadjiyianni (n 8) 240.

for the preservation of the environment in the context of the principle of sustainable development […].'[28]

Nonetheless, the Supreme Court of Cyprus has been particularly reluctant to adopt the case law of Greece's highest administrative court, which has gradually widened the *locus standi* of NGOs.[29]

Cyprus presents the particularity of being the only country in the European Union, and one of the few in the world, that is divided. An interesting issue here is the extent to which the regulation of rights to nature impacts the protection of rights in the northern part of the island, in the areas not under the effective control of the RoC. To date, no such impact has been documented. At the same time, there is information suggesting that the development taking place in the northern part of the island is insufficiently supervised, and this has adverse consequences for the environment.[30] Therefore, an important challenge for the RoC is not only to embed the rights to nature in its legal culture but also to effectively enforce those rights on the entire island. It is evident that the impact of environmental disasters cannot be stopped at the borders.

Regarding the impact of rights to nature on other jurisdictions, no such impact can be observed or has been documented to date.

In line with the aforementioned observations as to the status of the island, a preliminary question that should be addressed in this context concerns the enforceability of rights to nature upon the areas not under the effective control of the RoC. It is not possible for the RoC to enforce rights to nature that might be affected by acts or omissions of the *de facto* authority of the north. Notwithstanding that, the *Apostolidis v Orams*[31] judgment of the CJEU shows that individuals harmed by violations of environmental law could eventually seek enforcement of a Cypriot judgment granting damages in another Member State. This will be an option when an individual or a legal entity – such as a multinational company – with a domicile or assets in another Member State is involved in the violation concerned. The question is not purely hypothetical, given that in the last years there has been intense interest and activity regarding the exploration of hydro-

[28] Available at <www.hellenicparliament.gr/UserFiles/f3c70a23-7696-49db-9148-f24dce6a27c8/THE%20CONSTITUTION%20OF%20GREECE.pdf>.

[29] Kouspi (n 22).

[30] In that sense, see: Emine Eminel Sülün and Zehra Azizbeyli, 'The EU and Environmental Protection: Cyprus at the Crossroads of Europe' (2021) 20(1) Ankara Avrupa Çalışmaları Dergisi 293, esp. 294; according to the authors, '[a] new study has found that coastal areas in north Cyprus have the "second highest amount of microplastics among beaches studied across the world". This is no coincidence, as marine and coastal ecosystems are poisoned with toxic pollution every day due to uncontrolled growth of smaller towns' (in the northern part of the island).

[31] CJEU 28 April 2009, C-420/07, *Meletis Apostolides v David Charles Orams and Linda Elizabeth Orams*, ECLI:EU:C:2009:271.

carbons located offshore of Cyprus.[32] The RoC has granted offshore licenses to an Italian oil and gas company and to a South Korean company. Activities that create potential dangers for the environment could, however, also be initiated by the TRNC. The RoC might not be in a position to effectively block such activities, which could be in violation of environmental rules as well as of international law in general.

Apart from this very special situation of the Republic of Cyprus, one should also address concerns regarding the treatment of rights in a conflict-of-laws situation. It should be noted from the outset that there will hardly be a question of enforcement of the rights of nature under the Cypriot system outside of Cyprus. This is, first of all, because there are no specific rights of nature. As a result, one could hardly envisage a situation where such a question could actually arise.

Turning now to the enforcement of foreign rights of nature in the Cypriot legal order, it is first of all necessary to point out that the law applicable 'to a non-contractual obligation arising out of environmental damage or damage sustained by persons or property as a result of such damage' is governed by the Rome II Regulation (Regulation (EC) No 864/2007 on the law applicable to non-contractual obligations). Judgments of EU Member States that rule on compensation claims arising out of environmental damage are governed by the Brussels Ibis Regulation (Regulation (EU) 1215/2012 on jurisdiction and the recognition and enforcement of judgments in civil and commercial matters). National rules will apply as to the recognition of judgments emanating from countries outside of the EU.

The problem could be explained better through the formulation of certain hypothetical questions. For instance, if the Amazon forest did have legal personality recognized by certain States, would it still be able to found a company in Cyprus? Could a judgment awarding compensation to a chimpanzee in Hungary be recognized and enforced in Cyprus? If an ecosystem of the Dominican Republic, recognized as holder of rights, wanted to have a judgment in its favor recognized and enforced in Cyprus against a Cypriot company representing the interests of Spanish businesswomen, would that be possible? Would a cetacean recognized by Finland as having legal personality be permitted to appear before Cypriot courts?

Some initial answers can be summarized as follows: an element of nature would be precluded from founding a company. However, such a company established in another EU member state would be recognized in the RoC without any problem. As regards the right of an element of nature (an animal or otherwise) to appear before Cypriot courts, this issue would be governed by the *lex fori* in accord with the procedural nature of the question. In light of that, it is highly unlikely

[32] Ayla Gürel, Fiona Mullen and Harry Tzimitras, *The Cyprus Hydrocarbons Issue: Context, Positions and Future Scenarios* (Peace Research Institute Oslo (PRIO), PCC Report 1/2013); according to Charalampidou (n 1) 289, no 5: 'Developing offshore hydrocarbon resources, although only one commercially exploitable reserve was discovered in 2013, remains a critical component of the government's economic recovery efforts.'

that Cypriot judges would accept the legal standing of an element of nature, or of an animal, to bring a case before them. To date, it is impossible to appoint a guardian *ad litem* who would, for example, represent the rights of an animal or an ecosystem. As regards the recognition of a judgment of a third country awarding damages to holders of specific rights of nature, we believe, again, that courts will most likely refuse recognition on grounds of public policy. An outcome friendlier to recognition will be reached regarding a judgment emanating from other Member States, given the existing system of recognition and enforcement under the Brussels Ibis regime and the principle of mutual trust governing the administration of justice in the EU. Thus, we believe that enforcement of specific rights of nature that are progressively created in different Member States will in the long term affect the recognition of special rights which are more generously recognized outside of the EU.

VII. Efficacy

As explained above, it is mostly individuals and local Communities that have used the rights to nature and tried, more precisely, to explicitly rely on such rights. Cyprus' Technical Advisor of the State and Citizens (ETEK) has also tried to invoke the right to nature, but despite some initially positive signs as to the acceptance of its standing in cases challenging decisions of the administration, the relevant case law was later reversed.

As already explained, the implementation of a general right to nature has initially been limited, and it has been far from effective. On the other hand, and as opposed to the judiciary's restrictive reading of a right to nature, there are scholars that openly advocate for a different interpretation of the provisions of the Cypriot Constitution in the light of art 9(3) of the Aarhus Convention and in a way that will ensure that environmental NGOs will have standing to initiate judicial proceedings in order to generally challenge acts of the government and of the administration that may be contrary to the EU legislation on the protection of the environment.[33]

The main obstacle in the implementation of an effective human right to nature has been the narrow definition of persons having *locus standi* to initiate proceedings to challenge acts or omissions of private stakeholders and public authorities which are in violation of provisions relating to the protection of the environment. As a result of this approach of the Supreme Court, NGOs have been deprived, so far and to a great extent, of the possibility to actively engage in the field of protection of the environment. It is our hope that, following the constitutional enshrinement of a right to nature in 2024, the courts of the Republic of Cyprus will expand the legal standing of NGOs and other stakeholders to challenge decisions,

[33] Hadjiyianni (n 8) 239 (in Greek).

acts or omissions contravening environmental protection laws. This evolution is certain to result in a more robust and efficient protection of the environment.

On top of legal obstacles, account should be taken of the factual circumstances in Cyprus. The country is one of the most popular tourist destinations in the eastern Mediterranean; it relies heavily on tourism.[34] This leads to the existence of conflicting interests. On the one hand, local Communities and developers want to take advantage of and profit from the natural resources; on the other hand, a part of civil society and NGOs are fighting for an effective protection of the environment.[35] These conflicting interests make it difficult for governments to strike the right balance between the need to protect the environment and the interest of individuals to fully enjoy their rights to property.

VIII. Consequences

The evolution of the rights to nature and rights of nature in Cyprus has so far been modest. Environmental protection remains fragile and uncertain, although some steps have been taken in the right direction. Therefore, the answer to this question is partly prospective, in the sense that it takes into account the consequences of a broader and more generous recognition of rights of nature, both generally and specifically. In Cyprus, the incorporation of a general right to nature into the Cypriot Constitution in 2024 has the potential to significantly impact society as a whole. It could mark a shift from an excessive reliance of the economy on the construction industry to other activities that are more friendly to the environment. On the other hand, there is still a long way to go before the recognition of rights of nature becomes effective. For this to happen, much more is required than a mere changing of the legal framework. A change of mentality is far and foremost an indispensable requirement.

Nonetheless, European Union law – in combination with international law – has been and continues to be an accelerator for a change of attitude in all fields at cultural, political, and legal levels. The tools that EU law provides to actors of civil society could work as an encouragement and mobilize people's forces against actions in violation of environmental law which undermine the future and prospects of sustainable development on the island of Cyprus.

[34] Christina Hadjidemetriou, 'The Environment' in Constantin Stephanou (ed), *Cyprus and the EU. The Road to Accession* (Routledge 2005) 103.

[35] Hadjidemetriou (n 34) 106.

Czech Republic

An Anthropocentric Approach to Environmental Protection

Jakub Harašta[*]

I. Introduction

This report provides general information on the rights of nature under Czech law. The issue of the existence of rights of nature in the Czech Republic can be quickly summarized by the two following points:

There is no recognition of the rights of nature, ecosystems, or particular components of an ecosystem.

No overt activities, plans, or proposals exist to have these rights expressly recognized by law.

While these answers seemingly exhaust the questionnaire in response to which this report is prepared, the whole issue can be approached in a broader and more nuanced context.

Indeed, no legal provisions grant nature or its parts rights. No case law discusses this issue or is it used in a court's argumentation. Textbooks stay silent on the issue when explaining environmental law subjects.[1] However, rights of nature

[*] I would like to thank Vojtěch Vomáčka and Dominik Židek for their consultations. All mistakes remain mine. The manuscript was finalized in April 2024. The assessment of literature and case law, including internet materials, reflects the legal situation at that time. Translations are mine.

[1] See Vojtěch Vomáčka, 'Subjekty environmentálních právních vztahů' in Ilona Jančářová and others (eds), *Právo životního prostředí: obecná část* (Nakladatelství Munipress 2016) 192–195.

frequently appear in academic literature and commentaries on various acts within environmental law as a legitimate part of the discourse.

While this issue did not (and maybe never will) reach the mainstream, it frequently appears as a fringe but notable topic. For example, abuse of the environment was common during the communist era (1948–1989), and the environmental leitmotif was strongly present during the Velvet Revolution (November 1989). However, environmental protection was strongly anthropocentric. When discussing protecting the environment, the focus was mainly on the potential adverse influence on human health. To a certain extent, this focus remained in the legislation and case law through the 1990s and is only today being slowly modified (as is shown below).

On the other hand, critical musings on the rights of nature appeared already in 1981, when Petr Pithart[2] published his text *Stromy žalují* [Lawsuits by Trees] in the samizdat journal *Obsah* [Content].[3] Pithart reflected on Stone's seminal work *Should Trees Have Standing?*[4] and on the case of *Sierra Club v Morton* heard by the Supreme Court of the United States in 1971 and decided in 1972. Pithart described the development of legal subjectivity through the ages. He noted that the power of the Roman *pater familiae* over the life and death of family members (*ius vitae necisque*) is 'incomprehensible and absurd, cruel and unjust' by today's standards. Finally, he concluded that we might reach the point when the law breaks with its 'one-sided anthropocentric view of nature'. As a result, Pithart states, we might start protecting nature not for its usefulness but for its other values that we cannot convert to 'usefulness or even economic rentability'.

The moment of Czech law shifting away from the anthropocentric position has not yet been reached – and it may never be reached. However, there are tendencies to weaken the anthropocentric position and to introduce mechanisms that can be labelled as either biocentric or ecocentric correctives, or as weak anthropocentrism.[5] These discussions are mainly visible in academic literature but also in recent case law.

When reading this report, two of its limits must be considered.

Firstly, the author takes no stance on the desirability of legal recognition of rights of nature within the Czech legal system. The report is descriptive and, in some parts, tends towards the interpretation that the general rapporteur and broader audience might find interesting and useful for comparative and analytical purposes.

---

[2] After the Velvet Revolution, he became the prime minister (1990–1992) and later the chairman of the Senate (1996–1998 and 2000–2004).

[3] See Petr Pithart, 'Stromy žalují' (1981) 1 *Obsah* <https://denikreferendum.cz/clanek/4361-stromy-zaluji>.

[4] See Christopher D Stone, 'Should Trees Have Standing? – Toward Legal Rights for Natural Objects' (1972) 45 SCal LRev 450.

[5] The latter is the term Müllerová uses to describe her own approach. See Hana Müllerová, 'Perspektivy práva na příznivé životní prostředí: vzor přezkumu socioekonomických práv a test racionality' (2019) 158 Právník 533.

Secondly, while the text contains case law references, court decisions are generally not binding in the Czech Republic. Therefore, the author focuses on case law appearing in textbooks and commentaries to ensure that presented conclusions can be considered well established within the Czech legal system.

The predominantly negative answers to questions in the questionnaire have significantly affected the structure of this report. Following this introduction, the second section focuses on the rights of nature. It consists of three sub-sections, with the first describing the existing legal framework typical for its currently weakening anthropological reductionism. The second focuses on granting legal persons a right to (a healthy) environment, and the third sub-section reflects on the rights of nature in Czech academic literature. The third section captures the interesting academic effort (albeit one with limited impact) for a *Constitution for the Earth* and identifies some of the organizations engaged in protecting the environment for the public interest. Finally, the fourth section concludes this report.

II. Rights of Nature

1. State of Play: Anthropocentric Reductionism

The right to (a healthy) environment is guaranteed on a constitutional level. The existing constitutional framework consists of art 35 of the Charter of Fundamental Rights and Freedoms (referred to hereafter as the 'Charter'). This provision grants subjective rights (entitlements) to live in a favourable living environment (art 35(1)), procedural rights to timely and complete information about the state of the environment and natural resources (art 35(1)), and corresponding obligations of individuals to protect the environment (art 35(3)). In addition, art 7 of the Constitution obliges the state to protect the environment.

Article 35 of the Charter

'(1) Everybody has the right to live in a favourable environment.
(2) Everybody is entitled to timely and complete information about the state of the living environment and natural resources.
(3) In exercising his or her rights, nobody may endanger or cause damage to the living environment, natural resources, the wealth of biological species, and cultural monuments beyond the limits set by law.'

Article 7 of the Constitution – Environmental protection

'The state shall concern itself with the prudent use of its natural resources and the protection of its natural wealth.'

Act no 17/1992 Sb., on the environment (hereafter referred to as 'EnvAct') contains definitions of key terms which delineate the scope of what is protected. Section 2 defines the *environment* as everything providing natural conditions for the existence of organisms (including human beings). An *ecosystem* is defined in s 3 as a functional and interconnected set of living and inanimate components. Section 5 defines the term *acceptable regional load* as an effect of human activity not damaging the environment and its components. *Environmental protection* is defined in s 9 as activities preventing pollution or damage to the environment or reduction or elimination of pollution and environmental damage. Finally, s 11 stipulates that any territory must not be burdened by human activity over the level of acceptable regional load.

Section 2 of the EnvAct – Environment

'The environment is everything that provides natural conditions for the existence of organisms, including human beings, and is a precondition for their further development. Its major components are air, water, rocks, soil, organisms, ecosystems, and energy.'

Section 3 of the EnvAct – Ecosystem

'The ecosystem is the functional set of living and inanimate components of the environment, interlinked by metabolism, energy flow and the exchange of information, which influence one another and develop within a particular space and time.'

Section 5 of the EnvAct – Acceptable regional load

'The acceptable regional load is exerted upon a region by human activity at a level that does not damage the environment, particularly its components, the function of ecosystems or its ecological stability.'

Section 9 of the EnvAct – Environmental protection

'Environmental protection consists of activities that prevent pollution or damage to the environment or reduce or eliminate pollution and environmental damage. It includes protecting individual components of the environment, organisms or specific ecosystems and their interrelationship, as well as protecting the environment as a whole.'

Section 11 of the EnvAct

'The territory must not be burdened by human activity over the level of acceptable load.'

Such a set of definitions and basic obligations allows us to understand the scope of protection granted to maintain the fundamental human right to (a healthy) environment.

While ecosystems are generally recognized as leading components of the environment, the rights of ecosystems or other elements of the environment are not recognized. These terms appear in the EnvAct and related legislation as descriptive terms for an object of protection. The Czech legal framework recognizes the environment as consisting of living organisms, their communities, and whole ecosystems. It also explicitly includes inanimate components and the landscape for its aesthetic value and availability for human use.[6] Overall, this points us in the direction of focusing on protecting the environment not for its own sake but for humans.

However, even the nature of rights granted by art 35 of the Charter was unclear in the past. Initially, it was common to view the right to (a healthy) envi-

---

[6] See Vojtěch Vomáčka, '§2 Ochrana přírody a krajiny' in Vojtěch Vomáčka and others (eds), *Zákon o ochraně přírody a krajiny. Komentář* (CH Beck 2017) 13.

ronment 'not as a basic human right, but as a value judgment of the constitution-maker.'[7] The opinion shifted in time, and courts now generally recognize rights contained in art 35 of the Charter as part of fundamental human rights.[8] Such a shift allowed the protection of the environment to be taken more seriously in the public discourse. The stronger position of the right to (a healthy) environment was solidified by including the right to live in a favourable environment in art 81 of Act no 89/2012 Sb., the Civil Code.

Section 81 of the Civil Code

'(1) An individual's personality, including all his natural rights, is protected. Every person is obliged to respect the free choice of an individual to live as she pleases.
(2) Life and dignity of an individual, his health and the right to live in a favourable environment, his respect, honour, privacy and expressions of personal nature enjoy special protection.'

However, the core of rights guaranteed by art 35 of the Charter is still formed by the mid-1990s case law of the Constitutional Court (hereafter referred to as 'CC'). In its judgment of 9 October 1996, file no Pl. ÚS 16/95, the CC stated that the (healthy) environment is a public good. The court explicitly referred to Samuelson and Raz as an intellectual basis behind the understanding of the term.[9]

Recent commentary on the Charter by Tomoszek, Tomoszková, and Vomáčka labels the public-good approach as an instance of 'anthropocentric reductionism'. Authors specifically criticize the approach of the CC as too narrow to address the current 'global degradation of the environment caused by human activity.'[10] Authors further note that art 35 of the Charter, in its current interpretation, protects nature only for its 'usefulness' to humans, and not for its intrinsic value.[11]

The right to (a healthy) environment is guaranteed on a constitutional level. Therefore, the state and its courts are required to protect this right. Furthermore, when balancing different rights and freedoms in conflict, courts must protect the 'essential content' or 'essential core' of the right (s 4) of the Constitution).

Article 4 of the Charter

'(1) Duties are imposed only on the basis and within the bounds of law while respecting fundamental rights and freedoms.

---

[7] See Marian Kokeš, 'Čl. 35 (Právo na příznivé životní prostředí)' in Eliska Wagnerová and others (eds), *Listina základních práv a svobod. Komentář* (Wolters Kluwer 2012) 712–715.

[8] See Maxim Tomoszek, Veronika Tomoszková and Vojtěch Vomáčka, 'Čl. 35 (Životní prostředí)' in Faisal Husseini and others (eds), *Listina základních práva a svobod. Komentář* (CH Beck 2021) 977.

[9] See CC 9 October 1996, Pl. ÚS 15/96 citing to J Raz, 'Rights-Based Moralities' in Jeremy Waldron, *Theories of Rights* (OUP 1984) 187 and citing to Paul A Samuelson and William D Nordhaus, *Ekonomie* (Svoboda 1991) 770–771 (Public Goods and Externalities).

[10] See Tomoszek, Tomoszková and Vomáčka (n 9) 979.

[11] See Tomoszek, Tomoszková and Vomáčka (n 9). See also Dominik Žídek and Vojtěch Vomáčka, 'Ústavní zakotvení práva na životní prostředí' in Ilona Jančářová and others (eds), *Právo životního prostředí: obecná část* (Nakladatelství Munipress 2016) 153–154.

(2) Limitations may be placed upon fundamental rights and freedoms only by law and under the [Charter]'s conditions.
(3) Any statutory limitation upon fundamental rights and freedoms must apply similarly to all cases that meet the specified conditions.
(4) The essence and significance of these rights and freedoms must be preserved when employing the provisions concerning limitations upon the fundamental rights and freedoms. Such limitations shall not be misused for purposes other than those enacted.'

The call for protection of the essential core represents a strict prohibition against emptying any given right of its substance while engaging in the test of proportionality.[12] Given the composite nature of the right to (a healthy) environment, the definition of its core is complex and was firmly established in a comprehensive analysis published by Tomoszek and Tomoszková in 2016.[13] The authors identified general principles and components of the right to (a healthy) environment as consisting of several elements. First, it is the nature of the right, i.e., the definition of who can be the subject of the right, which in this case is an individual human being. Second, procedural aspects of the right also belong to its core, namely:

1. the right to defend oneself individually against interference,
2. the right to take part in decision-making processes,
3. the right to access data about the environment.

Third, substantive aspects belonging to the essential core consist of the obligation of the state to ensure such a quality of the environment that

1. allows individuals to meet their basic needs (sleep, food, water, etc) to the extent necessary to maintain their health,
2. does not lead to a restriction of other fundamental rights.[14]

In 2021, Tomoszek, Tomoszková, and Vomáčka, in their commentary, noted that the list is not exhaustive and might be further supplemented in literature and case law.[15] However they provided no specifics.

As a result, Czech law does not recognize nature as a subject but only as an object of rights. Furthermore, it is evident that the current regime is strictly anthropocentric and, at its core, focuses on protecting nature for its usefulness to human beings. The abovementioned critique is, to an extent, a fringe phenomenon which does not penetrate law-making and court decision-making. Therefore, the fate of nature essentially depends on whether there is a person nearby whom the law grants the possibility of defence against an imminent danger.[16]

---

[12] See Tomoszek, Tomoszková and Vomáčka (n 9) 991.

[13] See Veronika Tomoszková and Maxim Tomoszek, 'Esenciální obsah práva na příznivé životní prostředí' in Hana Müllerová and others (eds), *Právo na příznivé životní prostředí: Nové interpretační postupy* (Ústav státu a práva AV ČR 2016) 157.

[14] See Tomoszková and Tomoszek (n 14).

[15] See Tomoszek, Tomoszková and Vomáčka (n 9) 994.

[16] See Vojtěch Vomáčka, 'Země jako subjekt práv? Tendence v ekocentrickém pojetí právní ochrany životního prostředí' in Josef Šmajs, *Ústava Země* (Vydavate'stvo PRO 2015) 26.

However, that often might not be the case, as the current legal framework requires individual petitioners to be directly affected by environmental damage. For example, in 2010, the Supreme Administrative Court (hereafter the 'SAC') was petitioned to cancel the Visiting Rules of the National Park Šumava. Visiting Rules of the National Park were, in their legal nature, published as a *measure of a general nature* – an administrative act on the border between a legal act and an individual decision. Such an act pertains to a specific situation (it is not abstract as legal acts are) and is addressed generally (not specifically as an individual decision). Such a measure can be subjected to judicial review, as was the example of the Visiting Rules of the National Park Šumava. The petition was brought forth (among others) by an individual claiming the new Visiting Rules adversely affected the local population of protected Freshwater pearl mussels. However, the SAC refused to examine the case's merits and simply pronounced the lack of standing of the applicant. According to the SAC, the applicant was not entitled to bring an action because the potential negative effect had no direct impact on her quality of life.[17] In 2014, the SAC ruled similarly in a case involving the alleged adverse effect of the zoning plan for the population of birds. Again, the court asserted that the applicant lacked standing because the zoning plan did not have any direct impact on her quality of life.[18]

'[…] The Supreme Administrative Court is aware of the considerable variety of possible forms of interference with the right to the environment; typically, it could be, for example, pollution of air, water or soil, which would have an undeniable (adverse) impact on the environment even in an area outside the source of the pollution. However, the environmental impact in the case under consideration is very specific; the presence of pearl mussels in the upper reaches of the Vltava River does not directly affect the quality of life of appellant (b), and it is difficult to imagine the actual impact on her life from a reduction in the pearl mussel population in the section in question. The Supreme Administrative Court, therefore, concluded that the infringement of the appellant b)'s right to a favourable environment caused by the adoption of the contested measure of a general nature was not conceivable and that the appellant b) did not have active standing to bring an action for annulment of the measure of a general nature. […][19]

[…] However, in the appeal, the appellant only alleges procedural errors and the related interference with the Českobudějovické rybníky bird area, which may have a negative impact on specific bird species. However, the threat to the birds living in the bird area in question does not affect his quality of life. It is hardly possible to imagine the impact from the reduction of the bird population in the area on his life or property. The Supreme Administrative Court, therefore, finds that the defendant's misconduct challenged in the appeal does not constitute a violation of the appellant's right to a favourable environment or interference with his rights within the meaning of Article 101a(1) of the Code of Civil Procedure. […]'[20]

Therefore, the general approach to the right to (a healthy) environment requires an applicant to be affected by the adverse impact that human activity has on the

[17] See SAC 13 October 2010, 6 Ao 5/2010.

[18] See SAC 12 March 2014, 3 Aos 3/2013.

[19] See SAC 13 October 2010, 6 Ao 5/2010, para 34.

[20] See SAC 12 March 2014, 3 Aos 3/2013.

environment. Only then does the applicant have standing and an entitlement to sue. Müllerová, one of the leading Czech scholars on environmental law, noted the general unsuitability of the existing legal regime for protecting the environment 'as such'. She has contended that the ongoing shift towards a firm establishment of the right to (healthy) environment and environmental elements within existing human rights successfully broadens environmental protection.[21] However, this shift in interpretation, without the subsequent activity of lawmakers, faces an unbreakable barrier whereby environmental elements depend on existing human rights which usually revolve around human beings.[22] In conclusion, while it is an improvement, the shift in interpretation – firmly establishing the right to (a healthy) environment as a human right – can only go so far (especially in the continental system).

Therefore, nature is protected only due to human beings requiring favourable conditions. Hence, the environment is stripped of direct and complex legal protection because, ultimately, human beings are the sole beneficiaries of the remuneration for damages.[23] Scholars often strive to discuss the shortcomings of the strictly anthropocentric approach, but suggestions to replace it altogether and recognize the rights of nature (or its components) are non-existent.[24] Scholars note the existence of alternative models stemming from foreign literature or case law but do not discuss the application of these models in Czech law.[25] Hence, while alternative models are referenced in commentaries,[26] they have not gained traction in the decision-making or law-making processes.

While human beings are still at the centre of environmental protection, a noticeable shift away from anthropocentric reductionism is currently underway. The shift stems from courts granting legal persons procedural rights to the (healthy) environment and its substantive element. As a result, the prism is gradually shifting away from environmental protection exclusively for human beings.[27] The shift will be described in the next sub-section.

2. Progress: Rights of Legal Persons

In 1997, the CC stated that legal persons – especially associations whose mission as embedded in their statutes is the protection of nature – were entitled only to the procedural right to a (healthy) environment and not to substantive rights. As

---

[21] See Hana Müllerová, *Právo na životní prostředí: teoretické aspekty* (Ústav státu a práva AV ČR 2015) 11.

[22] See Müllerová (n 21).

[23] See Müllerová (n 21) 68.

[24] See Müllerová (n 21).

[25] See Müllerová (n 21) 115–116. See also Vojtěch Vomáčka, 'Za hranicí *Animal Welfare*: Matka Země a zvířecí superstar' in Hana Müllerová and others (eds), *Kapitoly o právech zvířat: 'my a oni' z pohledu filosofie, etiky, biologie a práva* (Academia 2016) 590–596.

[26] See Tomoszek, Tomoszková and Vomáčka (n 9) 975 and 978.

[27] See also Židek and Vomáčka (n 12) 154.

a result, legal persons were entitled to sue (to defend against interference) and to participate in decision-making processes. This decision can be understood as implicitly stating that only specific individuals are entitled to the substantive element, the right to the (healthy) environment itself. A year later, in 1998, the CC expressly concluded that only natural persons (human beings) were entitled to the right to (a healthy) environment because only they are 'biological organisms [...] subjected to possible negative environmental influences.'[28] While other courts did not always share this approach, the CC was firm to quickly uphold its original position as the ultimate judicial body.[29]

'[...] Article 35 of the Charter of Fundamental Rights and Freedoms, which enshrines the right to a favourable environment and timely and complete information on the state of the environment and natural resources, cannot be applied to legal entities. The rights relating to the environment belong only to natural persons since they are biological organisms which, unlike legal persons, are subject to possible adverse environmental effects. [...]'[30]

However, in 2014, the CC reconsidered this opinion. It ruled that environmental organizations are entitled to sue for the annulment of the zoning plan (as part of the judicial review of a measure of general nature). The CC stated that it would be 'absurd if a person fulfilling conditions [...] would not be entitled to sue [...] simply because they, together with other persons [...] have joined together and, on behalf of the association, are requesting the annulment of the zoning plan.'[31]

In this specific case, the CC ruled on a petition for the annulment of a measure of general nature. However, based on the CC's decision, administrative courts later extended the entitlement to sue also to proceedings against a decision of an administrative authority[32] and to proceedings against an unlawful intervention[33]. The outcome can be interpreted as allowing a claim of interference with the right to (a healthy) environment of legal persons even outside of associations whose focus is, per their statutes, on environmental protection.[34] As a result, legal persons are entitled to sue when they are at least marginally focused on environmental protection[35] or where the alleged interference has consequences preventing the legal person from achieving their objectives. For example, a gardening association or an association focusing on recreational use of a given location can now claim interference with their right to (healthy) environment.[36]

---

[28] See the resolution of the CC dated 6 January 1998, I. ÚS 282/97.

[29] See Tomoszek, Tomoszková and Vomáčka (n 9) 996.

[30] See the resolution of CC dated 6 January 1998, I. ÚS 282/97.

[31] See CC 5 May 2014, I. ÚS 59/14, para 22.

[32] See SAC 25 June 2015, 1 As 13/2015.

[33] See Municipal Court in Prague 9 March 2017, 45 A 31/2016. See also Tomoszek, Tomoszková and Vomáčka (n 9) 997.

[34] See Tomoszek, Tomoszková, and Vomáčka (n 9).

[35] See SAC 8 February 2018, 10 As 145/2017.

[36] See SAC 26 April 2017, 3 As 126/2016.

Municipalities are also entitled to sue to protect the right to (a healthy) environment. The case law of the SAC states that the municipality is a public corporation representing a territorial community of citizens. By its nature, it is called upon to represent and protect their rights and interests. Similar to associations, municipalities protect broader community interests.[37]

'[…] The Supreme Administrative Court […] held that, in terms of establishing an entitlement to sue for annulment of a measure of general nature, an application claiming the adverse effect of a measure of general nature is permissible. The applicant must therefore claim that certain subjective rights to which he is entitled are affected by the measure of a general nature. These findings are the alpha and omega of the applicant's entitlement to sue for the annulment of a measure of general nature. The categorical thesis […] according to which legal persons are not entitled to sue […] cannot hold.'[38]

'[…] The Supreme Administrative Court is aware that the case law [in question] concerns the review of measures of a general nature. However, the conclusions regarding the potential claimant's substantive sphere and the local element's necessity can also be applied to the question of the possible [entitlement to sue against the decision of an administrative authority]. […]'[39]

'[…] According to the established case law of the Supreme Administrative Court, a civil association (now an association) dealing with environmental protection does not hold substantive rights (the right to a favourable environment) and can effectively argue before the court only against the violation of its procedural rights in the previous proceedings […]. However, this legal opinion was corrected to a certain extent by the Constitutional Court in its decision of 30 May 2014, file no. I.ÚS 59/14, in which, when assessing the entitlement to sue for the annulment of a measure of a general nature […], it stated that the concept of 'impairment of rights' cannot be interpreted against associations established for nature and landscape protection as restrictively as the Supreme Administrative Court does, since the rights of a community may be affected more widely than in the case of an individual. The fact that a citizen prefers to promote his interest by associating with other citizens cannot be attributed to him. There is no reason not to apply those conclusions to assessing active standing in proceedings for protection against unlawful interference. […]'[40]

'[…] Although the associations' entitlement to sue in nature conservation matters is relatively broad, each claimant must be assessed individually, particularly regarding its relationship to the locality in question or its local context. […]'[41]

'[…] However, it should be taken into account that the Constitutional Court, in its ruling I. ÚS 59/14, did not limit the conclusion on the interference with substantive rights to the interference with the property or other substantive right of a member of the association who seeks protection of his rights through this legal entity. It also explicitly allowed, in the context of the Aarhus Convention, the interference with the right of members of the association to a favourable environment (without derivation from an existing property right in the regulated area), if the alleged interference has consequences for the achievement of the objectives pursued by the association in question. In addition to associations for the protection of nature and the countryside,

[37] See SAC 26 June 2013, 6 Aps 1/2013.
[38] See CC 5 May 2014, I. ÚS 59/14, para 21.
[39] See SAC 25 June 2015, 1 As 13/2015, para 79.
[40] See Municipal Court in Prague 9 March 2017, 45 A 31/2016.
[41] See SAC 8 February (n 36), para 31.

one can imagine, for example, gardening associations, associations organising the recreational use of a particular locality, etc.) […]'[42]

'[…] The complainant is a municipality, i.e., a public corporation, which, according to the Constitution, already constitutes a 'territorial community of citizens' […], i.e., the complainant is called directly by its nature to represent and protect the rights and interests of its constituent citizens. The Constitutional Court derives the municipality's competence to deal with environmental issues in the municipality's territory from [the law], according to which the municipality's independent competence includes environmental protection. […]'[43]

Concluding that such a shift grants any rights to nature would be a stretch. However, it brings forth tools that are more suitable for protecting nature 'as such' – it is shifting the prism of environmental protection away from being only for human beings.[44] By creating broader entitlement to sue – including legal persons (associations and municipalities), which typically have greater resources than natural persons – environmental law should become more efficient.[45] This comes as a natural conclusion, as human beings are more efficient when organized. Individual claims were often perceived as an impractical tool.

However, this does not mean that nature is granted legal subjectivity or that its rights are recognized. Neither associations nor municipalities are obliged to protect specific environmental interests. It falls within their discretion, capacity, and ability to do so.[46] While the environment is a public interest,[47] and several bodies can bring an action in the public interest (*actio popularis*) – specifically the Supreme State Prosecutor and the Ombudsman – they are not obligated to defend the public interest in terms of environmental protection. The law only allows them to defend the public interest,[48] as it, in a certain sense, now also allows legal persons.

3. *Reflecting on Rights of Nature*

The development of the rights of nature in other countries did not go unnoticed by the Czech academia. Vomáčka, while commenting on the Ecuadorian constitution, concluded that it is reasonable to expect efforts to conceptualize the Earth as a subject of rights. He believes the shift towards an ecocentric view of environmental protection will continue and gain traction among politicians, policymakers, and the wider public. Vomáčka criticized the Ecuadorian effort for an insufficient separation between nature and human culture, as he believes this issue weakens granted rights. Many provisions are, in his opinion, of a proclama-

---

[42] See SAC 26 April 2017, 3 As 126/2016.

[43] See SAC 26 June 2013, 6 Aps 1/2013, para 26.

[44] See Židek and Vomáčka, (n 12) 154.

[45] See Tomoszek, Tomoszková and Vomáčka (n 9) 1000.

[46] See Tomoszek, Tomoszková and Vomáčka (n 9).

[47] See Vomáčka (n 17) 27.

[48] See Vomáčka (n 17).

tory nature. Without further clarification, it is not easy to estimate their practical impact.[49]

A similar critique appears in literature even when discussing the possibility of granting rights (or legal subjectivity) to animals. For example, Müllerová notes that it is theoretically possible to grant legal subjectivity and rights to animals in the current postmodern situation. However, she states that this concept would require appointing human beings as guardians of those rights. Furthermore, this concept would have to be 'artificially created' by putting content (correct on its own) into a form that is not suitable for it.[50] Šejvl similarly notes that granting legal subjectivity to animals would require not only a modification of the system of legal subjectivity but also a thorough consideration of which animals should be the subjects of law, which rights should be recognized, and how should animals exercise them.[51] While the subjectivity of parts of nature was not expressly discussed, these conclusions could be extended to it.

III. Promotion of Rights of Nature

1. A Constitution for the Earth

In 2015, Josef Šmajs published the book *A Constitution for the Earth – Philosophical Concepts* in Czech, English, Germany, Slovak, and Russian.[52] Šmajs calls for the subjectivity of the Earth[53] in order to modify what he calls the 'predatory paradigm' of anthropocentric environmental protection.[54] Some organizations are listed as partners of the book and the idea it symbolizes.[55] Additionally, at the time of publication, the book was noted by some state-wide media. However, the further impact of this publication has been limited. It is not reflected in the sphere of politics and policy. The Green Party and the Pirate Party (as 'usual' suspects, given their overall political inclination) have not addressed this specific issue.

Therefore, the promotion of rights of nature in the Czech Republic is virtually non-existent. There is strong support in society for the protection of the environment. However, data suggest that most people believe the state should be the

[49] See Vomáčka (n 26) 590–596.

[50] See Hana Müllerová, 'Mohou mít zvířata práva? Vývoj konceptu právní subjektivity a právní postavení zvířat' (2012) 151 Právník 1074, 1103.

[51] See Michal Šejvl, 'Pojem subjektivních práv a právní subjektivity v kontextu práv zvířat', in Hana Müllerová and others (eds), *Kapitoly o právech zvířat: 'my a oni' z pohledu filosofie, etiky, biologie a práva* (Academia 2016) 436–437.

[52] See Josef Šmajs, *Ústava Země* (Vydavateľstvo PRO 2015).

[53] See Šmajs (n 53) 52–54.

[54] See Šmajs (n 53) 54–55.

[55] See list of partners, available at <https://ustavazeme.cz/cz/partneri>.

leading actor and driver of development in this domain,[56] and as of now, the state is not engaged in the topic of rights of nature.

2. Defending the Public Interest in Environmental Protection

There seems to be a consensus that the development of environmental law should be practically oriented. Therefore, more recognition has been granted to the right to (a healthy) environment of legal persons, which has deeply practical consequences, compared to efforts to establish rights of nature.

However, even within the current shift, there are many issues. Müllerová notes that the right to environment is poorly applied and scarcely enforced in front of courts despite its long-standing recognition. The existing case law is limited and is still struggling over the basic points of interpretation (as evidenced by the abovementioned shift in associations' entitlement to sue).[57]

Overall, no specific actor is focused on promoting the rights of nature in the Czech Republic. However, many different organizations are taking an interest in protecting the environment. The Czech Republic lacks a systematic effort to recognize the right of nature, but many different organizations focus on protecting nature in the public interest. Examples include *Egeria* (protection of voles and other rodents during construction of a highway),[58] *Voda z Tetčic* [Water of Tetčice] (protection of blind snakes and lizards affected by the construction of a train station),[59] or *Živé Kornice* [Kornice Alive] (protection for bumblebees and ants during construction of a highway).[60] The association *Děti Země* [Children of the Earth] is traditionally involved in environmental issues.

Additionally, in 2021, the association *Klimatická žaloba* [Climate Action] gained prominence by bringing a legal action claiming that the Czech government is not doing enough to promote sustainable protection of the environment. *Klimatická žaloba* contended that relevant administrative authorities (Government, Ministry of the Environment, Ministry of Industry and Trade, Ministry of Agriculture, and Ministry of Transport) have failed to act. Initially, the action succeeded when the Municipal Court in Prague in June 2022 ruled that the Ministries had failed to act (the petition against the Government was dismissed).[61] The Municipal Court ordered the Ministries not to 'continue violating the applicants' rights by failing to provide specific mitigation measures leading to [a reduction

---

[56] See Jan Krajhanzl, Tomáš Chabada and Renata Svobodová, *Vztah české veřejnosti k přírodě a životnímu prostředí: Reprezentativní studie veřejného mínění* (Nakladatelství Munipress 2018) 254.

[57] See Hana Müllerová, *Právo na příznivé životní prostředí: zkušenosti vybraných evropských zemí a návrhy pro budoucí uplatňování v ČR* (Ústav státu a práva AV ČR 2018) 133.

[58] See Regional Court in Brno 30 May 2018, 31 A 59/2017.

[59] See Regional Court in Brno 26 November 2015, 31 A 2/2015.

[60] See Regional Court in Hradec Králové – pobočka v Pardubicích 18 November 2020, 52 A 33/2020.

[61] See Municipal Court in Prague 15 June 2022, 14 A 101/2021.

in greenhouse gas emissions].' However, the Ministries appealed the ruling to the Supreme Administrative Court, which in February 2023 overruled the Municipal Court in Prague, citing the separation of powers as a critical issue which was omitted in the initial decision.[62] The Municipal Court in Prague, bound by the reasoning of the Supreme Administrative Court, ruled again in October 2023 and dismissed the action of *Klimatická žaloba* altogether.[63]

IV. Conclusion

The anthropocentrism of Czech environmental protection is well-noted and, at least within academia, discussed.

Section 2 of this report describes the existing legal framework. The legal framework is strictly anthropocentric. Not only it does not recognize any rights of nature, it also provides limited tools to protect nature 'as such', and not because of its value to human beings. When bringing forth a legal action, claimants must prove that an adverse effect on nature is closely tied to an adverse effect on themselves. However, the report also notes that a shift is underway. Since 2014, the entitlement to sue has been broadened. The shift did not happen through the activity of lawmakers but by an evolution of the case law of the Constitutional Court. Legal persons can now sue, shifting the focus away from anthropocentric reductionism. However, it is too soon to speak about ecocentric tendencies as the development aims more toward weak anthropocentrism. Czech academia has noted the developments in, for instance, Ecuador. However, authors seem to perceive little practical benefit in granting nature rights.

Section 3 touched on the issue of promoting the rights of nature. In 2015, Šmajs introduced the *Constitution of the Earth*. The work seems, however, to have had no practical impact. On the other hand, many legal persons (both on local and state levels) are willing to take a stand for the environment – even if none of them are promoting the rights of nature.

As a result, the push to have rights of nature established is non-existent. The Czech Republic, instead of an idealistic approach of 'allowing nature to sue', focuses on a pragmatic 'we can sue in nature's stead.'

[62] See SAC 20 February 2023, 9 As 116/2022.

[63] See Municipal Court in Prague 25 October 2023, 14 A 101/2021.

Denmark

Could Danish Human- and Animal-Rights based Environmental Protection Provide Rights to Nature?

Yvette Lind[*]

I. Background

In Denmark, the idea of rights of nature has been discussed,[1] yet it has so far not been formalized through law to the same extent that is the case in other countries such as New Zeeland and Bolivia. There is currently no constitutional nor national or local legislation which recognizes the rights of nature in Denmark. Nor are there ongoing legislative initiatives to introduce such rights. Subsequently, the rights of nature are not recognized as legally independent by any of the Danish governmental bodies or by the Danish courts.

Consequently, the legal preservation of nature in Denmark does not specifically concern nature and the development of awarding rights to nature itself. Instead, the Danish legal system is still, much like its Nordic neighbors, highly focused on the interests of humanity, ie how individuals or groups in society may gain

[*] This report was authored in December 2022. The present version notes some recent political occurrences but does not consider any subsequent legal developments from 2023 forward.

[1] This development is elaborated in: Nikolas Berg, Ingrid Berg and Martin Hultman, *Naturens Rättigheter*, (Universus Academic Press 2020); Katarina Hovden, Julia Gilvad Rasmussen and Anette Lytzen, *Danish Input to the UN Harmony with Nature Programme* <www.harmonywithnatureun.org/>; Mikkel Funder and Marie Ladkjær Gravesen, 'How to support a rights-based approach to nature-based solutions' *DIIS Policy Brief* (23 February 2022).

access to nature through, for instance, the so-called right to public access or, alternatively, how humanity may exploit the resources of nature in a sustainable fashion.

II. Danish (Animal and Nature-Oriented) Legislation in a Nutshell

Initially, it should be noted that Denmark – as an EU Member State – relies to a large extent on the environmental regulations that are either inspired by or passed down from EU level. The Danish legal system appears to be underpinned by the ambition of a healthy environment as a human right rather than nature having rights. This reasoning – and the core of environmental law – originated in the aftermath of the Second World War and has subsequently been fostered by the EU.[2] Modern environmental law is still largely built on the rights of humans rather than the rights of nature as has been increasingly noted and critiqued – both in legal scholarship generally and in this publication particularly.

To be able to discern the extent to which rights awarded to nature in Denmark may be developed in the future, there is a need to elaborate on already existing protection of nature as expressed through positive law.

Initially, within adjacent areas and most noticeably in animal welfare law, there has been a more far-reaching development. From 1 January 2020, there was an introduction of new legislation titled *Dyrevelfærdsloven*.[3] When introducing the animal welfare legislation, Denmark became the very first country to, through positive law, recognize that animals are sentient beings and not only living beings.[4] The new legislation replaced the more than 100-year-old predecessor legislation along with a sizeable amount of supplementary provisions. The new Animal Welfare Act is designed as a framework which may be expanded to include additional rights for animals.[5] An important but easily overlooked aspect of the new legislation is found within the name. The former legislation was named the Animal Protection Act, while the new legislation is named the Animal Welfare Act. The name change indicates the movement from merely protecting animals towards the goal of providing animals with safety and well-being as sentient beings.

Additionally, there is presently an extensive body of Danish environmental law:

---

[2] For further reading and support for this interpretation, see for instance Chapter 3 in Jan Darpö, *Can Nature Get It Right? A Study on Rights of Nature in the European Context*, Policy Department for Citizens' Rights and Constitutional Affairs Directorate General for Internal Policies, European Parliament, March 2021.

[3] LOV no 133, 25 February 2020 *Lov om dyrevelfærd.*

[4] § 1 LOV no 133 (n 3).

[5] LOV no 645, 8 June 2016 *Dyreværnsloven.*

- The Nature Preservation Act (Naturbeskyttelsesloven)[6]
- The International Habitats Preservation Act (Internationale beskyttelsesområder)[7]
- The Environmental Preservation Act (Miljøbeskyttelsesloven)[8]
- The Marine Environment Act (Havmiljøloven)[9]
- The Nature Act (Lov om Den Danske Naturfond)[10]
- The Marine Strategy Act (Havstrategiloven)[11]
- Hunting and Game Management Act (Jagt og vildtforvaltningsloven)[12]
- The Environmental Damage Act (Miljøskadeloven)[13]
- The Fisheries Act (Fiskeriloven)[14]

It is possible that Denmark could, similar to the previously described developments surrounding animal rights, introduce an Environmental Welfare Act which would replace the current Environmental Protection Act and the above-listed supplementary legislation. There have been no tangible discussions surrounding such a development, but given that the Animal Welfare Act is still fairly new, it may over time become a logical legal development.

The Danish Environmental Preservation Act provides the underpinning reasoning for the environmental protection you often find variants of within all the above cited supplementary environmental regulations:

'§ 1. The law must contribute to protecting nature and the environment, so that societal development can take place on a sustainable basis with respect for human living conditions and for the conservation of animal and plant life.

This Act is specifically intended:

1) to prevent and combat pollution of air, water, soil and subsoil as well as vibration and noise nuisance,
2) to provide important hygienic elements to the environment and to humans,
3) to limit the use and waste of raw materials and other resources,
4) to promote the use of cleaner technology, and
5) to promote recycling and reduce waste management problems.'[15]

This implies, as one can interpret from the above-quoted paragraph, that the enlarged legal preservation of nature is less about nature and the development of awarding rights to nature itself and is instead (still) substantially focused on the

---

[6] LBK no 240, 13 March 2019 under *Miljø- og Fødevareministeriet om Bekendtgørelse af lov om naturbeskyttelse* (Danish Implementation of Law for Nature's Preservation).

[7] BEK no 2091, 12 November 2021 *Habitatbekendtgørelsen.*

[8] LBK no 1218, 25 November 2019 *Miljøbeskyttelsesloven.*

[9] LBK no 1165, 25 November 2019 *Havmiljøloven.*

[10] LOV no 1518, 27 December2014 *Lov om Den Danske Naturfond.*

[11] LBK no 1161, 25 November 2019 *Havstrategiloven.*

[12] LBK no 265, 21 March 2019 *Jagt- og vildtforvaltningsloven.*

[13] LBK no 277, 27 March 2017 *Miljøskadeloven.*

[14] LBK no 764, 19 June 2017 *Fiskeriloven.*

[15] Translation by author.

interests of humanity and how humanity may exploit the resources of nature in a sustainable fashion.

This reasoning is supported by the fact that the Danish Parliament still holds the right to roll back nature conservation through legal proposals – a right which is utilized on occasion.

III. The *Amager* Case

One example of such action being taken was in 2019, when the Danish Parliament decided to give a natural habitat at Amager (a nature area in Copenhagen) the status of a city-zone when providing space for new residence areas.

Amager is a biodiverse area of over 200 hectares within the municipality of Copenhagen and includes several ecosystems in addition to providing a home and breeding ground for many species, including endemic species and species that are listed for preservation. The area was previously protected by the Nature Protection Act, but in 2019 there was a legislative reduction of these rights.[16] As a result, the local government permitted building projects to proceed by arguing that the projects would assist with the repayment of debts that were accrued through the previous expansion of the Metro infrastructure in the Copenhagen metropolitan area.

The parliamentary action to allow for commercial building projects on Amager induced a chain of events relevant when discussing the development of rights of nature: (1) complaints from the public and environmental NGOs, and (2) the introduction and framing of rights of nature in the public arena and in legal scholarship.

The first event concerned complaints from the public and environmental NGOs. Generally, the enforcement of preservation is done by the Board of Complaints for Environment and Food (*Miljø- og Fødevareklagenævnet*), in accordance with the Nature Preservation Act (*Naturbeskyttelsesloven*)[17] that specifies protected types of nature. §86 of the Nature Preservation Act specifies that anyone with an (1) individual and (2) significant interest in the outcome of a development plan decision is eligible to launch a formal objection. Subsequently, plans for developments accepted by municipalities or the Danish Environmental Ministry can be objected to by sending a formal written complaint to the Board of Complaints, which then assesses the complaint. It should be stated that the window for such a complaint is rather slim, merely four weeks.

The local population and several groups and organizations became involved in opposing the decision and the development plans through, for instance, demon-

[16] L 153 *Forslag til lov om ændring af lov om Metroselskabet I/S og Udviklingsselskabet By & Havn I/S.*

[17] LBK no 240 13 March 2019 under *Miljø- og Fødevareministeriet om Bekendtgørelse af lov om naturbeskyttelse* (Danish Implementation of Law for Nature's Preservation).

strations and petitioning.[18] Among the groups that have joined in supporting their efforts are the main environmental NGOs in the country, several local and activist groups, political parties, and many public figures. Nationally, local interest groups such as Friends of Amager Fælled (*Amager Fælleds venner*) and national interest groups such as Embassy of the Species (*Arternes Ambassadøre*) have mobilized resistance through formal written complaints, in which they often contend the lack of environmental assessments on of the behalf of project planners.[19]

IV. In the Aftermath of the *Amager* Case

Other key actors that have actively supported the development of enhanced environmental rights outside of the *Amager* case are:

- Forests of the World. An environmental NGO founded in Denmark in 1983 with the objective of conserving and managing the world's forests in a sustainable way.
- The Danish Society for Nature Conservation. Denmark's largest green organization with 130,000 members and over 1,500 volunteers. Additionally, it is the only association in Denmark that can raise conservation issues.
- Danish Nature Foundation (Danske Naturfond). A fund founded in 2015 by the Danish state and two large private conservation funds. It is the aim and purpose of the Fund to promote the protection of nature and the water environment in Denmark.

Consequently, on 7 July 2021, the Planning Complaints Board and the Environment and Food Complaints Board halted the building projects and called for further assessments of the potential negative impacts of the planned projects.[20] There was arguably support for such an assessment as the building construction risked destroying resting areas for the northern crested newt, a species which is listed under Annex IV in the EU Habitats Directive.[21] Subsequently, EU law provided legal support for the public complaints brought forward. The Board of Complaints for Environment and Food is expected to produce its final assessment by 2022.

The rights of nature framing came to resonate with defenders of the Amager Common. The rights of nature had at this time not been explicitly advanced in le-

[18] <www.skrivunder.net/vejledende_folkeafstemning_i_kobenhavns_kommune_om_byggerier_pa_amager_falled_og_stejlepladsen>.

[19] Jeres Klagenr. 1008139, Klagesag Vedr. Københavns Kommuneplan 2019, Lysholt Faber <https://blivhoert.kk.dk/sites/blivhoert.kk.dk/files/svar-materiale/bilag_1._2020-09-21_2._delreplik.pdf>.

[20] Nævnenes Hus, 'Planklagenævnet og Miljø- og Fødevareklagenævnet har tillagt klager over etablering og drift af bydel på Amager Fælled, Vejlands Kvarter, opsættende virkning' (published 7 July 2021) <https://naevneneshus.dk/>.

[21] Council Directive 92/43/EEC of 21 May 1992 on the conservation of natural habitats and of wild fauna and flora.

gal and democratic actions in Denmark. However, the Amager case introduced such discussions. For instance, in April 2021, Julia Gilvad Rasmussen gave a public talk on Earth Jurisprudence under the title 'Does the Amager Common have rights? Rights of nature and what they can be used for' ('Har Fælleden rettigheder? Naturens rettigheter og hvad de kan bruges til').

V. A Danish Green Constitution?

Politically, there has been some movement towards a Green Constitution in the aftermath of the Amager situation. For instance, former party leader of the Alternative (Alternativet) Uffe Elbæk addressed the development of global rights of nature in a 2018 a publication called "The Next Denmark".[22] The publication encompasses thirty-eight ideas that may, and possibly should, shape the future development of Denmark. Among these ideas we find the proposal for a Green Constitution. His political party, the Alternative, also included this proposal in its political program in 2018.[23]

In 2019, a public debate was initiated with a proposal to amend the constitution to recognize a human right to a healthy environment. Heads of large Danish companies took it upon themselves to advance the proposal for a Green Constitution. In 2021, there was an expansion of the coalition of supporters, and it now includes charities and organizations, such as the largest environmental organization in Denmark (The Danish Society for Nature Conservation).

Legal experts have made the connection between the cause for the protection of the Amager Common and the existence of potential rights of nature. This has instigated ongoing academic discussions on the introduction of a Green Constitution in Denmark. The initiative for such a Green Constitution is anthropocentric in orientation, with a vision to incorporate a human right to a healthy environment that focuses upon clean air, water, food, a sustainable climate, and diverse nature.

If such provisions were added to the Danish constitution, they would (1) introduce what is referred to as sympathy for nature as a positive legal right, (2) introduce access to information on nature's wellbeing, and (3) impose legal requirements on the state to prepare measures for preservation.[24] It should be emphasized

---

[22] Uffe Elbæk, *Det næste Danmark* (Alternativet 2018) <https://alternativet.dk/application/files/7515/2542/4885/Den_naeste_Danmark_low_res.pdf>.

[23] <https://alternativet.dk/application/files/2815/5800/0853/dereretalternativ_alternativet_nyversion.pdf>.

[24] Coalition for a Danish Green Constitution, *Hvorfor en Grøn Grundlov?*, <http://grongrundlov.dk/>; Mads Matzon, 'Forskere: Sådan Kan klimaet sikres I Grundloven' *Djøfbladet* (4 June 2020) <www.djoefbladet.dk/artikler/2020/6/forskere-saadan-kommer-klimaet-i-grundloven.aspx>; Kasper Kildegaard and Søren Domino, 'Topjurister i opråb: Klimaet er nu så afgørende, at det skal skrives ind i Danmarks vigtigste lov' *Berlinske Tidende* (18 October 2020).

that these proposed provisions are meant to support and enable a platform for sustainable governance and do not aim to enact rights for nature as a legal person.

However, these academic proposals have been met with resistance from both the Danish judiciary as well as some Danish politicians. Among the strongest arguments, we find an emphasis on the peril of enabling constitutional changes anchored in pure politics. The critics argue that such political matters should instead be brought forward in a political forum rather than a legal one. Furthermore, the naturally inherent ambiguity of what exactly constitutes the rights of nature is also emphasized as an apparent flaw.

Martin Henriksen, a representative of the Danish People's Party (Dansk Folkeparti) formulated it as such:

> 'The question about climate- and environment is a political matter, not a legal one. By moving the issue into the constitution, you move it away from the parliament and into the courts, in which judges without a popular mandate have the right to interpret to which extent this constitutional protection should be applied and that is absurd.'[25]

Similar concerns have been expressed by Danish Supreme Court judge Jens Peter Christensen on the matter of what constitutes appropriate constitutional safeguards for the preservation of nature. To him, this matter amounts to political considerations rather than legal ones.[26]

VI. Indigenous Rights as a Way of Giving Rights to Nature

Finally, the role and rights of indigenous people should also be acknowledged in the context of Denmark and the future development of rights of nature.

Greenland is home to Denmark's only recognized indigenous group, the Inuit. With the 2009 Act of Self-Government in Greenland, Greenland gained control of its natural resources.[27] In Greenland, there is currently a debate surrounding the extractive industry. Greenland holds a variety of precious natural resources, all of which are in high demand on the global market. These resources, including gas, oil, gold, coal, uranium, and rare earth elements, are particularly valuable in such great, untouched quantities. The extractive industry holds promises of (economic) independence for Greenland, thus the government of Greenland is creating pressure in favor of increased extraction activities in order to economically stabilize the island. However, extractive industry damages the environment, which the indigenous Inuit depend upon for their physical and cultural survival. Similar to many indigenous groups in this situation, there is, as such, a balancing act between the preservation of nature and culture, on the one hand, and potential financial independence on the other. Were the Inuit to gain additional inde-

[25] Translation and clarification by author, from: Kildegaard and Domino (n 24).

[26] Matzon (n 24).

[27] LOV no 473, 12 June 2009 *Lov om Grønlands Selvstyre*.

pendence – and/or extended legal protection as an indigenous group – it would not be unlikely for there to be a more tangible movement towards legalized rights of nature as we have witnessed in other jurisdictions around the globe. However, ongoing political developments linked to the governing of Greenland and its natural resources – largely induced by the U.S. under the Trump administration in 2024 – make this even less likely than before.

VII. Concluding Remarks

In conclusion, the Danish legal regulation of the rights of nature is arguably ambiguous. On the one hand, Denmark has been a trailblazer when it concerns animal welfare rights. The fact that Denmark was the first country in the world to acknowledge animals as sentient beings through positive law clearly underlines this.

However, at the same time there has been a noticeable reduction in environmental protection in Denmark as commercial interests have been allowed to take precedence. The described *Amager* case illustrates this development. Despite this negative development, there are still signs of an active movement towards the introduction of a Green Constitution if one considers the discussions within a variety of forums (public opinion, NGOs, political debate, and legal scholarship). The introduction of a Danish Green Constitution or an analogous development as in the case of the Danish Animal Welfare Act is not unlikely. Yet it is prudent to assume it may take some additional time before such a development takes place. It should also be noted that as of today (2024) the Danish ambition of introducing a green constitution is still centered on providing rights to humans rather than nature itself. More concretely, a Danish green constitution would extend human rights to clean air, clean drinking water, and clean food. It also been argued that it should contain the right to a sustainable climate and a diverse nature. These human rights could be seen as creating a bridge towards rights of nature, but it still appears as if Denmark has a long way to go before there is a Danish legal codification of the rights of nature.

Germany

Rights of Nature: A Paradigm Change for the German Legal System

Elena Ewering, Janina Reimann and Tore Vetter[*]

I. Introduction

Rights of Nature (RoN) are not recognized in German law, neither at the constitutional level nor in sub-constitutional law.[1] This means that natural entities or nature as a whole are to this day neither included in specific subjective rights nor

---

[*] The manuscript was finalized in February 2024. The assessment of literature and case law, including internet materials, reflects the legal situation at that time. The discussion on CJEU Preliminary Ruling and German Case Law in section III.4 and III.5 was supplemented in February 2025 to reflect more recent developments. This research was funded by the German Research Foundation (DFG), 421427080 – Die Natur als Rechtsperson, Andreas Fischer-Lescano.

[1] Jens Kersten, 'Natur als Rechtssubjekt' (2020) 11 APuZ 27; Klaus Ferdinand Gärditz, in Robert von Landmann and Gustav Rohmer (eds), *Umweltrecht* (CH Beck 2023) art 20a paras 23, 24; BVerfG 24 March 2021, 1 BvR 2656/18, 157 BVerfGE 30 para 112; English press release (29 April 2021) <www.bundesverfassungsgericht.de/SharedDocs/Pressemitteilungen/EN/2021/bvg21-031.html>.

recognized as legal personalities within the meaning of art 19(3) Basic Law. Therefore, the following cannot be about actually realized rights of nature. Nevertheless, it can be shown that even in a country unfamiliar with the concept of the rights of nature – like Germany – there is legal and social potential for the introduction of rights of nature.

II. Rights of Nature in Germany

The concept of rights of nature has been known among lawyers in Germany for years. Since the 1980s, the idea of rights of nature has been widely discussed by legal scholars.[2] In addition, the German courts were already engaged with the conception of subjectivity of natural entities as well. In the early 1980s, there was a mass mortality of seals in the North Sea, which was caused by high sea pollution.[3] In the famous 'seal case' ('Robbenklage'), a group of human activists supported by several environmental NGOs took legal action before the administrative court Hamburg in the name of the 'seals of the North sea'. They argued that the seals themselves had standing in the sense of the German law. The court, as expected at that time, dismissed the action as 'obviously non-admissible'.[4] It then stated that the seals, because they are animals, could not participate in an administrative court action in the sense of § 61 Code of Administrative Court Procedure (Verwaltungsgerichtsordnung, VwGO).[5] Because of this, the court did not take a stance on the question of possible infringements of subjective rights of the seals, which is necessary to have standing before the court, §42(2) VwGO. Nevertheless, the seal case was at the time considered a success by the claimants, as

---

[2] See in particular Klaus Bosselmann, 'Eigenrechte der Natur? Ansätze einer ökologischen Rechtsauffassung' [1986] KJ 1; Klaus Bosselmann, *Im Namen der Natur. Der Weg zum ökologischen Rechtsstaat* (Scherz 1992); Jörg Leimbacher, *Die Rechte der Natur* (Helbing & Lichtenhahn 1988). In the recent years eg: Andreas Fischer-Lescano, 'Nature as a Legal Person: Proxy Constellations in Law' (2020) 32(2) Law & Literature 237; Andreas Gutmann, *Hybride Rechtssubjektivität. Die Rechte der 'Natur oder Pacha Mama' in der ecuadorianischen Verfassung von 2008* (Nomos 2021); Andreas Fischer-Lescano and Andreas Gutmann, 'Nature's Day in Court. Rights of Nature and Access to Justice in Ecuador and Germany' (2021) 54(3) VRÜ 323; Elena Sofia Ewering and Andreas Gutmann, 'Landesgrundrechte für die Natur? – Eine Betrachtung des bayerischen Volksbegehrens „Rechte der Natur – Das Volksbegehren!"' [2021] DÖV 1118; Elena Sofia Ewering and Tore Vetter, 'Invisibilising Nature. Procedural Limits and Possibilities to Environmental Litigation in German Law' (2021) 54(3) VRÜ 376; Jens Kersten, *Das ökologische Grundgesetz* (CH Beck 2022); Kersten (n 1); Malte-Christian Gruber, *Rechtsschutz für nichtmenschliches Leben. Der moralische Status des Lebendigen und seine Implementierung in Tierschutz-, Naturschutz- und Umweltrecht* (Nomos 2006); Rejecting Rights of Nature in Germany: Gärditz (n 1) paras 23ff; Kersten Heinz, 'Eigenrechte der Natur' (1990) 29(3) Der Staat 415.

[3] With further reference: Gutmann (n 2) 87.

[4] VG Hamburg 22 September 1988, 7 VG 2499/88, [1988] NVwZ 1058.

[5] VG Hamburg 22 September 1988 (n 4).

it stirred up firm views on the representation of nature in German law. And for the first time it raised important questions regarding the legal personhood of non-humans in legal practice, questions which are once again present in today's discussions on rights of nature. In this sense, even if the seal case was widely considered grotesque by legal elites at that time, the concept of rights of nature was disseminated among German society and legal scholars alike and had a certain influence on the political decisions concerning the pollution of the North Sea.[6] As of today, the legal status of animals or of other non-humans remains unchanged, as shown by a recent constitutional complaint by the animal rights organization PETA.[7] The complainants objected to the possibility of castrating male piglets without anesthesia until the 31 December 2020, as provided for in §21(1) of the Animal Welfare Act.[8] In 2021, the Federal Constitutional Court rejected the complaint without giving reasons.[9] This indicated that the court did not consider the case to be of fundamental importance and thus presumably could not or did not want to contemplate a revision of the legal status of animals.[10]

1. Legal Status Quo for Rights of Nature

a) Constitutional status quo, environmental protection, art 20a Basic Law

The linchpin of German Law is the Grundgesetz (Basic Law), which entered into force on 24 May 1949. After World War II, the legal system in (West) Germany was fundamentally restructured with the constitution of the Federal Republic.[11]

---

[6] Gutmann (n 2) 87; Michael W Schröter and Klaus Bosselmann, 'Die Robbenklage im Lichte der Nachhaltigkeit' (2018) ZUR 195; Christopher D Stone, *Should Trees Have Standing? Law, Morality, and the Environment* (3rd revised edn, OUP 2010) 133.

[7] PETA, 'Grundrechte für Tiere' <www.peta.de/kampagnen/grundrechte-fuer-tiere/>.

[8] With an overview of the case: Cornelia Ziehm, 'Tiere als Rechtsperson' (2020) 18(1) EurUP 105.

[9] BVerfG 14 Mai 2021, 1 BvR 2612/19, [2021] BeckRS 13330.

[10] Jasper Mührel, 'Standing for Piglets: The Locus Standi of Non-Human Beings Under the German Constitution' (*Verfassungsblog*, 9 June 2021) <https://verfassungsblog.de/standing-for-piglets/>.

[11] And, of course, with the constitution of the German Democratic Republic in Eastern Germany, whose first constitution came into force on 7 October 1949 (the GDR reconstituted itself in 1968 and again in 1974, a 1990 draft constitution was never adopted). Subsequently, as a result of the unification treaty and the 2+4 treaty, the GDR was essentially absorbed into the Federal Republic and the (constitutional) law of the GDR no longer plays a de facto role in the legal system of today's Federal Republic. Its former territory is now divided into the new federal states of Mecklenburg-Western Pomerania, Brandenburg, Saxony, Saxony-Anhalt, Thuringia and as part of the united Berlin. The constitutions of the GDR did not contain any rights of nature, although art 15(2) of the 1974 constitution declared the 'preservation of the purity of waters and air' and 'the protection of the flora and fauna and the scenic beauty of the homeland' to be the responsibility of every citizen. The civil rights movement in eastern Germany before and after the reunion featured several innovative and progressive proposals for an ecological-social reconstruction both of the GDR-constitution and of a new Basic Law; see Jens

While civil[12] and criminal law[13] remained far more consistent with both pre-Weimar and Nazi law, this reconstruction was especially true for constitutional and administrative law.[14] The Basic Law was influenced both by contemporary human rights discourse in the tradition of the American Revolution and French Revolution as well as by the attempt to find a new intellectual-historical basis for the new constitution, one that was not encumbered by Nazi ideology.[15]

The Basic Law, therefore, already shows a clear humanistic, anthropocentric orientation in its structure. Unlike the Weimar Constitution, the Basic Law in its first 19 articles places fundamental rights in the foreground and not the norms of state organization typical of constitutions. The prominent feature of human dignity in art 1 Basic Law shows that the German legal system is anthropocentric. Art 1 Basic Law guarantees the inviolability of human dignity. Unlike the basic rights that follow, human dignity is, at its core, resistant to a balancing against other basic rights.[16] Art 1(1) is to be understood as a 'right to have rights'.[17] It is violated whenever the status of humans as legal subjects is put into question.[18] The guarantee of art 1(1) Basic Law is thus the basis for the legal subjectivity of every human being – a response to the dehumanization under the Nazi regime.[19] The guarantee of human dignity serves both as a legal safeguard – as under art 79(3)

Kersten, 'Die dritte Revolution – Plädoyer für ein ökologisches Grundgesetz' (2022) 67(6) Blätter für deutsche und internationale Politik 91, 95–96, with further references.

[12] With an overview on the history of the Civil Law Code (Bürgerliches Gesetzbuch, BGB) since 1945, see Franz Jürgen Säcker, in Franz Jürgen Säcker and others (eds), *Münchener Kommentar zum BGB* (9th edn, CH Beck 2021) Introduction paras 47–56.

[13] For continuities and differences in the German penal code in the post-war era, see the Third Criminal Law Amendment Act, 4 August 1953, Bundesgesetzblatt I, 735 <www.bgbl.de/xaver/bgbl/start.xav?startbk=Bundesanzeiger_BGBl&jumpTo=bgbl153s0735.pdf#__bgbl__%2F%2F*%5B%40attr_id%3D%27bgbl153s0735.pdf%27%5D__1654077683496>.

[14] For an overview of the development of German administrative law in the Federal Republic: Kyrill-Alexander Schwarz, in Michael Fehling, Berthold Kastner and Rainer Störmer (eds), *Verwaltungsrecht* (5th edn, CH Beck 2021) Introduction to the VwVfG paras 55, 56.

[15] Wolfram Höfling, in Michael Sachs (ed), *Grundgesetz* (9th edn, CH Beck 2021) art 1 para 2; Matthias Herdegen, in Günter Dürig, Roman Herzog and Rupert Scholz (eds), *Grundgesetz-Kommentar* (102th edn, CH Beck 2023) art 1 paras 14–16.

[16] BVerfG 10 October 1995, 1 BvR 1476/91, BVerfGE 93, 266 para 121; BVerfG 11 March 2003, 1 BvR 426/02, BVerfGE 107, 275 para 26.

[17] Jens Kersten, 'Die Rechte der Natur und die Verfassungsfrage des Anthropozän' in Jens Soentgen and others (eds), *Umwelt und Gesundheit* (Nomos 2020) 87, 109, with reference to Hannah Arendt, 'Es gibt nur ein einziges Menschenrecht' (1949) Die Wandlung 754, 759–760 <www.hannaharendt.net/index.php/han/article/download/154/273>. Arendt saw the realization that such a right to rights exists – and is also a right that one can be robbed of – as one of the most important lessons from the dehumanization and deprivation of rights that accompanied the Holocaust; Arendt (this note) 760–761, 769.

[18] BVerfG 20 October 1992, 1 BvR 698/89, BVerfGE 87, 209 para 113; BVerfG 15 February 2006, 1 BvR 357/05, BVerfGE 115, 118 para 121.

[19] Matthias Herdegen, in Günter Dürig, Roman Herzog and Rupert Scholz (eds), *Grundgesetz-Kommentar* (102nd edn, CH Beck 2023) art 1 para 21; Jens Kersten (n 17) 109.

Basic Law it will never be put into question – and as a basic principle for the political and legal culture of the Federal Republic, such that crimes like Auschwitz should never be allowed to happen again. Nevertheless, the guarantee of legal subjectivity for human beings under art 1(1) Basic Law does not exclude the granting of legal subjectivity to other entities.[20] Art 19(3) Basic Law supplements art 1(1) Basic Law and states that fundamental rights also apply to domestic legal entities insofar as they are 'by their nature' applicable to them. This is the constitutional basis for the legal personhood of, for example, companies, associations, or other legal persons, which, although they are not human individuals, can nevertheless assert violations of their rights in court.[21]

The central institution for the interpretation and further development of the Basic Law is the Federal Constitutional Court (Bundesverfassungsgericht, BVerfG), which is based in Karlsruhe. According to art 93 Basic Law, the Federal Constitutional court is responsible for disputes concerning the rights and duties of constitutional bodies and the constitutionality of sovereign acts. The Federal Constitutional Court Act (Bundesverfassungsgerichtsgesetz, BVerfGG) regulates the requirements for the proceedings before the Federal Constitutional Court.[22] The court's decisions are legally binding for the state organs of the Federation and the Länder (federal states), §31 BVerfGG. The central instrument of constitutional protection in cases of possible violations of individual subjective rights is the constitutional complaint (Verfassungsbeschwerde). According to art 93(1)(a) Basic Law, the complaint may be lodged by 'anyone' who claims a violation of his or her basic rights. A prerequisite for the admissibility of the complaint is that the complainant makes a prima facie case of a violation of his or her individual subjective rights – ie that he or she has standing to bring an action (BVerfGG, § 90(1)). In addition, the complainant must have exhausted the domestic legal remedies before the ordinary courts or the administrative courts (BVerfGG, § 90(2)).

Even if it does not (yet) contain specific subjective rights of nature, the Basic Law is not entirely silent regarding basic standards for the protection and treatment of nature. Art 20a, which was included in the Basic Law in 1994 and extended to include animals in 2002, now states that

> 'Mindful also of its responsibility towards future generations, the state shall protect the natural foundations of life and animals by legislation and, in accordance with law and justice, by executive and judicial action, all within the framework of the constitutional order.'[23]

---

[20] Jens Kersten (n 17) 109.

[21] For an overview also regarding the legal capacity of legal persons in civil matters: Säcker (n 12) Preface §21 BGB paras 41, 42.

[22] Christian Walter, in Günter Dürig, Roman Herzog and Rupert Scholz (eds), *Grundgesetz-Kommentar* (102th edn, CH Beck 2023) art 94 paras 34–37.

[23] Unless otherwise indicated, the English versions of German legislation reproduce the translations provided by the German Federal Ministry of Justice, <www.gesetze-im-internet.de/Teilliste_translations.html>.

In advance of the constitutional amendment, the wording of the new article was highly controversial. One part of the joint constitutional commission of the Bundestag and the Bundesrat was in favor of pronouncing an ecocentric perspective – protecting nature because of its own worth – another part wanted to protect nature as the foundation for human life and invoked the key role of human dignity under art 1 Basic Law.[24] Ultimately, art 20a Basic Law retains a certain degree of openness. Art 20a Basic Law tends to be understood as an expression of an anthropocentric view, but the provision is formulated so openly that it also permits protection of the environment for its own sake.[25]

Contrary to proposals by several political actors and legal scholars beforehand,[26] art 20a Basic Law was not introduced as a new 'environmental' basic right but rather as a state objective that should not be individually enforceable.[27] Representatives of the Green Party (Bündnis 90/Die Grünen) proposed to protect nature 'for its own sake', to expand the social liability of property under art 14(2) Basic Law toward an social-ecological liability and specifically to include standing for environmental organizations in a new art 19(5) Basic Law.[28] This proposal was rejected by the commission and received only three approving votes.[29]

While the scope and duties for the state deriving from art 20a Basic Law have been vividly discussed among legal scholars, it has because of its 'objective' character and the subsequent lack of relevant jurisdiction remained rather unnoticed and unapplied in legal and constitutional practice. Today, the prevailing opinion among scholars sees the provision as a purely state-objective provision, one which does not provide subjective, legally enforceable rights of any kind, neither to human applicants nor to nature itself.[30] As a state objective, art 20a Basic Law enshrined the environmental precautionary principle in the constitution.[31]

The judicial silence on the constitutional level was broken in March 2021, when the Federal Constitutional Court drew on art 20a Basic Law in its decision

---

[24] Report of the Joint Constitutional Commission, Deutscher Bundestag, 5 November 1993, BT-Drs. 12/6000, 65–66 <https://dserver.bundestag.de/btd/12/060/1206000.pdf>.

[25] Christian Calliess, in Günter Dürig and Roman Herzog and Rupert Scholz (eds), *Grundgesetz-Kommentar* (102th edn, CH Beck 2023) art 20a paras 60, 61.

[26] With further references to the discussion in the 1970s and 1980s: Rainer Wolf, 'Gehalt und Perspektiven des Art. 20a GG' (1997) 80 KritV 280 fn 4.

[27] Report of the Joint Constitutional Commission (n 24) 67.

[28] Jens Kersten proposes to pursue these proposals once again within the framework of an ecological reconstruction of the Basic Law. Especially proposals from the East German civil rights and environmental movement were far ahead of today's legal discussions on viable mechanisms to combat the climate crisis. Kersten (n 17) 97–98.

[29] Kersten (n 17) 150.

[30] Calliess (n 25) paras 30, 31; Kersten (n 1); Gärditz (n 1) paras 23–26; Dietrich Murswiek, in Michael Sachs (ed), *Grundgesetz* (9th edn, CH Beck 2021) art 20a paras 12–16.

[31] Stefan Huster and Johannes Rux, in Volker Epping and Christian Hillgruber (eds), *Beck'scher Online-Kommentar Grundgesetz* (57th edn, CH Beck 2024) art 20a paras 21–24; Wolf (n 26) 293–297; critical of the failure of the precautionary principle in the face of climate catastrophe: Kersten (n 12) 94–95.

declaring the Federal Climate Protection Act partially incompatible with basic rights. In its 'Climate Decision', the court argued that the pathway, as ruled out in the Federal Climate Protection Act, regarding the reduction of emissions by 2030 is insufficient for reaching the German climate targets and is therefore incompatible with the constitutional rights of future generations.[32]

The decision was viewed as groundbreaking, revolutionary even, by many observers.[33] The court clarified its view on the character and scope of art 20a Basic Law and for a first time – and extensively – took scientific knowledge into account[34] regarding the possibly devastating effects of the pending climate crisis (which climate scientists have warned about since the 1950s[35]). However, the court did not find a direct violation of the plaintiffs' (let alone nature's) rights under art 20a Basic Law. While the court made clear that art 20a Basic Law is indeed justiciable and not mere 'soft law', it upheld its view of the provision as a purely 'objective' one, which would not convey subjective rights of any sort.[36] The decision can certainly be seen as a pioneering re-interpretation of basic human rights, as the Federal Constitutional Court broke new ground in affirming a violation of 'almost all civil liberties' of future generations in their 'intertemporal dimension'.[37] However the Court did not find these interferences to be unjustifiable because the permissible emissions up to the year 2030 were per se incompatible with the climate protection duties of the state as contained in art 20a Basic Law[38]. Rather, they were unjustifiable because future generations post-2030 could be constitutionally obliged by art 20a Basic Law to tolerate far-reaching restrictions of their civil rights – today's shortcomings could justify a social 'full stop' in the future.[39] The court considered this inter-generational balancing of burdens disproportionate.

---

[32] BVerfG 24 March 2021 (n 1) paras 243–268. The decision was widely received among legal scholars, see eg the extensive debate on *Verfassungsblog* (including English-language contributions): <https://verfassungsblog.de/category/debates/der-klimabeschluss-des-bverfg/>.

[33] Felix Ekardt even called it 'probably the most far-reaching decision ever made by a supreme court worldwide on climate protection', Felix Ekardt, 'Climate Revolution with Weaknesses' (*Verfassungsblog*, 8 May 2021) <https://verfassungsblog.de/climate-revolution-with-weaknesses/>; Christian Calliess has emphasized what he sees as revolutionary step for the judicial review of the Paris climate goals, Christian Calliess, 'Das „Klimaurteil" des Bundesverfassungsgerichts: „Versubjektivierung" des Art. 20a GG?' (2021) 6 ZUR 355.

[34] See also Calliess (n 33) 355–356; Wolfgang Kahl, 'Klimaschutz und Grundrechte' [2021] JURA 117, 124; this consideration of scientific knowledge by courts had previously been held by some voices to be incompatible with the constitution, eg Thomas Voland, 'Zur Reichweite der Menschenrechte im Klimaschutz' [2019] NVwZ 114, 119.

[35] For an overview and further references, see the article on the history of climate change research in Wikipedia <https://en.wikipedia.org/wiki/History_of_climate_change_science>.

[36] BVerfG 24 March 2021 (n 1) paras 112, 205–207.

[37] BVerfG 24 March 2021 (n 1) paras 243–268.

[38] BVerfG 24 March 2021 (n 1) paras 208–210.

[39] BVerfG 24 March 2021 (n 1) para 192; see also Ewering and Vetter (n 2) 389–390; Calliess (n 33).

Yet the decision is far from being revolutionary in terms of being a (necessary) re-determination of the relation of humans to the rest of nature, and it proves to be no advantage in terms of rights of nature. The Federal Constitutional Court once again reached a decision from an exclusively anthropocentric – and sadly short-sighted – point of view. It rejected violations of the German state's duty to protect the rights to life (Basic Law, art 2(2)) and property (Basic Law, art 14(1)) of the German claimants on the grounds that, in addition to averting climate change by reducing emissions, the state could also, while not solely, turn to 'adaptation measures' to protect these rights.[40] This means that countermeasures against climate change need not be undertaken to the extent that the harmful consequences for people can be mitigated by adaptation measures, such as raising the height of dikes. With regard to the complainants from Bangladesh and Nepal, the court did not disavow the German state's duty to protect basic laws abroad under art 1(3) Basic Law in principle, but interestingly enough it rejected a violation on the grounds that the German state could not implement adaption measures abroad.[41]

All in all, the jurisdiction of the Federal Constitutional Court confirms the status of art 20a Basic Law as a state objective, which as of today does not contain subjective rights of (any) nature. In contrast, however, a new interpretation of art 20a Basic Law, contrary to the wording of the provision and its legislative history, would be at least unusual for the Federal Constitutional Court. Usually the court has taken a restrained approach in its further development of the law, and in the Climate Decision it also repeatedly emphasized the decision-making prerogative of the legislature. It might be that the Federal Constitutional Court would be more open to the concept of the rights of nature should they be implemented either by constitutional amendment or at the level of simple legislation. At any rate, this conclusion is suggested by the fact that the Federal Constitutional Court repeatedly discussed rights of nature with members of the Colombian Constitutional Court.[42]

b) Ordinary environmental law

In addition to the constitutional provisions of art 20a Basic Law, there are several 'simple' legal regulations concerning nature and the environment. Environmental law can be described as all legal provisions for the protection of humans and the environment which relate to environmental components, such as water, air, or natural habitats, and to environmental factors such as radiation, noise, or

[40] BVerfG 24 March 2021 (n 1) paras 164–172.

[41] BVerfG 24 March 2021 (n 1) paras 175–181.

[42] Press releases of the Federal Constitutional Court (BVerfG) no 43/2022 from 20 May 2022 <www.bundesverfassungsgericht.de/SharedDocs/Pressemitteilungen/DE/2022/bvg22-043.html> and no 65/2021 from 28 July 2021 <www.bundesverfassungsgericht.de/SharedDocs/Pressemitteilungen/DE/2021/bvg21-065.html>.

waste.[43] Environmental law is thus regulated in several laws, most of them being administrative law.[44]

(1) Overview on German administrative law

Specialized administrative courts are responsible for most areas of the public law including environmental law. As such, they are highly important in all legal matters concerning protective measures towards nature and violations thereof. The administrative courts are divided into the administrative courts (first instance), the higher administrative courts (second instance) and the Federal Administrative Court in Leipzig, as the highest 'simple' court beneath the Federal Constitutional Court. The prerequisites and procedural rules for all procedures at the administrative courts are regulated in the Administrative Court Order (VwGO). In environmental matters there are several types of action possible, depending on what the plaintiff wants to achieve, §88 VwGO.[45] Regardless of the particular type of proceeding, every action before the administrative courts is admissible only if the plaintiff can credibly claim that his or her rights have been violated by an administrative act or by its rejection or omission (VwGO, §42(2)). Plaintiffs are thus entitled to sue only if they can credibly claim a violation of their subjective rights. As such §42(2) VwGO, equivalent to the requirement of standing before the constitutional court, is an expression of the fundamentally (anthropocentric) subject-based German legal system of individual rights and acts as the major bottleneck for actions in the name of nature since a distinction is made between subjective rights protecting the individual (especially property under art 14(1) Basic Law[46]) and 'objective' rights protecting the general pub-

[43] Such is the definition in the Environmental Remedies Act §1(4) (Umweltrechtsbehelfsgesetz).

[44] Peter Christoph Storm, '§5 Umweltrecht' in Peter Christoph Storm (ed), *Umweltrecht* (11th edn, Erich Schmidt 2020) paras 52, 53.

[45] For a general overview on German administrative law in English, see: Florian Becker, 'The Development of German Administrative Law' (2017) 24 George Mason LRev 453, 466; specifically regarding environmental litigation options, see Ewering and Vetter (n 2) 383–387.

[46] Eg: BVerwG 19 September 1986, 4 C 8/84, [1987] NVwZ 409 para 11; in legal practice, rights of property are highly important in environmental protection cases. In the past, environmental activists often tried to gain legal standing by strategically obtaining ownership of land parcels where, for instance, major building or mining projects were planned. The Federal Administrative Court considered the acquisition of property of these 'Sperrgrundstücke' to be an abuse of rights and rejected the right to sue (eg BVerwG 27 October 2000, 4 A 10.99, 112 BVerwGE 135). For a discussion of such a case from an Ecuadorian perspective: María José Narváez Alvarez, 'Naturaleza, ecosistemas y acceso a la justicia: Estudio del caso Bosque de Hambach – Alemania' (2021) 54(3) VRÜ 352. – Property is, of course, also at the core of many civil actions involving the environment. One example is the lawsuit filed by the Peruvian Andean farmer Saúl Luciano Lliuya, who sued the German energy company RWE for damages in 2015. Due to its contribution to global warming, RWE was said to be at least partly responsible for the spread of the glacial lake above his hometown of Huaraz and for the rising threat of flooding of his property. After the Regional Court in Essen dismissed the lawsuit at first instance in 2016 due

lic.[47] On top of that, §61 VwGO states, that only 'natural' persons (eg humans) and legal persons (equivalent to art 19(3) Basic Law) can be participants in administrative actions, which is why the aforementioned requirements of administrative procedural law are of utmost importance in all matters of environmental law.

(2) Associative action: A window for rights of nature?

While there are no real rights of nature in simple legislation as well, there have been a number of modifications to the principles of standing and the general ban of popular actions enshrined in §42(2) VwGO. While the proposal for a constitutional implementation of associative actions during the consultations on the new art 20a Basic Law failed in 1994, there are nowadays several possibilities in environmental law allowing environmental protection organizations to bring legal action to court without being affected in terms of their own subjective rights in the sense of §42(2) VwGO. These legal mechanisms were heavily predetermined by both international as well as European law (on these inter-legal relations, see III.9, below). While they by no means convey rights of nature, procedural possibilities for associative action mark an important step toward a configuration of the legal system, which does not solely focus on the individual human subject and therefore can open up a door for rights of nature.

At first, §64 Federal Nature Conservation Act (Bundesnaturschutzgesetz, BNatSchG), which initially came into force in 2002 (then as §61),[48] provided that 'recognized nature conservation associations' could lodge appeals in accordance with the Administrative Court Code against specific decisions listed in §63 BNatSchG without having their own individual rights infringed.[49]

In 2006, this circle of 'legal persons' entitled to bring certain environmental actions to court was expanded, when the 'Environmental Remedies Act' (Umweltrechtsbehelfsgesetz, UmwRG) came into force. This Act implemented a corresponding Directive[50] of the European Community, which in turn responded to duties following from the Aarhus Convention.[51] Finally, in 2017 the Environmen-

---

to a lack of causality (LG Essen 15 December 2016, 2 O 285/15, [2017] NVwZ 734), the Higher Regional Court in Hamm is currently deciding on the plaintiff's appeal (OLG Hamm 30 November 2017, I-5 U 15/17). Recently, judges, together with experts, visited the hometown of Lliuya as part of the taking of evidence. An overview of the case can be found here: <https://rwe.climatecase.org/en/the-rwe-case>.

[47] Ewering and Vetter (n 2) 384.

[48] With further references, Ewering and Vetter (n 2) 385ff; Martin Gellermann, in Robert von Landmann and Gustav Rohmer (eds), *Umweltrecht* (CH Beck 2023) §64 BNatSchG para 1.

[49] Gellermann (n 48) para 4.

[50] Directive 2003/35/EC of the European Parliament and of the Council of 26 March 2003 providing for public participation in respect of the drawing up of certain plans and programmes relating to the environment and amending with regard to public participation and access to justice Council Directives 85/337/EEC and 96/61/EC [2003] OJ L156/17.

[51] Thomas Bunge, 'Die Verbandsklage im Umweltrecht' [2020] JuS 740, 741.

tal Remedies Act was revised again.[52] According to §2(1) UmwRG, 'recognized associations' can now file legal remedies under the Administrative Court Order, without having to have subjective standing, if one of the requirements named in §1 UmwRG is met. These are, for instance, admission decisions, which require an environmental impact assessment according to §1(1)(1) UmwRG, or permits for certain plants, §1(1)(2) and (3) UmwRG. These associations have standing if they can cumulatively claim that the permit violates legal provisions relevant to the decision and that the decision violates their statutory remit of promoting the objectives of environmental protection. In addition, the claiming association must have been entitled to participate in the administrative procedure.[53]

While these provisions for associative action mark an important step and were welcomed by environmental organizations, they contain high demands on the organizations entitled to take associative action,[54] and the UmwRG is applicable only in a limited field. This shows that the provisions for associative action can by no means be seen as a substitute for the fundamental changes rights of nature would bring in terms of procedural possibilities.[55]

2. Legal Proposals and Approaches for Rights of Nature in German Law

As already mentioned, there have been discussions on rights of nature among legal scholars, lawyers, and activists for years. Nowadays there are several approaches for rights of nature. While some propose to acknowledge rights of nature via the re-interpretation of existing constitutional provisions, others favor constitutional amendment and introduction of new specific rights of nature. Be-

[52] Bundesgesetzblatt I 2017, no 32, 1 June 2017, 1298 <www.bgbl.de/xaver/bgbl/start.xav?startbk=Bundesanzeiger_BGBl&jumpTo=bgbl117032.pdf#__bgbl__%2F%2F*%5B%40attr_id%3D%27bgbl117032.pdf%27%5D__1615546851359>.

[53] Ewering and Vetter (n 2) 386; Thomas Bunge, 'Die Verbandsklage im Umweltrecht' (2020) JuS 740, 743.

[54] §3 UmwRG contains the requirements for the recognition of organizations. They have to be formally organized, exist for at least three years, and pursue 'charitable purposes' within the meaning of German tax law. This excludes large parts of the environmental protection movement and less formal organizations. One example of this is the protests around the Hambach forest, which environmental activists have been fighting to preserve since 2012 at the latest. It was only after a lawsuit filed by the association BUND, which is recognized under the UmwRG, that the clearing of the forest was temporarily halted in 2018 by a decision of the OVG Münster (5 October 2018, 11 B 1129/18, [2019] openJur 4024). See also Christiane Gerstetter, 'Im Gerichtssaal und auf der Straße – unterschiedliche Wege, ein Ziel?' (2022) 1 juridikum 116.

[55] For a further overview on possible effects of RoN in this regard, see: Tore Vetter, 'Es macht einen Unterschied', short report on the initiative for constitutional amendment, (*Netzwerk Rechte der Natur*, 22 April 2022) <www.rechte-der-natur.de/files/opensauce/pdf/Es%20macht%20einen%20Unterschied.pdf>; Ewering and Vetter (n 2); for discussion of the decision of the administrative court of Cologne regarding the Hambach forest in Germany in the light of Ecuadorian rights of nature (in Spanish only with an English abstract), see: Narváez Alvarez (n 46).

cause of the federal design of the German Constitutional system, rights of nature are conceivable both on a state as well as on a federal level.

a) Rights of nature on the national level?

(1) Constitutional interpretation

Despite their history among legal scholars, rights of nature are still a conceptual novelty in practical constitutional law. There are currently no immediate plans to implement rights of nature in the Basic Law (for proposals on constitutional amendment see II.2.a)(2), below). Nevertheless, the lively scholarly discussion about the introduction of rights of nature into the German legal system provides some innovative proposals on how the Basic Law can be interpreted in a sense that already includes nature as legal person.

According to Andreas Fischer-Lescano, art 19(3) Basic Law, in conjunction with art 20a, provides a way to extend the scope of basic rights. Art 19(3) Basic Law opens up the protection of basic rights for legal persons, whereby the concept of legal person is much broader than the corresponding concept of civil law.[56] The purpose of art 19(3) Basic Law is to close gaps in the protection of basic rights with respect to collectives.[57] Up to now, the Federal Constitutional Court argued that 'legal persons' are capable of exercising basic rights, as long as they can be traced back to an expression of the general freedom of action of the human persons behind them – their legal status derives from the 'personnel substrate' of their human (natural person) members and representatives.[58] It is precisely nature as a collective of human as well as non-human entities that should be able to be considered as a legal person given that the free development of human beings depends on the free development of non-human entities.[59] More decisive for Fischer-Lescano, however, is the fact that art 19(3) Basic Law circularly determines that the person, who has basic rights and/or duties, has legal personality.[60] The most important aspect, however, seems to be that one should not proceed from the persons to certain rights, but from the rights to the persons. Even the question of whether legal persons under public law are entitled to basic rights is based on the construct of the 'endangerment situation typical of basic rights'. In view of the progressive destruction of entire ecosystems, a situation of 'endangerment typical of basic rights' in the sense of art 19(3) in conjunction with art 20a and, for instance, art 2(3) Basic Law could be asserted.[61]

56 Fischer-Lescano (n 2) 250ff; On the meaning of German Basic Law art 19(3), see Michael Sachs, in Michael Sachs (ed), *Grundgesetz* (9th edn, CH Beck 2021) art 19 paras 57ff.

57 Fischer-Lescano (n 2) 250.

58 Fischer-Lescano (n 2) 250; BVerfG 2 May 1967, 1 BvR 578/63, BVerfGE 21, 362 para 22; BVerfG 31 October 1984, 1 BvR 35/82, 356/82, 794/82, BVerfGE 68, 193 paras 42–46.

59 Fischer-Lescano (n 2) 251.

60 Fischer-Lescano (n 2) 250–252.

61 Fischer-Lescano (n 2) 250–252.

Malte-Christian Gruber sees a further possibility of granting constitutional protection to non-human entities de lege lata via the guarantee of human dignity from art 1(1) in conjunction with art 20a Basic Law.[62] The guarantee of human dignity is, in the overall view of art 20a Basic Law, not limited to human life since it is precisely the characteristic of human beings to put their own wishes and needs behind other interests. The ability to care for those who are weaker or in need of protection and to take a stand for them is part of human dignity.[63] This interpretation of art 1 Basic Law does not necessarily contradict the Federal Constitutional Courts view on the imponderability of human dignity as described initially. The guarantee of human dignity serves first and foremost the intangibility of the human status as a legal person. However, it does not prevent constitutional jurisprudence nor the legislature from expanding the circle of legal persons beyond human individuals.[64] This is also supported by the fact that the Federal Constitutional Courts view regarding the 'personnel substrate' of legal persons hardly reflects the legal practice of today's company law. For example, even large corporations and holdings might have the status of legal persons in the sense of art 19(3) Basic Law, without it being transparent on which 'natural persons' their personhood can be based.[65]

The possibilities outlined for reinterpreting the Basic Law in order to allow for nature's rights is a persuasive solution in this respect. Reinterpretation allows a constitutional amendment to be circumvented, which in practice might be much harder to accomplish, as is made clear in the following sub-section.

(2) Constitutional amendment

In contrast, the more obvious option for introducing rights of nature would be to explicitly incorporate them into the Basic Law. While this may be the most difficult alternative to achieve due to the constitutional and political hurdles associated with constitutional amendment, it might also be the most legally compliant way to enshrine rights of nature on a constitutional level and nip in the bud any differences of opinion on constitutional interpretation.

(3) Proposals

In his book 'Das ökologische Grundgesetz' (The Ecological Basic Law), Jens Kersten proposes a comprehensive constitutional amendment in detail. The elements of his proposal range from changing the preamble and the specified fundamental rights to amending the provisions on the organization of the state.[66] In

---

[62] Gruber (n 2) 212.

[63] Gruber (n 2) 212.

[64] Jens Kersten, 'Relative Rechtssubjektivität' (2017) 37(1) Zeitschrift für Rechtssoziologie 8, 11.

[65] Barbara Remmert, in Günter Dürig and Roman Herzog and Rupert Scholz (eds), *Grundgesetz-Kommentar* (102th edn, CH Beck 2023) art 19(3) paras 38–39.

[66] Kersten (n 2).

order to incorporate the rights of nature into the Basic Law, art 19(3) of the Basic Law should be amended to the effect that not only 'domestic legal persons' but also 'ecological persons' are entitled to fundamental rights to the extent that the fundamental rights are applicable to them.[67] Since art 19(3) of the Basic Law already regulates the transfer of fundamental rights to non-human entities, the addition of another type of non-human legal entity is not an absolute novelty; rather, it is merely a matter of familiarization.[68] Moreover, this would create an 'equality of arms' between social, economic, and ecological interests. The rights of nature would then be organized by parliamentary legislators.[69] It is important that 'nature' should not be authorized as a single entity but rather as various ecological actors, goods, and media.[70] Reference should be made to the proposed amendment of art 20a Basic Law. It should be stated there that nature and animals are to be protected. Biodiversity must be preserved and the oceans and climate protected.[71] But this list is not exhaustive. The legislature thus goes about finding a multitude of different non-human entities, which it is then supposed to transform normatively into ecological persons.[72]

The proposals of the initiative 'Rechte der Natur' (Rights of Nature) (see section III.1, below) go in a similar direction. The initiative also proposes to implement dignity and the integrity of every lifeform in art 20a Basic Law and to explicitly anchor the legal capacity of nature in a new art 19(4) Basic Law. Furthermore, they propose to implement rights of nature as limits to existing human basic rights, especially freedom of action (Basic Law, art 2(1)) and property (Basic Law, art 14(1)).[73] In proposing to expand art 14(2) by means of adding a social-ecological dimension, they are taking up proposals which were already discussed in advance of the 1994 constitutional amendment (see above).

(4) Procedure

Art 79 Basic Law is the relevant provision that governs the amendment of the Basic Law. Pursuant to art 79(2) Basic Law, the amendment must be carried by two-thirds of the members of the *Bundestag* (Parliament) and secure two-thirds of the votes of the *Bundesrat* (Federal Council). Para 3 further provides that amendments to the Basic Law affecting the division of the Federation into *Länder* (states), their participation in principle in the legislative process, or the principles laid down in arts 1 and 20 Basic Law shall be inadmissible.

---

[67] Kersten (n 2) 100.

[68] Kersten (n 2) 104.

[69] Kersten (n 2) 104.

[70] Kersten (n 2) 105.

[71] Kersten (n 2) 115.

[72] Kersten (n 2) 105–106.

[73] Initiative 'Rechte der Natur', proposal for a revision of the Basic Law <www.rechte-der-natur.de/files/opensauce/img/content/Pdf%20_%20Downloads/Netzwerk%20Rechte%20der%20Natur%20_%20Grundgesetzreform%2022%20April%202022-final.pdf>.

Opponents of the constitutional implementation of rights of nature and legal personhood for nature sometimes argue that such a step would not be compatible with the guarantee of human dignity under art 1(1) Basic Law and thus are not compatible with art 79(3) Basic Law.[74] However, as already explained above, human dignity does not guarantee an exclusive claim to legal subjectivity, an assertion which is also supported by art 19(3) Basic Law. An expansion of the circle of legal persons beyond 'traditional' domestic legal persons does not undermine the guarantee of human dignity any more than the status of already recognized legal persons.[75]

This makes it clear that an anchoring of the rights of nature in the constitution is in principle compatible with existing law. The most difficult hurdle will be to find a majority for such a constitutional amendment. As even the political party 'Bündnis 90/Die Grünen' (Alliance 90/The Greens) could not in 2020 bring themselves to include a demand for nature's rights in their basic program,[76] it can be concluded that a legislative majority is still a long way off in Germany. As of today, constitutional amendment is thus not a realistic political option.

b) An example from state constitutionalism: The Bavarian petition for a referendum on the rights of nature

Because of the high political and constitutional hurdles on the federal level, the level of the state constitutions and state law might prove an entrance point for rights of nature in the German legal system. The individual states (Bundesländer) act as sovereign states in their own right. The federal structure – one of the central principles of the German constitution– is covered and protected by art 20(1) Basic Law and concretized, for example, by the constitutional rights and duties of the states both in the executive branch (Basic Law, arts 30, 83–91) and in the legislative branch (Basic Law, arts 50–53, 70–82). Each federal state has its own constitution and, as a matter of principle, exercises state power itself. Art 30 Basic Law states: 'Except as otherwise provided or permitted by this Basic Law, the exercise of state powers and the discharge of state functions is a matter for the *Länder*.' Art 70 Basic Law concretizes this with regard to legislative power. It contains the basic principle, that 'The *Länder* shall have the right to legislate insofar as this Basic Law does not confer legislative power on the Federation'.[77] In addition, according to art 31 Basic Law: 'Federal law shall take precedence over *Land* law.' This means that even a federal law ordinance takes precedence

[74] Eg Gärditz (n 1) paras 23–24.

[75] See Kersten (n 1) 30.

[76] See Netzwerk Rechte der Natur, Die Grünen: Bundesarbeitsgemeinschaft Ökologie fordert Rechte der Natur <www.rechte-der-natur.de/de/aktuelles-details/87.html>; Bündnis 90/Die Grünen, Grundsatzprogramm <www.gruene.de/grundsatzprogrammprozess>.

[77] The German legislative branch is thus conceived as one of principle (states) and exception (federal). In fact, however, the federal legislative branch has made extensive use of its legislative competence (German Basic Law, arts 71–74).

over a state constitution. Nevertheless, the constitutions of the federal states are by no means of mere symbolic nature; rather, they are applicable law and thus constitute specific, important powers within the states. Their scope and integrity is monitored by the respective state constitutional courts, which act as integral parts of the system of judicative checks and balances.

A first step towards incorporating the rights of nature into a state level German law is currently being attempted in the south-eastern state of Bavaria.[78] Because of the mentioned status of the states and their constitutions, the implementation of rights of nature in the Bavarian constitution would be a very important step towards the acknowledgment of rights of nature in German legal practice. Their recognition would be a massive reform of Bavarian law, which would also call federal and other state law into question. Even if Bavaria's case law does not necessarily have an influence on the case law of other states or the federation, it can still be used as an argument in other proceedings. At present, the referendum is still in its initial phase. The activists are in the process of collecting the required 25,000 signatures. Once that figure is reached, the application for a review of admissibility can be filed.[79] Whether the referendum will be successful remains unclear.

The Bavarian popular initiative 'Rechte der Natur – Das Volksbegehren'[80] aims at the modification of art 101 of the Bavarian constitution, which provides currently to everyone 'the freedom to do whatever does not harm others within the limits of the laws and morality'.[81]

The initiative proposes to modify this passage to 'whatever does not harm others or the *connatural world* (*Mitwelt*)'.[82] This would recognize rights of nature for the first time in German law. The term 'connatural world' can be traced to Klaus Michael Mayer-Abich,[83] who tried to capture an enhanced understanding of nature by means of a term more inclusive of humans.[84] For him and also for other (legal) scientists like Klaus Bosselmann, this term underscores the need to abandon the dichotomy between human and nature. This is why the Bavarian popular initiative decided to use 'connatural world'[85] for the modification of art 101 Bavarian constitution.

---

78 See Rechte der Natur – Das Volksbegehren <https://gibdernaturrecht.muc-mib.de/>.

79 Pursuant to art 63(1) Landeswahlgesetz (Bavarian Election Law); in detail: Ewering and Gutmann (n 2) 1120.

80 Rechte der Natur – Das Volksbegehren <https://gibdernaturrecht.muc-mib.de/>.

81 See Constitution of the Free State of Bavaria <www.bayern.landtag.de/fileadmin/Internet_Dokumente/Sonstiges_P/BV_Verfassung_Englisch_formatiert_14-12-16.pdf>.

82 See Elena Sofia Ewering and Andreas Gutmann, 'Gibt Bayern der Natur Rechte?' (*Verfassungsblog*, 10 September 2021) <https://verfassungsblog.de/gibt-bayern-der-natur-rechte/>.

83 Martin Held and Hans-Jochen Luhmann, 'Von der Umwelt zur natürlichen Mitwelt' (2018) politische ökologie 142–143.

84 See Klaus Michael Meyer-Abich, *Aufstand für die Natur* (Carl Hanser 1990) 11, 35–55. English version: Klaus Michael Meyer-Abich, *Revolution for Nature* (University of North Texas 1994).

85 In German: *Mitwelt*.

Even though the concept of the connatural world is promising, the term 'nature' (or in German: 'Natur') seems to be sufficiently practicable. Nature is not to be understood as a gathering of non-human entities separated from human 'culture', but as a collective term, including all ecosystems as well as their human and non-human components. This perspective seems to be very common within the worldwide discussion of rights of nature.[86] To allow for a better understanding, the term 'nature' is used throughout the discussion of the Bavarian proposal as a synonym for 'connatural world'.

(1) Content of the proposed modification of art 101 Bavarian constitution

Currently, The Bavarian constitution – similar to art 20a Basic Law on the federal level – already contains a state objective in its art 141, which specifies, among other elements, the protection of the natural basis of life as a duty for individuals and the state community, the careful use of natural resources, and last but not least the enjoyment of natural beauty and recreation in the outdoors. With art 141 Bavarian constitution in mind, the new art 101 Bavarian constitution would implement at a minimum the right for nature to exist. Otherwise, the modification would be meaningless, as it would then merely be a repetition of art 141 Bavarian constitution.[87] Even if the modification of art 101 Bavarian constitution does not provide concretized rights of nature as is the case in the Ecuadorian constitution,[88] according to the initiators of the referendum, it guarantees their existence and thus also the legal subjectivity to be able to sue for these rights. Nevertheless, it remains the task of Bavarian legislation to concretize the rights of nature in simple legislation and to shape their enforceability.

By expanding the legal sphere to include nature, a new delimitation of already recognized legal subjects, especially that of humans, inevitably follows. However, this does not mean that all human use of nature must cease.[89] In fact, balancing the legal spheres of different legal subjects is always the task of German legislative and case law. In these procedures, nature would be helped to a stronger position through recognized legal subjectivity.

(2) Procedural opportunities

The wording of the Bavarian popular initiative does not contain any precise proposals for the enforcement of the new art 101 Bavarian constitution. To make nature's rights effective, however, it is necessary to broaden the legal process, especially recourse to the administrative courts. However, the court system's organi-

[86] See the United Nations program Harmony with Nature <www.harmonywithnatureun.org/>.

[87] Elena Sofia Ewering and Andreas Gutmann, 'Gibt Bayern der Natur Rechte?' (*Verfassungsblog*, 10 September 2021) <https://verfassungsblog.de/gibt-bayern-der-natur-rechte/>.

[88] See Gutmann (n 2) 213, 214.

[89] Exemplary in this regard: Alberto Acosta, *Buen Vivir. Vom Recht auf ein gutes Leben* (Oekom 2015) 118.

zation and procedures are subject to concurrent legislation pursuant to art 74 no (1) Basic Law, and, as explained, there are extensive regulations at the federal level. Notwithstanding this situation, §42(2) Code of Administrative Court Procedure allows states to enact deviating regulations at state level with regard to the right of action.[90]

With the introduction of art 101 Bavarian Constitution, subjective rights of nature would be introduced, so that nature could specifically assert its own rights within the meaning of §42 Code of Administrative Court Procedure.

But it could remain problematic that, according to the prevailing opinion, nature or natural entities are not eligible to participate in court proceedings pursuant to §61 Code of Administrative Court Procedure.[91] Consequently, the acknowledgement of subjective rights of nature in a state constitution should include recognition of the – at least partial – legal capacity of natural entities to bring violations of such subjective rights before court. The new subjective rights of nature in the Bavarian state constitution could therefore constitute legal personhood in the sense of §61 Code of Administrative Court Procedure. Yet it remains unforeseeable how the courts will assess the legal capacity of natural entities on the federal level on the basis of state rights.

(3) Legality of the new art 101 Bavarian constitution

Whether the referendum – if successful – recognizes the rights of nature in Bavaria will ultimately also depend on the compatibility of the new art 101 Bavarian constitution with the Basic Law. Pursuant to art 75(1) sent 2 Bavarian constitution, constitutional amendments must comply with the principle of democracy, at least at a minimum with art 28(1) Basic Law.

A first hurdle for the implementation of rights of nature in the Bavarian state constitution may arise in terms of legislative competence.[92] Arts 70–82 of the Basic Law contains provisions on whether the states or federation are responsible for legislation. What is disputed is whether the constitutional legislation of the states is also governed by arts 70–82 Basic Law or whether a legislative competence follows directly from the principle of constitutional autonomy of the states. In the specific case of the new art 101 Bavarian Constitution, the situation is made even more complicated by the fact that the rights of nature represent a fundamental innovation and are therefore difficult to assign to the catalogue of legislative powers. Due to its open wording, the new art 101 Bavarian Constitution has to be drafted in simple legislation. At the latest then, the question of legislative competence must be clarified.[93] As explained above, every federal law has primacy of application. However, since basic rights represent in particular the standard of legality for all state action, the new art 101 Bavarian Constitution must

[90] Ewering and Gutmann (n 2) 1123.

[91] As was the case in the *Robbenklage*, see section II, above.

[92] In detail, Ewering and Gutmann (n 2) 1121.

[93] Ewering and Gutmann (n 2) 1121.

above all be measured against these rights. The question of the compatibility of rights of nature with the basic rights of the Basic Law would arise not only at the state level but also at the federal level.

The new art 101 Bavarian Constitution could come into conflict with basic rights such as freedom of property (Basic Law, art 14), freedom of occupation (Basic Law, art 12), or the general freedom of action (Basic Law, art 2(1)). Especially in the case of freedom of property, it might seem paradoxical at first that natural entities such as animals, forests, etc. are also objects of property. However, since the rights of nature do not exclude any use of natural entities, but instead recognize their importance within the framework of the balancing process,[94] this does not pose a problem per se. The situation is similar to freedom of occupation and the general freedom of action. According to German legal academic literature, the new art 101 Bavarian Constitution will not initially and necessarily collide with these basic rights. Overall, however, the concrete implementation of laws will have to be measured against individual basic rights when these rights are drafted into simple legislation, and possible problems arising from the creation of new rights of nature which impinge on existing rights can be resolved at the stage of balancing.[95]

III. Legal and Political Operators

Rights of nature are still a marginal issue in Germany and have yet to be discovered as a possible viable legal instrument in re-defining the relation between humans and the rest of nature and in challenging the destruction and endangerment of ecosystems. Yet in recent years, and especially recently, interest seems to have increased particularly in civil society, but also in politics and case law, as several initiatives have formed around the claim for rights of nature. However, in the German federal election 2025 (Bundestagswahl) only one party of the parliament of the Federal Republic of Germany (Bundestag), namely 'Die Linke', explicitly incorporated rights of nature in their election manifesto.[96]

1. Network 'Rechte der Natur'

The national network 'Rechte der Natur' (Rights of Nature) is playing a pioneering role. In 2008, in a first attempt, the network 'Rechte der Natur und Ökonomie' (Rights of Nature and the Economy) was formed.[97] It is interesting to note that the members were mainly prestigious representatives of business and economics.

---

[94] Gutmann (n 2) 115.

[95] In detail, Ewering and Gutmann (n 2) 1124–1127.

[96] See: Programm der Partei Die Linke, <www.die-linke.de/fileadmin/user_upload/Wahlprogramm_Langfassung_Linke-BTW25_01.pdf> at 43.

[97] Netzwerk Rechte der Natur <www.rechte-der-natur.de/de/das-netzwerk.html>.

The recognition of the rights of nature was seen above all as a prerequisite for a sustainable economy. However, it was also emphasized that the rights of nature also require the successful realization of basic rights and human rights. The aim should be to advance the demand to endow nature with constitutional rights.[98]

Building on this network, the network 'Die Rechte der Natur – Initiative Grundgesetzreform' (The Rights of Nature – Basic Law Reform Initiative) was additionally founded in 2020. The amendment of the Basic Law is intended to oblige the German state to better protect nature for its own sake and to act on a responsibility for future generations.[99] In addition to many members of the original network, numerous legal scholars, academics, and civil society organizations are now represented as well.[100] In an intensive process, a position paper was drafted that contains concrete proposals for amendments to the law in order to include the rights of nature in the German constitution.[101] The proposal was presented to the general public in a press conference on 22 April 2022. The network plans to bring their proposal to the attention of political actors.[102] In May 2023 the network was formally founded and registered as an association (eingetragener Verein, e.V.) to enable a greater support for initiatives and further projects.[103]

At the level of the federal states, the Bavarian petition for the rights of nature in the state constitution, already described in detail above, was promoted for Bavaria within the framework of the Network 'Die Rechte der Natur – Initiative Grundgesetzreform'.[104]

2. Verein Rechte der Natur e.V.

A group of young lawyers founded the non-profit association 'Rechte der Natur' ('Rights of Nature') at the beginning of 2022.[105] The aim of the association is to introduce the legal concept of the rights of nature to the general public and to recruit members with different professional backgrounds.[106]

In order to achieve these goals, they want to carry out their own projects as well as serve as a platform capable of bundling the Germany-wide civil society

---

[98] Netzwerk Rechte der Natur (n 97).

[99] Netzwerk Rechte der Natur <https://ecojurisprudence.org/wp-content/uploads/2023/01/Germany-Proposal-to-Recognize-Rights-of-Nature-in-German-Constitution.pdf>.

[100] For a detailed list, see Netzwerk Rechte der Natur (n 97).

[101] Netzwerk Rechte der Natur, 'Rechte für die Natur – Initiative für eine Grundgesetzreform' <www.rechte-der-natur.de/de/initiative-grundgesetzreform.html>. See also the report by Tore Vetter, 'Rechte der Natur Biokratie' <www.rechte-der-natur.de/de/presse.html?file=files/opensauce/img/content/Pdf%20_%20Downloads/Kurzgutachten%20RdN%20Initiative%20Grundgesetzreform_Vetter_18.04.2022_final.pdf&cid=1602>.

[102] Netzwerk Rechte der Natur (n 97).

[103] Netzwerk Rechte der Natur, <www.rechte-der-natur.de/de/aktuelles-details/verein-netzwerk-rechte-der-natur-gegruendet.html>.

[104] Rechte der Natur, Das Volksbegehren <https://gibdernaturrecht.muc-mib.de/about-us>.

[105] Rechte der Natur e.V. <http://rechte-der-natur.org/>.

[106] Rechte der Natur e.V. <https://www.youtube.com/watch?v=F5HUqg2f3tw>.

engagement that already exists. There are already various civil society groups and associations that are committed to the protection of certain ecosystems. According to the understanding of the Rights of Nature Association, these commitments are basically already a concrete commitment to the rights of these specific ecosystems.

In addition, the idea of the rights of nature should be communicated to the public through concrete events in cultural as well as political and scientific fields.[107]

3. *Proposal of the Green Party's Federal Working Group on Ecology*

In 2020 the party 'Bündnis 90/Die Grünen' (Green Party) renewed its basic program. In this context, the Federal Working Group on Ecology has put forward a proposal to advocate a change of direction toward an ecological constitutional state in the future. According to this proposal, it is necessary to give nature rights because only in this way nature can be considered in court in opposition to the interests of the economy. In order to achieve this goal, the Basic Law and nature and animal protection law should be supplemented to the effect that nature's rights should be respected and protected.[108] Unfortunately, however, the proposal has not found its way into the new basic program.[109]

Even if the rejection of the rights of nature by a party whose core interest is climate and environmental protection is a defeat, it can nevertheless be emphasized that the rights of nature have at least become an issue also in political parties.

4. *CJEU Preliminary Ruling Concerning Rights of Nature*

In June and August 2021 as well as in April 2024 proceedings before the District Court of Erfurt (Landgericht Erfurt) were, pursuant to art 267 TFEU, suspended in order to make a referral to the Court of Justice of the European Union (CJEU).[110] The legal disputes were taking place in the context of the auto emissions scandal. Not only in Germany, but throughout Europe, claims for damages were made against manufacturers in cases where their vehicles or engines had an impermissible defeat device.[111] The CJEU was asked, inter alia, whether a claim for damages by vehicle buyers arises directly from EU law – irrespective of its character as a protective law. Among other things, the *effet utile* in connection with the

---

[107] Rechte der Natur e.V. <http://rechte-der-natur.org/>.

[108] BAG Ökologie <https://bagoekologie.wordpress.com/2020/07/10/die-rechte-der-natur-sind-zu-achten-und-zu-schuetzen/>.

[109] See Bündnis 90/Die Grünen, Grundsatzprogramm <www.gruene.de/grundsatzprogramm prozess>.

[110] LG Erfurt 9 August 2021, 8 O 481/21, juris; LG Erfurt 17 June 2021, 8 O 1130/20, juris; LG Erfurt 22 April 2024, 8 O 1045/18, juris.

[111] LG Erfurt 9 August 2021 (n 110) paras 7, 8.

rights of nature could be considered.[112] In brief, the submitting court reasons as follows:

The emission of more exhaust gases would also violate the rights of nature to a greater extent than permitted. Rights of nature could be derived from the Charter of Fundamental Rights of the European Union (CFREU) as well as from the EU treaties.[113] Many of these rights would by their nature or by analogy also apply to ecosystems, individual trees, or plants. With respect to the CFREU, it should be noted that the often-used term 'person' could also include nature or ecosystems.[114] Due to today's ecological challenge, the recognition of specific rights of nature through interpretation and the application of existing law would be required. The law of the European Union would be open for this, as Union law would also be open to international influences. Rights of nature are already recognized in other legal systems. The guarantee of human dignity found in art 1 CFREU also does not speak against but rather for the recognition of the rights of nature.[115] The same can be said regarding the responsibility for human communities and future generations as postulated in the preamble. In this way, the idea of equality enshrined in the CFREU could be taken into account.[116]

Certainly, these submissions by the District Court raised visionary questions on the interpretation of EU law, but eventually all three referrals were settled without a decision or statement by the CJEU.[117] For this reason the positioning of the CJEU regarding the recognition of rights of nature within the CFREU remains uncertain. However, the unanswered questions leave room for further influences and interpretations from member states in relation to the ongoing developments with regard to the rights of nature.

5. German Case Law Concerning Rights of Nature

Similar legal disputes were the subject of two proceedings of the District Court of Erfurt in August and October 2024.[118] Also in these cases the plaintiffs asked for compensation because their vehicles or engines had an impermissible defeat device. The rulings became the first ones by a German Court recognizing the rights of nature, the reasoning of the second judgement being more elaborate in this regard.

The District Court of Erfurt introduced in the two cases the rights of nature as 'objective values', which as such have an impact on the 'private enforcement'

---

[112] LG Erfurt 9 August 2021 (n 110) paras 36–40; LG Erfurt 17 June 2021 (n 110) para 22–26.

[113] LG Erfurt 9 August 2021 (n 110) paras 36–48; LG Erfurt 17 June 2021 (n 110) para 22–30.

[114] LG Erfurt 17 June 2021 (n 110) para 32; LG Erfurt 9 August 2021 (n 110) para 52.

[115] LG Erfurt 9 August 2021 (n 110) para 58; LG Erfurt 22 April 2024 (n 110) para 21.

[116] LG Erfurt 9 August 2021 (n 110) para 54.

[117] LG Erfurt 17 October 2024, 8 O 836/22, juris, para 69.

[118] LG Erfurt 2 August 2024, 8 O 1373/21, [2024] BeckRS 19541; LG Erfurt 17 October 2024 (n 117).

of rights.[119] Regarding the amount for compensation of the damages, the District Court found itself, in accord with the decision made by the German Federal Court of Justice (BGH),[120] able to decide 'on this issue at its discretion and conviction, based on its evaluation of all circumstances', §287 Code of Civil Procedure (Zivilprozessordnung, ZPO). Thereby, it introduced the rights of nature in favor of the plaintiffs. The District Court ruled that the rights of nature, specifically the rights of so-called 'ecological persons',[121] need to be considered within the presently existing European context and thus have a protection-reinforcing effect ('schutzverstärkende Wirkung') in the calculation of damage.[122] Especially arts 2, 3, and 37 CFREU are applicable to 'ecological persons', imparting them with the right to exist, regenerate, and preserve.[123]

The reasoning of the District Court is the subject of an ongoing academic discussion,[124] not least about the legal competence of the judge in charge of the proceedings.[125] It can be stated that the rulings opened up a space for legal dispute and offer an innovative and unusual idea for implementing rights of nature in German civil law cases. But also on the European level, the Erfurter legal cases potentially influence the further intercourse about rights of nature and promote their recognition within the member states.

---

[119] LG Erfurt 17 October 2024 (n 117) para 32; Andreas Gutmann and Jenny García Ruales, Von Flüssen und Dieselabgasen Rechte der Natur am LG Erfurt (*Verfassungsblog*, 18 June 2024), <https://verfassungsblog.de/rechte-der-natur-rio-machangara-dieselskandal/>.

[120] BGH 30 July 2024, VI ZR 122/23, [2024] NJW-RR 1350.

[121] See section III.4, above.

[122] LG Erfurt 2 August 2024 (n 118) para 29; LG Erfurt 17 October 2024 (n 117) paras 27, 56, 77.

[123] LG Erfurt 2 August 2024 (n 118) paras 30, 34; see section III.4, above.

[124] Mehrdad Payandeh, *Europarecht: Rechte der Natur*, [2024] JuS 989; Ralf Michaels, 'Rechte der Natur – ein erfolgreicher Reverse Legal Transplant?' [2024] Zeitschrift für Umweltrecht 643; Louis Johns, 'Anmerkung zum Urteil LG Erfurt, Urt. v. 17.10.2024, 8 O 836/22: Rechte der Natur II', [2025] Klima und Recht 26; Gutmann and García Ruales (n 119); Camilla Haake, 'Pioniertat oder Bärendienst? Das LG Erfurt über EU-Grundrechte für die Natur' (*Verfassungsblog*, 26 August 2024) <https://verfassungsblog.de/rechte-der-natur-erfurt/>; Christoph Degenhart, 'Irritierend distanzlos – Zu Andreas Gutmanns Kritik der Kritik am LG Erfurt zu Rechten der Natur' (*Verfassungsblog*, 5 November 2024) <https://verfassungsblog.de/rechte-der-natur-erfurt-kritik/>; Andreas Gutmann, 'Fruchtbare Irritationen – Das zweite Urteil des LG Erfurt zu Rechten der Natur' (*Verfassungsblog*, 28 October 2024) <https://verfassungsblog.de/rechte-der-natur-erfurt-2/>.

[125] Andreas Funke, 'Hier spricht der Richter Die Entscheidungen des LG Erfurt zu „Rechten der Natur“ und das richterliche Rollenverständnis' (*Verfassungsblog*, 31 October 2024) <https://verfassungsblog.de/rechte-der-natur-erfurt-unabhangigkeit/>; Christoph Degenhart, 'Irritierend distanzlos Zu Andreas Gutmanns Kritik der Kritik am LG Erfurt zu Rechten der Natur' (*Verfassungsblog*, 5 November 2024) <https://verfassungsblog.de/rechte-der-natur-erfurt-kritik/>.

6. Enforcement

It remains unclear how rights of nature would be enforced if they were to be acknowledged by any means.[126] As nature can neither speak for itself nor go to court on its own, the main challenge is how to articulate – interdependent and possibly internally conflicting[127] – interests of nature.

There are several conceivable ways of linking up with existing concepts in German law. One would be to institute representatives for natural entities, who could then enforce rights of nature in the name of nature in court. This would require an appropriate process to transfer power of representation, which in this case should be democratically legitimized and societally scrutinized to ensure an open debate regarding the interests of the represented natural entities and to prevent an abuse of representative powers.[128]

Another option would be the enforceability of rights of nature by any human being in the name of nature and abandonment of the requirement of individual standing on the whole. This comes close to the concept of 'Prozessstandschaft' (procedural status),[129] which means that someone sues for the rights of others in his own name (distinctive to representation, where the representative acts in the name of the represented). While voluntary procedural status ('gewillkürte Prozessstandschaft') is accepted under certain circumstances in civil law,[130] it is not considered permissible in administrative law.[131] Yet §42(2) VwGO opens up standing without the plaintiff being individually and subjectively affected in 'cases prescribed by law'. This means that legally prescribed procedural status for people to claim infringements of rights of nature in their own name would be compatible with existing law.[132] Nevertheless, the transferability of these concepts to rights of nature naturally involves considerable 'translation difficulties' with regard to questions in terms of the presumed 'interests' or 'will' of nature or of individual natural entities.

---

[126] For procedural obstacles and opportunities regarding the enforcement of rights of nature using the example of the Bavarian constitutional approach, see section II.2.b)(2), above.

[127] Cf Gutmann (n 2) 124–126, 244–263; Fischer-Lescano (n 2) 248–249.

[128] Gutmann (n 2) 198–203; Fischer-Lescano (n 2) 249; See also Carolin Raspé, *Die tierliche Person* (Duncker & Humblot 2013) 326; Bosselmann (n 2) 384.

[129] Cf Kersten (n 17) 113.

[130] Stephan Weth, in Hans Joachim Musielak and Wolfgang Voit (eds), *ZPO* (21th edn, Franz Vahlen 2024) §51 paras 25–32.

[131] Thomas Schmidt-Kötters, in Herbert Posser and Heinrich Amadeus Wolff (eds), *Beck'scher Online-Kommentar Verwaltungsgerichtsordnung* (61th edn, CH Beck 2019) §42 para 114; Michael Happ, in Erich Eyermann and Ludwig Fröhler (eds), *Verwaltungsgerichtsordnung* (16th edn, CH Beck 2022) §42 para 82; leaving the question undecided: VGH Munich 16 December 1993, 23 B 91.2659, [1995] NvWZ 617.

[132] Cf Happ (n 131).

7. *Theoretical Sources and Foundations*

The theoretical foundations for the current discussions on rights of nature in Germany are diverse. They are discussed as a consequence of a rearrangement of human-nature relations, the necessity of which has been made clear by climate change and the ongoing destruction of ecosystems. The approaches have in common that they require a rethinking of the ontological foundations of Western law, in which non-humans and nature are diminished to the role of objects. Of high influence in this matter are especially post-modern thinkers like Donna Haraway[133] or Bruno Latour, the latter of whom suggest an ontological-epistemological turn towards the recognition of 'non-human entities' as noteworthy participants in an Acteur-Network-Theory[134] and who has even proposed an inclusion of these entities in democratic negotiation processes.[135] Moreover, Global South thinking has also influenced the theoretical discourse on rights of nature in Germany. The reception of such thinking has led to controversy, especially due to the divergent assessments of indigenous influence on rights of nature in the Global South.[136] At the level of civil society, Global South thinking serves as a theoretical orientation for the formulation of demands by civil society actors.[137]

8. *Relationship between National Legal Systems*

Legal systems, where rights of nature have already been acknowledged or implemented, prove to be an important point of orientation for proposals to acknowledge rights of nature in German law. The Constitution of Rights of Nature in Ecuador has been increasingly received among legal scholars in Germany, as have acknowledgements regarding the rights of nature in other places around the world, including India, New Zealand, Spain, or Columbia. As regards the latter country, one sees a cautious but ongoing exchange initiated by the Federal Constitutional Court.[138] The development[139] of rights of nature in these countries and potentially resulting conflicts within their respective legal systems are carefully observed both by legal scholars and other jurists in Germany[140] and are gaining increasing

[133] Donna Haraway, *Unruhig bleiben* (Campus 2018) 72, 80.

[134] See Bruno Latour, *Eine neue Soziologie für eine neue Gesellschaft* (5th edn, Suhrkamp 2007) 185–189.

[135] See Bruno Latour, *Politics of Nature* (Harv UP 2004).

[136] Eg Gutmann (n 2) 132; Stefan Knauß, 'Pachamama als Ökosystemintegrität – Die Rechte der Natur in der Verfassung von Ecuador und ihre umweltethische Rechtfertigung' (2020) 7(2) Zeitschrift für Praktische Philosophie 221, 230.

[137] See Netzwerk Rechte der Natur (n 101).

[138] See II.1.a), above.

[139] Having reference to the work of Alberto Acosta, Andreas Gutmann understands the constitution of Ecuador, for example, as a 'constitution in the making', whose true constituting processes began only after formal adoption, Gutmann (n 2) 260–263.

[140] Eg Jula Zenetti, 'Natur und Recht – Eigenrechte der Natur: wegweisende Gesetze, Gerichtsentscheidungen und ausgewählte aktuelle Entwicklungen' (2022) 1 Natur und Landschaft 54.

public attention.[141] If rights of nature prove to be an efficient and mediating legal instrument in those countries, this might serve as a powerful token for the implementation of rights of nature in Germany as well. Recently, the Federal Constitutional Court has increasingly sought international dialogue. It published press releases in English[142] and in its Climate Decision explicitly referred to similar decisions in other countries, although unfortunately not to court decisions on rights of nature.[143]

9. International Law

As already mentioned, the German legal system is highly integrated in European and international law. Both have strongly influenced German environmental law. The Aarhus Convention and the EU Directives implementing it form the basis for the Environmental Appeals Act[144] and thus already exert considerable influence in changing German environmental and administrative law towards a less individual-centered understanding. In its Climate Decision in March 2021, the BVerfG once again committed itself to the validity and binding nature of the Paris climate targets.[145] The court assumed that by signing the Paris Agreement and the Climate Protection Act based upon it,[146] the Federal Republic of Germany had specified its constitutional mandate under art 20a Basic Law in a permissible manner and had thus committed itself to the Paris climate goals of limiting global warming to 1.5 degrees.[147]

Therefore, international and EU law could also play a key role for a possible implementation of rights of nature. Based on art 23 Basic Law, Germany participates in the development of the European Union and transferred sovereign rights

[141] Wolfgang Janisch, 'Klimakrise: Im Namen der Natur' *Süddeutsche Zeitung* (Germany, 2 June 2022); Michaela Haas, 'Der Fluss, der gegen seine Verschmutzung klagt' *Süddeutsche Zeitung Magazin* (Germany, 1 June 2022) <https://sz-magazin.sueddeutsche.de/die-loesung-fuer-alles/naturrechte-earth-laws-garn-91595>.

[142] BVerfG, press release no 31/2021 of 29 April 2021 <www.bundesverfassungsgericht.de/SharedDocs/Pressemitteilungen/EN/2021/bvg21-031.html>.

[143] Alexandra Kemmerer, 'Gericht halten über uns selbst – Klimaklagen als *Court Room Drama* und juristische Intervention' (2022) 1 juridikum 131, 133-134; Katja Gelinsky and Marie Christine Fuchs, 'Bitte noch mehr – Rechtsprechungsdialog im Karlsruher Klimabeschluss' (*Verfassungsblog*, 26 May 2021) <https://verfassungsblog.de/bitte-noch-mehr/>.

[144] See Ewering and Vetter (n 2) 385–387; Happ (n 131) Introduction to §1 UmwRG paras 2ff.

[145] Jannika Jahn, 'Internationaler Klimaschutz mithilfe nationaler Verfassungsgerichte? Erkenntnisse aus dem Klimabeschluss des BverfG' [2022] ZaöRV 47, 58.

[146] Art 2(1)(a) of the Paris Agreement obliges the convention states to hold 'the increase in the global average temperature to well below 2°C above pre-industrial levels and pursuing efforts to limit the temperature increase to 1.5°C above pre-industrial levels'. §1 of the Climate Protection Act is based on these objectives.

[147] BVerfG 24 March 2021(n 1) paras 208–247; For the important implications that the decision holds for the international integration of the Basic Law and its reception by the BVerfG, see Jahn (n 145).

to the Union. The relation between the German national legal system and European law, and especially between the case law of the Federal Constitutional Court and the case law of the CJEU, is subject both to cooperation and struggles for authority between both regimes. It cannot be said for sure how the Federal Constitutional Court would react to an implementation of rights of nature on the level of European law, but according to its previous rulings it would probably approve it, as long as basic rights of the Basic Law are not undermined.[148] Up to now there are no explicit rights of nature acknowledged under the CFREU; art 37 CFREU is mostly understood as a general principle that holds no specific, let alone subjective rights of nature.[149] The CJEU dismissed the 'People's Climate Case' in March 2021 as inadmissible on the grounds of the lack of individual concern. Having recourse to the notorious 'Plaumann formula', the court denied appellants standing before the court as they had not established that art 37 CFREU addressed them individually.[150]

Of particular relevance to German law is also the European Convention on Human Rights (ECHR). It is not only directly applicable in Germany through its ratification with the rank of simple federal legislation, art 59(2) Basic Law, but according to the jurisdiction of the Federal Constitutional Court it must also be observed in interpreting the Basic Law and the scope of basic rights.[151] While there are currently no rights of nature acknowledged by the ECtHR as well, the court regards the Convention as a 'living instrument',[152] and it might very well be that it opens the range of Convention rights toward rights of nature of some sort.

[148] BVerfG 22 October 1986, 2 BvR 197/83, BVerfGE 73, 339 – *Solange II*; More recently, the court decided to examine domestic law – which implements Union law without being fully determined by it – on the basis of the basic rights of the Basic Law, BVerfG 6 November 2019, 1 BvR 16/13, 152 BVerfGE 152, 152 – *Vergessen I*; in matters fully determined by Union law, it gave priority to the fundamental rights of the Union, but it decided to monitor them itself, BVerfG 6 November 2019, 1 BvR 276/17, 152 BVerfGE 152, 216 – *Vergessen II*.

[149] With further references, Steven Greer, Janneke Gerards and Rose Slowe, *Human Rights in the Council of Europe and the European Union: Achievements, Trends and Challenges* (CUP 2018) 353; Stefan Kadelbach, 'Charter of Fundamental Rights of the European Union (2000)' (2019) MPEPIL <https://opil.ouplaw.com/view/10.1093/law:epil/9780199231690/law-9780199231690-e1864>.

[150] CJEU 25 March 2021, C-565/19 P, *Armando Carvalho and Others v European Parliament and Council of the European Union*, ECLI:EU:C:2021:252 para 73.

[151] BVerfG 14 October 2004, 2 BvR 1481/04, BVerfGE 111, 307 paras 32–37; for an overview: Bedeutung der EMRK und der Entscheidungen des EGMR für die deutsche Gesetzgebung, Scientific Service of the German Bundestag, 22 June 2016 <www.bundestag.de/resource/blob/436800/8e288ab7acd739fcad2463017ad44b08/wd-3-162-16-pdf-data.pdf>.

[152] ECtHR 25 April 1978, 5856/72, *Tyrer v The United Kingdom*, para 31; George Letsas, 'The ECHR as a Living Instrument: Its Meaning and Legitimacy' in Andras Føllesdal, Birgit Peters and Geir Ulfstein (eds), *Constituting Europe: The European Court of Human Rights in a National, European and Global Context* (CUP 2013).

Most recently, the ECtHR's decisions on three climate cases in April 2024 have caused a stir. Although two of the three cases were dismissed,[153] the decision in the case of *Verein KlimaSeniorinnen Schweiz and Others v Switzerland*[154] is particularly important. The legal action sought to compel effective measures by Switzerland to prevent the temperature from rising by more than 1.5 degrees Celsius, as older women, for example, would suffer particularly during hot spells.[155] In the landmark decision of the ECtHR, the finding of a positive obligation of states to protect against the serious adverse effects of climate change under art 8 ECHR is of particular importance.[156] The ECtHR also found a violation of art 6 ECHR, as the national courts had failed to address the standing of the association.[157] Also revolutionary, and with potentially far-reaching consequences for the German legal system, is the strengthening of associative actions in relation to climate change.[158] Although the ECtHR rejected the victim status (ECHR, art 34) of the individual plaintiffs,[159] it affirmed it for the association.[160] If the ECtHR were to acknowledge rights of nature under the Convention, this could affect the Federal Constitutional Courts interpretation of the Basic Law. This applies in particular to art 20a Basic Law, the international dimension of which the Court particularly emphasized in its Climate Decision.[161] Because of the conflict rule in art 52(2) CFREU, a change in the ECtHR's position on this issue could also have an impact on European Union law and ultimately, via the latter, on German law.

10. Efficacy

It remains an open question to what extent rights of nature would be an efficient legal instrument in German legal practice. Since they have not yet been recognized, it can only be speculated what effects they might have. For example, the rights of nature would as a consequence of their recognition be included in the various balancing processes involving competing subjective rights. This does not mean that, for example, human interests must always take a back seat. How-

---

[153] The two dismissed cases were ECtHR 9 April 2024, 7189/21, *Carême v France*; ECtHR 9 April 2024, 39371/20, *Duarte Agostinho and Others v Portugal and 32 Others*.

[154] ECtHR 9 April 2024, 53600/20, *Verein KlimaSeniorinnen Schweiz and Others v Switzerland*.

[155] ECtHR 9 April 2024 (n 154) paras 10–21, 58–63.

[156] ECtHR 9 April 2024(n 154) para 519.

[157] ECtHR 9 April 2024 (n 154) paras 636–640.

[158] Sandra Arntz and Jasper Krommendijk, 'Historic and Unprecedented' (*Verfassungsblog*, 9 April 2024) <https://verfassungsblog.de/historic-and-unprecedented/>.

[159] ECtHR 9 April 2024 (n 154) paras 535, 536, 623.

[160] ECtHR 9 April 2024 (n 154) paras 526, 537, 623; The ECtHR states 'the standing of an association to act on behalf of the members or other affected individuals within the jurisdiction concerned will not be subject to a separate requirement of showing that those on whose behalf the case has been brought would themselves have met the victim-status requirements for individuals in the climate-change context' ECtHR 9 April 2024 (n 154) para 502.

[161] BVerfG 24 March 2021 (n 1) paras 199–204; Jahn (n 145) 71–72.

ever, the interests of nature could not be ignored, as is often the case at present. The experiences from countries that have already enshrined rights of nature are particularly valuable in assessing the potential efficacy of rights of nature in Germany.[162]

[162] For an estimation of how a German environmental law case would play out if it arose in Ecuador, see: Narváez Alvarez (n 46); for an assessment of the challenges an Ecuadorian case on rights of nature would face under German law: Ewering and Vetter (n 2).

Italy

Establishing the Rights of Nature in Italy: Between Tradition and Change

Domenico di Micco and Michele Graziadei[*]

I. Premise

Currently, several countries around the world recognize the 'rights of nature' in formal legal documents, ranging from constitutions to city ordinances, and in legal judgments. A similar tendency is gaining traction at the international level as well, through the activity of international organizations as well as through the initiatives of NGOs and community leaders. This contribution should therefore investigate whether and to what extent Italian law is participating in this movement and shares a similar approach. Before addressing this fundamental question, some preliminary observations on the subject are appropriate.

An investigation concerning the rights of nature is fraught with several initial difficulties, and they cannot simply be sidestepped by assuming that such an expression has a self-evident meaning for the present purposes.

[*] This paper is a collaborative work. Michele Graziadei wrote sections I–IV and X; Domenico di Micco is the author of sections V–IX. The manuscript was finalized in February 2025. The assessment of literature and case law, including internet materials, reflects the legal situation at that time.

To pick an illustrative example of problematic language, consider the preamble of the United Nations agreement on climate change (2015). This affirms 'the importance of ensuring the integrity of all ecosystems, including oceans, and the protection of biodiversity, *recognized by some cultures as Mother Earth*' and notes 'the importance for some of the concept of "*climate justice*" when taking action to address climate change'. (Emphasis added.)

What is the value of the recognition given by the preamble to the notion of 'Mother Earth' as a 'cultural' element? Is this language challenging the scientific language featured in the rest of the sentence, where the notion of 'climate' is advanced, or is it in continuity with it and with the concept of climate justice?

To enter a debate on the rights of nature without creating misunderstandings, it would first be necessary to clarify the relationship between that entity ('nature') and the concept of law, as a product of human cultures. The word 'nature' itself points in different directions in the various languages, and, at least in some cases, it is a notion that has been received from abroad rather than having been elaborated locally.[1] Secondly, one would have to consider what legal techniques are available to make any right ascribed to nature enforceable. If this is indeed desirable, a further step would be to decide what claims, immunities, privileges, and powers follow from the recognition of any such right.

Furthermore, the approach defended by the advocates of the rights of nature should be distinguished from the classical perspective of environmental law. The latter tends to focus on the notion of environmental damage and the need to (prevent and) remedy it. This perspective is dominant in market-oriented societies, whereas the rights of nature discourse, which has been developed in a number of contexts, squarely rejects a market-based approach toward the defense of nature. Lastly, it should be clear that the discussion over the rights of nature has little to do with all the legal and philosophical references to 'nature' in the context of the natural law approaches to a whole variety of topics, ranging from the protection of property to the manipulation of living entities through biotechnology. All these approaches draw upon the distinction between positive law and natural law. Such a distinction has a long tradition in Western legal thinking, but it is hardly relevant for present purposes. Natural law itself (as understood in this

[1] Frédéric Ducarme and Denis Couvet, 'What Does "Nature" Mean?' (2020) 6(1) Palgrave Communication 1; Frédéric Ducarme, Fabrice Flipo and Denis Couvet, 'How the Diversity of Human Concepts of Nature Affects Conservation of Biodiversity' (2021) 35(3) Conservation Biology 1019; Lissy Fehnker, Diane Pearson and Peter J Howland, 'Understanding Conceptions of "Nature" for Environmental Sustainability: A Case Study in Tāmaki Makaurau Auckland, Aotearoa New Zealand' (2021) 2(3) Earth 357. Similar definitional problems concern particular elements of nature: See, eg, Carmelo Dazzi and Giuseppe L Papa, 'A New Definition of Soil to Promote Soil Awareness, Sustainability, Security and Governance' (2022) 10(1) International Soil and Water Conservation Research 99. On the reception of the Western notion of 'nature' in Japan: Frederic Marcon, 'Without Nature: Thinking about the Environment in Tokugawa Japan' in Bonaventura Ruperti, Silvia Vesco and Carolina Negri (eds), *Rethinking Nature in Contemporary Japan From Tradition to Modernity* (Venice UP 2017).

tradition) is not innocent of the arguments that justify the exploitation that has endangered the rights of nature as understood by their contemporary defenders. It is clearly impossible to cover all this ground in the present essay. Furthermore, many of the issues just mentioned have already been discussed in existing literature. Therefore, in the following pages, we will on focus on those themes relating to the Italian legal tradition that may be of interest for a more general debate on the rights of nature.

II. Rights of Nature: A Game Changer?

No matter how acute the need for some clarifications as raised by the very idea of recognizing the 'rights of nature', for those who consult the literature on the topic, it is clear that all the developments linked to the emergence of the category 'rights of nature' are being received with increasing interest and excitement by parts of the legal community.[2] If we look not just at Italy, but at the whole world, these rights are surely considered as 'hot': they are part of a more general – by now global – set of concerns about the future of the planet and of the life on it, and they represent a set of emerging fundamental claims.

The rights of nature appear to many to be game changers, providing an appealing alternative to the market-led approach aiming for a minimization of environmental damage, an approach underlying environmental law since the 1960s.[3] Environmental law has neither halted the degradation of the natural world nor prepared a reassuring future for humanity. The proclamation of the rights of nature should instead foster a holistic approach to the regulation of the interaction between man and nature, driven by mankind's recognition of the necessity to live in harmony with nature.

The proclamation of the rights of nature is indeed unsettling deep-seated certainties on which modern law rests. This new approach targets the fundamental assumption that nature is not a subject but rather an object of rights, while rights can be held only by human beings, either as individuals or collectively. This belief is so deeply rooted in the law that advocates of recognizing the rights of nature were initially greeted with skepticism, if not ridicule, this being the same reaction provoked by the first proclamations of the rights of man in the eighteenth and nineteenth centuries. Now, however, the tide has turned, and the rights of nature are fully a part of the jurisprudential debate.

---

[2] For a global mapping of initiatives: Alex Putzer, Tineke Lambooy, Ronald Jeurissen and Eunsu Kim, 'Putting the Rights of Nature on the Map. A Quantitative Analysis of Rights of Nature Initiatives Across the World' (2022) 18(1) Journal of Maps 89.

[3] See, eg, Notre Affaire A Tous, *Les droits de la Nature: Vers un nouveau paradigme de protection du vivant* (Pommier 2022); Mumta Ito and Massimiliano Montini, 'Nature's Rights and Earth Jurisprudence. A New Ecologically Based Paradigm for Environmental Law' in Elia Apostolopoulou and Jose A Cortes-Vazquez (eds), *The Right to Nature: Social Movements, Environmental Justice and Neoliberal Natures* (Routledge 2018) 221–233.

The specificities of those bedrock certainties upon which modern law rests can be traced back to the specific cultural traits that first emerged in antiquity, in the Mediterranean–Middle East area, after the revolution of the Neolithic era, with an agrarian use of nature occurring within state-run structures organized as trade-facilitating entities. The technical revolutions that led to the mass production of fossil fuels in the 19th century, with the consequent explosion in the consumption of energy, exacerbated the impact of cultural attitudes confining nature to the reign of sheer matter, subject to free appropriation. The initial historical scene presented an ambivalent but somewhat balanced attitude toward nature, its being considered both a means of sustenance at the service of basic human needs as well as a cosmic force governing the fate of human communities. In the epoch of the industrial revolution this equilibrium was dramatically altered, and then destroyed. The industrial revolution and the advent of a new economy based on the large-scale production of unprecedented amounts of energy unleashed the devastating impact of ways of thinking that are still undermining the integrity of nature and the very existence of mankind on earth.

The debate over the rights of nature *eo nomine* in Italy has gained momentum in the last decades, thanks to the impulse provided by the comparison with the laws of those jurisdiction that first discussed or adopted the suggestion to protect the rights of nature as such[4] and following international developments such as the initiatives of the United Nations to proclaim Mother Earth Day (done for the first time in 2009) and to launch the Harmony with Nature initiative, the latter of which also includes the issue of the rights of nature.[5]

Recently, a team of Italian legal scholars working in collaboration with experts from the UK NGO *Nature's Rights* prepared the *Study Towards an EU Charter of the Fundamental Rights of Nature* (2020) for the European Economic and Social Committee.[6] This study could pave the way for the legal recognition of the rights of nature in the EU legal order, as a prerequisite for a different and improved relationship between human beings and nature. This could be achieved through the adoption of an EU Charter on the Fundamental Rights of Nature and a corresponding directive.

---

[4] For an excellent general treatment, with further bibliographic indications: Michele Carducci, 'Natura (diritti della)' in Raffaele Bifulco, Alfonso Celotto and Marco Olivetti (eds), *Digesto italiano delle Discipline Pubblicistе* VII supplement (Utet 2017); see, further, section X, below.

[5] See the Harmony with Nature website, maintained by the UN: <www.harmonywithnatureun.org/>. For the latest UN resolution on this initiative: UN General Assembly, 75th session, twelfth resolution on Harmony with Nature (A/RES/75/220).

[6] European Economic and Social Committee, Michele Carducci, Silvia Bagni and others, *Towards an EU Charter of the Fundamental Rights of Nature – Study*, European Economic and Social Committee (2020) <https://data.europa.eu/doi/10.2864/25113>. Participating Italian scholars were Michele Carducci, Silvia Bagni, Vincenzo Lorubbio, Elisabetta Musarò (UniSalento-CEDEUAM); Massimiliano Montini, Alessandra Barreca, Constanza Di Francesco Maesa (UniSiena).

The debate at the EU level does not point to a single direction, however. For example, the study *Can Nature Get It Right? A Study on Rights of Nature in the European Context* (2021),[7] commissioned by the European Parliament (Policy Department for Citizens' Rights and Constitutional Affairs), is much more cautious on this subject. This document, authored by to Prof. Jan Darpö, emeritus in environmental law at the Faculty of Law, Uppsala, does: '[…] not share the view that RoN entails a shift of paradigm in law that has the capacity to save the environment from the challenges we face today.' 'When deconstructing the RoN concept, no radical new instruments come to light compared with what we have today.' The author argues that although the RoN concept 'contains fresh insights in its critique of Western society and presents ideas that can be developed within our conventional legal notions', the failure to combat climate change and large-scale losses of biodiversity should not be ascribed to the circumstance that nature does not have rights, or to other basic flaws of the legal system, but to the lack of public support for a radical change and the lack of the political will to undertake such change.[8]

III. A Recent Constitutional Amendment in Italy

In 2022 the pressure to develop a response to environmental degradation led Italy to amend arts 9 and 41 of its Constitution.

Art 9 Constitution was originally dedicated to the development of culture and scientific and technical research, as well as to the protection of the landscape and the cultural heritage of the nation. The new version of this provision introduces the Republic's duty to protect '…the environment, biodiversity and ecosystems, also in the interest of future generations. The law of the State governs the ways and means of protection of animals.' The new text does not adopt the language of foreign constitutions that protect the 'rights of nature'. On the other hand, the textual references to biodiversity, the protection of ecosystems, and the protection of the interests of future generations all point to elements that feature prominently in the general discourse on the rights of nature and that reflect also the evolution of Italy's constitutional jurisprudence on environmental issues.[9]

---

[7] Jan Darpö, PE 689.328 (March 2021) <www.doi.org/10.2861/769001>.

[8] Darpö (n 7) 60. For further reflections on the developments in EU law, the essential references are: Hendrik Schoukens, 'Should Trees Have Standing – Also in the European Union?' (2018) 15(3–4) JEEPL 273; Katarina Hovden, 'The Best Is Not Good Enough: Ecological (Il) literacy and the Rights of Nature in the European Union' (2018) 15(3–4), JEEPL 281; Hendrik Schoukens, 'Granting Legal Personhood to Nature in the European Union: Contemplating a Legal (R)evolution to Avoid an Ecological Collapse (Part 1)' (2018) 15(3–4) JEEPL 309.

[9] On the preparatory works associated with this reform and the evolution of the Constitutional jurisprudence on environmental law in Italy, see: Senato della Repubblica – Ufficio Studi, *Dossier no 405/3: Modifiche agli articoli 9 e 41 della Costituzione in materia di tutela dell'ambiente*, Camera del Senato, Rome 2022.

The constitutional amendment was completed with references to the protection of the environment in art 41 Constitution, concerning the regulation of economic activities:

> 'Private economic initiative is free. It cannot be carried out in conflict with social utility or in such a way as to harm health, the environment, security, freedom and human dignity. The law determines the appropriate programmes and controls so that public and private economic activity can be directed and coordinated for social and environmental purposes.'[10]

The references to the environment were introduced in this text by the above-mentioned amendment.

Prior to this reform, a reference to the notion of 'ecosystem' was inserted in art 117 Constitution in 2001 in order to regulate the division of legislative competences belonging to the State and to the regional governments.

Any comparative law scholar knows that the enactment of a new legal text is not enough to draw conclusions about what the law in action will actually be. It is definitely too early to know whether this recent constitutional change will lead to better regulation and a more effective enforcement of rules protecting what would be the rights of nature in other countries.[11] Surely this amendment should have an impact in the resolution of cases such as the *ex Ilva* case (see section X., below). For the moment, the new amendment at least indicates a growing awareness of the necessity to change the course that has been followed so far and to adopt stricter measures and a more effective enforcement thereof to avoid further devastation. In any case, both national and local environmental associations in Italy are in most cases entitled to claim compensation of environmental damages.[12]

IV. Rights by Other Names

It follows from what has been said so far that the Italian legal system does not provide for any form of personification of nature in general, nor for the rights of nature in general terms.

When confronted with a similar question, an Italian lawyer could however point out that certain natural areas have been recognized as legal persons to protect their integrity, by recognizing them as parks under Italian law. Twenty-five national parks are currently included in the Official List of Protected Areas (EU-

---

[10] Our translation.

[11] As documented by a report published by one of Italy's leading environmental NGOs, the legislature that amended the Italian Constitution in 2022 has a very bad record on environmental issues: 'Unica nota positiva la tutela dell'ambiente in costituzione' (*WWF Italia*, 8 August 2022) <www.wwf.it/pandanews/societa/politica/ultima-legislatura-persi-5-anni-preziosi-per-clima-e-ambiente/>.

[12] For a full discussion of the standing requirements: Francesco Scalia, 'La giustizia climatica' (2021) 10 Federalismi 268.

APs), covering a total area of more than 16,000 km^2, which corresponds to approximately 5.3% of the national territory.[13]

It is very difficult to distinguish the institution of a park located on a certain territory from the environmental personhood now enjoyed by several natural entities around the world (eg the Whanganui River in New Zealand). Violations of the rules governing the park would surely trigger a legal response by the managers of the park in Italy. Analogous consideration could be formulated with regard to other protected areas such as those natural habitats that must be respected under the Habitat Directive (Council Directive 92/43/EEC of 21 May 1992).

Furthermore, Italian law no 168/2017 recognizes for the first time in the history of our country that certain original communities have normative primary powers concerning the management of the natural, economic, and cultural heritage whose territorial basis consists of the collective ownership of certain lands.[14]

These forms of collective landholding – these commons – are considered by the law to be original, primary legal orders governing a community.[15] Once more, these ancient forms of relations with nature indeed raise the question regarding the primacy of the (legal) order established in relation to certain natural elements over the law of the state, and they resonate with the idea that nature has an ontological status beyond the reach of the law of the state.

On the other hand, as mentioned above the general question concerning the personhood of nature has thus far not received a positive answer under our law.

Nonetheless, to do justice to the complexity of the problem, it must be observed that Italy historically responded to those same needs for protection and conservation that underlie the choice to grant nature a form of legal personhood through centuries-old legislation, with an innovative approach for those times. Hence, although neither the Italian legislature, nor the Italian courts, have so far ascribed personhood to natural elements in general as such, one can still argue that the Italian legal framework has nevertheless granted rights to nature through both protection and conservation principles, which have thus become the legal lever for nature's presence on the legal scene. One way in which this was done was to make intensive use of certain legal-administrative notions and legal devices such as the notion of 'natural reserve' (*riserva naturale*), namely the individuation of a certain territory as being subject to specific protective legal regimes established by private or public powers.[16]

---

[13] Elenco Ufficiale delle Aree Naturali Protette (EUAP), Supplemento ordinario no 115, Gazzetta Ufficiale no 125, 31 May 2010.

[14] Legge no 168, 20 November 2017, 'Norme in materia di domini collettivi'.

[15] On the historical debate over these forms of collective landholdings: Paolo Grossi, *An Alternative to Private Property: Collective Property in the Juridical Consciousness of the Nineteenth Century* (Lydia G Cochrane tr, Chicago UP 1981).

[16] Paolo Cendon, *Proprietà riserva e occupazione* (Jovene 1977).

With the purpose of offering as complete a reconstruction as possible of the Italian paradigm, the following pages will cover the main steps that led to this approach.

V. The 18th Century: The First Footprints of the Italian Approach to Rights of Nature

Looking back at the history of Italian law, some pre-unitary Italian states had already developed a significant legislative sensitivity toward nature at an early stage.[17] Most of them granted some protection to nature through legal provisions by which certain natural resources such as ancient trees, forests, and mountain areas found care and rights affirmed.

The Kingdom of Sicily and the Kingdom of Naples pioneered this type of legislation as early as the 18th century. This should not come as a surprise, since in the panorama of pre-unification Italian states, these two kingdoms were the main Italian gateway to the European Enlightenment.

In particular, as early as August 1745, the Kingdom of Sicily enacted measures to protect several territories and natural resources.[18] Those measures are generally regarded as among the world's oldest legal evidence of the public recognition and protection of nature by the State. By itself, this would justify an examination of these measure here. But the real reason why they are so valuable to us is that they already contain all the elements that historically constitute the Italian approach to these kinds of rights. Nearly two hundred years later, those same elements and that same approach would flow, almost unchanged, into the 1948 Republican Constitution. Let us then take a closer look at this original approach.

In 1745, the *Ordine del Real Patrimonio di Sicilia*[19] intended to protect the Greek theatre of Taormina, the 'Hundred Horse' Chestnut tree of Mascali, and the nearby ancient wood of Carpineto. This legal-administrative act was therefore aimed at protecting both a specific natural element – the Hundred Horses Chestnut tree – and an ecosystem too – the ancient wood of Carpineto. To this

---

[17] Until 1861, the Italian territory was divided between several sovereign states: the Sardinian-Piedmontese Kingdom was located in the north (Turin), with Lombardy (Milan) and Veneto (Venice) belonging to the Austrian Empire; the middle part of the Italian territory constituted the Papal States; while the Kingdom of the Two Sicilies ruled the south of Italy. The Italian Risorgimento culminated in the birth of the Kingdom of Italy in 1861. The Kingdom of Italy would last until the end of World War II, when, after the fall of Fascism, a referendum in 1946 sanctioned the choice of the Italian people to abandon the monarchy and establish the Italian Republic.

[18] *Ordini*, Tribunale del Real Patrimonio, Archivio di Stato di Palermo, vol atti 1744–1745.

[19] The *Ordine per il Real Patrimonio di Sicilia* was an administrative-judicial body that, in its centuries-long existence, had many areas of competence. It was a governmental body with judicial power. The mixture of these powers gave the Order a strong capacity to achieve results. The acts issued by this authority were called *Ordini*.

purpose, the *Ordine del Real Patrimonio di Sicilia* issued an *Ordine* – a legal-administrative act – identifying certain cultural and natural assets to be protected, stating reasons why they were to be protected and laying down sanctions for offenders. The Hundred Horse Chestnut is the largest and oldest known chestnut tree in the world. The tree's name originated from a legend. A queen of Aragon and her company of 100 knights are said to have found shelter under the tree during a severe thunderstorm. Located on the eastern slope of Mount Etna, it is generally believed to be 2,000 to 4,000 years old.[20]

The title of that old legal-administrative act makes clear the approach and the reasons for the protection it grants to these natural elements: 'Ordine Patrimoniale per la conservazione de' maravigliosi alberi nel bosco del Carpinetto sopra la città di Mascali'.[21]

The reason for granting legal protection to these trees – in particular to the Hundred Horse Chestnut tree – is their exceptional wonder, due to their exceptional age. In other words, the criterion that underlies this protection depends on the particular aesthetic value of these trees. In fact:

> 'Tra le cure pressanti, che tengono questo Sup.mo Patrim.le in continua attenzione non è la minore quella d'invigilare, in che si conservassero con pari diligenza ed oculatezza in questo Regno alcune meraviglie, che con le loro celebri rarità siccome appalesano i portenti della natura così ugualmente apportano lode e decoro al Regno, di cui elle ne sono la propagine, e lo germe; affinché con tale conservazione propria d'una buona Regenza, tramandar si potesse alla veggente posterità un monumento dell'insigne naturale portento'.[22]

The 1745 Sicilian order makes clear that protecting a marvelous specimen of nature also has inestimable value to the Kingdom. For this very reason, it must be preserved for posterity.

Once the intent and the reasons for this measure have been clarified, this legal-administrative text goes on to formulate the actual protective order:

> 'ordiniamo di dover con tutta diligenza, et ugual premura invigilare a che non fosse apportato ai cennati alberi di Castagno, o di altra sorte che siino, danno, o pregiudizio alcuno, o con tagli, o con fuoco, o con altra forma, e maniera che potesse andar da inferirgli il loro decadimento; ma

---

[20] The tree is now the property of the Sicilian Region; it is included in the Monumental Plants Register of Sicily (RPM) and in the Intangible Heritage Register (REI). In 2008 it was declared UNESCO's 'Messenger of Peace Monument'. Due to its extraordinary characteristics, this tree is included in the list of the Monumental Trees of Italy (*Alberi Monumentali d'Italia*), under national Legge no 10, 14 January 2013, *Norme per lo sviluppo degli spazi verdi urbani*, which requires Italian municipalities to list their veteran trees.

[21] Order for the preservation of the marvelous trees in the Carpinetto wood in Mascali [our translation].

[22] 'Among the pressing cares, which keep this Institution in constant attention, is not the least, that of guarantee that some marvels are preserved with equal diligence and care in this Kingdom, which with their famous rarity, as they show the portents of nature, so also bring praise and decorum to the Kingdom, of which they are the offshoot, and the germ; so that with such preservation, proper to a good Regency, a monument of the illustrious natural portent may be handed down to posterity' [our translation].

che venissero custoditi, e curati con tutt'attenzione, conforme ce lo persuadiamo dalla vostra buona condotta; imponendo delle pene pecuniarie, personali, carcerazioni, o altro a' Campieri, Guardiani, e Gabelloti, di esso Bosco, affin di accertarsi l'intento della conservazione di detti alberi, e mantenersi con ciò sempre più viva e recente la memoria di una tale naturale maraviglia'.[23]

The *Ordine* thus establishes a duty of protection. This is fully operative because it is sanctioned by pecuniary penalties and by imprisonment for offenders. Protection is granted by listing forbidden conduct (cutting down the tree, setting it on fire, etc.). The prohibitions, when read through the framework of rights, become the charter of tree's rights: the right not to be cut down, the right not to be set on fire, and the right not to be subjected to any other action that might prejudice it.

In other words, this 1745 Sicilian administrative provision affirms the will to protect and preserve these wonderful trees for the future generations; it reveals the aesthetic criterion according to which this protection is granted; it translates a mere principle into a precise series of legal commands that build up a protective regime around the tree, as if it were a right holder; finally, it provides for effective penalties and sanctions for transgressors. All this is done without elevating the tree to a legal person. This raises the inevitable question: Does it make any difference that the language of personhood is not used?

VI. The 19th Century: The Role of an Enlightened Political Elite

The path thus traced in the Italian 18th century continued in the 19th century. On closer inspection, the aforementioned Sicilian legal-administrative act represents the archetypal legal tool through which the Italian ruling class proceeded to create this kind of protection for what was to be preserved; whether statutes, decrees, administrative regulations, or bureaucratic practices – thus regardless of their exact legal nature – they all identically expressed a legal consciousness that was deeply ingrained in the political-legislative power itself.

All legal measures that were taken on this subject from the epoch of the pre-unitarian States up to the 1948 Republican Constitution are the legal product of a cultured bourgeoisie. Sitting in Parliament and holding governmental offices, this elite has historically found in such legal-administrative acts the tools to as-

[23] 'We order that no damage or harm is done to the aforementioned chestnut trees, or to any other kind of trees, either by cutting, or by fire, or in any other way that might lead to their decay; but that they be guarded and cared for with all due care, in accordance with our conviction of your good conduct; imposing pecuniary, personal, prison or other penalties on the Fieldsmen, Guardians, and Landlords of this Wood, so as to ascertain the intent of the conservation of these trees, and to keep the memory of such a natural marvel ever more alive and fresh' [our translation].

sert its vision of how to provide protection to certain natural elements that thus became part of the national heritage.

The Italian legislators who approved these measures could be described as 'cultured': they had their own sensibility toward nature; a *romantic* and *enlightened* sensibility. On the other hand, this elevated approach to the protection of 'beauty' in its various forms did not resonate with the popular elements of culture for the greatest part of the history we are considering. How is this to be explained?

The Italian cultural and religious tradition historically did not lead to considering nature as a possible object of veneration. Christianity posited a disconnect between the earthly and heavenly dimensions, with a clear preference for the latter. If the heavenly dimension was considered to be pure because it reflected God's perfection, all that belonged to the earthly dimension was considered impure, misguided, and unhealthy. All that formed part of natural reality was subject to decay, and ultimately to death.

This approach, which generally goes by the name of *contemptus mundi*, permeates the entire medieval worldview.[24]

The human being, too, was considered from this perspective. Man's soul was considered immortal and, for this reason, it received all the necessary attention and care, taking a central place in medieval thought; conversely, the human body was given no consideration other than as a weak point of human beings: the weak part that is subject to decay and death. Given these cultural and religious premises, it is not surprising that the lower classes were not interested praising nature nor in considering its protection a vital mission.

Within this cultural paradigm, some famous cases of attention to nature emerged nonetheless. Among these, St Francis of Assisi is the most famous. In his wonderful poem Canticle of the Creatures (1224), he thanks God for nature. Still, in St Francis's prayer, nature appears as something placed by God at man's disposal.[25] This vision, (later known as *anthropocentrism*) will be consolidated by Humanism, which will put man at the center of the world, with nature becoming, indeed, a resource at his disposal.[26]

This brief digression, which has brought us back to the Middle Ages and the Renaissance, highlights how a culture, strongly impregnated with Catholicism, had reasons to ignore nature or make a particular use of it. When it did not ignore it, it presented it as a gift from God to satisfy man's needs.

This approach would change, as we said before this digression, with the Enlightenment and Romanticism. But these two cultural movements affected only

---

[24] The *Contemptus mundi*, (contempt of the world) is a theme of the intellectual life of both classical antiquity and Christianity, one which cultivated a mystical vein and an ambivalence toward secular life. According to this vision, mankind would have to free itself from all interest in material things in order to achieve a state of grace and serenity.

[25] St Francis of Assisi, *Laudes Creaturarum*, 1224.

[26] See, eg, Pico della Mirandola, *Oratio de homine dignitatis*, 1486.

the elites (who, with their sensitivity, created the first pieces of legislation that granted certain rights to nature, such as the one of 1745 that we have just looked at). The development of a similar sensitivity among ordinary people was late to arrive.

Living in strongly anthropized territory, nature was hardly regarded by the population as an 'inviolable temple' but rather as part of an equilibrium that translates into coexistence between nature itself and human communities.

The proclamation of the oldest Chestnut tree as a national monument confirms this pattern. When the State catered for essential preservation needs, it did so by also taking into account the long-term interests of the human communities that lived in the area. Perhaps this is the reason why the Italian approach to the conservation of specific natural resources has been so successful in avoiding conflicts involving the local communities that were affected by preservation measures. This approach did not proclaim the rights of nature but rather affirmed them indirectly, by setting out duties rather than rights and by balancing all the other competing interests as well.

This approach was followed as well in the first decades of the 19th century. For example, in 1826, the Kingdom of the Two Sicilies – the absolute monarchy that ruled Southern Italy – decided to preserve the forests of Montecalvo, San Vito, and Calvi by granting them a special legal status.[27] These measures were the first legal protection in the world established in the form of a *natural reserve*.

After the pioneering initiatives of the previous century, the 19th century witnesses more institutionalized protective measures. The already mentioned establishment of the Montecalvo, San Vito, and Calvi forests as a 'conservation areas' were milestones in this change.

This idea of protecting nature and therefore of granting it certain rights – at least in the form of immunities, whereby it was not to be exploited without limits, not to be attacked by human activities, and not to be entirely subject to the will of the owner – become henceforth an ordinary way of proceeding. All those ingredients that we have already found, by analyzing the Sicilian Order of 1745, must now find a more stable and autonomous form of organization. This form will be finally the *Ente Parco*.

[27] Kingdom of the Two Sicilies Royal Rescript, 23 December 1826.

VII. The Early 20th Century: The Ente Parco Legal-administrative Figure

As early as 1905, Italy protected the ancient pinewood of Ravenna.[28] The initiative came from two eminent jurists and politicians of the time: Luigi Rava,[29] who would later become Minister of Agriculture, Industry, and Commerce, and Giovanni Rosadi,[30] who would later support several other laws on this subject.

Why did the Ravenna pine forest obtaine this special status?[31] This environment is remarkable both for its beauty and in the light of national history. This pinewood was one of Dante Alighieri's favorite places during his exile from Florence. It is also mentioned by many other great Italian poets, including Boccaccio and Pascoli.[32] Its preservation is strongly connected with key features of Italian culture, literature, and national history, thus demonstrating, once again, the aesthetic-cultural conception that permeated the Italian approach to this issue.

A missed opportunity, however, was law no 364/1909 on the protection of artistic beauties, in which Rava himself had included the protection of 'gardens, parks, waters and natural beauties'. The Senate removed this wording from the bill, fearing that it would hinder the country's development.[33]

A second very interesting aspect of this missed opportunity was the proposed introduction of an *actio popularis*, whereby any citizen could take legal action in the interest of the nation's archaeological, artistic, and historical heritage. Sometimes things that seem so modern to us are actually very ancient: this legal mechanism dates back to Roman law, where any citizen could bring judicial pro-

---

[28] Legge no 411, 16 July 1905, 'dell'inalienabilità dei relitti della pineta di Ravenna e della sua costa'. This law is taken as the beginning of Italy's contemporary efforts toward nature's conservationism by Paolo Passanti, *Il diritto cangiante: Il lungo novecento giuridico del paesaggio italiano* (Giuffrè 2019). The careful reconstruction of the historical context of this legislation by Passanti shows the complicated relationships between the raise of industrialism in Italy and attempts at nature's conservation as driven by an aesthetic sensibility in our country.

[29] Luigi Rava was a lawyer and professor of philosophy of law at the University of Bologna; the mayor of Rome (1920–1921); and he also served in the Chamber of Deputies and Senate of the Kingdom of Italy.

[30] Giovanni Rosadi was a lawyer who also served in the Chamber of Deputies and Senate of the Kingdom of Italy. Together with Benedetto Croce, he drafted the first national law on this subject in 1922.

[31] *Riserva Naturale Pineta di Ravenna* is a natural area protected by Emilia-Romagna with a surface of 709 hectares. The origins of the pinewood date back to the Roman Empire. The Romans planted it in order to produce the wood necessary for vessels. Luigi Rava, *La pineta di Ravenna: piccola storia di una grande bonifica* (Ente nazionale industrie turistiche 1926).

[32] The Ravenna pinewood is one of the natural places that have most influenced man's literary imagination. Dante Alighieri's *La Divina Commedia, Purgatorio*, XXVIII mentions it. Giovanni Boccaccio sets the novella of Nastagio degli Onesti in it. (Decamerone, 5th day, novella 8). Poet John Dryden made this pinewood known abroad, and Lord Byron galloped through it.

[33] Roberto Balzani, *Per le antichità e le Belle Arti. La legge n. 364 del 10 gennaio 1090 e l'Italia giolittiana* (Il Mulino 2003).

ceedings in defense of public interests, in particular common goods (*res communes omnium*) such as the air, the waters, the sea, and the coasts.

Rava and Rosadi, would once more take an historic initiative in 1922 when Italy passed its first national law on this subject. While the 1905 Italian act was still aimed at protecting a specific natural element (the Ravenna pine forest), the 1922 act achieved a more ambitious purpose: to establish a countrywide nature conservation paradigm. For this reason, law no 778/1922, (better known as 'Legge Croce' from the name of its proposer, the philosopher Benedetto Croce)[34] is an enormous step forward, crowning the ambition of a countrywide nature conservation project.

The 'Legge Croce' consists of only seven articles, which are clear, concise, and well drafted. The first one is perhaps the most significant for present purposes:

> 'Art. 1. Sono dichiarate soggette a speciale protezione le cose immobili la cui conservazione presenta un notevole interesse pubblico a causa della loro *bellezza naturale* o della loro particolare *relazione con la storia civile e letteraria*. Sono protette altresì dalla presente legge le *bellezze panoramiche*'.[35]

This article introduces the fundamental distinction between 'natural beauty', in the first paragraph, and 'panoramic beauty', in the second. In other words, from this moment on, it can no longer be said that the Italian legislator ignores 'nature' by dealing only with 'landscape'. On the contrary, through the distinction between 'natural beauty', (mountains, forests, a specific tree, a river, etc) and 'panoramic beauty' (which may also be referred to as landscape), lawmakers make explicit the will to protect both, in accordance with earlier practice.

This legislation therefore acknowledges the existence of this dual terminology and takes care to clarify its intention to protect everything the two linguistic formulae together may indicate.

This approach obviates the need to keep the two concepts separate: if the intention is to protect 'everything' that they can together indicate, they might as well be combined in a single term. This is essentially why, after the 'Legge Croce', subsequent legislation (primarily art 9 of the 1948 Constitution) abandoned such distinction and used a single encompassing term: 'landscape'.

For these reasons, the 'Legge Croce' was certainly a milestone in this process, but it also had a major limitation in that it was based solely on aesthetic criteria in deciding when to grant protection.[36]

The year 1922 was also a fateful year for Italy. On 28 October 1922, Benito Mussolini staged the 'March on Rome': the *coup d'état* that put an end to Italy's

---

[34] Legge no 778/1922, 11 June 1922, 'Legge Croce', Per la tutela delle bellezze naturali e degli immobili di particolare interesse storico.

[35] Legge Croce, art 1: 'Immovable objects whose conservation is of considerable public interest because of their *natural beauty* or their *particular relationship with civil and literary history* shall be declared subject to special protection. This law also protects *panoramic beauties*.' [our translation, emphasis added].

[36] Benedetto Croce was one of the most authoritative exponents of 'aesthetics'.

liberal political era and ushered in twenty years of Fascist dictatorship. During this period, the establishment of the great Italian National Parks was accomplished.

Just to mention some of the most notable parks, in 1922, King Vittorio Emanuele III donated to the Italian State 2,100 hectares of his hunting reserve for the creation of a national park: the Gran Paradiso National Park.[37] A few days later, on 11 January 1923, the Abruzzo National Park was created.[38] The Circeo National Park was established in 1934[39] and the Stelvio National Park the following year.[40]

Special consideration should be given to the creation of Circeo National Park. Mussolini had begun the reclamation of the Pontine Marshes in the same period. But given that marshes do not correspond to the traditional criterion of beauty, the decision to protect them as well reveals an awareness for their intrinsic environmental value, which today we would call 'biodiversity'.

All these examples bear witness to the political choice to institutionalize the protection of nature and conservation (which means at least a limited recognition of nature's right not to be exploited, consumed, and disfigured by human activities) through the *Ente Parco* legal-administrative institution. Far from pursuing this aim in a single-minded way, Fascism generally preferred economic and propagandistic initiatives over alternative approaches. The distinctive features of the conservationism it promoted were increasingly characterized by militarization, the use of violence, and administrative centralization.[41]

In addition to the twenty-five national parks mentioned above, Italy also has 105 regional parks, covering a total area of some 12,000 square kilometers, which include areas of land, sea, rivers and their banks, and lakes and their surroundings that are of environmental or naturalistic importance and that are often appreciated for their landscape features and for representing particular local traditions. Regional parks are regulated by Presidential Decree 616/77.[42] The official list of protected natural areas (abbreviated: EUAP) currently includes 134 regional parks, covering a total area of approximately 1.3 million hectares.[43]

National parks and regional parks are therefore the two expressions through which the *Ente Parco* legal-administrative institution takes shape. Whether national or regional, parks enjoy legal personhood. As such, they are for all intents and purposes legal entities, endowed with legal capacity and an ability to act. Hence the park can conclude contracts, take legal action, inherit, receive dona-

---

[37] Regio Decreto-Legge no 1584, 3 December 1922.

[38] Regio Decreto-Legge no 256, 11 January 1923.

[39] Circeo National Park was established by Legge no 285, 25 January 1934 by order of Benito Mussolini, under advice from Senator Raffaele Bastianelli, to preserve the last remains of the Pontine Marshes, which were being reclaimed in that period.

[40] Legge no 740, 24 April 1935.

[41] Roberta Biasillo, Marco Armiero and Wilko Graf von Hardenberg, *La natura del duce. Una storia ambientale del fascismo* (Einaudi 2022).

[42] Decreto del Presidente della Repubblica no 616/77, 24 July 1977.

[43] The list is drawn up and periodically updated by the Ministry for the Environment, Land and Sea.

tions, etc. The park is therefore also the entity that initiates legal action to secure the protection of the nature which it stands for. Which court is to be seized will obviously depend on the type of infringement or damage for which the action is brought. Each park has its own budget, management, and organizational capacity. Furthermore, a park also has, albeit limited, regulatory power. Each park can regulate many aspects relating to its internal life (including a wide range of aspects relating to resource management which also affect local communities' lives, such as wildlife control, controlled logging, etc.).

But where does the idea of a park come from? The concept of 'park' has evolved considerably through history.

The earliest forms of parks were those to which sacred and supernatural functions were attributed. In fact, many ancient cultures used to ascribe specific natural areas for magical purposes. For example, Nemi, a small town in central Italy, owes its name to the Latin *nemus aricinum*, meaning 'grove of Ariccia'. This grove was the site of one of the most famous Roman cults dedicated to Diana Nemorensis. This sacred use of nature was common among the Romans and other archaic Italic groups, but the advent of Christianity swept away this practice at the dawn of imperial Rome.[44]

From the Renaissance onwards a new form of park emerged throughout the Italian peninsula: the hunting reserve park. This new form of park was created for the exclusive use of noblemen and notables. The Bosco Fontana near Mantua,[45] the Boschi di Carrega near Parma,[46] and the numerous Savoy hunting reserves in various valleys of Piedmont and Valle d'Aosta (such as the Gran Paradiso National Park, mentioned above) belong to this typology.

In the 20th century, due to the changed historical and social conditions, these hunting reserves have been totally converted into national or regional parks.

However, even this most recent form of park has a certain number of characteristics that must be made explicit. Unlike other countries, which create parks as if they were inviolable sanctuaries, Italy opts instead for an 'integrated model', one capable of combining nature conservation with local communities' living and development needs. This approach is indeed typical of many other European countries, all of which have to deal with a densely populated territory. The main idea is that human activities can be compatible with the protection of those areas that are significant for their naturalistic constituents.

---

[44] The local cult was rendered famous by the classic study by James G Frazer, *The Golden Bough. A Study in Comparative Religion* (McMillan 1890).

[45] The first historical references to Bosco Fontana date back to the 14th century, when the Gonzagas, the future lords of Mantua, came into possession of this estate, which was covered by a vast and dense forest that immediately became their hunting reserve. It is thanks to the Gonzagas' hunting reserve that this untouched natural corner of the Po Valley has come down to us.

[46] The Boschi di Carrega regional park is a protected area covering an area of approximately 2600 hectares in the province of Parma.

VIII. 'Paesaggio': An Ambiguous Notion in the Italian Constitution

As motioned above, the Italian Constitution has been recently amended. First of all, it must be said that the Italian Constitution of 1948 was the first in the world to expressly protect the 'landscape'. The original text of art 9 Constitution provided: 'La Repubblica promuove lo sviluppo della cultura e la ricerca scientifica e tecnica. Tutela il paesaggio e il patrimonio storico e artistico della Nazione'.[47]

The first twelve articles of Italian Constitution, to which this article belongs, are defined by Constitution itself as 'fundamental principles'. Consequently, the framers' choice to insert among these twelve fundamental articles the protection of 'paesaggio' indicates that this element is one of the inalienable values on which Italian constitutional order is based. Accordingly, this represents the strongest recognition that Italy can give to that element.

The term 'paesaggio', however, needs some clarification. Should it be understood as that part of a territory on which the human gaze rests, or should it have a broader meaning, such as 'nature'? The preferred meaning is probably the latter; otherwise it would not really make sense to protect it in the founding articles of the Constitution.

However, since the most obvious things are often the most misunderstood, we will now look for evidence to support this statement. In this respect, although the Constitution is the founding document of the Italian Republic, which was born out of the resistance to the Fascist regime, it must also be said that, for a variety of purposes, the Constitution drew on aspects of the legislation enacted in previous decades, including legislation approved during Fascism. Art 9 of the Italian Constitution, until its recent amendment, possesses this genealogy. Its wording is in fact the 'efficient synthesis' of previous legislation on the subject.

Art 9 can thus be considered as the 'constitutionalization' of two acts enacted under Fascism, both called 'Leggi Bottai' in reference to the Minister who sponsored them in 1939.[48] They essentially mirror previous legislation on this subject: the so-called 'Legge Rosadi-Rava' (law no 364/1909), concerning cultural heritage conservation, and the previously mentioned the 'Legge Croce' (law no 778/1922). The term 'landscape' should therefore be properly understood in its more comprehensive sense, thus including 'nature'. The constitutional amendment of art 9, approved in 2022, supports this argument and sums up half a century of previous legislation and Constitutional court rulings on this topic. Even if the non-cultural dimension of nature had already previously been recognized, the aesthetic ap-

[47] Constitution of Italy, art 9: 'The Republic promotes the development of culture and scientific and technical research. It protects the *landscape* and the historical and artistic heritage of the Nation' [our translation, emphasis added].

[48] Legge no 1089/1939, 1 June 1939, 'Legge Bottai', concerning cultural heritage conservation ('sulla Tutela delle cose d'interesse artistico o storico'), and Legge no 1497/1939, 29 June 1939, 'Legge Bottai' concerning nature protection ('sulla Protezione delle bellezze naturali').

proach to its conservation prevailed until relatively recently.[49] The constitutional amendment of 2022 makes clear that a change of perspective occurred in Italy at least in this respect.

IX. The Second Half of the 20th Century: Affirmation of Nature's Biological Value and International Cooperation

In the last quarter of the 20th century the global discussion on all these issues gained momentum. Since that time, and as a consequence thereof, also Italy has been involved in the proliferation of international agreements, declarations of principles, and instruments of regulatory harmonization addressing all those aspects that, in general, have the purpose of advancing nature's conservation and protection.

The inordinate consumption of soil; the hydrogeological disasters that Italy has experienced over the years; the level of air and water pollution; the poor conservation of its rivers and costal resources and of many habitats; the clear signs that climate change is taking place now, rather than in the distant future: all of these things speak of the shortcomings and the flaws in what has been done so far to contain the degradation of nature.

In this context, the intrinsic biological value of nature is recognized more and more frequently, as international and European regulatory cooperation gradually overtakes national state policies in the 'design' of nature's protection.

The value of international cooperation for the protection of nature is, therefore, fundamental. Today, more than ever, nature must be seen as a global common good which cannot 'suffer' the costs arising from so many national administrations, which are very often inadequate.

Cultivating this perspective also inevitably leads to a rethinking of the basic assumption that underlies such massive, unsustainable exploitation of nature, namely the consideration of nature as an object rather than a subject, this in an ethical and legal world that still too often fails to conceive of obligations and rights toward any entity that is not recognized as a subject.

[49] The Constitutional Court marked a turning point with its judgment of 17 December 1987, no 641/1987, ECLI:IT:COST:1987:641: '[…] the environment is protected as a determining element of the quality of life. Its protection does not pursue abstract naturalistic or aesthetic purposes, but expresses the need for a natural habitat in which man lives and acts and which is necessary for the community and, for it, for citizens, according to values that are widely felt; it is imposed first and foremost by constitutional precepts (Articles 9 and 32 of the Constitution) for which it rises to a primary and absolute value.' [our translation].

X. The Rights of Nature in Italy: Inaugurating a New Trajectory

Italy's approach to the protection of nature is changing, as the constitutional amendment of 2022 shows. As part of this movement, over roughly the last two decades a few legal scholars have (as discussed in II., above) raised the question whether Italy could or should recognize nature's rights as such, as other countries have done in various ways. Some Italian scholars have not just dedicated research to this theme but have actually pushed for the recognition of such rights, as discussed above.

Currently, these debates are mostly confined to academic circles. So far no visible grassroots movement has taken up this theme to bring it to the forefront of public debate, although other related issues, eg climate change, are at the center of public discussions and gatherings in Italy.[50] No major Italian environmental NGO, such as *Lega ambiente*, has squarely put the issue on its agenda. Italy's Green Party – which represents a little less than 3% of the electorate – has not taken a position on this theme either. Italy is not the home of autochthonous peoples who would strongly advocate such a move. A rare exception in this landscape is the charter published by the Italian NGO *A Sud: Carta europea dei doveri umani per i diritti della natura e del clima*, which takes up the issue, but this document has not received much attention in Italy.[51]

The low level of mobilization of public opinion on the specific theme of the rights of nature is also reflected in the transcripts of the parliamentary debates on the 2022 reform leading to the constitutional protection of biodiversity, ecosystems, and the environment (see III., above). During these debates, nobody opened discussion on the issue or highlighted the positive contribution that the recognition of nature's rights could have for our country. The topic was simply not addressed or mentioned in Parliament.

Despite this situation, several scholarly contributions have opened the path toward a different future for the concept in our country as well.

First of all, the sheer number of Italian publications covering this subject with respect to foreign experiences around the world is remarkable. Over the years, comparative constitutional law scholars and other legal experts have studied the experience of the countries that have recognized the rights of nature in one way or another, with particular regard to Latin America.[52] The debate over those for-

---

[50] In 2018, one Italian municipality – Civita Castellana – adopted a resolution declaring the recognition of the rights of nature. We learned about this from a short video posted on a website that has a low number of views: <www.facebook.com/RightsofNatureEuropeItalia>. There is no other reporting in the media of similar initiatives elsewhere in Italy.

[51] The Charter is published as an appendix to the report: *Towards an EU Charter of the Fundamental Rights of Nature* (n 6). We have been unable to find this document on the website of the *A Sud* NGO.

[52] The important contribution by Carducci has already been cited in n 4: Some of the contributions published in recent years are: Luca Mezzetti, *I diritti della natura. Paradigmi di giuridificazione dell'ambiente nel diritto pubblico comparato* (Cedam 1997); Serena Baldin, 'I

eign experiences now involves other fields beyond the law, including, for instance, anthropology and philosophy.[53]

The attention paid to those foreign experiences is mostly due to the emerging conviction that an anthropocentric approach toward nature fails to meet vital challenges for the future of the entire humanity.[54]

Observing those foreign experiences and trying to learn from them is a form of cosmopolitanism or globalization *à rebours*.

Animated by the conviction that western constitutionalism has failed in this respect, the proposal coming from these contributions is mostly to draw upon 'theory from the South' to criticize the exploitative paradigm that the West has adhered to.[55] The academic literature on the rights of nature in Italy is by and large also acquainted with the US contributions on the topic, which also acted as a catalyst for the debate in Italy as well as in other EU countries (eg Germa-

---

diritti della natura: i risvolti giuridici dell'Etica Ambiental Exigente in America Latina' in Serena Baldin and Morenao Zago (eds), *Le Sfide della sostenibilità: Il buen vivir andino dalla prospettiva europea* (Filodiritto Editore 2014); Serena Baldin, 'I diritti della natura nelle costituzioni di Ecuador e Bolivia' (2014) 10 Visioni Latino Americane 25; Simonetta Fraudatario, 'Dai diritti dei popoli ai diritti della natura. La voce dei tribunali di opinione' (2018) 4 DPCE online 943; Silvia Bagni, 'Le voci (non del tutto inascoltate) della natura nella recente giurisprudenza colombiana e indiana' (2019) 37(4) DPCE online; Silvia Bagni and Giorgia Pavani, 'La comparación jurídica como instrumento epistemológico para el estudio y la investigación en el ámbito de los derechos de la naturaleza, la interculturalidad y los procesos de paz' in Manuel Alberto and Restrepo Medina, (eds), *Interculturalidad, protección de la naturaleza y construcción de paz* (Editorial Universidad del Rosario 2020); Silvia Bagni 'Back to the Future: Building Harmony with Nature in the European Union by Learning from Our Ancestors' in Serena Baldin and Sara De Vido (eds) *Environmental Sustainability in the European Union: Socio-Legal Perspectives*, (Edizioni Università di Trieste 2020); Emma A Imparato, 'I diritti della Natura e la visione biocentrica tral'Ecuador e la Bolivia' (2020) 41(4) DPCE online, 2455. Roberto Cammarata, 'La via latinoamericana ai beni comuni: Il buen vivir ei diritti della natura nel nuovo costituzionalismo andino' in Tecla Mazzarese (ed), *Ricucire il mondo: la necessaria sinergia fra diritti fondamentali e beni comuni* (Brixia UP 2021). With respect to Uganda: Laura A Nocera 'La Natura come soggetto di diritto nell'ordinamento dell'Uganda. Nuovi Autoritarismi e Democrazie: Diritto, Istituzioni, Società' (2021) 3(2) NAD-DIS 188; with respect to New Zealand: Pierre Brunet, 'Rights of Nature and Legal Personality of Natural Entities in New Zealand: Making a Commons' (2019) 38 Giornale di Storia Costituzionale 39.

[53] Flavia G Cuturi (ed) *La Natura come soggetto di diritti: prospettive antropologiche e giuridiche a confronto* (Franco Angeli 2020); Livio Perra, 'I guardiani dei diritti della natura e delle sue entità' (2021) 255(2) Il Politico 183. The entire December 2020 issue of the law journal *Diritto e questioni pubbliche* is dedicated to the topic.

[54] This argument is at the core of many contributions, including: Fritjof Capra and Ugo Mattei, *The Ecology of Law: Toward a Legal System in Tune with Nature and Community* (Berrett-Koehler 2015); see further: Ugo Mattei and Alessandra Quarta, *The Turning Point in Private Law* (Edward Elgar 2019). Both books are available in Italian, and both Mattei and Quarta are Italian.

[55] Serena Baldin, *Il buen vivir nel costituzionalismo andino* (G Giappichelli editore 2019); Lucio Pegoraro, 'Blows against the empire. Contra la híper-constitución colonial de los derechos fundamentales, en búsqueda de un núcleo intercultural compartido' (2021) 47 Teoría y realidad constitucional 97.

ny, France). This is indeed one of the areas of the law in which movements from the North and from the South join forces for a common cause. Within this framework, one can take good notice of, for example, collaborative efforts such as those that led a US non-governmental organization fighting for the recognition of nature's rights to assist in the drafting of Ecuador's constitutional provision recognizing Nature's rights.[56]

Although the debate in Italy on the recognition of the rights of nature has for the moment not really gone beyond the walls of university institutions, a paradigm shift toward the protection of nature is currently being advocated also in the country as a whole. The visible effects of the destruction of nature and its impact on everyday life are, additionally, a real concern in Italy. This is attested to in a number of surveys. The results of these surveys are interesting: between the need to improve the environment and increase employment, concern for the environment for the first time outweighs concern for the employment situation – for 54% of Italians.[57]

A sign of the times is the encyclic *Laudato sì* issued by Pope Francis in 2005, which makes the point: 'Never have we so hurt and mistreated our common home as we have in the last two hundred years'. Not surprisingly, this text has been welcomed by quite a few legal scholars who have taken it as a positive sign of a change of mentality in these matters.[58]

The 20th century is witnessing the crisis resulting from the classical private law paradigm drawing a sharp distinction between subjects and objects.[59] Legal personality is now considered mostly as a mere technique, and legal personality

---

[56] For a detailed account: Mihnea Tănăsescu, 'The Rights of Nature in Ecuador: The Making of an Idea' (2013) 70(6) International Journal of Environmental Studies 846 (noting that the actors involved in the process of constitution making provide somewhat different accounts of the actual role of grassroots movements in this context).

[57] This is the result of an SWG survey carried out by the British embassy in Italy for the UK government: <www.gov.uk/government/news/italians-the-climate-crisis-is-the-greatest-concern>. The survey reveals that the climate crisis is the top concern for the vast majority of Italians, especially among the youngest.

[58] See, eg, Silvia Bagni and Giorgia Pavani, 'La comparación jurídica como instrumento epistemológico para el estudio y la investigación en el ámbito de los derechos de la naturaleza, la interculturalidad y los procesos de paz' in Restrepo Median and Manuel Alberto (eds), *Interculturalidad, protección de la naturaleza y construcción de paz* (Editorial Universidad del Rosario 2020); Luigi Colella, 'Ecologia integrale e «diritto ambientale del cambiamento». L'attualità dell'Enciclica «Laudato Si» (cinque anni dopo)' (2020) 3 Quaderni amministrativi 82; Rodrigo Míguez Núñez, 'La vocazione giuridica di un'enciclica ecologica: note civilistiche a proposito della Laudato si' (2017) 2 Politica del diritto 267; Giovanni Razzano, 'La sfida della responsabilità ambientale nel sistema costituzionale alla luce dell'enciclica Laudato si' (2017) 11 Federalismi.it.

[59] With specific regard to the debate over the rights of nature, see, eg: Rodrigo Míguez Núñez, 'Soggettività giuridica e natura. Spunti per una riflessione civilistica' (2020) 2 Diritto e questioni pubbliche 34; Rodrigo Míguez Núñez, 'Né persone né cose: lineamenti de costruttivi per un rinnovamento concettuale della «summa divisio»' [2021] Rivista critica del diritto privato 359.

can be granted to serve a variety of purposes, including the protection of nature. Nothing in principle could then prevent the recognition in Italy of 'nature' as such as a legal subject, endowed with legal capacity, although, as Guido Alpa recently noted, many will still think that this is a rather fanciful move.[60] Perhaps the hardest thing to change is the deep rooted cultural assumption, one shared by all secularized monotheistic societies, embracing a separation between sacred space and profane space – the former being the temple and the latter the space of politics and the market, a space.

Some of the Italian legal scholars who have devoted substantial research to the rights of nature have thus joined forces in preparing the European Commission Study *Towards an EU Charter of the Fundamental Rights of Nature* (2021), working in cooperation with the UK-based NGO 'Rights of Nature'.[61] As far as Italy is concerned, this study deals with one of the greatest ongoing environmental disasters that is still unfolding in our country, namely the large scale environmental pollution caused by the steelworks of the former ILVA industry based in Taranto. The focus it has received is justified, as the European Court of Human Rights has twice condemned Italy in relation to the plant. Specifically, the Strasbourg Court held that the applicants suffered a violation of their human rights because, inter alia, the Italian legal framework does not provide effective judicial remedies to react to disasters of this nature.[62]

The above mentioned study draws several important conclusions and outlines several initiatives to make a better future possible for us all, including the recognition of nature's rights in formal EU legal documents. These initiatives would help, inter alia, to overcome the fragmented approach to nature's protection that has prevailed so far in Europe. Such recognition would also help avoiding a fundamental error, namely further regulation based on balancing:

> '[…] between the "value" of Nature as "good" or "function" and other human values (political, social, economic values). This balance is regarded as "legitimate", only because it conforms to the plurality of human interests (procedural and substantive) and not to the functioning rules of ecosystems.'[63]

The study makes it clear that the rights of nature label brings with it, above all, a new legal method of approaching the conservation of the Earth's ecosystems, one which also involves the development of new moral values and sensitivities. A new path is thus open to us, not only as lawyers but also as citizens.

---

[60] Guido Alpa, 'Dalla tutela dell'ambiente al riconoscimento della «natura» come soggetto di diritto. Una rivisitazione delle categorie del diritto civile?' (2020) 11 Rivista Italiana per le Scienze Giuridiche 34, Spanish translation in (2022) 43 Revista de Derecho Privado 73.

[61] See (n 6) for a listing of the involved Italian authors; collaborating authors from Rights of Nature were Mumta Ito, Lindsey Spinks, and Paul Powlesland.

[62] ECtHR 5 May 2022; 4642/17, *Ardimento and Others v Italy*, ECtHR 24 January 2019, 54414/13 and 54264/15, *Cordella and Others v Italy*; ECtHR 25 January 2025, 51567/14, *Cannavacciuolo and Others v Italy*.

[63] European Economic and Social Committee and others (n 6) 166.

Japan

Bringing the Alternative View to Fruition

Keisuke Mark Abe[*]

I. Introduction

Surrounded by ocean and adorned with lakes and rivers, respect for nature has always been a part of Japanese culture. Thousands of mountains and forests have since antiquity been considered to be gods incarnate; numerous animals, including foxes, deer, and monkeys, have been enshrined and are worshipped across the country. In the city of Nara, the ancient capital, the primeval forest of Mount Kasuga, covering some 298 hectares [approximately 736 acres], sits barely touched by humans, as hunting and logging have been forbidden since the Emperor Ninmyo designated the entire mountain as a sacred area in the ninth century. Roughly sixty kilometers off the northern coast of Kyushu lies the island of Okinoshima, a UNESCO World Heritage site, which rises sheer from the sea and where no one is allowed to set foot without the permission of the Munakata Taisha Shrine, the landowner of the island. Nothing, not even a leaf of grass, may be removed from this pristine island where gods are said to dwell. Similar examples abound.

The roots of these intrinsic and fundamental beliefs in nature, in something of a contrast to the adopted Western legal system, may be fertile enough to allow the concept of rights of nature to bear fruit in the near future. Yet at present, one may think of it as cultivating seeds which are just beginning to sprout. Although various creative legal theories have been crafted and asserted, the Japanese judiciary, generally disinclined to take independent innovative action, has so far adhered to a traditional formalistic approach in dealing with litigation involving

---

[*] The manuscript was finalized in October 2025. The assessment of literature and case law, including internet materials, reflects the legal situation at that time.

environmental issues. As a result, rights of nature are not currently recognized in the Japanese legal system; however, one can point to a few small but notable developments relevant thereto.

II. *Toshogu v Minister of Construction*[1]

Although there have been a few instances where judicial decisions placing emphasis on the interest of the general public in enjoying fine landscapes led to the suppression of bridge or building construction, probably the only successfully-litigated Japanese case which approaches the concept of rights of nature in substance, if not in form, is one involving the giant cedar tree named 'Taro' in the precincts of Toshogu, the location where Ieyasu Tokugawa, the founder of the Tokugawa Shogunate [the Edo period], is enshrined. Litigated during an era of rampant environmental destruction, it is a case that was decided by the court of first instance in 1969 and affirmed by the Tokyo High Court in 1972.

In 1964, the Governor of Tochigi Prefecture applied to the Minister of Construction for certification of a business plan to widen a national highway, which the Minister issued shortly thereafter, in order to accommodate the increased number of tourists during the first Tokyo Olympics. The governor then requested a judgment of the regional administrative commission, a quasi-judicial body, to acquire a piece of land owned by Toshogu, a religious organization registered as a corporation. The commission approved the acquisition as proposed by the governor. This public construction project was scheduled to be conducted pursuant to the Expropriation of Land Act.[2] However, Toshogu immediately challenged the widening of the highway in question because the work involved the logging of fifteen cedar trees, including Taro, estimated to be over 500 years old. The plaintiff Toshogu sought revocation of the ministerial certification, maintaining that it would be unlawful to lumber these trees.

The focus of the litigation was on whether the development plan was such that it would contribute to an 'appropriate and reasonable use of land' as provided in the Expropriation of Land Act. The court of first instance, the Utsunomiya District Court, held that it would not, pointing out that the necessity of widening the highway was highly questionable in respect of the damage to the scenic, religious, and historical value of Toshogu, which is located in the protected Nikko National Park. The cultural value of the land concerned was irreplaceable in the sense that, once lost, it would be impossible to restore. The intended construction project, on the other hand, was one of four possible plans to inexpensively cope with expected traffic congestion; the need in question could be equally met by any of the other plans, all of which the governor himself had drawn up and

---

[1] Tokyo Koto Saibansho [Tokyo High Court] 13 July 1973, *Toshogu v Minister of Construction*, 710 Hanrei Jiho 23 (1973) <https://www.courts.go.jp/assets/hanrei/hanrei-pdf-18123.pdf>.

[2] Law No 219 of 1951.

examined before reaching the conclusion that the plan involving the logging of Taro and other trees would be easiest to carry out. Expropriating the land and damaging its cultural value for the purpose of widening a road, when in fact there were other options, could not be seen as contributing to an 'appropriate and reasonable use of land', the district court concluded.

The Tokyo High Court [subordinate only to the Supreme Court of Japan] unanimously affirmed this ruling, stating that, notwithstanding the government's discretion in making administrative decisions on matters such as land appropriation, the exercise of its discretion is unlawful if there is an error in the decision-making process. This applies, for example, when the government disregards the most significant matters and takes into consideration inconsequential ones. Here, this was clearly the case because the Minister had unduly made light of the irreplaceable cultural value of the land in question and the preservation of the surrounding environment, while he had overestimated the importance of widening the road in dealing with the surging traffic during the Tokyo Olympics. With words of stern reproof for the Minister's inadequate decision-making, the High Court invalidated his certification of the business plan and also overturned the administrative commission's judgment allowing the acquisition to move forward, effectively sparing Taro and the other cedar trees from destruction. Due to the unequivocal decision of the appellate court, the case was not taken to the Supreme Court.

III. *Amami Black Rabbits v Governor of Kagoshima Prefecture*[3]

Notwithstanding its practical importance, *Toshogu v Minister of Construction* was, of course, not a case recognizing the rights of nature. The dispute being between the landowner Toshogu and the government, there was no discussion as to whether Taro should be granted legal personhood. Neither the district court nor the appellate court suggested that the cedar trees could possess inherent rights distinct from those of the property owner. This was because the idea that an ecosystem or a part thereof could be a holder of rights on its own, capable of suing for its own preservation, was almost unknown in legal circles at that time.

In the 1970s, however, the efforts of environmentally minded citizens and organizations started to take shape. One of the most symbolic developments was the foundation of the National Union for Conservation of Nature, a network of over seventy environmental organizations across Japan. The Union's management was administered by the secretariats of two major conservation groups; the Nature Conservation Society of Japan, incorporated in 1960 as the first of its kind in the nation, and the Wild Bird Society of Japan, formed in 1934 by ornithologists. The inaugural meeting of the Union in 1971 was attended by the Secretary

[3] Kagoshima Chiho Saibansho [Kagoshima District Court] 22 January 2001, *Amami Black Rabbits v Governor of Kagoshima Prefecture*, <https://www.courts.go.jp/assets/hanrei/hanrei-pdf-15675.pdf>.

of the Environment Agency, also created that same year [a precursor to the Ministry of Environment]. Since its inception, the Union has been a zealous advocate for the rights of nature, raising people's awareness about the criticality of environmental conservation. Although the coalition of politically diverse groups disbanded shortly thereafter, numerous entities previously or currently under the umbrella of the Union have continued to take action to prevent nature-destructive business and government activities over the past several decades. For instance, the Nature Conservation Society of Japan held the Beech Symposium in 1985 to protect the virgin forest biosphere of the Shirakami mountain range, following which citizen letters calling for the halt of the forest road construction therein flooded the office of the Governor of Aomori Prefecture, who ultimately decided to suspend the construction in 1987; the Shirakami Mountains were later designated as one of Japan's first UNESCO World Heritage sites in 1993. To take another example, the Nature Conservation Society of Japan has recently met with the Ministries of Environment and Defense to discuss the construction of a United States Marine Corps base in Henoko, Okinawa, which might interfere with the presence of rare marine life (discussed further in Part IV, below).

Researchers in various fields began to investigate the concept of rights of nature as well, put forward most notably by the American philosopher Christopher Stone, whose article, 'Should Trees Have Standing? – Toward Legal Rights for Natural Objects',[4] influenced the dissenting opinion of Justice William O. Douglas in *Sierra Club v Morton*,[5] decided by the United States Supreme Court in 1972, in which Douglas argued for the right of 'environmental objects to sue for their own preservation'. Both of their arguments have been frequently cited by Japanese commentators exploring ways to overcome technical legal barriers to achieve conservationist goals. Takemichi Hatakeyama, who has pioneered the field of comparative environmental law, proposed that, in view of effective judicial control over unbridled development as well as the high level of public participation in forest management in the United States, Japan should urgently establish procedures to incorporate 'diverse voices' into environmental law and politics.[6] On a similar but different note, Hitoshi Aoki, an animal law specialist, elaborated on Jean-Pierre Marguénaud's thesis, 'La personalité juridique des animaux', in which the French law professor had suggested that animals be treated as legal persons, and concluded that it is, in fact, theoretically possible to extend the concept of personhood to animals under Japanese law, although this is arduous to put into practice.[7]

---

[4] Christopher Stone 'Should Trees Have Standing? – Toward Legal Rights for Natural Objects' (1972) 45 SCalLRev 450.

[5] US Supreme Court *Sierra Club v Morton* 405 US 727 (1972).

[6] Takemichi Hatakeyama, *Amerika no kankyo hogo ho* [Environmental Protection Law in the United States] (Hokkaido University Press 1992) 309.

[7] Hitoshi Aoki, *Dobutsu no hikaku ho bunka* [A Comparative Perspective on Animal Law and Culture] (Yuhikaku Publishing Co. 2002) 268.

It is then little wonder that lawsuits began to be filed by grassroots organizations on behalf of environmental proxies such as endangered species over the following decades. By far, the most well-known Japanese case relating to the rights of nature – a case which was decided by the district court in 2001 but which still reflects the state of Japanese law today – is *Amami Black Rabbits v Governor of Kagoshima Prefecture*, usually referred to as the *Amami Black Rabbits* case. There, environmental protection groups and local residents sued the governor over the planned construction of two golf courses on Amami Oshima, the largest island in the Amami Archipelago in southwestern Japan, alleging that the right to life of Amami black rabbits, as well as that of Amami thrushes, Lidth's jays, and Amami woodcocks (all threatened or near-threatened and officially protected species native to Amami Oshima) would be infringed by the construction. The plaintiffs sought revocation of the prefectural permission to develop the parcels of land in question; after originally filing in the name of these animals, following the dismissal of the first complaint as not being in compliance with a standing court order to state the name and address of a specific individual or corporation, they subsequently filed in their own names on behalf of the animals. They did so on the grounds that, under the Japanese Forest Law,[8] the governor is not authorized to approve the development application if the development significantly degrades the environment in the area.

The Kagoshima District Court dismissed the suit, holding that the plaintiffs had no legal interest and therefore no standing to sue under the Administrative Case Litigation Act,[9] as their rights or otherwise legally protected interests were not infringed upon or on the verge of being so. Seeing the threshold question to be whether those conducting nature observation, conservation, or recreation activities on the planned golf course sites had standing to challenge the governor's approval of the forest land development, the District Court noted that, even if the plaintiffs had an interest in such activities, they were activities that could be freely conducted by any member of the public, and therefore the plaintiffs were not in any special legal relationship to the forest concerned; they were thus not in any position to contest the governor's decision. Accordingly, the lawsuit was terminated without any examination as to the legality of the development permit for the golf course construction.

In a final observation, however, the three judges of the District Court highlighted the significance of rights of nature. They admitted that the Japanese legal system limited the subjects of rights to individuals and corporations, meaning that nature itself, or animals and plants, which the plaintiffs claimed were valuable and precious to humankind, could only be the objects of rights, and reiterated that the plaintiffs were not entitled to bring this lawsuit in accordance with the existing legislation. Having said that, however, the judges clearly pointed out that the concept of rights of nature as raised by the plaintiffs 'has presented an

[8] Japanese Forest Law, Law No 249 of 1951.

[9] Administrative Case Litigation Act, Law No 139 of 1962.

extremely difficult but inescapable question of whether it will continue to be acceptable to adhere to the current legal framework, which is focused on remedies for invasions of personal and corporate interests.'

This case attracted extensive media attention, particularly due to the plaintiffs' attempt to bring a lawsuit on behalf of endangered native animals. As major newspapers picked up the story, the public began to understand the momentousness of this unusual litigation. Bipartisan support began to emerge among politicians, who put pressure on the Ministry of the Environment and the Agency for Cultural Affairs. In response to this groundswell of attention, the Agency for Cultural Affairs ordered a reexamination of the tracts of land in question on the basis of the Law for the Protection of Cultural Properties,[10] which revealed the existence of significantly more burrows inhabited by Amami black rabbits than previously thought. Due to vehement public opposition, the development of the golf courses was halted; 'we have in essence prevailed', commented Takaaki Kagohashi, the executive director of the plaintiffs' legal team.[11] Kagohashi emphasized that, despite technically a defeat, the litigation was meaningful, as it played the role of a rallying point facilitating the gathering of sympathizers from around the nation under the flag of protecting indigenous animals.

At the same time, however, the *Amami Black Rabbits* case exposed the limits of Japanese law: The threshold that a plaintiff must satisfy in order to be granted standing in administrative disputes is set at a higher level than in many other countries. This has been noted by academics and practicing attorneys specializing in environmental law. Suggestions have been made to graft the American theory of standing, under which a number of threatened or endangered species have been accorded de facto standing to sue in their own right, onto the existing Japanese legal system, but these efforts have been unsuccessful so far. Numerous lawsuits have been brought by animals and plants, including pikas [whistling hares], flying squirrels, goshawks, sea turtles, sweetfish, mudskippers, fiddler crabs, and great burnets [herbs of the rose family], as well as by the Isahaya Bay in western Japan, but the courts have dismissed every one of them on the grounds that the current law does not confer animals and plants, or nature itself, the capacity to be a party in court proceedings. In other words, they can neither sue nor be sued. Additionally, conservation groups and residents of areas affected by development projects have been generally denied standing to pursue relief in court because of the lack of legally cognizable interests.

Bean Geese v Governor of Ibaraki Prefecture[12] is another district court decision illuminating this point. When a construction plan for a new highway connecting the outskirts of Tokyo arose in the 1990s, the population of bean geese, mi-

---

[10] Law for the Protection of Cultural Properties, Law No 214 of 1950.

[11] *Amami no Kuro Usagi, Henoko Dugong* [Amami Black Rabbits and Dugongs in Henoko], 20 September 2020, <https://www.bengo4.com/c_1017/n_11754/>.

[12] Mito Chiho Saibansho [Mito District Court] 20 February 1996, *Bean Geese v Governor of Ibaraki Prefecture*, 957 Hanrei Taimuzu 194 (1998).

gratory birds that arrived on the southern coast of Lake Kasumigaura in Ibaraki Prefecture, was a meager fifty. In a lawsuit brought by bean geese, local residents, and a non-profit association, it was maintained that the governor was acting unreasonable by failing to designate the bean geese destination as a wildlife sanctuary. The Mito District Court dismissed the bean geese's complaint as they did not have the requisite capacity to be a party to a court proceeding, while also dismissing the claims of the other plaintiffs, stating that the governor's failure to designate the area as a wildlife sanctuary was not extremely unreasonable or unlawful. The Tokyo High Court affirmed.

However, the manner in which the environmental assessment was conducted in this case was found by the district court to have 'lacked adequacy' given that a biologist who had cautioned that bean geese would be unlikely to use alternative sites as foraging and resting areas was refused an opportunity to be heard at the prefectural urban planning council. In addition, the district court stressed the importance of preserving the flocks in question, pointing out that biodiversity is better ensured by maintaining local animal populations and ecosystems as a whole, rather than just looking at each individual species.

The judiciary's reluctant approval of land development affecting wildlife habitats not only vividly reveals that there remain serious technical hurdles to be crossed before rights of nature can be protected in Japan, such as amending rules and procedures governing litigation practice, but also shows that a growing number of judges are skeptical of the traditional legal theory which falls short of acknowledging that nature and natural objects have inherent standing to assert their rights. Reflecting on the *Amami Black Rabbits* case and the *Bean Geese* case, Tadashi Otsuka, one of the foremost authorities on Japanese environmental law, concludes that granting *locus standi*, or the capacity to bring an action to court, to nature or natural objects is almost impossible under the current law of Japan, although he hastens to add that the awareness surrounding the concept of rights of nature may prove strategically beneficial in the long run by helping environmental groups to persuasively assert standing to sue while at the same time acting as a catalyst for new environmental protection legislation.[13]

IV. *Center for Biological Diversity v Esper*[14]

In 2008, the Japanese parliament enacted the Basic Act on Biodiversity.[15] Sponsored by a cross-party group of lawmakers, the bill passed both houses unanimously. Although there is no reference to the rights of nature, it is a landmark piece of legislation in that it is the first comprehensive statute aimed at conserv-

---

[13] Tadashi Otsuka, *Kankyo ho* [Environmental Law] (3rd ed, Yuhikaku Publishing Co. 2010) 64–65.

[14] *Center for Biological Diversity v Esper* 958 F.3d 895 (9th Cir. 2020).

[15] Basic Act on Biodiversity, Law No 58 of 2008.

ing biodiversity, including wildlife, their habitats, and their connections to a variety of ecosystems. It has provisions promoting public participation in both the design and evaluation stages of environmental policies and, based on the proposals of various NGOs, it also features provisions directing the government to implement strategic environmental assessments beginning in the planning phase of potentially harmful development projects, a point of time earlier than in preceding environmental assessments.

More specifically, the Basic Act on Biodiversity can serve as a vehicle to strengthen the enforcement of numerous already existing laws related to conservation, such as the Wildlife Protection and Hunting Management Law,[16] the Law for the Conservation of Endangered Species of Wild Fauna and Flora,[17] and the Invasive Alien Species Act,[18] thereby providing citizens and organizations with a basis to request the government's revision of these laws when necessary. It provides for the sustainable use of natural resources and also highlights the importance of adopting precautionary measures prior to the initiation of environment-threatening projects. Through ensuring that public opinion is taken into account, this act increases the chances of effectively implementing important policies that are practiced widely internationally but which have not yet been introduced in Japan; it may well be an essential first step towards the realization of rights of nature. Despite this, the Basic Act on Biodiversity remains a declaration of principles, with the exception that it obligates the national government to take measures to formulate a 'National Biodiversity Strategy' and requires local governments to make efforts to prepare 'Regional Biodiversity Strategies'.

Marine life habitat destruction also illustrates the inadequacy of the present legal framework, strongly indicating that many more legislative efforts are needed before the rights of nature paradigm can be embraced. The circumstances are as follows: Japan's coastal ecosystems have been severely damaged by land reclamation and other development activities over the past several decades. The loss of seaweed beds and coral reefs has deprived aquatic organisms of their habitats and the resources necessary for survival, reducing the self-cleansing capacity of marine environments. The Aichi Targets, agreed to at the Conference of the Parties to the Convention on Biological Diversity held in Nagoya, Japan, in 2010 and attended by 180 countries, included a commitment to minimize anthropogenic pressures on coral reefs and other vulnerable ecosystems affected by climate change or ocean acidification, but this has not yet been fully achieved, at least not in Japan.

Against this background, and pursuant to a decision jointly made with the United States Department of Defense, the Japanese government is presently carrying out reclamation work on roughly 160 hectares [approximately 395 acres]

---

[16] Wildlife Protection and Hunting Management Law, Law No 88 of 2002.

[17] Law for the Conservation of Endangered Species of Wild Fauna and Flora, Law No 75 of 1992.

[18] Invasive Alien Species Act, Law No 78 of 2004.

in Henoko Oura Bay in Okinawa Prefecture in order to build a replacement facility for the aging US Marine Corps base. Located on the coast of a subtropical island, the bay is known to be home to over 5,300 species, including 262 endangered species, one of which, the dugong, a herbivorous marine mammal similar to a manatee, is critically endangered. The reclamation plan has therefore been strongly opposed from the beginning by citizens, environmental groups, and numerous academic societies.

Amidst Japan's uninspiring judicial climate described in Part III, above, conservation-minded individuals and organizations filed a unique case in the United States District Court for the Northern District of California in 2003, apparently driven by hope that American courts might be more likely to listen to their claims with open ears. They challenged the construction of aircraft runways on landfill in Henoko Oura Bay, as such construction was likely to damage or even destroy feeding grounds and habitat for the dugong. The plaintiffs sought declaratory and injunctive relief based on a provision in the National Historic Preservation Act [NHPA],[19] a US statute stipulating that, prior to the approval of any federal undertaking outside the United States that may adversely affect protected property, the head of the federal agency having jurisdiction over the undertaking must 'take into account' any adverse effect on the property.

The plaintiffs' expectation turned out to not be baseless. The first stage of their litigation was a success, because, after a lengthy process involving a court order in 2008 directing the Department of Defense to comply with the NHPA and the Department of Defense's eventual report in 2014 specifying that it had discharged its NHPA obligations, the United States Court of Appeals for the Ninth Circuit, the intermediate appellate court, ultimately sided with the plaintiffs as to the justiciability of their claims. The Court of Appeals held that the plaintiffs did indeed have standing to pursue declaratory and injunctive claims and that the issues were not political questions outside the realm of judicial review. This means that they were adjudged to have met the threshold to move on to the merits, which was a result that could not likely have been achieved in a Japanese court. The citizens' and organizations' wishes were fulfilled up to this point.

The plaintiffs, however, ultimately lost the case because the Department of Defense's action was judged to be within the bounds of its discretion. Rather than focusing on technical procedural issues, the United States District Court considered the case on its merits, but its 2018 ruling found that the Department of Defense had complied with the procedural requirement to 'take into account' the potential adverse effects of its proposed action on a historic property, the dugong, and that its decision was not arbitrary or capricious as there was sufficient evidence that the presence of the dugong in the area was sporadic. The Court of Appeals subsequently upheld the District Court's conclusion that the Department of Defense had adequately taken into account the environmental impact of the reclamation and runway construction project in question. According to the Court of

[19] NHPA, 54 USC. §307101(e).

Appeals, the Department of Defense's decision to consult with the Japanese national government, but not with the plaintiffs and local community members, was not unreasonable, as the NHPA delegates to federal agencies 'the specific decisions of which organizations, individuals, and/or entities to consult (or not consult) and the manner in which such consultation occurs'.

Numerous lawsuits have been filed in Japanese courts as well over the relocation of this US military base, raising various issues such as whether the approval of the landfill work by the Governor of Okinawa Prefecture was duly granted or revocable by succeeding governors, but all have so far been dismissed. As of 2022, more than five years after the project began, only about thirty percent of the planned landfill area has been reclaimed. Furthermore, with the discovery of soft ground on the seafloor having necessitated large-scale soil stabilization, the construction period has now been re-estimated at twelve years instead of the original eight. The protection of dugong habitat in Henoko Oura Bay is expected to continue to be a major point of contention in every local election in Okinawa for the foreseeable future.

V. Conclusion: Breaking Away from Fossilized Jurisprudence

Let us return to the story of the Amami black rabbit, a relict species that arrived and settled in Japan millions of years ago, when it was part of the Eurasian continent. This living fossil now only inhabits Japan's Amami Islands, long after closely related species died out in mainland Asia. A phenomenon similar to this occasionally takes place in the world of law as well.

The idea that only humans and entities with legal personhood can be subjects of rights was imported from Western jurisdictions when Japan embarked on its modernization drive in the 1860s. Additionally, transplanted continental doctrines that confer broad discretion to public authorities to approve or disapprove development plans, as well as associated complex litigation procedures, seem to be remnants of an industrial past, embodying the essence of nineteenth-century legal formalism, disconnected from the traditional Japanese outlook on nature. It may very well be the case that the useful life of those rules and doctrines has long since passed.

In light of the recent trend that sees more and more countries around the world beginning to recognize rights of nature, it appears to be high time for Japanese lawyers and legislators to squarely face the 'inescapable question' of whether or not to continue to adhere to the existing legal framework as noted by the judges in *Amami Black Rabbits v Governor of Kagoshima Prefecture*; otherwise Japanese law may soon be a specimen of an archaic mindset that is growing extinct in other jurisdictions, one characterized by an inflexible dichotomy between humans and nature. Fortunately, the various litigation attempts of private individuals and organizations examined in this report seem to provide a strong indication that the underlying legal consciousness of the Japanese general public is developing towards an acknowledgement of rights of nature. There is, therefore, hope that courts will begin to entertain lawsuits brought by nature or natural objects in the not too distant future.

However, in order to actually realize rights of nature, including the right to life of endangered species, the Japanese legal system should not be so outdated that it may be perceived as 'relict' in the world of comparative law. A legal system that is not able to protect living fossils, let alone nature as a whole, is a system that itself must remain a living fossil.

The Netherlands

Rights of Nature, Globalization, and Legal Pluralism

Phillip Paiement[*]

I. Ongoing Initiatives to Introduce Rights of Nature

There are currently no recognized rights of nature in the Dutch legal system. They have not been introduced through legislation, constitutional provisions, or through case law. As far as the author is aware, they have never been pursued in litigation within the Netherlands in a noteworthy manner.[1] In other words, rights of nature do not exist in the *lex lata*. As such, the following report will, instead, as-

[*] The manuscript was finalized in March 2024. The assessment of literature and case law, including internet materials, reflects the legal situation at that time. I would like to thank Sophie van Wijk for her research support in the preparation of this report. This publication has been made possible by the TransLitigate project (ERC-2021-STG 101039648), funded by the European Union. Views and opinions expressed are, however, those of the author(s) only and do not necessarily reflect those of the European Union or the European Research Council. Neither the European Union nor the granting authority can be held responsible for them.

[1] By contrast, in Belgium, a legal system that shares many common characteristics with the Dutch legal system and with which there is considerable exchange of legal professionals and scholars, recent climate change litigation attempted (unsuccessfully) to have the personality and rights of 'an Italian alder and 81 other trees' recognized through their petition as intervening parties in a lawsuit against the Belgian state. See Francophone First Instance Court of Brussels, 4th chamber, 17 June 2021, *Klimaatzaak ASBL v. Belgium*, no 2015/4585/A June. To the author's knowledge, this was the first attempt to recognize rights of nature via litigation in the Belgian legal system.

sess ongoing initiatives and campaigns to introduce rights of nature to the Dutch legal system.

There is an ongoing campaign to develop a rights framework for the Dutch portion of the Wadden Sea,[2] as well as more minor appeals for rights to be created for smaller natural areas, such as the Maas River,[3] the Biesbosch wetlands,[4] and the Amelisweerd forest on the outskirts of the city Utrecht.[5] Interestingly, these campaigns have been inspired by the institutional model of 'waterships' (*waterschappen*, sometimes translated as 'water boards'), democratically-formed administrative bodies responsible for water management in the Netherlands with historical origins tracing back to at least the 13th century.[6] Building on this concept, the campaigns seek the development of 'natureships' (*natuurschappen*) to function as public-law legal persons that are statutorily tasked with protecting the rights of the natural areas covered by the natureship (eg the Dutch portion of the Wadden Sea).[7]

It is very unlikely that these campaigns will be successful in enacting legislation to create rights of nature for these various natural areas. So far, only one national political party, the Party for the Animals (*Partij van de Dieren*) has sup-

---

[2] Tineke Lambooy, Jan van de Venis and Christiaan Stokkermans, 'A Case for Granting Legal Personality to the Dutch Part of the Wadden Sea' (2019) 44 Water International 786; Jan van de Venis, Tineke Lambooy and Arjan Berkhuysen, 'De Waddenzee Heeft Ook Rechten En Die Horen in de Grondwet Te Staan' *Trouw* (Amsterdam, 20 January 2018); Laura Burgers and Jessica den Outer, *Rights of Nature: Case-studies from Six Countries* (Embassy of the North Sea 2021) 52.

[3] Tineke Lambooy, 'Rechten van de natuur wereldwijd' (*Lab Toekomstige Generaties*, 18 December 2020) <https://labtoekomstigegeneraties.nl/2020/12/18/rechten-van-de-natuur-wereldwijd/>; Jessica den Outer, 'A Plea to Recognize the Legal Rights of the Maas River' *Earth Island Journal* (16 April 2021) <www.earthisland.org/journal/index.php/articles/entry/a-plea-to-recognize-the-legal-rights-of-the-maas-river/>.

[4] Tineke Lambooy, 'Nature Needs Legal Rights to Really Protect Biodiversity' (*Open Access Government*, 28 January 2021) <www.openaccessgovernment.org/protect-biodiversity/102506/>.

[5] Jan van de Venis, Tineke Lambooy and Arjan Berkhuysen, 'Amelisweerd is de ideale proeftuin om de werking van rechten voor de natuur te verkennen' *de Volkskrant* (Amsterdam, 22 April 2021).

[6] For a basic introduction to the Dutch *waterschappen* in English, see Herman Havekes and others, *Water Governance: the Dutch Waterschap Model* (Unie van Waterschappen, Drukkerij Opmeer 2008). Tineke Lambooy, Jan van de Venis and Christiaan Stokkermans, proponents of integrating Rights of Nature for the Wadden Sea via the institutional structure of *waterschappen* institutions, acknowledge that the idea originates from a 1976 advisory report on the future governance of the nature area: Commissie Toekomstige Bestuursstructuur Waddengebied (Commissie Staatsen), *Bestuur en Beheer van het Waddengebied* (Harlingen, January 1976). As such, this specific Rights of Nature project builds on existing legal and policy discussions rooted in a longer, pre-Rights of Nature discourse on how to best institutionalize the protection of environmental interests in the Netherlands. This example offers the most specific vision of how Rights of Nature concepts might be adapted to 'fit' among existing legal institutions in the Netherlands.

[7] Lambooy, Venis and Stokkermans (n 2) 786.

ported the initiative, but they hold just six of the 150 seats in the lower chamber. The Party for the Animals' proposal to create legal personality for the Wadden Sea was discussed during a 2018 parliamentary session with the Minister of Infrastructure and Water Management, but it is yet to gain further political support.[8]

It is important to note the timing of the campaign to establish personality and rights for the Wadden Sea. This campaign coincides with the 2017 Dutch government's decision to create a new public management authority for the Wadden Sea. The pursuit of legal personality and rights for the Wadden Sea represents a strategic attempt to influence the nature of this new public management authority, and these suggestions were debated by the lower chamber of Parliament with respect to this issue. Ultimately, in 2020, the new authority (*Beheerautoriteit Waddenzee*) was established as an entity without individual legal personality and which is to function as a coordinative forum between the two ministries that work most closely on governance of the Wadden Sea (the Ministry of Agriculture, Nature, and Food Quality and the Ministry of Infrastructure and Water Governance) as well as the three provinces whose coasts include the Wadden Sea (North Holland, Friesland, and Groningen).[9] Without legal personality, a primary responsibility to protect the ecological interests of the Wadden Sea area, or a role for representatives of environmental NGOs or local residents, the current organizational format fails to reflect what the campaigns for rights of nature had hoped to achieve.

Given the unlikelihood of a legislative pathway for creating rights of nature in the Netherlands, a separate initiative has been spearheaded by The New Institute (*Het Nieuwe Instituut*), a cultural center focused on architecture, design, and planning. Its 'zoöp' initiative – a portmanteau of the Greek '*zoë*' (meaning 'life') and 'cooperation' – offers an organizational model that would functionally create rights of nature in the private sphere of an economic or social organization.[10] Operating according to the logic of a certification scheme, The New Institute's Zoönomic Foundation seeks to certify a diverse range of types of organizations – for-profit, non-profit, public, etc. – that wish to become zoöps. Conditions for certification primarily include: (i) installing a board member in the zoöps governance body, selected from a list of individuals compiled by the Zoönomic Foundation, whose role is to represent the interests of non-human life within the 'operational sphere' of the zoöp and (ii) conducting an annual cycle of evaluations and improvements of the ecological integrity of the zoöp. In effect, the zoöp initiative seeks to pragmatically create mechanisms whereby the interests of non-human

[8] Kamerstukken II 16 February 2018, *Waddenzee Beleid: Verslag van een algemeen overleg*, no 158, 2017/2018, 29658, 9, 18, 21.

[9] 'Bestuursovereenkomst Beheerautoriteit Waddenzee, Ministerie van Landbouw, Natuur en Voedselkwaliteit' *Staatscourant* (no 36033, 8 July 2020).

[10] Het Nieuwe Instituut, 'Zoöp' <https://zoop.hetnieuweinstituut.nl/>. Details concerning the zoöp model presented in this report were acquired from the website of *Het Nieuwe Instituut* as well as in private conversation with the initiator of the zoöp project. See also Burgers and den Outer (n 2) 52. Further public information about the initiative is not available at this time.

lifeforms are represented in the governance bodies of various types of organizations, in acknowledgment of the unlikelihood of legislative pathways for securing legal personality at the national level. It constitutes a private law approach to securing rights of nature.[11] Importantly, it is limited to securing the representation of ecological interests in a future-oriented manner, as zoöps do not incur liability for historical damage caused to the environment.[12] Currently a half dozen organizations (proto-zoöps) are pilot testing the governance model, and The New Institute is investigating the possibility of transplanting the model to other national jurisdictions in Europe.

II. Legal and Political Operators

The campaigns for creating rights of nature in the Netherlands have largely been raised by civil society organizations and academic researchers. Civil society organizations that are actively campaigning for the establishment of rights of nature in the Netherlands include the Embassy of the North Sea,[13] the Dutch branch of Greenpeace,[14] and the Institute for Nature Education and Sustainability *(Instituut voor Natuureducatie en Duurzaamheid (IVN))*.[15] A number of legal scholars hold prominent positions in these campaigns, including Prof. Tineke Lambooy[16] (Nyenrode Business University), Dr. Laura Burgers[17] (University of Amsterdam),

---

[11] The zoöp model also seems to avoid the pitfalls of using private rights to protect environmental interests under Dutch private law as identified by Anne De Vries-Stotijn, Ilon Van Ham and Kees Bastmeijer, 'Protection through Property: From Private to River-Held Rights' (2019) 44 Water International 736.

[12] Burgers and den Outer (n 2) 52.

[13] Embassy of the North Sea, 'About' <www.embassyofthenorthsea.com/over/>.

[14] The Dutch branch of Greenpeace does not explicitly campaign for the creation of rights of nature, but it does use this language at times to describe its legal and political interventions. For example, in announcing an action against the Dutch government with regards to the ongoing nitrogen pollution crisis in the Netherlands ('de stikstof crisis'), they describe their actions as acting on behalf of nature to 'let the right of nature prevail', see Hilde de Vries, 'Nederland schendt Europese natuurregelgeving, wij stappen naar de rechter' (*Greenpeace*, 20 May 2021) <www.greenpeace.org/nl/natuur/46203/nederland-schendt-europese-natuurregelgeving/>.

[15] The Institute's support for rights of nature manifests itself in its actions to restore and preserve the Maas River in the Netherlands. See, for example, 'Maas in de Wet' (*IVN*, 25 March 2022) <www.ivn.nl/provincies/limburg/nieuws/maas-in-de-wet>; Paul de Vries, 'Moeder Aarde in de grondwet' (*IVN*) <www.ivn.nl/mens-en-natuur/nieuws/moeder-aarde-in-de-grondwet>.

[16] Lambooy, Venis and Stokkermans (n 2) 786; Venis, Lambooy and Berkhuysen (n 2); Venis, Lambooy and Berkhuysen (n 5).

[17] Burgers and den Outer (n 2); see also an interview with Laura Burgers in Frank Straver, 'Groene activisten zoeken de juridische confrontatie. Is het recht de laatste reddingsboei voor klimaat en natuur?' *Trouw* (Amsterdam, 27 March 2022).

Dr. Daphina Misiedjan[18] (Erasmus University), and Prof. Kees Bastmeijer[19] (Tilburg University). Environmental lawyers outside of academia are also prominent in these campaigns, including Jessica den Outer[20] and Jan van de Venis (who also serves as an unofficial 'Ombudsman for Future Generations' in the Netherlands[21]). It is apparent that legal professionals play an essential role in the campaigns to establish rights for nature in the Netherlands. It is also noteworthy that one of the most prominent law firms in the Netherlands – De Brauw Blackstone Westbroek – collaborated with The New Institute in developing their standard contract to govern the certification relationship between the Zoönomic Foundation and the participating zoöps, giving credence to the model's legal integrity.[22]

In addition to civil society and the academic sector, a small number of public-sector actors are involved in the campaign for rights of nature in the Netherlands. As noted in Section I, the Party of the Animals has explicitly endorsed the idea of creating legal personality for the Wadden Sea. In addition, the municipality of Noardeast-Fryslân passed a motion in 2019 endorsing the idea of creating legal personality and associated rights for the Wadden Sea.[23] The Dutch judiciary has been notably absent from these initiatives, as have other levels of political institutions (municipalities, provinces, etc.).

In terms of strategy, campaigns to establish rights of nature have: (i) sought to leverage a legislative reform moment, (ii) linked rights of nature campaigns onto existing environmental grievances, and (iii) tied rights of nature to youth environmental politics and awareness. The most robust proposal to establish a formal rights of nature regime in the Netherlands arose in conjunction with the 2017–2020 process of establishing a new public authority for governing the Wadden Sea (see discussion in Section I, above). Legal and environmental scholars used this reform moment to suggest institutional structures that would embed representatives for the living environment in the new public law authority. In addition, a more minor proposal to create rights for the Amelisweerd forest near the city of Utrecht was linked to a long-existing environmental dispute between local residents and the national government. For decades, the national government has been pursuing an expansion of the city's highway system which would require cutting down substantial portions of the forest. By proposing that the Utrecht municipal council establish rights for the Amelisweerd forest, campaigners seized

[18] Netherlands Institute for Advanced Studies, 'Daphina Misiedjan' <https://nias.knaw.nl/fellow/misiedjan-daphina/>.

[19] de Vries-Stotijn, Van Ham and Bastmeijer (n 11) 736; see also Bastmeijer's statement in Jean-Pierre Geelen, 'De roep klinkt steeds luider: "Wil je de natuur echt beschermen, geef haar dan rechten"' *de Volkskrant* (Amsterdam, 16 April 2021).

[20] Burgers and den Outer (n 2).

[21] Lab Toekomstige Generaties, 'Ombudspersoon' <https://labtoekomstigegeneraties.nl/ombudspersoon/>.

[22] Het Nieuwe Instituut (n 10).

[23] *Gemeenteraad Noardeast-Fryslân*, 'Motie: Bijzonder rechten UNESCO Werlederfgoed Waddenzee' (2019/Z235986, 11 July 2019).

on existing environmental disputes to illustrate how a rights of nature paradigm could provide an enduring resolution and relocate governance of the forest to the local (municipal) level and away from a national government that, in their eyes, undervalues the importance of the forest.[24] Finally, the campaign to establish rights for the Maas River has been organized in collaboration with IVN Natuureducatie, an organization devoted to offering nature education to children as well as adults. Notably, the Maas River campaign included a traineeship for students who researched the possibility of establishing rights for the river.[25] Ultimately, however, all campaigns which are premised on legislative actions for establishing rights of nature face the considerable obstacle of widespread political opposition, with only the Party for the Animals (six of 150 seats in the lower chamber) having expressed support for establishing rights of nature (see discussion in Section I, above).

III. Enforcement

Enforcement plays a key role in discussions about potential rights of nature in the Netherlands. The idea to create legal personality for nature areas, and to give them associated rights, stems from frustration with the State's failure to comply with existing obligations for nature conservation and protection under European Union law as well as under international environmental law.[26] De Vries-Stotijn, Van Ham, and Bastmeijer describe the weakness of existing conservation law frameworks as resulting from 'the difficulty of addressing diffuse sources of pollution, the role of politics in implementing strict licensing systems, the difficulty of ending historic rights to pollute, and weak supervision and enforcement.'[27] The failure to meet these obligations is bound up in the conflicting interests at stake in public decisions about environmental conservation. According to Bastmeijer, the Dutch courts give too much deference to the State in its appraisal of conflicting interests when environmental conservation is at stake, which frequently results in economic interests outweighing competing environmental concerns. The installation of rights of nature would not remove the need for these evaluations, but it would reverse the burden of proof by requiring the State to prove that the environmental harm of an action is less than it appears or is proportionate to the other competing interests at stake.[28]

Under the proposed 'natureship' model of creating rights for nature in the Netherlands, the rights of nature would be encompassed in the public 'natureship' entities that are statutorily empowered to govern activities that effect the respective

[24] Venis, Lambooy and Berkhuysen (n 5).

[25] den Outer (n 3).

[26] See Bastmeijer's statement in Geelen (n 19).

[27] de Vries-Stotijn, Van Ham and Bastmeijer (n 11) 736.

[28] See Bastmeijer's statement in Geelen (n 19).

nature area. The Wadden Sea Natureship would constitute the public legal person of the Wadden Sea, and it would be given the legal responsibility 'to protect and support the ecological integrity of a specified geographical area [...] and to regulate human activity in it so that the area's ecological integrity is not jeopardized.'[29] It could enforce environmental protections through the establishment of tax measures on polluting or harmful activities, through the pursuit of civil claims where harm has been done to the protected natural area, or through the issuance of penalties for violating regulatory measures protecting the natural area.[30] In sum, the natureship model seeks to remove responsibility for environmental conservation from ministries which are simultaneously pursuing competing (economic, social) goals and to instead create a new public law entity that combines representatives from the ministries, affected municipalities and provinces, environmental NGOs, and local citizens.

Enforcement under the zoöp model takes a far more private-law approach. Fundamentally, in the zoöp model there are no formal rights created for natural entities, nor is legal personality given to nature. Instead, representation of ecological interests is incorporated into the governance bodies of the zoöp organizations through a designated board member. Should a zoöp fail to sufficiently protect ecological interests within its sphere of activity, enforcement would ultimately result in termination of the certification agreement between the Zoönomic Foundation and the zoöp organization and, correspondingly, the inability of the latter to publicly identify itself with the zoöp title. Although this enforcement mechanism is not rigorous, the zoöp model relies more on the soft enforcement of environmental protection commitments within the internal decision-making of the zoöp, as guaranteed by the board member charged with overseeing these commitments.

IV. Theoretical Issues and Foundation

There is no single philosophical, religious, or cultural tradition that particularly underpins the campaigns to establish rights for nature in the Netherlands, but rather a loose collection of partially convergent and partially divergent foundations. One of the prominent philosophical defenses of these campaigns is found in landscape architect Erik Kaptein's *Rechtsgelijkheid voor de Natuur* ('Legal Equality for Nature'), which emphasizes the similarities between the current environmental crisis and the crisis of the French Revolution, comparing the extension of rights to citizens with the prospect of extending rights to nature.[31] This Enlightenment-based underpinning of rights of nature is linked with the ultimate

[29] Lambooy, van de Venis and Stokkermans (n 2) 798.

[30] Lambooy, van de Venis and Stokkermans (n 2) 798.

[31] Erik Kaptein, *Rechtsgelijkheid voor de Natuur: Waarom niet-menselijk leven rechten verdient* (ISVW Uitgevers 2021) 96.

well-being and happiness of humanity, according to Kaptein.[32] Other scholars present the foundation of rights of nature in terms of addressing the shortcomings of the West's anthropocentricism and its strict distinction between humanity and nature, parameters which have traditionally defined the Dutch and other Western legal systems. They pose this negative argumentation without offering a specific, positive foundation as an alternative.[33] The Embassy of the North Sea explicitly acknowledges French social theorist Bruno Latour's notion of the 'parliament of things,' with the flat ontology and relational metaphysics on which it is based, as inspiration for the Embassy's foundation.[34] Notably, the Latourian foundation shares a common critique of the West's distinctions between human-nature and subject-object, distinctions which, ironically, are often associated with the Age of Enlightenment. Others imply that the justification of establishing rights of nature may arises out of a pragmatic interest in securing greater compliance with the Dutch state's existing environmental obligations under national, European, and international law.[35] They suggest that the establishment of legal personality and associated rights to natural areas would be procedurally important in influencingg the weight given to environmental interests in public decision-making. In sum, the philosophical or theoretical foundations of the campaigns for rights to nature are best characterized as a bricolage of different ontologies and value systems rather than as a coherent worldview.

V. Relationship between National Legal Systems

Foreign legal systems have been often named as a source of inspiration for the pursuit of rights of nature in the Netherlands. Campaigns in the Netherlands often point to national constitutions that have explicitly enshrined rights of nature or that have rights to a healthy environment. Specifically, the following national constitutions have been named as inspirations: Belgium (art 23),[36] Norway (art 112),[37]

---

[32] 'By extending and sharing the individual rights of liberty with all life on earth, humanity increases its happiness.' (Original: '[D]oor het recht op individuele vrijheid uit te breiden en te delen met al het leven op aarde, vergroot de mens zijn geluk.'), see Kaptein (n 31) 11.

[33] For example, the IVN presents rights of nature as based on 'insights that we in the West have forgotten,' such as *buen vivir* and the intrinsic value of nature. See de Vries (n 15).

[34] Embassy of the North Sea (n 13). Latour introduces his notion of the 'Parliament of Things' in Bruno Latour, *We Have Never Been Modern* (Harv UP 1991) 142–145.

[35] de Vries-Stotijn, Van Ham and Bastmeijer (n 11) 736.

[36] van de Venis, Lambooy and Berkhuysen (n 2).

[37] van de Venis, Lambooy and Berkhuysen (n 2); Lambooy, van de Venis and Stokkermans (n 2) 786.

Germany (art 20a),[38] Ecuador (arts 10, 71–74),[39] Bolivia (preamble),[40] and Colombia (arts 79–80).[41] In addition, foreign legal systems that have used legislation to establish rights for nature have also been explicitly named as inspirations for the campaigns in the Netherlands, including both New Zealand's legislative actions concerning the Te Awa Tupua (Whanganui River)[42] and the City of Pittsburgh's (USA) 2017 enshrinement of rights of nature in the city charter (which has since been invalidated by the courts).[43] Additionally, the High Court of Uttarakhand in India has been pointed to as a source of inspiration following its judicial decisions establishing personality and associated rights for various rivers and natural areas.[44] Finally, indigenous peoples' legal systems have provided inspiration as well, specifically the Ponca Tribe of Oklahoma's 2017 legislation establishing rights of nature.[45]

Reflecting on these foreign sources of inspiration for the Dutch campaigns, two notable observations arise. First, there is a particular focus on foreign examples of rights of nature in respect of rivers and other aquatic areas, regarding in particular the examples of India, New Zealand, and Pittsburgh (USA). This is logical considering not only the prominence of rivers and coastlands in the Dutch landscape and environment but also the concentration of campaigns around the establishment of personality and rights for the Wadden Sea, Maas River, and Biesbosch wetlands.

Secondly, the influence of foreign examples that are connected to indigenous peoples and their worldviews (eg the Ponca Tribe example and Pachamama in the Ecuador legal system) are prominent in the Dutch campaigns. This is counterintuitive insofar as there are not any indigenous communities in the territory of the Netherlands which have experienced the (post-)colonial processes which are apparent in these examples. One proponent of rights of nature in the Netherlands notes that the municipal body (Noardeast-Fryslân) supporting the establishment of rights for the Wadden Sea comes from the province of Friesland, which is home to a sub-national community that speaks its own language (Frisian), thus making it an 'indigenous' call to action.[46] It is important to note, however, that the Frisian people are not commonly identified as an indigenous peoples community, and

[38] Lambooy, van de Venis and Stokkermans (n 2) 786.

[39] van de Venis, Lambooy and Berkhuysen (n 2); Lambooy, van de Venis and Stokkermans (n 2) 786.

[40] Lambooy, van de Venis and Stokkermans (n 2) 786.

[41] Lambooy, van de Venis and Stokkermans (n 2) 786.

[42] van de Venis, Lambooy and Berkhuysen (n 2); Lambooy, van de Venis and Stokkermans (n 2) 786; den Outern (n 3); Het Nieuwe Instituut (n 10).

[43] Lambooy, van de Venis and Stokkermans (n 2) 786.

[44] van de Venis, Lambooy and Berkhuysen (n 2); Lambooy, van de Venis and Stokkermans (n 2) 786.

[45] van de Venis, Lambooy and Berkhuysen (n 2); Lambooy, van de Venis and Stokkermans (n 2) 786.

[46] Lambooy, van de Venis and Stokkermans (n 2) 794.

they have not been subjected to a colonial process of exploitation. Furthermore, the rights of nature campaign in the Frisian portion of the Wadden Sea is not connected to any philosophical, religious, or cultural values unique to the Frisian community. One potential explanation for the inspiration from indigenous peoples is an implicit recognition of the incompatibility of rights of nature with the historically predominant Western and capitalist cultural values in the Netherlands, although such an explanation has not been explicitly offered by the campaigns as far as the author is aware.

Finally, to the author's knowledge there have not been any attempts to have foreign rights of nature enforced under private international law proceedings within the Dutch courts. However, the Dutch courts have increasingly become a forum for complex, multijurisdictional disputes regarding environmental harm (eg the *Shell Nigeria* cases).[47] Should such claims be brought before the Dutch courts, there are no evident principles that would categorically prevent the Dutch courts from enforcing such rights.

VI. International Law

International law has not been invoked as either an explicit justification or as a reason in opposition to the creation of rights of nature in the Netherlands. Proponents for establishing legal personality for the Wadden Sea have emphasized its status as a UNESCO World Heritage site with considerable biological diversity, but they do not suggest that this status under international law requires the creation of legal personality for the site.[48] Instead, as stated above, the campaigns seeking to establish rights of nature believe that personality status and rights of nature would be instrumental in achieving greater compliance with the Dutch state's conservation obligations under international and EU law. It is notable, however, that the extensive existence of Dutch, European and international environmental protections is also the justification offered by the Dutch government as to why the establishment of legal personality and rights for the Wadden Sea is unnecessary. In the government's view, a sufficient normative framework for protecting the ecological interests of the Wadden Sea already exists.[49]

---

[47] Gerechtshof den Haag 29 January 2021, *Oguru, Efanga & Milieudefensie v Shell*, ECLI:NL:GHDHA:2021:132; Gerechtshof den Haag 29 January 2021, *Dooh & Milieudefensie v Shell*, ECLI:NL:GHDHA:2021:133; Gerechtshof den Haag 29 January 2021, *Akpan & Milieudefensie v Shell*, ECLI:NL:GHDHA:2021:134.

[48] See, for instance, the extensive discussion of the Wadden Sea's UNESCO World Heritage status in Lambooy, van de Venis and Stokkermans (n 2) 786.

[49] Kamerstukken II 25 July 2019, *Waddenzee Beleid: Verslag van een algemeen overleg*, no 187, 2018/2019, 29648, 15.

VII. Efficacy

Under the natureship model, separate public bodies would be statutorily created for each natural area that is given personality and rights.[50] These natureship bodies would be the legal persons entrusted with the protection of the natural areas. Formally speaking, the natural areas themselves would not be understood to have legal personality. The natureship would have the legal capacity to issue public regulatory decisions, to collect tax revenue on activities within that natural area under its responsibility, and to file civil claims for damage caused to that natural area. The natureships would be directed by a board or committee composed of representatives from relevant ministries, provinces, municipalities, environmental NGOs, and local residents. In contrast, zoöps are premised on the representation of the interests of non-human life as embedded at the level of a private organization, with a board member dedicated to that representative role. They are, therefore, are not dependent on the creation of public bodies by the legislature. The efficacy of the zoöp model rests on the integration of rights of nature interests into the regular decision-making and governance of an organization. However, they are ultimately a voluntary tool, and should a zoöp decide that the protection of non-human life in its organization has become undesirable, it would retain the possibility of terminating its status as a zoöp. In this manner, the efficacy of zoöps reflects the efficacy of a 'soft', voluntary private governance, whose binding nature depends more on the socio-economic environment of the organization than on the formal character of the governance norms.[51]

VIII. Consequences

In conclusion, rights of nature do not exist as *lex lata* in the Netherlands. While there are prominent ongoing campaigns to have them established for particular natural areas, these campaigns are premised on legislative pathways that are very unlikely in the currently political context. Aside from these campaigns for legislative action, there are no evident campaigns for a constitutional amendment to establish rights of nature, nor are there any litigation-based attempts to establish them. In this political context, perhaps the most interesting development in the Netherlands – vis-à-vis a comparative study of the rights of nature – is the Zoönomic Foundation's development of the zoöp model and the de facto integration of rights of nature into the internal decision-making processes of organizations (private or public, for-profit or non-profit) that agree to become zoöps. As a private-law based model for incorporating ecological interests into an organization's

---

[50] Lambooy, van de Venis and Stokkermans (n 2) 786.

[51] See, for example, Gunther Teubner, 'Self-Constitutionalizing TNCs? On the Linkage of "Public" and "Private" Codes of Conduct' (2011) 18 IJGLS 617; Phillip Paiement, *Transnational Sustainability Standards* (CUP 2017) 44–66.

governance, it holds considerable potential for transplantation into other legal systems and for being taken up as an alternative in political contexts that do not seem to offer legislative or constitutional pathways for establishing rights of nature.

Norway

Rights of Nature in the Norwegian Legal Culture: To Be or Not to Be?

Sören Koch, Esmeralda Colombo and Catalina Vallejo Piedrahíta[*]

I. Introduction

In a global context of pervasive air, water, and soil pollution, climate change, and the loss of biodiversity – all backlit by evolving progressive interpretations of socio-ecological justice – new ontological approaches are emerging for Western law: the legal relationship between humans and nature is changing.[1] From a legal system where nature was objectified as either property or a set of ecosystem services, legal orders are beginning to recognize ecosystems as subjects with rights

[*] The manuscript was finalized in June 2022. The assessment of literature and case law, including internet materials, reflects the legal situation at that time.

[1] Kristina Lyons, 'Mejorar los conflictos: Derechos de la Amazonía en mundos cosmopolíticos' (2021) 23 Revista de Antropología y Sociología: Virajes 105.

and obligations of their own, a new step in environmental constitutionalism.[2] As the constitutional incorporation of substantive and procedural environmental rights, duties, and remedies, environmental constitutionalism has been deployed to protect the natural environment in new ways.[3] Nonetheless, questions linger on whether protecting nature via constitutional stipulations is sufficient or whether nature itself should be subject to rights.[4]

In Norway, rights of nature (RoN), rights to nature, and environmental rights have been discussed for over a decade.[5] To better understand the role of RoN in Norwegian law, this article's research question concerns the current status of the rights of nature in Norway, connecting existing debates on RoN internationally with ongoing climate and biodiversity litigation in Norway. Our approach combines legal culture with ethical considerations that have germinated in Norway, notably Deep Ecology.

To discuss the research question, we divide the article into three main parts. Firstly, we explore the conceptualization and expansion of RoN globally and review its treatment in the literature. Secondly, we offer a general view of Norway, its legal culture, and the state of legal recognition of RoN therein. Thirdly, building on the results from the previous sections, we discuss the notion of legal duties toward nature as either an alternative or potential predecessor of RoN.

II. Zooming Rights of Nature In: Global and Norwegian Perspectives

Worldwide, RoN exist in at least 40 countries, where they are recognized in the form of constitutional provisions, treaty agreements, statutes, local ordinances, or court decisions.[6] RoN conceptualize nature in various ways, spanning specific ecosystems and nature as a whole, and can involve a range of legal mechanisms, such as rights of nature, legal or juristic persons or subjects, or models that recognize ecosystems as living entities.[7]

[2] Elizabeth Macpherson and others, 'Where Ordinary Laws Fall Short: "Riverine Rights" and Constitutionalism' (2021) 30 GLR 438.

[3] Erin Daly, 'Environmental Dignity Rights Primer. New Frontiers in Environmental Constitutionalism', Widener University Delaware Law School Legal Studies Research Paper Series 17-08; Erin Daly and James R May, *Implementing Environmental Constitutionalism: Current Global Challenges* (CUP 2018).

[4] Louis Jacobus Kotzé and others, 'Earth System Law: Exploring New Frontiers in Legal Science' (2022) 11 Earth System Governance 100126.

[5] See, eg, the conference organized by the University of Oslo in 2012 on the rule of law for nature (University of Oslo, 2022).

[6] Alex Putzer and others, 'Putting the Rights of Nature on the Map. A Quantitative Analysis of Rights of Nature Initiatives Across the World' (2022) 18(1) Journal of Maps 89.

[7] Odd Harald Eidsmo Barder, *Naturen som rettssubjekt. Kan miljøproblemene løses ved å gi naturen rettigheter?* [Master Thesis, University of Oslo] (2015) 13ff., <https://www.duo.uio.no/handle/10852/49618>; Macpherson and others (n 2) 439.

The reasons for affording rights to nature directly – in contrast to protecting the right of humans to enjoy a safe and healthy environment – vary.[8] First, there are ecocentric reasons, which highlight the need to limit human power over animals as they are sentient beings, subject to abuse, and possessing the capacity to experience suffering.[9] Second, protecting the natural environment directly, rather than through human values and human rights – such as the right to health and a healthy environment – seems a proper counterweight to the failures of traditional environmental laws and regulations.[10] Advocates of environmental constitutionalism argue that RoN should be recognized at a constitutional level, thus protecting the intrinsic value of nature more efficiently.[11] Third, granting nature legal personhood would not only give natural entities standing in court but would also allow courts to consider scientific evidence on the consequences of human actions on the natural environment.[12] Fourth, RoN have been identified as a means for Indigenous peoples to uphold their rights to the traditional use of natural resources.[13]

The first legal manifestation of the rights of nature is often traced to the 2008 Constitution of Ecuador, which incorporated protection of the rights of *Pacha Mama* (Mother Nature) alongside the protection of the human right to a healthy environment.[14] These stipulations maintain the rights-based approach that characterizes environmental constitutionalism 'but argues for the expansion of rights-holders beyond humans to include the natural environment as a whole, or specific parts of it, what Tănăsescu calls "the expanding circle of moral concern."'[15] Despite different factual, legal, and cultural preconditions, those driving the recognition of RoN in different countries seek to elevate particular fundamental and inviolable interests – both of nature and sometimes of peoples – above the every-

---

[8] Barder (n 7) 3; Erling Fjeldaas, 'Norges Lover – av, og for, mennesker' in Marit M Simonsen and others (eds), *Da mennesker regjerte kloden. Etikk, dinosaurer og juss for en verden i krise* (Fagbokforlaget 2020) 19.

[9] Eg, *Proposal for an Act on Animal Welfare (dyrevelferdslov)* p 23, in which the justice department proposes to provide sentient animals with 'value independent from their economic or affective value for humans' <www.regjeringen.no/contentassets/4c83935a183e45ea92761d8b864383dd/no/pdfs/otp200820090015000dddpdfs.pdf>.

[10] Joshua C Gellers, 'Earth System Law and the Legal Status of Non-humans in the Anthropocene' (2021) 7 Earth System Governance 100083.

[11] Jan Darpö, 'Can Nature Get It Right? A Study on Rights of Nature in the European Context' (Policy Department for Citizens' Rights and Constitutional Affairs, EU, PE 689.328, 2021) 49; Gellers (n 10).

[12] Mauricio Guim and Michael A Livermore, 'Where Nature's Rights Go Wrong' (2021) 107 Va L Rev 1347, 1352, 1381.

[13] Guillaume Chapron, Yaffa Epstein and José Vicente López-Bao, 'A Rights Revolution for Nature' (2019) 363 Science 1392; Darpö (n 11) 13.

[14] Macpherson and others (n 2) 439.

[15] Macpherson and others (n 2), 439; Mihnea Tănăsescu, *Understanding the Rights of Nature* (transkript 2022).

day business of environmental management; for this, they appeal to higher-level norms, which have the potential to transcend managerial regulatory approaches.[16]

The fundamental challenge for environmental law is, as Bugge emphasizes, the incredible complexity of its object: nature itself.[17] It can be added that addressees of environmental law are always and exclusively humans, making a thorough recognition of RoN through legal frameworks per se challenging.

Remarking on this limitation of RoN, the very idea of RoN and its theoretical foundation have also received criticism.[18] Some disapprove of RoN because, in some cases, they have been recognized before necessary reforms in civil and procedural law have occurred; for example, changes in the legal person or legal subject regimes or in procedural law, allowing natural entities to stand in court, or defining who will speak on their behalf, among other considerations.[19] From the perspective of Indigenous cultures, the rights of nature theory and practice are considered valuable but not entirely unproblematic.[20] This problem particularly emerges in models that seek to recognize or protect the rights of the totality of 'nature' from 'all' humanity.[21] Rights of nature advocates emphasize the alignment between Indigenous worldviews and their ecocentric approaches. However, it is critical to note that the RoN movement has Western and non-Indigenous origins[22] 'and can at times exist in detriment of Indigenous agency and difference'.[23]. The support for the rights of nature may vary from one Indigenous group to another, depending on a myriad of contextual factors.

In the case of ecosystems such as rivers, mountains, glaciers, lakes, or, significant for Norway, fjords, it is argued that their intrinsic value and quality demand the highest level of legal protection from harmful human intervention.[24] Legally enforceable rights are regarded as the most powerful tool to limit human action that trespasses the sphere of another.[25] However, what would be the foundation of those rights?

Ethics can offer a theoretical approach to analyze and propose RoN or alternative ideas. Surveying ethical approaches to this topic in Norway, Ragnhild Freng

---

[16] Macpherson and others (n 2) 439.

[17] Hans C Bugge, *Klimarett. Internasjonal, europeisk og norsk klimarett mot 2030* (Universitetsforlaget 2021) 52ff.

[18] Darpö (n 11) 46ff.

[19] Darpö (n 11) 51.

[20] Erin O'Donnell and others, 'Stop Burying the Lede: The Essential Role of Indigenous Law(s) in Creating Rights of Nature' (2020) 9 TEL 403.

[21] Tănăsescu (n 15).

[22] Macpherson and others (n 2).

[23] Macpherson and others (n 2) 446.

[24] Barder (n 7) 15ff.; see also Natural Diversity Act (*Nature mangfold loven*) para 1, which acknowledges that nature might have some intrinsic value but needs to be protected for the sake of humans (Environmental and Natural Diversity Act 2009).

[25] Barder (n 7) 2.

Dale argues for a new, modified social contract in the form of a nature contract.[26] Similarly, Barder advocates for revising the legal and philosophical foundations of the human–nature relationship,[27] including through meta norms to underpin legal reforms.[28] With a similar preoccupation, Næss proposes focusing on duties toward nature within a 'Deep Ecology' approach.[29] Despite these theoretical debates, environmental lawyers and political analysts demand institutional reforms such as implementing specific review organs that will guarantee more efficient protection over nature or substantive changes to environmental administrative law in Norway.[30]

However, neither politicians, activists, philosophers, legal scholars, nor courts have applied or even seriously discussed recognizing the rights of nature as legally enforceable standards. Explaining this reluctant approach toward RoN requires a brief presentation of Norway's geography and the Norwegian and Sámi legal cultures, their specific features, and their mutual relationship. These features constitute the legal and theoretical framework that influences the degree to which rights or duties toward nature could be implemented and enforced in the future and why they are not recognized today.

III. Rights of Nature in Norwegian Legal Culture

1. Some Facts about Norway

Norway is a country on the northern periphery of the European continent. It stretches across the west coast of the Scandinavian peninsula from Lindesnes in the South over 2,500 kilometers to Hammerfest in the North. According to an international convention, Norway also governs the Arctic archipelago Spitzbergen.[31] Norway's population density is relatively low compared to many other European countries. On average, there are approximately 15 inhabitants per square kilometer, with only 2.2% of the land under cultivation.[32]

---

[26] Ragnhild Freng Dale, 'En ny naturkontrakt? Natursyn i mennesets tidsalder og noen mulige veier videre' in Simonsen and others (eds) (n 8).

[27] Odd Harald Eidsmo Barder, 'Tingene omkring oss – kan rettigheter for dyr og natur redde kloden?' in Simonsen and others (eds) (n 8) 115.

[28] Barder (n 27) 119ff.; Norith Eckbo, 'Life Finds a Way' in Simonsen and others (eds) (n 8) 107f.; Dale (n 26) 65ff.

[29] Arne Næss, 'En fisks iboende verdi' in Andreas Follesdal (ed), *Dyreetikk* (Fagbokforlaget 2000).

[30] Esmeralda Colombo and Eivind Hoff-Elimari, 'En stemme for naturen: Slik kan vi forbedre naturens rettssikkerhet i Norge' (2022) 1 Pan – Foreningen Grunnloven § 112 – Report 74ff.

[31] Treaty Concerning the Archipelago of Spitsbergen, 1920, <www.jus.uio.no/english/services/library/treaties/01/1-11/svalbard-treaty.html>.

[32] Statistics Norway (2021) <www.ssb.no/en>.

Norway's geography is dominated by mountains, rising up to 2,400 meters above sea level, partly covered by glaciers, high plains, and thousands of lakes and rivers, descending into numerous fjords, the longest of which reach more than 200 km inland. Forests cover most of the coastline and stretch out to the East toward the Swedish border. Norway's rivers and fjords were once famous for their diversified wildlife, such as wild salmon, whales, and deep-sea corals. Some rivers and fjords still maintain some of the most precious underwater ecosystems in Europe. However, salmon farming and grand-scale pollution have significantly impaired these biological subsystems. Wild salmon has grown rare, and the surviving populations are in constant decline, while industrial salmon farming has soared.[33]

The exclusive economic zone (EEZ) assigned to Norway by the law of the sea stretches 200 nautical miles from the outer limits of its territorial sea,[34] which is also inhabited by important fish populations. Rich petroleum resources are located on the continental shelf, which the law of the sea defines as the seabed and subsoil of the submarine areas beyond the territorial sea within the natural prolongation to the outer edge of the continental margin – or a distance of 200 nautical miles from the baselines whenever the outer edge of the continental margin does not extend up to that distance.[35] Effective control is also exercised over the aquatic resources on the continental shelf, where Norway's rich oil and gas reserves are located.

Due to the exploration of these natural resources on the continental shelf, Norway is one of the world's wealthiest countries and at the top of the Worldwide Governance Index prepared by the World Bank, also leading the Human Development Index.[36] Notwithstanding ambiguities,[37] the literature indicates a positive association between strong governance indicators and environmental protection.[38] Nevertheless, Norway seems to dispel this positive narrative with respect to its ranking in the Environmental Performance Index (EPI). Developed by Yale University and Columbia University, the EPI scores sustainability at the national scale

---

33 Statistics Norway, 'Export of Salmon' (2021) <www.ssb.no/en/utenriksokonomi/statistikker/laks/uke/2021-06-30>.

34 United Nations Convention on the Law of the Sea, Pub L No 1833 UNTS 3 (1982), art 57 (LOS Convention 1982).

35 LOS Convention 1982, art 76.

36 D Kaufmann and A Kraay, WGI 2022 Interactive <https://info.worldbank.org/governance/wgi/>; United Nations Human Development Index Ranking, in Human Development Reports (United Nations 2021) <https://hdr.undp.org/data-center/country-insights>.

37 Oskar Rydén and others, 'Linking Democracy and Biodiversity Conservation: Empirical Evidence and Research Gaps' (2020) 49 Ambio 419.

38 Shouro Dasgupta and Enrica De Cian, 'The Influence of Institutions, Governance, and Public Opinion on the Environment: Synthesized Findings from Applied Econometrics Studies' (2018) 43 Energy Research & Social Science 77; Robert T Deacon, 'Public Good Provision Under Dictatorship and Democracy' (2009) 139 Public Choice 241; Quan Li and Rafael Reuveny, 'Democracy and Environmental Degradation' (2006) 50 International Studies Quarterly 935.

using 40 performance indicators across 11 issue categories. In EPI's ranking, from 2020 to 2022, Norway moved from position nine to position 20,[39] making this research on RoN particularly momentous to increase our understanding of Norway's legal stance toward nature.

2. The Norwegian Legal Culture

Legal culture is often used as a catch-all expression with unclear content. Conversely, this paper applies an operationalized concept of legal culture, defined as legal ideas and expectations of law made operational by institutional(-like) practice.[40] In our view, the interplay between the intellectual and institutional structures of a legal culture can provide interesting insights and explanations for specific features or concepts. Based on this assumption, Koch and Sunde have developed an analytical tool to compare different legal cultures based on six structural elements that underpin all modern legal cultures.[41] We will use this Legal Cultural Model (LCM) to explain why there has been reluctancy to recognize RoN in Norway.

Despite political unions with the other Nordic kingdoms, notably Denmark and Sweden between 1397–1905, the Norwegian legal culture possesses unique characteristics that can be traced back to the late medieval period.[42] Its characteristics include a strong egalitarian structured society with a comparably low degree of social hierarchies, a rather short distance between the elite and the common people, and a high degree of confidence in the state, which have contributed to a pragmatic and realistic approach to law and its application.[43] Social welfare is the dominant policy promoted by almost all political parties. The court system

[39] Daniel C Esty and others, *Environmental Performance Index* (Yale Center for Environmental Law and Policy 2008); Zachary A Wendling and others, *Environmental Performance Index* (Yale Center for Environmental Law and Policy 2020).

[40] Regarding the concept of legal culture and the specific understanding of the operationalized concept of legal culture transferred into a legal cultural model with six elements that can be identified in all modern legal cultures, see Sören Koch, 'Legal Culture and Comparative Law' in Sören Koch and Jørn Øyrehagen Sunde (eds), *Comparing Legal Cultures* (Fagbokforlaget 2020) 42; Jørn Øyrehagen Sunde, 'Managing the Unmanageable' in Koch and Sunde (eds) (this note) 25. The Legal Cultural Model (LCM) is an analytical tool that splits legal cultures up into the subcategories norm production, conflict resolution, ideal of justice, legal method, attitude toward legal professionalization, and internationalization.

[41] Koch and Sunde (eds) (n 40).

[42] For a more concise overview of the Norwegian legal culture, see Marius Mikkel Kjølstad, Sören Koch and Jørn Øyrehagen Sunde, 'An Introduction to Norwegian Legal Culture' in Koch and Sunde (eds) (n 40); Koch (n 40). See for an updated version of this chapter Marius Mikkel Kjølstad, Sören Koch and Jørn Øyrehagen Sunde, 'An Introduction to Norwegian Legal Culture' in Sören Koch and Marius Mikkel Kjølstad (eds), *Handbook on Legal Cultures, A Selection of the World's Legal Cultures* (Springer Nature 2022) vol II, 801–849.

[43] Anna Nylund, 'Klima, miljø og domstoler i et komparativt perspektiv' in Ole Kristian Fauchald and Eivind Smith (eds), *Mellom jus og politikk: Grunnloven* § 112. (Fagbokforlaget 2019) 104.

is structured in a simple three-tier hierarchy with the Supreme Court at the top. Like all Scandinavian countries, Norway has no separate constitutional court. In contrast to Sweden and Finland, however, no specialized administrative courts have been established in Norway. This situation has contributed to a comparably low degree of specialization both in the judiciary and in the education and professionalization of lawyers. Constitutional review is acknowledged (in Norwegian Constitution, §89) but applied somewhat reluctantly by the courts.[44]

In accordance with the Western European legal tradition, Norwegian law is based on an anthropocentric approach.[45] Accordingly, the protection of nature and environmental law have become separate branches of administrative law.[46] In this sense, the 2009 Nature Diversity Act (NDA) is the most important framework for biodiversity protection. The NDA aims

> 'to protect biological, geological, and landscape diversity and ecological processes through conservation and sustainable use, and in such a way that the environment provides a basis for human activity, culture, health, and well-being, now and in the future, including a basis for Sámi culture'.[47]

The noted provision suggests that biodiversity shall be promoted per se and across policy sectors. However, the NDA stops short of articulating enforceable duties placed upon government authorities, for instance, the coverage quality and restrictions on resource harvesting in protected areas (eg NDA 2009, sections 33 and 48).[48] Norway is not part of the European Union, but through the European Economic Area Agreement (EEA), it is connected to the European market and its underlying legal frameworks. Duties toward nature are often construed based on EEA law,[49] but two of the most important EU directives concerning 'duties toward nature', namely the Habitat Directive and the Birds Directive, are not included in the EEA agreement and thus do not apply to Norway.

If we look more closely at the elements of contemporary Norwegian legal culture, we can provide the following explanation for the solidly anthropocentric approach in Norway: In terms of institutionalized forms of norm production and

---

[44] Sören Koch, 'Den tyske klimadommen viser at domstolene kan hjelpe politikerne å tenke langsiktig om klimaet' (*Juridika*, 3 June 2021) <https://juridika.no/innsikt/den-tyske-klimadommen>; Malcolm Langford and Beate Kathrine Berge, 'Norway's Constitution in a Comparative Perspective' (2019) 6 Oslo LRev 198.

[45] Barder (n 27) 118; Thea Rølsåsen, 'This is How You Play God' in Simonsen and others (eds) (n 8) 33ff.

[46] The first legal institutions concerning the protection of the environment can be traced back to the enactment of the Nature Protection Act and the establishment of a Department of Environmental Matters in 1972. However, the operationalization of environmental protection on a wider scale had not taken place before the early 2000s.

[47] NDA 2009, §1.

[48] See also Esmeralda Colombo, 'A Paper Tiger in the Fog of Governance: Norway's Riddle in Biodiversity Matters' (2024) 33(1) RECIEL 109.

[49] Gunnhild Storbekkrønning Solli, 'Rammer for domstolskontroll ved brudd på EØS-rettslige forpliktelser på miljøforvaltningsrettens område' (2021) 60(4) Lov Og Rett 225.

conflict resolution, there is a tendency to balance environmental protection against other relevant interests. Almost all legislative acts to protect nature contain provisions providing government administrators with wide discretion. For example, the NDA, which binds only public agencies, establishes a reasonability test in central sections 4 and 8.[50] The concerns of nature need to be considered only to the extent this is regarded as reasonable in the eyes of public authorities, who often prioritize economic considerations when granting permission to interfere in natural habitats.[51]

Albeit problematic concerning the conservation and promotion of biodiversity, balancing interests is the fundament of the Norwegian legal method and pragmatic approach to law.[52] This feature again correlates with an ideal of justice that focuses more on the individual case than the long-term consequences of legal measures in terms of planetary boundaries.[53] Furthermore, the lack of specialized administrative courts or environmental boards, as well as the high costs of accessing civil justice,[54] makes it challenging to address and legally enforce the interests of nature in Norway. Environmental protection is just one of many relevant *reelle hensyn* (translated from Norwegian as 'policy considerations') that the administration and courts can take into consideration. In legal education, environmental law is not a mandatory subject, and few elective courses are offered at Norwegian universities. The degree of legal specialization in environmental law is comparably low. The predominant dualistic approach to international law allows the legislature to amend international instruments in the implementation process or reject the incorporation where the EEA agreement allows. Therefore, courts can relatively easily interpret national and international legislation restrictively or extensively where judges conclude that a literal application of legislation would be inconsistent with the anthropocentric and right-based approach to law or the economic and welfare policies facilitated by the human-centered foundation of Norwegian legal culture. In what follows, we will provide specific examples of the legal-cultural determination of the Norwegian approach toward RoN.

3. The Sámi Legal Culture and its Relationship to Norwegian and International Law

The Sámi people are an Indigenous ethnic minority living in Finland, Norway, Russia, and Sweden. Throughout history, Sámis have been living close to nature in their day-to-day occupations, such as 'reindeer husbandry, fishing-farming,

---

[50] Norwegian Nature Diversity Act, sect 4: 'The objective is also to maintain ecosystem structure, functioning and productivity to the extent this is considered to be reasonable.'

[51] Fjeldaas (n 8) 21ff.

[52] Ingvell Helland and Sören Koch, 'Introduction to Norwegian Legal Method' in Ingvell Helland and Sören Koch (eds), *Nordic and Germanic Legal Methods* (Mohr Siebeck 2014).

[53] Guillaume Chapron and others, 'Bolster Legal Boundaries to Stay Within Planetary Boundaries' (2017) 1 Nature Ecology & Evolution 1.

[54] Colombo and Hoff-Elimari (n 30) 71ff.

hunting, herding, and harvesting for food supplies'.[55] Out of a total of 81,000 Sámis, roughly 50,000–65,000 live in Norway.[56] Of these, only 2,500 Sámis conduct reindeer husbandry due to the 'Norwegianization process' from the 1850s on, which has threatened the Sámi cultures.[57] Still, since Norway is the sole country that ratified the ILO 169 Convention on Indigenous Peoples, Sámis in Norway have stronger collective rights.[58]

In contrast to the Norwegian (state) legal culture, the Sámi people have for centuries cultivated a rather different approach to law and conflict resolution. Based on customary law and religious beliefs, conflicts were resolved without establishing a strictly institutionalized mechanism.[59] Since Norway signed and implemented the ILO 169 Convention on Indigenous peoples, the International Covenant on Civil and Political Rights (ICCPR), and the Nordic Saami-Convention,[60] the rights and customs of the Sámi people have been protected by national and international law. In this regard, the Supreme Court has several times considered Sámi customs, especially their rights to explore and use natural resources in accordance with their cultural practices.[61]

One source of concern in this regard is that, notwithstanding such recognition, the government continues to refuse to address its lack of enforcement of a 2021 decision where Norway's Supreme Court declared that public authorities had breached the ICCPR, art 27, by violating the right to culture for Sámi Indigenous peoples by installing windmills in their traditional lands through administrative abuse.[62] Compounding this problem, Sámi representatives have shown limited interest in RoN, with only the Sámi Parliament in Sweden endorsing the Universal Declaration of the Rights of Mother Earth (UDRME) in 2018, a gesture that was considered symbolic by legal literature.[63] Both lawyers and scholars dealing with

[55] Susann Funderud Skogvang, *Samerett* (3rd edn, Universitetsforlaget 2017).

[56] Skogvang (n 55). See also <https://nordnorge.com/en/topic/the-sami-are-the-indigenous-people-of-the-north/#:~:text=The_Sami_people_live_in,around_half_live_in_Norway>.

[57] Skogvang (n 55); see a different approach in The Sámi Act 1987 (2007), §2-6.

[58] Skogvang (n 55) 40ff.

[59] Jørn Øyrehagen Sunde, *Speculum legale: Rettsspegelen: Ein introduksjon til den norske rettskulturen si historie i eit europeisk perspektiv* (Fagbokforlaget 2005) 327ff.

[60] ILO Convention no 169, C169 – Indigenous and Tribal Peoples Convention, 1989 (2007) <www.ilo.org/jakarta/whatwedo/publications/WCMS_124013/lang--en/index.htm>; International Covenant on Civil and Political Rights, Pub. L. No. 999 UNTS 171 (1996); Nordic Saami Convention, draft in English translation: <www.regjeringen.no/globalassets/upload/aid/temadokumenter/sami/sami_samekonv_engelsk.pdf>.

[61] Supreme Court of Norway 11 October 2021, HR-2021-1975-S, English translation <https://www.domstol.no/globalassets/upload/hret/decisions-in-english-translation/hr-2021-1975-s.pdf>.

[62] Kjell M Derås, 'Debatt: To høyesterettsdommer, men svært ulik politisk oppfølging/På tide å ta ansvar, statsminister Støre?' (*Ságat*, 10 October 2022) <www.sagat.no/to-hoyesterettsdommer-men-svart-ulik-politisk-oppfolging/19.34587>.

[63] Darpö (n 11) 52.

Sámi rights prefer a discourse based on human- and property rights rather than RoN.[64]

More generally, Sámi are Norwegian citizens and must abide by Norwegian law. In terms of institutional relations, the district of the inner part of Finnmark, the Sis-Finnmárkku diggegoddi, holds a unique position, protecting the Sámi language and giving special priority to Sámi customs. For appeals, the Hålogaland Appellate Court and the Supreme Court apply the same privileges, still these are all ordinary courts applying Norwegian law and procedure. However, further analysis of Sámi law and legal culture exceeds the scope of the article and the knowledge of the authors.

4. Rights of Nature in Norway

To date, rights of nature have not been recognized in Norwegian law. Moreover, the very concept of making nature or natural entities legal subjects, rather than mere objects of property, is regarded as foreign in the view of Norwegian lawyers and legal scholars.[65] The prevailing theoretical approach to law is anthropocentric. The Norwegian Dispute Procedure Act 2005 (*Tvisteloven*) §2-1 contains a closed list of potential litigants. This list does not include nature or natural entities. Leading textbooks and commentaries on civil procedure simply state that neither nature nor natural objects have legal standing ('hverken dyr eller naturobjekter har partsevne').[66] The same applies to administrative cases, which the same courts handle.

In 2020, one proposal to the Norwegian Parliament (*Stortinget*) suggested evaluating whether the rights of nature could be used to preserve certain culturally important mountains and other holy places of the Sámi.[67] However, the proposal was voted down and dismissed.[68]

The Norwegian Constitution does not contain a specific provision that provides nature or natural entities with rights, nor does environmental legislation.[69] However, nature is not entirely unprotected under Norwegian constitutional law. Like many other countries, Norway has been influenced by 'constitutional environmentalism'.[70] After 20 years of parliamentary debate and many different approaches,

---

[64] Darpö (n 11) 52.

[65] Barder (n 7) 12; Torstein Eckhoff, 'Kan elver og skoger ha rettigheter?' (1980) 6(1) Hefte for Kritisk Juss 2; Fjeldaas (n 8) 20.

[66] Inge Lorange Backer, *Rettslig interesse for søksmål, skjønn og klage, særlig ved naturinngrep* (Universitetsforlaget 1984) 138.

[67] Per Espen Stoknes, Representantforslag om verneprosess for utvalgte hellige fjell i samisk kultur (14 December 2020), Stortinget, Familie- og kulturkomiteen, Innst. 300 S (2020–2021) <www.stortinget.no/no/Saker-og-publikasjoner/Saker/Sak/?p=82539>.

[68] Innst. 300 S (2020–2021) sak no 23, Stortinget 2020.

[69] Environmental and Natural Diversity Act 2009; Anti-Pollution Act 1981.

[70] Joshua C Gellers *The Global Emergence of Constitutional Environmental Rights* (Routledge 2017); Jérémie Gilbert, *Natural Resources and Human Rights: An Appraisal* (OUP 2018).

in 1992, a provision was enshrined to guarantee everyone the right to an environment that is conducive to health and to a natural environment.[71] The provision was slightly amended in connection with a significant reform of the Norwegian Constitution in 2014 and now features the following wording:

> 'Every person has the right to an environment that is conducive to health and to a natural environment whose productivity and diversity are maintained. Natural resources shall be managed on the basis of comprehensive long-term considerations, which will safeguard this right for future generations as well.
>
> In order to safeguard their right in accordance with the preceding paragraph, citizens are entitled to information on the state of the natural environment and on the effects of any encroachment on nature that is planned or carried out.
>
> State authorities shall take measures for the implementation of these principles.'

The framing of the provision indicates that the norm contains a legally enforceable subjective right. This reading was further supported by the choice of constitutional legislators to relocate the provision in the new section on human rights in the Norwegian Constitution in connection with the 2014 reform.[72]

Nature itself has not been granted any rights, but the wording seems to indicate that private parties, be they citizens or interest groups, might be competent to act on behalf of those claiming that specific conduct infringes upon their human rights to a healthy environment. Some scholars have even argued that the provision gives rise to the rights of nature.[73] However, as noted above, the wealth of Norway mainly depends on the exploration of oil and gas and on the fishing industry. Consequently, social welfare based on extractive economic growth is the guiding policy principle.

Against this background, the government and the majority of Parliament ruled out any interpretation of the constitutional provision as an enforceable judicial standard extending discretionarily. In a landmark climate decision, the Norwegian Supreme Court supported the government's point of view.[74] In contrast to the courts of first and second instance,[75] the judges unanimously ruled that no environmental review is warranted whenever the Parliament has taken a position – any position – on a particular topic. Exceptionally, Norwegian Constitution §112 could be invoked as a 'safety valve' only when the legislature has 'grossly neglected' its obligations pursuant to the third paragraph of §112, which the Supreme Court did not expound on by way of examples. In other words, the scope of application of the constitutional environmental protection norm was severely

---

[71] Norwegian Constitution 1814, §112, see <www.nhri.no/en/report/climate-and-human-rights/4-section-112-of-the-norwegian-constitution/>.

[72] Norwegian Constitution 1814, n 51, ch E.

[73] Barder (n 27) 121.

[74] Supreme Court of Norway 11 October 2021 (n 61).

[75] Oslo Court of Appeals 23 January 2020, No 18-060499ASD-BORG/03; Oslo District Court 4 January 2018, No 16-166674TVI-OTIR/06.

limited, making it virtually impossible to establish subjective and enforceable rights to protect nature.[76]

The right to information pursuant to para 2 of §112 might force public authorities to be more transparent regarding the reasons for their decisions. However, as long as Parliament documents that they have considered implications for nature, no procedural rights toward protecting nature and the environment can be deduced from this section either. Consequently, applicable Norwegian law provides the government and Parliament with the utmost discretion on how to discharge their environmental duties, as per §112.[77] Overall, this backward reading of §112 would make a judicial recognition of the rights of nature or a nature-protective interpretation of existing law particularly difficult in Norway. The same applies to ordinary environmental legislation, in which Parliament has granted the administration broad discretion, particularly when balancing environmental with economic considerations and interests.[78] Empirical studies have proven that (as a rule) the latter will prevail.[79]

Several intertwined legal-cultural factors have contributed to and facilitated this outcome. International studies have shown that Norwegian citizens have a relatively high degree of confidence in the authorities, especially the government.[80] This belief system promotes an understanding that sensitive political topics should be decided by a democratically elected Parliament as proposed by the government rather than courts. The courts acknowledge and respect such political prerogatives. Although granted the competence to conduct a constitutional review (Norwegian Constitution, §89), judges at all levels of the court hierarchy are reluctant to use this competence, and if they do, they are wary of interfering with politics or economic policy considerations.[81]

Supreme Court judges have argued that such a reluctance or self-limitation is a necessary precondition for the perceived legitimacy of the judiciary and for confidence in courts.[82] In the famous Kløfta ruling, the Supreme Court established a theory that the intensity of review depends on the object of regulation.[83] While a rather strong demand for judicial review is recognized for the violation of hu-

---

[76] Marius Gulbranson Nordby, *Det store klimasøksmålet: Saken som tok livet av Grunnlovens miljøbestemmelse* (Cappelen Damm akademisk 2021) 126.

[77] Esmeralda Colombo, 'People v. Arctic Oil and Its Discontents: The Norwegian Paradox in Global Perspective on Climate and Energy Justice' (2021) 51 ELR 10457.

[78] Fjeldaas (n 8) 21ff.

[79] Fjeldaas (n 8) 23; cf examples provided below.

[80] Council of Europe, *European Judicial Systems: Efficiency and Quality of Justice*. (CEPEJ Studies 26, 2018).

[81] Koch (n 44) 111; Eivind Smith, *Konstitusjonelt demokrati: Statsforfatningsretten i prinsipielt og komparativt lys* (Fagbokforlaget 2017) 317.

[82] Jørn Øyrehagen Sunde, *Høgsteretts historie 1965–2015: At dømme i sidste instans* (Fagbokforlaget 2015) 66.

[83] Norwegian Supreme Court 27 January 1976, Rt. 1976 s1.

man rights, courts are reluctant to review the work of state powers in other fields.[84] In particular, the courts are obliged to respect the policy decisions of Parliament in the sense that their role is to assess the legality of the decision, not whether it is appropriate politically.[85] Indeed, the majority of lawyers in Norway are rather skeptical of any attempts to establish a constitutional court or specialized administrative courts.[86] This skepticism has consequences for the legal protection of nature. Nylund has comparatively analyzed climate litigation and concludes that the procedural preconditions in a legal system without administrative courts, such as in Norway, make it particularly difficult to enforce duties toward and rights of nature.[87] The decision of the Norwegian Supreme Court on the climate change cases (cited above) seems to prove such a thesis right.

If the Kløfta ruling were read in light of the entwinements between climate change and human rights violations, case law on environmental judicial review might progressively change toward more protective stances regarding nature's promotion. Human rights violations can be conceptualized, for instance, as abuses of future generations' freedoms.[88] Importantly, Norwegian Constitution, §112 protects the rights and freedoms of future generations. However, in the above-cited climate decision, the Supreme Court did not engage with the consequences of regulatory action for future generations but instead virtually excluded environmental policy from constitutional and more general judicial review. The courts legitimize this stance by dismissing the participation of the youth in democratic processes so long as they do not vote for popularly elected bodies. In fact, the court stated:

> '[…] decisions involving basic environmental issues often require a political balancing of interests and broader priorities. Democracy considerations also suggest that such decisions should be made by popularly elected bodies, and not by the courts.'[89]

In addition, the Supreme Court held that §112 of the Norwegian Constitution only protects the environment in Norway.[90] Consequently, the emission of greenhouse gases outside of Norway would fall into the scope of constitutional protection only if they directly affect Norwegian territory. In other words, if Norwegian oil and gas companies cause harm to nature outside Norway, there is no constitutional protection or remedy.

---

[84] E Smith (n 81) 317.

[85] E Smith (n 81) 5.

[86] See Norwegian Public Reports (*Norges Offentlige Utredninger*) NOU 2019:17, 12ff.; NOU 2017: 8, 51ff.; NOU 2020:11, 290ff.; containing further references.

[87] Nylund (n 43).

[88] Louis Jacobus Kotzé, 'Neubauer et al. Versus Germany: Planetary Climate Litigation for the Anthropocene?' (2021) 22 GLJ 1423.

[89] Norwegian Supreme Court 22 December 2020, HR-2020-2472-P (English translation, courtesy of the Court <www.domstol.no/globalassets/upload/hret/translated-rulings/hr-2020-2472-p.pdf> para 141.

[90] Norwegian Supreme Court 22 December 2020 (n 89) paras 149 and 155.

This self-inflicted idea of separation of powers based on considerations of democratic prerogative in combination with a restrictive interpretation of Norwegian Constitution, §112 has contributed to limiting the court's competence to review environmental legislation.[91] Specifically, it poses very high obstacles to judicial review of specific administrative decisions or actions regarding environmental considerations. The conclusion of the Supreme Court is telling:

> 'When a production license follows directly from the *Storting*'s endorsement of the opening of relevant areas, there is little left for the Supreme Court to control.'[92]

This summary positions Norwegian environmental constitutionalism as an aspiration with little to no applicability *so far*. In such a political and legal cultural climate, the recognition of enforceable rights of nature is more than unlikely, at least in the near future.

5. Political and Legal Operators

Environmental policy is the domain of the political branches of power in Norway. However, environmental constitutionalism and nature's interests have traditionally had a weak standing in Norwegian politics compared to other Western European countries. Significantly, in the 2021 national elections, the environmental party (Miljøparti de Grønne – MDG) did not manage to win the necessary 4% of votes required to be seated in Parliament. As far as we can evaluate, politicians have not yet actively invoked the idea that nature or natural entities can be provided with rights, have legal personality, or be subjects of law. Even the main policy principles of the MDG, which are based on an ecological approach, are vague regarding RoN.[93]

Legal scholars have not shown considerable interest in RoN either. In 1980 one of Norway's most prominent and influential legal scholars posed the following question: 'Do rivers and mountains have rights?'.[94] Eckhoff was deeply impacted by legal realism in the United States and operationalized a realistic approach to the law, especially in his works on legal methodology.[95] His interest in the topic was triggered by Stone's article 'Should Trees Have Standing',[96] which he calls thought-provoking but not entirely convincing.[97] Consequently, natural entities have no subjective rights under applicable Norwegian law. This highly authoritative point of view has remained unchallenged until recently.

---

[91] Koch (n 44).

[92] Norwegian Supreme Court 22 December 2020 (n 89) para 157.

[93] Miljøpartiet De Grønne, *Main Policy Principles* (2021) (in English) <www.mdg.no/main_policy_principles_in_english>.

[94] Eckhoff (n 65).

[95] Torstein Eckhoff, *Rettskildelære* (Tanum 1971).

[96] Christopher D Stone, 'Should Trees Have Standing?: Toward Legal Rights for Natural Objects' (1972) 45 SCal LRev 450.

[97] Backer (n 66) 19; Eckhoff (n 95) 5.

The climate change debate has contributed to reviving the idea of RoN.[98] In 2019, law professors Fauchald and Smith published an anthology concerning §112 of the Norwegian Constitution.[99] While many contributors to this book and other legal scholars have argued that the provision would provide enforceable rights,[100] even encompassing the rights of nature,[101] others, like Smith, took a strict legal positivistic position, denying ordinary courts any competence to use the provision to review the legislature or any administrative act on the grounds of environmental considerations.[102] He argues that only the semi-political organ of the 'Riksrett' (a court staffed by laymen and five Supreme Court judges) has the competence to apply the provision to review acts of Parliament. The underlying assumption is that only Parliament can decide how to use and protect nature; ordinary courts have no such competence.

Even environmental lawyers seem to prioritize alternative mechanisms to reform Norwegian law in concordance with the 'green turn' (*grønt skifte*). Winge proposes a general prohibition for all projects that will cause considerable and irreversible damage to nature.[103] In addition, he claims that the principle of legality should also be established for any administrative act impairing environmental protection. This situation would mean that any infringement of nature or the environment would require an explicit foundation in legislation. Furthermore, Winge advocates specialized environmental courts.[104] The Norwegian Human Rights Institute (NIM) has recommended the establishment of an independent and permanent climate commission with surveillance and review competence.[105] From a comparative perspective, both Fauchald and Bragdø-Ellenes conclude that only an efficient review of administrative decisions in courts of justice would guarantee a proper independent consideration of environmental aspects.[106]

All of these proposals do not aim to directly implement the rights of nature but might well serve as a first step toward creating institutional preconditions for providing a forum for and foundation upon which the idea of the rights of nature might prosper. The underlying problem is that, as Bugge states, '[e]nvironmen-

---

98 Gjermund Aasbrenn, 'Litteratur' (2020) 59 Lov Og Rett 185.

99 Fauchald and Smith (eds) (n 43).

100 Simonsen and others (eds) (n 8).

101 Barder (n 27) 121.

102 Eivind Smith, 'Miljøparagrafen – kritisk lest' in Fauchald and Smith (eds) (n 43) 151.

103 Nicolai K Winge, 'Miljørettens evolusjon-Veien mot et grønt skifte i jussen' (2016) 32(3) Praktisk Økonomi & Finans 282, 287.

104 Winge (n 103) 289.

105 Norwegian National Human Rights Institution, *Report. Climate and Human Rights. Summary* (2021) <https://www.nhri.no/en/report/climate-and-human-rights/summary/>.

106 Sunniva Cristina Bragdø-Ellenes, 'Overprøving av vedtak i miljøsaker – i domstoler og klagenemnd?' in Fauchald and Smith (eds) (n 43) 122, 129; Ole Kristian Fauchald, 'Environmental justice in courts–a case study from Norway' (2010) 1 Nordisk Miljörättslig Tidskrift 49, 58ff.

tal law is more politics than law' in Norway.[107] Creating an institutional legal or procedural framework for protecting nature and the environment would limit the political prerogative. This change, in turn, would require a major paradigm shift in the Norwegian legal culture. The enactment of the NDA can be regarded as a step in this direction. However, as discussed above, its practical impact on the ecosystem and biodiversity protection is limited.

Interestingly, NGOs such as Greenpeace and the Association of Nature and Youth have not actively argued based on an ecocentric point of view in Norway. At first glance, this situation might seem surprising, not least because these same organizations are fighting against invasive intrusions into local environments with devastating effects on nature all over Norway. One example is the establishment of an underwater waste disposal site in the Førdefjord. This fjord is nationally protected due to its favorable habitat for wild salmon, and the approval process illustrates the general weakness of Norwegian environmental law, with few strict limitations and wide discretion for authorities. In the view of the relevant authorities, the disposal site did not violate the Pollution Control Act (PCA), as economic profit and employment are relevant factors when assessing such applications pursuant to §11, the final paragraph of the PCA.[108]

The waste disposal site was approved by the government against the recommendation of the Institute of Marine Research and significant resistance from the local population. Norway is one of only a few countries worldwide that allow mining waste to be dumped directly into the ocean.[109] According to experts, this will have devastating effects on the ecosystem of the fjord and probably the migration of fish along the Norwegian coast.[110] In a preliminary ruling, the European Surveillance Authority (ESA) has supported the view of several NGOs that Norway has incorrectly implemented the Directive on Mining Waste.[111] Nevertheless, it also seems that the current center-left government has no ambition to stop the project. Quite the opposite, the current government has been the first worldwide to officially open its continental shelf to deep-sea mining, specifically in the areas circumscribing a nature reserve in the Arctic, short of a fully-fledged environmental impact assessment and against the recommendations of the Nor-

---

[107] Hans Christian Bugge, 'Grunnlovsbestemmelsen om miljøvern: Hvordan ble den til?' in Fauchald and Smith (eds) (n 43) 436; Hans Christian Bugge, *Lærebok i miljøforvaltningsrett* (5th edn, Utg. Universitetsforlaget 2019) 39.

[108] Anti-Pollution Act 1981 No. 6 Concerning Protection Against Pollution.

[109] Tor Bjarne Christensen, *Krever blankt avslag til giftutslipp i Førdefjorden* (15 September 2020) <https://naturvernforbundet.no/krever-blankt-avslag-til-giftutslipp-i-fordefjorden/>.

[110] Institute of Marine Research of Norway, *Report of the Institute of Marine Research 4.10.2019 to the Department of Administration of Resources*, <www.hi.no/Horing-Driftskonsesjonen-til-Nordic-Mining-i-Fordefjorden.pdf>.

[111] The Green Warriors of Norway (NMF), *Claim to ESA on 14 June 2021*, <www.eftasurv.int/cms/sites/default/files/documents/gopro/Paragraph%2048%28b%29%20letter%20-%20Mining%20operating%20licenses%20%28complaint%29%20-publication%20-%20Website%20Version.pdf>.

wegian Institute of Marine Research and independent academics.[112] Apparently, NGOs fighting both projects consider it a more efficient strategy to apply to supranational standards and institutions than arguing that the fjord or inner ecosystems have subjective rights. Non-representative interviews with selected representatives of involved NGOs confirm this observation. Again, the impression remains that even NGOs consider the proclamation of RoN as too radical or unrealistic, given the prevailing perceptions of the Norwegian legal system and its underlying legal culture.

Resistance against the expansion of wind- and water-based electricity infrastructure serves as another example. As mentioned previously, the Supreme Court found that the establishment of two wind parks violated Sámi rights to use the land in concordance with their tradition as range land for their reindeer herds.[113] The judges found that the authorities had unlawfully neglected the interests of the Indigenous people in the administrative approval process. As the wind parks have already been built and are operational, the licensee and the state were sentenced to pay compensation, but the decision is subject to different interpretations.[114]

In the same case, the Sámi claimants did not argue based on an ecocentric or ecological approach to law but applied principles of property law and expropriation as well as national and international guarantees for the protection of their cultural heritage.[115] Interestingly for our context, the Supreme Court has, to some extent, lowered the threshold toward the rather strict criterium of a 'denial' of [minority and Indigenous] rights. By conducting a concise comparative analysis of international sources and case law,[116] the Court concludes that for a violation of art 27, it is sufficient to prove a considerable negative impact on the natural preconditions necessary to exercise a cultural practice.[117] Furthermore, the infringement of the mere opportunity to exercise these cultural practices ('vesentlige negative konsekvenser for muligheten til kulturutøvelse') can establish such a violation of minority rights.

In other words, the convention and Constitution pose an absolute duty that all state authorities must respect the natural resources for the preservation of pro-

---

[112] Ministry of Petroleum and Energy, Impact assessment – Exploration and extraction of seabed minerals on the Norwegian continental shelf, 27 October 2022, available at <www.regjeringen.no>.

[113] Supreme Court of Norway 11 October 2021 (n 61).

[114] Derås (n 62).

[115] Based on a violation of art 27 of the Convention on Civil and Political Rights, incorporated into Norwegian law by the Human Rights Act (Menneskerettsloven) LOV-1999-05-21-30, attachment 6, part III.

[116] Human Rights Committee 26 October 1994, CCPR-1992-511, *Ilmari Länsman and others v Finland*; 30 October 1996, CCPR-1995-671, *Jouni E. Länsman and others v Finland*; 27 March 2009, CCPR-2006-1457, *Ángela Poma Poma v Peru*; 26 March 1990, CCPR-1984-167, *Lubicon Lake Band v Canada.*

[117] Supreme Court of Norway 11 October 2021 (n 61) para 119.

tected cultural practices, such as reindeer husbandry.[118] These duties (owed indirectly toward nature) can be balanced against other considerations only when they are of outstanding public importance. As a specific example of such an interest, the Court refers to considerations of environmental protection ('the green turn').[119] By allowing a proportionality test, even under narrow preconditions, the duty to protect natural habitats was once again made dependent on human interests in the utilization of nature. In this case, the economic interests of the Sámi were regarded as more important than the sustainable production of energy, even though the latter should be considered a limiting factor for the rights of the Sámi according to art 27 of the ICCPR. However, the interests of the Sámi in this case go beyond merely economic concerns, as they are intertwined with indigenous ethnic and territorial rights, as well as social, economic, and cultural rights. Those who have argued for a wider scope of rights – such as Stoknes in his proposal concerning holy mountains and Sámi territories – claim that the advocates of these rights should be Sámi political institutions, particularly their assembly (*Sametinget*).[120] Once again, a political organ representing the common interests of the people is regarded as the best and most efficient guarantor of such rights.

Consequently, if RoN were recognized in Norway, these rights most likely would be represented and enforced by state or local authorities and not by individuals or interest groups. This reality would make them vulnerable to political interference and hardly guarantee independent legal protection. By contrast, in most countries that have recognized RoN today, individuals, NGOs, and Indigenous peoples are regarded as legal guardians of natural entities, while public authorities are regarded the main addressees for claims of violation of these rights. According to this view, it makes little sense to impose the duty to protect RoN on the very institutions most likely to violate them. In Norway, however, the state is regarded a guarantor for due process and balanced and well-informed decisions based on scientific evidence and sustainability principles. Despite the mentioned examples disproving this assumption, lawyers' and citizens' high confidence in public authorities is a legal cultural feature that – as positive as it might be from a political perspective – might prove an obstacle to a more sustainable or critical approach to the protection of the environment.

Be that as it may, any journey toward recognizing RoN in Norway should start from the fundamental task of protecting and promoting biodiversity.

---

[118] Supreme Court of Norway 11 October 2021 (n 61) para 130.

[119] Supreme Court of Norway 11 October 2021 (n 61) para 131.

[120] Proposal by Stoknes (n 67).

IV. Alternatives that May Increase the Legal Protection of Nature in Norway

1. A Moderate Approach to Changes

We have seen that the attribution of rights to nature has not (yet) impacted the Norwegian legal or philosophical discourse because it has been regarded as incompatible with the anthropocentric foundation of its legal culture. The minimal impact of even the rights-based approach of environmental constitutionalism in Norway can be regarded as both a symptom and a contributing factor.

Nevertheless, legal cultures are not static but constantly and dynamically changing.[121] An important feature of the Norwegian legal culture is that it has changed rather evolutionarily or organically. Radical changes are the exception. One of the more radical changes occurred around the 1260s and 1270s, triggered by the legislative activities of King Magnus the Law Mender, which has been labeled a legal-cultural revolution.[122] An important trigger of this development in the late medieval period was the prevailing ideas and expectations of the law, which fell out of step with social needs and realities. The radical changes of the legal culture were manifested in the Code of the Realm from 1274. However, the king carefully presented them as amendments to existing law to prevent resistance from political and social actors. This strategy has been a common feature of the history of the Norwegian legal culture,[123] indicating that necessary changes to institutional and intellectual structures with radical consequences for the legal culture will have a greater chance of success if they are presented as minor adjustments. Based on ongoing theoretical debates, we have identified three possible ways to transform the Norwegian legal culture to suit the current need to protect nature and the environment.

First, the determination of absolute (in contrast to relative and fragmented) duties toward nature, instead of the attribution of rights of nature, may help reconcile Norwegian legal culture with the need for an improved rule of law for nature. The rule of law for nature is not a commonly used term among lawyers and politicians in Norway, notwithstanding a conference organized in 2012 on the subject and the resulting edited volume.[124] Recently, however, public debate and specific reports have helped highlight the rule of law for nature as a collective term when discussing the legalization of the environmental struggle and its potential for safe-

---

[121] Sunde (n 40) 28.

[122] Jørn Øyrehagen Sunde, 'Daughters of God and Counsellors of the Judges of Men: Changes in the Legal Culture of the Norwegian Realm in the High Middle Ages' in Stefan Brink and Lisa Collinson (eds), *New Approaches to Early Law in Scandinavia* (Brepols 2014) 144ff.

[123] Sören Koch, 'Consequences of Changing Expectations to Law and its Institutions: Illustrated by the Role and Background of Judges in Norway' (2015) 83(3–4) Tijdschrift Voor Rechtsgeschiedenis/Revue d'histoire Du Droit/The Legal History Review 461.

[124] Christina Voigt (ed), *Rule of Law for Nature: New Dimensions and Ideas in Environmental Law* (CUP 2013).

guarding national interests.[125] Such interests include soil protection, beach zones, children's and young people's interests, reindeer husbandry, biodiversity, public health, and avalanche danger.[126] In the words of Bugge, the

> '[r]ule of law for nature means predictability, security and the absence of arbitrariness and bias in decisions that affect nature and the full accounting of environmental values in decision-making – be it by private interests or public authorities.'[127]

The advantage of emphasizing duties toward nature, compared to the status quo of a managerial approach, is that public authorities would be under the obligation of not only mapping the natural environment but also of accounting for it in their decision-making processes. Such procedural protection of nature would be complemented by the substantive protection of wildlife thresholds (for instance, by minimizing the environmental consequences of development plans). Overall, a duty-based paradigm would have ripple effects on private parties, who would be required to provide alternatives to the most invasive development plans they aim to carry out and compensate for detrimental effects on the natural environment.

Second, the Norwegian Constitution contains concepts that point to a potential for increasing the importance of the rule of law for nature in Norway. The first sentence of Norwegian Constitution §112 refers to the term 'diversity' of the natural environment, which expresses an ecocentric perspective, and further mentions its 'productivity', which can be interpreted as both an anthropocentric and ecocentric concept.[128] Furthermore, Norwegian Constitution §112 allows for the 'encroachment on nature that is planned or carried out', which is an anthropocentric aspect.

§112 of the Constitution thus enshrines a spectrum of perspectives, from ecocentric to anthropocentric. In this sense, it is unsurprising that §112 asks many questions and responds with few answers.[129] This situation still applies after the Supreme Court ruling in the climate lawsuit, which marked the first time the section was judicially applied in court. The Court confirmed that Norwegian Constitution §112 has legal significance but concluded that there is a very high threshold for the courts to legally challenge a government decision to which the Parliament has given consent.[130] What this means for all other administrative acts and decisions not expressly sanctioned by Parliament remains unclear.

---

125 Hallvard Surlien, 'Naturen trenger rettssikkerhet' (*Agenda Magasin*, 3 January 2017) <https://agendamagasin.no/debatt/naturen-trenger-rettssikkerhet/>.

126 Riksrevisjonens undersøkelse av behandling av innsigelser i plansaker. Dokument 3:6 (2018–2019), 2018.

127 Hans Christian Bugge, 'Twelve Fundamental Challenges in Environmental Law: An Introduction to the Concept of Rule of Law for Nature' in Voigt (ed) (n 124) 8.

128 Ole Kristian Fauchald, '§ 112 i Grunnloven' in Ola Mestad and Dag Michalsen (eds), *Grunnloven: Historisk kommentarutgave 1814–2020* (Universitetsforlaget 2021).

129 Bugge (n 107) 40.

130 Norwegian Supreme Court 22 December 2020 (n 89).

Third, rights and duties toward future generations enshrined in Norwegian Constitution §112 might offer another possible way to strengthen the rule of law perspective toward the sustainable protection of nature. A decision on the Norwegian climate case by the European Court of Human Rights will likely highlight this uncommunicated aspect because considerations concerning future generations are essential from a human rights perspective.[131] The climate ruling of the German Federal Constitutional Court (FCC) from March 2021 has proven the persuasive power of this perspective, allowing the judicial review of environmental policies based on the generation-comprehensive consequences of government actions. The FCC argued that in conjunction with the international obligations resulting from the Paris Agreement, the human rights protection of future generations could limit the room for political action (*Handlungsraum*) and even the prerogative of the legislature.[132] Decisive for the competence for constitutional review of a parliamentary policy decision was that harm to nature, particularly greenhouse gas emissions, is often irreversible. The same will then apply if one recognizes RoN. An encroachment of these rights with irreversible effects would require substantial constitutional legitimation. Hence, through the proxy of future generations, the rights of nature could be protected indirectly. However, in the Norwegian Supreme Court climate judgment, this aspect of intergenerational fairness was barely mentioned, which will limit the binding effect of the judgment on later cases.

Finally, in terms of philosophical sources, the most relevant philosopher to mention is Arne Næss (1912–2009). As one of the most prominent Norwegian philosophers of the 20th century, Næss believed that the root causes of environmental crisis were to be found in utilitarian pragmatism – which idolized managerial approaches to nature – and anthropocentrism, which for its part was a detrimental force for other forms of life and the life quality of humans themselves.[133] In response to existing 'shallow' ecology, which has joined in a devil's pact with anthropocentric approaches to nature, Næss proposed 'Deep Ecology', a term that he coined to pose and answer deep questions, rejecting all types of rationality based on economic growth.[134] In his critique of the utilitarian tenets underlying law and public policy, however, Næss did not advocate legal frameworks enshrining the rights of nature; rather, he interspersed his writings with the enucleation of duties toward nature that public policy should incorporate.[135] It is thus fair to hold that Deep Ecology deems it possible to radically reform law toward non-an-

---

[131] Application no 34068/21.

[132] German Federal Constitutional Court 24 March 2021, BvR 2656/18.

[133] Arne Naess, 'The Shallow and the Deep, Long-Range Ecology Movement. A Summary.' (1973) 16 Inquiry 95.

[134] Arne Næss, 'The Deep Ecological Movement: Some Philosophical Aspects' in David R Keller (ed), *Environmental Ethics: The Big Questions* (Wiley-Blackwell 2010) 240.

[135] Esmeralda Colombo, 'Law & Ethics: Deep Ecology, Climate Change, and Norway's Wolf Policy' (2021) 67 Scandinavian Studies in Law 273.

thropocentric directions within the existing rights paradigm that excludes the rights *of* nature[136] but promotes collective and individual duties *toward* nature.

Such a paradigm would promote the establishment and enforcement of duties toward nature, which is recognized for its intrinsic value rather than as an instrument for growth.[137] The effects of a duty paradigm based on Deep Ecology are consequential, as Deep Ecology would limit unnecessary encroachments of natural entities and ecosystems and protect the animals therein. In this regard, an application of Deep Ecology has already cast light on the intrinsic incoherence of Norway's wolf policy while proposing specific changes to existing legislation to avoid an unjustifiable loss of biodiversity.[138] The argument resonates with broader critiques of the rights of nature movement and a preference for emphasizing and specifying duties toward nature.[139]

Further, intermediate concepts to protect nature can provide an incremental process for more protective stances toward nature, albeit not in the form of rights. Lastly, the duties paradigm deployed to evaluate Norwegian law and policy can be a viable alternative in legal cultures that currently do not seem to acknowledge the rights of nature. Thus, the duties paradigm can eventually become a predecessor of fully-fledged rights.

2. The Impact of EU and International Law

The current anthropocentric focus, in combination with a dualistic approach toward international law, seems to facilitate economic considerations at the expense of biodiversity and environmental protection. With the notable limits of protecting birds and habitats, as previously mentioned, Norwegian environmental law has strong roots in EEA law.[140] Article 73 of the EEA agreement binds Norway's environmental policy to EU law and, to some extent, to the case law of the European Court of Justice. Nevertheless, the Norwegian Supreme Court has been criticized for not sufficiently considering European environmental protection standards. For example, Solli argues that the Court was obliged to interpret Norwegian legislation in concordance with the applicable EU directives and Norway's international obligations, but on several occasions it failed, willingly or negligently, to meet European standards or even consider them.[141]

Similarly, in the Norwegian climate case, the ruling majority of the Supreme Court found that the government was under no obligation to evaluate the envi-

[136] Colombo (n 135); Arne Næss, 'Den dypøkologiske bevegelse: Aktivisme ut fra helhetssyn' in Svein Gjerdåker and others (eds), Den uoverstigelige grense: tanke om handling i miljøkampen (JW Cappelens Forlag 1991).

[137] Næss (n 29) 13.

[138] Colombo (n 135) 295.

[139] Luc Ferry, *Le Nouvel Ordre écologique. L'arbre, l'animal et l'homme* (Éditions Grasset 1992) 245.

[140] Solli (n 49) 225.

[141] Solli (n 49) 226.

ronmental impacts of petroleum licenses in the Barents Sea in light of future exported emissions.[142] Conversely, the Supreme Court's minority and some scholars maintained that the climate effects of petroleum licenses had been inadequately analyzed and that both the government and the Supreme Court's majority had misinterpreted Directive 2001/42/EC (the Strategic Environmental Assessment Directive).[143] Even EU initiatives toward an EU Charter of the Fundamental Rights of Nature, or a Draft Directive on the Rights of Nature,[144] have not generated discussions in Norway and would probably be disregarded as they exceed EEA relevance.

Considering international law, in line with the Constitution and the general presumption principle, Norwegian law is assumed to be in accordance unless there are signs pointing otherwise.[145] Norway has undertaken several international obligations for the protection of nature, and mentioning just a few may be clarificatory: the United Nations Convention on Biological Diversity,[146] the Ramsar Convention, which is the global agreement for protecting wetlands,[147] the Bern Convention on the Conservation of European Wildlife and Natural Habitats,[148] and the Bonn Convention on the Protection of Migratory Species of Wild Animals.[149] In addition, the Bern Convention obliges cooperation between member states to decide how to implement nature conservation. The Aarhus Convention contains procedural rules to ensure that the consequences of a specific measure or a development plan are studied and easily accessible to the public to allow for access to justice.[150]

Notwithstanding Norway's wide participation in such international conventions for the protection of nature and the environment, a 'Norwegian paradox' can be observed. In environmental matters, Norway presents itself as an environmental champion abroad, although it continues to be an environmental laggard domestically.[151] The impact of international law is particularly controversial in

[142] Norwegian Supreme Court 22 December 2020 (n 89) para 193ff.

[143] Henrik Bjørnebye, 'Spørsmålet om mangelfull utredning av klimavirkninger i HR-2020-2472-P' (2021) 60 Lov Og Rett 159.

[144] Darpö (n 11) 51.

[145] Geir Ulfstein, 'Internasjonal miljøretts stilling i norsk rett' (1999) 38 Lov Og Rett 402.

[146] Convention on Biological Diversity (1992), (CBD).

[147] Convention on Wetlands of International Importance especially as Waterfowl Habitat (1971).

[148] Convention on the Conservation of European Wildlife and Natural Habitats (1979), (The Bern Convention).

[149] Convention on the Conservation of Migratory Species of Wild Animals (1979).

[150] Convention on Access to Environmental Information, Public Participation in Decision-Making and Access to Justice in Environmental Matters (1998), arts 6–9.

[151] Elise Åsnes, 'Norges dobbeltmoral' (*Dagsavisen*, 17 October 2021) <www.dagsavisen.no/debatt/2021/10/17/norges-dobbeltmoral/>; David R Boyd, 'Report of the Special Rapporteur on the Issue of Human Rights Obligations Relating to the Enjoyment of a Safe, Clean, Healthy and Sustainable Environment on His Visit to Norway' [Report to the UN General Assembly], UN A/HRC/43/53/Add.2, 3 January 2020; Esmeralda Colombo, 'Alle mann på dekk:

climate change matters. In the sole climate case that has unfolded in Norwegian courts, the state's representative criticized the idea that Norwegian law could or should be interpreted in accordance with Norway's international obligations, especially the Paris Agreement.[152] Further, the value of comparative considerations has been questioned.[153]

In the first trial of the climate case, the lawyer representing the State claimed that reviewing the allocation of petroleum licenses in the Norwegian Arctic in light of the Norwegian Constitution, §112, as the environmental organizations demanded, was an attempt to 'Americanize' Norwegian legal culture.[154] The 'Americanization argument' of the State lawyer was twofold: First, the plaintiffs used the law to raise awareness of the government's climate and oil policy in courts and society, even though climate policy allegedly belongs to the political sphere. Secondly, the environmental organizations used 'liberal legal argumentation' (*frirettslig argumentasjon*, a derogatory term among lawyers), especially as regards the constitutional provision that protects the right to the environment.[155] In terms of the role played by other international elements in the Norwegian climate case, building blocks such as the precautionary principle or the international climate regime were sidelined and had little impact on the decision.[156]

Conclusively, it is clear that international law fails to offer a decisive anchor to enshrine the rights of nature in Norway. Therefore, in terms of actual change, advocating for the rights of nature may be counterproductive in Norway. Instead, focusing on duties toward nature, as based on international law and domestic law, may delimit state power over nature and decrease the level of government discretion vis-à-vis the natural environment. Duties toward nature can be specified as cooperation duties, due diligence duties, and minimum threshold protection within international law sources. For instance, under the 1979 Bern Convention, which Norway ratified in 1986, Norway should cooperate with Sweden to protect the Scandinavian wolf. However, cooperation is nonexistent because Norwegian leg-

---

Klimasøksmålet og amerikaniseringen av norsk rettskultur' (*Harvest Magazine*, 4 November 2020) <www.harvestmagazine.no/pan/alle-mann-pa-dekk-klimasoksmalet-og-amerikaniseringen-av-norsk-rettskultur>; Somini Sengupta, 'Both Climate Leader and Oil Giant? A Norwegian Paradox' (*The New York Times*, 17 June 2017) <www.nytimes.com/2017/06/17/world/europe/norway-climate-oil.html>.

[152] Norwegian Supreme Court 22 December 2020 (n 89) para 145.

[153] Advocating for 'traditional law', where the courts would not meddle with the legislative and executive branches, see Christian Reusch, 'Kvitt eller dobbelt – Høyesteretts dom i klimasøksmålet' (*Rett24*, 5 January 2021) <https://rett24.no/articles/kvitt-eller-dobbelt-hoyesteretts-dom-i-klimasoksmalet>.

[154] Cathrine Hambro, 'Retorikk i retten' (*Wahl-Larsen Advokatfirma*, 10 December 2020) <https://wla.no/klimasaken-retorikk-i-retten/>.

[155] Esmeralda Colombo, '«Det norske paradokset» i Klimasøksmål Arktis: Ein rettsleg analyse av økonomiske vurderingar' (2020) 2 Retfærd 47; Ole Kristian Fauchald, 'Klimarettssaken og «amerikanisering» av norske domstoler' (2018) 57 Lov Og Rett 158.

[156] Colombo (n 155); Christina Voigt 'The First Climate Judgment Before the Norwegian Supreme Court: Aligning Law with Politics' (2021) 33 JEL 697.

islation does not set any strict duty to protect wolves or their habitats. Even the Norwegian Environmental Agency[157] has maintained that Norwegian policy overrode Sweden's ability to maintain a viable wolf population. Moreover, the Norwegian Parliament advised the government in 2017 to revise the interpretation of the NDA, openly neglecting its obligations following the Bern Convention, so as to allow the killing of more wolves.[158]

3. *Institutional Considerations*

A paradigm based on duties toward nature would thus set a more ambitious agenda for the rule of law for nature in Norway compared to the status quo. Currently, the rule of law for nature is defined by predictability, security, and the absence of arbitrariness and bias in decisions that affect nature, which means a full accounting of environmental values in the decision-making process. This approach aims to fulfill, to a considerable degree, similar functions to those entailed by RoN but by applying alternative legal mechanisms and a different theoretical approach. Implementing duties toward nature directly deduced from the highly valued standard of the rule of law can thus be regarded as a functional equivalent of RoN but better suited for the specific preconditions of the Norwegian legal culture.

In the following section, we propose some avenues on how the seeds of legal duties toward nature can be planted in existing Norwegian legal frameworks. In particular, we identify the institutional settings and tools that best adapt existing law to the duty paradigm without contextually causing a radical rupture with the prevailing features of the institutional and intellectual structures of the Norwegian legal culture, which were surveyed earlier. Duty-paradigmatic approaches and tools need to be operationalized institutionally and substantively to provide efficient environmental protection.

Institutionally, the consideration of duties toward nature is currently missing in both administrative and judicial bodies but can be relatively easily phased in. Notably, the Norwegian legal culture features generalist judges who are routinely called to oversee civil and criminal cases that vary widely. Thus, they have no opportunity to specialize in environmental matters. Indeed, almost 80% of all cases treated at first-instance courts (*Tingrett*) are criminal cases. Of the remaining cases, 19% are civil law matters, mostly family law, and only 0.5% are related to environmental issues, mostly environmental criminal cases.[159] As a matter of consequence, Norwegian judges tend to decide environmental cases without recourse to socio-ecological or specialized legal knowledge. Instead, they employ

---

[157] Norwegian Environmental Agency, *Stevning og begjæring om midlertidig forføyning til Oslo Tingrett* (2017).

[158] Eirik Holmøyvik, 'Ulvestrid mellom statsmaktene' (*Rett24*, 19 February 2020) <https://rett24.no/articles/ulvestrid-mellom-statsmaktene>.

[159] Fauchald (n 155).

a high threshold of review by affording public authorities broad discretion,[160] not even limited by judicial protection of the human right to a clean and safe environment whenever such right would infringe on political prerogatives.

Moreover, the constellation of roughly 40 administrative review bodies that offer alternative dispute resolution to court proceedings does not have competence in environmental matters, leaving it to the original decision-maker, albeit the superior body in the administrative hierarchy, to review its own decision, with obvious impartiality issues.[161] In contrast, decisions are routinely reviewed by specialized environmental courts and administrative bodies in similar legal cultures, such as Sweden and Denmark.[162] Establishing such organs, therefore, poses no insuperable legal-cultural obstacles but would suit a mediation-friendly approach to conflict resolution in the Nordic countries and Norway in particular.[163]

To phase in the consideration of duties toward nature, Fauchald suggests establishing an impartial and independent administrative body[164] that would exclusively be devoted to environmental matters. This possibility has appeared feasible and would adhere to a legal culture that relies on avoiding litigation as far as possible,[165] particularly because accessing courts is extremely expensive in Norway. On average, expenditures for completing litigation at the first instance total circa 10,000 EUR (app. 100,000 NOK), putting Norway at loggerheads with the Aarhus Convention.[166] Nonetheless, the generalist character of the judiciary and the lack of specialized judges should be tackled by requiring at least one mandatory course on environmental law and the rule of law for nature as a prerequisite for obtaining a university law degree.

Based on similar considerations, several legal scholars have suggested establishing a specialized court that might handle interests for nature better than ordinary courts of justice.[167] One of Winge's arguments is that environmental law is mostly based on international and supranational sources and thus requires a spe-

---

160 Ole Kristian Fauchald, 'Er det behov for miljøombud' in Helle Tegner Anker and Birgitte Egelund Olsen (eds), *Miljøretlige emner: Festskrift til Ellen Margrethe Basse* (Djøf Forlag 2008).

161 Sølvi Maria Stordalen Blindheim, *Kontroll av enkeltvedtak med betydning for miljø: I hvilken grad overholder norske kontrollinstanser kravene i Århuskonvensjonen artikkel 9.4?* [Master Thesis] (Norwegian University of Life Sciences, Ås 2018) <https://nmbu.brage.unit.no/nmbu-xmlui/handle/11250/2503570Y>; Bragdø-Ellenes (n 106) 122ff.

162 Helle Tegner Anker and Annika Nilsson, 'The Role of Courts in Environmental Law-Nordic Perspectives' (2010) 3 Journal of Court Innovation 111, 116.

163 Anna Nylund, 'The Many Ways of Civil Mediation in Norway' in Laura Ervo and Anna Nylund (eds), *The Future of Civil Litigation. Access to Court and Court Annexed Mediation in the Nordic Countries* (Springer 2014).

164 Fauchald (n 160) 209ff.

165 Anker and Nilsson (n 162).

166 International Commission of Jurists, *Ikke uoverkommelig dyrt? By Den internasjonale juristkommisjon* (2020) <https://issuu.com/deninternasjonalejuristkommisjon-norge/docs/icj-rapport_print>; Convention on Biological Diversity (n 146).

167 Bragdø-Ellenes (n 106) 129; Winge (n 103) 289.

cialized and professionalized review organ to efficiently protect the environment in accordance with the applicable law. This approach might be regarded as a middle way between a cosmopolitan and a domestic account of duties toward nature, which eventually can develop into full-fledged RoN. However, in 2017, the commission on the establishment of specialized courts (*Særdomstolsutvalget*)[168] refused to consider establishing a specialized court for environmental administrative cases. Moreover, the commission was skeptical about transferring any competencies of administrative dispute resolution organs (which are provided with court-like competencies but still governed by the executive branch of government) to specialized administrative courts or specialized chambers in ordinary courts.[169]

Based on a specific instruction by the government, on the other hand, the Courts Reform Commission discussed a higher degree of specialization in courts for the protection of children's interests.[170] Both children and nature have in common a limited ability to defend themselves in court, but Norway is not seriously considering specialized environmental courts. This interesting empirical observation can easily be explained by the lack of *administrative* courts. By looking at neighboring countries Sweden and Finland, where administrative courts handle thousands of environmental cases in specialized chambers and at comparably low costs for the plaintiff, it appears that the low number of environmental cases in Norway is instead caused by the non-existence of specialized courts and other access-to-justice conundrums, not a lack of demand.[171] It therefore seems highly questionable to argue that specialization in environmental law is redundant in Norway because there is no legal demand.[172]

This argument substantiates the anthropocentric point of departure promoted by Norwegian legal pragmatism. In the most recent proposal from 2020, the suggested 'moderate specialization' of courts for several areas of law, such as Sámi interests, did not explicitly mention environmental protection. The proposal gives rise to developing increasing special competence within ordinary courts, but as we have seen, this is unlikely to have a considerable effect on a more efficient legal protection of nature or the environment.[173] Recent institutional reforms in Norway have therefore proven insufficient.

In addition to efficient environmental review bodies, duties toward nature can be anchored in an Ombudsman for the environment or for future generations, which would be a step away from RoN and a step toward rights-based approaches. As mentioned above, the protection of the environment for future generations

---

[168] Norwegian Public Reports (*Norges Offentlige Utredninger*), NOU 2017:8, 47ff. Norwegian Public Reports (*Norges Offentlige Utredninger*), NOU 2020:11, 290ff.

[169] NOU 2017:8 (n 168) 47ff.

[170] Norwegian Public Reports (*Norges Offentlige Utredninger*), NOU 2019:3.

[171] Nylund (n 43) 101ff.

[172] See the Courts Reform Commission's report applying this argument in NOU 2017:8 (n 168) 49.

[173] NOU 2020:11 (n 168) 295ff. Regarding Sámi interests, see p 298.

is enshrined in Norwegian Constitution §112. As the first country in the world, Wales recently appointed a commissioner for future generations.[174] The role of the commissioner in Wales is to ask difficult questions and force the authorities to justify their decisions. Such comments focus, for instance, on planning policies that are particularly detrimental to nature and the interests of future generations. However, such comments are for guidance only.

When it comes to Norway's Parliamentary Ombud, the Ombud is nominated by Parliament and consequently not entirely impartial. Such an official does not provide a viable alternative to administrative and legal review by specialized alternative dispute resolution organs or courts. In fact, the Ombud's competence lies in safeguarding the rights of individuals in their dealings with public administrators, including human rights. However, they are not appointed to advocate for nature's interests. Consequently, a debate has recently flared up among academics on whether the Ombud could consider a case brought by an environmental organization to challenge a municipality dispensation and the relevant permit to build in the 100-meter belt along the sea (the beach zone), which is by law covered by a building ban.[175] An Ombudsman for nature was evaluated in the process of discussing Norway's Biodiversity Act, short of being approved.[176]

4. Substantive Aspects

Having considered institutional changes that might contribute to strengthening the rule of law for nature in Norway, some substantive aspects are also noteworthy. Substantively, the rule of law for nature would increase with specific changes in Norway's regulatory frameworks on spatial planning. Notably, Section 3-1, the first paragraph of the Planning and Building Act (2008), emphasizes considerations that must be included in the decision-making process to protect nature and the environment. For example, planning authorities – notably municipalities – must account for soil resources, landscape quality, the protection of valuable landscapes and cultural environments, the natural basis for Sámi culture, and the climate through the reduction of greenhouse gas emissions and adaptations to expected climate change. However, such considerations are not the equivalent of providing natural entities with rights. The subject of the duty is municipalities and the state and its organs. Furthermore, the consideration of environmental protection is not outlined hierarchically or given priority over other conflicting considerations such as economic growth. They lack specificity and enforceability. Particularly, there is no enforceable duty on a public authority to apply the Planning and Building Act consistently with the Environmental and Natural Diver-

---

[174] The Future Generations Commissioner for Wales, *Well-being of Future Generations (Wales) Act 2015*, <www.futuregenerations.wales/about-us/future-generations-act/>.

[175] Hanne Harlem and Maria Bakke, 'Hvem skal Sivilombudet beskytte?–en kommentar til Erik Magnus Boe og Erik Keiserud' (2021) 60 Lov Og Rett 447.

[176] Fauchald (n 160) 218.

sity Act (2009), although the authorities must consider the general principles laid down in chapter II of the NDA.

To increase the rule of law for nature, it has been proposed to amend the Planning and Building Act by organizing relevant considerations hierarchically and by imposing specific requirements on how environmental interests must be considered to avoid results that would be detrimental to the aim of protecting nature and securing the survival of crucial ecosystems.[177]

Importantly, three mechanisms can be correlated to the proposed amendment of the Planning and Building Act: spatial budgeting, spatial neutrality, and compensation for an ecological loss, all of which have been recently discussed by academics.[178] Further discussions would include practitioners and NGOs and would mandate that public authorities account for and fulfill specific duties toward nature.[179] Through spatial budgeting (*arealregnskap*), public authorities, eg municipalities, would be obligated to account for the loss of wildlife and areas untouched by human settlements. In Norway, there is no obligation to develop areas already partly touched by human settlement. This situation results in pristine environments being chosen for private and public buildings or roads without accounting for all the cumulative consequences of the decision or a duty to consider alternatives. As a result, buildings and roads are currently the primary cause of the loss of crucial ecosystems such as wetlands in Norway.[180] The obligation to at least account for the effects of such loss would significantly increase the rule of law for nature.

As a prerequisite, municipalities should gather more resources, notably from the state, to map the natural environment. Accordingly, municipalities would be obliged to issue plans that specifically concern biodiversity (*kommunedelplaner for naturmangfold*), constituting a relevant repository of knowledge for further planning concerning the natural environment. Notwithstanding some experimental programs in this direction,[181] there is a striking lack of knowledge, funding, and incentives for municipalities to map the ecosystems they manage.[182] As academics and practitioners have elucidated, the resource-sharing system in Norway makes it profitable for municipalities to allow for building and planning in pris-

---

177 Colombo and Hoff-Elimari (n 30).

178 Anna M Muhrman, Per Kristian Rørstad and Jonathan E Colman, Økokontoer og arealbanker i kommunal natur-og arealforvaltning (2021) 114(1–2) Kart Og Plan 55, 60; Winge (n 103).

179 Bjørn Asle Nord, 'Frykter naturens rettsvern svekkes' (NRK Nettavis, 3 July 2020) <www.nrk.no/klima/frykter-naturens-rettsvern-svekkes-1.15074131>.

180 Klima og miljødepartementet, Naturstrategi for våtmark [Plan] (23 June 2021). <www.regjeringen.no/nn/dokumenter/naturstrategi-for-vatmark/id2863261>.

181 Miljødirektoratet, Kommunedelplan for naturmangfold. M-1676 (2020) <https://www.miljodirektoratet.no/publikasjoner/2020/april-2020/evalueringsrapport-pilotprosjektet-kommunedelplan-for-naturmangfold/>.

182 Nicolai K Winge, *Arealregnskap – i skjæringsfeltet mellom juss og politikk* (November 2021).

tine areas since municipalities have jurisdiction to levy taxes on productive activities occurring in those areas.[183]

Connected to spatial budgeting, the second mechanism that can increase the rule of law for nature is spatial neutrality. Spatial neutrality would ensure that construction outside of building zones, namely in the pristine natural environment, should be minimized and, if permitted, fully compensated.[184] However, ecological compensation should be a last resort after avoiding, mitigating, and restoring affected natural areas. Spatial neutrality is no magic wand if the conditions and criteria of ecological compensation fall short of being specific and enforceable.

In this sense, the third and final mechanism for increasing the rule of law for nature in Norway is compensation for the loss of the natural environment. There is no general requirement (yet) to compensate for nature lost to development projects under Norwegian law. Paragraph 48 of the Biodiversity Act is the sole provision discussing this point, although it only concerns dispensation from already-decided protective measures. Paragraph 48 requires ecological compensation when the valuable natural environment has been impaired by development projects, meaning that compensatory measures for biodiversity would physically replace the loss of valuable nature. In practice, however, small areas and very few projects trigger the claim for compensation under the Biodiversity Act.[185]

By carrying out an interview-based qualitative study, Muhrman, Rørstad, and Colman find that, among biodiversity experts and public authorities in charge of biodiversity tasks in Norway, two alternative concepts are of particular interest. Accordingly, the authors recommend a basis for amendments to the Biodiversity Act: eco-accounts and area banks, which can achieve zero net loss of biodiversity and increase the protection of nature.[186] A zero net loss means that measures should eliminate the deterioration of nature, with the result being that biodiversity either exceeds the loss or is compensated beyond the loss.[187] When using an eco-account, monetary compensation is paid into a public account, and the municipality decides how to use those funds for compensation measures in line with some basic legislative principles. In area banks, the developer would be a prime actor, becoming responsible – together with the municipality – for compensation measures. Valuable natural areas should be protected and would not be compensated if they cannot be restored or belong to ecosystems that do not have enough protection at present, such as wetlands and millennial forests. If there are alter-

---

183 Eivind Hoff-Elimari, 'Sp: Krev stans i «veksttilskuddet» som sentraliserer Norge' (*Nationen*, 2 October 2021) <www.nationen.no/motkultur/kronikk/sp-krev-stans-i-veksttilskuddet-som-sentraliserer-norge/>.

184 Kommuneplan Nordre Follo kommune 2019–2030 Arealdel Oppegård (2019).

185 Muhrman, Rørstad and Colman (n 178) 60.

186 Muhrman, Rørstad and Colman (n 178) 59.

187 Muhrman, Rørstad and Colman (n 178) 56.

natives to destroying an ecosystem, a compensation scheme would not be triggered, and the municipality would be obliged to change its spatial planning.

Such a mechanism seems feasible in a Norwegian context, where an academic study has evaluated its applicability in light of comparative experience, especially from Germany.[188] In Germany, such instruments have worked for 20 years, with good evaluation and affordable schemes. An example is the development of motorways, where the compensation, on average, accounts for 5.4% of the total project costs with effective restoration measures.[189] If a project continues to impact the natural environment, even when the damage to the environment is minimized, the unavoidable impact must be compensated for, or offset by, nature conservation or landscaping measures.

According to the Supreme Federal Administrative Court of Germany, the definition of compensation is situational. It must not coincide with the impact, but there must be a functional connection between the impact and compensatory measures.[190] This functional context means that the compensation measure applies to the site of impact and restores the ecological function disturbed by it. Compensation measures must achieve the goals within a 'reasonable deadline'. In practice, ecological functions can be compensated only if they can be restored within a maximum of 25 years. If the adverse effects cannot be avoided or compensated within a reasonable time, the authorities are prohibited from allowing the activity.[191] As mentioned, similar measures and mechanisms are now being discussed and proposed in Norway.[192]

According to current Norwegian law, destroying nature appears to bear no economic or liability consequences for public authorities or private investors. In practice, public authorities have little incentive to protect and preserve valuable natural habitats or ecosystems: profit is privatized, while the loss of nature is socialized because ecological compensation is never compulsory. Above, three mechanisms – spatial budgeting, spatial neutrality, and compensation for ecological loss – were described to show how amendments to applicable law may increase consideration of the natural environment in public authorities' everyday decision-making processes. Such consideration would not amount to bestowing full-fledged rights of nature. However, in line with theoretical approaches discussed in Norwegian culture, notably Deep Ecology, the proposed amendments would ramp up duties toward nature at the local and central government levels. Such duties may even become predecessors of rights of nature in the long run.

---

[188] Muhrman, Rørstad and Colman (n 178).

[189] Juliane Albrecht, Jochen Schumacher and Wolfgang Wende, 'The German Impact-Mitigation Regulation – A Model for the EU's No-net-loss Strategy and Biodiversity Offsets?' (2014) 44(3) EPL 317, 323.

[190] Federal Administrative Court 17 February 1997, 4 VR 17.96, NuR 1998, 305.

[191] Albrecht, Schumacher and Wende (n 189) 319.

[192] Muhrman, Rørstad and Colman (n 178).

V. Concluding Remarks

The consequences that recognition of the rights of nature would have for politics, law, and culture in Norway can be discussed on a theoretical basis only and remain, to some degree, speculative. Nevertheless, reviewing the development and current features of the Norwegian legal culture allows us to provide some 'educated guesses'.

First, to recognize the rights of nature would require a considerable change in what Tuori calls the deep structure of the legal culture.[193] Both politicians and lawyers would need to adapt to a more ecologic or biocentric approach to law. As we have argued, 'Deep Ecology' and a stronger emphasis on the rule of law for nature might offer more easily digestible solutions for actors that are based on an anthropocentric view of politics and law. Thus, such an approach would be better suited for the existing features of Norwegian legal culture.

Second, the acceptance of the Norwegian climate litigation case by the ECtHR offers an opportunity to reassess domestic environmental principles following a less parochial approach, although the outcome remains to be seen.[194]

Third, a stronger focus on duties toward nature, ecosystems, and natural entities will enforce and facilitate a broader interdisciplinary scientific and public debate on the rights of nature, which again might trigger changes in the deep structure of the legal culture. A radical re-evaluation of fundamental values, which could restructure the legal and political discourse and underlying mentality, is a precondition for any sustainable change in Norway as in any other legal culture. Deep Ecology offers one way to calibrate the underlying value of nature without altogether abandoning the established institutional and intellectual structures of the current legal and political culture.

However, before we can start discussing these questions, the deep structure of the Norwegian legal culture needs to change considerably. This change is not limited to law and legal principles but must impact the underlying values and mentality of politicians and lawyers alike. A new natural ideology and philosophy supported by at least a broad popular majority would be required. Simply recognizing RoN will not solve current problems.

---

[193] Kaarlo Tuori, *Critical Legal Positivism* (Routledge 2002) 12.

[194] ECtHR, Application no 34068/21 Greenpeace Nordic and Others against Norway from 15 June 2021. See also the resulting 'Questions posed to the Norwegian Government' (2022). See Sabin Center for Climate Change, Law, *Greenpeace Nordic and Others v Norway* (2022) <http://climatecasechart.com/non-us-case/greenpeace-nordic-assn-v-ministry-of-petroleum-and-energy-ecthr/>.

Québec (and Canada)

Nature, Culture, Rupture: Tools for Environmental Protection in a Liberal Anthropocentric Legal System

Stéphanie Roy, Gaële Gidrol-Mistral and Alexandra Popovici[*]

I. Existence and Contents

Neither Québec nor Canada has formally recognized any rights to nature envisaged as an autonomous entity, as has been done in Ecuador and Bolivia, where the Constitution, in the first case, and a federal law, in the second, now recognize rights in Mother Earth (*Pasha Mama* in Ecuador and *Madre Tierra* in Bolivia).[1] Canadian law's liberal values and other obstacles to the development and implementation of rights of nature are at the core of this silence. Despite the absence

[*] The manuscript was finalized in April 2023. The assessment of literature, case law, and other materials, reflects the legal situation at that time.

[1] Constitution of the Republic of Ecuador 2008; Ley No 300 of 15 October 2012 (Bolivia).

of a fundamental shift toward the protection of rights of nature, some local initiatives have been the ground of interesting manifestations. Indeed, a local community of the province has granted legal personality to the Magpie River – *Mutehekau Shipu* in Innu – an important and untouched river at the heart of the traditional territory of the Innu community of Ekuanitshit.

1. Granted Rights

Indeed, on 16 February 2021, the Innu Council[2] of Ekuanitshit and the Regional County Municipality of Minganie adopted mirror resolutions granting legal personhood to the river and recognizing nine fundamental rights upon the Magpie River: the right to live, to exist and to flow; the right to respect its natural cycles; the right to evolve naturally, to be preserved and protected; the right to maintain its natural biodiversity; the right to perform essential functions within its ecosystem; the right to maintain its integrity; the right to be free from pollution; the right to regeneration and restoration; and the right to take legal action.[3] These resolutions also provide that the municipality and the Innu Council may appoint guardians whose mission will be to ensure respect for the rights and interests of the river.[4] The appointment of guardians to protect the rights and interests of the river is reminiscent of Aotearoa New Zealand's regime surrounding the protection of the Whanganui River, which was recognized as a law by the *Te Awa Tupua (Whanganui River Claim Settlement) Act* in 2017.[5] It should be noted, however, that the concepts of 'guardian' and 'representative' of nature are not synonyms, as will be explained in Section III 'Enforcement', below.

The Québec provincial legislature has been silent on the matter and did not pass any such legislation, which significantly diminishes the effects of these resolutions. The jurisdiction of the Regional County Municipality of Minganie is limited to flow regulations[6] and that of the Ekuanitshit Innu Council – a band coun-

---

[2] The plural of Innu in this language is 'Innuat' and the Council does not spell its name with an 's'. Note that this section reproduces some passages from Stéphanie Roy, 'Traverser le pont une fois à la rivière : réflexions sur la mise en œuvre des droits de la nature et l'évolution du droit de l'environnement' (2022) 19(1) McGill Journal of Sustainable Development Law 32.

[3] Morgan Lowrie and La Presse canadienne, 'La rivière Magpie obtient le statut de personnalité juridique' (*Le Devoir*, 27 February 2021), <www.ledevoir.com/societe/environnement/596053/la-riviere-magpie-obtient-le-statut-de-personnalite-juridique>; Innu Council of Ekuanitshit, Résolution No 919-082 (2021) 12; Municipalité régionale de comté de Minganie, Résolution No 025-21 – Reconnaissance de la personnalité juridique et des droits de la rivière Magpie – Mutehekau Shipu (2021).

[4] See Innu Council of Ekuanitshit's Resolution (n 3) 12.

[5] Te Awa Tupua (Whanganui River Claim Settlement) Act 2017 (New Zealand).

[6] Municipal Powers Act, CQLR, c C-47.1, ss 103–106.

'103. A regional county municipality has jurisdiction over continuously or intermittently flowing watercourses, including those artificially created or modified, except

(1) watercourses or parts of watercourses that the Government determines, after consultation with the Minister of Sustainable Development, Environment and Parks, by an order in

cil under the *Indian Act* – to the by-laws relating to the maintenance of watercourses and the protection of fur-bearing animals, fish, and other game on the reserve.[7] Although their powers allow them to adopt certain conservation measures vis-à-vis nature (the Council) and water flow management (the municipality), one can imagine that a court, adopting a restrictive interpretation of their powers, would conclude that neither the municipality nor the Council can change the legal status of the river, and they would consider that only the province has jurisdiction to do so under its power to legislate property and civil rights.[8] The question of whether the legal personality and rights of the Magpie River will be formally recognized in Québec law therefore remains open.[9] The Magpie River, despite the vagueness surrounding the validity of its regime, constitutes the only example of nature rights protection in Canada.

The Montréal-based International Observatory on the Rights of Nature and the St. Lawrence River Alliance/Magtogoek are also advocating for the recognition of legal personality in the St. Lawrence River, the province's largest river.[10] On

---

council that comes into force on the date of its publication in the *Gazette officielle du Québec* or any later date specified in the order;

(2) a ditch along a public or private road;

(3) a common ditch within the meaning of article 1002 of the Civil Code; and

(4) a drainage ditch

(a) used solely for drainage or irrigation;

(b) that was artificially created; and

(c) the watershed of which has an area of less than 100 hectares.

The part of a watercourse used as a ditch remains under the jurisdiction of the regional county municipality.'

[7] Indian Act, RSC 1985, c I-5, s 81(1)(f) and (o).

'81(1) The council of a band may make by-laws not inconsistent with this Act or with any regulation made by the Governor in Council or the Minister, for any or all of the following purposes, namely,

[…]

(f) the construction and maintenance of watercourses, roads, bridges, ditches, fences and other local works;

[…]

(o) the preservation, protection and management of fur-bearing animals, fish and other game on the reserve;

[…]'

[8] Constitution Act, 1867 (UK), 30 & 31 Vict C 3, s 92(13).

'92. In each Province the Legislature may exclusively make Laws in relation to Matters coming within the Classes of Subjects next hereinafter enumerated; that is to say,

[…]

13. Property and Civil Rights in the Province.

[…].'

[9] See Yenny Vega Cárdenas and Daniel Turp (eds), *A Legal Personality for the St. Lawrence River and the Rivers of the World* (JFD 2021).

[10] Observatoire international des droits de la Nature, 'Fleuve Saint-Laurent (Québec, Canada)' <www.observatoireNature.org/fleuve-saint-laurent>; Yenny Vega Cárdenas, 'Introduction'

5 May 2022, MP Alexandre Boulerice of the New Democratic Party (NDP) of Canada introduced a bill in the Federal Parliament that proposed the conferral of legal personhood upon the St. Lawrence River, alongside the establishment of a framework comprising *interprovincial* guardians for its enforcement.[11] Concurrently, on the same date, MP Émilie Lessard-Therrien of Québec Solidaire presented a similar bill in the Québec Provincial Parliament advocating for the attribution of legal personhood to the river, supplemented by a system of *provincial* guardians tasked with its enforcement.[12] Subsequent to their introduction, neither bill underwent further deliberation in Parliament, primarily due to a lack of endorsement from a majority party. Nevertheless, the Assembly of First Nation Québec-Labrador (AFNQL), comprising the Assembly of Chiefs representing the 43 First Nations communities situated in Québec and Labrador, unanimously endorsed a resolution on 19 April 2023 for the conferment of legal personhood upon the St. Lawrence River. This resolution was subsequently submitted to the United Nations, thereby amplifying the political significance of the initiative.[13] However, notwithstanding these developments, the actualization of legal personhood for the St. Lawrence River remains largely speculative in nature.

On the other side of the country, in British Columbia, the First Nation ?Esdilagh (Alexandria), a member of the Tsilhqot'in Nation, passed legislation – the *Elhdaqox Dechen Ts'edilhtan* (Sturgeon River Law) – in favor of a river that lies on its traditional territory. Without granting rights to the river directly, this law recognizes the ?Esdilagh First Nation's stewardship *responsibilities* toward the Fraser River to curb its degradation as caused by industrial projects.[14] The law requires that 'any proposed activities that could impact the Fraser (Sturgeon) River have the nation's free, prior and informed consent.'[15] As with the Magpie River, the scope that the courts will assign to this Act emanating from the Aboriginal community itself, particularly, the obligation of the federal and provincial governments to justify any legislative measure that adversely affects an Aborigi-

in Yenny Vega Cárdenas and Daniel Turp (eds) (n 9) 33; Observatoire international des droits de la Nature, 'St. Lawrence River Alliance/Magtogoek' <https://observatoirenature.org/observatorio/en/st-lawrence-river-alliance-2/>.

[11] Bill C-271, An Act to give legal capacity to the St. Lawrence River and to provide for measures respecting its protection, 1st Sess, 44th Parl, 2022 (First Reading).

[12] Bill 990, An Act to confer rights on the St. Lawrence River, 2nd Sess, 42nd Leg, 2022 (First Reading).

[13] Joe Lofaro, 'First Nations Chiefs Adopt Resolution Declaring St. Lawrence River a Legal Person' (*CTV News*, 24 April 2023) <https://montreal.ctvnews.ca/first-nations-chiefs-adopt-resolution-declaring-st-lawrence-river-a-legal-person-1.6369335>. The resolution was submitted to the United Nations on the Twelfth Interactive Dialogue of the General Assembly on Harmony with Nature in New York.

[14] Rebecca Dyok, 'B.C. First Nation Adopts Historic Law to Protect Fraser River' (*The Star*, 28 May 2020) <www.thestar.com/news/canada/2020/05/28/bc-first-nation-adopts-historic-law-to-protect-fraser-river.html>.

[15] Tsilhqot'in Nation, 'Elhdaqox Dechen Ts'edilhtan (Sturgeon River Law)' from 27 May 2020, s 10 <http://files.harmonywithNatureun.org/uploads/upload995.pdf>.

nal right protected by s 35 of the *Constitution Act 1982* in the absence of the consent of the peoples concerned, remains whole.[16] In Québec, as in other Canadian provinces, the laws on natural resources and land use planning, and the *Civil Code of Québec*, do not reflect Aboriginal rights, which would be beneficial to the development of rights of nature. These laws – which are not in line with international law, specifically the 2007 *United Nations Declaration on the Rights of Indigenous Peoples*, which was introduced into Canadian law by federal law on 21 June 2021 – should be amended to respect Indigenous rights and traditions.[17]

Indigenous communities around the world view humans as part of nature and therefore maintain a relationship of interdependence with it, contrary to the dominance relationship that prevails in Western liberal philosophy.[18] This relationship has inspired the *guardianship* regimes surrounding the protection of the former Te Urewera Park and the Whanganui River in Aotearoa New Zealand, as well as the protection of Mother Earth's rights in Ecuador and Bolivia.[19] Canada is no exception to this trend, which probably explains why the initiatives to protect the rights of nature in two out of the three previous examples emerge from Indigenous communities, namely the Ekuanitshit Innu community in Québec and the First Nation ?Esdilagh (Alexandria) in British Columbia.

2. Other Legal Provisions Conferring Special Status to Nature

Some environmental protection mechanisms in various provincial statutes provide increased protection for nature by recognizing its special status or intrinsic value, but without granting it any rights. For example, in Québec, the *Charter of Human Rights and Freedoms* – a quasi-constitutional document – states, in s 46.1,

---

[16] Constitution Act 1982, Appendix B of the Canada Act 1982 (UK), 1982, c 11, s 35; Constitution Act 1982, Schedule B to the Canada Act 1982, 1982, c 11 (UK), s 35; *R v Sparrow*, [1990] 1 SCR 1075 at 1110; Jean Leclair and Michel Morin, 'Peuples autochtones et droit constitutionnel, feuilles mobiles (consultées le 3 mars 2020)' [2011] JurisClasseur Québec (Collection Droit public – Droit constitutionnel) 15, 144–145.

Section 35. Recognition of existing aboriginal and treaty rights

'(1) The existing aboriginal and treaty rights of the aboriginal peoples of Canada are hereby recognized and affirmed.

Definition of aboriginal peoples of Canada

(2) In this Act, "aboriginal peoples of Canada" includes the Indian, Inuit and Métis peoples of Canada.

Land claims agreements

(3) For greater certainty, in subsection (1) treaty rights includes rights that now exist by way of land claims agreements or may be so acquired.

Aboriginal and treaty rights are guaranteed equally to both sexes

(4) Notwithstanding any other provision of this Act, the aboriginal and treaty rights referred to in subsection (1) are guaranteed equally to male and female persons.'

[17] United Nations Declaration on the Rights of Indigenous Peoples Act, SC 2021, c 14.

[18] Catherine Iorns Magallanes, 'Maori Cultural Rights in Aotearoa New Zealand: Protecting the Cosmology that Protects the Environment' (2015) 21(2) Widener L Rev 273, 277–279.

[19] Iorns Magallanes (n 18) 283f; Vega Cárdenas (n 10) 74.

that every person has the right 'to live in a healthful environment in which biodiversity is preserved, to the extent and according to the standards provided by law.'[20] This article, invoked in conjunction with fundamental rights and freedoms, such as the right 'to life, and to personal security, inviolability and freedom' of s 1, could allow such an interpretation.[21] However, it is more difficult to invoke autonomously, since the Charter does not confer on it the special status it confers to ss 1 to 38. As provided by s 52 of the Charter, '[n]o provision of any Act [...] may derogate to the Charter,' which means that any Act or provision can be invalidated by a court when contrary to any section between 1 and 38, which is not the case of subsequent provisions, including s 46.1.[22] In practice, the right to a healthy environment has to be invoked jointly with another right protected by ss 1 to 38 to have any effect.

Also in Québec, the *Act to Affirm the Collective Nature of Water Resources and to Promote Better Governance of Water and Associated Environments* provides that water is the 'common heritage of the Québec nation' and that 'water is for everyone's use and must be accessible in the quantity and quality required to meet every individual's essential needs.'[23] Above all, its preamble confirms that '[t]he State, as custodian of the interests of the nation in water resources, must be invested with the necessary powers to ensure their protection and management.' The law also gives the Attorney General the power to take legal action in the event of damage to water 'in the name of the State as custodian of the interests of the nation in water resources.'[24] However, this power of action, as well as the law in general, has not been applied and has rather fulfilled an interpretative function in case law.[25] It also imposes few real duties on the State with regard to the protection of water resources.[26]

---

[20] Charter of Human Rights and Freedoms, CQLR, c C-12, s 46.1. See below in section VIII for a bilingual version.

[21] Sophie Thériault and David Robitaille, 'Les droits environnementaux dans la Charte des droits et libertés de la personne du Québec: Pistes de réflexion' (2012) 57(2) McGill LJ 211, 225.

[22] Charter of Human Rights and Freedoms (n 20) s 52.

'52. No provision of any Act, even subsequent to the Charter, may derogate from ss 1 to 38, except so far as provided by those sections, unless such Act expressly states that it applies despite the Charter.'

[23] Act to affirm the collective nature of water resources and to promote better governance of water and associated environments, CQLR, c C-6.2 preamble and ss 1–2 (Water Act). This section includes excerpts from Stéphanie Roy, 'La théorie de l'État fiduciaire et le contrôle judiciaire des actes de l'Administration en matière environnementale: vers une redéfinition des obligations de l'État' (LL.D Thesis, Laval University 2021) 255–261. See also Alexandre Lillo, The Shape of Water – Building a Legal Framework for Water Management in Canada (University of Ottawa Thesis Collection 2020) <https://ssrn.com/abstract=3698036>.

[24] Water Act, s 8.

[25] *Wallot v Quebec (City of)*, 2011 QCCA 1165 paras 25–26.

[26] On this subject, see Paule Halley and Christine Gagnon, 'Les réformes du droit de l'eau au Québec: un patrimoine commun et son État gardien' in Paule Halley and Julia Sotousek

Finally, the laws of some jurisdictions explicitly state that the government must act as an environmental trustee.[27] This relates in particular to the *Northwest Territories Environmental Rights Act* and the *Yukon Environment Act*. The first enshrines the right of every adult citizen residing in the Northwest Territories to take legal action 'to protect the environment and the public trust,' which it defines as 'the collective interest of the people of the Northwest Territories in the quality of the environment and the protection of the environment for future generations.'[28]

The Act further provides that the Executive Council 'shall prepare a draft statement of environmental values that explains how environmental considerations, including the right to a healthy environment, will be integrated into decisions that might have a significant impact on the environment' (s 17). Ministers or deputy heads of public bodies covered by such a statement must take reasonable steps to ensure that the statement of environmental values is taken into account (s 18). Finally, the Act requires the government to 'report publicly on the state of the environment in the Northwest Territories' (s 20(1)). The purpose of the report is:

'(a) to provide an assessment of environmental status and trends in the Northwest Territories;
(b) to provide data and information for territorial, national and international state of the environment initiatives; and
(c) to provide an early warning system of possible impacts resulting from environmental change.'[29]

Specifically, as required by the Act, the report has to:

'(a) present baseline information on the environment;
(b) incorporate traditional, local and scientific knowledge;
(c) establish and report on indicators of impairment of or improvement to the environment;
(d) identify and present analyses of trends and changes in indicators established as required by paragraph (c);

(eds), *L'environnement, notre patrimoine commun et son État gardien: aspects juridiques, nationaux, transnationaux et internationaux* (Yvon Blais 2012) 245f.

[27] Raphaëlle Groulx-Julien, 'Les obligations fiduciaires de l'État pour la protection de l'environnement' in Paule Halley and Julia Sotousek (eds) (n 26) 313–319.

[28] The relevant articles in the Environmental Rights Act, SNWT 2019, c 19, read as follows:

'1. In this Act,

[…]

"public trust" means the collective interest of the people of the Northwest Territories in the quality of the environment and the protection of the environment for future generations.

[…].'

'13.(1) Subject to subsection (3), every adult resident in the Northwest Territories has the right to protect the environment *and the public trust*, by commencing an action in the Supreme Court against any person for any act or omission that the resident believes on reasonable grounds has caused or is likely to cause significant harm to the environment. […].'

[29] Environmental Rights Act, s 20(2).

(e) identify links between changes in the environment and causal factors such as development activity, climate change and natural wildlife population trends;
(f) identify emerging problems for the environment, especially those involving long-term and cumulative effects; and
(g) address any other matter that the Minister considers appropriate.'[30]

This law has been in force since 2019, and the first report was filed on 3 June 2022. The report summarizes current environmental issues in the Northwest Territories and those linked to climate change as required by the Act. It also discusses future actions to be taken by the government on various themes, such as the implementation of the *NWT Water Stewardship Strategy* to ensure water quality and abundance, and the implementation of three new waste diversion programs.[31] According to s 20(3) of the Act, a new report has to be submitted every four years subsequent to the publication of the first report. Although the law imposes few substantive duties on the government, the procedural obligations it entails closely resemble the duties that one would typically expect of a trustee.

The *Yukon Environment Act*, by contrast, states in its preamble that the Government of Yukon 'is the trustee of the public trust and is therefore responsible for the protection of the collective interest of the people of the Yukon in the quality of the natural environment'.[32] It defines 'public trust' in s 2 as 'the collective interest of the people of the Yukon in the quality of the natural environment and the protection of the natural environment for the benefit of present and future generations.' In addition to providing, as in the *Northwest Territories Act*, that any resident of Yukon (including the government) may take action against a person who degrades the environment, it clearly provides that the Government of Yukon may be held liable if it 'has failed to meet its responsibilities as trustee of the public trust to protect the natural environment from actual or likely impairment.'[33]

In addition, the Act broadens legal standing by providing that any interested person may initiate an action even if they are unable to show 'any greater or different right, harm, or interest than any other person' or 'any pecuniary or proprietary right or interest in the subject matter of the proceeding' (s 10(1)(a) and (b)). This distinguishes this regime from 'traditional' legal mechanisms that generally require a personal harm to a person's rights or property for a remedy to be pos-

---

30 Environmental Rights Act, s 20(4).

31 Government of Northwest Territories, 'NWT State of Environment Report 2022' (3 June 2022) <www.gov.nt.ca/ecc/sites/ecc/files/resources/wt_state_of_the_environment_report_2022.pdf>.

32 Environment Act, RSY 2002, c 76, preamble.

33 Environment Act, s 8(1): 'Every adult or corporate person resident in the Yukon who has reasonable grounds to believe that
(a) a person has impaired or is likely to impair the natural environment; or
(b) the Government of the Yukon has failed to meet its responsibilities as trustee of the public trust to protect the natural environment from actual or likely impairment
may commence an action in the Supreme Court. […]'

sible. The *Yukon Environment Act* also provides that '[a]ny two persons resident in the Yukon, who believe on reasonable grounds that an activity is impairing or is likely to impair the natural environment may apply to the Minister for an investigation.' (s 14(1)). The Minister will then be required to prepare a report on the progress of the investigation (s 16(1)).

As a substantive duty, the Act provides that:

'*38 Public trust*
(1) The Government of the Yukon is the trustee of the public trust.
(2) The Government of the Yukon shall, subject to this Act or a schedule 1 enactment, conserve the natural environment in accordance with the public trust.'

To ensure that this objective is achieved, the Act provides that the members of the Executive Council shall take the following measures:

'*39 Responsibilities of the Members of the Executive Council*
(1) For the purpose of achieving the objectives of this Act, the members of the Executive Council shall
(a) identify the implications for the environment of their decisions and, if a decision may have significant implications for the environment or sustainable development, provide a public report;
(b) ensure that consideration of the environment is integrated in all decisions of the Government of the Yukon, including policies, programs, budgeting, funding, regulatory initiatives, plans, developments, and projects; and
(c) ensure that standards for conservation of the environment and sustainable development are incorporated in the purchasing policies and practices of the Government of the Yukon. [...]'

The members of the Executive Council must prepare a report tabled in the Legislative Assembly every three years to ensure that these responsibilities are carried out effectively and fairly (s 39(2) and (3)). Under the Act, the Minister may also establish an all-party council to advise on sustainable development, which is currently non-existent (ss 40–41). From a procedural perspective, Yukon's legislation also requires the government to produce a public report on the state of the environment every three years (ss 47(1) and 48(1)). The goal is, among other things, 'to allow the public to monitor progress toward the achievement of the objectives of this Act.'[34] Interestingly, in addition to providing information on the impairment or improvement of the environment, it must 'incorporate the traditional knowledge of Yukon First Nation members as it relates to the environment.'[35] Finally, the Act provides that the Minister of the Environment *may* take measures to promote understanding and a sense of responsibility for the environment, such as a public information and awareness program (s 51(1)(a)).

These laws are interesting in that they enshrine the right of citizens to take legal action against any person who may have caused harmful effects on the environment, including the government that is considered to be the 'trustee of the public trust' in environmental matters. It must be noted that, these provisions have

[34] Environment Act, s 47(2)(b).
[35] Environment Act, s 48(2).

only seldom been raised in court and never successfully. What is more, the duties they impose are broad and have undefined contours ('protect the natural environment,' 'determine the impact of their decisions on the environment,' 'ensure that the issue of the environment is taken into account'), which makes their implementation more difficult. On the other hand, the procedural duties of reporting to the public are articulated more clearly and constitute a step forward.

In Ontario, Canada's most populous province, the *Environmental Bill of Rights* provides that any resident of Ontario may bring an action against a 'person [who] has contravened or will imminently contravene an Act, regulation or instrument prescribed for the purposes of Part V and the actual or imminent contravention has caused or will imminently cause significant harm to a public resource of Ontario,' except where the damage is 'from odour, noise or dust resulting from an agricultural operation.'[36] This provision therefore also broadens citizens' standing to bring proceedings in favor of the environment.

Another interesting initiative can be found in British Columbia, where Rachel Garret and Stepan Wood proposed to either amend the existing law or change the law by giving distinctive rights to nature itself.[37] Their text argues that 'protecting the rights of Indigenous peoples, including rights to self-government and to free, prior and informed consent to activities affecting them and their territories, can be an effective way to protect the rights of nature.'[38]

More symbolically, the Inuit people of Nunavut passed the *Nunatsiavut Constitution Act*, in which they provide that every Inuk has the 'right to an environment that is not harmful to his or her health or well-being and to have the environment protected for the benefit of present and future generations.'[39] However, it is a law that has only a symbolic effect in the community, since it is the *Nunavut Act* that governs the administration of this territory under federal law.[40]

II. Legal and Political Operators

In the case of the Magpie River, citizen groups have long been engaged in its protection to counter the threat of a potential hydroelectric dam, ultimately leading to the creation of the Muteshekau-shipu Alliance.[41] The latter is composed of the

[36] Environmental Bill of Rights 1993, 1993 SO, c 28 (Ontario), s 84(1) and (4).

[37] Rachel Garrett and Stepan Wood, 'CLE Working Paper No 1/2020 – Rights of Nature Legislation for British Columbia: Issues and Options' (2020) Centre for Law and the Environment 4, <https://commons.allard.ubc.ca/cle/4>. Preliminary draft model legislation is suggested at 17.

[38] Garret and Wood (n 37) at 11.

[39] Nunatsiavut Constitution Act (CIL 31-12-2012 N-3, Schedule A, c 2), s 2.4.20.

[40] Nunavut Act, SC 1993, c 28.

[41] Pier-Olivier Boudreault, 'Dans les coulisses de la reconnaissance des droits de la rivière Magpie' (*SNAP*, 15 March 2021) <https://snapquebec.org/dans-les-coulisses-de-la-reconnaissance-des-droits-de-la-riviere-magpie/>.

Canadian Parks and Wilderness Society (CPAWS), the Association Eaux Vives, the Regional County Municipality of Minganie, and the Innu Council of Ekuanitshit.[42] They first proposed a protected area project, which was set aside by the government, probably because of the hydroelectric potential of the river.[43] That is why the Alliance sought a way to protect the river without the assistance of the provincial government, which led to the adoption of resolutions by the Ekuanitshit Innu Council and the Minganie Regional County Municipality, aimed at protecting the intrinsic value of the river. For the time being, initiatives to protect the rights of nature are the result of local actors and not government actors. Except for the cases already mentioned, the media have not been vocal about other citizen initiatives related to rights of nature, which makes it difficult to identify other political operators precisely.

III. Enforcement[44]

Regarding the implementation and enforcement of potential rights of nature, it is important to highlight the essential distinction in law between custodianship or guardianship and representation or agency. Indeed, in Québec civil law, the distinction is clear. Representation presupposes the existence of subjective rights that are exercised by the representative in the name of the represented, who has legal personality.[45] The mandate is its paradigm.[46] The mandator is the one supervising the mandatory. The mechanism implemented under guardianship is quite different; it is based not on the mechanism of representation but on the notion of power exercised in the interest of others or for a particular purpose, such as the protection of nature.[47] Here, the guardian is considered an 'administrator of the property of others,' as defined by the legislature in the *Civil Code of Québec*, and

42 'La rivière Magpie est reconnue comme une «personne juridique», une première au Canada' (*SNAP*, 23 February 2021) <https://snapcanada.org/la-riviere-magpie-est-reconnue-comme-une-personne-juridique-une-premiere-au-canada/>.

43 Alexandre Shields, 'L'industrie avant la protection du territoire?' (*Le Devoir*, 11 February 2021) <www.ledevoir.com/societe/environnement/595022/environnement-l-industrie-avant-la-protection-du-territoire>; according to Quebec's Natural Heritage Conservation Act, CQLR, c C-61.01, s 2, a protected area 'means a "protected area" within the meaning of the United Nations Convention on Biological Diversity and as interpreted by the IUCN in the Guidelines for Applying Protected Area Management Categories (2008)'; it is a status that restricts development activities and increases conservation measures in a particular area.

44 See also section VIII., below.

45 Madeleine Cantin Cumyn, *L'administration du bien d'autrui*, 2nd edn with the collaboration of Michelle Cumyn (Yvon Blais 2014).

46 Adrian Popovici, *La couleur du mandat* (Thémis 1995).

47 Madeleine Cantin Cumyn, 'The Legal Power' (2009) 17(3) ERPL 345.

their duties can extend to both persons and things.[48] This stewardship regime does not impose obligations per se, but fiduciary duties on the person vested with powers, and any interested party can enforce these duties. Although in some circumstances, for example in the case of the Magpie River, the language used seems to attribute legal personality to nature – which would mean we are in the realm of representation[49] – it is rather a situation of guardians who are vested with powers and made responsible for protecting nature, and this regardless of the legal status of nature.[50]

IV. Theoretical Sources and Foundations

There is no school of thought specific to Québec and Canada. The theoretical movements that are used come, for example, from Earth Jurisprudence[51] or from ethical movements such as the Land Ethic.[52] Christopher Stone, with his article *Should Trees Have Standing? Toward Legal Rights for Natural Object* published in 1973, as well as the 2017 book *The Rights of Nature: a Legal Revolution That Could Save the World* from the Canadian lawyer David R. Boyd, Special Rapporteur on Human Rights and The Environment at the United Nations, have both contributed to the spread of these ideas and ethics in Canadian legal literature.[53] While the influence of Indigenous thought and practice should also be mentioned (see section I, above), it should not be idealized, as explained by Professor Kirsten Anker: 'idealized stereotypes replace attention to individual specificities and erase Indigenous voices – and their heterogeneity – in the process.'[54]

---

[48] Madeleine Cantin Cumyn, 'De l'administration des biens à la protection de la personne d'autre' in *Obligations et recours contre un curateur, tuteur ou mandataire défaillant 2008*, vol 283 (Yvon Blais 2008).

[49] In the common law, Garret and Wood propose to have an 'Office of Nature' (n 37) 22.

[50] On the new regime imposed by the law on guardianship of animals as sensitive beings, see Alexandra Popovici 'Chercher la petite bête: l'animal dans le Code civil du Québec' in Nathalie Vézina, Pascal Fréchette and Louise Bernier (eds), *Mélanges Robert P. Kouri – L'humain au cœur du droit* (Yvon Blais 2021).

[51] Peter Burdon, *Earth Jurisprudence: Private Property and the Environment* (Routledge 2015).

[52] Aldo Leopold, 'The Land Ethic' (2012) 14(1) Emergence: Complexity and Organization 59.

[53] Christopher D Stone, 'Should Trees Have Standing – Toward Legal Rights for Natural Objects' (1972) 45 SCal LRev 450; David R Boyd, *The Rights of Nature: A Legal Revolution That Could Save the World* (ECW Press 2017). They are both cited in the Innu Council of Ekuanitshit's Resolution (n 3).

[54] Kirsten Anker, 'Ecological Law and Indigenous Relational Ontologies: Beyond the "Ecological Indian"?' in Kirsten Anker and others (eds), *From Environmental to Ecological Law* (Routledge 2021) 111 <https://papers.ssrn.com/sol3/papers.cfm?abstract_id=3703825>.

V. Relationship Between National Legal Systems

The resolution of the Innu Council of Ekuanitshit expressly mentions having drawn inspiration from municipalities (Mexico City, Santa Monica, Crestone, and San Francisco), provinces (Victoria in Australia and Colima in Mexico), and countries (Ecuador, Bolivia, and Aotearoa New Zealand) that have recognized rights and legal status in nature or certain natural entities.[55] It also refers to the jurisprudence of many courts, including the Inter-American Court of Human Rights, and constitutional courts (including those of Colombia, Ecuador, India, and Bangladesh) that have recognized the legal personality of nature.[56] The Council also invokes sources of international law to justify its competence to legislate in this area, referring, for example, to the right to self-determination recognized in art 3 of the *United Nations Declaration on the Rights of Indigenous Peoples*.[57] The right to self-determination is thus used to justify their measure to protect a river that is significant to the community. Therefore, the Magpie River case, the only rights of nature initiative in Québec, was clearly influenced by other national and international legislation.

VI. International Law

The rights of nature are not integrated into the binding legal instruments of international law, even if they have been mentioned in Québec and Canadian scholarship since the 1970s. Despite the absence of a binding legal instrument at the international level, the developments related to the guardianship of the Whanganui River in Aotearoa New Zealand have influenced the resolutions adopted to protect the Magpie River in Québec. Indeed, as mentioned above, the Ekuanitshit Innu Council and the Regional County Municipality of Minganie's resolutions provided for the appointment of guardians to protect the river, citing Aotearoa New Zealand's case.[58] If the province were to adopt a statute to implement this regime, it would be interesting to see if provincial lawmakers would also follow Aotearoa New Zealand's example.

---

[55] See references cited in the Innu Council of Ekuanitshit's Resolution (n 3) 7 (fn xxxiv).

[56] See other references cited in the Innu Council of Ekuanitshit's Resolution (n 3) 7 (fn xxxv).

[57] United Nations Declaration on the Rights of Indigenous Peoples, A/Res/61/295 2007, s 3; Innu Council of Ekuanitshit's Resolution (n 3) 10.

[58] See Innu Council of Ekuanitshit's Resolution (n 3) (fn xxxiv).

VII. Efficacy

1. Potential Actors for Invoking Rights of Nature

As the development of the rights of nature is only embryonic, these rights have not yet been invoked before the courts, which have therefore not been able to confirm their application in provincial law. In the case of the Magpie River, we have seen that it is citizen groups (the Association Eaux Vives and the Regional County Municipality of Minganie), an Indigenous community (the Innu Council of Ekuanitshit), and an NGO (CPAWS Québec) that have worked to implement the regime. We can therefore imagine that it will be the same groups that will invoke these rights to defend nature.

2. Obstacles to the Development of Rights of Nature

Why have the rights of nature not developed further? To answer this question, it is first necessary to address the broader issue of obstacles to the development and implementation of environmental law. Despite the development of environmental law globally and in Canada since the 1970s, ecosystems have increasingly continued to degrade. For example, wild vertebrate species experienced a 60% decline between 1970 and 2014 due to excessive pressures exerted by humanity on planetary resources.[59] Environmental law has therefore clearly failed to fulfil its objective. Several factors explain this phenomenon. In particular, environmental law has not called into question the liberal paradigm in which it is embedded, and which has led to the degradation of ecosystems. While liberal assumptions and values have indeed greatly influenced Western society and propelled its economic development, they have also led to destructive behaviors of ecosystems.[60] The Canadian legal system is based on these liberal assumptions and therefore shares common theoretical obstacles to the implementation of nature's rights with other Western countries.

The philosophers of the Enlightenment, at the origin of the liberal school of thought, were interested in human rights and society in opposition to the arbitrariness of the monarchical authorities. They wanted to put forward the protection of individual rights and their freedom to counterbalance monarchical power. They therefore valued individual autonomy, the higher moral status of the human being (for their ability to think), and private ownership.[61] The importance

[59] Monique Grooten and Rosamunde Almond (WWF-NL), *Rapport Planète Vivante 2018: Soyons ambitieux* (WWF 2018) 3, 18.

[60] Michael M'Gonigle and Louise Takeda, 'The Liberal Limits of Environmental Law: A Green Legal Critique' (2013) 30(3) Pace Envtl L Rev. 1005, 1060. Note that this section reproduces some passages from Roy (n 2).

[61] Laura S Underkuffler, 'On Property: An Essay' (1990) 100(1) Yale LJ 127, 142; Philippe Boudreau and Claude Perron, *Lexique de science politique* (3rd edn, Chenelière éducation 2011) 111–112; Hugues Puel, 'Le néolibéralisme, modèle économique dominant' (2005) 1(233)

given to individual autonomy and the superior value of human beings generates an anthropocentric world view, where individual and human interests are at the center of considerations and the legal system. The law therefore operates in a world separate from nature.[62]

The Canadian and Western conception of private property as conferring 'absolute' rights on owners over their land, also stems from liberalism. According to the English philosopher John Locke, one of the leading figures of Judeo-Christian liberalism,[63] an acre of cultivated land has a value ten times higher than wasteland – the etymology of the word is telling here –, since it produces (according to him) ten times more fruit. Wasteland, on the other hand, has no use since it cannot be profited from.[64] This ignores any form of intrinsic value or ecosystem services.[65] According to Locke, an individual may appropriate property or land only to the extent that it ensures their own subsistence – or that of others by selling the fruits of their labour – while also leaving for others an adequate quantity of equally good property or land.[66]

His reasoning assumes an infinite abundance of natural resources, as he argues that individuals can appropriate land even when there may not be enough for everyone. Relying on the notion of the 'vacant lands of America,' his theory ignores the presence of First Nations.[67] Ecosystems, unfortunately, are not infinite, making the infinite abundance of natural resources a mistaken premise. His thinking, however, greatly influenced the notion of common law ownership, the different rights attached to it, and the virtual absence of any obligation for the owner to their community. It should be noted that the French Civil Code also adopted this liberal posture and, in 1804, gave the owner the 'most absolute' use of their property; it thus aimed to avoid a return to feudalism as well as the separation of ownership and the use of the property that accompanied the system of royalties due to the lords.[68]

Although Canadian laws – environmental and otherwise – today limit the exercise of property rights in various ways (think of urban planning rules, those gov-

---

Rev Déthique Théologie Morale 29, 31; on property, see eg John Locke, *Deux traités du gouvernement*, translated by Bernard Gilson (J. Vrin 1997).

[62] Aaron Mills, 'The Lifeworlds of Law: On Revitalizing Indigenous Legal Orders Today' (2016) 64(1) McGill LJ 847, 864–865.

[63] John Locke, *Second Treatise of Government: An Essay Concerning the True Original, Extent and End of Civil Government* (John Wiley & Sons 2014).

[64] John Locke, 'Deuxième traité du gouvernement – Essai sur l'origine, les limites et les fins véritables du gouvernement civil' in John Locke, *Deux traités du gouvernement* (J Vrin 1997) 137 at paras 36–42.

[65] Eric Zencey, 'Fixing Locke: Civil Liberties on a Finite Planet' in Peter N Goggin (ed), *Rhetorics, Literacies, and Narratives of Sustainability* (Routledge 2009) 186–187.

[66] Locke (n 64) at paras 31–33.

[67] Zencey (n 65) 181–182.

[68] Jean-Louis Halpérin, *Histoire du droit des biens* (Economica 2008) 196.

erning neighborhood annoyance and non-contractual liability,[69] or prohibitions on polluting),[70] owners traditionally have the right to use their property 'freely and completely.'[71] The search for individual interest in a capitalist economy thus leads to the commodification of nature. By considering the land as a commodity, however, its natural production capacity is ignored and subjected to transformations – drainage, levelling, paving, deforestation, watering of chemicals, etc. – which have the effect of destroying the natural attributes it possesses for food production or ecosystem support, to the detriment of the needs of the community.[72]

Also derived from liberalism is the idea that the market can act as a regulator of individual actions, placing rationality, individualism, and free competition at the center of social life.[73] State intervention, in Canada and elsewhere, must therefore be limited to putting in place the tools to exercise individual autonomy, including the application and respect of private property rights and contractual relations.[74] However, the market does not redistribute wealth but, rather, works to increase it globally.[75] Moreover, capitalism also ignores the finitude of natural resources and their intrinsic value, so that the search for infinite growth leads to the overexploitation of these resources.[76] The tool that must act as a regulator of individual actions therefore increases inequalities and the destruction of ecosystems without being able to overcome them.

The liberal democratic State – such as Canada – therefore also has the function of alleviating the inequalities generated by the economic system and protecting those who suffer them. This function – and, by extension, the legitimacy of the State – is also contingent on economic growth, as the taxation of such growth provides the State with revenue that can be redistributed to fund social programs.[77]

---

[69] CCQ, arts 7, 976 and 1457; Yaëll Emerich and Alexis Hudon, 'Les assises conceptuelles du droit de l'environnement en droit des biens: entre patrimoine commun et relation fiduciaire' (2017) 47(2) Rev Générale Droit 519, 527.

[70] Emerich and Hudon (n 69) 527–534.

[71] CCQ, art 947.

« La propriété est le droit d'user, de jouir et de disposer librement et complètement d'un bien, sous réserve des limites et des conditions d'exercice fixées par la loi.	'Ownership is the right to use, enjoy and dispose of property fully and freely, subject to the limits and conditions for doing so determined by law.
Elle est susceptible de modalités et de démembrements. »	Ownership may be in various modalities and dismemberments.'

[72] Mary Christina Wood, *Nature's Trust: Environmental Law for a New Ecological Age* (CUP 2013) 312–317.

[73] Puel (n 61) 32. We are talking here about economic liberalism, which should be distinguished from political liberalism.

[74] M'Gonigle and Takeda (n 60) 1060–1061. David Harvey, *Brève histoire du néolibéralisme*, translated by Antony Burlaud and others (Les Prairies ordinaires 2014) 16.

[75] M'Gonigle and Takeda (n 60) 1064–1065.

[76] James Gustave Speth, *The Bridge at the End of the World: Capitalism, the Environment, and Crossing from Crisis to Sustainability* (Yale UP 2008) 53.

[77] M'Gonigle and Takeda (n 60) 1065–1066.

The State thus finds itself in a contradictory position where it must, on the one hand, promote economic growth and, on the other, lessen its negative effects on both the population and its environment. This context has thus influenced the development of environmental law, the goal of which Canadian professor and author Micheal M'Gonigle and Louise Takeda summarize as follows:

> 'The goal is not to challenge the liberal economic order, but to make it work in harmony with the environment. In other words, environmental law works as a kind of *law of mitigated production*, the primary goal of which is to improve existing processes and structures of production but without fundamentally challenging the context in which they operate.'[78]

Therefore, the objective of protection pursued by environmental law must necessarily be accompanied by a reflection on the – liberal – causes of ecological destruction, and measures must be put in place to tackle these causes; otherwise it will not have the desired effect.

Nature's rights have the potential to change the relationship humans have with their environment. The granting of rights to the whole of nature or to a natural entity – when accompanied by clear responsibilities imposed on humans, as was done in Aotearoa New Zealand by defining the duties of guardians to protect the former Te Urewera National Park and the Whanganui River (and as desired with the Magpie River) – recognizes the interdependence of humans and nature as well as the intrinsic value of nature.[79]

But is this the way to go in order to instigate the paradigm shift necessary for better environmental protection? In other words, is changing the status of nature – from an *object* of law to a *subject* of law – by granting it legal personality one of the *sine qua non* conditions of this paradigm shift? The granting of rights to nature (as was done for human rights after the Second World War) is based on liberal logic and does not call into question the premises underlying the legal system: individual autonomy, the higher status of the human being (the human/nature division; subject/object dichotomy), the extent of the rights attributed to landowners (and their control over ecosystems), and the lack of any intrinsic value of nature. However, the mechanisms that stem from liberalism perpetuate the practices that lead to the degradation of ecosystems. For instance, adopting the monetization of ecosystem services as the principal strategy would fail to account for the complexity of ecosystems while perpetuating the inequitable distribution of access to these services characteristic of the capitalist economy.[80]

Recognizing legal rights for nature can be beneficial to ecosystems in Canada and other Western countries only if it is embedded within a broader reform of the legal system in which it operates. Such an approach would restore the intrinsic value of the natural world and acknowledge the interdependent relationships

---

[78] M'Gonigle and Takeda (n 60) 1067–1068. Emphasis in the original.

[79] Te Urewera Act 2014, New Zealand, ss 16 and 18; Te Awa Tupua (Whanganui River Claim Settlement) Act 2017 (n 5) ss 18 and 19; Iorns Magallanes (n 18) at 327.

[80] M'Gonigle and Takeda (n 60) 1092.

among living beings and the elements that constitute it. Granting rights to a natural entity such as the Magpie River is nevertheless likely to bring about a change in mentality, particularly for the communities involved in its conservation if correlative responsibilities have been granted. However, it may be insufficient.

VIII. Nature, Objectively Protected: How to Protect Nature in Québec Private Law

If the tools of public law leave us helpless in our desire to protect nature, it seems that Québec private law and more particularly its Civil Code offer interesting perspectives.

The preliminary provision of the *Civil Code of Québec* is particularly interesting in that it incorporates general principles of law as well as the Québec Charter into the Québec *jus commune*:

Disposition Préliminaire	*Preliminary Provision*
«Le Code civil du Québec régit, en harmonie avec la Charte de la langue française (chapitre C-11), la Charte des droits et libertés de la personne (chapitre C-12) et les principes généraux du droit, les personnes, les rapports entre les personnes, ainsi que les biens. Le code est constitué d'un ensemble de règles qui, en toutes matières auxquelles se rapportent la lettre, l'esprit ou l'objet de ses dispositions, établit, en termes exprès ou de façon implicite, le droit commun. En ces matières, il constitue le fondement des autres lois qui peuvent elles-mêmes ajouter au code ou y déroger.» 1991, c 64; 2022, c 14, s 123	'The Civil Code of Québec, in harmony with the Charter of the French language (chapter C-11), the Charter of human rights and freedoms (chapter C-12) and the general principles of law, governs persons, relations between persons, and property. The Civil Code comprises a body of rules which, in all matters within the letter, spirit or object of its provisions, lays down the jus commune, expressly or by implication. In these matters, the Code is the foundation of all other laws, although other laws may complement the Code or make exceptions to it.' 1991, c 64; 2022, c 14, s 123

Accordingly, the new section of the Québec Charter of Human Rights and Freedoms, s 46.1, mentioned above, which recognizes the right to a biodiverse and healthy environment, is understood to be part of Québec private law.

Article 46.1	*Section 46.1*
«Toute personne a droit, dans la mesure et suivant les normes prévues par la loi, de vivre dans un environnement sain et respectueux de la biodiversité.» 2006, c 3, s 19	'Every person has a right to live in a healthful environment in which biodiversity is preserved, to the extent and according to the standards provided by law.' 2006, c 3, s 19

It should be emphasized that while s 46.1 recognizes biodiversity and the right to live in a healthy environment, it does not have the effect of a fundamental right, as it is envisaged as an economic and social right creating neither a right to an injunction remedy nor even a right to information. However, because of the preliminary provision of the Civil Code, it does insert the concept of biodiversity as a general principle into private law.[81]

The Civil Code thus becomes a conceptual toolbox that is particularly relevant when it comes to identifying solutions to protect nature. Some of these mechanisms are already well known and have been used by nature conservation associations and citizens. The best known mechanisms are neighborhood annoyance,[82] social utility trust,[83] long-term indivision,[84] and conservation easements.[85] These instruments, while offering a private form of environmental protection,[86] do not recognize nature as a distinct entity requiring protection. This is perhaps why the solution most frequently invoked is that of conferring legal personality

---

[81] See *Coalition Climat Montréal c Couillard*, 2018EXP-338 (CS), 2017 QCCS 5623 appel rejeté à 2018 QCCA 2013; *Alarie c Propriétés Sommet prestige Inc.*, J.E. 2012-1895, EYB 2012-211054 (CS); Dominique Amyot-Bilodeau, 'La constitutionnalisation du droit à l'environnement: fondements théoriques et mise en oeuvre québécoise' (2006–2007) 19 RJEUL 3; David Robitaille and Sophie Thériault, 'Le droit à un environnement sain: normativité et limites d'un principe phare de la charte québécoise des droits' in A Braën (ed), *Droits fondamentaux et environnement* (Wilson & Lafleur 2013) 203; Thériault and Robitaille (n 21).

[82] CCQ, art 976:

«Les voisins doivent accepter les inconvénients normaux du voisinage qui n'excèdent pas les limites de la tolérance qu'ils se doivent, suivant la nature ou la situation de leurs fonds, ou suivant les usages locaux.»	'Neighbours shall suffer the normal neighbourhood annoyances that are not beyond the limit of tolerance they owe each other, according to the Nature or location of their land or local usage.'

See also *St. Lawrence Cement v Barrette*, 2008 CSC 64 and annotations made by Élise Charpentier on art 976f of *Code civil du Québec, Annotations – Commentaires* (6th edn, Yvon Blais 2021) at 1040f.

[83] See CCQ, arts 1260–1261 and 1270; see also Tiess, 'Le Guide sur les diducies d'utilités sociale à l'usage des idealistes' (2021) <https://tiess.ca/download/documents/TIESS-FUS-Guide-fiducies-utilites-sociales.pdf>.

[84] See CCQ, art 1030; Gaële Gidrol-Mistral, 'L'affectation à un but durable, vers une nouvelle forme d'appropriation des biens communs? Réflexions autour de l'article 1030 du Code civil du Québec' (2016) 46(1) RGD 95; Gaële Gidrol-Mistral and Alexandra Popovici, 'Duguit Appropriated: The Case of Trusts and Collective Ownership in Québec' in Paul Babie and Jessica Viven-Wilksch (eds), *Léon Duguit and the Social Obligation Norm of Property* (Springer 2019).

[85] Gaële Gidrol-Mistral, 'L'environnement à l'épreuve du droit des biens' (2017) 62(3) McGill LJ 687; Gaële Gidrol-Mistral, 'Quebec Private Law, Destined to Preserve the Environment?' in Siel Demeyere and Vincent Sagaert (eds), *Contract and Property with an Environmental Perspective* (Intersentia 2020). See also Jean-François Girard, 'La vraie nature de la servitude de conservation: analyse d'un outil juridique méconnu' in Service de la Formation continue du Barreau du Québec, *Développements récents en droit de l'environnement*, vol 352 (Yvon Blais 2012) 158; Jean Trudelle, *La servitude de conservation et la protection durable des milieux naturels au Québec: constats et recommandations* (M Env Essay, University of Sherbrooke 2014).

[86] Gaële Gidrol-Mistral, 'La destinée environnementale du droit des biens' in Max Falque (ed), *Environmental Property Rights/Property Rights for the Environment* (Bruylant 2022) 285.

to nature or one of its elements, such as the Magpie River or the St. Lawrence River,[87] which in turn allows a form of sacralization of nature.

As Yaëll Emerich explains, this method of protecting nature may appear effective, as it grants subjective rights to the protected entity, similar to those of a person (human or legal). However, the fact remains that this legal technique operates within the anthropocentrism framework it aims to dismantle.[88] Thinking outside the subjectivist mold requires revisiting the foundations of private law in light of its own theories. Indeed, recourse to doctrines of ecocentrism and biocentrism[89] does not help to de-center the anthropocentrism inherent in private law. The American legal tradition in which these doctrines are inscribed does not enter easily into dialogue with that of Québec civil law.

Yet Québec is a fertile alter-subjective terrain that, in terms of its architecture, techniques, and concepts, already possesses effective, and above all, a-subjective tools to protect nature.[90] Several instruments are present in the Québec legal arsenal and make it possible to move from the legal fiction of personification to the legal reality of an objectively protected nature. *Res communis* is one of the most promising concepts within Québec civil law for addressing the protection of nature, as it provides a framework that safeguards nature without resorting to personification, while granting it a distinct status.

If the *res communis* can claim such a place, it is because it benefits from the support of the law on water (1). Reinvigorated, it can therefore rely on existing mechanisms of civil law, such as the preventive action of article 982 (2), the ordinary law regime of reparation in kind (3), and the general regime of private administration (4).

1. The res communis of Article 913 of the Civil Code, Upgraded by the Water Act

While the *Water Act*[91] is often presented as having a purely symbolic significance, it is an important moment for private law. Through the public concepts of the common heritage of the nation and the custodian State, it puts the concept of the commons back at the center of the private law debate on the protection of nature.[92]

[87] Daniel Turp and Yenny Vega Cárdenas, *A Legal Personality for the St. Lawrence River and Other Rivers of the World* (JFD editions 2021).

[88] Yaëll Emerich, 'Vers une reconceptualisation du droit des biens face aux défis environnementaux' (2017) 119(2) La Revue du Notariat 321.

[89] In the case of Emerich, he considers the Earth Jurisprudence theory as advanced by Thomas Berry and Peter Burdon and also the Ethics of the Earth movement articulated by Aldo Leopold.

[90] Alexandra Popovici, *Êtres et Avoirs – Les droits sans sujet en droit privé actuel* (Yvon Blais 2019).

[91] See note 23.

[92] Madeleine Cantin Cumyn, Michelle Cumyn and Claire Skrinda, 'L'eau, chose commune: un statut juridique a confirmer' (2000) 79(3) The Canadian Bar Review 398; Madeleine Cantin Cumyn, 'La notion de chose commun et les conflits d'usage' (2007) 12(2) Lex Electronica

Indeed, the real contribution of this statute is to refocus the ideas of nature protection on the 'reality' of the *res*.

Article 913	*Article 913*
«Certaines choses ne sont pas susceptibles d'appropriation; leur usage, commun à tous, est régi par des lois d'intérêt général et, à certains égards, par le présent code. L'air et l'eau qui ne sont pas destinés à l'utilité publique sont toutefois susceptibles d'appropriation s'ils sont recueillis et mis en récipient.»	'Certain things may not be appropriated; their use, common to all, is governed by general laws and, in certain respects, by this Code. However, water and air not intended for public utility may be appropriated if collected and placed in receptacles.'

Art 913 CCQ defines the *res communis* as something whose use is common to all. Civil law usually allows some things to be absolutely removed from the relationship of appropriation. If, among these things, the legislator expressly names water and air, it is not impossible to conceive that other natural elements could claim this status and thus escape the logic of subjective rights and appropriation. Building on this mechanism, each citizen can use (*usus*) the common resource for their personal service. Unlike the traditional ownership system, the limit imposed on users is abuse, understood here not as an abuse of subjective rights[93], but as *abusus*. Still, despite paragraph 1 imposing a strong principle, paragraph 2 allows the users to find a way back to ownership. The collection and containerization of the common resource instantly transform it into an appropriated property. Thus, these things, which by their nature should have been definitively removed from the mechanism of acquisition by occupation, are treated as *res nullius*, a thing without an owner. This change in nature, which erases the mechanism of the *res communis*, therefore prevents us from thinking of this category as an effective one.

However, the question regarding the real power of art 913 arises all the more forcefully as the Québec Charter's reference to biodiversity would make it possible to escape the proprietary grip of paragraph 2. Yet, even if the legal nature of the *res communis* were to be affirmed, the fact remains that its regime would continue to be elusive: What actions could non-owner users take? Would they

---

<www.lex-electronica.org/s/649>; Madeleine Cantin Cumyn, 'Le régime juridique de l'eau, chose commune' in Catherine Choquette and Alain Létourneau (eds), *Vers une gouvernance de l'eau au Québec* (MultiMondes 2008); Madeleine Cantin Cumyn, 'L'eau, une ressource collective: portée de cette désignation dans la Loi affirmant le caractère collectif des ressources en eau et visant à renforcer leur protection' (2010) 51(3/4) Les Cahiers de Droit 595. See also the annotations made by Élise Charpentier to arts 913 and 982 of the *Code civil du Québec, Annotations – Commentaires* (n 82) 1002f and 1050ff.

[93] CCQ, art 7; see Pierre-Emmanual Moyse, 'L'abus de droit: l'anténorme, Partie 1' (2012) 57(4) McGill LJ 859; Pierre-Emmanual Moyse, 'L'abus de droit: l'anténorme, Partie 2' (2012) 58(1) McGill LJ 1.

have some obligations, and, if so, who would be their creditors? If art 913 alone does not provide the answer, it is possible to find one in art 982.

2. The Preventive Protection of Private Law

Although art 982 CCQ is concerned specifically with the protection of the use of water resources, it is possible to imagine a general system of preventive protection of all common things.

Article 982	*Article 982*
« À moins que cela ne soit contraire à l'intérêt général, celui qui a droit à l'usage d'une source, d'un lac, d'une nappe d'eau ou d'une rivière souterraine, ou d'une eau courante, peut, de façon à éviter la pollution ou l'épuisement de l'eau, exiger la destruction ou la modification de tout ouvrage qui pollue ou épuise l'eau. »	'Unless it is contrary to the general interest, a person having a right to use a spring, lake, sheet of water, underground stream or any running water may, to prevent the water from being polluted or depleted, require the destruction or modification of any works by which the water is being polluted or depleted.'

According to this article, any person who uses a thing in common may act in such a way as to avoid its qualitative (pollution) or quantitative (depletion) damage. If we take seriously the integration into the *Civil Code of Québec* of a *res communis* that is truly distinct from the *res nullius* that is appropriable (CCQ, art 913(1)), then art 982 allows us to conclude that any person, even one without a traditional legal interest, can take action as a preventive measure.[94] Such an understanding would make it possible to recognize that there exist in Québec civil law objective remedies that are detached from the classic figure of subjective remedies.[95] Nature, if considered as a *res communis*, could then benefit from an a-subjective and therefore anthropo-decentric protective regime. In this vision, preventive action is detached from personal interest in order to crystallize an interest external to the person, emancipating nature and civil law from the yoke of personification. This new way of considering action and reparation, which is set out in art 982 of the Code, gives concrete form to the notion of 'diffuse interest'[96] in Québec civil law.

---

[94] Usually, to have a right to act in court, one has to be personally interested, and the judgment will also make reparation to the injured party personally. But the CCQ has paved the way for actions of 'uninterested parties.' See Alexandra Popovici, 'Droits de regard: la fiducie dans le Code civil du Québec' in Christine Morin and Brigitte Lefebvre (eds), *Mélanges en l'honneur du professeur Jacques Beaulne* (Wilson & Lafleur 2018); see also Robert Godin, 'L'intérêt général – commentaire sur l'article 982 du *Code civil du Québec* ou "JE PUISE MAIS N'ÉPUISE"' [2010] Revue du Barreau 137.

[95] In French law, this type of situation has been called 'un contentieux objectif'. See eg. H Motulsky, 'Le droit subjectif et l'action en justice', in *Écrits: Études et notes de procédure civile* (Dalloz 1973).

[96] According to Judith Rochfeld and others, diffuse interests are interests of an undefined group of persons. See Judith Rochfeld and others, *Rapport final de recherche – L'Échelle de communalité* (IRJS 2021) 244 <www.gip-recherche-justice.fr/publication/lechelle-de-communalite/>.

Unfortunately, to our knowledge, no such action has yet been brought. Moreover, this autonomization of the action would place exorbitant powers in the hands of the courts, breaking with the interpersonal structure of private law. Finally, this also raises the question of the type of remedy that would be appropriate for such damages detached from any personal interest: monetary compensation would be useless and only a remedy in kind would permit a real reparation of nature.

3. Reparation in Kind

Although these 'objective actions' are still struggling to find their basis and their methods of financing,[97] remedies in kind are included in the range of remedial tools within civil law, serving as a fundamental legal means of reparation.[98] Thus, it seems to us that the actions undertaken in matters of environmental damage should better favor remedies in kind insofar as the principal objective of recourse is precisely not only to put an end to the disturbance, but also to repair nature.

For example, if several authors consider neighborhood disturbances as the epitome of private environmental action, one must still imagine a remedy that is not limited to the mere personal financial compensation of the property owner, without guaranteeing the actual implementation of a restoration of the property in kind by the latter. This way of thinking is interesting because it highlights the intrinsic integrity of nature, which thereby acquires a legal reality that can do without the myth of personification to obtain reparation in kind.

In light of this prospective analysis, the absence of ecological damage or real environmental obligation seems less problematic in Québec private law than one might spontaneously think, especially since the Québec Civil Code has a special regime allowing the implementation of objective actions.

4. Private Administration

The new regime of private administration[99] makes it possible to imagine ways of objectively protecting nature. Indeed, the *res communis* could perfectly well benefit from the regime established in Title 7 of the Book on Property, that of the administration of the property of others, which establishes a general regime applicable to any situation where a person acts furtherance of an external interest in private law.[100]

The title chosen by the legislator for this regime in the Civil Code is, however, misleading. As Professor Madeleine Cantin-Cumyn clearly demonstrated, this regime affects both persons, subject to a protective regime, and things, understood

[97] Judith Rochfeld and others explore the idea of a 'dedicated fund' (n 96) at 415f.

[98] Whereas execution in kind may previously have been thought of as an exceptional measure, it is today the principal remedy: *Group HRC Canada Inc. v Beauregard*, 2018 QCCA 1063, para 78.

[99] Cantin Cumyn and Cumyn (n 45) 573.

[100] Cantin Cumyn and Cumyn (n 45).

widely, administered by a third party.[101] Thus, there is no need to personify nature to allow it to benefit from this protective regime.[102] Centered around the notion of private power, this autonomous regime parallels the central regime of subjective rights. Although the distinction between subjective rights and powers is too often overlooked,[103] Madeleine Cantin-Cumyn has clearly highlighted this difference.

A power is a legal prerogative conferred on a person in the interest of others or for the achievement of a purpose. The interest of others or the goal to be achieved, to which the exercise of power is always subordinated, distinguishes it essentially from subjective rights. A subjective right is a legal prerogative that confers on its holder an advantage in their own interest. In principle, the holder of a right exercises it freely. On the other hand, the holder of a power must, in its exercise, take into account the purpose for which the prerogative was granted.

The potential of these powers, subject to the regime of a-subjective administration, could benefit the *res communis*, which being non-appropriable, must be administered objectively. Under this regime, the users would be subject to the rules of administration found in arts 1308 et seq. CCQ, the most relevant of which are the following:

Article 1308
« L'administrateur du bien d'autrui doit, dans l'exercice de ses fonctions, respecter les obligations que la loi et l'acte constitutif lui imposent; il doit agir dans les limites des pouvoirs qui lui sont conférés.

Il ne répond pas de la perte du bien qui résulte d'une force majeure, de la vétusté du bien, de son dépérissement ou de l'usage normal et autorisé du bien. »

Article 1308
'The administrator of the property of others shall, in carrying out his duties, comply with the obligations imposed on him by law and by the constituting act. He shall act within the powers conferred on him.

He is not liable for loss of the property resulting from superior force or from its age, its perishable nature or its normal and authorized use.'

Article 1309
« L'administrateur doit agir avec prudence et diligence.

Il doit aussi agir avec honnêteté et loyauté, dans le meilleur intérêt du bénéficiaire ou de la fin poursuivie. »

Article 1309
'An administrator shall act with prudence and diligence.

He shall also act honestly and faithfully in the best interest of the beneficiary or of the object pursued.'

Article 1310
« L'administrateur ne peut exercer ses pouvoirs dans son propre intérêt ni dans celui d'un tiers; il ne peut non plus se placer dans une situation de conflit entre son intérêt personnel et ses obligations d'administrateur.

Article 1310
'No administrator may exercise his powers in his own interest or that of a third person or place himself in a position where his personal interest is in conflict with his obligations as administrator.

[101] Cantin Cumyn (n 48) 203–221.

[102] See the argument made by Popovici (n 50).

[103] Madeleine Cantin Cumyn, 'Le pouvoir juridique' (2007) 52 McGill LJ 215; see also Madeleine Cantin Cumyn, 'The Legal Power' (2009) 17(3) ERPL 345.

S'il est lui-même bénéficiaire, il doit exercer ses pouvoirs dans l'intérêt commun, en considérant son intérêt au même titre que celui des autres bénéficiaires.»

If the administrator himself is a beneficiary, he shall exercise his powers in the common interest, giving the same consideration to his own interest as to that of the other beneficiaries.'

Article 1313
«L'administrateur ne doit pas confondre les biens administrés avec ses propres biens.»

Article 1313
'No administrator may mingle the administered property with his own property.'

Article 1316
«L'administrateur peut ester en justice pour tout ce qui touche son administration; il peut aussi intervenir dans toute action concernant les biens administrés.»

Article 1316
'An administrator may sue and be sued with respect to anything connected with his administration; he may also intervene in any action respecting the administered property.'

Recalling fiduciary duties, this regime would make it possible to require any user to act in the sole interest of nature. The complexity of the situation lies in the fact that each user is simultaneously both the beneficiary and the administrator of the common resource. According to art 1310 CCQ, the user would therefore have to take into account not only the interest of nature but also the common interest.

In civil law, these 'fiduciary obligations' are akin to duties insofar as they objectively bind the user, who has the duty to protect the resource in the name of the purpose itself and in the common interest. This duty becomes a justiciable obligation only in the event of misuse of power, thus opening the way to an action that can be implemented by any interested party. In this case, since it is a *res communis*, any person who may be a user may bring an action before the courts to seek redress. In order to truly repair the damage caused, reparation in kind should be the general principle, but it does not exclude an action for personal liability in the case of prejudicial fault.

It is interesting to note that this system of powers offers, in addition to traditional sanctions, a sanction of its own that does not apply to strictly selfish subjective prerogatives, such as the right of ownership or a claim. The user can indeed have their powers taken away. This forfeiture of the prerogative, which can be found in art 1361(2) CCQ, is a punitive sanction that the law applies to situations involving an interest superior to one own or a collective interest.[104] Of course, this 'death penalty' must be pronounced by a judge.[105]

Echoing s 44 of the Québec Charter, this regime also imposes, through art 1354 CCQ, a right of access to information which confers on any user the right to demand an account from the administrator.[106]

---

[104] This sanction is found in dismemberments, such as usufruct or servitude, or in the exercise of rights for others, such as parental authority.

[105] *Droit de la famille – 1914*, 2019 QCCA 20, para 24 and *Droit de la famille – 151074*, 2015 QCCS 2120, para 29.

[106] Although the users have not yet seized upon this regime of private administration, they have used the right of access to information as a means of ensuring protection, see Patrice Bergeron, 'Des organismes souhaitent lever le secret sur les prélèvements d'eau' (*La Presse*

IX. Conclusion

Despite the local initiative surrounding the Magpie River in Québec or the 'public trust' of the natural environment set forth by the Northwestern Territories and the Yukon environmental statutes, it seems difficult to claim that nature can at present assert subjective rights or legal personality in Québec or Canadian law. However, the fact remains that the general arsenal of Québec private law confers upon nature an objective protection. To this end, it is necessary to take seriously both *res communis* and private powers as well as the human right to a biodiverse and healthy environment enshrined in the Québec Charter, as these mechanisms could be used in court to ensure a better protection of nature detached from personal interests. Although they do not confront the liberal causes of ecological destruction, these tools could allow to move away from the anthropocentric paradigm without waiting for legislative intervention.

Canadienne, 14 February 2022) <www.lapresse.ca/actualites/environnement/2022-02-14/des-organismes-souhaitent-lever-le-secret-sur-les-prelevements-d-eau.php>.

Romania

Unfertile Ground for the Rights of Nature

*Diana Botău**

I. Existence and Contents

The Romanian legal system does not recognize any legal personality for nature as a whole, for ecosystems, or for particular components of an ecosystem. Therefore, the so-called rights of nature are not included as such in any official document that may be considered a source of law, be it the Constitution, statues, decrees, or even bylaws.

A legislative initiative aimed at recognizing the legal status of dolphins as non-human persons was put forward in 2014 by one independent member of the legislature, based on arguments related to the extraordinary intelligence and empathy of dolphins. However, the Parliament rejected the bill, mainly on the ground that legal subjects are always faced with both rights and obligations, whereas non-human legal subjects would contradict such a principle since they could only be granted rights but would never be bound to obligations.[1]

---

* The manuscript was finalized in May 2025. The assessment of literature and case law, including internet materials, reflects the legal situation at that time.

[1] Legislative Proposal for Recognizing Dolphins as Non-human Persons, available at <http://www.cdep.ro/pls/proiecte/uplpck2015.proiect?cam=2&idp=14254>.

Further, since Romania is a civil law country, whose laws can be created only by the legislature, there is no case law recognizing such rights, and there is no bureaucratic practice in this regard.

Civil society may join forces and seek such legislation, but going through the necessary procedure will be both long and uncertain;[2] moreover, the legislature is not under the obligation to pass a certain statute related to such an initiative, instead merely being required to consider the opportunity and to discuss such a proposal within the regular legislative process, which might or might not end with the passing of a statute. However, the narrative regarding the rights of nature is absent from the public debate or from public concern; Romanian society does not, at the current time, indicate any awareness regarding this topic, and the cultural and economic advance of the country demands a focus on what are considered to be more urgent matters.

Romania is not a party to any international law treaties or soft law instruments that recognize such rights of nature. Occasionally, Romania does have representatives at events that are related to such rights, but no real importance is attached to such gestures, no publicity surrounds them, and no consequences follow; they are but diplomatic courtesies, at the most.

Certain natural sites, however, are registered in the UNESCO universal cultural heritage and are protected as such. The Danube Delta is such an example. Also, the mining site of Roșia Montană has been the subject matter of an ongoing debate regarding its registration as a universal cultural heritage site; since there are competing interests regarding this area, the official application for registering Roșia Montană on the UNESCO list has been delayed several times by national officials who are colluding with the mining industry. Popular support has so far hindered mining in the area; however, the entire movement supporting Roșia Montană was not framed around the idea of the rights of nature but was rather focused on the negative consequences that would follow from such exploitation and on environmental arguments.

There have been some domestic associations or individuals who have sometimes vaguely mentioned nature's rights, but generally, there is little to almost no talk about such rights in official and/or unofficial discourse. Scholarly literature that addresses the subject make no reference whatsoever to Romania, not even in hypothetical terms; commentators explain only the meaning of the term, the philosophy behind it,[3] its conceptual evolution, and they offer some relevant examples from South America,[4] or emphasize the strategic relation of the concept of

---

[2] Law no 189/1999 on the citizens' legislative initiative, Monitorul Oficial al României (Official Gazette of Romania), part I no 516, 8 June 2004.

[3] Ana Maria Moraru, 'Considerații istorice și etico-morale asupra relației om-natură/Historical and Ethical Considerations upon the Realtionship Human-Nature' (2010) 2 Revista de Științe Juridice 83.

[4] Andrei Duțu-Buzura, 'Natura ca subiect de drept: experiențe antropo-culturale necesare ori/și progres juridic?/Nature as Subject of Law: Necessary Anthropo-Cultural Eexperiences or/and Legal Progress?' (2020) 6 Pandectele Române 49.

rights of nature to Indigenous emancipatory politics.[5] Other papers do not at all tackle the rights' of nature issue and refer only to the broad issue of environmental protection, or they refer to the idea of solidarity rights (eg, the right to peace, to a healthy environment, or to development),[6] to natural heritage that needs preservation,[7] or to the criminal law concept of ecocide as related to international law principles.[8]

Therefore, since no national importance is attributed to the rights of nature, they are mostly neglected and only occasionally and unofficially mentioned on special occasions like Earth Day or the International Day of Biological Diversity. Even then, the notion is not referred to in terms of the rights of nature but in broader concepts like the importance of nature for the overall wellbeing, etc.

In Romania, such lack of interest in rights of nature reflects a complex interplay of the civil law system's reliance on legislative codification, the anthropocentric nature of the existing legal framework, historical priorities focused on post-communist transition and economic development, and a less developed public and advocacy discourse specifically on this novel legal concept. The traditional civil law framework being inherently anthropocentric, it views nature primarily as property or as a resource for human use. Legal rights and duties are typically assigned to humans and moral persons. While there might be instances where environmental protection is considered in judicial decisions, these are typically framed within existing anthropocentric laws (eg, pollution regulations), rather than nature being recognized as a rights-bearing entity. Shifting to an 'ecocentric' view, where nature itself holds rights, would require a significant and deliberate departure from consecrated philosophical and legal traditions.

Moreover, unlike common law systems where case law plays a crucial role in developing and evolving legal principles, in civil law countries like Romania, law is primarily created by the legislature. Therefore, new legal concepts, especially those as paradigm-shifting as the rights of nature, require explicit legislative enactment to gain national importance and legal enforceability.

Furthermore, Romania's recent history, marked by a transition from communism to democracy and a market economy, has shaped its legislative priorities. Much of the focus has been on establishing democratic institutions, promoting economic growth, and aligning its legal framework with European Union law. For a legislature grappling with immediate socio-economic challenges, the focus

---

[5] Mihnea Tănăsescu, 'Rights of Nature, Legal Personality, and Indigenous Philosophies' (2020) 9(3) TEL 429.

[6] Mădălina Virginia Antonescu, 'Conceptul de "securitate a biosferei". Spre o diplomație globală de securizare/The Concept of "Biosphere Security". Towards a Global Diplomacy of Biosphere Security' (2017) 62(1) Impact Strategic 76.

[7] Andrei Duțu-Buzura, 'Cultural and Natural Heritage and Its Law' (2020) RJPA 17.

[8] Anca Ileana Dușcă, 'Natura ca subiect de drept/Nature as Subject of Law' (2020) 11 Dreptul 33.

has been mostly on tangible issues and less on those that may be perceived as mainly abstract or philosophical.

A lack of public discourse and awareness regarding the rights of nature adds to the explanation for the legislative vacuum in this regard. Environmental NGOs often focus on practical conservation, advocate for stricter enforcement of existing laws, or raise awareness about specific environmental threats. Moreover, unlike some countries where indigenous movements or severe environmental disasters have served as catalysts for recognizing the rights of nature, Romania has not experienced an impact event able to galvanize public and political will specifically around this concept.

Accordingly, the domestic discourse is mostly centered on the right *to* nature as opposed to the rights *of* nature. There is an entire body of law that aims at protecting nature, ie the environment, which falls into the broader environmental law domain and does not comprise nature's rights in themselves. And, truth be said, even the environmental laws lack proper implementation within Romania; there is much to be done in order to have them function properly, and it is likely to take a very long time and perhaps repeated sanctions from the European Commission before a real change will be seen. Therefore, it is highly unlikely that such debatable rights as the rights of nature could find fertile soil for implementation in Romania.

II. Legal and Political Operators

Except for a failed bill aiming at recognizing dolphins as non-human persons, the rights of nature have not been promoted in Romania by any domestic legal, political, or civil society institution or group. A Romanian branch of the *Pachamama* organization does exist in Romania and has been created by one passionate individual;[9] however, its activities are mainly related to organizing sports events in natural settings and fostering the relationship between humans and nature. Likewise, no international or foreign institutions has played any similar roles regarding the rights of nature. NGOs like *Greenpeace* do have local branches in Romania as well, but such entities do not plead for the rights of nature as they are understood in Bolivia, New Zealand, or other parts of the globe where they are given actual legal force.

Since there is no relevance attached to the rights of nature in Romania, there is no tension between domestic and international foreign institutions. In terms of environmental law, however, there is a certain tension between the EU rules regarding the protection of the environment and the lack of efficacy on the part of the national authorities in implementing such regulations. However, none of these EU rules are framed within rights of nature theory.

---

[9] <www.pachamamaromania.ro/>.

III. Enforcement

Since there are no rights of nature recognized within the Romanian jurisdiction, the topic is to be related to the broader area of environmental law. However, not even within the environmental body of laws is there any regulation regarding a human right to a healthy environment; the so-called 'right' to a healthy environment is at its most political discourse, lacking a robust and clearly defined legal framework for its practical implementation and enforcement. This creates a challenging landscape for individuals seeking to assert this right.

Article 35 of the Romanian Constitution does state, although in limited detail, that '[t]he right to a healthy, well-preserved and balanced environment is guaranteed. The State provides the legislative framework for the exercise of this right.' This constitutional provision serves as the bedrock of the political discourse and politicians frequently reference it to demonstrate their commitment to environmental protection. There is often a general rhetoric about the importance of environmental protection, sustainable development, and a 'green future', aligned with European Union directives and the broader global push for environmental responsibility. However, this discourse often remains at a high level of abstraction, focusing on broad goals rather than concrete mechanisms for individual redress.

When environmental problems arise (eg, illegal deforestation, pollution incidents, waste management crises), the political discourse frequently devolves into a blame game. Different political parties or government entities may accuse each other of inaction or corruption, rather than focusing on systemic solutions or empowering citizens to effectively exercise their environmental rights. Rather than proactive policy-making, the political discourse tends to be reactive, emerging strongly in response to specific environmental crises. During such times, politicians may make pledges, initiate investigations, or propose ad-hoc measures, which often fade from the public eye once the immediate pressure subsides.

Moreover, since a significant part of the political discourse is driven by the necessity to transpose and implement EU environmental legislation, this often leads to the adoption of laws and regulations that are technically compliant but may lack effective enforcement mechanisms or broad public awareness and participation. The discourse thus focuses on fulfilling obligations to the EU rather than genuinely empowering citizens to claim their environmental rights.

Environmentally related actions or omissions of public authorities can be challenged in court by individuals or organizations, but only when the claimant can prove a personal interest in doing so. Most of such procedures address the formal requirements of the law and do not reach the merits of the case.

In the absence of a specific and comprehensive legal framework for the right to a healthy environment, proving a personal interest in Romanian courts becomes a significant hurdle. Typically, in administrative law (where most environmental disputes would fall), personal interest requires demonstrating a direct and actual prejudice or harm. Such prejudice is mostly linked to proving direct harm to health or property (eg, respiratory issues due to air pollution, water contamination leading to illness, agricultural land ruined by industrial discharge), or at least impact on quality of life (eg, constant noise pollution from an unauthorized construction, loss of access to a recreational natural area due to illegal dumping, persistent noxious odours).

However, while the Romanian Constitution provides an aspirational statement about the right to a healthy environment, the absence of detailed legislation and clear procedural rules means it largely remains a political talking point rather than a truly enforceable individual right. This void places a heavy burden on individuals to navigate a complex legal landscape without the necessary tools or pathways for effective redress.

IV. Theoretical Sources and Foundations

Most of the Romanian legislation adopts an anthropocentric view of nature and natural resources, regarding them mainly as property and/or objects. Nature as a whole is fragmented and compartmentalized into different natural resources or it is divided into one particular medium at a time, such as soil, air, water, plants, trees, or animals and birds; environmental protection encompasses nature conservation and heritage law. Also, environmental concerns are to be balanced with other rights protected under legislation, and the outcome is not always environmentally friendly.

A foundation for rights of nature has not been formulated in Romanian literature.

V. Relationship between National Legal Systems

Other legal systems have not influenced the Romanian view on the rights of nature topic. Most of the jurisdictions that have acknowledged such rights have done so under specific cultural and anticolonial circumstances. Such circumstances are not to be found within the Romanian context. Therefore, the concept does not have traction in this country.

Foreign rights of nature cannot be enforced in the Romanian system.

VI. International Law

1. European Convention on Human Rights

The European Convention on Human Rights (ECHR), to which Romania is a party, does not include a human right to a healthy environment. Over time, its provisions have been expanded towards a 'greener' interpretation in the case-law of the European Court of Human Rights (ECtHR), but the Convention is still firmly rooted in the idea of a 'direct victim' whose right to life and respect for private and family rights should be protected.[10]

The issue of environmental rights has been raised throughout the years in different cases, but the ECtHR has consistently held that the Convention cannot be read to include an environmental right in itself.[11] Over the years, the ECtHR has adopted an evolutionary approach, showing an increased awareness of the relationship between human rights and the environment. However, in the environment-related cases that the ECtHR has ruled on so far, it has done so mainly under the rights to life (art 2), respect for private and family life (art 8), peaceful enjoyment of property (art 1 of Protocol no 1), and a fair hearing (art 6).[12]

Accordingly, the ECtHR has consistently rejected any idea of an *actio popularis*,[13] while showing an increasing awareness of the role that NGOs play in a democratic society. Individuals can work via NGOs 'by proxy', but only when their individual rights under the ECHR are affected.[14]

2. *Aarhus Convention*

Up to now, the 1998 Aarhus Convention, adopted within the United Nations Economic Commission for Europe, is one of the most advanced instruments for environmental democracy.[15] The EU and all its Member States, Romania included, are Parties to this Convention. Its Preamble emphasizes the importance of a close relationship between environmental rights and human rights. Nearly all the provisions in the Convention are implemented in the EU by various directives.

3. *EU Law*

Although rights of nature are not recognized as such in the EU, the preservation and protection of the environment, as well as the notion of sustainable development, has a strong constitutional position in the Treaties of the European Union

---

[10] Alan Boyle, 'Human Rights or Environmental Rights? A Reassessment' (2007) 18 Fordham Env L Rev 471, 505.

[11] ECtHR 22 May 2003, App no 41666/98, *Kyrtatos v Greece*; ECtHR 9 June 2005, App no 55723/00, *Fadeyeva v Russia*) or ECtHR 25 September 2014, App no 29878/09, *Karin Andersson ao v Sweden*.

[12] See Council of Europe, 'Protecting the Environment Using Human Rights Law' <www.coe.int/en/web/portal/human-rights-environment>.

[13] ECtHR 6 September 1978, App no 5029/71, *Klass ao v Germany*; ECtHR 9 November 2004, App no 22494/93, *İlhan v Turkey*.

[14] ECtHR, 12 June 2007, App no 75218/01, *Collectif Stop Melox et Mox v France*; ECtHR 24 February 2009, App no 49230/07, *L'Erablière asbl v Belgium*.

[15] The Aarhus Convention (Convention on Access to Information, Public Participation in Decision-making and Access to Justice in Environmental Matters) was adopted in Aarhus, Denmark, on 25 June 1998, by the United Nations Economic Commission for Europe (2161 UNTS 447), available at <https://unece.org/DAM/env/pp/documents/cep43e.pdf>.

and in the Charter of Fundamental Rights of the European Union (CFREU). Moreover, in secondary EU law, all kinds of environmental law issues are covered by directly applicable regulations and directives that leave room for Member State implementation: eg the Industrial Emissions Directive[16] and the Regulation concerning the Registration, Evaluation, Authorization and Restriction of Chemicals.[17] Regulations and directives cover a wide variety of environmental issues, from the soil and water, to the air, atmosphere, and climate. From a rights of nature perspective, the most prominent instruments are the two 'nature directives', that is, the Birds Directive[18] and the Habitats Directive.[19] These directives require Member States (Romania included) to designate areas for the conservation of natural habitat types of European interest and for the protection of rare and threatened species, thus establishing the Natura 2000 network. Romania has entered several hundred national sites on the list in accord with the provisions of these directives.[20]

The intrinsic value of nature is also reflected in the case law of the Court of Justice of the European Union (CJEU). According to the CJEU, the public concerned, including NGOs, must have the possibility to challenge in court any administrative decisions (actions or omissions) in environmental matters. The national courts are called upon to disregard any national law in breach of European law provisions containing clear, precise, and unconditional obligations concerning nature conservation and environmental protection.

4. Soft Law

Several soft law instruments that Romania has endorsed also acknowledge that the Earth and its ecosystems are our common home and that is it necessary to promote harmony with nature in order to achieve a balance among the economic, social, and environmental needs of present and future generations. Such soft law instruments include the Universal Declaration of the Rights of the Mother Earth adopted in 2010 at the First World People's Conference on Climate Change

---

[16] Council Directive 2010/75/EU of 24 November 2010 on industrial emissions (integrated pollution prevention and control) [2010] OJ L334/17.

[17] Council and Parliament Regulation (EC) 1907/2006 of 18 December 2006 concerning the Registration, Evaluation, Authorisation and Restriction of Chemicals (REACH), establishing a European Chemicals Agency, amending Directive 1999/45/EC and repealing Council Regulation (EEC) No 793/93 and Commission Regulation (EC) No 1488/94 as well as Council Directive 76/769/EEC and Commission Directives 91/155/EEC, 93/67/EEC, 93/105/EC and 2000/21/EC [2006] OJ L396/1.

[18] Council and Parliament Directive 2009/147/EC of 30 November 2009 on the conservation of wild birds [2009] OJ L20/7.

[19] Council Directive 92/43/EEC of 21 May 1992 on the conservation of natural habitats and of wild fauna and flora [1992] OJ L206/7.

[20] The list can be consulted at: <http://ananp.gov.ro/ariile-naturale-protejate-ale-romaniei/>.

and the Rights of Mother Earth, held in Cochabamba, Bolivia; the 1992 Rio Declaration;[21] and the 2000 Earth Charter.[22]

VII. Efficacy

Due to its membership in the European Union, Romania does have quite an extensive body of environmental legislation, in terms of both substance and procedure. However, environmental objectives are often in competition with economic and social ones. More often than not, the environmental regulations fail to compete with the legislation supporting economic growth, private property, and state sovereignty.

To a certain extent, environmental protection agencies and environmental NGOs are awarded standing in the court, allowing them to challenge decision-making under environmental protection laws. They can either initiate a case against, for instance, a polluter, or join an existing case already begun. Administrative decisions can be challenged and remedial actions can be requested. However, the protection offered by the national legislation is rather weak and court decisions are often not properly enforced.

Environmental protection under the ECHR is indirect since the beneficiaries of the duty of States to regulate and control sources of environmental harm are only those individuals whose rights are affected. Therefore, the duty of States is not so much about protecting the environment as it is about protecting individuals from harmful environmental effects. Relevant cases in this regard deal with natural disasters, city dumps, incineration plants, industrial installations, mining facilities, and similar activities generating severe environmental pollution.

Moreover, the ECHR gives States a wide margin of appreciation/discretion in such matters, considering that national authorities are best placed to make decisions on environmental issues which often have difficult social and technical aspects. That is why even the ECHR falls short of guaranteeing the right to a healthy environment when the concept is understood in broader terms, unrelated to any impact on humans.

In Romania, one of the most notable cases that ended up on the ECtHR's docket is *Tătar v Romania*.[23]

The applicants, Vasile Gheorghe Tătar and Paul Tătar, father and son, are Romanian nationals. At the relevant time, they lived in Baia Mare (Romania). The company S.C. Aurul S.A., afterwards operating as S.C. Transgold S.A., had obtained a licence in 1998 to exploit the Baia Mare gold mine. The company's ex-

---

[21] Rio Declaration on Environment and Development, UN General Assembly (12 August 1992) <www.un.org/en/conferences/environment/rio1992>.

[22] Earth Charter 2000 <https://earthcharter.org/>.

[23] ECtHR 27 January 2009, App no 67021/01, *Tătar v Romania*.

traction process involved the use of sodium cyanide. Part of its activity was located in the vicinity of the applicants' home.

On 30 January 2000, an environmental accident occurred at the site. A United Nations study reported that a dam had breached, releasing about 100,000 cubic meters of cyanide-contaminated tailings water into the environment. The report stated that S.C. Aurul S.A. had not halted its operations following the accident.

After the accident, Vasile Gheorghe Tătar filed various administrative complaints concerning the risk incurred by him and his family as a result of the use of sodium cyanide by S.C. Aurul S.A. in its extraction process. He also brought criminal proceedings, in 2000, complaining that the mining process was a health hazard for the inhabitants of Baia Mare, that it posed a threat to the environment, and that it was aggravating his son's medical condition, namely asthma.

By an order of 20 November 2001, the Romanian courts discontinued the criminal proceedings concerning the accident of 30 January 2000 on the ground that the facts complained of did not constitute offences. No judicial order or decision concerning the other complaints has been issued to date.

In its admissibility decision of July 2007, the ECtHR ruled that there had been a violation of art 8 of the ECHR (right to respect for private and family life), on account of the Romanian authorities' failure to protect the right of the applicants, who lived in the vicinity of a gold mine, to enjoy a healthy and protected environment.

The Court observed that pollution could interfere with a person's private and family life by harming his or her well-being and that the State had a duty to ensure the protection of its citizens by regulating the authorization, setting-up, operation, safety, and monitoring of industrial activities, especially activities that were dangerous for the environment and human health.

The Court further noted that the company had been able to continue its industrial operations after the January 2000 accident, in breach of the precautionary principle, according to which the absence of certainty with regard to current scientific and technical knowledge could not justify any delay on the part of the State in adopting effective and proportionate measures.

The Court concluded that the Romanian authorities had failed in their duty to assess, to a satisfactory degree, the risks that the company's activity might entail and to take suitable measures in order to protect the rights of those concerned to respect for their private lives and homes, within the meaning of art 8, and, more generally, to ensure their right to enjoy a healthy and protected environment.

Despite its evolutionary and greening character, the ECHR still offers only indirect environmental protection. This position is likely to be challenged by climate cases that are being brought in Europe today, under the tenet of climate change.[24]

---

[24] For instance, a group of six Portuguese minors, under the name of *Youth for Climate Change*, sued all Member States of the EU (Romania included) and six other Parties to the ECHR for breaches of art 2 and art 8 due to their failure to undertake necessary steps to limit climate change, according to their commitments as Signatory States to the 2015 Paris Agree-

VIII. Conclusions

Rights of nature appear to be a legal transplant that is unlikely to find fertile soil in Romania. A very large section of the existing environmental law comes from binding European law, and authorities are struggling to cope with the demands of this law, often unsuccessfully. Rather than adopting a new label, efforts need to be made to improve the effectiveness of already existing environmental law.

In Romania, the rights of nature paradigm is unlikely to rescue the environment from the deficits of current legislation and, above all, from its insufficient effectiveness and weak enforcement. Therefore, the notion is mainly conceived as a symbolic concept.

ment (limiting global temperature increase to 1.5–2°C). The applicants claimed the States' failure to meet Paris Agreement targets on global temperature rise contributes to climate change and impacts their health and living conditions through heatwaves. They argued for shared State responsibility and sought exemption from exhausting domestic remedies due to the urgent need for climate action. See ECtHR 9 April 2025, App no 393/71/20, *Duarte Agostinho v Austria and Parties to ECHR*; the application was deemed inadmissible.

Switzerland

Mapping the Scene

Laura Affolter and Sian Affolter[*]

I. Some Introductory Remarks

As is happening around the globe, in recent years there have been several attempts to introduce rights of nature in Switzerland. Switzerland provides an interesting example for studying nature's rights because, on the one hand, the interconnectedness of nature's rights advocates working on a global level can be nicely shown. On the other hand, the parliamentary attempts to endow nature with rights have met with much opposition. The grounds for this opposition will be explored in this report.

The report is a product of the collaboration between a legal scholar with expertise in Swiss and European environmental law and a social anthropologist specializing in legal anthropology. Whilst Sian Affolter identified and interpreted the existing law according to the method of legal hermeneutics, Laura Affolter

[*] The manuscript was finalized in January 2023. The assessment of literature and case law, including internet materials, reflects the legal situation at that time. The interviews that this report draws on were conducted in July and August 2022. The discussions on the judgment of the ECtHR on the 'climate seniors' in section II. 1. b) and of the Association *Rechtsperson Reuss* in section II. 2. d) were supplemented in June 2024 to reflect more recent developments. The part on the Appel du Rhône in section II. 2. c) was also updated in June 2024.

conducted empirical research on the rights of nature movement in Switzerland. Her research was carried out as part of the project 'Nature's Rights in Europe: frictions, collisions and transformative potentials' that was funded by the Amsterdam Centre for European Studies.[1] As part of this research, Laura Affolter conducted interviews with Marionna Schlatter, member of the Swiss parliament; Doris Ragettli, co-founder of Rights of Mother Earth; Thomas Egli, founder of Objectif Sciences International and CEO of the Geneva Forum; as well as with a member of id·eau, the NGO behind the *Appel du Rhône*. These interviews provide the empirical basis for this report, along with publicly available texts such as campaign websites, minutes of parliamentary debates, written parliamentary motions, and press statements. The interviews were all conducted in July and August 2022 and reflect the state of affairs at that time. As academics, we were also granted access to the minutes of a parliamentary debate that took place behind closed doors. Due to confidentiality issues, however, we are not allowed to quote from this debate. Finally, as part of her fieldwork, Laura Affolter participated in the World Ethic Forum in Pontresina in August 2022, where the proposition of nature's rights was discussed.

The report is structured as follows: Part II, after discussing the way existing law in Switzerland regulates the relationship between humans and nature and environmental protection, presents three recent initiatives that on different levels attempt or attempted to introduce nature's rights in Switzerland. Part III then outlines how these initiatives came about, who the people and organizations behind them are, and what the connections between these actors and other actors working on a more global level are. In Part IV, we discuss the ways in which other legal systems influenced the different proposals, which directly leads into Part V, in which we tentatively suggest what the possible theoretical arguments underlying the different initiatives might be. Finally, in Part VI we set forth the arguments that were brought against the initiatives as well as the impact these initiatives have had so far on Swiss law and politics – or that they might have in the future.

II. Nature and its Rights in Switzerland: Status Quo and Recent Initiatives

1. Existing Law

Unlike some other legal orders, Swiss law does not (yet) contain any rights of nature as such. However, the relationship between humans and nature is regulated by the Swiss legal order in multiple ways. First and foremost, the federal constitution lays down the guiding principles and aims which are to be followed by the federal legislator. The latter has in turn enacted various legal acts regarding

[1] <www.uva.nl/en/profile/d/i/m.c.dietrich/m.c.dietrich.html?cb#Research>.

the protection of the environment. Finally, Switzerland is a federal country, meaning that law is not only set at a federal level but also at a cantonal level. In the text that follows, it shall be shown to what extent Swiss law at these three different levels contains provisions that are relevant to the issue at hand.

a) (Federal) Constitutional law

On a constitutional level, the framework regarding the human relationship with nature is quite elaborate. It starts with the preamble, stating that the constitution is adopted with the Swiss people and the cantons being mindful of their responsibility towards creation and conscious of their responsibility towards future generations. It is widely recognized that the preamble thus includes a commitment towards the principle of sustainability as well as to the existence of an 'ecological responsibility'[2] of the state without, however, precisely defining its content.[3]

Art 2 Constitution, in enshrining the so-called aims of the Swiss confederation, again takes up the relationship with nature. According to art 2 para 2 Constitution, the Swiss Confederation is to, inter alia, promote sustainable development.[4] Thus, the Swiss constitution requires the state to respect the 'ecological boundaries' whilst at the same time demanding that the adopted measures consider social and economic developments.[5] Apart from the reference to the principle of sustainability, art 2 also states that the Swiss Confederation is committed to the long-term preservation of natural resources (para 4). It has been shown elsewhere that the constitutional aim of preserving natural resources requires two things.[6] On the one hand, it entails that the state must ensure that nature as a whole does not deteriorate (*Verschlechterungsverbot*). On the other hand, it also implies a certain 'improvement imperative' (*Verbesserungsgebot*), thus requiring the state

---

[2] Frank Haldemann, *Verantwortung als Verfassungsprinzip. Die schweizerische Verfassungsordnung im Spannungsfeld der Verantwortungsethik* (Schulthess 2003) 41.

[3] Bundesrat, Botschaft über eine neue Bundesverfassung, BBl 1997 I, p. 124; Alexandre Flückiger, 'Le développement durable en droit constitutionnel suisse' (2006) 20(5) Umweltrecht in der Praxis 471, 484.

[4] It would go far beyond the scope of this report to analyze the numerous discussions regarding the content of the principle of sustainability. See in this regard Klaus Bosselmann, *The Principle of Sustainability. Transforming Law and Governance* (2nd edn, Routledge 2017); Klaus Mathis, *Nachhaltige Entwicklung und Generationengerechtigkeit. Eine interdisziplinäre Studie aus rechtlicher, ökonomischer und philosophischer Sicht* (Mohr Siebeck 2017); regarding Switzerland, Flückiger (n 3).

[5] Sian Affolter, *Der Umgang der Landwirtschaft mit der natürlichen Umwelt – de lege lata und de lege ferenda* (Schulthess 2021) 67–68 and 38–43; Flückiger (n 3) 490–496.

[6] See S Affolter (n 5) 31–33, essentially – due to an absence of such discussions in Swiss scholarship – drawing on the discussions in German literature on art 20a GG; cf. Astrid Epiney, 'Commentary on Art. 20a' in Hermann von Mangoldt, Friedrich Klein and Christian Starck (eds), *Kommentar zum Grundgesetz* (7th edn, CH Beck 2018) marg nos 65–71; Dietrich Murswiek, 'Commentary on Art. 20a' in Michael Sachs (ed), *Grundgesetz* (9th edn, CH Beck 2021) marg nos 43–48.

to take positive measures to improve the current state of nature. These aspects are to be taken into account by the legislator. However, art 2 is not precise enough to prescribe what measures need or need not be taken. It remains up to the legislator to decide on the specific instruments. Therefore, while art 2 Constitution defines the principle of sustainability and the protection of nature as guiding principles for state action,[7] explicit, concrete requirements may not and cannot be derived from art 2 Constitution.[8]

The role and protection of the environment is further laid down in art 73–78 Constitution. Art 73 requires the state to 'endeavour to achieve a balanced and sustainable relationship between nature and its capacity to renew itself and the demands placed on it by the population'. According to legal commentators, the provision postulates the principle that the Confederation and the cantons are obliged to ensure, within the framework of their actions, that the regenerative capacity of nature is guaranteed insofar as the natural resources are to be preserved as undiminished as possible for future generations.[9] We can see here again that the Swiss constitution requires the legislator as well as any state entity applying the law to take into account that natural resources are finite and that 'ecological boundaries' are to be respected.

Art 74 Constitution states that the 'Confederation shall legislate on the protection of the population and its natural environment against damage or nuisance' (para 1). 'It shall ensure that such damage or nuisance is avoided. The costs of avoiding or eliminating such damage or nuisance are borne by those responsible for causing it' (para 2). The basic message of art 74 Constitution is, thus, to give the Confederation a mandate to protect the environment and to lay down the principles for this. The latter includes, in particular, the precautionary principle and the polluter pays principle. At the same time, it also becomes clear that the concept of Swiss environmental law is based on an anthropocentric rather than an ecocentric approach, since art 74 Constitution is concerned with the protection

---

[7] Bernhard Ehrenzeller, 'Commentary on Art. 2' in Bernhard Ehrenzeller, Benjamin Schindler and Rainer J Schweizer (eds), *Die schweizerische Bundesverfassung. St. Galler Kommentar* (4th edn, Dike 2023) marg nos 11–15; Eva Maria Belser 'Commentary on Art. 2' in Bernhard Waldmann, Eva Maria Belser and Astrid Epiney (eds), *Bundesverfassung* (Helbing & Lichtenhahn 2015) marg no 7; Giovanni Biaggini, *Bundesverfassung der Schweizerischen Eidgenossenschaft* (2nd edn, Orell Füssli 2017) art 2 marg no 4.

[8] Cf. Bundesrat, Botschaft über eine neue Bundesverfassung, BBl 1997 I, p. 126; Martin Bertschi and Thomas Gächter, 'Schöne Worte?: Zur Eignung der Präambel, des Zweckartikels und des Appells an die Verantwortung als Leitlinien staatlichen Handelns' in Thomas Gächter and Martin Bertschi (eds), *Neue Akzente in der ‚nachgeführten' Bundesverfassung* (Schulthess 2000) 19–21; Biaggini (n 7) art 2 marg no 4.

[9] See Alain Griffel, 'Commentary on Art. 73' in Bernhard Waldmann, Eva Maria Belser and Astrid Epiney (n 7) marg no 6; Flückiger (n 3) 498.

of humans and their environment and not primarily with the protection of the environment for its own sake.[10]

Arts 75 to 80 Constitution contain specific legislative mandates in connection with individual aspects of nature, such as forests, water, or natural habitats.

Finally, art 120 Constitution also appears noteworthy in the present context. It regulates non-human gene technology and states that the Confederation, when legislating on the use of reproductive and genetic material from animals, plants, and other organisms, is to take into account the dignity of living beings. What exactly the dignity of living beings comprises, however, remains to be concretized. According to the case law of the Federal Supreme Court, the concept of the dignity of living beings is close to that of human dignity.[11] Accordingly, it can be concluded that living beings may not be degraded to mere 'means to an end'.[12] However, the dignity of living beings – in contrast to human dignity – is not absolute but can be restricted vis-à-vis other public interests.[13] It is noteworthy that the Federal Supreme Court has stated that the dignity of living beings is to be taken into account also outside the scope of art 120 Constitution, thus not solely when non-human gene technology is at issue.[14] This means that there is a general constitutional recognition that animals, plants, and other organisms have an intrinsic value which must be respected.[15]

After having laid down the constitutional framework regarding the human relationship with nature, it can be concluded that the current constitution foresees numerous obligations for the state to protect the environment. However, none of the dispositions laid out confer subjective rights.[16] Individuals, let alone nature as such, cannot derive any rights for themselves from these provisions which could then be judicially enforced. Hence, the constitution limits itself to regulating the responsibilities of the state towards nature without foreseeing mechanisms for nature to demand compliance with the constitutional requirements. Yet even if the Swiss constitution were to foresee such possibilities – respective proposals are discussed below – judicially enforcing the respect of such constitutional provisions could still prove difficult. This is due to art 190 Constitution, which states

---

[10] Alain Griffel, 'Commentary on Art. 74' in Bernhard Waldmann, Eva Maria Belser and Astrid Epiney (n 7) marg no 24.

[11] BG 7 October 2009, 2C_422/2008, BGE 135 II 384 cons. 4.6.1; BGE 135 II 405 cons. 4.3.4.

[12] Bundesrat, 'Botschaft zu einer Änderung des Bundesgesetzes über den Umweltschutz (USG) vom 1. März 2000', BBl 2000, 2391 ff, 2405; Bernhard Waldmann, 'Commentary on Art. 120' in Bernhard Waldmann, Eva Maria Belser and Astrid Epiney (n 7) marg no 17.

[13] Waldmann (n 12) marg no 17.

[14] BGE 135 II 384 (n 11) cons. 3.1.

[15] See Johan Rochel, 'Des invités dans la Constitution cantonale?' (2021) 2 Schweizerisches Zentralblatt für Staats- und Verwaltungsrecht 67, 72.

[16] See Luzius Mader, 'Die Umwelt in neuer Verfassung? Anmerkungen zu umweltschutzrelevanten Bestimmungen der neuen Bundesverfassung' (2000) 2 Umweltrecht in der Praxis 105, 115.

that the Federal Supreme Court and other judicial authorities apply federal acts and international law. This means that if a federal act proves to be contrary to constitutional law, it still remains applicable despite the constitution being higher up in the hierarchy of norms.[17] However, the fact that the constitutional norm may not be enforceable in court does not, from a conceptual point of view, change the legally binding nature of constitutional law. The respective norm remains binding for its addressee and must be respected.[18]

b) Statutory (federal) law

The constitutional provisions presented above are implemented in different federal and cantonal acts. It would go far beyond the scope of this paper to describe their content in detail. However, with regard to the underlying question of this study, it appears interesting to analyze just what statutory law provides with regards to access to judicial review of acts or rulings allegedly infringing environmental protection law. Two aspects are worth highlighting. First, individuals can demand judicial review of an act or omission allegedly infringing environmental protection law when they are particularly affected by the act in question and have an 'interest worthy of protection'.[19] This essentially means that they must be more affected by the act than just anyone.[20] As a result, in general only people living in the proximity of where environmental protection law is being infringed are entitled to bring an action before court.[21] Second, Swiss law foresees a so-called organizations' right of appeal. This is contained in certain federal acts, namely in the Environmental Protection Act (EPA) and the Federal Act on the Protection of Nature and Cultural Heritage (NCHA). It enables designated organizations to appeal against individual rulings which fall into the scope of application of the respective act.[22] Hence, the concept of certain people or organizations being en-

[17] See Astrid Epiney, 'Commentary on Art. 190' in Bernhard Waldmann, Eva Maria Belser and Astrid Epiney (n 7) marg no 22.

[18] Astrid Epiney (n 17) marg no 22.

[19] See Administrative Procedure Act (APA), art 25a, in regards to acts committed by the state, and BG 1C_165/2009, 3 November 2009, cons 2.2–2.4, regarding acts committed by individuals.

[20] For an overview of the respective case law, see Beatrice Wagner Pfeifer, *Umweltrecht. Allgemeine Grundlagen* (Dike 2017) 351–355.

[21] Cf BG 5 May, 1C_37/2019, BGE 146 I 145, where the Federal Supreme Court concluded that the appellants, all elderly women who claimed that the state's inaction with regard to the climate crises infringed their right to life, were not sufficiently affected in their rights in the sense of art 25a APA. The court referred to the Paris Agreement, according to which global warming is to be limited to well below 2 degrees. It then argued that even without further measures, this point would only be reached in the more or less distant future. Hence, the appellants were at this point, according to the court, not sufficiently affected in their rights to be able to judicially force the government to take action.

[22] See, for example, art 12 para 1 NCHA: 'A right of appeal against rulings of the cantonal authorities or the federal authorities shall be accorded to […] the organizations concerned with nature protection […] or related objectives, subject to the following requirements: (1.) the or-

titled to defend nature's interests is in no way foreign to Swiss law. The organizations' right to appeal has proven to play a central role in ensuring effective protection of nature. It especially proves to have a substantial preventive effect. If an appeal by an organization is imminent, authorities and project managers are deemed to be more willing to take environmental protection laws into account from the outset and to take the organizations seriously as negotiating partners.[23] As a result, many appeals never actually have to be submitted because disputes could already be settled by negotiation. Where an appeal has been submitted, they have proven to be quite successful. In 2021, 49.2% of the appeals submitted by organizations were fully or partially upheld by the Federal Supreme Court whilst only 26.2% were dismissed or not admitted.[24]

In this regard, it is also worth noting the recent and widely discussed judgement of the ECtHR against Switzerland regarding the so-called climate seniors.[25] In this case, an association (Verein KlimaSeniorinnen, whose members are all elderly women living in Switzerland) and four individual women appealed a judgment of the Swiss Federal Supreme Court. The latter had held that neither the individual women nor the association were affected by Swiss federal climate policy (or rather the lack thereof) in a sufficiently particular manner to ask a court to condemn the federal administration for infringing their human rights.[26] The ECtHR equally held that the case of the individual women was inadmissible. However (and rather surprisingly), the ECtHR came to the conclusion that

> 'the special feature of climate change as a common concern of humankind and the necessity of promoting intergenerational burden-sharing in this context [...], speak in favour of recognising the standing of associations before the Court in climate-change cases. [...] The Court therefore considers it appropriate in this specific context to acknowledge the importance of making allowance for recourse to legal action by associations for the purpose of seeking the protection of the human rights of those affected, as well as those at risk of being affected by the adverse effects of climate change, instead of exclusively relying on proceedings brought by each individual on his or her own behalf.'[27]

It thus concluded that the association in this case had standing before the Court and that Switzerland had, in fact, violated human rights by not acting in a timely

---

ganization is active throughout Switzerland, (2.) it pursues solely non-profit-making objects; any commercial activities must serve the achievement of its non-profit-making objects.'

23 Regina Meier, *Das ideelle Verbandsbeschwerderecht. Eine Darstellung der Regelungen auf Bundesebene* (Schulthess 2015) 58; Alexandre Flückiger, Charles A Morand and Thierry Tanquerel, *Evaluation du droit de recours des organisations de protection de l'environnement* (Federal Office for the Environment 2000) 137–140.

24 Bundesamt für Umwelt, 'Statistik und Evaluation des Verbandsbeschwerderechts' <www.bafu.admin.ch/bafu/de/home/themen/recht/fachinformationen/verbandsbeschwerderecht/statistik-und-evaluation-des-verbandsbeschwerderechts.html>.

25 Cf the numerous commentaries on <https://verfassungsblog.de/tag/klimaseniorinnen/>.

26 BGE 146 I 145 (n 21).

27 ECtHR 9 April 2024, App no 53600/20, *Verein KlimaSeniorinnen Schweiz and others v Switzerland*, marg no 499.

and appropriate way to comply with its duty to ensure effective protection from the serious adverse effects of climate change on lives, health, well-being, and quality of life.[28] Therefore, it appears that the judgment has significantly strengthened the organizations' right of appeal, at least in matters relating to climate change. The effects the judgment will have on Swiss law, however, remain to be seen.

c) Cantonal law

Being a federal country, Switzerland's cantons all have their own constitution. Of these 26 constitutions, none confers rights to nature. However, the canton of Geneva's does contain a (human) right to a healthy environment. Art 19 Constitution of Geneva states that 'Everyone has the right to live in a healthy environment'. To our knowledge, the provision has not yet been extensively concretized by case law.[29] It thus remains to be seen what role it will play in legal disputes. According to Francesca Magistro, art 19 Constitution of Geneva aims to protect the environment through a subjective right of the individual to demand a healthy environment, on the one hand, and to protect the individual through a healthy environment on the other hand.[30] She contends that art 19 Constitution of Geneva essentially presents the individual with a means to combat inaction by the state in environmental matters, for example when the canton inaptly implements federal provisions or when the cantonal legislator refuses to legislate.[31]

With regard to cantonal law, it should also be pointed out that the Federal Supreme Court was recently confronted with the question of whether a canton could grant fundamental rights to entities other than the ones already recognized by Swiss law.[32] This was due to a popular initiative in the canton of Basel-Stadt which foresaw the granting of fundamental rights to primates. The cantonal government had ruled that the initiative was contrary to federal law and could thus not be submitted to a popular vote. The decision was appealed, and at last instance the Federal Supreme Court concluded that the granting of fundamental rights to primates was possible as long as those rights were binding only for the state and there was no binding effect for individuals. It upheld the reasoning of the lower court, which

[28] *Verein KlimaSeniorinnen Schweiz and others v Switzerland* (n 27) marg no 573.

[29] In ACST/25/2020 the Cour de Justice, the court of last instance in the canton drew on art 19 Constitution of Geneva – to our knowledge for the first and only time – to argue that the prohibition forbidding certain classes of vehicles from operating within a certain perimeter in the event of peaks in air pollution was a justified restriction of the economic freedom.

[30] Francesca Magistro, *Le droit à un environnement sain revisité* (Schulthess Editions Romandes 2017) 229; regarding the introduction of a right to a healthy environment on a federal level, see also Alexandre Flückiger, 'Art. 10a. Droit à un environnement sain' in Sophie Weerts and others (eds), *Révision imaginaire de la Constitution fédérale. Mélanges en hommage au prof. Luzius Mader* (Helbing Lichtenhahn 2018) 54–57.

[31] Magistro (n 30) 229; see also Eva Maria Belser, Thea Bächler and Sandra Egli, *Recht auf Umwelt. Eine Untersuchung der geplanten Anerkennung eines Rechts auf Umwelt durch die UN und ihrer Folgen für die Schweiz* (SKMR 2021) 89–90 <https://boris.unibe.ch/162996/>.

[32] BG 16 September 2020, 1C_105/2019, BGE 147 I 183.

had argued that the relationship between individuals was conclusively regulated by federal legislation, thus prohibiting cantons from deviating in this regard. Therefore, the canton could not introduce rights for 'new' legal subjects which were binding for individuals. However, as long as the rights would bind only the state, the court saw no reason to deny the canton the possibility of granting rights to primates.[33] The initiative was eventually submitted to a popular vote but was rejected by a majority of 75%.[34] For the issue at hand, the ruling of the Federal Supreme Court would mean that the cantons would be able to grant rights to nature but only to the extent that these would have to be respected by the state.

2. *Initiatives*

As shown above, nature or aspects of nature are not (yet) recognized as subjects of law in Switzerland. However, there have been efforts to change this. In this section, we present three recent proposals: a parliamentary postulate that was presented by a member of parliament in 2017, a parliamentary initiative submitted by five members of parliament in 2021, and a citizens' initiative called the *Appel du Rhône*. In updating this article for publication in June 2024, we came across another initiative that we very briefly introduce at the end: an initiative to amend the constitution of the canton of Lucerne in order to recognize the legal personality of the Reuss River.

a) Parliamentary Postulate 17.4312 (2017)

In 2017, a member of the Green Party and then member of the National Council, Lisa Mazzone, submitted a postulate, which in Swiss parliamentary politics is a means through which the Federal Council, hence the government, is mandated 'to examine and report on whether to submit a bill to the Federal Assembly or to take [some other] measure' (Parliament Act, art 123). In her postulate, Lisa Mazzone mandated

> '[t]he Federal Council […] to submit a report on the question of whether it might be opportune to endow Swiss glaciers with legal personality and to establish legal channels that would allow the violation of these legal persons' rights to be asserted before the courts.'[35]

---

[33] BGE 147 I 183 (n 32); for discussions on the rulings by the cantonal and the Federal Supreme Court, see Andreas Glaser, 'Entscheidbesprechung Bundesgericht, I. öffentlich-rechtliche Abteilung, Urteil 1C_105/2019 vom 16. September 2020' (2021) 1 Aktuelle Juristische Praxis 88; Andreas Glaser and Irina Lehner 'Entscheidbesprechung Appellationsgericht des Kantons Basel-Stadt als Verfassungsgericht, Urteil VG.2018.1 vom 15. Januar 2019, A., B. und C. gegen Grosser Rat des Kantons Basel-Stadt, Beschwerde gegen einen Beschluss des Grossen Rates vom 10. Januar 2018 betreffend rechtliche Zulässigkeit der kantonalen Volksinitiative „Grundrechte für Primaten"' (2019) 7 Aktuelle Juristische Praxis 2019 724.

[34] Kanton Basel-Stadt, 'Wahl- und Abstimmungsresultate (Archiv)' <www.bs.ch/regierungsrat/staatskanzlei/politische-rechte/wahlen-und-abstimmungen/wahl-und-abstimmungsresultate-archiv>.

[35] Lisa Mazzone, 'Die Gletscher mit Rechtspersönlichkeit ausstatten und Rechtswege einrichten. Eine gute Gelegenheit für unser Land? – Postulat 17.4312' <www.parlament.ch/de/

In its response to the postulate, the Federal Council argued that

'[f]rom a legal point of view, Switzerland's glaciers are considered "ownerless objects" within the meaning of Article 664 of the Swiss Civil Code (CC). As "public objects", they are subject to cantonal law (Art. 664 para 3 CC and Art. 78 of the Federal Constitution). According to the Swiss legal system, existence alone does not establish legal personality. The latter is inseparably linked to the natural person as an individual and to the exercise of his or her civil rights, i.e. to his or her social life (see Art. 11–19 CC). Granting glaciers a form of legal personality [comparable to legal entities] is also not a viable option, because legal entities are a delimited legal category (Art. 52 CC). Legal entities always pursue an inherent purpose. This is also the only reason for their personification. A glacier or other objects, on the other hand, do not have such an inherent purpose. Therefore, it would run counter to our understanding of the law to give them a legal personality. The question of whether objects can be given a legal personality was already debated in 2002 when the new Article 641a CC was introduced in relation to animals. However, this possibility was rejected because it is not compatible with the Swiss legal system. Moreover, it is difficult to imagine how a legal personality would actually improve the protection of glaciers.'[36]

Because the postulate did not make it onto the agenda of the National Council within two years, it was, following formal requirements, disregarded.[37]

b) Parliamentary Initiative 21.436–21.440 (2021)

Four years later, the next attempt was launched by members of the Swiss parliament. Initiated by Marionna Schlatter from the Green Party, a group of five parliament members representing five different political parties ranging from the left to the middle-right of the political spectrum presented five identically worded parliamentary initiatives on 19 March 2021.[38] Their initiatives requested that '[t]he protection of the environment and nature […] be strengthened in the Federal Constitution'. This should be done by pursuing two different directions in the revision of the constitution. First, 'the right of humans to a healthy environment should be enshrined as a fundamental right' and, second, 'nature should be given at least partial status as a legal subject'.[39] The five members of parliament grounded their initiative as follows:

'The intensive usage of natural resources is increasingly putting nature under pressure. The decline in biodiversity and climate change are consequences of a lack of equilibrium between humans and nature. As part of the natural environment, humans are directly affected by the environmental crisis: through the scarcity of resources, through increasing environmental burdens and the consequences of climate heating. This is why the United Nations General Assembly

---

ratsbetrieb/suche-curia-vista/geschaeft?AffairId=20174312>, translation from German to English by the authors.

[36] Mazzone (n 35), translation from German to English by the authors.

[37] Mazzone (n 35).

[38] Items of business 21.436–21.440, Marionna Schlatter, 'Recht auf gesunde Umwelt und Rechte der Natur – Parlamentarische Initiative 21.436' <www.parlament.ch/de/ratsbetrieb/suche-curia-vista/geschaeft?AffairId=20210436>.

[39] Schlatter (n 38), translation from German to English by the authors.

adopted Resolution 75/220 "Harmony with Nature" on 21 December 2020. In Switzerland, new constitutional foundations are needed. The right to a healthy environment should become an enforceable fundamental right of all people. And nature, understood as a unified ecosystem, is to be given at least partial status as a legal subject. When elaborating these rights in detail, it will have to be clarified who will be entitled to assume the legal representation of nature.'[40]

The initiative thus sought to change the constitution, adding rights of nature to the already rather elaborate constitutional framework when it comes to environmental protection as it was laid out above.

As is the procedure for parliamentary initiatives, the initiative was first debated by the Legal Affairs Committee of the National Council, who subsequently made the recommendation to the National Council to reject the initiative on the following grounds:

'In the Commission's view, the term "healthy environment" is not sufficiently defined in order to be enshrined in the Federal Constitution as a fundamental right. The Commission also considers the term "nature" to be too imprecise for the description of a legal subject. The implementation of the parliamentary initiatives would raise further difficult questions, such as the procedural representation of nature. It is unclear how representatives could claim to know the legitimate interests of nature. Furthermore, the Commission recalls that in our legal system, a legal subject is not only entitled to rights, but also has duties. For these reasons, the Commission rejects the parliamentary initiative.'[41]

On 12 December 2022, the parliament followed the recommendation of the Committee with a majority of 101 out of 189 members of parliament voting to dismiss the initiative. As Marionna Schlatter told us in her interview, she and her colleagues expected this to happen. At the time, she told us that she was planning to relaunch the proposal through different means, possibly including a popular initiative. A group of people we met at the World Ethic Forum was also exploring this same possibility. To what extent this has been ongoing, we do not know.

c) *Appel du Rhône*

In addition to the above-discussed parliamentary initiative and postulate which aimed to enshrine some sort of rights of nature on a national level in Switzerland, there is another initiative (the term not being used here in a strict political sense) working at a binational French-Swiss level that is, however, possibly looking to adopt a broader transnational approach linked to other developments around the world. This is the *Appel du Rhône* – literally the 'appeal of the (river) Rhône' – launched by the NGO id·eau.[42] The *Appel du Rhône* demands the following:

---

[40] Schlatter (n 38), translation from German to English by the authors.

[41] Kommission für Rechtsfragen des Nationalrates, 'Bericht der Kommission für Rechtsfragen vom 19. Mai 2022' <www.parlament.ch/centers/kb/Documents/2021/Kommissionsbericht_RK-N_21.436_2022-05-19.pdf>, tranlation from German to English by the authors.

[42] Association id-eau, 'Activités' <www.id-eau.org/activités>, translation from French to English by the authors.

'In the context of the current ecological crisis and climate emergency, in view of the gaps in existing law which exclude the possibility of taking into account our interdependence with other living systems and species whose good health is vital for our survival, and in order to equip ourselves with the means to fight for the preservation of the Rhône, we demand the recognition of a legal personality for the Rhône in order to defend its rights to exist, to be preserved, to regenerate, to evolve in order to maintain and guarantee its environment and its biodiversity for us and for future generations.

The *Appel du Rhône* aims to raise awareness of the issues involved in preserving the river from an ecological point of view and to fill the gaps in legal provisions in order to enable the communities and people living in vicinity to the river to:

- Have the Rhône recognized as a natural entity with a legal personality;
- Guarantee its legal status in transnational law;
- Provide it with a legal framework which will regulate human and industrial activity and enable it to protect itself against destructive activities, while promoting a circular and regenerative economy;
- Give everyone, all individuals and communities concerned, the means to preventively ensure its preservation and anticipate any irreversible damage to the river and its ecosystems;
- Create enforceable rights to defend this status effectively and in a binding manner in the event of an attack on its integrity;
- Establish the primacy of water as a universal resource essential to our survival and that of future generations.'[43]

How exactly the recognition of the legal personality of the river Rhône is supposed to happen in this transnational setting was from the beginning left unspecified. This was purposefully done, or as one member of id·eau explained to us:

'That is the process. We are not giving a ready-made legal object that will be used. It is co-construction of something that we want to be community-based and that we want to come from all the inhabitants in France and in Switzerland that are living along the river and that share a common concern and interest in the Rhône.'[44]

In order for the *Appel du Rhône* to work in this community-based, bottom-up way, id·eau set up a Popular Assembly of the Rhône. The Popular Assembly consisted of twenty-five members that were recruited by an independent agency hired by id·eau. Based on social and demographic information available to them and following certain diversity criteria, the agency contacted different people living on the banks of the river Rhône both in Switzerland and in France. The idea was that this Popular Assembly would meet five times in the course of two years and that, at the end, they would make recommendations as to what (legal and political) steps should be taken.

'The Popular Assembly is supervised by a Governance Committee independent from id·eau. The Committee was set up to ensure the support of the Assembly, to preserve its independence and to respect its will. To prepare the random selection process, draw up the agenda and invite experts and facilitators, the Committee appoints an Advisory Board whose members are com-

---

[43] Association id-eau, 'Appel du Rhône' <www.appeldurhone.org/_files/ugd/6963cc_274fe3817316413e9135f89144352232.pdf>.

[44] Interview with a member of id·eau, 10 August 2022.

mitted and expert in their respective fields of activity: rights of nature, social ecology, bioregionalism, institutional design, participatory democracy, the economic and social field, planetary limits, arts, […].'[45]

This meant that id·eau itself was not involved in the community-based, bottom-up process other than providing the framework and the infrastructure of the assembly meetings.

In March 2024, the Popular Assembly together with id·eau published a report summarizing its discussions, findings, and outcomes. The report outlines a new 'governance system for the Rhône and its tributaries'.[46] There is to be a group of guardians for the Rhône. In order to become such a guardian, the Popular Assembly has defined a set of criteria that need to be met:

'[Guardians must be volunteers and actively engaged; respect the mission and values of the project; not have economic interests linked to the Rhône; live in the catchment area; have completed an initial training course on planetary boundaries; [and] be over 16 years old.'[47]

Other stakeholders, such as 'professionals working for and on the Rhône, experts in various fields […], local associations, as well as lawyers specializing in the issues at stake', are to provide advice and, as far as we are able to tell, also be included in decision-making.[48] The guardians will provide authorities and local residents with information, suggest preventive measures, engage in awareness-raising and also be entitled to take legal action if need be. There will be a hotline and an online platform for people to contact guardians. Decision-making is to happen by consensus unless a large number of participants is involved, in which case decisions will be taken by majority vote.[49]

d) Association Rechtsperson Reuss

In September 2023, the association *Rechtsperson Reuss* (translating roughly to: Legal Personality for the Reuss River) was founded. Together with other environmental, animal, and nature rights organizations, the association is planning to launch a popular initiative aiming to amend the constitution of the canton of Lucerne (and possibly also of the cantons Uri, Zug, Zurich, and Aargau) in order to recognize the legal personality of the Reuss River and grant it basic rights.[50] The proposition would be to include a new article in the cantonal constitution which could read:

---

[45] Assemblée Populaire du Rhône, 'Gouvernence et organisation' <www.assembleepopulairedurhone.org/organisation>, translation from French to English by the authors.

[46] Association id-eau, 'Donner une voix au Rhône' (2024) 14 <www.assembleepopulairedurhone.org/_files/ugd/098f20_42bf530b83db4f879565dbefe3ba94f5.pdf>, translation from French to English by the authors.

[47] Association id-eau (n 46).

[48] Association id-eau (n 46).

[49] See Association id-eau (n 46) 15.

[50] See Verein Rechtsperson Reuss, 'Unser Ziel' <www.r-reuss.ch/index.html>.

'§ Section 10 para 3 Const./LU (new)
The rights of non-human natural persons must be respected and protected.

§ Section 10[bis] Const./LU (new) Protection of the non-human natural person "Reuss"

1 The Reuss has the right to represent its own interests as a hybrid personality in the legal system.
2 The Reuss has the right to physical integrity, in particular to clean water.
3 Any cruel or degrading treatment of any person of nature (*Naturperson*) in and on the Reuss is prohibited. Their right to physical integrity must be respected.'[51]

III. Initiators, Key Brokers, and Travelling Ideas

After having briefly introduced the three main initiatives to endow nature with rights in Switzerland, we will now outline who the actors are behind these proposals and who or what motivated them to take action. This section is mainly based on the interviews we conducted; as we unfortunately did not interview Lisa Mazzone and only later included her parliamentary postulate in this report – and as the initiative to recognize the legal personality of the Reuss River did not exist at the time of our initial research – the main focus is on the parliamentary initiative presented by Marionna Schlatter and her colleagues and on the *Appel du Rhône*.

1. Parliamentary Initiative 21.436–21.440 (2021)

As mentioned above, the parliamentary initiative was presented by five members of parliament from five different political parties: Marionna Schlatter from the Green Party,[52] Beat Flach from the Green Liberal Party,[53] Anna Giacometti from The Liberals,[54] Niklaus-Samuel Gugger from the Evangelical People's Party,[55] and Jon Pult from the Social Democratic Party.[56] The main initiator behind this initiative was Marionna Schlatter. Knowing that left-wing and green initiatives normally don't stand a chance, she set out to find members of other political parties – particularly from those positioned in the middle and middle-right of the po-

---

[51] *Luzerner Zeitung*, 23 November 2023, 23, translation from German to English by the authors.

[52] Schlatter (n 38).

[53] Beat Flach, 'Recht auf gesunde Umwelt und Rechte der Natur – Parlamentarische Initiative 21.437' <www.parlament.ch/de/ratsbetrieb/suche-curia-vista/geschaeft?AffairId=20210437>.

[54] Anna Giacometti, 'Recht auf gesunde Umwelt und Rechte der Natur – Parlamentarische Initiative 21.438' <www.parlament.ch/de/ratsbetrieb/suche-curia-vista/geschaeft?AffairId=20210438>.

[55] Niklaus-Samuel Gugger, 'Recht auf gesunde Umwelt und Rechte der Natur – Parlamentarische Initiative 21.439' <www.parlament.ch/de/ratsbetrieb/suche-curia-vista/geschaeft?AffairId=20210439>.

[56] Jon Pult, 'Recht auf gesunde Umwelt und Rechte der Natur – Parlamentarische Initiative 21.440' <www.parlament.ch/de/ratsbetrieb/suche-curia-vista/geschaeft?AffairId=20210440>.

litical spectrum – who would present the initiative with her, hoping that this might help gain the support of a majority of voters. For this reason too, the wording of the initiative, which was carefully chosen, reads as quite moderate. The initiative speaks of climate change and not of climate crisis, and, all in all, the initiative is very vaguely phrased, possibly even more so than initiatives usually are.

Originally, the idea to launch the above-described parliamentary initiative apparently emerged out of a meeting organized by the founders of the World Ethic Forum in November 2020, to which Marionna Schlatter had been invited. Swiss activist Doris Ragettli, who is the co-founder of Rights of Mother Earth, an organization dedicated to getting the UN to adopt the Declaration of the Rights of Mother Earth, and part of the so-called 'feedback circle' of the World Ethic Forum, also participated in that meeting. If we look at the organizational structure of the World Ethic Forum, the global connections in rights of nature advocacy are clearly visible. Doris Ragettli is one of the key figures acting as an intermediary between different activists and organizations in the sense that she puts different actors in touch with each other. She and her organization Rights of Mother Earth collaborate with other organizations such as the Global Alliance for the Rights of Nature (GARN), the Community Environmental Legal Defense Fund (CELDF), the Australian Earth Laws Alliance (AELA), and the Earth Law Center, amongst many others. All of these organizations were identified in an extensive study by Craig Kauffman as being amongst the most central nodes in the rather lose network of rights of nature advocacy, meaning 'that most other organizations depend on [them] to make connections with other organizations', thus making these organizations what Kauffman calls 'key brokers'.[57]

In an interview, Doris Ragettli told us that she had the idea for Rights of Mother Earth in 2009.[58] That was before she knew about the developments in Ecuador and before the World People's Conference on Climate Change and the Rights of Mother Earth in Cochabamba (Bolivia) drafted such a declaration in April 2010. She learned about those already existing developments when going on trips to the rainforest in Ecuador that were organized by the Pachamama Alliance. It was on those trips that she met Robin Milam, who is a member of GARN, and together they co-founded Rights of Mother Earth and started collecting 'signatures for a Declaration of Rights of Mother Earth to be adopted by the UN'.[59] They handed over the first 117,000 signatures to the then Secretary-General of the UN, Ban Ki-moon, at the Rio+20 conference. One of the people contributing signatures was the Indian physicist, environmental activist, and ecofeminist Vandana Shiva, who is also the 'matroness' and one of the so-called 'firekeepers' of the World

---

[57] Craig M Kauffman, 'Mapping Transnational Rights of Nature Networks & Laws: New Global Governance Structures for More Sustainable Development' (International Studies Association Annual Conference, Toronto 2020) 12 <http://files.harmonywithnatureun.org/uploads/upload924.pdf>.

[58] Interview with Doris Ragettli from Rights of Mother Earth, 25 July 2022.

[59] <www.rightsofmotherearth.com/>.

Ethic Forum. Vandana Shiva is also on the Executive Committee of GARN and a member of the World Future Council.

The initiator of the World Ethic Forum is the Swiss activist and singer-songwriter Linard Bardill. The first forum under the overall topic of 'radically shared aliveness' took place in August 2022 in Pontresina (Switzerland). The stated proposition of the World Ethic Forum on their website is to 'radically change the relationship between humans and other living subjects', which amongst other things requires that 'we create a legal system that treats non-humans as full subjects'.[60] The forum brought together (eco-)activists from all over the world or, as the website puts it, 'engaged people from fields such as art, law, ethics, natural medicine, agriculture, politics, transformation [and] peace work' to discuss this and other propositions.[61] The aim of the World Ethic Forum thus goes beyond enshrining rights of nature in the Swiss constitution. Yet some members of the forum are working specifically towards this aim. They worked together with the five members of parliament in preparing the parliamentary initiative and, at least at the time of our research, were considering launching a popular referendum now that the parliamentary initiative had failed. Support for the parliamentary initiative was furthermore provided by the Center for Democratic and Environmental Rights (CDER), an organization that was started by the original founders of CELDF.[62] In an online communique, CDER writes that it 'has been working closely over the past year with Swiss activist Doris Ragettli of the organization Rights of Mother Earth, and Swiss author and activist Linard Bardill, and others with the World Ethic Forum, on this project'.[63] In the same report, Linard Bardill is quoted as thanking CDER 'for their consulting, research, and fine-tuning'.[64]

2. *Appel du Rhône*

The main support for the *Appel du Rhône* has come from Valérie Cabanes, with the initial call for the *Appel du Rhône* already being inspired by her work. Valérie Cabanes, an international lawyer specializing in human rights and humanitarian law, is also part of the Executive Committee of GARN as well as an expert of the UN Harmony with Nature program and legal advisor for the Stop Ecocide Foundation. Moreover, she is co-founder of the NGO Notre Affaire à Tous (which works for climate justice and was one of the NGOs that filed the successful lawsuit in France against TotalEnergies) and also of the Wild Legal Program, a school

---

60 World Ethic Forum, 'For a Culture of Radically Shared Aliveness – Principles and Initial Appeal of the WorldEthicForum' <www.worldethicforum.com/en/proposition/>.

61 <www.worldethicforum.com/en/home-en/>.

62 See Mihnea Tănăsescu, *Understanding the Rights of Nature: A Critical Introduction* (transcript Verlag 2022) 108.

63 Center for Democratic Environmental Rights, 'Switzerland Parliament – First Step to Protect the Rights of Nature-3/26/2021' <www.centerforenvironmentalrights.org/news/switzerland-parliament-first-step-to-protect-the-rights-of-nature>.

64 Center for Democratic Environmental Rights (n 63).

for the rights of nature. All of these organizations are important collaboration partners and supporters of the initiative, as is Doris Ragettli's Rights of Mother Earth. Id·eau has now itself become a member organization of GARN and was also invited to become part of the UN Harmony with Nature network.

3. *Geneva Forum*

Both these examples nicely show the interconnectedness of the different actors working both on a local and global level. Nature's rights advocates all over the world collaborate with one another, making reference to and taking inspiration from case law and legislative advances elsewhere. Ideas about nature's rights travel through these networks, creating new political and legal subjectivities through the forming of people's legal consciousness.[65] In what ways other legal systems served as a source of inspiration in Switzerland will be discussed in more detail in the next section. Before doing so, however, it seems worth pointing out that beyond the above-discussed initiatives, Switzerland is also a site of exchange for rights of nature advocates from all over the world.

Every year in December, a conference is organized by the NGO Objectif Sciences International (OSI) with the support of mapas, End Ecocide on Earth, Rights of Mother Earth, the Earth Law Alliance, and the UN Harmony with Nature program, all organizations that are amongst the key brokers in global rights of nature advocacy.[66] This conference, which is called the Geneva Forum, is organized in addition to the Interactive Dialogue of the UN Harmony with Nature program that takes place on Earth Day every year in New York. The aim of the Geneva Forum is to bring together

- Local and regional actors of different countries
- Thematic actors …
- Regional or national federations
- Thematic federations …
- Large institutions of rights of nature
- Associations of defence and protection of nature
- Government departments (environment, education, research, sustainable development …) and international associations of Ministries
- Specialized journalists (law, science, environment, education, sustainable development …)
- UN agencies (UNDP, UNEP …)[67]

[65] See Julia Eckert and others, 'Introduction: Law's Travels and Transformations' in Julia Eckert and others (eds), *Law Against the State: Ethnographic Forays into Law's Transformations* (CUP 2012); Patricia Ewick and Susan S Silbey, *The Common Place of Law: Stories from Everyday Life* (Chi UP1998); Karen Ann Faulk and Sandra Brunnegger, 'Introduction. Making Sense of Justice' in Sandra Brunnegger and Karen Ann Faulk (eds), *A Sense of Justice. Legal Knowledge and Lived Experience in Latin America* (Stanford UP 2016); Alexandra Huneeus, Javier Couso and Rachel Sieder, 'Cultures of Legality: Judicialization and Political Activism in Contemporary Latin America' in Javier Couso, Alexandra Huneeus and Rachel Sieder (eds), *Cultures of Legality. Judicialization and Political Activism in Latin America* (CUP 2010).

[66] See <www.osi-genevaforum.org/rights-of-nature-for-peace-and-development.html?lang=en>.

[67] See <www.osi-genevaforum.org/rights-of-nature-for-peace-and-development.html?lang=en>.

to discuss ongoing international initiatives on the rights of nature. Thomas Egli, the founder of OSI and CEO of the Geneva Forum, is one of the experts of the UN Harmony with Nature program, and his NGO OSI has special consultative status with the UN's Economic and Social Council. The main aim of OSI is to 'empower individuals to take responsibility for the sustainable development of our planet […] through science education, scientific reasoning and (re)connecting with Nature'.[68] It does so through the organization of participatory research camps for adolescents and young adults. OSI offers such science camps specifically on the topic of the rights of nature.

IV. Sources of Inspiration: Other Legal Systems

For all the above-discussed initiatives in Switzerland, other legal systems were an important source of inspiration. The parliamentary initiative makes direct reference to Resolution 75/2020 'Harmony with Nature' of the United Nations General Assembly,[69] and in her interview, Marionna Schlatter explained to us:

> 'The impetus for us was international: On the one hand, the UN campaign and, on the other hand, the court rulings regarding the diesel [emissions] scandal and these kinds of international efforts.'[70]

She told us that she had studied various cases from South America and New Zealand – she deemed the latter particularly interesting due to the similarities regarding the democratic system – and that she had been in touch with key actors involved in these cases as well as the people behind the popular initiative in Bavaria, Germany. In offering support for the initiative during the plenary discussion of the National Council and in the debate of the Legal Affairs Committee, members of parliament furthermore referred to existing laws in Ecuador and Bolivia; the example of a river that was granted the status of a legal subject in Florida was also mentioned as well as Sweden as a place where the introduction of nature's rights is currently being discussed.

Similarly, in her parliamentary postulate, Lisa Mazzone made reference to the Whanganui case in New Zealand and to the cases in Uttarakhand in India, where 'rivers, watercourses, streams, lakes, the air, the grasslands, valleys, the jungle, forest wetlands, springs, even waterfalls' were recognized as legal subjects with legal personality, so as to stress the timeliness of her request.[71] She thus stated that

[68] Objectif Sciences International, 'What is OSI, what OSI does' <www.osi-ngo.org/qui-nous-sommes/article/ce-qu-est-osi-ce-que-fait-osi?lang=en>.

[69] See General Assembly resolution 75/2020, *Harmony with Nature*, A/RES/75/2020 (21 December 2020), at <https://undocs.org/en/A/RES/75/20>.

[70] Interview with Marionna Schlatter, 12 August 2022, translation from Swiss German to English by the authors.

[71] Mazzone (n 35), translation from German to English by the authors.

it was 'time to consider whether it would not be appropriate to grant legal personality to Swiss glaciers' too.[72]

Finally, in their description of the *Appel du Rhône*, id·eau makes reference to the *Vilcabamba* case in Ecuador as well as to other cases where rivers were assigned legal status: the *Atrato* case in Colombia, the *Whanganui* case in New Zealand, and the *Turag* case in Bangladesh. Overall, the work of id·eau is guided by the nine planetary boundaries.[73]

When asked about the main lessons learned from studying cases from other countries, the member of id·eau we interviewed said:

'When we looked at different case studies, we saw that where rights of nature succeeded, it was often indigenous people who managed that or communities close to nature and we started thinking: "Well, who are the indigenous people in Switzerland and in France that could bear this and push it forward?" We think that [the Popular Assembly of the Rhône] is the closest we can come to really having the roots of the Rhône in the process.'[74]

By contrast, Marionna Schlatter explained to us, one of the main things that she learned was that the 'Mother Earth discourse would probably not work in Switzerland', that ascribing an essence to a river like indigenous people sometimes did, would probably 'put people off in Switzerland', leading her to the conclusion that 'in Switzerland, you have to be very technocratic about it'.[75] Marionna Schlatter and the member of id·eau concurred in their view that it was important to revise the current legal system because 'as it is today, it is not up to the environmental crisis'.[76] However, they differed in their belief as to how this was to be achieved. Marionna Schlatter has opted for the institutional path, believing in change from within the system, while id·eau is more of the idea

'that we don't expect governments to do the whole thing. It's a sort of branch of mobilization that really relies on citizens more than on the authorities to move these issues forward.'[77]

Both our interview partners again agreed that what was needed was a different conceptualization of nature. In the words of Marionna Schlatter, we need to

'get away from the dominance of man over weak nature. [We need to understand that] the purpose of nature is not [...] to provide resources for people, but that it has the right to regenerate itself. And that it has the right to exist per se.'[78]

---

[72] Mazzone (n 35), translation from German to English by the authors.

[73] As developed by Johan Rockström and others, 'A Safe Operating Space for humanity' (2009) 461 Nature 472, and as updated by Will Steffen and others, 'Planetary Boundaries: Guiding human development on a changing planet' (2015) 347(6223) Science.

[74] Interview member id·eau (n 44).

[75] Interview Schlatter (n 70), translation from Swiss German to English by the authors.

[76] Interview Schlatter (n 70), translation from Swiss German to English by the authors.

[77] Interview member id·eau (n 44).

[78] Interview Schlatter (n 70), translation from Swiss German to English by the authors.

This is, of course, a very common position in rights of nature advocacy. Furthermore, her wording reminds us of nature's rights in Ecuador, where art 71 of the constitution grants

'[n]ature, or Pacha Mama, where life is reproduced and occurs, […] the right to integral respect for its existence and for the maintenance and regeneration of its life cycles, structure, functions and evolutionary processes.'[79]

The phrasing of the *Appel du Rhône* is also similar:

'[I]n order to equip ourselves with the means to fight for the preservation of the Rhône, we demand the recognition of a legal personality for the Rhône in order to defend its rights to exist, to be preserved, to regenerate, to evolve in order to maintain and guarantee its environment and its biodiversity for us and for future generations.'[80]

These similarities might stem from Marionna Schlatter and her colleagues as well as id·eau studying existing Ecuadorian constitutional law and cases, but it also might be due to the same intellectual sources being at work in different cases and legislative processes, as Mihnea Tănăsescu shows in his analysis.[81] CELDF, for instance, played an important role in the constitution-building process in Ecuador,[82] and it has, indirectly, also been playing an influential role in Switzerland: through CDER, the organization created by the original founders of CELDF, through GARN, through Doris Ragettli and Rights of Mother Earth, and through the many cases influenced by their work globally.

V. Theoretical Foundations

In the interviews, we did not discuss in detail the theoretical (philosophical, biological, religious, cosmological, etc.) arguments that gave rise to the initiatives in Switzerland. However, drawing on Mihnea Tănăsescu's careful analysis, we would like to tentatively suggest what might be possible theoretical arguments underlying the different initiatives, mostly implicitly. Tănăsescu criticizes that in rights of nature advocacy an 'orthodox view' has emerged that draws on what he calls an 'ecotheological view', a tradition that can be traced from Godofredo Stutzin to Thomas Berry to Cormac Cullinan. He writes that

'[f]or Stutzin, as well as for his followers, the moral standing of nature obviously demands legal standing, the two being inseparable. Furthermore, once the law catches up with the sup-

---

[79] <https://pdba.georgetown.edu/Constitutions/Ecuador/english08.html>.

[80] Association id-eau (n 43), translation from French to English by the authors.

[81] See Tănăsescu (n 62).

[82] See Christina Espinosa, 'Interpretive Affinities: The Constitutionalization of Rights of Nature, Pacha Mama, in Ecuador' (2019) 21(5) Journal of Environmental Policy & Planning 608, 611; Andreas Gutmann, *Hybride Rechtssubjektivität. Die Rechte der ‚Natur oder Pacha Mama' in der ecuadorianischen Verfassung von 2008* (Nomos 2021) 35; Tănăsescu (n 62) 53.

posedly obvious fact of nature's moral standing, ecological problems can be solved, because of this alignment of the law with moral sensibility.'[83]

He goes on to argue that this

'ecotheological history […] doesn't seem to be primarily interested in the pragmatism of given situations, but rather in advancing a framework that subsumes any given situation under the Great Work, the totality that imposes, as if on its own, a series of "fundamental rights" that have to be recognized (as opposed to granted).'[84]

Such rights are, according to Cormac Cullinan in his book 'Wild Law', 'the right to exist, to have a habitat, and to evolve as part of the earth community',[85] hence, the same rights the *Appel du Rhône* and Marionna Schlatter indirectly draw on. Here we can also see similarities with Jörg Leimbacher's book on the rights of nature (in Switzerland) published in 1988. He concludes that nature should be granted a fundamental right to exist. Such a right of nature to exist would have to protect the 'being-there', the 'being-so', and the development opportunities of nature.[86] He argues that only by enshrining a fundamental right of nature can it be prevented that nature can simply be disposed of. In his view, only a fundamental right of nature gives it immediate and direct claims that can be enforced – without the intervention of the legislator – with the help of the courts.

One of the things Tănăsescu criticizes about the ecotheological view is the idea that a shift from so-called anthropocentric law to ecocentric law will quasi automatically lead to the fixing of environmental problems. The absoluteness of such a view is probably not shared by the members of parliament behind the parliamentary initiative nor by id·eau. Thus, as Marionna Schlatter explained:

'I think nature is diversity and our political system will never be able to cope with this diversity. We will never be able to map it into the legal system so that it is fair for nature, because our human system always simplifies. We try to make the system so that it is somewhat right. […] You have to see it as a project. The legal system is a project that must evolve with society. It is not set in stone.'[87]

Yet we do also see this idea of a quasi-automatic solution to the environmental crisis in this statement by the member of id·eau:

'What we're convinced of is that if you give a legal status to nature, then it doesn't need us to renature it. It will regenerate and it will have the right to regenerate. […] We really have faith in a law that would be preventive and not reparationist. […] What we say is that law as it is will not be the solution. So, it's not faith in the legal solution as such, but in a law that prevents something from happening and makes sure that we have tools that assess the damages, the potential damages before anything is allowed to happen, […] meaning that you will have to prove that you will not cause damage.'[88]

---

[83] Tănăsescu (n 62).
[84] Tănăsescu (n 62) 30.
[85] Tănăsescu (n 62) 29.
[86] Jörg Leimbacher, *Die Rechte der Natur* (Helbing & Lichtenhahn 1988) 107–108.
[87] Interview Schlatter (n 70), translation from Swiss German to English by the authors.
[88] Interview member id·eau (n 44).

The shift towards prevention can also be observed elsewhere in connection with rights of nature.[89] Rights of nature advocates criticize that existing environmental legislation focuses too much on reparation of past damage, mostly through financial punitive means. In turn, rights of nature and the recognition of natural entities with legal personality are believed to work preventively, especially if established through a transnational framework.

VI. A Change in Sight?

Despite recent efforts, not much has changed in terms of nature's rights in Switzerland. Both the parliamentary postulate and initiative failed. Yet this does not mean, of course, that the debate is off the table. As already briefly mentioned above, Marionna Schlatter and her colleagues never expected the initiative to be passed at first go. For them, the initiative was more a means of starting a new political discussion. In the interview, Marionna Schlatter told us that she was happy that in the run-up to the parliamentary debate, the initiative gained quite a bit of media attention.[90] Similarly, the member of id·eau that we interviewed was pleased about the public interest they were receiving. Apparently, id·eau is often invited by other organizations pursuing similar aims to present their project, and both political parties and municipalities have collectively signed the *Appel du Rhône*, which had not been id·eau's original intention.[91] Nevertheless, it seems to us that rights of nature have not yet made it into mainstream environmental debates and activism in Switzerland; rather, they are, at least at a political level, met with fierce resistance. This can be well shown by taking a closer look at the debates within the Legal Affairs Committee as well as the debate in the plenary session of the National Council. Both were remarkably short, the latter discussion lasted only fifteen minutes. Other than the member of parliament representing the so-called majority vote of the Legal Affairs Committee, only the initiators spoke. Furthermore, as far as we are able to tell, this debate did not get much media coverage subsequent to the parliamentary debate.

The debate in the Legal Affairs Committee which preceded the plenary debate started with a member of the federal administration presenting a view as to how feasible the application would be. This was followed by members of the different parliamentary groups presenting their positions. In the end, it was a very political vote, with only members of the two green parties and the Social Democratic Party voting in favor of the initiative.[92] This is, of course, hardly surprising for parliamentary proceedings.

---

[89] See Laura Affolter, 'The Responsibility to Prevent Future Harm: Anti-Mining Struggles, the State, and Constitutional Lawsuits in Ecuador' (2020) 4(2) JLA 78.

[90] Interview Schlatter (n 70).

[91] Interview member id·eau (n 44).

[92] See Kommission für Rechtsfragen des Nationalrates (n 41).

In the opening statement, the administration representative claimed that introducing a new (human) right to a healthy environment would not create any major legal problems as such and whether to accept this constitutional change or not was therefore entirely dependent on the political will of the parliament.[93] This was asserted mainly on the basis that several countries have already adopted such a right and that, on 8 October 2021, the Human Rights Council of the UN passed a resolution encouraging states 'to adopt policies for the enjoyment of the right to a safe, clean, healthy and sustainable environment'.[94] This position was confirmed by the UN General Assembly on 28 July 2022 when it declared 'access to [a] clean and healthy environment a universal human right'.[95] With regard to the second part of the parliamentary initiative, namely to grant nature legal personality, it was claimed that this would pose more difficulties in that it would mean a substantial change for Swiss law. The proposition was thus made by the administration representative to envisage, as an alternative to granting nature rights, a broadening of the already existing organizations' right of appeal. During the discussion, the opponents of the initiative focused mainly on the arguments that the term 'healthy environment' was not sufficiently defined to be enshrined as a fundamental right and that the term 'nature' was also too vague to be granted legal personality. Furthermore, they argued that granting nature legal personality was not possible because having rights, in their view, also entails having obligations. In the plenary session of the National Council, Yves Nidegger from the right-wing Swiss People's Party made the following statement on behalf of the majority of the Legal Affairs Committee voting against further consideration of the initiative:

'The majority's reasons for saying that there is no need to legislate are essentially that there is no possibility to legislate. Regarding the first part of the proposal, namely a fundamental right to a healthy environment, the Federal Constitution, in Article 73, speaks of sustainable development as a principle, but in our legal system, the notion of a right to a healthy environment, which can be enforced in court, is so undefined that, in the opinion of the majority of the Committee, it cannot be made a constitutional right of the same rank as the protection of property, freedom of expression or other such rights, since this is what it is about. The second part the proposal, [...] to make nature a subject of law, runs counter to our modern legal order. It is obviously very refreshing and interesting to re-enchant nature by giving it rights – a bit like the shamanism that was our way of thinking for a long time in neolithic and ancient times, when animals, trees and rivers had a soul, had a spirit, could demand justice, could have claims – but modern times,

---

93 See Nationalrat, 'Wintersession 2022 – Achte Sitzung vom 12.12.22 – Votum 3' <https://www.parlament.ch/de/ratsbetrieb/amtliches-bulletin/amtliches-bulletin-die-verhandlungen?SubjectId=59236#votum3>.

94 UN, 'Human Rights Council adopts four resolutions on the right to development, human rights and indigenous peoples, the human rights implications of the COVID-19 pandemic on young people, and the human right to a safe, clean, healthy and sustainable environment' (8 October 2021) <www.ohchr.org/en/press-releases/2021/10/human-rights-council-adopts-four-resolutions-right-development-human-rights>.

95 UN, 'UN General Assembly declares access to clean and healthy environment a universal human right' (28 July 2022) <https://news.un.org/en/story/2022/07/1123482>.

with positive law, have created a legal order that is totally impervious to this order of things. There are only subjects of law that have rights but also obligations, hence, human beings. And even though the Middle Ages still saw trials in which animals were brought to justice because locusts or rats had destroyed crops […] those times have passed, and since the introduction of positive law, we cannot conceive of the possibility of re-enchanting nature to the point of giving it rights of action. We can only end up in the situation we have elsewhere where more or less self-proclaimed defenders of nature can sue. This already exists in our legal system: when there are ecological and environmental interests, associations have rights of action. There is even an international convention, the Aarhus Convention, which gives an extremely broad definition of environmental issues and which offers, in all the countries that have adopted it, including Switzerland, extremely broad rights of action to nature conservation groups. In other words, there is no need to legislate. Rather, it is a romantic proposal. However sympathetic it may seem, this parliamentary initiative does not meet a need and it leads to an impossibility to legislate, which is why the majority of the committee recommends that you do not proceed with it.'[96]

In our view, this statement is problematic in various respects. First, it proposes a hierarchization of knowledge, portraying philosophies that recognize the spirit of animals, trees, and rivers in the past as a less developed form of knowledge. Second, the argument that holders of rights must also have obligations was already dismissed by Jörg Leimbacher in 1988, who argued that if this were the case, people who were no longer capable of judgment for any (mental and/or somatic) reason, and who thus were no longer capable of entering into obligations, would no longer be considered legal subjects.[97] The supporters of the initiatives countered the opponents by arguing that it was feasible to recognize both the right to a healthy environment and to grant nature or aspects of nature legal personality. As shown above, they did this by drawing on existing laws in other countries. A strong emphasis was put on the above-mentioned UN resolution,[98] and reference was also made to the judgment of the Federal Supreme Court[99] mentioned above, according to which granting rights to primates is, from a legal point of view, feasible, albeit only to a certain extent. The proponents pointed out that the existing law creates the, in their view, absurd situation where an individual can easily bring an action before court when they feel the color of the neighbor's roof is wrong but not when a river is being polluted, because, as it was laid out above, there is no interest worthy of protection.

Finally, the debate brought to the surface concerns as to who would represent nature if it were to be granted rights. To some extent, this discussion was antici-

---

[96] See Nationalrat, 'Wintersession 2022 – Achte Sitzung vom 12.12.22 – Votum 6' <www.parlament.ch/de/ratsbetrieb/amtliches-bulletin/amtliches-bulletin-die-verhandlungen?SubjectId=59236#votum6>, translation from French to English by the authors.

[97] Leimbacher (n 86) 50; see also Saskia Stucki, *Grundrechte für Tiere. Eine Kritik des geltenden Tierschutzrechts und rechtstheoretische Grundlegung von Tierrechten im Rahmen einer Neupositionierung des Tieres als Rechtssubjekt* (Nomos 2016) 273–279.

[98] UN, 'Resolution adopted by the Human Rights Council on 8 October2021' <https://documents-dds-ny.un.org/doc/UNDOC/GEN/G21/289/50/PDF/G2128950.pdf?OpenElement>.

[99] BGE 147 I 183 (n 32).

pated by Marionna Schlatter, who connected this particular challenge to the functioning of Swiss democracy:

> '[W]e have a democracy of interest representatives, which works very differently than in other countries. The fact that we have a militia system, so to speak, where everyone is dependent on the interest representatives having such a direct influence on the political system, makes it much more politically charged. And that means that it would be an enormous challenge for Switzerland to find representatives who are really, how shall I say, not prejudiced. [...] Who would represent these interests? I think there would have to be an association that would be organized on a grassroots level and represent the interests. But then it would quickly become incredibly complicated. I think that will be a challenge for the Swiss system to translate.'[100]

It is to be expected that these discussions will continue in the future. For Marionna Schlatter and her colleagues the parliamentary initiative was conceived as a first step to initiate a new political discussion, which has now begun. What she suspects will ultimately happen is that other European countries will first introduce rights of nature and then, as has occurred in numerous other legislative matters, Switzerland will eventually follow suit. Due to its geographical location and economic situation, Switzerland is factually obliged to ensure that its domestic law is compatible with the law of its neighboring states and of the European Union.

The hope of id·eau is even more ambitions, namely an internationally binding law that states could not afford to go against; or as the member that we interviewed explained:

> 'It's that universally there is a right of nature that supersedes everything and that you cannot go against because it would be against humanity and against survival.'[101]

Whatever the outcome will be, what we can already observe is that new legal subjectivities are emerging and that new legal topics are being brought to the fore. Taking into account international developments, it is a trend that we expect to continue.

---

[100] Interview Schlatter (n 70), translation from Swiss German to English by the authors.

[101] Interview member id·eau (n 44).

Appendix

Questionnaire

Rights of Nature, Globalization and Legal Pluralism
International Academy of Comparative Law
General Congress – Asunción 2022

I. Description

During the last two decades, the connections between nature, human beings and law have become a relevant object of study around the globe. The recognition of rights to domestic animals in India and Pakistan, the constitutionalization of the rights of nature in Bolivia and Ecuador, and the recognition of rights to a river in New Zealand, for example, have attracted the attention of legal scholars both in the Global North and South. The description, analysis, and evaluation of the intersections between nature, persons, and legal systems, have also generated discourses that complement, contradict, and interweave. These discourses examine different dimensions of the problem from diverse political, theoretical, and disciplinary perspectives. Thus, for example,

(i) environmental law scholars have focused on the way in which law intersects with biocentric and anthropocentric perspectives of nature to protect or undermine ecosystems;
(ii) constitutional law scholars have focused on the constitutionalization of the rights of plants, minerals, and animals in countries such as Ecuador and Bolivia, as well as the advantages and disadvantages generated by these processes of juridification of nature;
(iii) legal theorists have focused on the way in which modern law intersects with moral, political, and legal traditions usually marginalized such as those of Indigenous groups; and
(iv) civil law scholars have focused in the way that traditional ways of conceptualizing and regulating property are challenged by the idea that nature has rights.

The diversity and richness of these political and epistemological processes, however, are not only valuable for the different ways in which they contribute to the understanding and criticizing the relationships between nature, human beings, and law. These processes are valuable as well because they are a privileged space for understanding and questioning key aspects of contemporary globalization and legal pluralism. The rights of nature allow us to examine four problems that are of particular interest to comparative law and for global legal pluralism: the processes of cross-fertilization between national legal systems; the interaction, clash, and influence of national legal systems and international law; the relationships between transnational theoretical discourses, international law, and national legal systems; and the geopolitics of legal knowledge.

This project, which is part of the XXI General Congress of the International Academy of Comparative Law, therefore, pursues the following four objectives:

(i) To examine how the dialogue between legal systems of both the Global South and North has contributed to the creation of national and international concepts, rules, and principles related to the rights of nature.
(ii) To explore the interactions between national legal systems with international law that have allowed the articulation of both a transnational discourse on the rights of nature and the creation of national and international legal regimes that aim to regulate the relationship between human beings and nature.
(iii) To analyze the way in which transnational theoretical discourses on the rights of nature have interacted with and have been used by very diverse legal actors and levels: from local and transnational non-governmental organizations to governments, international institutions, and universities, passing through political parties and ordinary citizens.
(iv) To study the processes of production, exchange, use, and legitimation of legal knowledge that cut across the discourses on the rights of nature.

To achieve these ends, national chapters of the International Academy of Comparative law have chosen a group of national rapporteurs. These rapporteurs are in charge of drafting national reports on the object of study of this project. Ralf Michaels and Daniel Bonilla, the general rapporteurs of the project, have also chosen a group of special rapporteurs who will write reports on aspects related to the international and theoretical dimensions of the rights of nature. Michaels and Bonilla will also be in charge of drafting a general report that puts national and special reports in dialogue. In this general report, the rights of nature will be in dialogue with globalization and legal pluralism. The former will not simply be an example to illustrate the latter, nor the latter a pretext for studying the former. In the general report, the particular objects of study will intertwine with the general objects of study.

II. Questions

National rapporteurs are asked to write their reports as responses to the following questionnaire in the structure given. Your reports should not only consist of brief answers to the question: we hope you will write elaborate and referenced responses, focusing not just on statutes and results of court opinions (law in the books) but also the law in action. Also, although we urge you to stick with the structure of the questionnaire, we would much prefer your reports to be in the form of an independent text, rather than simply answers to individual questions. There is no need to repeat the questions in your reports.

Special rapporteurs have some more freedom in their texts, given that many of the questions will not fit their particular topic. Nonetheless, we believe the questionnaire can serve as a source of information for what to expect from other reports, and as a good guideline of questions that may be relevant for you as well.

We ask that your reports not extend thirty pages.

1. Existence and Contents

a) Are rights of nature (nature as a whole, ecosystems, or particular components of an ecosystem) recognized in your legal system? (We do not ask for rights of animals here.)

   If the answer to this question is positive, when, in what form (eg Constitution, statutes, decrees, case law, bureaucratic practice), and through what processes were these rights recognized? Are there proposals to abolish or limit them or their effects? Have they been criticized?

   If negative, are there any proposals to recognize the rights of nature in your legal system? How are they being discussed, how likely is it that they will be adopted?

b) What specific rights, if any, were recognized to nature (or are proposed to be recognized to nature) in your legal system, eg, legal personality, life, standing, ability to enter into contracts, ability to inherit, etc? How do these rights relate to broader legal frameworks, eg, on environment, nature conservation, land ownership, or natural resources? Are they parts of such larger areas, or do they stand alone?

c) What are the central characteristics of the rights of nature that your jurisdiction recognizes (or that are proposed), eg, main legal concepts and key procedural elements? Or What are the central characteristics of the rights of nature that are proposed to be recognized in your legal system?

d) With what other rights can the rights of nature collide in your legal system, eg, resource extraction or property rights? How are such conflicts resolved? Be specific – report existing case law.

2. Legal and political operators

a) What *domestic* legal, political, or civil society institutions or groups promoted the recognition of or effectively recognized the rights of nature in your jurisdiction, eg, congress, courts, administrative agencies, NGOs, or Indigenous groups? What motivated those institutions to do it? What legal, political, or civil society institutions or groups opposed the recognition of rights of nature? What motivated them?

b) What *international or foreign* institutions or groups played similar roles?

c) As between domestic and international/foreign influences, are they in tension and, if so, how is that tension resolved? What is their relative grade of importance?

3. Enforcement

a) Who can exercise the rights of nature eg, the State, any citizen, specific groups of citizens?

b) In what courts, if any, or before what state authorities, can such rights be enforced?

c) What procedure(s) must be followed to exercise these rights?

d) How are rights of nature enforced?

4. Theoretical sources and foundations

What are the theoretical (philosophical, biological, religious, cosmological, etc) arguments, sources and foundations that gave rise to and ground the rights of nature that were recognized or are intended to be recognized in your legal system? Are there any tensions between the theoretical sources and foundations of the rights of nature recognized in your jurisdiction and traditional sources and foundations of your legal system?

5. Relationship between national legal systems

a) Did other legal systems (foreign or substate) influence, eg, were an inspiration or source for the recognition (or the recognition proposals) of the rights of nature in your jurisdiction?

b) Have the rights of nature that were recognized by your legal system had any impact on other national jurisdictions?

c) How are rights of nature treated in conflict of laws? Can foreign rights of nature be enforced in your system? Can your system's rights be enforced abroad?

6. International law

a) Did international law influence in any way the recognition (or the proposals for the recognition) or the rejection/limitation of the rights of nature in your jurisdiction?

b) Have the rights of nature that were recognized by your legal system, or their rejection, had any impact on international law?

c) Have there been, or could there be, any conflict of laws issues between international law and the rights of nature recognized by your legal system? If any, how have they been or would they be resolved?

7. *Efficacy*

a) What individuals, groups, or institutions, if any, are recognized as legal guardians of nature in your jurisdiction? Who has used rights of nature to defend nature in your jurisdiction, eg, NGOs, state, citizens, Indigenous groups, universities?

b) Has the implementation of the rights of nature that were recognized by your legal system been effective (please state the indicators that you are using to evaluate the effectiveness of rights of nature)? What legal, theoretical, or empirical factors have been an obstacle for the implementation of these rights?

8. *Consequences*

What political, legal, environmental, economic, or cultural consequences has the recognition of the rights of nature had in your country?

III. Criteria for writing the reports

In writing your report, please keep the following points in mind:

1. Write the answers to the questionnaire in English (preferred), French, or Spanish.
2. Cite the relevant literature and include a list of the bibliography used at the end of the questionnaire.
3. Cite the case law, statutes, decrees, etc completely. Include the full texts of the relevant legal standards in the footnotes.
4. Use technical terms consistently.
5. Explain concepts that are typical of your legal system or legal culture and unknown or little known in other legal systems.